Springer Undergraduate Texts in Philosophy

The Springer Undergraduate Texts in Philosophy offers a series of self-contained textbooks aimed towards the undergraduate level that covers all areas of philosophy ranging from classical philosophy to contemporary topics in the field. The texts will include teaching aids (such as exercises and summaries) and will be aimed mainly towards more advanced undergraduate students of philosophy.

The series publishes:

- All of the philosophical traditions
- Introduction books with a focus on including introduction books for specific topics such as logic, epistemology, German philosophy etc.
- Interdisciplinary introductions – where philosophy overlaps with other scientific or practical areas

This series covers textbooks for all undergraduate levels in philosophy particularly those interested in introductions to specific philosophy topics.

We aim to make a first decision within 1 month of submission. In case of a positive first decision the work will be provisionally contracted: the final decision about publication will depend upon the result of the anonymous peer review of the complete manuscript. We aim to have the complete work peer-reviewed within 3 months of submission.

Proposals should include:

- A short synopsis of the work or the introduction chapter
- The proposed Table of Contents
- CV of the lead author(s)
- List of courses for possible course adoption

The series discourages the submission of manuscripts that are below 65,000 words in length.

For inquiries and submissions of proposals, authors can contact Ties.Nijssen@Springer.com

More information about this series at http://www.springer.com/series/13798

Harrie de Swart

Philosophical and Mathematical Logic

Springer

Harrie de Swart
Faculty of Philosophy
Erasmus University Rotterdam
Rotterdam, The Netherlands

Department of Philosophy
Tilburg University
Tilburg, The Netherlands

ISSN 2569-8737   ISSN 2569-8753   (electronic)
Springer Undergraduate Texts in Philosophy
ISBN 978-3-030-03253-1   ISBN 978-3-030-03255-5   (eBook)
https://doi.org/10.1007/978-3-030-03255-5

Library of Congress Control Number: 2018960381

© Springer Nature Switzerland AG 2018
This work is subject to copyright. All rights are reserved by the Publisher, whether the whole or part of the material is concerned, specifically the rights of translation, reprinting, reuse of illustrations, recitation, broadcasting, reproduction on microfilms or in any other physical way, and transmission or information storage and retrieval, electronic adaptation, computer software, or by similar or dissimilar methodology now known or hereafter developed.
The use of general descriptive names, registered names, trademarks, service marks, etc. in this publication does not imply, even in the absence of a specific statement, that such names are exempt from the relevant protective laws and regulations and therefore free for general use.
The publisher, the authors and the editors are safe to assume that the advice and information in this book are believed to be true and accurate at the date of publication. Neither the publisher nor the authors or the editors give a warranty, express or implied, with respect to the material contained herein or for any errors or omissions that may have been made. The publisher remains neutral with regard to jurisdictional claims in published maps and institutional affiliations.

This Springer imprint is published by the registered company Springer Nature Switzerland AG
The registered company address is: Gewerbestrasse 11, 6330 Cham, Switzerland

Logic is to improve human thinking in order to improve human existence.
[Andrzej Grzegorczyk]

However, this same [mathematical] form of thinking, this same kind of concept analysis, is also applicable to many other areas that are directly related to the immediate reality of our daily lives. And such a broader application of the mathematical form of thought seems to me to be of the highest importance. After all, the unparalleled development of the technique in a narrow sense, of the technical technique, one could say, is followed by a hardly less important development of the psychological technique, of the advertising technique. propaganda technique, in short, of means to influence people. However, we have failed to strengthen our defense equipment against belief and suggestion attempts by others by improving our *thinking technology*. [...] In this tangle of questions and sham questions we can find a guide in the conceptual analysis, demonstrated in the mathematical way of thinking. Against all these known and unknown psychic influences we can forge a weapon by *improving our thinking technique*. And that such a reinforcement of our spirit is required, urgently needed, is my deepest conviction. [David van Dantzig, 1938, inaugural lecture, Delft, the Netherlands; translated from Dutch]

This book is dedicated to Johan J. de Iongh (1915 - 1999)

My friend and teacher

It is the main task of a philosopher to show people that things do not have to be the way they are, that they might be different and that in some cases they should be different. [Johan de Iongh]

Johan de Iongh (1915 - 1999) was a student of L.E.J. Brouwer (1881 - 1966), the founding father of intuitionism. He was convinced of the soundness of the intuitionistic view of mathematics. He also had a great affinity with the signific position, represented by Gerrit Mannoury (1867 - 1956).

He became professor in Nijmegen in 1961, where he was teaching the course on analysis for first-year students. Later de Iongh devoted most of his teaching to courses on logic, the foundations and the philosophy of mathematics, and in particular intuitionistic mathematics. He was very careful in giving an accurate presentation of Brouwer's views. He took a great interest in the well-being of his students and found it important to know them personally.

Johan de Iongh was as much a philosopher as a mathematician. He shared Plato's view that the study of mathematics is the correct introduction to philosophy. He has published very little. His Platonic distrust towards the written word was great; his tendency to share his thoughts and ideas with friends, rather than to write them down, much greater. Yet some texts from him have been preserved, and many of his ideas have been worked out in Ph.D. theses and papers by his students.

His broad scholarship was impressive. He read Greek and Latin authors in the original. His interest in science reached far beyond mathematics and he was widely read in world literature.

He was a convinced Catholic and his thinking on mathematics and philosophy has developed in continuing discussion with St Augustine, St. Thomas Aquinas, St. Thomas More and Nicholas of Cusa. He always started his lectures with a short prayer in Latin: Spiritus sancti gratia illuminet sensus et corda nostra [May the grace of the Holy Spirit illuminate our senses and our hearts]. And he always finished his lectures with the following prayer: Gratias tibi agimus, Domine, pro omnibus beneficiis tuis [We thank you, my Lord, for all your blessings].

It was a privilege to be his student, his PhD student, his assistant and his friend.

Foreword

The following quotation is from Lewis Carroll, *Symbolic Logic and The Game of Logic*; Introduction.

> The learner, who wishes to try the question *fairly*, whether this little book does, or does not, supply the materials for a most interesting recreation, is *earnestly* advised to adopt the following Rules:
>
> (1) Begin at the *beginning*, and do not allow yourself to gratify a mere idle curiosity by dipping into the book, here and there. This would very likely lead to your throwing it aside, with the remark 'This is *much* too hard for me!', and thus losing the chance of adding a very *large* item to your stock of mental delights. ... You will find the latter part *hopelessly* unintelligible, if you read it before reaching it in regular course.
>
> (2) Don't begin any fresh Chapter, or Section, until you are certain that you *thoroughly* understand the whole book *up to that point*, and that you have worked, correctly, most if not all of the examples which have been set. So long as you are conscious that all the land you have passed through is absolutely *conquered*, and that you are leaving no unsolved difficulties *behind* you, which will be sure to turn up again later on, your triumphal progress will be easy and delightful. Otherwise, you will find your state of puzzlement get worse and worse as you proceed, till you give up the whole thing in utter disgust.
>
> (3) When you come to any passage you don't understand, *read it again*: if you *still* don't understand it, *read it again*: if you fail, even after *three* readings, very likely your brain is getting a little tired. In that case, put the book away, and take to other occupations, and next day, when you come to it fresh, you will very likely find that it is *quite* easy.
>
> (4) If possible, find some genial friend, who will read the book along with you, and will talk over the difficulties with you. *Talking* is a wonderful smoother-over of difficulties. When *I* come upon anything - in Logic or in any other hard subject - that entirely puzzles me, I find it a capital plan to talk it over, *aloud*, even when I am all alone. One can explain things so *clearly* to one's self! And then, you know, one is so *patient* with one's self: one *never* gets irritated at one's own stupidity!
>
> If, dear Reader, you will faithfully observe these Rules, and so give my little book a really *fair* trial, I promise you, most confidently, that you will find Symbolic Logic to be one of the most, if not *the* most, fascinating of mental recreations!
>
> ...
>
> Mental recreation is a thing that we all of us need for our mental health; and you may get much healthy enjoyment, no doubt, from Games, such as Back-gammon, Chess, and the new Game 'Halma'. But after all, when you have made yourself a first-rate player at any

one of these Games, you have nothing real to *show* for it, as a *result*! You enjoyed the Game, and the victory, no doubt, *at the time*; but you have no *result* that you can treasure up and get real *good* out of. And, all the while, you have been leaving unexplored a perfect *mine* of wealth. Once master the machinery of Symbolic Logic, and you have a mental occupation always at hand, of absorbing interest, and one that will be of real *use* to you in *any* subject you may take up. It will give you clearness of thought - the ability to *see your way* through a puzzle - the habit of arranging your ideas in an orderly and get-at-able form - and, more valuable than all, the power to detect *fallacies*, and to tear to pieces the flimsy illogical arguments, which you will so continually encounter in books, in newspapers, in speeches, and even in sermons, and which so easily delude those who have never taken the trouble to master this fascinating Art. *Try it*. That is all I ask of you!

[From Lewis Carroll, *Symbolic Logic and The Game of Logic*. Introduction; Dover Publications, Mineola, NY, 1958.]

Preface

Having studied mathematics, in particular foundations and philosophy of mathematics, it happened that I was asked to teach logic to the students in the Faculty of Philosophy of the Radboud University Nijmegen. It was there that I discovered that logic is much more than just a mathematical discipline consisting of definitions, theorems and proofs, and that logic can and should be embedded in a philosophical context. After ten years of teaching logic at the Faculty of Philosophy at the Radboud University Nijmegen, thirty years at the Faculty of Philosophy of Tilburg University and nine years at the Faculty of Philosophy of the Erasmus University Rotterdam, I got many ideas how to improve my LOGIC book which was published twenty five years ago in 1993 by Verlag Peter Lang. Although the amount of work was enormous, I felt I should do it. It is like working on a large painting where you put some extra color in one corner, add a little detail at another place, shed some more light on a particular face, etc.

This book was written to serve as an introduction to *logic*, with special emphasis on the interplay between logic and *mathematics*, *philosophy*, *language* and *computer science*. The reader will not only be provided with an introduction to *classical propositional and predicate logic*, but to *philosophical* (modal, deontic, epistemic) and *intuitionistic logic* as well. *Arithmetic and Gödel's incompleteness theorems* are presented, there is a chapter on the *philosophy of language* and a chapter with applications: *logic programming*, *relational databases* and SQL, and *social choice theory*. The last chapter is on *fallacies* and *unfair discussion methods*.

Chapter 1 is intended to give the reader a first impression and a kind of overview of the field, hopefully giving him or her the motivation to go on.

Chapter 2 is on (classical) *propositional logic* and Chapter 4 on *predicate logic*. The notion of *valid consequence* is defined, as well as three notions of (formal) *deducibility* (in terms of logical axioms and rules, in terms of tableaux and in terms of rules of natural deduction). A procedure of searching for a formal deduction of a formula B from given premisses $A_1, \ldots, A_n$ is given in order to show the equivalence of the notions of valid consequence and (formal) deducibility: *soundness and completeness*. This procedure will either yield a (formal) deduction of B from $A_1, \ldots, A_n$

xi

– in which case B is deducible from $A_1, \ldots, A_n$ and hence also a valid consequence of these premisses – or (in the weak, not necessarily decidable sense) if not, one can immediately read off a counterexample – in which case B is not a valid consequence of $A_1, \ldots, A_n$ and hence not deducible from these premisses.

Chapter 3 contains the traditional material on *sets* treated informally in such a way that everything can easily be adapted to an axiomatic treatment. A sketch of the axioms of Zermelo-Fraenkel is given. The notions of *relation* and *function* are presented, since these notions are useful instruments in many fields. From a philosophical point of view *infinite sets* are interesting, because they have many properties not shared by finite sets. The notion of *enumerable* set is needed in the Löwenheim-Skolem theorem in predicate logic, reason why the chapter on sets is presented before the chapter on predicate logic.

At appropriate places *paradoxes* are discussed because they are important for the progress in philosophy and science. Chapter 5 presents a discussion of formal number theory (arithmetic). Peano's axioms for *formal number theory* are presented together with an outline of Gödel's *incompleteness theorems*, which say roughly that arithmetic truth cannot be fully captured by a formal system.

Chapter 6 deals with modal, deontic, epistemic and temporal logic, frequently called *philosophical logic*. It has several applications in the *philosophy of language* whose major topics are discussed in Chapter 7.

It is interesting to note that traditional or classical logic silently is presupposing certain philosophical views, frequently called Platonism. L.E.J. Brouwer (1881 - 1966) challenged these points of view, resulting in a completely different and much more subtle *intuitionistic logic* which we present in Chapter 8.

Interestingly, both logic and set theory have applications in computer science. In Chapter 9 we discuss *logic programming* and the programming language PROLOG (PROgramming in LOGic), which is a version of the first-order language of predicate logic. To illustrate the role of set theory in the field of computer science, we discuss the logical structure of *relational databases* and the *query language* SQL. In this chapter we also discuss *social choice theory* which deals with elections and voting rules. Finally, in Chapter 10 we discuss a number of *fallacies* and *unfair discussion methods*.

I have tried to give the reader some impressions of the *historical development of logic*: Stoic and Aristotelian logic, logic in the Middle Ages, and Frege's Begriffsschrift, together with the works of George Boole (1815 - 1864) and August De Morgan (1806 - 1871), the origin of modern logic.

Since 'if ..., then ...' can be considered to be the heart of logic, throughout this book much attention is paid to *conditionals*: material, strict and relevant implication, entailment, counterfactuals and conversational implicature are treated and many references for further reading are given.

At the end of most sections are exercises; the solutions can be found at the end of the chapter in question. Starred items are more difficult and can be omitted without loss of continuity. The expression := is used as an abbreviation for 'is by definition'.

Tilburg, Rotterdam, summer 2018   *H.C.M. (Harrie) de Swart*

Acknowledgements

It was Johan de Iongh (1915 – 1999) in Nijmegen who introduced me to mathematics, foundations and philosophy of mathematics, logic in particular, history of mathematics, Plato and other philosophers. I had the privilege of studying and working under his guidance from 1962 till 1980. We became friends forever. I also owe much to my collegues in the group around prof. de Iongh: Wim Veldman and Wim Gielen in particular.

The influence of Kleene's books, *Introduction to Metamathematics* and *Mathematical Logic*, is noticeable throughout.

I spent the academic year 1976 – 1977 at the department of History and Philosophy of Science of the Faculty of Philosophy of Princeton University, with a grant of the Niels Stensen Foundation. It is here that I attended courses by John Burgess (philosophy of language), David Lewis (modal logic, counterfactuals) and had conversations with Saul Kripke. The chapters on the philosophy of language and modal logic are to a high degree influenced by these lectures.

The subsection on relational databases and SQL is the result of taking a course given by Frans Remmen at the Technical University of Eindhoven. I am grateful to Luc Bergmans and Amitabha Das Gupta for their contributions on G. Mannoury and L. Wittgenstein respectively.

I am most grateful to the Faculty of Humanities of Tilburg University and to the Faculty of Philosophy of the Erasmus University Rotterdam for providing me with the facilities of office space, computer, etc. In particular, I like to thank Willy Ophelders, who was instrumental in my appointment in Rotterdam; without this appointment this book would not have appeared.

I am happy that Springer Verlag is willing to publish this work. I thank Ties Nijssen and Christi Lue, who were extremely helpful in the preparation of this book.

Most of all I owe a lot to the many students who attended my courses and even were willing to pay for that. Their critical questions and remarks helped enormously to shape this book. It is a privilege that people are willing to listen to you, even when they have troubles with understanding what you are trying to say. I only realized this when I was a member of the local city council, where almost nobody was willing to listen to anybody.

Contents

1 Logic; a First Impression 1
   1.1 General 1
   1.2 Propositional Logic 2
   1.3 Sets; Finite and Infinite 8
   1.4 Predicate Logic 8
   1.5 Arithmetic; Gödel's Incompleteness Theorem 12
   1.6 Modal Logic 13
   1.7 Philosophy of Language 13
   1.8 Intuitionism and Intuitionistic Logic 14
   1.9 Applications 15
      1.9.1 Programming in Logic: Prolog 15
      1.9.2 Relational Databases 16
      1.9.3 Social Choice Theory 16
   1.10 Fallacies and Unfair Discussion Methods 17
   1.11 Solutions 19
   References 20

2 Propositional Logic 21
   2.1 Linguistic Considerations 21
   2.2 Semantics; Truth Tables 29
      2.2.1 Validity 33
   2.3 Semantics; Logical (Valid) Consequence 37
      2.3.1 Decidability 40
      2.3.2 Sound versus Plausible Arguments; Enthymemes 41
   2.4 Semantics: Meta-logical Considerations 44
   2.5 About Truthfunctional Connectives 51
      2.5.1 Applications in Electrical Engineering and in Jurisdiction .. 53
      2.5.2 Normal Form*; Logic Programming* 55
      2.5.3 Travelling Salesman Problem (TSP)*; NP-completeness* ... 58
   2.6 Syntax: Provability and Deducibility 62
   2.7 Syntax: Meta-logical Results 71

		2.7.1	Deduction Theorem; Introduction and Elimination Rules ...	73
		2.7.2	Natural Deduction*	78
	2.8	Tableaux		82
	2.9	Completeness of classical propositional logic		93
	2.10	Paradoxes; Historical and Philosophical Remarks		97
		2.10.1	Paradoxes	97
		2.10.2	Historical and Philosophical Remarks	102
	2.11	Solutions		112
	References			128
3	**Sets: finite and infinite**			129
	3.1	Russell's Paradox		129
	3.2	Axioms of Zermelo-Fraenkel for Sets		132
	3.3	Historical and Philosophical Remarks		140
		3.3.1	Mathematics and Theology	140
		3.3.2	Ontology of mathematics	140
		3.3.3	Analytic-Synthetic	141
		3.3.4	Logicism	142
	3.4	Relations, Functions and Orderings*		144
		3.4.1	Ordered pairs and Cartesian product	144
		3.4.2	Relations	146
		3.4.3	Equivalence Relations	149
		3.4.4	Functions	151
		3.4.5	Orderings	156
		3.4.6	Structures and Isomorphisms	158
	3.5	The Hilbert Hotel; Denumerable Sets		162
	3.6	Non-enumerable Sets		168
	3.7	Solutions		175
	References			180
4	**Predicate Logic**			181
	4.1	Predicate Language		181
		4.1.1	Quantifiers, Individual Variables and Constants	182
		4.1.2	Translating English into Predicate Logic, Intended and Non-intended Interpretation	185
		4.1.3	Scope, Bound and Free Variables	188
		4.1.4	Alphabet and Formulas	189
	4.2	Semantics: Tarski's Truth Definition; Logical (Valid) Consequence		194
	4.3	Basic Results about Validity and Logical Consequence		204
		4.3.1	Quantifiers and Connectives	204
		4.3.2	Two different quantifiers	208
		4.3.3	About the axioms and rules for $\forall$ and $\exists$	209
		4.3.4	Predicate Logic with Function Symbols*	211
		4.3.5	Prenex Form*	212
		4.3.6	Skolemization, Clausal Form*	213

Contents

  4.4 Syntax: Provability and Deducibility 216
    4.4.1 Natural Deduction 221
    4.4.2 Tableaux 222
  4.5 Completeness, Compactness and Löwenheim-Skolem 228
    4.5.1 Undecidability 230
    4.5.2 Compactness and Löwenheim-Skolem Theorems 232
    4.5.3 Second-order Logic 234
    4.5.4 Skolem's Paradox 235
  4.6 Predicate Logic with Equality 237
  4.7 About the Relation of Logic with other Disciplines 242
    4.7.1 Logic and Philosophy of Language 242
    4.7.2 Logic and Philosophy of Science 244
    4.7.3 Logic and Artificial Intelligence; Prolog 246
    4.7.4 Aristotle's Organon 247
  4.8 Solutions 248
  References 260

5 **Arithmetic: Gödel's Incompleteness Theorems** 261
  5.1 Formalization of Elementary Number Theory 261
  5.2 Gödel's first Incompleteness Theorem 266
    5.2.1 Gödel-numbering 268
    5.2.2 Provability predicate for $\mathscr{P}$ 270
  5.3 Gödel's second Incompleteness Theorem 271
    5.3.1 Implications of Gödel's Incompleteness Theorems 272
  5.4 Non-standard Models of Peano's Arithmetic 272
    5.4.1 Second-order Logic (continued) 273
  5.5 Solutions 275
  References 276

6 **Modal Logic** 277
  6.1 Modal Operators 277
  6.2 Different systems of Modal Logic 279
  6.3 Possible World Semantics 282
  6.4 Epistemic logic 287
    6.4.1 Muddy Children Puzzle; Reasoning about Knowledge 288
  6.5 Tableaux for Modal Logics 290
  6.6 Applications of Possible World Semantics 296
    6.6.1 Direct Reference 296
    6.6.2 Rigid Designators 298
    6.6.3 De dicto - de re distinction 299
    6.6.4 Reasoning about Knowledge 300
    6.6.5 Common Knowledge 302
  6.7 Completeness of Modal Propositional Logic 303
  6.8 Strict Implication 309
  6.9 Counterfactuals 310

	6.10	Weak and Relevant Implication; Entailment*	313
	6.11	Modal Predicate Logic	315
		6.11.1 Modal Predicate Logic and Essentialism	316
	6.12	The Modal Logic *GL*	318
	6.13	Solutions	320
		References	327
7	**Philosophy of Language**	329
	7.1	Use and Mention	329
	7.2	Frege's Sinn und Bedeutung (Sense and Reference)	331
	7.3	Mannoury (1867-1956), Significs	335
	7.4	Speech Acts	340
	7.5	Definite Descriptions	342
	7.6	Berry's and Grelling's Paradox	344
	7.7	The Theory of Direct Reference	346
	7.8	Analytic - Synthetic	349
	7.9	Logicism	350
	7.10	Logical Positivism	351
	7.11	Presuppositions	353
	7.12	Wittgenstein on meaning	357
	7.13	Syntax - Semantics - Pragnatics	363
	7.14	Conversational Implicature	365
	7.15	Conditionals	366
	7.16	Leibniz	367
	7.17	De Dicto - De Re	369
	7.18	Grammars	370
	7.19	Solutions	374
		References	376
8	**Intuitionism and Intuitionistic Logic**	379
	8.1	Intuitionism vs Platonism; basic ideas	379
		8.1.1 Language	381
		8.1.2 First Steps in Intuitionistic Reasoning	382
	8.2	Intuitionistic Propositional Logic: Syntax	385
	8.3	Tableaux for Intuitionistic Propositional Logic	387
	8.4	Intuitionistic Propositional Logic: Semantics	393
	8.5	Completeness of Intuitionistic Propositional Logic	397
	8.6	Quantifiers in Intuitionism; Intuitionistic Predicate Logic	402
		8.6.1 Deducibility for Intuitionistic Predicate Logic	404
		8.6.2 Tableaux for Intuitionistic Predicate Logic	406
		8.6.3 Kripke Semantics for Intuitionistic Predicate Logic	407
		8.6.4 Soundness and Completeness	409
	8.7	Sets in Intuitionism: Construction Projects and Spreads	411
	8.8	The Brouwer Kripke axiom	416
	8.9	Solutions	417

Contents

References 426

9 Applications: Prolog; Relational Databases and SQL; Social Choice Theory 427
9.1 Programming in Logic 427
9.1.1 Recursion 430
9.1.2 Declarative versus Procedural Programming............. 432
9.1.3 Syntax 434
9.1.4 Matching versus Unification 436
9.1.5 Lists, Arithmetic 438
9.1.6 Cut 440
9.1.7 Negation as Failure................................ 443
9.1.8 Applications: Deductive Databases and Artificial Intelligence 444
9.1.9 Pitfalls 447
9.2 Relational Databases and SQL............................ 451
9.2.1 SQL 459
9.3 Social Choice Theory; Majority Judgment 462
9.3.1 Introduction..................................... 463
9.3.2 Plurality Rule (PR): most votes count 464
9.3.3 Majority Rule (MR): pairwise comparison 466
9.3.4 Borda Rule (BR).................................. 467
9.3.5 Outcome depends on the Voting Rule 468
9.3.6 Arrow's Impossibility Theorem 468
9.3.7 Domination 470
9.3.8 Majority Judgment (MJ) 471
9.3.9 Properties of Majority Judgment 473
9.3.10 Point Summing and Approval Voting.................. 474
9.3.11 Majority Judgment with many Voters 475
9.3.12 Presidential Elections in the USA.................... 475
9.3.13 Presidential Elections in France 477
9.3.14 Elections for Parliament in the Netherlands 479
9.4 Solutions 482
References 487

10 Fallacies and Unfair Discussion Methods 489
10.1 Introduction 489
10.2 Fallacies 491
10.2.1 Clichés and Killers 491
10.2.2 Improper or hasty Generalizations 493
10.2.3 Thinking simplistically 494
10.2.4 Appeal to ignorance 496
10.2.5 Speculative Thinking 497
10.2.6 Incredulity 499
10.2.7 The use of Terms with a vague Meaning 500
10.2.8 The Danger of Words with more than one Meaning 502

        10.2.9 Aprioristic Reasoning 504
        10.2.10 Circular Reasoning 504
        10.2.11 Applying double Standards 505
        10.2.12 Rationalizing 506
        10.2.13 After this, therefore because of this 508
    10.3 Unfair Discussion Methods 510
        10.3.1 Pushing someone into an extreme corner 510
        10.3.2 Straw man argument 511
        10.3.3 Diversion maneuvers 512
        10.3.4 Suggestive Methods 515
        10.3.5 Either/Or Fallacy 524
        10.3.6 The treacherous paradox 524
        10.3.7 Ad Hominem Arguments 525
        10.3.8 Argumentum ad baculum 527
        10.3.9 Secrecy 527
        10.3.10 The Retirement Home's Discussion 528
    10.4 Summary 529
    References 530

Index 531

Chapter 1
Logic; a First Impression

H.C.M. (Harrie) de Swart

Abstract In this introductory chapter the topic of the book is explained: distinguishing valid patterns of reasoning from invalid ones. The validity may depend on the meaning of connectives like 'if ..., then ...', 'and', 'or' and 'not', in which case one speaks of propositional logic. But the validity may also depend on the meaning of the quantifiers 'for all' and 'for some', in which case one speaks of predicate logic. If we extend the logical language with symbols for addition and multiplication of natural numbers, Gödel's famous incompleteness theorems show up. In order to have meaning, logical formulae presuppose a universe of discourse, or a set, which may be finite or infinite. In particular infinite sets have peculiar properties. If the validity of a reasoning pattern also depends on the meaning of modalities, like 'necessary' and 'possible', one speaks of modal logic. Modal logic helps to clarify or solve certain issues in the philosophy of language. It turns out that validity of an argument is also dependent on philosophical presuppositions. Changing the philosophical point of view may result in intuitionistic logic. The language of logic may be used as a programming language: Prolog (Programming in Logic); and the theory of sets is the basis for relational databases and the query language SQL; another application of logic is social choice theory. Fallacies and unfair discussion methods are abundantly present in daily discourse and hence deserve attention too.

1.1 General

The study of logic is the study of reasoning. The basic question in this book is what conclusions can be drawn with absolute certainty from a particular set of premises. To illustrate what we mean by this, let us consider Euclid's geometry.

Euclid (c. 330 B.C.) began his geometry books, called the 'Elements', with a precise formulation of the geometrical axioms (postulates, premises) on which he wanted to found his geometry. For instance, one of the axioms says that it is possible to draw a straight line through any two points. Next, Euclid used (informal) reasoning to deduce theorems from the geometrical axioms, for instance, the theorem that any triangle which is equiangular also is isosceles.

premisses (postulates, axioms)

reasoning (studied in logic)

conclusion (theorem)

In this book deductive logic is studied and not probabilistic logic which studies the question what conclusions can be drawn from a set of premisses with a certain probability. An example of the latter is, for instance, the question how likely it is that a person gets a certain disease when he has been in touch with other people having the disease.

Logic has a long history: it was studied by the Stoics (see [1, 5, 10, 12]), by Aristotle (see [1, 10, 11]) and by many medieval philosophers (see [1, 2, 10, 13]); the study of logic was greatly advanced by the works of Boole (1847, 1854) [3, 4], Frege (1879) [6, 7] and Russell (1910) [14], becoming a full-fledged discipline with the work of Gödel (1930-1931) [9, 15].

In addition to the term 'logic', one also encounters in the literature the expressions 'mathematical logic', 'philosophical logic' and 'formal (or symbolic) logic', which are used to stress one of the many aspects of logic.

1.2 Propositional Logic

Below we give some concrete simple arguments from different fields.

Example 1.1.

a1)   If $1 = 2$, then I am the Pope of Rome.
I am not the Pope of Rome.
Therefore: not $1 = 2$.

a2)   If $1 = 2$, then I am the Pope of Rome.
Not $1 = 2$.
Therefore: I am not the Pope of Rome.

b1)   If triangle ABC is equiangular, then it is isosceles.
Triangle ABC is not isosceles.
Therefore: Triangle ABC is not equiangular.

b2)   If triangle ABC is equiangular, then it is isosceles.
Triangle ABC is not equiangular.
Therefore: Triangle ABC is not isosceles.

1.2 Propositional Logic

c1) If it snows, then it is cold.
It is not cold.
Therefore: It does not snow.

c2) If it snows, then it is cold.
It does not snow.
Therefore: It is not cold.

Note that all the arguments above consist of two *premisses* and one (putative) conclusion. Further note that all arguments a1, b1 and c1 in Example 1.1 have the same structure, namely, the following *pattern of reasoning*:

1. if P_1, then P_2 $P_1 \to P_2$
    not P_2 $\neg P_2$
    therefore: not P_1 $\neg P_1$

Using $\to$ for 'if ..., then ...' and $\neg$ for 'not', this pattern of reasoning can be represented by the schema to the above right. This pattern of reasoning is called *Modus Tollens*.

The arguments a2, b2 and c2 in Example 1.1 also have the same pattern, namely,

if P_1, then P_2 $P_1 \to P_2$
not P_1 $\neg P_1$
therefore: not P_2 $\neg P_2$

The first pattern of reasoning is *valid*, i.e., it is impossible to replace P_1, P_2 by such propositions that the premisses $P_1 \to P_2$ and $\neg P_2$ result in true propositions and that at the same time the conclusion $\neg P_1$ results in a false proposition. For suppose P_1, P_2 are interpreted as propositions P_1^* (e.g., it snows) and P_2^* (e.g., it is cold) respectively and suppose that

'if P_1^*, then P_2^*' (if it snows, then it is cold) and 'not P_2^*' (it is not cold) are both true.

Then 'not P_1^*' (it does not snow) must be true too. For suppose that P_1^* (it snows) would be true; then – by the first premiss – P_2^* (it is cold) would be true too. This is a contradiction with the second premiss 'not P_2^*' (it is not cold).

Note that this insight does not depend on the particular choice of P_1^* and P_2^*. P_1^* and P_2^* may be any propositions from number theory, geometry, economics, philosophy, from daily life, and so on.

Concrete arguments which have an underlying pattern of reasoning which is valid are called *correct arguments*. Thus the arguments a1, b1 and c1 in Example 1.1 are correct, since they are particular instances of the valid pattern 1:

$$\frac{P_1 \to P_2 \quad \neg P_2}{\neg P_1}$$

We say that $\neg P_1$ is a logical (or valid) consequence of $P_1 \to P_2$ and $\neg P_2$. Notation: $P_1 \to P_2, \neg P_2 \models \neg P_1$.

We know that it is impossible for the premises of a correct argument to be true and at the same time its conclusion to be false. Whether the premises and the conclusion of a concrete argument are true or false is not the business of the logician, but of the mathematician, the economist, the philosopher, the physicist, and so on, depending on what these propositions are about. The logician is not concerned with the truth or falsity of the axioms of geometry. Given a concrete argument, he is only concerned with the validity or invalidity of the underlying pattern of reasoning and if this is valid, he can only say that *if* the premises of the concrete argument in question are true, *then* the conclusion must likewise be true.

Warning: If a pattern of reasoning is valid, a concrete argument with that pattern does not imply that the premises are true, nor that the conclusion is true.

Counterexample Pattern 1 $\quad\dfrac{P_1 \to P_2 \quad \neg P_2}{\neg P_1}\quad$ is a valid pattern of reasoning.

Now take P_1^*: Bill Gates is wealthy.
$\quad\quad P_2^*$: Bill Gates owns all the gold in Fort Knox.
Then we get the following concrete argument:
$\quad\quad$ If Bill Gates is wealthy, then he owns all the gold in Fort Knox.
$\quad\quad$ Bill Gates does not own all the gold in Fort Knox.
$\quad\quad$ Therefore: Bill Gates is not wealthy.
So, we have a correct argument, since the underlying pattern is valid, with a false conclusion. This is only possible if at least one of the premises is false. And indeed, the first premise is actually false. Correctness of a concrete argument means that it is impossible that all the premises are true and at the same time the conclusion false, in other words: *if* all premises are true (which actually may not be the case), *then* the conclusion must be true too.

From the definition of validity it follows that a pattern of reasoning is *invalid* if it is possible to interpret $P_1, P_2, \ldots$ in such a way that all premises result in true propositions while at the same time a false one results from the conclusion. An example of an invalid pattern is the following one:

$$\dfrac{P_1 \to P_2 \quad \neg P_1}{\neg P_2}$$

underlying the concrete arguments a2, b2 and c2 in Example 1.1. Taking
$\quad\quad P_1^*$: Bill Gates owns all the gold in Fort Knox,
$\quad\quad P_2^*$: Bill Gates is wealthy,
results in the following concrete argument :
$\quad\quad$ If Bill Gates owns all the gold in Fort Knox, then he is wealthy.
$\quad\quad$ Bill Gates does not own all the gold in Fort Knox.
$\quad\quad$ Therefore: Bill Gates is not wealthy.
So, all the premises are true, while the conclusion is false.

We say that $\neg P_2$ is not a logical (or valid) consequence of $P_1 \to P_2$ and $\neg P_1$. Notation: $P_1 \to P_2, \neg P_1 \not\models \neg P_2$.

1.2 Propositional Logic

Concrete arguments with an underlying pattern of reasoning which is invalid are called *incorrect*. So, the arguments a2, b2 and c2 in Example 1.1 are incorrect.

Warning: A concrete argument with an underlying pattern of reasoning which is invalid does not necessarily imply that the conclusion is false; the conclusion may be true, but in that case the truth of the conclusion does not depend on the truth of the premises.

Counterexample: The pattern $\dfrac{\begin{array}{c} P_1 \to P_2 \\ \neg P_1 \end{array}}{\neg P_2}$ is an invalid pattern of reasoning.

Taking P_1^*: I own all the gold in Fort Knox,
    P_2^*: I am wealthy,
we obtain the following concrete incorrect argument with true premises and a true conclusion:

  If I own all the gold in Fort Knox, then I am wealthy.
  I do not own all the gold in Fort Knox.
  Therefore: I am not wealthy.

Below is a non exhaustive list of valid patterns of reasoning frequently used in practice:

Example 1.2 (some valid patters of reasoning).

1. if P_1, then P_2 $\dfrac{\begin{array}{c} P_1 \to P_2 \\ \neg P_2 \end{array}}{\neg P_1}$ *Modus Tollens*
    not P_2
    therefore: not P_1

2. if P_1, then P_2 $\dfrac{\begin{array}{c} P_1 \to P_2 \\ P_1 \end{array}}{P_2}$ *Modus Ponens (MP)*
    P_1
    therefore: P_2

3. P_1 if and only if (iff) P_2 $\dfrac{\begin{array}{c} P_1 \rightleftarrows P_2 \\ \neg P_1 \end{array}}{\neg P_2}$
    not P_1
    therefore: not P_2

4. not (P_1 and P_2) $\dfrac{\begin{array}{c} \neg(P_1 \land P_2) \\ P_1 \end{array}}{\neg P_2}$
    P_1
    therefore: not P_2

5. P_1 or P_2 $\dfrac{\begin{array}{c} P_1 \lor P_2 \\ \neg P_2 \end{array}}{P_1}$
    not P_2
    therefore: P_1

We have introduced above $\rightleftarrows$ for 'if and only if (iff)', $\land$ for 'and', $\lor$ for the inclusive 'or', i.e., $P_1 \lor P_2$ stands for 'P_1 or P_2 or both P_1 and P_2'. The reader should verify that all patterns in Example 1.2 are valid.

The following two patterns of reasoning are frequently used in practice, although they are invalid:

if P_1, then P_2 $\qquad\qquad P_1 \to P_2$
not P_1 $\qquad\qquad\qquad\quad \neg P_1$
therefore: not P_2 $\qquad\overline{\qquad\neg P_2\qquad}$

if P_1, then P_2 $\qquad\qquad P_1 \to P_2$
P_2 $\qquad\qquad\qquad\qquad\;\; P_2$
therefore: P_1 $\qquad\quad\overline{\qquad P_1\qquad}$

So, the following concrete arguments are not correct:

> If it rains, then the street becomes wet.
> It does not rain.
> Therefore: The street does not become wet.

> If it is raining, then the street becomes wet.
> The street becomes wet.
> Therefore: It is raining.

It should now be clear that the expressions in patterns of reasoning are built from $P_1, P_2, P_3, \ldots$ using the connectives $\rightleftarrows, \to, \wedge, \vee$ and $\neg$. In fact, we have introduced a new language for representing patterns of reasoning, the alphabet of which consists of the symbols:

$P_1, P_2, P_3, \ldots$ $\qquad\qquad$ called atomic formulas
$\rightleftarrows, \to, \wedge, \vee, \neg$ $\qquad\qquad$ called connectives
$(\,,\,)$ $\qquad\qquad\qquad\quad$ called parentheses.

Of course, $\wedge P_1 P_2 \neg$ is not a well-formed expression of this language. Let us define how the well-formed expressions or formulas of this language are built up.

Formulas:
1. $P_1, P_2, P_3, \ldots$ are formulas. In other words, if P is an atomic formula, then P is a formula.
2. If A and B are formulas, then $(A \rightleftarrows B)$, $(A \to B)$, $(A \wedge B)$ and $(A \vee B)$ are formulas.
3. If A is a formula, then $(\neg A)$ is a formula too.

Example 1.3. P_1, P_3 and P_5 are formulas.
$(\neg P_1)$ and $(P_3 \to P_5)$ are formulas.
$((\neg P_1) \vee (P_3 \to P_5))$ is a formula.

We can minimize the need for parentheses by agreeing that we leave out the most outer parentheses in a formula and that in

$$\rightleftarrows, \to, \wedge, \vee, \neg$$

any connective has a higher rank than any connective to the right of it and a lower rank than any connective to the left of it. According to this convention $\neg P_1 \to P_2 \vee P_3$ means $(\neg P_1) \to (P_2 \vee P_3)$, because $\to$ has a higher rank than $\neg$ and $\vee$, and it does not mean $((\neg P_1) \to P_2) \vee P_3$ nor $\neg((P_1 \to P_2) \vee P_3)$. According to the convention just mentioned the expression $\neg P_1 \vee P_3 \to P_5$ stands for the formula $((\neg P_1) \vee P_3) \to P_5$,

1.2 Propositional Logic

because $\rightarrow$ has the highest rank and $\vee$ has a higher rank than $\neg$. Notice that the formula $(\neg P_1) \vee (P_3 \rightarrow P_5)$ is a different formula with a quite different meaning.

It is important to notice that the validity or invalidity of the reasoning patterns above does not depend on the content of the P_1, P_2, but solely on the meaning of the connectives $\rightleftarrows$, $\rightarrow$, $\wedge$, $\vee$ and $\neg$. In *propositional logic* one studies the (in)validity of reasoning patterns of which the (in)validity is completely determined by the meaning of the connectives between the propositions in question.

In Chapter 2 a characterization of validity is given both in semantic and in syntactic terms, and it is shown that these two characterizations are equivalent, which gives us confidence that we have given an adequate definition of the notion in question.

In logic we study the validity or invalidity of patterns of reasoning. The expressions in these patterns are formulas of the language specified above. This language is called the *object-language*, because it is the object of study. The language used in studying the object-language is called the *meta-language* or the *observer's language*. In our case the meta-language will be part of English. The situation is similar to the one where an English speaking person is studying Russian, in which case Russian is the object language and English is the meta-language. It is important to keep in mind this distinction between the object-language and the meta-language; otherwise, one may get involved in paradoxes like the antinomy of the liar.

That intuition is not always a reliable guide in judging correctness of a given argument will become clear from a few examples. At the end of this section are a few exercises in which the reader is challenged to judge on intuitive grounds whether the argument given is correct. Although the arguments are simple, they are sufficiently complex to puzzle an untrained intuition. When the reader has finished Chapter 2 he or she will be able to judge the correctness of these arguments with certainty!

Exercise 1.1. Check whether the following argument is correct by translating the propositions in the argument into the language of propositional logic and by determining whether the corresponding pattern of reasoning is valid.
If Socrates did not die of old age [$\neg O$], then the Athenians sentenced him to death [D].
The Athenians did not sentence Socrates to death.
If Socrates died from poison [P], then he did not die of old age.
Therefore: Socrates did not die from poison.

Exercise 1.2. Check whether the following argument is correct.
If the weather is nice [N], then John will come. [J].
The weather is not nice.
Therefore: John will not come,

Exercise 1.3. Check whether the following argument is correct.
John comes [J] if the weather is nice [N].
John comes.
Therefore: the weather is nice.

Exercise 1.4. Check whether the following argument is correct.
John comes [*J*] *only if* the weather is nice [*N*].
John comes.
Therefore: the weather is nice.

Exercise 1.5. Check whether the following argument is correct.
It is not the case that John gets promotion [*P*] and at the same time not a higher salary [¬*S*].
John does not get promotion or he is not diligent [¬*D*].
John is diligent.
Therefore, John will not get a higher salary.

1.3 Sets; Finite and Infinite

The quantifiers ∀ (for all *x*) and ∃ (for some *x*) in (the language of) predicate logic are ranging over a certain domain: the set of all persons, the set of all natural numbers, the set of all real numbers, etc. In fact, there are many possible domains, where a domain is just a set of objects. These sets may be finite, like the set consisting of Ann, Bob and Coby, or the set {1, 2, 3} consisting of the numbers 1, 2 and 3, but they may be also infinite, like the set $\mathbb{N}$ of all natural numbers. We will study these sets more closely in Chapter 3 with particular attention for the properties of infinite sets. As we shall see, infinite sets have properties quite different from the properties of finite sets. For instance, a proper part of a finite set will be smaller than the original set. But as we shall see in Chapter 3, this property does not hold for infinite sets: a proper part of an infinite set may be equally large as the original set. A simple example is the set $\mathbb{N}_{even} = \{0, 2, 4, 6, \ldots\}$ of the even natural numbers which is a proper subset of the set $\mathbb{N} = \{0, 1, 2, 3, 4, 5, 6, \ldots\}$ of all natural numbers. That these sets are equally large may be seen as follows: there is a one-one correspondence between the elements of both sets.

$$\begin{array}{c} \mathbb{N}: \ 0 \ \ 1 \ \ 2 \ \ 3 \ \ 4 \ \ \ldots \\ | \ \ | \ \ | \ \ | \ \ | \ \ \ldots \\ \mathbb{N}_{even}: \ 0 \ \ 2 \ \ 4 \ \ 6 \ \ 8 \ \ \ldots \end{array}$$

1.4 Predicate Logic

An example of a simple argument which we cannot adequately analyse with the means developed in propositional logic, is the following:

    All men are mortal.
    Socrates is a man.
    Therefore, Socrates is mortal.

1.4 Predicate Logic

If we translate this argument in the formal language of propositional logic, we find as the underlying pattern of reasoning:

$$\frac{\begin{array}{c}P_1\\P_2\end{array}}{P_3}$$

and we know this pattern is invalid since we can substitute true propositions for P_1 and P_2 and at the same time a false one for P_3. On the other hand, it seems to us that the argument above, about Socrates, is correct.

The point is that in the translation of the premises into P_1 and P_2 and of the conclusion into P_3, the internal structure of the sentences is lost: P_1, P_2 and P_3 are unrelated atomic formulas. But the premises and the conclusion of the argument are not unrelated; in fact, it is this relationship which causes the argument to be correct. We have to exhibit the internal subject-predicate structure of the premises and the conclusion in order to make visible that these three sentences are related and in order to see that the underlying pattern of reasoning is valid.

The structure of the argument above is the following pattern:

For all objects x, if x is a person, then x is mortal. $\forall x[P(x) \to M(x)]$
Socrates is a person. $\dfrac{P(c)}{M(c)}$
Therefore: Socrates is mortal.

Using $\forall x$ for 'for all x', $P(x)$ for 'x has the property P (to be a Person)', $M(x)$ for 'x has the property M (to be Mortal)' and c for 'Socrates', this pattern of reasoning can be represented by the schema to the above right.

Notice that the following arguments have the same underlying pattern of reasoning:

All philosophers are smart   All natural numbers are positive
John is a philosopher   5 is a natural number
Therefore, John is smart   Therefore, 5 is positive

The pattern just mentioned is *valid*, i.e., it is impossible to choose a domain of individuals and to give to P, M and c appropriate meanings such that from the premises $\forall x[P(x) \to M(x)]$ and $P(c)$ true propositions result and at the same time from the conclusion $M(c)$ a false proposition.

But, for instance, the pattern

$$\frac{\forall x[P(x) \to M(x)]}{P(c)}\;M(c)$$

is *invalid*, since it is possible to choose a domain, to interpret the symbols P, M as predicates P^*, M^* over the domain chosen and to interpret the symbol c as an element c^* in the domain, such that true propositions result from the premises and a false proposition from the conclusion. For instance, take as domain the set of all persons, let P^* be the predicate 'is a man', M^* the predicate 'is mortal' and let c^* be the element 'Queen Maxima'. Then $\forall x[P(x) \to M(x)]$ yields the true proposition: For every person x, if x is a man, then x is mortal. $M(c)$ yields the true proposition: Queen Maxima is mortal. But $P(c)$ yields the false proposition: Queen Maxima is a man.

Next consider the following elementary argument:

> John is ill
> Therefore: someone is ill.

In order to exhibit the structure of this argument, we need one more symbol: $\exists x$, for 'there is at least one x such that ...'. Then the underlying pattern of reasoning of this argument is the following:

$$\frac{I(c)}{\exists x[I(x)]}$$

This pattern of reasoning is again valid: it is impossible to take a domain D and to interpret the symbol I as a predicate I^* over D and the symbol c as an individual c^* in D such that a true proposition (c^* has the property I^*) results from the premiss $I(c)$ and at the same time a false proposition (there is at least one individual which has the property I^*) from the conclusion $\exists x[I(x)]$.

Note that the following arguments also have the same (valid) underlying pattern of reasoning and hence are correct.

> 5 is odd
> Therefore: some natural number is odd

> Peter is rich
> Therefore: someone is rich

In order to be able to exhibit the internal subject-predicate structure of atomic sentences and the mutual relationships between them, we need the following symbols:

SYMBOLS	NAME	MEANING
$x,\ldots$	individual variables	individuals in a given domain
$P, M, I, \ldots$	predicate symbols	predicates over the given domain
$c,\ldots$	individual constants	concrete individuals in the given domain
$\rightleftarrows; \rightarrow; \wedge; \vee; \neg$	connectives	iff; if ..., then ...; and; or; not
$\forall, \exists$	quantifiers	for all; there exists
$[,],(,)$	parentheses	

In fact, we have introduced a new (subject-) predicate language, richer than the former propositional language, in which we can translate the subject-predicate structure of concrete arguments, exhibiting the underlying pattern of reasoning.

Of course, $P\exists\forall\neg$ is not a well-formed expression of this language and we have to define precisely what the well-formed expressions or formulas of this language are. We shall do so in Chapter 4; for the moment it is sufficient to work with a not precisely defined notion of formula.

It turns out that one can select a few elementary steps of reasoning, among which

$$\frac{A \quad A \rightarrow B}{B} \text{ called Modus Ponens,} \quad \frac{A \wedge B}{B}, \quad \frac{A}{A \vee B}, \quad \frac{\forall x[A(x)]}{A(t)},$$

such that every valid pattern of reasoning, no matter how complex, can be built up from these elementary steps. This is Gödel's *Completeness Theorem*, 1930.

1.4 Predicate Logic

For instance, the following correct argument can be built up from the elementary steps just specified.

> John loves Jane and John is getting married.
> If John is getting married, then he is looking for another job.
> Hence: John is looking for another job or he does not love Jane.

The underlying pattern of reasoning is:
$$\frac{P_1 \wedge P_2}{P_2 \to P_3}$$
$$P_3 \vee \neg P_1$$

And indeed, this pattern can be built up from the elementary steps specified above as follows:

$$\frac{\dfrac{\text{premiss}}{P_1 \wedge P_2}\quad \dfrac{\text{premiss}}{P_2 \quad P_2 \to P_3}}{\dfrac{P_3}{P_3 \vee \neg P_1}}$$

And the four elementary steps of reasoning specified above can be supplemented by a few more elementary steps to form what is called Gentzen's [8] system of *Natural Deduction* – to be discussed in Subsection 2.7.2 – such that every correct argument can be simulated by an appropriate combination of the elementary steps in Gentzen's system (1934-5). We shall prove Gödel's completeness theorem in Chapter 2 for propositional logic and in Chapter 4 for predicate logic.

Another example: the argument above about Socrates which has as its underlying pattern of reasoning

$$\frac{\forall x[P(x) \to M(x)]}{P(c)}$$
$$M(c)$$

can be built up from the elementary steps in the system of Natural Deduction as follows:

$$\frac{\text{premiss}}{P(c)} \quad \frac{\dfrac{\text{premiss}}{\forall x[P(x) \to M(x)]}}{P(c) \to M(c)}$$
$$M(c)$$

The schema above is called a logical *deduction* (in the system of Natural Deduction) of $M(c)$ from the premisses $\forall x[P(x) \to M(x)]$ and $P(c)$. We say that $M(c)$ is logically *deducible from* $\forall x[P(x) \to M(x)]$ and $P(c)$, since such a logical deduction exists.

In Chapter 4 a characterization of validity is given both in semantic and in syntactic terms, and it is shown that these two characterizations are equivalent, which gives us confidence that we have given an adequate definition of the notion in question.

Exercise 1.6. Check whether the following argument is correct by translating the propositions in the argument into the language of predicate logic and by determining

whether the corresponding pattern of reasoning is valid.
All gnomes have a beard or a conical cap.
Therefore: all gnomes have a beard or all gnomes have a conical cap.

Exercise 1.7. Check whether the following argument is correct.
All gnomes with a beard have a conical cap.
All gnomes have a beard.
Therefore: all gnomes have a conical cap.

Exercise 1.8. Check whether the following argument is correct.
There is a gnome with a beard.
There is a gnome with a conical cap.
Therefore: there is a gnome with a beard and a conical cap.

Exercise 1.9. Check whether the following argument is correct.
There is at least one gnome such that he has no beard or he has a conical cap.
There is at least one gnome who has a beard.
Therefore: there is at least one gnome who has a conical cap.

1.5 Arithmetic; Gödel's Incompleteness Theorem

In Chapter 2 we shall see that it is possible to fully capture the meaning of the logical connectives in terms of certain logical axioms. For instance, the meaning of the connective $\wedge$ can be fully captured by the following logical axioms: $A \wedge B \to A$, $A \wedge B \to B$ and $A \to (B \to A \wedge B)$. In other words, the propositional connectives can be characterized by appropriate logical axioms. This is expressed by the completeness theorem for propositional logic.

This result can be extended to predicate logic. In Chapter 4 we shall see that the meaning of the quantifiers $\forall$ and $\exists$ may also be fully captured by certain logical axioms. For instance, the meaning of $\forall$ is fully captured by the logical axioms $\forall x[A(x)] \to A(t)$, where t is either an individual variable or an individual constant, and $A(y) \to \forall x[A(x)]$, assuming there are no restrictions on the individual variable y. Gödel's completeness theorem for predicate logic (1930) expresses that the propositional connectives and the quantifiers can be characterized by appropriate logical axioms and rules.

Now, if we add to the logical language symbols $+$ and $\times$ to render addition and multiplication of natural numbers, naturally the question arises whether we may fully capture the meaning of these symbols in terms of certain arithmetical axioms, like $x + 0 = x$ and $x + sy = s(x+y)$, where sy denotes the successor of y. Amazingly, Kurt Gödel [9] proved in 1931 that it is impossible to fully capture the meaning of $+$ and $\times$ by arithmetical axioms. This is his famous Incompleteness theorem. This result has far reaching philosophical consequences.

We shall present Gödel's result and its philosophical implications in Chapter 5.

1.6 Modal Logic

The language of propositional and predicate logic may be further extended with a symbol $\Box$ for modalities, like necessary, obligatory, knowing that, etc. Depending on the precise meaning of the modality one may add several logical axioms for these modalities. For instance, $\Box A \to A$, in case $\Box$ stands for 'necessary' or for 'knowing that'. But for the modality 'obligatory' the axiom $\Box A \to A$ seems to be inappropriate: it is obligatory to stop for a red traffic light, but that does not imply that one actually does so. Since these modalities are used in several philosophical arguments, it is worthwhile to give a logical analysis of them.

By defining $\Diamond A$ by $\neg \Box \neg A$ we get modalities like 'possibly': $\neg A$ is not necessary, in other words, A is possible.

In Chapter 6 we will adapt the notions of validity and deducibility to modal logic and show that these two notions are again equivalent, just as in propositional and predicate logic. However, the notion of validity is now more complicated, since it is given in terms of possible worlds. $\Box A$ (A is necessary, or knowing A) is true in a given world means that A is true in all worlds imaginable from that given world. And $\Diamond A$ (A is possible) is true in a given world means that A is true in at least one world imaginable from that given world.

1.7 Philosophy of Language

In Chapter 7 we shall see that several problems in the philosophy of language are better understood or may be clarified by using the notion of possible world.

For instance, the *de re - de dicto* distinction in a sentence like 'it is possible that a republican will win' may be made clear by giving two different logical translations of this sentence:

de re: $\exists x[R(x) \land \Diamond W(x)]$: there is an individual x in the actual world w such that x is a Republican in world w and such that there is a world w' (imaginable from the actual world w) in which x wins.

de dicto: $\Diamond \exists x[R(x) \land W(x)]$: there is a world w' (imaginable from the actual world w) in which an individual x exists who is a Republican in that world w' and who wins in that world w'.

In the *de re* version the modality $\Diamond$ is within the scope of the existential quantifier $\exists$, while in the *de dicto* version the existential quantifier $\exists$ is within the scope of the modality $\Diamond$.

Another example is the difference between a name like 'Aristotle' and the corresponding description, like 'the most well known student of Plato'. Traditionally these two expressions were identified. But that causes the problem that a sentence like 'Aristotle is the most well known student of Plato' would be nothing more than a logical truth, or, using Kant's terminology, an analytic statement. Kripke proposed to solve this problem by conceiving proper names like 'Aristotle' as a rigid designator, i.e., as referring in all possible worlds to the same object. While the name 'Aristotle' refers in all possible worlds to the same object, also in the world in which he actually was a carpenter instead of a philosopher, the description 'the most well known student of Plato' may refer to different objects in different worlds. The description 'the most well known student of Plato' may help us to pick the proper reference of the name 'Aristotle', but it should not be identified with the name 'Aristotle'.

1.8 Intuitionism and Intuitionistic Logic

A *classical mathematician* studies the properties of mathematical objects like an astronomer, who studies the properties of celestial bodies. From a classical point of view, mathematical objects are like celestial bodies in the sense that they exist independently of us; they are created by God.

An *intuitionist* creates the mathematical objects himself. According to *Brouwer's intuitionism*, mathematical objects, like 5, 7, 12 and +, are mental constructions. A proposition about mathematical objects (like $5 + 7 = 12$) is true if one has a proof-construction that establishes it. Such a proof is again a mental construction.

> Mathematics is created by a free action, independent of experience [L.E.J. Brouwer, *Collected Works*, Vol. 1, p. 97].

Since, intuitionistically, the truth of a mathematical proposition is established by a proof – which is a particular kind of mental construction –, the meaning of the logical connectives has to be explained in terms of proof-constructions.

A proof of $A \wedge B$ is anything that is a proof of A and of B.

A proof of $A \vee B$ is, in fact, a proof either of A or of B, or yields an effective means, at least in principle, for obtaining a proof of one or other disjunct.

A proof of $A \rightarrow B$ is a construction of which we can recognize that, applied to any proof of A, it yields a proof of B. Such a proof is therefore an operation carrying proofs into proofs.

Intuitionists consider $\neg A$ as an abbreviation for $A \rightarrow \bot$, postulating that nothing is a proof of $\bot$ (falsity).

It follows that from an intuitionistic point of view it is reckless to assume $A \vee \neg A$. The validity of $A \vee \neg A$ means, intuitionistically, that we have a method adequate in principle to solve any mathematical problem A. However, consider *Goldbach's conjecture*, G, which states that each even number is the sum of two odd primes: $2 = 1+1, 4 = 3+1, 6 = 5+1, 8 = 7+1, 10 = 7+3, 12 = 7+5, 14 = 7+7, 16 = 13+3, 18 = 13+5, \ldots$. One can check only finitely many individual instances, while Goldbach's Conjecture is a statement about infinitely many (even) natural numbers. So far neither Goldbach's Conjecture, G, nor its negation, $\neg G$, has been proved. An intuitionist is therefore not in a position to affirm $G \vee \neg G$. A person who claims that he or she can provide a proof either of G or of $\neg G$ is called *reckless*.

Notice that from a classical point of view $A \vee \neg A$ is valid, since A is a statement about mathematical objects created independently of us, for which either A or $\neg A$ holds, although we may not know which one.

1.9 Applications

In Chapter 8 we will elaborate on Brouwer's intuitionism and see that his different philosophical point of view about the nature of mathematical objects results in a logic which is much more fine-grained, but also more difficult, than classical logic.

1.9 Applications

1.9.1 Programming in Logic: Prolog

Since Gödel's completeness theorem expresses that every valid pattern of reasoning can be built up from a certain small collection of logical rules in a logical proof-system (such as the system of Natural Deduction), the idea to equip a computer with these logical rules is quite natural. If we do so, the computer will be able to simulate reasoning and hence disposes of *Artificial Intelligence*. By adding to such a computer-program a number of data $A_1, \ldots, A_n$, concerning a small and well-described subject, the so-called *knowledge base*, the computer is able to draw conclusions from those data. If $A_1, \ldots, A_n$ represent someone's expertise, one speaks of an *expert system*. And if the knowledge base consists of Euclid's axioms for geometry or Peano's axioms for number theory or of axioms for some other part of mathematics, one speaks of *automated theorem proving*.

It was only in the early 1970's that the idea emerged to use the formal language of logic as a programming language. An example is PROLOG, which stands for PROgramming in LOGic. A logic program is simply a set of formulas (of a particular form) in the language of predicate logic. The formulas below constitute a logic program for kinship relations. The objects are people and there are two binary predicates 'parent of' (p), and 'grandparent of' (g).

- A_1: p(art, bob).
- A_2: p(art, bud).
- A_3: p(bob, cap).
- A_4: p(bud, coe).
- A_5: $g(x,z)$:- $p(x,y)$, $p(y,z)$.

'art', 'bob', 'bud', 'cap' and 'coe' are individual constants and A_5 stands for $p(x,y) \wedge p(y,z) \rightarrow g(x,z)$. Now if we ask the question

$$?\text{-} g(\text{art}, \text{cap})$$

the answer will be 'yes', corresponding with the fact that g(art, cap) can be logically deduced from the premises or data $A_1, \ldots, A_5$.

But if we ask the question

$$?\text{-} g(\text{art}, \text{amy})$$

the answer will be 'no', corresponding with the fact that g(art, amy) cannot be logically deduced from $A_1, \ldots, A_5$. Note that this does not mean that $\neg g$(art, amy) logically follows from $A_1, \ldots, A_5$.

And if we ask the question

$$?\text{-} g(\text{art}, X)$$

the answer will be $X = \text{cap}$, $X = \text{coe}$.

Once we have observed that data can be translated into formulas in the formal language of logic and that queries concerning the objects in the data – again translated into formulas – can be answered with 'yes' or 'no', depending on whether the putative conclusion can or cannot be logically deduced from the given data, it becomes clear that there is an interesting connection between logic and databases.

In Chapter 9 we shall study more closely how the language of logic may be used as a programming language in the context of artificial intelligence.

1.9.2 Relational Databases

The theory of finite sets is the basis for *relational databases*, which we shall present in Chapter 9. In fact, the query language SQL formulates questions to the database in terms of sets. To illustrate, suppose we have a table P with patients containing their number (nmb), name (nm), address (addr), residence (res) and gender (gen).

	nmb	nm	addr	res	gen
t	$t(\text{nmb})$	$t(\text{nm})$	$t(\text{addr})$	$t(\text{res})$	$t(\text{gen})$

Each row in the table, called a tuple t, represents one patient. Mathematically, a tuple t assigns to every attibute nmb, nm, addr, res, gen a value $t(\text{nmb})$, $t(\text{nm})$, $t(\text{addr})$, $t(\text{res})$, $t(\text{gen})$ in a predefined domain. Then

$$\{\, t(\text{nm}) \mid t \in \mathsf{P} \mid t(\text{res}) = \text{'Princeton'} \wedge t(\text{gen}) = \text{'male'} \,\}$$

is the set of all names of patients in table P who live in Princeton and are male.

This set is generated by the Structured Query Language SQL as follows:
SELECT t.nm
FROM P t
WHERE t.res = 'Princeton'
AND t.gen = 'male'

1.9.3 Social Choice Theory

In social choice theory one studies how individual preferences or evaluations should be aggregated to a common (group or social) preference or evaluation respectively. That this is problematic may be demonstrated by the following simple example. Suppose there are nine voters (or judges) who have the following preferences over

three candidates or alternatives a, b and c:
4: $a\,b\,c$
3: $b\,c\,a$ That is, the first
2: $c\,b\,a$

four voters prefer a to b, b to c and also a to c; similarly for the other voters.

If we apply Plurality Rule (PR) or 'most votes count', only the most preferred candidate is taken into account. So, a has four first votes, b three and c only two. Consequently, the common or social ranking under PR will be: $a\,b\,c$.

If we apply Majority Rule (MR) or pairwise comparison, we see that $3 + 2 = 5$ voters, hence a majority, prefer b and c to a; and that $4 + 3 = 7$ voters prefer b to c. So, under Majority Rule the common or social ranking will be: $b\,c\,a$.

Many other voting rules exist, which will all lead to different outcomes. But already at this stage we see that the outcome depends on the aggregation rule, rather than on the preferences of the voters.

Another problem is that all familiar voting rules may yield an outcome which is counter-intuitive. For instance, Plurality Rule makes a the winner, while a for a majority of the voters is the least preferred candidate. And Majority Rule in some cases does not even yield a winner, for instance, when there are three voters with the following preferences 1: $a\,b\,c$; 1: $b\,c\,a$ and 1: $c\,a\,b$. So the question arises whether there exists a voting rule that has only nice properties. This question was answered negatively by K. Arrow in 1951: there cannot exist a voting rule, which takes individual preferences as input, that satisfies certain desirable properties among which being non-dictatorial. This *impossibility theorem* has puzzled the social choice community, consisting of political scientists, economists, mathematicians and philosophers, ever since.

However, in 2010 Balinski and Laraki pointed out that the framework of Arrow, in which voters are supposed to give a preference ordering, is ill conceived. Voters should be asked to give evaluations of the candidates, for instance in terms of 'excellent', 'good', 'acceptable', 'poor' and 'reject'. Notice that evaluations are much more informative than preference orderings. Next, Balinksi and Laraki present a voting rule, called *Majority Judgment* (MJ), which takes evaluations of the candidates by the voters as input and yields a social ranking of the candidates as output. This Majority Judgment does satisfy the desired properties.

In Section 9.3 we shall discuss Plurality Rule, Majority Rule and the Borda Rule and show that they all violate one or more of the desired properties. Also a version of Arrow's theorem will be proved. Next we present Balinski and Laraki's Majority Judgment and show that it does satisfy the desired properties.

1.10 Fallacies and Unfair Discussion Methods

For many discussions and meetings it holds that they are led perfectly from a formal, procedural and technical point of view, but that the quality of the in-depth discussion is poor. The cause of poor thinking should be sought in the weakness of human nature, rather than in the restrictions of our intelligence. Among the weaknesses of human nature are ambitions, emotions, prejudices and laziness of thinking. The goal of a discussion is not to be right or to overplay or mislead the other, but to discover the truth or to come to an agreement by common and orderly thinking.

Ideally, an argument consists of carefully specified premises or assumptions and a conclusion which logically follows from the premises. Logical correctness of an argument means that *if* the premises are true, *then* the conclusion must also be true. In Section 1.2 we have already seen that logical correctness of an argument does not mean that the premises are true, neither that the conclusion is true. We may have a logically correct argument with a false conclusion when at least one of the premises is false. And a logically incorrect argument may have a conclusion that is true, when its truth is not based on the given premises but on other grounds. One should also realize that from contradicting premises one may conclude anything one wants: ex falso sequitur quod libet; a principle popular among many politicians.

In real life premises and even the conclusion may be tacit, in which case one speaks of *enthymemes*. Premises may not be explicitly stated for practical reasons or because the speaker is not aware of them himself, but also to mislead the audience.

One may distinguish formal and informal fallacies. A *formal fallacy* is an incorrect argument which may be represented in a formal logical system such as propositional logic. A simple example is: A implies B ($A \rightarrow B$) and B; hence A. For instance: if the weather is nice, then John will come. John comes; hence the weather is nice. That this argument is incorrect may become clear from the following example which has exactly the same structure: if Bill Gates owns all the gold in Fort Knox, then he is rich. Bill Gates is rich; hence Bill Gates owns all the gold in Fort Knox. However, a doctor frequently has to reason this way: a patient comes with a certain complaint B that may have several causes A; $A \rightarrow B$ and B, so the doctor will start with treating the most likely cause A.

An argument is an *informal fallacy* when the putative conclusion is not supported by the content of the premises, but is based on the ambitions, emotions, prejudices and/or laziness of thinking of the people involved. In real life, ambitions, emotions, prejudices and laziness of thinking play a major role in argumentation, debating and discussions. A speaker may be too proud to admit that he is wrong, he may be irritated by his opponent and consequently say more than he can justify, he may have prejudices which he does not want to give up and/or he may be too lazy to study an issue carefully and for that reason oversimplify it.

So, in real life discussions and debating it is important that one is aware of all kinds of tricks which are used, consciously or unconsciously, by one's opponent to suggest that you are wrong, while in fact your opponent is wrong. In Chapter 10 we discuss a dozen different fallacies and a dozen unfair discussion methods.

1.11 Solutions

Solution 1.1. The pattern of reasoning is the following one:

$$\begin{array}{ll}(1) & \neg O \to D \\ (2) & \neg D \\ (3) & \underline{P \to \neg O} \\ & \neg P\end{array}$$

This pattern is valid; hence the argument is correct. Suppose (1), (2) and (3) and P. Then by (3) $\neg O$. Then by (1) D, contradicting (2). Therefore, if (1), (2) and (3), then $\neg P$. Note that both the conclusion and the second premiss in this argument are false.

Solution 1.2. The pattern of reasoning is the following one:

$$\begin{array}{ll}(1) & N \to J \\ (2) & \underline{\neg N} \\ & \neg J\end{array}$$

This pattern is invalid and hence the argument is not correct. It may well be that John comes, while the weather is not nice. In that case J is true and hence also the premisses (1) and (2) are true, while the conclusion $\neg J$ is false.

Another counterexample: take for N the proposition 'Bill Gates owns all the gold in Fort Knox' and for J the proposition 'Bill Gates is rich'. Then all premisses are true, while the conclusion is false.

Solution 1.3. The pattern of reasoning is the following one:

$$\begin{array}{ll}(1) & N \to J \\ (2) & \underline{J} \\ & N\end{array}$$

This pattern is equivalent to the former one, since $\neg N \to \neg J$ is equivalent to $J \to N$, and hence invalid.

Solution 1.4. The pattern of reasoning is the following one:

$$\begin{array}{ll}(1) & J \to N \\ (2) & \underline{J} \\ & N\end{array}$$

This pattern is valid and hence the argument is correct. The first premiss may be expressed by $\neg N \to \neg J$ or equivalently by $J \to N$. If both premisses (1) and (2) are true, then the conclusion must be true too.

Solution 1.5. The pattern of reasoning is the following one:

$$\begin{array}{ll}(1) & \neg(P \land \neg S) \\ (2) & \neg P \lor \neg D \\ (3) & \underline{D} \\ & \neg S\end{array}$$

This pattern is not valid and hence the argument is incorrect. If P is false and S and D are true, then the premisses are all true, while the conclusion is false.

Solution 1.6. The pattern of reasoning is the following one:

$$\frac{\forall x[B(x) \lor C(x)]}{\forall x[B(x)] \lor \forall x[C(x)]}$$

using $B(x)$ for 'x has a beard' and $C(x)$ for 'x has a conical cap'. This pattern is not valid; hence the argument is not correct: taking natural numbers as domain, interpreting $B(x)$ as 'x is even' and $C(x)$ as 'x is odd' yields a true premiss and a false conclusion.

Solution 1.7. The pattern of reasoning is the following one:
$$\frac{\forall x[B(x) \to C(x)]}{\forall x[B(x)]}$$
$$\overline{\forall x[C(x)]}$$

This pattern is valid; hence the argument is correct: if all objects with the property B also have the property C, and all objects have the property B, then all objects must have the property C, no matter what the objects or what the properties B and C are.

Solution 1.8. The pattern of reasoning is the following one:
$$\frac{\exists x[B(x)]}{\exists x[C(x)]}$$
$$\overline{\exists x[B(x) \wedge C(x)]}$$

This pattern is not valid and hence the argument is not correct: taking natural numbers as domain, interpreting $B(x)$ as 'x is even' and $C(x)$ as 'x is odd', yields true premises and a false conclusion.

Solution 1.9. The pattern of reasoning is the following one:
$$\frac{\exists x[\neg B(x) \vee C(x)]}{\exists x[B(x)]}$$
$$\overline{\exists x[C(x)]}$$

This pattern is not valid and hence the argument is not correct: taking natural numbers as domain, interpreting $B(x)$ as 'x is even' and $C(x)$ as 'x is negative', yields true premises and a false conclusion.

References

1. Bochénski, I.M., *A History of Formal Logic*. University of Notre Dame Press, 1961.
2. Boehner, P., *Medieval Logic*. Manchester, 1952.
3. Boole, G., *The mathematical analysis of logic, being an essay toward a calculus of deductive reasoning*. Cambridge and London, 82 pp., 1847. Reprinted by Basil Blackwell, Oxford and the New York (Philos Libr.), 1948.
4. Boole, G., *An investigation of the laws of thought, on which are founded the mathematical theories of logic and probabilities*. Walton and Maberly, London, 1854. Reprinted as vol. 2 of George Boole's collected works, Ph.E.B. Jourdain (ed.), Chicago & London, 1916. Reprinted by Dover Publications, New York, 1951.
5. Frede, M., *Die stoïsche Logik*. VandenHoeck & Ruprecht, Göttingen, 1974.
6. Frege, G., *Begriffsschrift und andere Aufsätze*. I. Angelelli (ed.), Olms, Hildesheim, 1964.
7. Frege, G., *Philosophical Writings*. Translated by P. Geach and M. Black. Basil Blackwell, Oxford, 1970.
8. Gentzen, G., Untersuchungen über das logische Schliessen. *Mathematische Zeitschrift*, Vol. 39, pp. 176-210, 405-431, 1934-1935.
9. Gödel, K., Über formal unentscheidbare Sätze der Principia Mathematica und verwandter Systeme, *Monatshefte für Mathematik und Physik* 38: 173-98, 1931.
10. Kneale, W. & M., *The Development of Logic*. Clarendon Press, Oxford, 1962.
11. Łukasiewicz, J., *Aristotle's Syllogistic*. From the standpoint of modern formal logic. Clarendon Press, Oxford, 1957.
12. Mates, B., *Stoic Logic*. University of California. Publication in Philosophy, Vol. 26, Berkeley and Los Angeles, 1953 and 1961.
13. Moody, E., *Truth and Consequence in Medieval Logic*. North-Holland Publ. Co., Amsterdam, 1953.
14. Whitehead, A.N. and B. Russell, *Principia Mathematica*. Vol. 1, 1910 (2nd ed. 1925). Vol. 2, 1912 (2nd ed. 1927). Vol. 3, 1913 (2nd ed. 1927). Cambridge University Press, England.
15. Smullyan, R.M., *Gödel's Incompleteness Theorems*. Oxford University Press, Oxford, 1992.

Chapter 2
Propositional Logic

H.C.M. (Harrie) de Swart

Abstract In this chapter we analyse reasoning patterns of which the validity only depends on the meaning of the propositional connectives 'if ..., then ...', 'and', 'or' and 'not'. By giving a precise description of the meaning of these propositional connectives one is able to give a precise definition of the notion of logical or valid consequence. Two such definitions are given: a semantic one, in terms of truth values and hence in terms of the meaning of the formulas involved, and a syntactic one in terms of logical axioms and rules of which only the form is important. The semantic and the syntactic definition of logical consequence turn out be equivalent, giving us confidence that we gave a proper characterization of the intuitive notion of logical consequence. We prove or disprove all kinds of statements *about* the notion of logical or valid consequence, which is useful in order to get a good grasp of this notion. The last section treats a number of paradoxes which have been important for the progress in science and philosophy; it also contains a number of historical and philosophical remarks.

2.1 Linguistic Considerations

Logic is such a rich, broad and varied discipline that it is necessary to approach it by picking a small and manageable portion to treat first, after which the treatment can be extended to include more. In this Chapter we restrict our study of reasoning to what is called *propositional logic* or the *propositional calculus*.

A *proposition* is the meaning of a declarative sentence, like 'John is ill', 'Coby goes to school', etc., where a sentence has been obtained from letters or words from a given alphabet according to certain grammatical rules. So, a sentence is just a combination of letters or words, while the corresponding proposition is the meaning of the sentence in question. One says that a sentence *expresses* a proposition. This explains the term '*sentential calculus*' instead of 'propositional calculus'. A proposition is either true or false, although we do not have to know which of the two.

Besides declarative sentences one can distinguish interrogatory sentences which ask questions and imperative sentences which express commands. These latter sentences do not express propositions: it does not make sense to ask whether they express something true or something false. Note that different declarative sentences may express the same proposition. Thus the same proposition is expressed by 'John reads the book' and 'the book is read by John'. $3^2 + 4^2 = 5^2$ and $5^2 = 3^2 + 4^2$ also express the same proposition (which happens to be true); but $5 + 7 = 12$ expresses a different proposition (also true).

By means of connectives one may construct more complex propositions from more elementary ones. For instance, 'John is ill and Coby goes to school' has been composed from the two more elementary propositions by means of the connective 'and'. The most important *connectives* or *propositional operations* are: 'if and only if (iff)', 'if ..., then ...', 'and', 'or' and 'not'. In propositional logic one uses the symbols $\rightleftarrows$, $\rightarrow$, $\wedge$, $\vee$ and $\neg$ for these connectives, respectively.

We distinguish *atomic propositions*, like 'John is ill' and 'Coby goes to school' on the one hand and *composite propositions* on the other hand. *Atomic propositions* are those propositions which cannot be composed of yet more simple propositions by means of propositional operations. If a proposition has been composed from more elementary propositions by means of one or more propositional operations we call it a composite proposition. Thus, 'John is ill and Coby goes to school' is a composite proposition.

In propositional logic one uses letters $P_1, P_2, P_3, \ldots$ to denote atomic propositions. For instance, 'John is ill' may be translated by P_1, while 'Coby goes to school' may be translated by P_2. The composite proposition 'John is ill and Coby goes to school' is then translated by $P_1 \wedge P_2$.

In *propositional logic* one studies the (in)validity of reasoning patterns of which the (in)validity is completely determined by the meaning of the connectives 'if and only if' ($\rightleftarrows$), 'if ..., then ...' ($\rightarrow$), 'and' ($\wedge$), 'or' ($\vee$) and 'not' ($\neg$) between the propositions in question. A simple example is the following reasoning pattern, called *Modus Ponens* (MP):

It snows (1)	P_1
If it snows, then it is cold. (2)	$P_1 \rightarrow P_2$
Therefore: it is cold. (3)	P_2

The pattern to the above right is *valid*, i.e., no matter what propositions the formulas P_1, P_2 stand for, if the resulting two premises are both true, then also the conclusion must be true; in particular, if (1) and (2) are true, then (3) must be true too. Notice that the validity of this pattern only depends on the meaning of the connective $\rightarrow$ and not on the meaning of the formulas P_1, P_2. We call the concrete argument about snow and being cold *correct*, because the underlying reasoning pattern is valid.

But, for instance, the validity of the reasoning pattern

For all x, if x is a person, then x is mortal	$\forall x[P(x) \rightarrow M(x)]$
Socrates is a person	$P(c)$
Therefore: Socrates is mortal	$M(c)$

2.1 Linguistic Considerations

not only depends on the meaning of the connective $\rightarrow$, but also on the meaning of the universal quantifier $\forall$ (for all). Notice that P_1, P_2 above stand for propositions, while $P(x), M(x)$ stand for the predicates 'x is a person' and 'x is mortal' respectively. In predicate logic, to be treated in Chapter 4, we study reasoning patterns of which the validity also depends on the meaning of the quantifiers $\forall$ (for all) and $\exists$ (for at least one).

The study of propositional logic was initiated by the Stoics (see Subsection 2.10.2), some 300 years before Aristotle developed his theory of the syllogisms (see Subsection 4.7.4).

Let us start by considering some examples of propositions about numbers, some of which are true and some of which are false. We give their translation into the language of propositional logic, and their translation into the language of predicate logic.

Proposition	*prop. formula*	*pred. formula*
1. All numbers are positive (≥ 0)	P_1	$\forall x[P(x)]$
2. All numbers are negative (≤ 0)	P_2	$\forall x[N(x)]$
3. All numbers are positive or negative	P_3	$\forall x[P(x) \vee N(x)]$

Here $\forall$ is the *universal quantifier* expressing 'for all', $P(x)$ stands for the predicate 'x is positive', $N(x)$ for the predicate 'x is negative' and $\vee$ stands for the connective 'or'. It is important to notice that the propositional translation of sentence 3 cannot be rendered by $P_1 \vee P_2$, because this formula expresses the proposition 'all numbers are positive or all numbers are negative' which happens to be false, while sentence 3 is true. Also notice that in $P_1 \vee P_2$ the connective $\vee$ stands between two propositions, while in $\forall x[P(x) \vee N(x)]$ the connective $\vee$ stands between two predicates.

Proposition	*prop. formula*	*pred. formula*
4. There is at least one even number	P_4	$\exists x[E(x)]$
5. There is at least one odd number	P_5	$\exists x[O(x)]$
6. There is a number that is both even and odd	P_6	$\exists x[E(x) \wedge O(x)]$

Here $\exists$ is the *existential quantifier* expressing 'there is at least one', $E(x)$ stands for the predicate 'x is even', $O(x)$ for the predicate 'x is odd' and $\wedge$ stands for the connective 'and'. It is important to notice that the propositional translation of sentence 6 cannot be rendered by $P_4 \wedge P_5$, because this formula expresses the proposition 'there is at least one even number and there is at least one odd number' which happens to be true, while sentence 6 is false. Also notice that in $P_4 \wedge P_5$ the connective $\wedge$ stands between two propositions, while in $\exists x[E(x) \wedge O(x)]$ the connective $\wedge$ stands between two predicates.

Proposition	*prop. formula*	*pred. formula*
7. There is a number x such that $x > 0$	P_7	$\exists x[x > 0]$
8. There is a number x such that not $x > 0$	P_8	$\exists x[\neg(x > 0)]$

$x > 0$ is not a proposition, but a predicate, while $5 > 0$, for instance, is a proposition. Similarly, 'not $x > 0$' is not a proposition, but the negation of a predicate, while 'not $5 > 0$' is a proposition. It is important to notice that proposition 8 is not the negation of proposition 7; the negation of 7 is 'there is no number x such that $x > 0$', $\neg \exists x[x > 0]$, which is equivalent to 'for all numbers x, not $x > 0$'. This latter proposition is false, while proposition 8 is true. In the negation of sentence 7, the negation stands in front of the existential quantifier, while in sentence 8 the negation stands in front of the predicate $x > 0$.

Proposition	prop. formula	pred. formula
9. All persons have a mother	P_9	$\forall x \exists y[M(x,y)]$
10. There is one mother of all persons	P_{10}	$\exists y \forall x[M(x,y)]$

$\forall x \exists y[M(x,y)]$ says: for every person x there is a person y such that x stands in the child-mother relation $M(x,y)$ with y. But by changing the order of the quantifiers one obtains $\exists y \forall x[M(x,y)]$ which says: there is at least one person y such that for all persons x, x stands in the child-mother relation $M(x,y)$ with y. Notice that sentence 9 is true, while sentence 10 is false.

Proposition	prop. formula	pred. formula
11. For every number there is a larger one	P_{11}	$\forall x \exists y[x < y]$
12. There is a largest number	P_{12}	$\exists y \forall x[x < y]$

$\forall x \exists y[x < y]$ says: for every number x there is a number y such that x is smaller than y. But changing the order of the quantifiers one obtains $\exists y \forall x[x < y]$ which says: there is a number y such that for all numbers x, x is smaller than y. Notice that sentence 11 is true, while sentence 12 is false. So, the order of the quantifiers $\forall$ and $\exists$ does matter!

Let us have a closer look at proposition 9: 'all persons have a mother', or equivalently:

> For every person x there is some person y such that y is the mother of x.

$$\underline{}$$
$$\text{I}$$
$$\underline{}$$
$$\text{II}$$
$$\underline{}$$
$$\text{III}$$

I, 'y is the mother of x', does not express a proposition, but a binary predicate or relation; neither does 'Mary is the mother of x', which expresses a unary predicate. However, 'Mary is the mother of John' does express a proposition.

II, 'there is some person y such that y is the mother of x', does not express a proposition, but a unary predicate, which may become more clear if we formulate II as follows: someone is the mother of x or, equivalently, x has a mother. However, 'someone is the mother of John' does express a proposition.

III does express the proposition 'every person has a mother'. Note that all variables x, y occurring in III also occur in the context 'for every' or 'there is'.

In *propositional logic* one ignores the internal subject-predicate structure of the atomic propositions. The atomic propositions can have the form 'for all x, x has

2.1 Linguistic Considerations

a certain property P', like the propositions 1 up to 3 inclusive, or the form 'there is at least one x such that x has the property P', like the propositions 4 up to 8 inclusive, or the form 'for every x there is a y such that x is in relation $R(x,y)$ to y', like proposition 9 and 11, and so on. In the *propositional calculus* we restrict ourselves to arguments like Modus Ponens and the arguments a), b) and c) in Chapter 1, the correctness of which only depends on how the different propositions are composed of more elementary ones by means of operations like 'iff', 'if ..., then ...', 'and', 'or' and 'not'. In the propositional calculus the internal subject-predicate structure of the elementary propositions is not taken into consideration. However, the argument above about Socrates makes it clear that the correctness of an argument may also depend on this subject-predicate structure. Therefore, the propositional calculus has to be extended to the predicate calculus, which is treated in Chapter 4.

Below we list the symbols we are using for the propositional operations, mentioning their name and alternative symbols which may be used in the literature.

name	symbol	alternatives	meaning
equivalence	$\rightleftarrows$	$\leftrightarrow, \sim, \equiv$	(is) equivalent (to); if and only if; iff
(material) implication	$\rightarrow$	$\supset$	if ..., then ...; implies
conjunction	$\wedge$	&	and
disjunction	$\vee$		or; and/or
negation	$\neg$	-	not

Instead of the atomic propositions considered above, being about numbers, and the propositions that can be built from them by the propositional connectives, we can of course consider different atomic propositions, for instance of geometry, physics or of some other sharply circumscribed part of natural language, together with the composite propositions that can be built from them. So, in order to retain flexibility for the applications, we shall simply assume, throughout this chapter, that we are dealing with an object language in which there is a class of (declarative) sentences consisting of certain building blocks

$$P_1, P_2, P_3, \ldots$$

called *atomic formulas*, from which *composite formulas* can be built by means of the propositional connectives. By a *formula* we mean either an atomic or a composite formula. So, throughout this chapter our object language will be the following symbolic language:

	symbols	names
Alphabet:	$P_1, P_2, P_3, \ldots$	atomic formulas or propositional variables
	$\rightleftarrows, \rightarrow, \wedge, \vee, \neg$	connectives
	(,)	parentheses

Definition 2.1 (Formulas).

1. Each atomic formula is a formula. In other words, if P is an atomic formula, then P is a formula.

2. If each of A and B is a given formula (i.e., either an atomic formula or a composite formula already constructed), then $(A \rightleftarrows B)$, $(A \rightarrow B)$, $(A \wedge B)$ and $(A \vee B)$ are (composite) formulas.
3. If A is a given formula, then $(\neg A)$ is a (composite) formula.

This language is the *formal language of propositional logic*. It consists of the formulas built from the given alphabet according to Definition 2.1.

$P_1, P_2, P_3, \ldots$ are symbols to be interpreted as atomic propositions from arithmetic, geometry, physics, any other science or daily life. The first four connectives are binary connectives, the last one is unary. The connectives are symbols whose meanings are the respective propositional operations; in Section 2.2 we will fix and stylize these meanings by truth tables. The parentheses are punctuation marks. In $A \rightarrow B$ we call A the *antecedent* and B the *succedent*.

Example 2.1. Here are some examples of formulas:

$$P_1, P_2, P_3, P_4$$
$$(\neg P_2), (\neg P_3)$$
$$(P_1 \vee (\neg P_2)), ((\neg P_3) \rightarrow P_4)$$
$$((P_1 \vee (\neg P_2)) \wedge ((\neg P_3) \rightarrow P_4)).$$

Notice that the number of left parentheses must be equal to the number of right parentheses.

if, only if and **iff**:
'B if A' is translated by $A \rightarrow B$, which may also be read as 'if A, then B'.
'B only if A' is translated by $B \rightarrow A$, and 'B if and only if (iff) A' is translated by $(A \rightarrow B) \wedge (B \rightarrow A)$, or, equivalently, by $A \rightleftarrows B$.

Convention: When we want to state something about arbitrary natural number, the letters n, m are used to stand for any of the natural numbers 0, 1, 2, 3, ... For instance, when we state that for all natural numbers n, m: $n + m = m + n$. Similarly, the letters P, Q and R are used to stand for any atomic formulas $P_1, P_2, P_3, \ldots$ and the letters $A, B, C, A_1, A_2, \ldots, B_1, B_2, \ldots, C_1, C_2, \ldots$ are used to stand for any formulas, not necessarily atomic. For instance, the letters A and B may stand for any of the formulas in Example 2.1. Distinct such letters need *not* represent distinct formulas, in contrast to $P_1, P_2, P_3, \ldots$ which are distinct atomic formulas.

Parentheses in formulas are essential: they indicate which parts belong together. Leaving them out may cause ambiguity. For instance, $A \wedge B \rightarrow C$ might mean:

- $(A \wedge B) \rightarrow C$, which is an implicational formula with $A \wedge B$ as antecedent, and
- $A \wedge (B \rightarrow C)$, which is a conjunction of the formulas A and $B \rightarrow C$.

'If John wins the lottery and is healthy, then he will go to the Bahamas' is a proposition of the first form, while 'John wins the lottery and if he is healthy, then he will go to the Bahamas' is a proposition of the second form. Only in the second proposition it is stated that John wins the lottery.

2.1 Linguistic Considerations

Convention We can minimize the need for parentheses by agreeing that we leave out the most outer parentheses in a formula and that in

$$\rightleftarrows, \rightarrow, \wedge, \vee, \neg$$

any connective has a higher rank than any connective to the right of it and a lower rank than any connective to the left of it.

According to this convention, $A \wedge B \rightarrow C$ should be read as $(A \wedge B) \rightarrow C$, because $\rightarrow$ has a higher rank than $\wedge$, and not as $A \wedge (B \rightarrow C)$, which has a different meaning. The formula $\neg A \vee B$ should be read as $(\neg A) \vee B$, because by convention $\vee$ has a higher rank than $\neg$, and not as $\neg (A \vee B)$, which means quite something else. And $C \rightleftarrows A \wedge B \rightarrow C$ should be read as $C \rightleftarrows ((A \wedge B) \rightarrow C)$.

It is interesting to notice that the build-up of formulas is very similar to the build-up of natural numbers. Formulas are generated by starting with atomic formulas $P_1, P_2, P_3, \ldots$ and successively passing from one or two formulas already generated before to another formula by means of the connectives. Natural numbers are generated by starting with one initial object 0 and successively passing from a natural number n already generated before to another natural number $n+1$ or n' (the *successor* of n).

Since natural numbers are built up from 0 by repeated application of the successor operation, the theorem of *mathematical induction* follows immediately from the definition of natural numbers:

Theorem 2.1 (Mathematical induction). *Let Φ be a property of natural numbers such that*

1. *(induction basis:) 0 has property Φ, and*
2. *property Φ is preserved when going from a natural number n to its successor n', i.e., for all natural numbers n, if n has property Φ (induction hypothesis), then also n' has property Φ.*

Then all natural numbers have property Φ.

Using mathematical induction, one can prove, for instance, that for all natural numbers n, $1 + 2 + \ldots + n = \frac{1}{2}n(n+1)$. See Exercise 2.5.

The *induction principle* for formulas is similar to mathematical induction for natural numbers. Since (propositional) formulas are built up from atomic formulas $P_1, P_2, P_3, \ldots$ by successive applications of connectives to formulas already generated before, the following theorem, called the *induction principle* (for propositional formulas), follows immediately from the definition of formulas.

Theorem 2.2 (Induction principle). *Let Φ be a property of formulas, satisfying*

1. *(induction basis:) every atomic formula has property Φ and*
2. *property Φ is preserved in building more complex formulas by means of the connectives, i.e., if A and B have property Φ (induction hypothesis), then $(A \rightleftarrows B)$, $(A \rightarrow B)$, $(A \wedge B)$, $(A \vee B)$ and $(\neg A)$ also have property Φ.*

Then every formula (of the propositional calculus) has property Φ.

Using Theorem 2.2 one can prove, for instance, that every formula contains as many left parentheses as right parentheses (see Exercise 2.6.) Another application is Theorem 2.18 which says that every formula can be written in normal form.

Notice that we have introduced a logical (propositional) language such that English sentences may be translated into this logical language and conversely one may translate the logical formulas into the corresponding English sentences. What holds for English sentences of course also holds for German, French, Spanish and all other sentences. With this in mind one might build for each natural language a machine that translates the sentences of the language in question into logical formulas and back. By combining these machines with logic as the intermediate language, one obtains an automatic translation of, for instance, English to, for instance, German: automatically translate the English sentences into logical formulas and next automatically translate the resulting logical formulas into German sentences. This was roughly the Rosetta translation project of the European Union.

Exercise 2.1. Let P_1 stand for 'John works hard',
$\quad\quad\quad\quad P_2 \quad$ for 'John is going to school', and
$\quad\quad\quad\quad P_3 \quad$ for 'John is wise'.
Translate the following sentences into the language of propositional logic, using the least possible number of parentheses.

i) If John works hard and is going to school, then John is not wise.
ii) John works hard and if John is going to school, then he is not wise.
iii) John works hard, or if John is going to school, then he is wise.
iv) If John is going to school or works hard, then John is wise.
v) If John works hard, then John is not wise, at least if he is going to school.

Exercise 2.2. Translate the following formulas into English sentences, reading P_1, P_2 and P_3 as indicated in exercise 2.1.
i) $(P_1 \to P_2) \to \neg P_3$ $\quad\quad\quad$ iv) $\neg P_2 \wedge P_3$
ii) $\neg P_1 \vee P_3$ $\quad\quad\quad\quad\quad\quad\quad$ v) $\neg(P_2 \wedge P_3)$
iii) $\neg(P_1 \vee P_3)$

Exercise 2.3. Translate the following propositions into propositional formulas and into predicate formulas:
1. Every gnome has a beard.
2. All gnomes have no beard.
3. Not every gnome has a beard.

Exercise 2.4. Which of the following expressions are formulas (of the language of propositional calculus)? P_1, P, $\neg P_8$, $\neg Q$, $P_1 \wedge \neg P_8$, $P \wedge \neg Q$, A, B, $A \wedge \neg B$, $(P_1 \wedge P_2) \to \neg P_3$, $(P_1 \wedge P_2) \to Q$, $(P_1 \wedge P_2) \to B$, $A \wedge B \to C$.

Exercise 2.5. Use mathematical induction (Theorem 2.1) to prove that for all natural numbers n, $1 + 2 + \ldots + n = \frac{1}{2}n(n+1)$.

Exercise 2.6. Use the induction principle (Theorem 2.2) to show that every formula of propositional logic contains as many left parentheses '(' as right parentheses ')'.

2.2 Semantics; Truth Tables

In the first section of this chapter a logical (propositional) language was introduced in which we can translate the premisses and the conclusion of an argument, resulting in a reasoning pattern. We have indicated the meaning of the atomic formulas: atomic propositions which are either true or false. And we have indicated the meaning of the propositional connectives $\rightleftarrows, \rightarrow, \wedge, \vee,$ and $\neg$: 'if and only if', 'if ..., then ...', 'and', 'or', and 'not', respectively.

In this section the meaning of the atomic formulas and the propositional connectives is made more precise, where we restrict ourselves (in this chapter) to classical logic. Owing in part to different analyses of implication, the heart of logic, there are different systems of logic: classical logic, intuitionistic logic, relevance logic and so on. Although we will treat the latter logic systems in other chapters, in this chapter we shall concern ourselves primarily with *classical logic*, because it is the simplest and most commonly used system of logic. In classical logic we assume that each proposition is either *true*, indicated by 1, or *false* indicated by 0. We do not, however, suppose that one always *knows* whether a particular proposition is true or false.

To start with, the atomic formulas $P_1, P_2, P_3, \ldots$ stand for (or are interpreted as) atomic propositions, such as 'John is ill', the 'weather is nice', etc. These atomic propositions may be true, indicated by 1, or false, indicated by 0. We standardize this in the so-called *truth table* of the atomic formulas $P_1, P_2, P_3, \ldots$. So, by definition the truth table of an atomic formula P, where P stands for any of the atomic formulas $P_1, P_2, P_3, \ldots$, is the following one:

$$\begin{array}{c} P \\ \hline 1 \\ 0 \end{array}$$

For two atomic formulas P and Q there are four different assignments of the values 1 (true) and 0 (false), schematically rendered as follows:

$$\begin{array}{cc} P & Q \\ \hline 1 & 1 \\ 1 & 0 \\ 0 & 1 \\ 0 & 0 \end{array}$$

In the first line the atomic formulas P and Q are both interpreted as true atomic propositions, in the fourth line both as false atomic propositions.

For three atomic formulas P, Q and R there are eight different assignments of the values 1 and 0. Notice that the number of different assignments of the values 1 and 0 to P, Q and R is two times as many as for the two atomic formulas P and Q, since for each of the four different assignments of the values 1 and 0 to P and Q, one may assign a 1 or a 0 to R: More generally:

Lemma 2.1. *For n atomic formulas $P_1, \ldots, P_n$, $n = 1, 2, \ldots$, there are 2^n different assignments of the values 1 and 0.*

If a formula A, for instance $A = P_1 \to (P_2 \to P_3)$, has been built from three atomic formulas, there are $2^3 = 8$ different assignments of the values 1 and 0 to the atomic formulas P_1, P_2 and P_3. But the formula A itself can have at most two different values: 1 and 0.

Next a precise meaning has to be given to the propositional connectives. This is done in the so-called *truth tables for the propositional connectives*, where it is specified how the truth value of the composite formulas $A \rightleftarrows B$, $A \to B$, $A \land B$, $A \lor B$ and $\neg A$ is completely determined by the truth values of the components A and B.

Two different formulas A and B can have at most four different values of truth (1) and falsity (0), represented by the four rows in the table below. Each column in the table below indicates how the truth or falsity of the composite formula heading that column depends on the truth values of its immediate components A and B.

A	B	$A \rightleftarrows B$	$A \to B$	$A \land B$	$A \lor B$		A	$\neg A$
1	1	1	1	1	1		1	0
1	0	0	0	0	1		0	1
0	1	0	1	0	1			
0	0	1	1	0	0			

Thus $A \rightleftarrows B$ is true exactly when A and B have the same truth value; hence, the reading 'equivalent', i.e., 'equal valued', for $\rightleftarrows$.
$A \to B$ is false exactly when A is true and B is false.
$A \land B$ is true exactly when A and B are both true.
$A \lor B$ is false exactly when both A and B are false.
And $\neg A$ is true exactly when A is false.

The truth tables for the propositional connectives may also be presented in the following way:

		$\rightleftarrows$	$\to$	$\land$	$\lor$			$\neg$
1	1	$1 \rightleftarrows 1 = 1$	$1 \to 1 = 1$	$1 \land 1 = 1$	$1 \lor 1 = 1$		1	$\neg 1 = 0$
1	0	$1 \rightleftarrows 0 = 0$	$1 \to 0 = 0$	$1 \land 0 = 0$	$1 \lor 0 = 1$		0	$\neg 0 = 1$
0	1	$0 \rightleftarrows 1 = 0$	$0 \to 1 = 1$	$0 \land 1 = 0$	$0 \lor 1 = 1$			
0	0	$0 \rightleftarrows 0 = 1$	$0 \to 0 = 1$	$0 \land 0 = 0$	$0 \lor 0 = 0$			

The truth tables for $\rightleftarrows$, $\land$, $\lor$ and $\neg$ are self evident and give little or no reason for discussion. However, the table for $\to$ was already disputed by the Stoics, see Subsection 2.10.2. Nevertheless, it is the only one of the 16 possible columns of length 4 consisting of 1's and 0's which is tenable; any other proposal can easily be rejected as unreasonable.

First, let us notice that the propositional connectives $\rightleftarrows$, $\to$, $\land$, $\lor$ and $\neg$ as defined in the truth tables are *truthfunctional*, i.e., the truth values of $A \rightleftarrows B$, $A \to B$, $A \land B$, $A \lor B$ and $\neg A$ are completely determined by the truth values of its components A and B. This is not always the case for the connective 'if ..., then ...' from daily language, as may be illustrated by the following two sentences:
1. If I would have jumped out of the window on the 10th floor, then I would have been injured.

2.2 Semantics; Truth Tables

2. If I would have jumped out of the window on the 10th floor, then I would have changed into a bird.

Although in both sentences the components have the same truth value 0 (I have not jumped out of the window, I have not been injured and I have not changed into a bird) the first sentence is held to be true, while the second sentence is held to be false. In other words, in sentence 1, the combination 'if 0, then 0' gives a 1, while in sentence 2 the same combination 'if 0, then 0' gives a 0. So, the 'if ..., then ...' from daily language is not truthfunctional. Consequently, the $\rightarrow$ may be different from the 'if ..., then ...' from daily language.

Nevertheless, in daily life the 'if ..., then ...' is frequently, although not always, used precisely as described in the truth table of $\rightarrow$. We may illustrate this with the following example:

For all integers n and m, if $n = m$, then $n^2 = m^2$.

Why is this proposition true? Simply because it is impossible that for some integers n and m the proposition $n = m$ has truth value 1, while the proposition $n^2 = m^2$ has truth value 0. In other words, the combination 1 for $n = m$ and 0 for $n^2 = m^2$ does not occur. Only the combinations 1 - 1, 0 - 1 and 0 - 0 may occur and these give the value 1, just as in the truth table of $\rightarrow$:

	$n = m$	$n^2 = m^2$	if $n = m$, then $n^2 = m^2$
$n = 2, m = 2$	1	1	1
$n = 2, m = -2$	0	1	1
$n = 2, m = 3$	0	0	1

From the table for $\rightarrow$ one sees that $A \rightarrow B$ is true (has value 1; is 1) if and only if A is false ($\neg A$ is true) or B is true (has value 1); in other words, it is easy to check that $A \rightarrow B$ and $\neg A \vee B$ have the same truth table. The truth table of $A \rightarrow B$ is also the same as the one of $\neg(A \wedge \neg B)$, which corresponds with our intuitions:

A B	$\neg A$	$\neg A \vee B$	$\neg B$	$A \wedge \neg B$	$\neg(A \wedge \neg B)$
1 1	$\neg 1 = 0$	$0 \vee 1 = 1$	$\neg 1 = 0$	$1 \wedge 0 = 0$	$\neg 0 = 1$
1 0	$\neg 1 = 0$	$0 \vee 0 = 0$	$\neg 0 = 1$	$1 \wedge 1 = 1$	$\neg 1 = 0$
0 1	$\neg 0 = 1$	$1 \vee 1 = 1$	$\neg 1 = 0$	$0 \wedge 0 = 0$	$\neg 0 = 1$
0 0	$\neg 0 = 1$	$1 \vee 0 = 1$	$\neg 0 = 1$	$0 \wedge 1 = 0$	$\neg 0 = 1$

Warning One frequently is inclined to read $A \rightarrow B$ as: A and hence B. But this is wrong! If I assert $A \rightarrow B$, I do not assert A, neither B. Consider, for instance, the sentence: if I win the lottery, then I will give you a Cadillac. This does not mean that I win the lottery and hence will give you a Cadillac.

Why is $A \rightarrow B$ true (1) in case A is false (0)? Consider the following example. Suppose I am determined never to play in a lottery; in this case I can truthfully state: If I win the lottery, then I will give you a Cadillac. Assuming I never play in a lottery, this is an empty statement, without content, and hence this statement cannot be false.

And why is $A \to B$ true (1) if B is true (1)? Suppose B stands for 'I give you a Cadillac' and suppose this is true (1). Then the sentence 'if I win the lottery, then I will give you a Cadillac' is certainly true (1) too.

The reader should also verify that the truth table for $A \rightleftarrows B$ is the same as the one of $(A \to B) \land (B \to A)$, which also corresponds with our intuition:

A B	$A \rightleftarrows B$	$A \to B$	$B \to A$	$(A \to B) \land (B \to A)$
1 1	1	1	1	1
1 0	0	0	1	0
0 1	0	1	0	0
0 0	1	1	1	1

If one constructs the truth tables for $A \land B$ and for $B \land A$, one will find that these two truth tables are the same:

A B	$A \land B$	$B \land A$
1 1	$1 \land 1 = 1$	$1 \land 1 = 1$
1 0	$1 \land 0 = 0$	$0 \land 1 = 0$
0 1	$0 \land 1 = 0$	$1 \land 0 = 0$
0 0	$0 \land 0 = 0$	$0 \land 0 = 0$

However, a sentence like 'Ann had a baby and got married' will leave another impression than the sentence 'Ann got married and had a baby'. In this example the order of the two atomic propositions suggests a temporal or causal succession. Also in the sentence 'John fell into the water and drowned' one cannot easily change the order of the atomic components. These examples show that the connectives from daily language may have shades of meaning which are lost in their translation to the corresponding propositional connectives. Notice that the expression 'A but B' has nuances of meaning not possessed by 'A and B' and lost in the translation $A \land B$: 'I love you and I love your sister almost as well' will leave another impression than 'I love you but I love your sister almost as well'.

In daily life, the connective 'or' is sometimes used in an exclusive way. For instance, when the dinner menu says 'tea or coffee included', we do not expect to get both. But in 'books can be delivered at school or at church' the connective 'or' is used in an inclusive way: we may deliver books at school and/or at church. Notice that the symbol $\lor$, coming from the Latin 'vel', corresponds with the inclusive 'or' and that $A \lor B$ has the same truth table as $B \lor A$.

Analysing the use of the propositional operations 'iff', 'if ..., then ...', 'and', 'or', and 'not' in arithmetic, calculus and more generally in mathematics, it turns out that these operations are used precisely as described in the truth tables of $\rightleftarrows, \to, \land, \lor$ and $\neg$ respectively. This should make it clear that our propositional connectives and *material implication* $A \to B$ in particular are useful and natural forms of expression. In natural language the propositional operations are frequently, but not always, used as described in the truth tables above.

No disagreement exists that 'if A, then B' is false if A is true and B is false. Problems arise with the claim that 'if A, then B' is false *only* if A is true and B is false, and is true in all other cases. 'If these three chairs cost 6 dollars (A), then

2.2 Semantics; Truth Tables

one chair costs 2 dollars (B)' is true, because it is impossible that A is true and B is false, due to the causal relation between A and B; in this example both A and B are supposed to be false. Problems arise if there is no connection of ideas between A and B, like in 'if I would have jumped out of the window, then I would have changed into a bird', which is true under our table. $A \to B$ is called a *conditional* or a *material implication*; the latter name because the truth of 'if A, then B' in general depends on matters of empirical fact.

Example 2.2. Let us illustrate the repeated use of the truth tables by computing the one for $P_1 \to (P_2 \to P_3)$ and the one for $(P_1 \wedge P_2) \to P_3$:

P_1	P_2	P_3	$P_2 \to P_3$	$P_1 \to (P_2 \to P_3)$	$P_1 \wedge P_2$	$(P_1 \wedge P_2) \to P_3$
1	1	1	$1 \to 1 = 1$	$1 \to 1 = 1$	$1 \wedge 1 = 1$	$1 \to 1 = 1$
1	1	0	$1 \to 0 = 0$	$1 \to 0 = 0$	$1 \wedge 1 = 1$	$1 \to 0 = 0$
1	0	1	$0 \to 1 = 1$	$1 \to 1 = 1$	$1 \wedge 0 = 0$	$0 \to 1 = 1$
1	0	0	$0 \to 0 = 1$	$1 \to 1 = 1$	$1 \wedge 0 = 0$	$0 \to 0 = 1$
0	1	1	$1 \to 1 = 1$	$0 \to 1 = 1$	$0 \wedge 1 = 0$	$0 \to 1 = 1$
0	1	0	$1 \to 0 = 0$	$0 \to 0 = 1$	$0 \wedge 1 = 0$	$0 \to 0 = 1$
0	0	1	$0 \to 1 = 1$	$0 \to 1 = 1$	$0 \wedge 0 = 0$	$0 \to 1 = 1$
0	0	0	$0 \to 0 = 1$	$0 \to 1 = 1$	$0 \wedge 0 = 0$	$0 \to 0 = 1$

Notice that $P_1 \to (P_2 \to P_3)$ has the same truth table as $(P_1 \wedge P_2) \to P_3$, which corresponds with our intuition: $P_1 \to (P_2 \to P_3)$ is read as 'if P_1, then (if - in addition - P_2, then P_3), which is equivalent to 'if P_1 and P_2, then P_3'.

2.2.1 Validity

Atomic formulas have (by definition) two truth values, 1 and 0. However, it is easy to see that some composite formulas have only one truth value. For instance, the formula $P_1 \to P_1$ can only have the truth value 1, no matter what the truth value of P_1 is. And the formula $P_1 \wedge \neg P_1$ can only have the truth value 0, no matter what the truth value of P_1 is:

P_1	$P_1 \to P_1$	$\neg P_1$	$P_1 \wedge \neg P_1$
1	$1 \to 1 = 1$	$\neg 1 = 0$	$1 \wedge 0 = 0$
0	$0 \to 0 = 1$	$\neg 0 = 1$	$0 \wedge 1 = 0$

Other formulas with only the truth value 1 are $P_1 \vee \neg P_1$, $P_1 \wedge P_2 \to P_1$, $P_1 \to (P_2 \to P_1)$ and $P_1 \to P_1 \vee P_2$. These formulas are called *always true* or *valid*. Wittgenstein (1921) called these formulas *tautologies*.

Definition 2.2 (Valid; Consistent; Contingent). Let A be a formula.
A is *always true* or *valid* := the truth table of A – entered from the atomic formulas from which A has been built – contains only 1's. **Notation:** $\models A$.
A is *consistent* or *satisfiable* := the truth table of A contains at least one 1; that is, the formula A may be true.

A is *contingent* := the truth table of *A* contains at least one 1 and at least one 0; that is, *A* may be true and it may be false.
A is *inconsistent* or *always false* or *contradictory* := the truth table of *A* contains only 0's; that is, *A* cannot be true, in other words, is always false.

Notice that a valid formula is consistent, but not contingent; that a contingent formula is by definition also consistent; and that an inconsistent formula is by definition not contingent.

So, for instance, the formula $P_1 \to P_1$ is valid and hence also consistent, the formula $P_1 \to P_2$ is contingent and consistent, but not valid, and the formula $P_1 \wedge \neg P_1$ is inconsistent or always false.

On the one hand, valid formulas are uninteresting because they give no information. On the other hand, since valid formulas are always true regardless of the truth or falsity of their atomic components, they may be used in reasoning as may be illustrated by the following example.

Example 2.3. Consider the following argument:

$$\begin{array}{ll} \text{John is lazy } [L]. & L \\ \text{If John is ill } [I] \text{ or lazy, he stays at home } [H]. & \dfrac{I \vee L \to H}{H} \\ \text{Therefore: John stays at home.} & \end{array}$$

In this valid reasoning pattern we use silently that $\models L \to I \vee L$. The argument might be simulated as follows:
$$\dfrac{\dfrac{L \quad L \to I \vee L}{I \vee L} \quad I \vee L \to H}{H}$$

Note that there are infinitely many valid formulas. Although it is not exhaustive (for instance, $P \vee \neg P$ does not occur in it), the following list enumerates infinitely many valid formulas.

$P \to P$
$P \to (P \to P)$
$P \to (P \to (P \to P))$
$\vdots$

Warning: While the symbol *A* stands for any formula, like $P_1 \to P_2$, $P_2 \wedge \neg P_3$, etc., the expression $\models A$ is not a formula, but a statement *about* the formula *A*, namely, that the truth table of *A* contains only 1's. The symbol $\models$ does not occur in the logical alphabet, and '$\models A$' is shorthand for '*A* is valid' or '*A* is always true', which clearly is not a logical formula. In other words, the symbol *A* indicates a formula from the logical language, our object language, while the expression $\models A$ belongs to the meta-language, in which we make statements *about* formulas of the object language.

Notation: If a particular formula *A* is not valid, this is frequently written by $\not\models A$ instead of 'not $\models A$'. For instance: $\models P_1 \to P_1 \vee P_2$, but $\not\models P_1 \to P_1 \wedge P_2$.

Definition 2.3 (Interpretation; Model). Let *A* be a formula built from the atomic formulas $P_1, \ldots, P_n$. An *interpretation i* of *A* assigns a value 1 or 0 to all the atomic

2.2 Semantics; Truth Tables

components of A; so, an interpretation i of A corresponds with a line in the truth table for A and interprets each atomic formula in A as either a true or a false proposition. Interpretation i of A is a *model* of A := i assigns to A the value 1, in other words, $i(A) = 1$. In this terminology the definition of 'A is valid' can be reformulated as follows: every interpretation i of A is a model of A.

Example 2.4. Thus, if A has been built from only two atomic formulas P and Q, then there are four different interpretations of A: i_1, i_2, i_3, i_4.

	P	Q	$P \to Q$	
i_1	1	1	1	$i_1(P) = 1, i_1(Q) = 1, i_1(P \to Q) = 1$
i_2	1	0	0	$i_2(P) = 1, i_2(Q) = 0, i_2(P \to Q) = 0$
i_3	0	1	1	$i_3(P) = 0, i_3(Q) = 1, i_3(P \to Q) = 1$
i_4	0	0	1	$i_4(P) = 0, i_4(Q) = 0, i_4(P \to Q) = 1$

For instance, i_1, i_3 and i_4 are a model of $P \to Q$, but i_2 is not a model of $P \to Q$.

Definition 2.4. Let Γ be a (possibly infinite) set of formulas and i an interpretation, assigning the values 0 or 1 to all the atomic components of the formulas in Γ.
i a *model* of Γ := i is a model of all formulas in Γ, i.e., i makes all formulas in Γ true.
Γ is *satisfiable* := there is at least one assignment i which is a model of Γ.

Example 2.5. If Γ consists of $P_1 \to P_2$ and $P_1 \vee P_2$, then i_1 and i_3 are models of Γ.

	P_1	P_2	$P_1 \to P_2$	$P_1 \vee P_2$
i_1	1	1	1	1
i_2	1	0	0	1
i_3	0	1	1	1
i_4	0	0	1	0

Theorem 2.3 (Compactness theorem). * *Let Γ be a (possibly infinite) set of formulas such that every finite subset of Γ has a model. Then Γ has a model.*

Proof. Let Γ be a (possibly infinite) set of formulas such that every finite subset of Γ has a model. We will define an interpretation i of the atomic propositional formulas $P_1, P_2, P_3, \ldots$ such that for every natural number n, $\Phi(n)$, where $\Phi(n)$:= every finite subset of Γ has a model in which $P_1, P_2, \ldots, P_n$ take the values $i(P_1), i(P_2), \ldots, i(P_n)$.

Once having shown this, it follows that $i(A) = 1$ for every formula A in Γ. For given a formula A in Γ, take n so large that all atomic formulas occurring in A are among $P_1, \ldots, P_n$. Since $\{A\}$ is a finite subset of Γ and because of $\Phi(n)$, A has a model in which $P_1, \ldots, P_n$ take the values $i(P_1), \ldots, i(P_n)$. So, $i(A) = 1$.

Let $i(P_1) = 0$ and suppose $\Phi(1)$ does not hold. That is, there is a finite subset Γ' of Γ which has no model in which P_1 takes the value $i(P_1) = 0$. Then we define $i(P_1) = 1$ and show that $\Phi(1)$, i.e., every finite subset of Γ has a model in which P_1 takes the value $i(P_1) = 1$. For let Δ be a finite subset of Γ. Then $\Delta \cup \Gamma'$ is a finite subset of Γ and hence has a model i. Since i is a model of Γ', $i(P_1) = 1$.

Suppose we have defined $i(P_1), \ldots, i(P_n)$ such that $\Phi(n)$. Then we can extend the definition of i to P_{n+1} such that $\Phi(n+1)$. For suppose that $\Phi(n+1)$ does not hold if $i(P_{n+1}) = 0$. That is, there is a finite subset Γ' of Γ which has no model in which $P_1, \ldots, P_n, P_{n+1}$ take the values $i(P_1), \ldots, i(P_n), 0$. Then we define $i(P_{n+1}) = 1$ and show that $\Phi(n+1)$, i.e., every finite subset of Γ has a model in which $P_1, \ldots, P_n, P_{n+1}$ take the values $i(P_1), \ldots, i(P_n), 1$. For let Δ be a finite subset of Γ. Then $\Delta \cup \Gamma'$ is a finite subset of Γ and hence, by the induction hypothesis, $\Delta \cup \Gamma'$ has a model in which $P_1, \ldots, P_n$ take the values $i(P_1), \ldots, i(P_n)$. Since i is a model of Γ', $i(P_{n+1}) = 1$. □

For applications of the compactness theorem in mathematics see Exercises 2.16, 2.17 and 2.18.

Exercise 2.7. Show that the formulas in the pairs below have the same truth table:
a) $\neg(A \wedge B)$ and $\neg A \vee \neg B$.
b) $\neg(A \vee B)$ and $\neg A \wedge \neg B$.
c) $\neg A \vee B$ and $A \to B$.
d) $\neg(A \to B)$ and $A \wedge \neg B$.
e) $A \to B$ and $\neg B \to \neg A$.
f) $A \to B$ and $\neg(A \wedge \neg B)$.

Exercise 2.8. Compute and compare the truth tables for:
a) $P_1 \wedge P_2 \to \neg P_3$ and $P_1 \wedge (P_2 \to \neg P_3)$ (see Exercise 2.1).
b) $P_1 \vee (P_2 \to P_3)$ and $P_1 \vee P_2 \to P_3$ (see Exercise 2.1).
c) $P_1 \to (P_2 \to \neg P_3)$ and $(P_1 \to P_2) \to \neg P_3$ (see Exercise 2.1 and 2.2).
d) $\neg P_1 \vee P_3$ and $\neg(P_1 \vee P_3)$ (see Exercise 2.2).
e) $\neg P_2 \wedge P_3$ and $\neg(P_2 \wedge P_3)$ (see Exercise 2.2).

Exercise 2.9. Prove that a) $(A \vee \neg A) \to B$ has the same truth table as B,
b) $(A \vee \neg A) \wedge B$ has the same truth table as B, and
c) $(A \wedge \neg A) \vee B$ has the same truth table as B.

Exercise 2.10. Prove that $A \vee B$, $(A \to B) \to B$ and $(B \to A) \to A$ all have the same truth table.

Exercise 2.11. Verify that the following formulas are valid by showing that it is impossible that at some line in the truth table they have the value 0.
a) $\neg\neg A \to A$
b) $(A \to B) \vee (B \to A)$
c) $(P \to Q) \to (\neg Q \to \neg P)$.

Exercise 2.12. Show that the following formulas are not valid by computing just one suitable line of the table: a) $P \vee Q \to P \wedge Q$ b) $(P \to Q) \to (Q \to P)$.

Exercise 2.13. Which of the following alternatives applies to the following formulas?
1. $P_1 \to \neg P_1$
2. $P_1 \rightleftarrows \neg P_1$
3. $P_1 \to P_1 \wedge P_2$
4. $P_1 \to P_1 \vee P_2$
5. $P_1 \to P_2$
6. $(P_1 \to P_2) \rightleftarrows (\neg P_1 \vee P_2)$
7. $\neg(P_1 \to P_2) \rightleftarrows (P_1 \wedge \neg P_2)$
8. $\neg(P_1 \wedge P_2) \rightleftarrows (\neg P_1 \vee \neg P_2)$
9. $\neg(P_1 \vee P_2) \rightleftarrows (\neg P_1 \wedge \neg P_2)$
10. $(\neg P_1 \vee P_2) \rightleftarrows (P_1 \to P_2)$

Alternative A: not satisfiable (inconsistent).
B: satisfiable (consistent), but not valid.
C: valid, and hence satisfiable.

Exercise 2.14. Show that each formula built by means of connectives from only one atomic formula P has the same truth table as either $P \wedge \neg P$, P, $\neg P$ or $P \to P$.

2.3 Semantics; Logical (Valid) Consequence

Exercise 2.15. Consider the following truth table for the *exclusive* 'or', $\underline{\vee}$.

A	B	$A \underline{\vee} B$
1	1	0
1	0	1
0	1	1
0	0	0

a) Verify that $A \underline{\vee} B$ has the same truth table as $(A \vee B) \wedge \neg(A \wedge B)$ and as $\neg(A \rightleftarrows B)$.
b) Verify that $(A \underline{\vee} B) \underline{\vee} C$ and $A \underline{\vee} (B \underline{\vee} C)$ have the same truth table and in particular that these formulas have the value 1 in the first line of the truth table (where A, B and C are 1). Note that this does not correspond with the intended meaning of 'A or B or C', if the 'or' is used exclusively.

Exercise 2.16. * (Kreisel-Krivine [18]) A group G is said to be *ordered* if there is a total ordering < of G (see Chapter 3) such that $a \leq b$ implies $ac \leq bc$ and $ca \leq cb$ for all c in G. Show that a group G can be ordered if and only if every subgroup of G generated by a finite number of elements of G can be ordered.

Exercise 2.17. * (Kreisel-Krivine [18]) A graph (a non-reflexive symmetric relation) defined on a set V is said to be *k-chromatic*, where k is a positive integer, if there is a partition of V into k disjoint sets $V_1, \ldots, V_k$, such that two elements of V connected by the graph do not belong to the same V_i. Show that for a graph to be k-chromatic it is necessary and sufficient that every finite sub-graph be k-chromatic.

Exercise 2.18. * Suppose that each of a (possibly infinite) set of boys is acquainted with a finite set of girls. Under what conditions is it possible for each boy to marry one of his acquaintances? It is clearly necessary that every finite set of k boys be, collectively, acquainted with at least k girls. The *marriage theorem* says that this condition is also sufficient. More precisely, let B and G be sets (of Boys and Girls respectively) and let $R \subseteq B \times G$ be such that (i) for all $x \in B$, $R_{\{x\}}$ is finite, and (ii) for every finite subset $B' \subseteq B$, $R_{B'}$ has at least as many elements as B', where $R_{B'} := \{y \in G \mid \text{for some } x \text{ in } B', R(x, y)\}$. Then there is an injection $f : B \to G$ such that for all $x \in B$ and $y \in G$, if $f(x) = y$, then $R(x, y)$. In **The Marriage Problem** (*American Journal of Mathematics*, Vol. 72, 1950, pp. 214-215) P. Halmos and H. Vaughan prove first the case in which the number of boys is finite. Using this result prove the marriage theorem for the case that B is infinite.

2.3 Semantics; Logical (Valid) Consequence

Consider the following concrete argument:
    John is intelligent [I] or John is diligent [D].
    If John is intelligent, then he will succeed [S].
    If John is diligent, then he will succeed (too).
    Therefore: John will succeed.

We may translate the propositions in this argument into formulas:
$$\frac{\begin{array}{l} I \vee D \\ I \to S \\ D \to S \end{array}}{S}$$

To help our memory, for convenience we have used the symbols I, D and S instead of P_1, P_2, P_3. Intuitively, this pattern of reasoning is valid: no matter what propositions I, D, S stand for, if all premises are true, the conclusion must be true too; in other words, it is impossible that the premises are all true and at the same time the conclusion false. Now we have given in Section 2.2 a precise meaning to the atomic formulas and to the connectives in terms of truth tables, we can make the notion of *valid* or *logical consequence* precise: in the truth table starting with I, D and S, at each line in which all of $I \vee D$, $I \to S$ and $D \to S$ have the value 1, also S must have the value 1; in other words: there is no line in the truth table starting with I, D, and S in which the premises $I \vee D$, $I \to S$, $D \to S$ are all 1 and the conclusion S is 0.

I	D	S	$I \vee D$	$I \to S$	$D \to S$	S
1	1	1	1	1	1	1
1	1	0	1	0	0	0
1	0	1	1	1	1	1
1	0	0	1	0	1	0
0	1	1	1	1	1	1
0	1	0	1	1	0	0
0	0	1	0	1	1	1
0	0	0	0	1	1	0

In this example there are three lines, line 1, 3 and 5, in which all premises are true and, as we can see, in each of these lines also the conclusion is true. So, in each case that all premises are true, the conclusion is true too. We say that S is a *valid* or *logical consequence* of the premises $I \vee D$, $I \to S$ and $D \to S$.

Definition 2.5 (Logical or valid consequence).
a) B is a *logical* or *valid consequence* of premises $A_1, \ldots, A_n :=$ in each line of the truth table for $A_1, \ldots, A_n$ and B in which all premises $A_1, \ldots, A_n$ are 1, also B is 1; in other words, there is no line in the truth table in which all premises $A_1, \ldots, A_n$ are 1 and at the same time B is 0. **Notation:** $A_1, \ldots, A_n \models B$.
b) Let Γ be a (possibly infinite) set of formulas. B is a *logical* or *valid consequence* of $\Gamma :=$ for each interpretation i, if $i(A) = 1$ for all formulas A in Γ, then also $i(B) = 1$. In other words, each interpretation which is a model of all formulas in Γ is also a model of B. **Notation:** $\Gamma \models B$.

The notion of logical (or valid) consequence is a *semantical* notion: it concerns the truth or falsity, and hence the meaning, of the formulas in question. Notice that in case $n = 0$, i.e., there are no premises, the definition of $A_1, \ldots, A_n \models B$ reduces to the definition of $\models B$: there is no line in the truth table for B in which B is 0.

Next consider the following argument.
      If the weather is nice [N], then John will come [C].
      The weather is not nice.

2.3 Semantics; Logical (Valid) Consequence

Therefore: John will not come.

We may translate these propositions into the following formulas:
$$\frac{\begin{array}{c} N \to C \\ \neg N \end{array}}{\neg C}$$

Again, for convenience, we have used the symbols N and C instead of the atomic formulas P_1, P_2 in order to help our memory.

Intuitively, this argument is not correct: John may also come when the weather is not nice; for instance, because someone offers John to bring him by car. So, the premisses may be true, while the conclusion is false. We see this clearly in the truth table for the formulas in question:

N	C	$N \to C$	$\neg N$	$\neg C$
1	1	1	0	0
1	0	0	0	1
0	1	1	1	0
0	0	1	1	1

There are two lines in the truth table in which both premisses are 1 (true): line 3 and line 4. In line 4 the conclusion $\neg C$ is 1 too, but in line 3 the conclusion is 0! Line 3 is the case that N is 0 and C is 1, i.e., the weather is not nice, while John does come; in this case both premisses $N \to C$ and $\neg N$ are true, while the conclusion $\neg C$ is false. So, there is a line in the truth table, in which all premisses are true, while the conclusion is false; in other words, $\neg C$ is not a logical consequence of $N \to C$ and $\neg N$. Therefore, not $N \to C, \neg N \models \neg C$ or $N \to C, \neg N \not\models \neg C$.

Notation: Instead of 'not $A_1, \ldots, A_n \models B$' one usually writes: $A_1, \ldots, A_n \not\models B$.

Another intuitive counterexample is the following one; Suppose Berta is a cow and interpret N as 'Berta is a dog' and C as 'Berta has four legs'. Then we have the situation of line 3 in the table: N is 0, C is 1, $N \to C$ is 1, $\neg N$ is 1, but $\neg C$ is 0.

Theorem 2.4.
a) $A \models B$ if and only if (iff) $\models A \to B$.
More generally,
b) $A_1, A_2 \models B$ if and only if (iff) $A_1 \models A_2 \to B$
     if and only if (iff) $\models A_1 \to (A_2 \to B)$
     if and only if (iff) $\models A_1 \wedge A_2 \to B$.
Even more generally,
c) $A_1, \ldots, A_n \models B$ if and only if (iff) $A_1, \ldots, A_{n-1} \models A_n \to B$
     if and only if (iff) $\models (A_1 \wedge \ldots \wedge A_n) \to B$.

Proof. a) $A \models B$ iff there is no line in the truth table in which A is 1 and B is 0. This is equivalent to: there is no line in the truth table in which $A \to B$ is 0. In other words, equivalent to: $\models A \to B$.
b) $A_1, A_2 \models B$ iff there is no line in the truth table in which A_1 and A_2 are both 1 and B is 0. This is equivalent to: there is no line in the truth table in which A_1 is 1 and $A_2 \to B$ is 0, i.e., $A_1 \models A_2 \to B$. This is - in its turn - equivalent to: there is no line in the truth table in which $A_1 \to (A_2 \to B)$ is 0, i.e., $\models A_1 \to (A_2 \to B)$. Or equivalently, there is no line in the truth table in which $(A_1 \wedge A_2) \to B$ is 0, i.e., $\models (A_1 \wedge A_2) \to B$.
c) Similarly.   □

It is important to notice that $A \to B$ is a formula of the logical language, while $\models A \to B$, or equivalently $A \models B$, is a statement in the meta-language *about* the formulas A and B, namely, that there is no line in the truth table in which A is 1 and B is 0. The symbol $\models$ does not occur in the logical language, but is just an abbreviation from the metalanguage.

Definition 2.6. In the statement $A_1,\ldots,A_n \models B$ we call $A_1,\ldots,A_n$ the *premisses* and B the (putative) *conclusion*. In particular, in $A \models B$ we call A the premiss and B the conclusion. However, in the formula $A \to B$ we call A the *antecedent* and B the *succedent*.

Theorem 2.5. * *Let Γ be a (possibly infinite) set of formulas. B is a valid consequence of Γ ($\Gamma \models B$) if and only if there are finitely many formulas $A_1,\ldots,A_n$ in Γ such that B is a valid consequence of $A_1,\ldots,A_n$ ($A_1,\ldots,A_n \models B$).*

Proof. The 'if' part is evident. To show the 'only if' part, suppose that $\Gamma \models B$, that is, $\Gamma \cup \{\neg B\}$, i.e., the set consisting of $\neg B$ and of all formulas in Γ, does not have a model. Then, according to the Compactness Theorem 2.3, there is a finite subset $\Gamma' = \{A_1,\ldots,A_n\}$ of formulas in Γ such that $\{A_1,\ldots,A_n\} \cup \{\neg B\}$ does not have a model, which means that $A_1,\ldots,A_n \models B$. □

2.3.1 Decidability

The notion of validity (for the classical propositional calculus) is clearly *decidable*, i.e., there is an algorithm (an effective computational procedure), also called a *decision procedure*, to determine for any formula A in a finite number of steps (depending on the complexity of A) whether it is valid or not. Namely, in order to determine whether A is valid or not, we simply have to compute the truth table of A, entered from its atomic components, and see whether it has 1 in all its lines or not. Computing a truth table of a given formula A and checking whether it has 1 in all its lines can be carried out by a machine and yields an answer 'yes' or 'no' in finitely many steps, the number of steps depending on the complexity of A. Because $A_1,\ldots,A_n \models B$ is equivalent to $\models A_1 \wedge \ldots \wedge A_n \to B$, also the notion of valid consequence (of a finite number of premisses) is clearly decidable.

One of Leibniz' ideals was to develop a *lingua philosophica* or *characteristica universalis*, an artificial language that in its structure would mirror the structure of thought and that would not be affected with ambiguity and vagueness like ordinary language. His idea was that in such a language the linguistic expressions would be pictures, as it were, of the thoughts they represent, such that signs of complex thoughts are always built up in a unique way out of the signs for their composing parts. Leibniz (1646 - 1716) believed that such a language would greatly facilitate thinking and communication and that it would permit the development of mechanical rules for deciding all questions of consistency or consequence. The language, when it is perfected, should be such that 'men of good will desiring to settle a controversy on any subject whatsoever will take their pens in their hands and say *Calculemus* (let us calculate)'. If we restrict ourselves to the propositional calculus, Leibniz' ideal has been realized: the classical propositional calculus is decidable, more precisely, given premises $A_1, \ldots, A_n$ and a putative conclusion B, one may decide whether B is a logical consequence of $A_1, \ldots, A_n$ by simply calculating the truth tables of $A_1, \ldots, A_n, B$. However, A. Church and A. Turing proved in 1936 that the predicate calculus is undecidable, i.e., there is no mechanical method to test logical consequence (in the predicate calculus), let alone philosophical truth.

For more information the reader is referred to W. & M. Kneale [16], *The Development of Logic* and to B. Mates [20], *Elementary Logic*, Chapter 12.

Now, if A has been built from n atomic formulas, the truth table of A has 2^n lines. So, a formula built from 10 atomic formulas has a truth table with $2^{10} = 1024$ lines. And if $n = 20$, the truth table of A has $2^{20} = 2^{10} \times 2^{10} = 1024 \times 1024$, so more than a million lines. Hence, the number of steps needed to decide whether a given formula A is valid or not grows fast if A becomes more complex. In fact, if A has been built from 64 atomic formulas, it will take many lifetimes in order to compute whether A is valid or not, even with very futuristic computers, the number of lines being $2^{64} = 2^4 \times (2^{10})^6 \approx 16 \times (10^3)^6 = 16 \times 10^{18}$. In Subsection 2.5.3 we will construct such a formula, built from 64 atomic formulas, to describe a particular travelling salesman problem. Supposing a computer computes $16000 = 16 \times 10^3$ lines per second, in one human lifetime it can compute about 100 (years) × 365 (days) × 24 (hours) × 60 (minutes) × 60 (seconds) × 16000 (lines) ≈ 16×10^{13} lines. So, in order to compute a truth table of a formula built from 64 atomic formulas, our computer needs about $(16 \times 10^{18}) / (16 \times 10^{13}) = 10^5$ human lifetimes, supposing it can compute 16000 lines per second. This means that our decision procedure to determine whether a given formula A (of the propositional calculus) is valid or not, is a rather theoretical one if the complexity of A is great, more precisely, if A has been built from say 64 atomic components.

One may wonder whether there are more effective or more realistic decision procedures to determine validity, other than making the truth table and checking whether it has 1 in all its lines. No such procedure is known, although for many concrete formulas ad hoc solutions can give a quick answer to the question whether they are valid or not. But no (general) procedure is known, other than making truth tables, to determine the validity of an arbitrary formula.

2.3.2 Sound versus Plausible Arguments; Enthymemes

A concrete argument consists of a number of premises and a (putative) conclusion. The atomic propositions of the argument are translated into atomic formulas $P_1, P_2, \ldots$ and the composite propositions of the argument are translated into composite formulas which are composed by the logical connectives from the atomic formulas. The result is a logical reasoning pattern:

premisses
|
logical | reasoning
|
conclusion

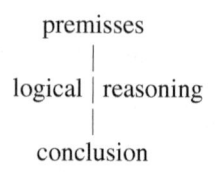

A reasoning pattern is *valid* if it is impossible that the premisses are true and at the same time the conclusion false. A concrete argument is *correct* if the underlying reasoning pattern is valid, otherwise it is incorrect.

The correctness of a concrete argument is not determined by the content or meaning of the atomic propositions in question, but by the meaning of the propositional connectives (and in predicate logic also by the meaning of the quantifiers) which occur in the argument. That is why one abstracts from the content of the atomic propositions in question by translating them into $P_1, P_2, \ldots$, as pointed out by Frege [8] in his Begriffsschrift (1879).

The atomic formulas may be interpreted as true or false propositions, denoted by 1 and 0 respectively, and the meaning of the logical connectives is specified precisely in the truth tables. Validity of a reasoning pattern means that for every interpretation of the atomic formulas it is impossible that the premisses become true propositions while the conclusion becomes a false proposition.

In his Begriffsschrift [8] of 1879 Gottlob Frege compares the use of the logical language with the use of a microscope. Although the eye is superior to the microscope, for certain distinctions the microscope is more appropriate than the naked eye. Similarly, although natural language is superior to the logical language, for judging the correctness of a certain argument the logical language is more appropriate than natural language. Since the content or meaning of the atomic propositions does not matter for the correctness of the argument, it is more convenient to abstract from this content by replacing the atomic propositions by atomic formulas $P_1, P_2, \ldots$.

It is possible that the study of logic does not augment our native capacity to *discover* correct arguments; but it certainly is of value in *checking* the correctness of given arguments. However, the reader should realize that at this stage we are not yet able to give an adequate logical analysis of, for instance, the following argument.

>All men are mortal.
>Socrates is a man.
>Therefore: Socrates is mortal.

In order to see the correctness of this argument one has to take into account the internal subject-predicate structure of the atomic propositions involved, and this is precisely what is ignored in the propositional calculus and what we shall study in the predicate calculus; see Chapter 4. Using only the means of the propositional calculus, all we can say is that the foregoing argument is of the form $P, Q \models R$, which does not hold, because we may interpret P and Q as true propositions and R as a false one; in other words, P and Q may have the value 1, while R may have the value 0. In order to see the correctness of the argument above, one has to analyse the internal subject-predicate structure of the atomic formulas P, Q and R; but this is

2.3 Semantics; Logical (Valid) Consequence

beyond the scope of the propositional calculus. In the propositional calculus we can adequately analyse only those arguments the correctness of which depends on the way the composite propositions are composed of the atomic propositions by means of the propositional operations.

Arguments are frequently used to persuade the hearer of the truth of the conclusion on the grounds that (i) the conclusion logically follows from the premises and in addition (ii) the premises are true. Let us use $A_1, \ldots, A_n :: B$ to denote
(i) $A_1, \ldots, A_n \models B$, and
(ii) $A_1, \ldots, A_n$ are true; and therefore B is true.

When both (i) and (ii) hold, we call the argument not simply 'valid', but *sound*. And we call an argument *plausible*, when it is valid, but we can only say that $A_1, \ldots, A_n$ are plausible.

It frequently happens that speakers in giving an argument do not explicitly mention *all* their premises; in some cases they even leave the conclusion tacit. For instance, if someone offers me coffee, I might respond as follows:

If I drink coffee [C], I can't get to sleep early [$\neg S$]. So please don't pour me any.

The argument given is of the form $C \to \neg S :: \neg C$, which is clearly an abbreviation for $C \to \neg S,\ S :: \neg C$.

I might even leave out the conclusion; if I have just been offered a cup of coffee, simply $C \to \neg S$ might be sufficient not to let the hostess pour me any coffee.

Arguments in which one or more premises or the conclusion is tacit are called *enthymemes*. Premises may not be explicitly stated for practical reasons, but also to mislead the audience.

Exercise 2.19. Translate the propositions in the following argument into formulas of the language of propositional logic and check whether the (putative) conclusion is a logical (or valid) consequence of the premises:
If the government raises taxes for its citizens, the unemployment grows.
The unemployment does not grow or the income of the state decreases.
Therefore: if the government raises taxes, then the income of the state decreases.

Exercise 2.20. Translate the propositions in the following argument into formulas of the language of propositional logic and check whether the putative conclusion is a logical (or valid) consequence of the premises:
Europe may form a monetary union only if it is a political union.
Europe is not a political union or all European countries are member of the union.
Therefore: If all European countries are a member of the union, then Europe may form a monetary union.

Exercise 2.21. Verify by making truth tables:
a) $A, A \to B \models B$ b) $A \to B, \neg B \models \neg A$ c) $A, \neg A \models B$
d) $A \to B \not\models B \to A$ e) $A \to B, \neg A \not\models \neg B$ f) $A \to (B \lor C) \models (A \to B) \lor (A \to C)$
g) $A \lor B, \neg A \models B$ h) $\neg(A \land B), A \models \neg B$

Exercise 2.22. Translate the propositions in the following argument into formulas of the language of propositional logic and check whether the putative conclusion is

a logical (or valid) consequence of the premisses:
John does not win the lottery or he makes a journey [J].
If John does not make a journey, then he does not succeed for logic.
John wins the lottery [W] or he succeeds for logic [S].
Therefore: John makes a journey.

Exercise 2.23. Translate the propositions in the following argument into formulas of the language of propositional logic and check whether the putative conclusion is a logical (or valid) consequence of the premisses:
If Turkey joins the EU [T], then the EU becomes larger [L].
It is not the case that the EU becomes stronger [S] and at the same time not larger.
Therefore: Turkey does not join the EU or the EU becomes stronger.

2.4 Semantics: Meta-logical Considerations

In this section we will prove results *about* the notions of validity and valid consequence of the type: if certain formulas are valid, then also some other formulas are valid.

Suppose we want to determine whether the formula $(P_3 \land \neg P_4) \land (\neg P_4 \lor P_5 \lor P_6) \to (P_3 \land \neg P_4)$ is valid. Making the truth table of this formula, starting with the atomic formulas P_3, P_4, P_5, P_6 occurring in it, will yield a positive answer. But this table contains $2^4 = 16$ rows and the chance of making a computational mistake is considerable. However, notice that the formula has the form $P_1 \land P_2 \to P_1$ with P_1 replaced by $A_1 = (P_3 \land \neg P_4)$ and P_2 replaced by $A_2 = (\neg P_4 \lor P_5 \lor P_6)$. Although the table for $A_1 \land A_2 \to A_1$ may consist of many lines, 16 in our example, there cannot be more than 4 different combinations of 1 and 0 for A_1 and A_2. In our example the second row, in which $A_1 = P_3 \land \neg P_4$ has value 1 and $A_2 = \neg P_4 \lor P_5 \lor P_6$ has value 0, will even not occur, because if $\neg P_4$ is 1, then also $A_2 = \neg P_4 \lor P_5 \lor P_6$ is 1.

A_1 A_2	$A_1 \land A_2 \to A_1$
1 1	$(1 \land 1) \to 1 = 1$
1 0	$(1 \land 0) \to 1 = 1$
0 1	$(0 \land 1) \to 0 = 1$
0 0	$(0 \land 0) \to 0 = 1$

All four possible combinations of 1 and 0 for A_1 and A_2 will yield for $A_1 \land A_2 \to A_1$ the value 1. So, from the fact that the formula $P_1 \land P_2 \to P_1$ is valid, we may conclude that also the formula $A_1 \land A_2 \to A_1$ is valid for any formulas A_1 and A_2; in particular, that the formula $(P_3 \land \neg P_4) \land (\neg P_4 \lor P_5 \lor P_6) \to (P_3 \land \neg P_4)$ is valid. What we have won is that the table for $P_1 \land P_2 \to P_1$ requires only the computation of 4 instead of 16 rows.

The substitution theorem below reduces the amount of work needed to establish the validity of certain formulas.

2.4 Semantics: Meta-logical Considerations

Theorem 2.6 (Substitution theorem). *Let $E(P_1,P_2)$ be a formula containing only the atomic formulas P_1,P_2, and let $E(A_1,A_2)$ result from $E(P_1,P_2)$ by substituting formulas A_1,A_2 simultaneously for P_1,P_2, respectively.*

$$\text{If} \models E(P_1,P_2), \text{ then } \models E(A_1,A_2).$$

More generally: if $\models E(P_1,\ldots,P_n)$, then $\models E(A_1,\ldots,A_n)$, where the latter formula results from the former one by replacing the atomic formulas $P_1,\ldots,P_n$ by the (composite) formulas $A_1,\ldots,A_n$.

So, since $\models P_1 \to P_1$, the substitution theorem tells us that
$\models P_2 \wedge \neg P_3 \to P_2 \wedge \neg P_3$ $\quad (A_1 = P_2 \wedge \neg P_3)$
$\models (P_3 \to P_5 \wedge \neg P_7) \to (P_3 \to P_5 \wedge \neg P_7)$ $\quad (A_1 = P_3 \to P_5 \wedge \neg P_7)$
and so on. So, the purpose of the substitution theorem is to reduce the amount of work needed to establish the validity of certain formulas.

Proof. Suppose $E = E(P_1,\ldots,P_n)$ contains only the atomic formulas $P_1,\ldots,P_n$ and $\models E$, i.e., the truth table of E entered from the atomic formulas $P_1,\ldots,P_n$ is 1 in each line.

P_1	...	P_n	...	E
1	...	1	...	1
:		:		:
0	...	0	...	1

Now $E^* = E(A_1,\ldots,A_n)$ results from E by substituting the formulas $A_1,\ldots,A_n$ for the atomic formulas $P_1,\ldots,P_n$ in E. Let us suppose that the formulas $A_1,\ldots,A_n$ and hence also E^* are built from the atomic formulas $Q_1,\ldots,Q_k$. Then the computation of the truth table of E^* is as follows.

Q_1	...	Q_k	...	A_1	...	A_n	...	E^*
1	...	1	...					
:		:		:		:		:
0	...	0	...					

Since the construction of E^* from $A_1,\ldots,A_n$ is the same as the construction of E from $P_1,\ldots,P_n$, the truth table of E^* is computed from those of $A_1,\ldots,A_n$ in precisely the same manner as the truth table of E is computed from those of $P_1,\ldots,P_n$. Hence, because by assumption the computation of the values of E from the values for $P_1,\ldots,P_n$ only yield 1's, also the computation of the values of E^* from the values for $A_1,\ldots,A_n$ will only yield 1's. I.e., $\models E^*$.

Note that it may happen that some combinations of 0's and 1's for $A_1,\ldots,A_n$ do not occur. For instance, if $A_1 = Q_1 \vee \neg Q_1$, then A_1 will have the value 1 in all lines and the value 0 for A_1 will not occur. □

Remark 2.1. : The converse of the substitution theorem, if $\models E^*$, then $\models E$, does not hold. For instance, let $E(P_1) = P_1$ and let $A_1 = P_2 \to P_2$. Then $E^* = E(A_1) = P_2 \to P_2$ is valid, but $E(P_1) = P_1$ is not valid.

In the next theorem the validity of many formulas is shown by means of the substitution theorem. For example, 5b) says that for any choice of formulas A and B, $B \to A \vee B$ is valid. Taking $A = P_1 \wedge \neg P_2$ and $B = P_2 \to P_3$, we find that $(P_2 \to P_3) \to ((P_1 \wedge \neg P_2) \vee (P_2 \to P_3))$ is valid. This method of proving the validity of the latter formula is much more economical than proving the validity directly from its definition by making the truth table of the latter formula entered from the atomic components P_1, P_2 and P_3; this table would consist of eight lines!

Theorem 2.7. *For any choice of formulas A, B, C:*

1	$\models A \to (B \to A)$	or $A \models B \to A$ or $A, B \models A$
2	$\models (A \to B) \to ((A \to (B \to C)) \to (A \to C))$	or $A \to B, A \to (B \to C), A \models C$
3	$\models A \to (B \to A \wedge B)$	or $A, B \models A \wedge B$
4a	$\models A \wedge B \to A$	or $A \wedge B \models A$
4b	$\models A \wedge B \to B$	or $A \wedge B \models B$
5a	$\models A \to A \vee B$	or $A \models A \vee B$
5b	$\models B \to A \vee B$	or $B \models A \vee B$
6	$\models (A \to C) \to ((B \to C) \to (A \vee B \to C))$	or $A \to C, B \to C \models A \vee B \to C$
7	$\models (A \to B) \to ((A \to \neg B) \to \neg A)$	or $A \to B, A \to \neg B \models \neg A$
8	$\models \neg\neg A \to A$	or $\neg\neg A \models A$
9	$\models (A \to B) \to ((B \to A) \to (A \rightleftarrows B))$	or $A \to B, B \to A \models A \rightleftarrows B$
10a	$\models (A \rightleftarrows B) \to (A \to B)$	or $A \rightleftarrows B \models A \to B$
10b	$\models (A \rightleftarrows B) \to (B \to A)$	or $A \rightleftarrows B \models B \to A$

Proof. The statements in the right column, after the 'or', are according to Theorem 2.4 equivalent to the corresponding statements in the left column, before the 'or'. The statements in the left column follow from the substitution theorem. For instance, to show 1, $\models A \to (B \to A)$, it is easy to verify that $\models P_1 \to (P_2 \to P_1)$, from which it follows by the substitution theorem that for any formulas A, B, $\models A \to (B \to A)$. □

The student is not expected to learn the list in Theorem 2.7 outright now. In the course of time he or she will become familiar with the most frequently used results.

Later in Section 2.9 it will be shown that all valid formulas may be obtained (or deduced) by applications of Modus Ponens to formulas of the ten forms in Theorem 2.7; this is the so-called *completeness theorem* for propositional logic. For that reason formulas of the form 1, ..., 10 in Theorem 2.7 are called logical *axioms for (classical) propositional logic*. Notice that the formulas in 1 and 2 concern $\to$, the formulas in 3 and 4 concern $\wedge$, the formulas in 5 and 6 concern $\vee$, the formulas in 7 and 8 concern $\neg$ and the formulas in 9 and 10 concern $\rightleftarrows$. For instance, the formulas in 1 and 2 would not be valid if the $\to$ were replaced by any other connective. The completeness theorem says essentially that formulas of these ten forms together characterize the meanings of $\rightleftarrows$, $\to$, $\wedge$, $\vee$ and $\neg$: every valid formula may be obtained by applications of Modus Ponens to formulas of these ten forms.

Paradoxes of Material Implication $\models A \to (B \to A)$, or, equivalently, $A \models B \to A$, and $\models \neg A \to (A \to B)$, or, equivalently, $\neg A \models A \to B$, have been called *paradoxes of*

2.4 Semantics: Meta-logical Considerations

material implication. This has been illustrated by examples like the following ones: $A \models B \rightarrow A$: I like coffee; therefore: if there is oil in my coffee, I like coffee. $\neg A \models A \rightarrow B$: I do not break my legs; therefore: if I break my legs, I will go for skying. This sounds very strange indeed. However, Paul Grice [10] has pointed out that in conversation one is supposed to take social rules into account, such as being relevant and maximally informative. And although $B \rightarrow A$ is true when A is true, it is simply misleading to say $B \rightarrow A$, or equivalently $\neg B \vee A$, when one knows that A is true, because A is clearly more informative than $B \rightarrow A$ or, equivalently, $\neg B \vee A$. Similarly, although $A \rightarrow B$ is true when $\neg A$ is true, it is misleading to say $A \rightarrow B$, or, equivalently, $\neg A \vee B$, when one has the information $\neg A$, because $\neg A$ is clearly more informative than $A \rightarrow B$ or, equivalently $\neg A \vee B$.

Also the proof of the next theorem is by showing that one obtains valid formulas if one replaces A, B, C by the atomic formulas P_1, P_2, P_3; next application of the substitution theorem yields the desired result.

Theorem 2.8. *For any formulas A, B, C:*

11	$\models \neg\neg A \rightleftarrows A$	*law of double negation*
12	$\models A \vee \neg A$	*law of excluded middle*
13	$\models \neg(A \wedge \neg A)$	*law of non-contradiction*
14	$\models \neg A \rightarrow (A \rightarrow B)$ or $\neg A, A \models B$	*ex falso sequitur quod libet*
15	$\models (A \rightarrow B) \rightarrow ((B \rightarrow C) \rightarrow (A \rightarrow C))$	or $A \rightarrow B, B \rightarrow C \models A \rightarrow C$

From the table for $\rightleftarrows$ follows immediately the next theorem.

Theorem 2.9. *Let A, B be any formulas. $\models A \rightleftarrows B$ if and only if A and B have the same truth table.*

Proof. Suppose $\models A \rightleftarrows B$. Then from the table for $\rightleftarrows$ it follows that it is impossible that in some line of the truth table one of A, B is 1 while the other is 0. Conversely, suppose A and B have the same truth table. Then in every line of the truth table both formulas are 1 or both formulas are 0. In either case $A \rightleftarrows B$ is 1. Since this holds for every line in the truth table, $\models A \rightleftarrows B$. □

Theorem 2.10. *For any formulas A, B, C:*

16	$\models (A \rightarrow B) \rightleftarrows (\neg B \rightarrow \neg A)$	*contraposition*
17a	$\models \neg(A \vee B) \rightleftarrows \neg A \wedge \neg B$	*De Morgan's laws 1847*
17b	$\models \neg(A \wedge B) \rightleftarrows \neg A \vee \neg B$	
18	$\models \neg(A \rightarrow B) \rightleftarrows A \wedge \neg B$	
19	$\models (A \rightleftarrows B) \rightleftarrows (A \rightarrow B) \wedge (B \rightarrow A)$	
20	$\models A \rightarrow B \rightleftarrows \neg(A \wedge \neg B)$	
21	$\models A \rightarrow B \rightleftarrows \neg A \vee B$	
22	$\models A \wedge (B \vee C) \rightleftarrows (A \wedge B) \vee (A \wedge C)$	*distributive law*
23	$\models A \vee (B \wedge C) \rightleftarrows (A \vee B) \wedge (A \vee C)$	*distributive law*
24	$\models A \rightarrow (B \rightarrow C) \rightleftarrows B \rightarrow (A \rightarrow C)$	
25	$\models A \rightarrow (B \rightarrow C) \rightleftarrows A \wedge B \rightarrow C$	

Proof. One easily verifies that $A \to B$ and $\neg B \to \neg A$ have the same truth truth table. Hence, by theorem 2.9 it follows that $\models (A \to B) \rightleftarrows (\neg B \to \neg A)$. Another way of showing this is to verify that $\models (P_1 \to P_2) \rightleftarrows (\neg P_2 \to \neg P_1)$, simply by computing the truth table. Next the substitution theorem 2.6 yields the desired result.
The other items are shown similarly. □

A reasoning rule like Modus Ponens or Modus Tollens should, of course, be *sound*, i.e., if its premises are true (1), then its conclusion must be true (1) too. In other words, these rules should preserve truth. One easily verifies that Modus Ponens and Modus Tollens are sound.

Theorem 2.11. *(a) For every line in the truth table: if A is 1 and $A \to B$ is 1 in that line, then B is also 1 in that line. In other words: $A, A \to B \models B$.*
We say that the rule of Modus Ponens (MP) is sound. Consequently:
(b) For all formulas A and B, if $\models A$ and $\models A \to B$, then $\models B$. In other words: for all formulas A and B, if $\models A \to B$, then if (in addition) $\models A$, then $\models B$.
(c) However, not for all formulas A and B, if (if $\models A$, then $\models B$), then $\models A \to B$.

Proof. (a) follows immediately from the truth table for $\to$.
From (a) follows: if A is 1 in all lines and $A \to B$ is 1 in all lines of the truth table, then B is 1 in all lines of the truth table. In other words, if $\models A$ and $\models A \to B$, then $\models B$. This proves (b).
(c) 'if $\models A$, then $\models B$' means: if A is 1 in all lines of the truth table, then B is 1 in all lines of the truth table (*). $\models A \to B$ means: in every line in which A is 1, B must be 1 too. Notice that this does *not* follow from (*). For suppose that A is 1 in *some* line of the truth table, we do not know whether A is 1 in *all* lines of its truth table. In fact, there are formulas A and B such that 'if $\models A$, then $\models B$' holds, while $\models A \to B$ does not hold. For example, take $A = P_1$ (it is cold) and $B = P_2$ (it is snowing). Since $\not\models P_1$ (not always it is cold) and $\not\models P_2$ (not always it is snowing), 'if $\models P_1$, then $\models P_2$' holds, while $\models P_1 \to P_2$ (always if it is cold, then it is snowing) does not hold. □

Theorem 2.12. *(a) For all formulas A, if $\models \neg A$, then not $\models A$.*
However, the converse does not hold:
(b) Not for all formulas A, if not $\models A$, then $\models \neg A$.

Proof. (a) Suppose $\models \neg A$, i.e., $\neg A$ is 1 in all lines of its truth table. Equivalently: A is 0 in all lines of its truth table. So, for sure, it is not the case that A is 1 in all lines of its truth table, i.e., not $\models A$.
(b) 'Not $\models A$' means that not in all lines of its truth table A is 1, in other words, A is 0 in *some* line of its truth table. This does not mean that $\models \neg A$, or equivalently, that A is 0 in *all* lines of its truth table. In fact, there are formulas A such that not $\models A$, while $\models \neg A$ does not hold. For instance, take $A = P_1$ (it is raining). Then not $\models P_1$ (not always it is raining), while $\models \neg P_1$ (always it is not raining; it never rains) does not hold. □

Warning One might be inclined to write: for all formulas A, if not $\models A$, then *not* $\models \neg A$. However, this is false. For instance, taking $A = P_1 \wedge \neg P_1$ we have not $\models P_1 \wedge \neg P_1$, but also $\models \neg(P_1 \wedge \neg P_1)$. The expression

2.4 Semantics: Meta-logical Considerations

$$\text{if not} \models A, \text{then} \models \neg A \qquad (*)$$

does hold for some formulas, for instance, for $A = P_1 \wedge \neg P_1$, but it does not hold for other formulas, for instance, not for $A = P_1$.

A formula that refutes (*) is called a *counterexample* to the statement (*). So, P_1 is a counterexample to (*).

Theorem 2.13. *(a) For all formulas A and B, if $\models A$ or $\models B$, then $\models A \vee B$. However, the converse does not hold:*
(b) Not for all formulas A and B, if $\models A \vee B$, then $\models A$ or $\models B$.

Proof. (a) Suppose $\models A$ or $\models B$. Consider the case that $\models A$, i.e., A is 1 in all lines of its truth table. Then clearly, also $A \vee B$ is 1 in all lines of its truth table, i.e., $\models A \vee B$. The case that $\models B$ is treated similarly.
(b) $\models A \vee B$ means: $A \vee B$ is 1 in all lines of its truth table, i.e., in each line of the truth table A is 1 or B is 1. So, there might be lines in which A is 1 and B is 0, while there might be other lines in which A is 0 and B is 1. So, this does not mean that A is 1 in all lines, i.e., $\models A$, nor that B is 1 in all lines, i.e., $\models B$. In fact, there are formulas A and B, such that $\models A \vee B$, while neither $\models A$ nor $\models B$. For instance, take $A = P_1$ (it is raining) and $B = \neg P_1$. Then $\models P_1 \vee \neg P_1$ (always it is raining or not raining), while neither $\models P_1$ (always it is raining), nor $\models \neg P_1$ (always it is not raining; it never rains). □

Warning One might be inclined to write: for all formulas A and B, if $\models A \vee B$, then not $\models A$ and not $\models B$. However, this is false. For instance, take $A = P_1 \to P_1$ and B arbitrary, then $\models (P_1 \to P_1) \vee B$, but also $\models P_1 \to P_1$ holds. The expression

$$\text{if} \models A \vee B, \text{then} \models A \text{ or } \models B \qquad (*)$$

does hold for some formulas, for instance, for $A = P_1 \to P_1$ and B arbitrary, but it does not hold for other formulas, for instance, not for $A = P_1$ and $B = \neg P_1$. So, $A = P_1$ and $B = \neg P_1$ is a *counterexample* to the statement (*).

Notice that, for instance, $A = P$ and $B = Q$ with P, Q atomic, is not a counterexample against (*), because such a counterexample should consist of formulas A and B such that '$\models A \vee B$' does hold, while '$\models A$ or $\models B$' does not hold; and $\models P \vee Q$ is not the case.

Theorem 2.14. *For all formulas A and B, $\models A \wedge B$ if and only if $\models A$ and $\models B$.*

Proof. $\models A \wedge B$ means: in all lines of its truth table, $A \wedge B$ is 1, i.e., in all lines, A is 1 and B is 1. This is equivalent to: in all lines A is 1 and in all lines B is 1, i.e., $\models A$ and $\models B$. □

In order to be able to formulate the replacement theorem, we first have to define the notion of *subformula*.

Definition 2.7 (Subformula). 1. If A is a formula, then A is a subformula of A.
2. If A and B are formulas, the subformulas of A and the subformulas of B are subformulas of $A \rightleftarrows B$, $A \to B$, $A \wedge B$, and $A \vee B$.
3. If A is a formula, then the subformulas of A are subformulas of $\neg A$.

Example 2.6. The subformulas of $\neg P \vee Q \to (P \to \neg P \vee Q)$ are: $\neg P \vee Q \to (P \to \neg P \vee Q)$, $\neg P \vee Q$, $P \to \neg P \vee Q$, $\neg P$, Q and P. Notice that $P \vee Q$ is not a subformula of $\neg P \vee Q \to (P \to \neg P \vee Q)$.

Theorem 2.15 (Replacement theorem). *Let C_A be a formula containing A as a subformula, and let C_B come from C_A by replacing the subformula A by formula B. If A and B have the same truth table, then C_A and C_B have the same truth table too.*

Proof. Assume A and B have the same table. If, in the computation of a given line of the table for C_A, we replace the computation of the specified part A by a computation of B instead, the outcome will be unchanged. Thus, C_B has the same table as C_A. □

Corollary 2.1 (Replacement rule). *If $\models C_A$ and A and B have the same table, then $\models C_B$.*

Warning: do not confuse object- and meta-language The reader should realize that the symbol '$\models$' does not occur in the alphabet of the propositional calculus and that consequently any expression containing $\models$ is not a formula. '$\models A$' is a statement *about* formula A, saying that A is valid, i.e., A is 1 in all lines of its truth table (always true). 'A' stands for a formula in the *object-language*, i.e., the language of propositional logic, but '$\models A$' is an expression in the *meta-language* about formula A, saying that A is always true.

$\models A \rightleftarrows B$ means $\models (A \rightleftarrows B)$; it cannot mean $(\models A) \rightleftarrows B$, because '$\models A$' belongs to the meta-language, while '$\rightleftarrows$' and 'B' belong to the object language. So, '$\models$' stands outside every formula.

Because '$\models \neg A$' is an expression of the meta-language and '$\to$' is a symbol of the object language, we are not allowed to write 'if $\models \neg A$, then not $\models A$' in Theorem 2.12 as '$\models \neg A \to \neg \models A$'; '$\to$' should connect formulas and '$\models \neg A$' and 'not $\models A$' are not formulas.

We can compare '$\models A$' with for instance "'Jean est malade' is a short sentence". This is not a sentence in French (the object language), but a statement in English (the meta-language) about a sentence ('Jean est malade', 'A') of the object language.

Below we have listed a number of expressions on the left hand side and the language they belong to on the right hand side.

$P \wedge \neg P$: Formula of the object-language.
$\models P \wedge \neg P$: Statement in the meta-language about the formula $P \wedge \neg P$.
'$\models P \wedge \neg P$' is false: Statement in the meta-meta-language about $\models P \wedge \neg P$.

Because our meta-language is a natural language (English), the meta-meta-language coincides with the meta-language itself.

Exercise 2.24. Show that for all formulas A and B,
1) if $\models A \rightleftarrows (A \to B)$, then $\models A$ and $\models B$;
2) if $A \models \neg A$, then $\models \neg A$.
3) if $A \to B \models A$, then $\models A$.

Exercise 2.25. Prove or refute: for all formulas A and B,

2.5 About Truthfunctional Connectives

a) if not $\models A \to B$, then $\models A$ and $\models \neg B$. b) if $\models \neg(A \to B)$, then $\models A$ and $\models \neg B$.
c) if not $\models A \wedge B$, then $\models \neg A$ or $\models \neg B$. d) if $\models \neg(A \wedge B)$, then $\models \neg A$ or $\models \neg B$.
e) if not $\models A \vee B$, then $\models \neg A$ and $\models \neg B$. f) if $\models \neg(A \vee B)$, then $\models \neg A$ and $\models \neg B$.

Exercise 2.26. Establish the following.
(a1) $A_1, A_2, A_3 \models A_1$, $A_1, A_2, A_3 \models A_2$, $A_1, A_2, A_3 \models A_3$.
(a2) More generally: $A_1, \ldots, A_i, \ldots, A_n \models A_i$ for $i = 1, \ldots, n$.
(b1) If $A_1, A_2, A_3 \models B_1$ and $A_1, A_2, A_3 \models B_2$ and $B_1, B_2 \models C$, then $A_1, A_2, A_3 \models C$.
(b2) More generally, for any $n, k \geq 0$: if $A_1, \ldots, A_n \models B_1$ and ... and $A_1, \ldots, A_n \models B_k$ and $B_1, \ldots, B_k \models C$, then $A_1, \ldots, A_n \models C$.

Exercise 2.27. Show directly from the definition of valid consequence:
1) if $A \models B$ and $A \models \neg B$, then $\models \neg A$. (*Reductio ad absurdum*)
2) if $A \models C$ and $B \models C$, then $A \vee B \models C$. (*Proof by cases*)

Exercise 2.28. Which of the following statements are right and which are wrong, and why is that the case? For all formulas A, B, C,
(a) $A \to B \vee C \models (A \to B) \vee (A \to C)$.
(b) if $\models (A \to B) \vee (A \to C)$, then $\models A \to B$ or $\models A \to C$.
(c) if $A \models B$, then $B \to C \models A \to C$.

Exercise 2.29. Prove: if $T \wedge A \wedge B \models P$, then $\neg P \models \neg T \vee \neg A \vee \neg B$.
Interpreting T as a Theory, A as Auxiliary hypotheses, B as Background hypotheses and P as Prediction, this is actually the *Duhem-Quine thesis*. In 1906 Pierre Duhem argued that the falsification of a theory is necessarily ambiguous and therefore that there are no crucial experiments; one can never be sure that it is a given theory rather than auxiliary or background hypotheses which experiment has falsified. [See S.C. Harding, [11], *Can theories be refuted?* p. IX.]

Exercise 2.30. Prove or refute: for all formulas A, B and C,
a) if $A \models B$, then $\neg B \models \neg A$.
b) if $A \models B$ and $A, B \models C$, then $A \models C$.
c) if $A \vee B \models A \wedge B$, then A and B have the same truth table.

2.5 About Truthfunctional Connectives

One may wonder if the object-language of propositional logic may be enriched by adding some new truthfunctional connectives, for instance, the connective $\uparrow$, called the *Sheffer stroke*, to be read as 'neither ..., nor ...' and defined by the following truth table.

A	B	$A \uparrow B$
1	1	0
1	0	0
0	1	0
0	**0**	**1**

In this case we see immediately that ↑ may be defined in terms of ¬ and ∧: $A \uparrow B$ has the same truth table as $\neg A \wedge \neg B$. But maybe there are other *binary* (i.e., with two arguments A and B) *truthfunctional connectives* which cannot be defined in terms of the ones we already have: $\rightleftarrows$, →, ∧, ∨ and ¬.

Now it is easy to see that there are $2^4 = 16$ possible binary truthfunctional connectives, each of them corresponding with a table of length 4:

A	B			...		
1	1	1	1	...	0	0
1	0	1	1	...	0	0
0	1	1	1	...	0	0
0	0	1	0	...	1	0

It is not difficult to see that each of these 16 truthfunctional connectives may be expressed in terms of ∧, ∨ and ¬. Consider, for instance, the three truthfunctional connectives corresponding with the following truth tables:

A	B	
1	1	0
1	**0**	**1**
0	1	0
0	0	0

A	B	
1	1	0
1	0	0
0	**1**	**1**
0	0	0

A	B	
1	1	0
1	**0**	**1**
0	**1**	**1**
0	0	0

The left truth table is precisely the table of $A \wedge \neg B$, the truth table in the middle is precisely the table of $\neg A \wedge B$, and the right truth table is precisely the truth table of $(A \wedge \neg B) \vee (\neg A \wedge B)$. So, the following Theorem is evident:

Theorem 2.16. *Each binary (i.e., having two arguments A and B) truthfunctional connective may be expressed in terms of* ∧, ∨ *and* ¬.

We say that the set $\{\wedge, \vee, \neg\}$ is a *complete set of truthfunctional connectives*: each binary truthfunctional connective may be expressed in terms of these three connectives. We have already seen earlier that → and $\rightleftarrows$ can be expressed in terms of ∧, ∨ and ¬: $A \to B$ has the same truth table as $\neg A \vee B$, and also as $\neg(A \wedge \neg B)$; and $A \rightleftarrows B$ has the same truth table as $(A \to B) \wedge (B \to A)$.

Theorem 2.16 can easily be generalized to truth tables entered from more than two formulas. Consider, for instance, the truth table below entered from three atomic formulas P, Q and R:

P	Q	R	
1	1	1	0
1	**1**	**0**	**1**
1	0	1	0
1	0	0	0
0	**1**	**1**	**1**
0	1	0	0
0	0	1	0
0	0	0	0

2.5 About Truthfunctional Connectives

The formula corresponding with this table is clearly: $(P \wedge Q \wedge \neg R) \vee (\neg P \wedge Q \wedge R)$. More generally, we see that for every formula A there is a formula A' which is a disjunction of conjunctions of *literals*, i.e., atomic formulas or negations of atomic formulas, such that A and A' have the same truth table. We shall say that A' is in *disjunctive normal form*. By applying the de Morgan's laws (Theorem 2.10), we may conclude that for every formula A there is also a formula A'' in *conjunctive normal form*, i.e., which is a conjunction of disjunctions of literals, and which has the same truth table as A. See Theorem 2.18.

Next we shall show that each truthfunctional connective may be expressed in terms of only one connective: the Sheffer stroke $\uparrow$.

Theorem 2.17. *Every binary truthfunctional connective may be expressed in terms of the Sheffer stroke $\uparrow$.*

Proof. In order to prove this, by Theorem 2.16 it suffices to prove that $\wedge$, $\vee$ and $\neg$ may be expressed in terms of the Sheffer stroke $\uparrow$.

a) $\neg A$ has the same truth table as $\neg A \wedge \neg A$, and hence as $A \uparrow A$ (neither A, nor A).
b) $A \wedge B$ has the same truth table as $\neg(\neg A) \wedge \neg(\neg B)$, hence as $\neg A \uparrow \neg B$ (neither $\neg A$, nor $\neg B$) and therefore as $(A \uparrow A) \uparrow (B \uparrow B)$.
c) $A \vee B$ has the same truth table as $\neg(\neg A \wedge \neg B)$, hence as $\neg(A \uparrow B)$ and therefore as $(A \uparrow B) \uparrow (A \uparrow B)$. □

2.5.1 Applications in Electrical Engineering and in Jurisdiction

There are many situations in which there are two opposites analogous to the case of truth and falsity of propositions. For example, in electrical engineering: on (lit, 1) and off (unlit, 0); and in jurisdiction: innocent and guilty. In all such situations one can work with truth tables in a similar way as in propositional logic.

Suppose we have two switches A and B, both with a 0- and a 1- position, a bulb and a battery and that we want the bulb to burn (1, lit) precisely if both switches are in the 1-position. So, the corresponding table is the one for $A \wedge B$:

switch A	switch B	bulb
1	1	1
1	0	0
0	1	0
0	0	0

The following electric circuit will satisfy our wishes.

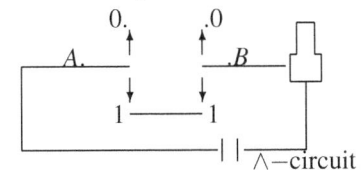

$\wedge$−circuit

If we want the bulb to burn if at least one of the two switches A and B is in the 1-position, then we find a table corresponding with the one for $A \vee B$ and the corresponding electric circuit is as follows.

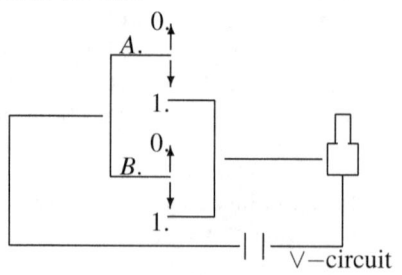

∨–circuit

And if we want the bulb to burn if switch A is in the 0-position, then we find a table corresponding with the one for $\neg A$ and the corresponding electric circuit is the following one.

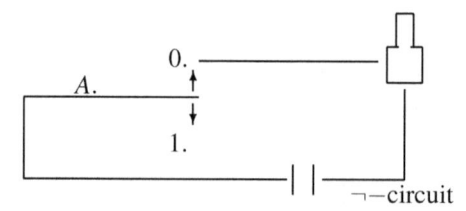

¬–circuit

Theorem 2.16 formulated in terms of electric circuits now tells us that each electric circuit can be built from the electric circuits for $\wedge$, $\vee$ and $\neg$, and the proof of Theorem 2.16 provides us with a uniform method to build any circuit we want from the circuits for $\wedge$, $\vee$ and $\neg$. We shall consider some examples below. However, the circuits resulting from our uniform method in the proof of Theorem 2.16 will not always be the simplest ones and for economic reasons one may in practice use circuits other than the ones found by this uniform method.

Example 2.7. Suppose we want our bulb to burn in all cases except one: if switch A is in position 1 and switch B is in position 0. So the corresponding table is the following one.

switch A	switch B	bulb
1	1	1
1	0	0
0	1	1
0	0	1

We see that this table corresponds with the one for $A \rightarrow B$. The proof of Theorem 2.16 tells us that the circuit corresponding with $(A \wedge B) \vee (\neg A \wedge B) \vee (\neg A \wedge \neg B)$ will satisfy our wishes. However, a much simpler, and hence less expensive circuit, doing the same job, can be found if we realize that $A \rightarrow B$ has the same truth table as $(\neg A) \vee B$. So in order to achieve our purpose, we can take the ∨-circuit described above with instead of switch A the circuit for $\neg A$.

2.5 About Truthfunctional Connectives

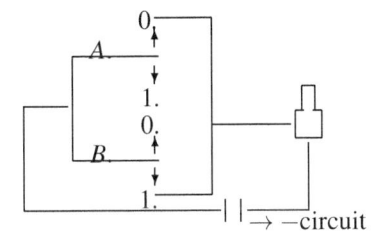

[Diagram of $\rightarrow$-circuit with switches A and B]

Example 2.8. Suppose we want to build a two-way switch: a switch A at the foot of the stairs and a switch B at the top of the stairs such that we can turn the light on and off both at the foot and at the top of the stairs by changing the nearest switch over into another position.

[Diagram of staircase with switches A and B]

We can achieve our purpose by making the electric circuit such that the light is on when both switches are in the same position and off when both are in a different position. The corresponding table is the following one.

switch A	switch B	light
1	1	1
1	0	0
0	1	0
0	0	1

This table corresponds with the one for $A \rightleftarrows B$. Applying the proof of Theorem 2.16, we shall find that the circuit corresponding with $(A \wedge B) \vee (\neg A \wedge \neg B)$ will satisfy our requirements. So we can take the $C \vee D$-circuit described above with the circuit for $A \wedge B$ instead of switch C and the circuit for $\neg A \wedge \neg B$ instead of switch D. And this latter circuit is obtained by replacing in the $E \wedge F$-circuit described above switch E by the circuit for $\neg A$ and switch F by the circuit for $\neg B$.

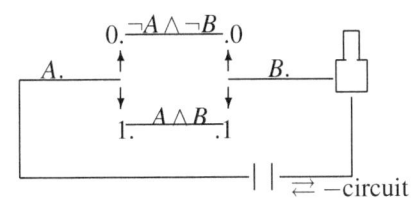

[Diagram of $\rightleftarrows$-circuit]

For an application of truth tables in jurisdiction we refer the reader to Exercise 2.31.

2.5.2 Normal Form*; Logic Programming*

Definition 2.8 (Normal form). A *literal* is by definition an atomic formula or the negation of an atomic formula.

A formula B is in *disjunctive normal form* if it is a disjunction $B_1 \vee \ldots \vee B_k$ of formulas, where each B_i ($1 \leq i \leq k$) is a conjunction $L_1 \wedge \ldots \wedge L_n$ of literals.

A formula B is in *conjunctive normal form* if it is a conjunction $B_1 \wedge \ldots \wedge B_k$ of formulas, where each B_i ($1 \leq i \leq k$) is a disjunction $L_1 \vee \ldots \vee L_n$ of literals.

Example 2.9. So, P_2 and $\neg P_1$ are examples of literals. $(\neg P_1 \wedge P_2) \vee \neg P_3$ is a formula in disjunctive normal form, and $(\neg P_1 \vee \neg P_3) \wedge (P_2 \vee \neg P_3)$ is a formula in conjunctive normal form.

Theorem 2.18 (Normal form theorem). *For each formula A (of classical propositional logic) there are formulas A' and A'' in disjunctive or conjunctive normal form respectively, which have the same truth table as A. In other words, each formula A of classical propositional logic may be written in disjunctive, respectively, conjunctive, normal form.*

Proof. We will use the induction principle (Theorem 2.2) to show that every formula A has the property Φ: there are formulas A' and A'' in disjunctive or conjunctive normal form respectively, which have the same truth table as A. Since all truth-functional connectives can be expressed in terms of $\neg$, $\wedge$ and $\vee$, we may assume that all formulas are built from atomic formulas by means of these three connectives.

1. If A is an atomic formula P, then $A = P$ itself is both in disjunctive and in conjunctive normal form.
2. Suppose $A = \neg B$ and (induction hypothesis) that there are formulas B' and B'' which are in disjunctive or conjunctive normal form respectively, and which are equivalent to B. Then $A = \neg B$ has the same truth table as $\neg B'$, which by the De Morgan's laws, Theorem 2.10, 17, can be rewritten as a conjunction of disjunctions of literals. And $A = \neg B$ has the same truth table as $\neg B''$, which by the De Morgan's laws, Theorem 2.10, 17, can be rewritten as a disjunction of conjunctions of literals.
3. Suppose $A = B \wedge C$ and (induction hypothesis) that there are formulas B', C' and formulas B'', C'' which are in disjunctive or conjunctive normal form respectively and which are equivalent to B, respectively C. Then $A = B \wedge C$ has the same truth table as $B'' \wedge C''$, which is again a conjunction of disjunctions of literals. And $A = B \wedge C$ has the same truth table as $B' \wedge C'$, which by the distributive laws, Theorem 2.10, 22 and 23, can be rewritten in disjunctive normal form.
4. Suppose $A = B \vee C$ and (induction hypothesis) that there are formulas B', C' and formulas B'', C'' which are in disjunctive or conjunctive normal form respectively and which are equivalent to B, respectively C. Then $A = B \vee C$ has the same truth table as $B' \vee C'$, which is again a disjunction of conjunctions of literals. And $A = B \vee C$ has the same truth table as $B'' \vee C''$, which by the distributive laws, Theorem 2.10, 22 and 23, can be rewritten as a conjunction of disjunctions of literals.

Example 2.10. $A = P \rightarrow \neg(\neg Q \vee P)$ has the same truth table as, subsequently, $\neg P \vee \neg(\neg Q \vee P)$, $\neg P \vee (\neg\neg Q \wedge \neg P)$, $\neg P \vee (Q \wedge \neg P)$, which is in disjunctive normal form, and $(\neg P \vee Q) \wedge (\neg P \vee \neg P)$, which is in conjunctive normal form.

2.5 About Truthfunctional Connectives

Knowledge Representation and Logic Programming The language of logic may be used to represent knowledge. For instance, suppose a person has the following knowledge at his disposal:
(1) John buys the book if it is about logic and interesting.
(2) The book is about logic.
(3) The book is interesting if it is about logic.
Using P to represent 'John buys the book',
Q to represent 'the book is about logic', and
R to represent 'the book is interesting',
the person's knowledge can be represented by the following logical formulas:
(1a) $Q \wedge R \to P$,
(2a) Q,
(3a) $Q \to R$.

In the programming language Prolog (Programming in Logic), which will be treated in Chapter 9, these formulas are rendered as follows:
(1b) P :- Q, R. (to be read as: P if Q and R)
(2b) Q.
(3b) R :- Q. (to be read as: R if Q)

(1b) and (3b) are called *rules* and (2b) is called a *fact*. Using logical reasoning 'new' knowledge can be deduced from the knowledge already available. For instance, from (2a) and (3a) follows R (4a), and from (2a), (4a) and (1a) follows P, i.e., 'John buys the book'.

(1b), (2b) and (3b) together can be considered to form a *knowledge base* from which new knowledge can be obtained by logical reasoning or deduction.

The programming language Prolog, to be treated in Chapter 9, has a built in logical inference mechanism. When provided with the database consisting of (1b), (2b) and (3b), Prolog will answer the question '?- P.' with 'yes', corresponding to the fact that P is a logical consequence of (1b), (2b) and (3b).

The following definition introduces some terminology which is used in logic programming and which is needed in Chapter 9.

Definition 2.9 (Literal). a) A *positive literal* is an atomic formula. A *negative literal* is the negation of an atomic formula.
b) A *clause* is a formula of the form $L_1 \vee \ldots \vee L_m$, where each L_i is a literal.

Because clauses are so common in logic programming, it will be convenient to adopt a special clausal notation. In logic programming the clause $\neg P_1 \vee \ldots \vee \neg P_k \vee Q_1 \vee \ldots \vee Q_n$, where $P_1, \ldots, P_k, Q_1, \ldots, Q_n$ are atomic, is denoted by

$$Q_1, \ldots, Q_n \text{ :- } P_1, \ldots, P_k \ (k \geq 0).$$

which stands for $P_1 \wedge \ldots \wedge P_k \to Q_1 \vee \ldots \vee Q_n$, which has the same truth table as $\neg P_1 \vee \ldots \vee \neg P_k \vee Q_1 \vee \ldots \vee Q_n$.

Theorem 2.18 says that each formula of (classical) propositional logic may be written as a finite conjunction of clauses.

For reasons of efficiency, to be explained in Chapter 9, in Prolog only *Horn clauses* are used, i.e., clauses which contain at most one positive literal, in other words, which are of the form Q :- $P_1, \ldots, P_k$ or of the form :- $P_1, \ldots, P_k$.

(1b), (2b) and (3b) above are examples of Horn-clauses. Q_1, Q_2 :- P_1, P_2, P_3. or equivalently $P_1 \wedge P_2 \wedge P_3 \rightarrow Q_1 \vee Q_2$, is not a Horn clause.

Definition 2.10 (Horn clause).

a) A *definite program clause* is a clause of the form

$$Q :- P_1, \ldots, P_k \ (k \geq 0, P_1, \ldots, P_k, Q \text{ atomic})$$

which contains precisely one atomic formula (viz. Q) in its consequent. Q is called the *head* and $P_1, \ldots, P_k$ is called the *body* of the program clause.

b) A *unit clause*, also called a *fact*, is a clause of the form

$$Q :-$$

that is, a definite program clause with an empty body.

c) A *definite program* is a finite set of definite program clauses.

d) A *definite goal* is a clause of the form

$$:- P_1, \ldots, P_k$$

that is, a clause which has an empty consequent. Each P_i ($i = 1, \ldots, k$) is called a *subgoal* of the goal.

e) A *Horn clause* is a clause which is either a definite program clause or a definite goal. So, a Horn clause is a clause with at most one positive literal.

Example 2.11. The following is an example of a definite program:

$P :- Q, R.$
$Q :-.$
$R :- Q.$

This program corresponds with the formula $(P \vee \neg Q \vee \neg R) \wedge Q \wedge (R \vee \neg Q)$, which is in conjunctive normal form, and where each conjunct contains precisely one positive literal (and hence is a Horn clause). Note that this formula has the same truth table as $(Q \wedge R \rightarrow P) \wedge Q \wedge (Q \rightarrow R)$.

Given this program, in logic programming the goal ':- P.' will be answered with 'yes', corresponding with the fact that P logically follows from $(P \vee \neg Q \vee \neg R) \wedge Q \wedge (R \vee \neg Q)$.

The goal ':- S' will be answered with 'no', corresponding with the fact that S does not logically follow from the given program.

Logic programming in general and Prolog in particular will be treated in Chapter 9. However, this treatment also presupposes familiarity with classical predicate logic, which will be treated in Chapter 4.

2.5.3 Travelling Salesman Problem (TSP)*; NP-completeness*

The Traveling Salesman Problem is the problem of computing the shortest itinerary, when a number, n, of cities with given distances has to be visited, each city to be

2.5 About Truthfunctional Connectives

visited only once. From a theoretical point of view there is no problem at all: if there are n cities to be visited, there are $(n-1)!$ itineraries; compute the total distance of each of them and take the shortest. However, from a practical point of view there are problems: if 10 cities are to be visited, there are $9! = 362{,}880$ itineraries; and if a sales-representative has to visit 30 cities, there are 29! itineraries and 29! is larger than 10^{29}. Supposing that a computer could calculate the distances of $1000 = 10^3$ itineraries per second, in one human lifetime it could compute about 100 (years) × 365 (days) × 24 (hours) × 60 (minutes) × 60 (seconds) × 10^3 (itineraries) ≈ 10^{13} itineraries. So, in order to compute the distances of 29! itineraries, our computer would need more than $10^{29} / 10^{13} = 10^{16}$ human lifetimes! Thus, like the validity problem for formulas of propositional logic, also the Travelling Salesman Problem is solvable in theory, but no realistic solution is known.

We will see below how the following *Traveling Salesman Problem* can be reduced to a *satisfiability problem* in the propositional calculus. In the map, the vertices are towns and the lines are roads, each 10 miles long. This example is from A. Keith Austin [1].

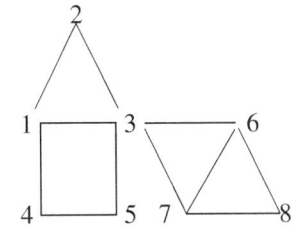

PROBLEM: Can the salesman start at 1 and visit all the towns in a journey of only 70 miles?

Theorem 2.19. *There is a formula E of the propositional calculus such that there is a journey of only 70 miles starting at 1 if and only if E is satisfiable.*

CONSTRUCTION of E: To express the problem in propositional logic, we introduce the atomic formulas P_t^m, for $m = 0, 1, \ldots, 7$, $t = 1, 2, \ldots, 8$, the intended meaning of P_t^m being: after $10 \times m$ miles the salesman is at town t. Given any journey of 70 miles, each P_t^m is either true or false. We now express the conditions of the problem as logical formulas.

i) If the salesman is at 5 after 30 miles, then he is at 3 or 4 after 40 miles, i.e., if P_5^3 is true, then either P_3^4 or P_4^4 is true. Let $J_5^3 := P_5^3 \to P_3^4 \vee P_4^4$ be the formula in our propositional language expressing this. Similarly we have $P_5^m \to P_3^{m+1} \vee P_4^{m+1}$, and $P_3^m \to (P_1^{m+1} \vee P_2^{m+1} \vee P_5^{m+1} \vee P_6^{m+1} \vee P_7^{m+1})$ for $m = 0, 1, \ldots, 6$, and so on for each town. Denote each of these by the corresponding J_y^x. All these have to be true and so we write

$$J := J_1^0 \wedge J_2^0 \wedge \ldots \wedge J_8^0 \wedge J_1^1 \wedge J_2^1 \wedge \ldots \wedge J_8^1 \wedge \ldots \wedge J_8^6.$$

ii) Another condition is that each town has to be visited. That town 1 has to be visited can be expressed as $P_1^0 \vee P_1^1 \vee P_1^2 \vee \ldots \vee P_1^7$ and similarly for the other towns. Let

$$V := (P_1^0 \vee \ldots \vee P_1^7) \wedge (P_2^0 \vee \ldots \vee P_2^7) \wedge \ldots \wedge (P_8^0 \vee \ldots \vee P_8^7).$$

iii) Also the salesman is only at one town at any one time, so we have, e.g., $P_3^5 \to \neg P_1^5$. Let $N_3^5 := P_3^5 \to \neg P_1^5 \wedge \neg P_2^5 \wedge \neg P_4^5 \wedge \ldots \wedge \neg P_8^5$. And let

$$N := N_1^0 \wedge N_2^0 \wedge \ldots \wedge N_8^7.$$

iv) Finally, he has to start at 1, so we require P_1^0 to be true.

Now let $E := J \wedge V \wedge N \wedge P_1^0$. Then E has the required property: there is a journey of only 70 miles starting at 1 if and only if E is satisfiable.

Theorem 2.19 reduces the Traveling Salesman Problem for eight cities to a satisfiability problem in the propositional calculus. However, the formula E constructed in the proof of Theorem 2.19 is built from $8^2 = 64$ atomic formulas P_t^m. So, in order to check whether E is satisfiable, we have to compute a truth table entered from 2^{64} lines. We have already seen in Subsection 2.3.1 that making truth tables with so many entries does not yield a practical or realistic decision method to decide whether arbitrary formulas are satisfiable or not. Since the original problem can be solved by computing the distances of $(8 - 1)!$ itineraries, the reduction of the Traveling Salesman Problem to the satisfiability problem for propositional logic has not helped us to find a practical or realistic solution for the former. We have to wait for a realistic solution of the satisfiability problem or for a proof that no such solution exists.

Of course, in order to see whether a given formula E is satisfiable, i.e., has at least one 1 in its truth table, one might *non-deterministically* choose a line in the truth table and compute whether E is 1 in that line. The computation of one line in the truth table can be done in a realistic way: the time required to do so is a *polynomial* of the complexity of the formula in question. If it turns out that E is 1 in the chosen line, one knows that E is satisfiable, but when it turns out that E is 0 in the chosen line, one does not know whether E is satisfiable or not. And we have seen in Subsection 2.3.1 that it is not realistic to compute all lines in the truth table of E if E has been built from many, say 64, atomic formulas. For that reason, the satisfiability problem for propositional calculus is said to belong to the class *NP* of all problems which may be decided Non-deterministically in Polynomial time.

In 1971, S. Cook showed that not only the Traveling Salesman Problem, but also all other problems in the class *NP*, can be reduced to a satisfiability problem in the propositional calculus. For that reason the satisfiability problem for propositional logic is called *NP-complete*.

Exercise 2.31. [Keisler; appearance in S.C. Kleene [14], p. 67] Brown, Jones and Smith are suspected of income tax evasion. They testify under oath as follows.
BROWN: Jones is guilty and Smith is innocent.

2.5 About Truthfunctional Connectives

JONES: If Brown is guilty, then so is Smith.
SMITH: I'm innocent, but at least one of the others is guilty.
Let B, J, S be the statements 'Brown is innocent', 'Jones is innocent', 'Smith is innocent', respectively. Express the testimony of each suspect by a formula in our logical symbolism, and write out the truth tables for these three formulas (in parallel columns). Now answer the following questions.
a) Are the testimonies of the three suspects consistent, i.e., is the conjunction of these testimonies consistent?
b) The testimony of one of the suspects follows from that of another. Which from which?
c) Assuming everybody is innocent, who committed perjury?
d) Assuming everyone's testimony is true, who is innocent and who is guilty?
e) Assuming that the innocent told the truth and the guilty told lies, who is innocent and who is guilty?

Exercise 2.32. [W. Ophelders] The football clubs Pro, Quick and Runners play a football tournament. The trainers of these clubs make the following statements.
Trainer of Pro: If the Runners win the tournament, then Quick does not.
Trainer of Quick: We or the Runners win the tournament.
Trainer of the Runners: We win the tournament.
Express the three statements by formulas in our logical symbolism and write out the truth tables for these three formulas. Next answer the following questions, supposing there can be at most one winner.
a) Assuming everyone's statement is true, which club wins the tournament?
b) Assuming only the trainer of the winning club makes a true statement, which club wins the tournament?

Exercise 2.33. Find formulas composed from P, Q, R, $\wedge$, $\vee$ and $\neg$ only, whose truth tables have the following value columns:

P	Q	R	(a)	(b)	(c)	(d)
1	1	1	0	1	0	0
1	1	0	0	0	0	1
1	0	1	0	0	0	1
1	0	0	0	0	0	1
0	1	1	1	1	0	1
0	1	0	0	0	0	0
0	0	1	0	1	0	1
0	0	0	0	0	0	1

Exercise 2.34. Let $A \downarrow B$ be defined by the following truth table:

A	B	$A \downarrow B$
1	1	0
1	0	1
0	1	1
0	0	1

$A \downarrow B$ may be read as 'not A or not B'. Prove that $\neg$, $\vee$ and $\wedge$, and hence each of the 16 binary truthfunctional connectives, can be expressed in terms of $\downarrow$.

Exercise 2.35. A set of binary truthfunctional connectives is *independent* iff none of the members of the set can be expressed in terms of the other members of the set.
i) Show that $\{\wedge, \vee, \neg\}$ is not independent.
ii) Show that $\{\wedge, \neg\}$, $\{\vee, \neg\}$ and $\{\rightarrow, \neg\}$ are independent and complete sets of truthfunctional binary connectives.

Exercise 2.36. Show that there are only two binary connectives, namely, $\uparrow$ (the Sheffer stroke) and $\downarrow$ (see Exercise 2.34) such that every binary truthfunctional connective can be expressed in it.

Exercise 2.37. Construct formulas in conjunctive normal form which have the same truth table as the following formulas:
i) $(P \rightarrow (Q \rightarrow P)) \wedge (P \rightarrow Q \vee P)$
ii) $(P \rightarrow \neg(Q \rightarrow P)) \wedge (P \rightarrow Q \wedge P)$
iii) $(P \rightarrow \neg(Q \rightarrow P)) \vee (P \rightarrow Q \wedge P)$

2.6 Syntax: Provability and Deducibility

By now it will be clear that there are a great many, in fact even infinitely many, valid formulas. And given premisses $A_1, \ldots, A_n$, there are infinitely many valid consequences of those premisses. The question now arises whether it is possible to select a few valid formulas, to be called *logical axioms*, together with certain rules – which applied to valid formulas produce (or generate) new valid formulas – such that any valid formula can be obtained (or generated) by a finite number of applications of the given rules to the selected logical axioms. This question can be answered positively, which means that in a certain sense we have reduced the big collection of valid formulas to a surveyable subset: any formula in the big collection of valid formulas can be generated by the given rules from formulas in the subset.

There are several possibilities for choosing the logical axioms and rules such that the desired goal is accomplished. In this section one of them is presented, namely, a system for propositional logic developed by Frege, and adapted by Russell and Hilbert. Henceforth, we shall speak of a *Hilbert-type* system. In Section 2.8 two other, more recent, systems will be treated which achieve the same goal.

One may design production methods satisfying the following two conditions:

(I) the production method produces in the course of time only formulas which are valid, and, more generally,

(II) the production method if applied to certain formulas given as premisses, only produces formulas which are a valid consequence of those premisses.

There are in fact many such production methods, each of them consisting of (i) a set of valid formulas, and (ii) a set of rules of inference. One such production method satisfying (I) and (II) can be obtained by taking:

2.6 Syntax: Provability and Deducibility

(i) All formulas of any of the forms $A \to (B \to A)$ and
$$(A \to B) \to ((A \to (B \to C)) \to (A \to C));$$
We have seen in Theorem 2.7, 1 and 2, that such formulas are valid. We call these formulas (logical) *axioms for the connective* $\to$.

(ii) As the sole rule of inference, called the $\to$-*rule* or *Modus Ponens (MP)*, we take the operation of passing from two formulas of the respective forms D and $D \to E$ to the formula E, for any choice of formulas D and E.

$$\text{Modus Ponens } (MP): \quad \frac{D \qquad D \to E}{E}$$

In an inference by this rule, the formulas D and $D \to E$ are the *premisses*, and E is the *conclusion*. The following statements can easily be checked:

(α) Any interpretation that makes the premisses of the rule true, also makes the conclusion of the rule true. For our particular rule MP: for any interpretation i, if $i(D) = 1$ and $i(D \to E) = 1$, then $i(E) = 1$, and consequently

(β) If all premisses of the rule are valid, then also the conclusion of the rule is valid. For our particular rule MP: if $\models D$ and $\models D \to E$, then $\models E$ (Theorem 2.11). Our rule of inference may be applied zero, one, two or more times to formulas of the form mentioned in (i) or to formulas which we have already generated earlier.

Example 2.12. This production method yields, among other things, the following formulas for any choice of the formula A:

1. $A \to (A \to A)$ This is a formula of the form $A \to (B \to A)$, taking $B = A$.
2. $(A \to (A \to A)) \to ((A \to ((A \to A) \to A)) \to (A \to A))$ This is a formula of the form $(A \to B) \to ((A \to (B \to C)) \to (A \to C))$, taking $B = A \to A$ and $C = A$.
3. $(A \to ((A \to A) \to A)) \to (A \to A)$ This formula is obtained by an application of Modus Ponens to 1 and 2.
4. $A \to ((A \to A) \to A)$ This formula is of the form $A \to (B \to A)$, taking $B = A \to A$.
5. $A \to A$ This formula is obtained by an application of Modus Ponens to 3 and 4.

Schematically:
$$\frac{A \to (A \to A) \quad (A \to (A \to A)) \to ((A \to ((A \to A) \to A)) \to (A \to A))}{(A \to ((A \to A) \to A)) \to (A \to A)} \; MP$$
$$\frac{A \to ((A \to A) \to A)}{A \to A} \; MP$$

This schema is called a (logical, Hilbert-type) *proof* of the formula $A \to A$ and $A \to A$ is called (logically) *provable*, because there exists such a schema using only logical axioms and Modus Ponens. Note that each of the formulas in this schema, and $A \to A$ in particular, is produced by our production method and that each of these formulas is valid, since we started with valid formulas and since Modus Ponens applied to valid formulas only yields formulas which are valid (Theorem 2.11 or (β) above).

Example 2.13. The production method described above applied to the formulas $A \to B$ and $B \to C$, for instance, yields the following formulas:

1. $A \to B$ This formula is a given premiss.
2. $(A \to B) \to ((A \to (B \to C)) \to (A \to C))$ This formula is of the appropriate form.
3. $(A \to (B \to C)) \to (A \to C)$ Obtained by applying Modus Ponens to 1 and 2.
4. $B \to C$ This formula is a given premiss.
5. $(B \to C) \to (A \to (B \to C))$ This is a formula of the form $A \to (B \to A)$, taking $A = B \to C$ and $B = A$.
6. $A \to (B \to C)$ This formula is obtained by applying Modus Ponens to 4 and 5.
7. $A \to C$ This formula is obtained by an application of Modus Ponens to 6 and 3.

Schematically:

$$\dfrac{\begin{array}{cc}\text{premiss} & \text{axiom 2}\\ A \to B & (A \to B) \to ((A \to (B \to C)) \to (A \to C))\end{array}}{(A \to (B \to C)) \to (A \to C)}$$

$$\dfrac{\begin{array}{cc}\text{premiss} & \text{axiom 1}\\ B \to C & (B \to C) \to (A \to (B \to C))\end{array}}{A \to (B \to C)}$$

$$A \to C$$

This schema is called a (logical, Hilbert-type) *deduction* of $A \to C$ from the premisses $A \to B$ and $B \to C$ and $A \to C$ is said to be *deducible* from the premisses $A \to B$ and $B \to C$, using only these premisses, logical axioms and Modus Ponens. Note that each of the formulas in this schema, and $A \to C$ in particular, is produced by our production method applied to the premisses $A \to B$ and $B \to C$, and that each of these formulas is a valid consequence of the premisses $A \to B$ and $B \to C$, since we started with valid formulas, the premisses $A \to B$ and $B \to C$ only, and because of (α) above.

It will be clear now that any production method, consisting of (i) a set of valid formulas and (ii) a set of rules of inference satisfying (α) and (β), will satisfy the conditions (I) and (II), mentioned in the beginning of this section.

One can prove (see Exercise 2.44) that Peirce's law, $((A \to B) \to A) \to A$, although it contains only the connective $\to$, is not generated by the production method consisting of the two logical axioms for $\to$ and Modus Ponens. This raises the question whether there is a *complete* production method satisfying I and II, i.e., a production method which in the course of time generates *all* valid formulas and, more generally, which generates, if applied to certain formulas, given as premisses, *all* valid consequences of those premisses. The answer to this question is affirmative. In Section 2.9 we shall prove that the production method consisting of all formulas of any of the forms shown after the symbol $\models$ in Theorem 2.7, and of the sole rule of inference, Modus Ponens, is complete. For convenience these formulas are again listed below and will be called (logical) *axioms for (classical) propositional logic*.

1. $A \to (B \to A)$
2. $(A \to B) \to ((A \to (B \to C)) \to (A \to C))$
3. $A \to (B \to A \wedge B)$
4a $A \wedge B \to A$

2.6 Syntax: Provability and Deducibility

4b $\quad A \wedge B \to B$
5a $\quad A \to A \vee B$
5b $\quad B \to A \vee B$
6. $\quad (A \to C) \to ((B \to C) \to (A \vee B \to C))$
7. $\quad (A \to B) \to ((A \to \neg B) \to \neg A)$
8. $\quad \neg\neg A \to A$
9. $\quad (A \to B) \to ((B \to A) \to (A \rightleftarrows B))$
10a $\quad (A \rightleftarrows B) \to (A \to B)$
10b $\quad (A \rightleftarrows B) \to (B \to A)$

Numbers 1 and 2 concern axioms for the connective $\to$, numbers 3 and 4 concern axioms for $\wedge$, numbers 5 and 6 concern axioms for $\vee$, numbers 7 and 8 concern axioms for $\neg$ and numbers 9 and 10 concern axioms for $\rightleftarrows$. Notice that in a sense they describe the typical properties of the connective in question; for instance, the axioms for $\wedge$ will not hold if we replace $\wedge$ by $\vee$.

These forms themselves will be called *axiom schemata*. Each schema includes infinitely many axioms, one for each choice of the formulas denoted by A, B, C. For example, corresponding to 1 in Theorem 2.7, we have as Axiom Schema 1: $A \to (B \to A)$. Particular axioms in this schema are $P \to (P \to P)$, $P \to (Q \to P)$, $Q \to (P \to Q)$, $\neg P \to (Q \wedge R \to \neg P)$, $(P \to (\neg Q \to P)) \to (R \to (P \to (\neg Q \to P)))$, etc.

The choice of the logical axioms is a subtle matter. For instance, if one would replace axiom schema 8, $\neg\neg A \to A$, by its converse, $A \to \neg\neg A$, then the resulting system would not be complete, in particular, the resulting system would not be able to generate Peirce's law, $((A \to B) \to A) \to A$. Also, if one replaces axiom 8, $\neg\neg A \to A$, by $\neg A \to (A \to B)$ one obtains intuitionistic propositional logic, which is completely different from classical logic; see Chapter 8. Small changes may have far reaching consequences!

Example 2.14. For illustration, let us show that from the premises
$P \to W$: I will pay them for fixing our TV [P] only if it works [W].
$\neg W$: Our TV still does not work.
the logical consequence $\neg P$ (I will not pay) can be generated by using the logical axioms 1 and 7 and by three applications of Modus Ponens.

$$
\frac{\begin{array}{cc}\text{prem} & \text{axiom 1} \\ \neg W & \neg W \to (P \to \neg W)\end{array}}{P \to \neg W} \quad \frac{\begin{array}{cc}\text{prem} & \text{axiom 7} \\ P \to W & (P \to W) \to ((P \to \neg W) \to \neg P)\end{array}}{(P \to \neg W) \to \neg P}
$$
$$\overline{\neg P}$$

The schema above is called a (logical, Hilbert-type) *deduction* of $\neg P$ from the premisses $P \to W$ and $\neg W$ and we say that $\neg P$ is (logically) *deducible from $P \to W$ and $\neg W$*, meaning that there exists a (logical, Hilbert-type) deduction of $\neg P$ from $P \to W$ and $\neg W$.

Definition 2.11 (Deduction; Deducible). Let $B, A_1, \ldots, A_n$ be formulas.

1. A (logical, Hilbert-type) *deduction* of B from $A_1, \ldots, A_n$ (in classical propositional logic) is a finite list $B_1, \ldots, B_k$ of formulas, such that
   (a) $B = B_k$ is the last formula in the list, and
   (b) each formula in the list is either one of $A_1, \ldots, A_n$, or one of the axioms of propositional logic (see Theorem 2.7), or is obtained by an application of Modus Ponens to a pair of formulas preceding it in the list.
2. B is *deducible from* $A_1, \ldots, A_n$:= there exists a (logical, Hilbert-type) deduction of B from $A_1, \ldots, A_n$.
   Notation: $A_1, \ldots, A_n \vdash B$, where the symbol $\vdash$ may be read 'yields'. If there does not exist a deduction of B from $A_1, \ldots, A_n$ this is written as $A_1, \ldots, A_n \not\vdash B$ as shorthand for: not $A_1, \ldots, A_n \vdash B$.
3. In case $n = 0$, i.e., in case there are no premisses, these definitions reduce to:
   A (logical, Hilbert-type) *proof* of B is a finite list of formulas with B as last formula in the list, such that every formula in the list is either an axiom of propositional logic or obtained by Modus Ponens to formulas earlier in the list.
   B is (logically) *provable* := there exists a (logical, Hilbert-type) proof of B.
   Notation: $\vdash B$
4. For Γ a (possibly infinite) set of formulas, B is *deducible from* Γ, if there is a finite list $A_1, \ldots, A_n$ of formulas in Γ such that $A_1, \ldots, A_n \vdash B$.
   Notation: $\Gamma \vdash B$.

Example 2.15. We have seen in Example 2.13 that $A \to B, B \to C \vdash A \to C$ and in Example 2.14 that $P \to W, \neg W \vdash \neg P$. And also in Example 2.12 that $\vdash A \to A$.

So, $A_1, \ldots, A_n \vdash B$, in words: B is *deducible from* $A_1, \ldots, A_n$, if and only if there exists a finite schema of the form

$$\frac{A_1 \quad \cdots \quad A_n \qquad \text{axiom} \qquad \text{axiom}}{\dfrac{D \quad D \to E}{\dfrac{E}{B}}}$$

And in case there are no premisses $A_1, \ldots, A_n$, i.e., $n = 0$, we say that $\vdash B$, in words: B is (logically) *provable* or *deducible*.

Example 2.16. Consider the following sequence of formulas:

$$\frac{\text{premiss}}{A \wedge B} \quad \frac{4\text{b}}{A \wedge B \to B} \text{ MP} \quad \frac{\dfrac{\text{premiss}}{A \wedge B} \quad \dfrac{4\text{a}}{A \wedge B \to A}}{A} \text{ MP} \quad \frac{\text{premiss}}{A \to (B \to C)}$$

$$\frac{B \qquad \qquad \qquad B \to C}{C} \text{ MP}$$

2.6 Syntax: Provability and Deducibility

For each choice of formulas A, B, C, this sequence of formulas is a deduction of C from $A \to (B \to C)$ and $A \wedge B$. Hence, C is deducible from $A \to (B \to C)$ and $A \wedge B$; i.e., $A \to (B \to C), A \wedge B \vdash C$.

The notion of logical consequence, $A_1, \ldots, A_n \models B$, is in terms of the truth or falsity and hence in terms of the *meaning* of the formulas involved. Therefore, this notion of logical consequence is a semantic notion. But the notion of (logical) deducibility, $A_1, \ldots, A_n \vdash B$, is in terms of the *forms* of the formulas involved. One does not have to know the meaning of the connectives, one only has to distinguish the form of the formulas involved. Therefore, this notion is a syntactic notion.

In $A_1, \ldots, A_n \vdash B$ one may think of the premises $A_1, \ldots, A_n$ as being the (non-logical) axioms of Euclid ($\pm$ 300 B.C.) for geometry, the axioms of Peano for arithmetic (see Chapter 5), the axioms of Zermelo - Fraenkel for set theory (see Chapter 3) or the laws of Newton for classical mechanics.

The premises $A_1, \ldots, A_n$, formulated in an appropriate formal language, constitute what one calls a (formal) *theory*: Euclid's geometry, Peano's arithmetic, the set theory of Zermelo - Fraenkel, Newton's mechanics, and so on. Each science is continually trying to re-adjust its foundations, as formulated in its premises. For instance, Cantor's naive set theory had to be replaced by the set theory of Zermelo - Fraenkel (see Chapter 3) and Newton's (classical) mechanics by Einstein's theory of relativity.

Of course, we want that our production method, consisting of the (logical) axioms for propositional logic and Modus Ponens, is *sound*, that is, when applied to given premises $A_1, \ldots, A_n$, it should generate only formulas which are a logical (or valid) consequence of $A_1, \ldots, A_n$. This is indeed the case, as stated in the following *soundness theorem*.

Theorem 2.20 (Soundness theorem).
(a): If $A_1, \ldots, A_n \vdash B$, then $A_1, \ldots, A_n \models B$, or, equivalently,
(a') if $A_1, \ldots, A_n \not\models B$, then $A_1, \ldots, A_n \not\vdash B$.
(b): In case $n = 0$, i.e., there are no premises: if $\vdash B$, then $\models B$.
(c): If $\Gamma \vdash B$, then $\Gamma \models B$.

Proof. Suppose $A_1, \ldots, A_n \vdash B$, i.e., there is a finite schema of the form

$$
\begin{array}{ccccc}
\underline{A_1} & \cdots & \underline{A_n} & \underline{\text{axiom}} & \underline{\text{axiom}} \\
 & & & & \\
 & \underline{D} & & \underline{D \to E} & \\
 & & E & & \\
 & \underline{} & & \underline{} & \\
 & & \underline{B} & &
\end{array}
$$

Note the following:
i) Each axiom of propositional logic has the value 1 in each line of the truth table.

ii) For all lines of the truth table, given that in an application of Modus Ponens the premisses D and $D \to E$ have the value 1, the conclusion E has the value 1 as well.

We have to show that $A_1,\ldots,A_n \models B$. So, suppose that the premisses $A_1,\ldots,A_n$ are 1 in a given line of the truth table. Then it follows from i) and ii) that, going from top to bottom in the deduction of B from $A_1,\ldots,A_n$, every formula in the deduction has value 1 in the given line. Hence, in particular, B has value 1 in that same line of the truth table. □

One may illustrate this proof by a concrete example, for instance, for the case that $A \to (B \to C)$, $A \wedge B \vdash C$.

Corollary 2.2 (Simple consistency).
There is no formula B such that both $\vdash B$ and $\vdash \neg B$.

Proof. Suppose $\vdash B$ and $\vdash \neg B$ for some B. Then according to the soundness theorem 2.20, $\models B$ and $\models \neg B$. Contradiction. □

We hope that the production method, consisting of the (logical) axioms for propositional logic and Modus Ponens, is *complete*, that is, that every valid consequence of given premisses $A_1,\ldots,A_n$ may be logically) deduced from these premisses. This is indeed the case, as is stated in the following theorem, which will be proved in Section 2.9 and in Exercise 2.59.

Theorem 2.21 (Completeness theorem).
(a): If $A_1,\ldots,A_n \models B$, then $A_1,\ldots,A_n \vdash B$, or, equivalently,
(a') if $A_1,\ldots,A_n \nvdash B$, then $A_1,\ldots,A_n \nvDash B$.
(b): In case $n = 0$, i.e., there are no premisses: if $\models B$, then $\vdash B$.
(c): If $\Gamma \models B$, then $\Gamma \vdash B$.

By the *soundness* of the axiomatic-deductive system for (classical) propositional logic we mean that *at most* certain formulas are provable, namely only those which are valid; by the *completeness* we mean that *at least* certain formulas are provable, namely, all which are valid. By the end of Section 2.9 we shall have proved the completeness theorem and hence (combining completeness and soundness) have shown the following equivalences:

$$A_1,\ldots,A_n \models B \text{ iff } A_1,\ldots,A_n \vdash B$$
$$\Gamma \models B \text{ iff } \Gamma \vdash B$$
$$\models B \text{ iff } \vdash B$$

There are a number of arguments underscoring the *philosophical meaning of the completeness theorem*, which justify taking the trouble to prove this theorem.

1. The completeness theorem tells us that any correct argument (in the object language) has a rational reconstruction which has the standard form described in the definition of $A_1,\ldots,A_n \vdash B$. Arguments in science and in daily life usually do not proceed in the way described in the definition of $A_1,\ldots,A_n \vdash B$, but according to the completeness theorem for any such correct argument there is a rational reconstruction which does.

2.6 Syntax: Provability and Deducibility

2. Note that whether B is deducible from $A_1, \ldots, A_n$ or not only depends on the *form* of the formulas $A_1, \ldots, A_n$ and B. Hence, the question whether B is a valid consequence of $A_1, \ldots, A_n$ or not has been reduced to a question about the form of the formulas $A_1, \ldots, A_n$ and B.
3. We have defined the intuitive notion of 'B is a logical consequence of $A_1, \ldots, A_n$' in two completely different ways; we have given a semantic definition in terms of truth values ($A_1, \ldots, A_n \models B$) and a syntactic one in terms of logical axioms and the rule Modus Ponens ($A_1, \ldots, A_n \vdash B$). That these two notions turn out to be equivalent suggests that our definitions indeed capture the corresponding intuitive notion.
4. We have given a mathematically precise definition of the intuitive notion of logical consequence in order to make this notion mathematically manageable, which is necessary if one wants to prove in a precise way certain statements about this notion. Now it is safe to assume that

   a) if B is intuitively a logical consequence of $A_1, \ldots, A_n$, then $A_1, \ldots, A_n \models B$.
   According to the completeness theorem,
   b) if $A_1, \ldots, A_n \models B$, then $A_1, \ldots, A_n \vdash B$.
   An analysis of the axioms and rules of propositional logic indicates that
   c) if $A_1, \ldots, A_n \vdash B$, then B is intuitively a logical consequence of $A_1, \ldots, A_n$.

   (a), (b) and (c) show that the intuitive notion of logical consequence and the mathematical notions of $A_1, \ldots, A_n \models B$ and of $A_1, \ldots, A_n \vdash B$ coincide extensionally.
5. In Chapter 4 we shall extend the notion of valid or logical consequence and of (logical) deducibility to (classical) *predicate logic*. Then we shall prove that these notions are again equivalent (soundness and completeness). On that occasion we shall further elaborate on the meaning of the completeness theorem in the case of predicate logic.

In Example 2.14 we have constructed a logical deduction of $\neg P$ from the premises $P \rightarrow W$ and $\neg W$, hence, $P \rightarrow W, \neg W \vdash \neg P$, where P and W were atomic formulas. More generally, in the same way one can show that for arbitrary formulas A and B, $A \rightarrow B, \neg B \vdash \neg A$. That is, the rule *Modus Tollens*

$$\frac{A \rightarrow B \quad \neg B}{\neg A}$$

is a derived rule, that from now on may be used in the construction of (logical) deductions. There are many more derived rules, for instance, see Exercise 2.39.

Exercise 2.38. Translate the following arguments in logical terminology and check whether the (putative) conclusion is *deducible* from the premises. If so, give a deduction, using the logical axioms $K \rightarrow (R \rightarrow K)$ and $(R \rightarrow K) \rightarrow ((R \rightarrow \neg K) \rightarrow \neg R)$. If not, then why not?
a) If it rains [R], then John will not come [$\neg C$]. John will come. Therefore: it does not rain.
b) Only if it rains [R], John will not come [$\neg C$]. John will come. Therefore: it does not rain.

Exercise 2.39. By constructing appropriate deductions, show that
(a) $A, A \rightarrow B \vdash B$ (f) $B \vdash A \vee B$
(b) $A, B \vdash A \wedge B$ (g) $\neg\neg A \vdash A$
(c) $A \wedge B \vdash A$ (h) $A \rightarrow B, B \rightarrow A \vdash A \rightleftarrows B$
(d) $A \wedge B \vdash B$ (i) $A \rightleftarrows B \vdash A \rightarrow B$
(e) $A \vdash A \vee B$ (j) $A \rightleftarrows B \vdash B \rightarrow A$

Hence, from now on, the following derived rules may be used in the construction of (logical) deductions:

$$\frac{A \quad B}{A \wedge B} \quad \frac{A \wedge B}{A} \quad \frac{A \wedge B}{B} \quad \frac{A}{A \vee B} \quad \frac{B}{A \vee B} \quad \frac{\neg\neg A}{A}$$

Exercise 2.40. Prove that $A, \neg A \vdash B$ by using the following axioms:
axiom 1 (a): $A \rightarrow (\neg B \rightarrow A)$ axiom 1 (b): $\neg A \rightarrow (\neg B \rightarrow \neg A)$
axiom 7: $(\neg B \rightarrow A) \rightarrow ((\neg B \rightarrow \neg A) \rightarrow \neg\neg B)$ axiom 8: $\neg\neg B \rightarrow B$

Exercise 2.41. By using the soundness theorem show that
(a) not $P \vee Q \vdash P \wedge Q$,   (c) not $P \vdash Q$,
(b) not $P \rightarrow Q \vdash Q \rightarrow P$,   (d) not $P \rightarrow Q \vdash P \wedge Q$.

Note that in order to show that $A \vdash B$, it suffices to exhibit at least one logical deduction of B from A; but in order to show that not $A \vdash B$, one has to prove that no logical deduction of B from A can exist, in other words, that any deduction is not a deduction of B from A. In order to prove the latter, it suffices – according to the soundness theorem – to show that $A \not\models B$.

Exercise 2.42. Prove or refute: $P \rightarrow Q, P \vdash R \vee Q$ either by giving a deduction of $R \vee Q$ from $P \rightarrow Q$ en P, using the logical axiom $B \rightarrow A \vee B$, or by showing that such a deduction cannot exist.

Exercise 2.43. Translate the following argument in logical terminology and check whether the (putative) conclusion is *deducible* from the premisses. If so, give a deduction, using the logical axioms $A \rightarrow (B \rightarrow A)$, $(A \rightarrow B) \rightarrow ((A \rightarrow \neg B) \rightarrow \neg A)$ and $\neg\neg A \rightarrow A$. If not, why not?
If John succeeds [S], then John works hard [H].
If John is not intelligent [$\neg I$], then John does not succeed.
Therefore: if John is intelligent, then John works hard.

Exercise 2.44. Consider a system of three truth values, 0, 1 and 2, of which 0 is the only designated truth value, and let the truth table of $\rightarrow$ be as follows.

A	B	$A \rightarrow B$
0	0	0
0	1	1
0	2	2
1	0	0
1	1	0
1	2	2
2	0	0
2	1	0
2	2	0

Show that for any choice of formulas A, B, C

a) for every interpretation i, $i(A \to (B \to A)) = 0$,
b) for every interpretation i, $i((A \to B) \to ((A \to (B \to C)) \to (A \to C))) = 0$,
c) for every interpretation i, if $i(A) = 0$ and $i(A \to B) = 0$, then $i(B) = 0$,
d) for some interpretation i, $i(((A \to B) \to A) \to A) \neq 0$.

Conclude that Peirce's law, $((A \to B) \to A) \to A$, is independent of $A \to (B \to A)$ and $(A \to B) \to ((A \to (B \to C)) \to (A \to C))$, in other words, that Peirce's law is not generated by the production method consisting of only the two axioms for $\to$ and Modus Ponens.

2.7 Syntax: Meta-logical Results

In this section (logical) proofs and deductions in the object-language will be studied, using (of necessity) informal proofs and deductions in the meta-language. The main results are the Deduction theorem, and the Introduction and Elimination rules. Given premisses $A_1, \ldots, A_n$ and given a formula B, these theorems are crucial in facilitating the search for a logical deduction of B from $A_1, \ldots, A_n$, if there is one. Next Gentzen's system of Natural Deduction is presented. It is shown that any formula which is logically provable in this system is also provable in the proof-system of Section 2.6, and conversely.

In Section 2.6 we defined a (logical) deduction of B from premisses $A_1, \ldots, A_n$ as being a finite sequence of formulas which satisfies certain conditions. It is important to realize that whether a given sequence of formulas is a (logical) deduction or not only depends on the *form* of the formulas in the sequence. In other words, whether a given sequence of formulas is a (logical) deduction can be checked mechanically; one can write a computer program to check the correctness of a given putative (logical) deduction. An example is *Automath*, developed by N.G. de Bruijn [3] and others at Eindhoven University.

It is also important to distinguish between logical deductions (of formulas) in the object language and informal proofs of certain statements *about* logical deductions. For instance, in Theorem 2.22 (b1) we will prove informally that if $A_1, A_2, A_3 \vdash B_1$ and $A_1, A_2, A_3 \vdash B_2$ and $B_1, B_2 \vdash C$, then $A_1, A_2, A_3 \vdash C$. This theorem is *about* logical proofs and deductions in the object-language; however, the formulation and the (informal) proof of this theorem are given in the meta-language. Notice that this Theorem is the syntactic counterpart of Exercise 2.26.

Theorem 2.22.
(a1) $A_1, A_2, A_3 \vdash A_1$, $A_1, A_2, A_3 \vdash A_2$, $A_1, A_2, A_3 \vdash A_3$.
(a2) More generally: $A_1, \ldots, A_i, \ldots, A_n \vdash A_i$ for $i = 1, \ldots, n$.
(b1) If $A_1, A_2, A_3 \vdash B_1$ *and* $A_1, A_2, A_3 \vdash B_2$ *and* $B_1, B_2 \vdash C$, *then* $A_1, A_2, A_3 \vdash C$.
(b2) More generally, for any $n, k \geq 0$: *if* $A_1, \ldots, A_n \vdash B_1$ *and* $\ldots$ *and* $A_1, \ldots, A_n \vdash B_k$ *and* $B_1, \ldots, B_k \vdash C$, *then* $A_1, \ldots, A_n \vdash C$.

Proof. *(a1)* For each $i, 1 \leq i \leq 3$, A_i itself is a (logical) deduction of A_i from A_1, A_2, A_3. In the definition of a logical deduction it is not required that all the premisses are actually used; they *may* be used, but not necessarily so.
(a2) is shown similarly.
(b1)

$$(\beta_1)\begin{cases} \underline{A_1, A_2} \quad \underline{A_3} \quad \underline{\text{axiom}} \\ \underline{} \\ \underline{B_1} \end{cases} \quad (\beta_2)\begin{cases} \underline{A_1, A_2} \quad \underline{A_3} \quad \underline{\text{axiom}} \\ \underline{} \\ \underline{B_2} \end{cases} \Bigg\} (\gamma)$$

$$\underline{}$$
$$C$$

Assume $A_1, A_2, A_3 \vdash B_1$, $A_1, A_2, A_3 \vdash B_2$ and $B_1, B_2 \vdash C$. That is, there are deductions (β_1) and (β_2) of B_1 and B_2 respectively, from A_1, A_2, A_3 and there is a deduction (γ) of C from B_1, B_2. By replacing the premisses B_1 and B_2 in (γ) by the deductions (β_1) and (β_2), we obtain a (logical) deduction of C from A_1, A_2, A_3. Hence, $A_1, A_2, A_3 \vdash C$.
(b2) is shown similarly. □

If we take in Theorem 2.22 (b1) $B_1 = B_2 = A_3 = A$, we obtain the following result.

Corollary 2.3. *If $A \vdash C$, then $A_1, A_2, A \vdash C$.*
More generally: If $A \vdash C$, then $A_1, \ldots, A_{n-1}, A \vdash C$.

Proof. In the definition of $A_1, \ldots, A_{n-1}, A \vdash C$ it is not required that each of the assumption formulas $A_1, \ldots, A_{n-1}$ actually occur in the deduction. □

Theorem 2.22 can be reformulated in set-theoretic terms: let $L(A_1, \ldots, A_n)$, called the *logic of* $A_1, \ldots, A_n$, be the set of all formulas that are deducible from $A_1, \ldots, A_n$. Then Theorem 2.22 says that i) for each $i, 1 \leq i \leq n$, A_i is in $L(A_1, \ldots, A_n)$, and ii) if each of $B_1, \ldots, B_k$ is in $L(A_1, \ldots, A_n)$ and $B_1, \ldots, B_k \vdash C$, then C is in $L(A_1, \ldots, A_n)$.

Since in Corollary 2.3 the premisses $A_1, \ldots, A_{n-1}$ are not relevant to C, Corollary 2.3, which just has been shown for classical logic, does not hold for the so-called *relevance logic*; see Section 6.10.

Let us consider the following four expressions:
(i) $\models A \rightarrow B$ i.e., $A \rightarrow B$ is valid,
(ii) $A \models B$ i.e., B is a valid consequence of A,
(iii) $\vdash A \rightarrow B$ i.e., $A \rightarrow B$ is (logically) provable,
(iv) $A \vdash B$ i.e., B is (logically) deducible from A.
(i) and (ii) are *semantic notions*, i.e., they are concerned with the *meaning* of the formulas in question; (iii) and (iv) are *syntactic notions*, i.e., they are concerned with the *form* of the formulas in question.

In Theorem 2.4 we have already shown that (i) and (ii) are equivalent. In Theorems 2.23 and 2.24 we will prove that (iii) and (iv) are equivalent.

2.7 Syntax: Meta-logical Results

In the soundness theorem (Theorem 2.20) we have shown that (iii) implies (i) and that (iv) implies (ii). The converses of these results, (i) implies (iii) and (ii) implies (iv), will be shown in Section 2.9.

So, by the end of Section 2.9 we shall have proved that (i), (ii), (iii) and (iv) are equivalent. But remember that 'if $\models A$, then $\models B$' is a weaker statement than (ii), $A \models B$ (see Theorem 2.11). Consequently, 'if $\vdash A$, then $\vdash B$' is a weaker statement than (iv), $A \vdash B$.

Theorem 2.23. *(a) If $\vdash A \to B$, then $A \vdash B$. (b) More generally, for any $n \geq 1$, if $A_1, \ldots, A_{n-1} \vdash A \to B$, then $A_1, \ldots, A_{n-1}, A \vdash B$.*

Proof. (b) Suppose $A_1, \ldots, A_{n-1} \vdash A \to B$, i.e., there is a deduction (α) of $A \to B$ from $A_1, \ldots, A_{n-1}$.

$$\left. \begin{array}{c} \dfrac{A_1 \qquad\qquad A_{n-1} \qquad\qquad \text{axiom}}{} \\ \\ \dfrac{A \qquad\quad \overline{A \to B}}{B}\, MP \end{array} \right\} (\alpha)$$

By adding one more premiss, A, to this deduction and one more application of Modus Ponens, one obtains a deduction of B from $A_1, \ldots, A_{n-1}, A$. □

2.7.1 Deduction Theorem; Introduction and Elimination Rules

In order to establish an implication 'if A, then B', one often assumes A and then continues to conclude B. The following theorem, called the deduction theorem, which is the converse of Theorem 2.23, captures this idea in a precise form: in order to establish that $A_1, \ldots, A_{n-1} \vdash A \to B$, it suffices to show that $A_1, \ldots, A_{n-1}, A \vdash B$.

That the deduction theorem is a very useful tool may be seen from the following. In order to show that $\vdash A \to ((A \to B) \to B)$, it suffices by the deduction theorem to show that $A \vdash (A \to B) \to B$. Likewise, in order to show the latter statement it suffices to prove $A, A \to B \vdash B$; and this is very easy (one application of Modus Ponens suffices), while to show that $\vdash A \to ((A \to B) \to B)$ directly is much more complicated.

Theorem 2.24 (Deduction theorem, Herbrand 1930).
(a) If $A \vdash B$, then $\vdash A \to B$. More generally,
(b) If $A_1, \ldots, A_{n-1}, A \vdash B$, then $A_1, \ldots, A_{n-1} \vdash A \to B$.

Proof. (b) Suppose $A_1, \ldots, A_{n-1}, A \vdash B$, i.e., there is a (logical) deduction (α) of B from the premisses $A_1, \ldots, A_{n-1}, A$. Below we shall change (α) step by step into a (logical) deduction (γ) of $A \to B$ from $A_1, \ldots, A_{n-1}$, hence showing that $A_1, \ldots, A_{n-1} \vdash A \to B$.

$$\left.\begin{array}{c} \cfrac{A_1 \qquad A_{n-1} \qquad A \qquad \text{axiom}}{} \\ \\ \cfrac{C \quad C \to D}{D} \\ \\ \cfrac{}{B} \end{array}\right\} \ (\alpha)$$

The first step consists in prefixing the symbols $A \to$ to each formula occurring in (α). This results in the schema (β).

$$\left.\begin{array}{c} \cfrac{A \to A_1 \qquad A \to A_{n-1} \qquad A \to A \qquad A \to \text{axiom}}{} \\ \\ \cfrac{A \to C \qquad A \to (C \to D)}{A \to D} \\ \\ \cfrac{}{A \to B} \end{array}\right\} \ (\beta)$$

Although the last formula in (β) is $A \to B$, (β) itself is not a deduction of $A \to B$ from $A_1, \ldots, A_{n-1}$ for the following reasons:
(i) (β) does not start with logical axioms or premises $A_1, \ldots, A_{n-1}$, and (ii)

$$\cfrac{A \to C \qquad A \to (C \to D)}{A \to D}$$

is not an application of Modus Ponens.

However, by inserting appropriate formulas into (β), one can transform (β) into a (logical) deduction (γ) of $A \to B$ from $A_1, \ldots, A_{n-1}$ as follows.

1. For $1 \le j \le n-1$ replace $A \to A_j$ at the top in (β) by the following:

$$\cfrac{A_j \qquad \overset{\text{axiom 1}}{A_j \to (A \to A_j)}}{A \to A_j} \ \text{MP}$$

2. Replace $A \to A$ at the top in (β) by the (logical) proof of $A \to A$, given in Section 2.6.

3. Replace $A \to$ axiom at the top in (β) by the following:

$$\cfrac{\text{axiom} \qquad \overset{\text{axiom 1}}{\text{axiom} \to (A \to \text{axiom})}}{A \to \text{axiom}} \ \text{MP}$$

4. Replace

$$\cfrac{A \to C \qquad A \to (C \to D)}{A \to D}$$

by the following:

2.7 Syntax: Meta-logical Results

$$\frac{A \to C \qquad \overset{\text{axiom 2}}{(A \to C) \to ((A \to (C \to D)) \to (A \to D))}}{(A \to (C \to D)) \to (A \to D)} \text{ MP} \qquad A \to (C \to D)$$

$$\overline{A \to D}$$

Each formula of the resulting sequence (γ) either is one of $A_1, \ldots, A_{n-1}$ or is a logical axiom or comes from two preceding formulas in the sequence by Modus Ponens, and the last formula of the sequence is $A \to B$. So (γ) is a deduction of $A \to B$ from $A_1, \ldots, A_{n-1}$. □

In Exercise 2.58 the proof of the deduction theorem is applied to a deduction of $Q \vee R$ from $P \to Q$ and P in order to obtain a deduction of $P \to Q \vee R$ from $P \to Q$.

Example 2.17. In Example 2.16 we have seen that $A \to (B \to C)$, $A \wedge B \vdash C$. By the deduction theorem it follows that $A \to (B \to C) \vdash A \wedge B \to C$. And, again by the deduction theorem, it also follows that $\vdash (A \to (B \to C)) \to (A \wedge B \to C)$. The reader would find it a difficult exercise to construct in a direct way (i.e., without applying the deduction theorem or using its method of proof) a logical proof of $(A \to (B \to C)) \to (A \wedge B \to C)$.

In general, it is much easier to show that $A_1, \ldots, A_{n-1}$, $A \vdash B$ than to show that $A_1, \ldots, A_{n-1} \vdash A \to B$. The deduction theorem is a simple way to show the existence of certain (logical) deductions without having to exhibit those logical deductions explicitly. It is easy to write down a logical deduction of C from $A \to (B \to C)$ and $A \wedge B$; so, $A \to (B \to C)$, $A \wedge B \vdash C$. Then, by two applications of the deduction theorem, one knows that $\vdash (A \to (B \to C)) \to (A \wedge B \to C)$, without having to write down a logical proof of the latter formula, which would be a rather complicated job. Following the proof of the deduction theorem one is able in principle to exhibit such a logical proof, but in most cases we are not interested in writing down this (logical) proof explicitly.

It is possible to derive additional results which make it easy to show that certain deductions exist without having to write down those deductions explicitly. One result is called *Reductio ad absurdum*; it says that in order to deduce $\neg A$ (from Γ, where Γ is a finite list of zero or more formulas) it suffices to deduce a contradiction (B and $\neg B$) from the assumption A (together with Γ). Another result is called $\vee$-*elimination*: in order to deduce C from $A \vee B$ (and Γ), it suffices to deduce C from A (and Γ) and to deduce C from B (and Γ).

The proof system of Section 2.6 contains only one rule, Modus Ponens. However, many other rules can be derived, for example, the rule called $\wedge$-*introduction*: from the two formulas A and B one can deduce the one formula $A \wedge B$. This result is obtained by using the axiom $A \to (B \to A \wedge B)$ and two applications of Modus Ponens. The next theorem contains the results just mentioned and a number of related similar results.

Theorem 2.25 (Introduction and Elimination Rules). *For any finite list Γ of (zero or more) formulas, and for any formulas A, B, C:*

	INTRODUCTION	ELIMINATION
$\rightarrow$	If $\Gamma, A \vdash B$, then $\Gamma \vdash A \rightarrow B$	$A, A \rightarrow B \vdash B$
$\wedge$	$A, B \vdash A \wedge B$	$A \wedge B \vdash A$ $A \wedge B \vdash B$
$\vee$	$A \vdash A \vee B$ $B \vdash A \vee B$	If $\Gamma, A \vdash C$ and $\Gamma, B \vdash C$, then $\Gamma, A \vee B \vdash C$
$\neg$	If $\Gamma, A \vdash B$ and $\Gamma, A \vdash \neg B$, then $\Gamma \vdash \neg A$ (reductio ad absurdum)	$\neg\neg A \vdash A$ (double negation elimination) $A, \neg A \vdash B$ (weak negation elimination)
$\rightleftarrows$	$A \rightarrow B, B \rightarrow A \vdash A \rightleftarrows B$	$A \rightleftarrows B \vdash A \rightarrow B$ $A \rightleftarrows B \vdash B \rightarrow A$

Proof. $\rightarrow$-*introduction* is the deduction theorem.

$\rightarrow$-*elimination*, $\wedge$-*introduction*, $\wedge$-*elimination*, $\vee$-*introduction*, *double negation elimination* and the three $\rightleftarrows$-*rules* are done in Exercise 2.39.

$\vee$-*elimination*: Suppose $\Gamma, A \vdash C$ and $\Gamma, B \vdash C$. Then by the deduction theorem $\Gamma \vdash A \rightarrow C$ and $\Gamma \vdash B \rightarrow C$. The following schema shows that $A \rightarrow C, B \rightarrow C, A \vee B \vdash C$:

$$\cfrac{A \rightarrow C \quad \cfrac{\text{axiom 6}}{(A \rightarrow C) \rightarrow ((B \rightarrow C) \rightarrow (A \vee B \rightarrow C))}}{\cfrac{B \rightarrow C \quad (B \rightarrow C) \rightarrow (A \vee B \rightarrow C)}{\cfrac{A \vee B \rightarrow C \quad A \vee B}{C} \text{ MP}} \text{ MP}} \text{ MP}$$

Hence, $\Gamma, A \vee B \vdash C$.

Weak negation elimination: Evidently, (1) $A, \neg A, \neg B \vdash A$, and (2) $A, \neg A, \neg B \vdash \neg A$. From (1) and (2) it follows by $\neg$-introduction that (3) $A, \neg A \vdash \neg\neg B$. And, by double negation elimination, also (4) $\neg\neg B \vdash B$. From (3) and (4) it follows that $A, \neg A \vdash B$. By this rule, from a contradiction $A, \neg A$, any formula B can be deduced.

$\neg$-*introduction* (reductio ad absurdum): Suppose $\Gamma, A \vdash B$ and $\Gamma, A \vdash \neg B$. Then by the deduction theorem $\Gamma \vdash A \rightarrow B$ and $\Gamma \vdash A \rightarrow \neg B$. Let (α) be a deduction of $A \rightarrow B$ from Γ and let (β) be a deduction of $A \rightarrow \neg B$ from Γ. Then the schema below is a deduction of $\neg A$ from Γ.

2.7 Syntax: Meta-logical Results

$$(\beta)\begin{cases} \begin{array}{cc} \Gamma & \Gamma \\ \rule{1cm}{0.4pt} & \rule{1cm}{0.4pt} \\ \rule{1cm}{0.4pt} & \rule{1cm}{0.4pt} \\ \rule{1cm}{0.4pt} & \rule{1cm}{0.4pt} \\ & A \to B \end{array} \Bigg\} (\alpha) \\ \rule{3cm}{0.4pt} \\ A \to \neg B \end{cases} \quad \dfrac{\quad \dfrac{\text{axiom 7}}{(A \to B) \to ((A \to \neg B) \to \neg A)}\quad}{\dfrac{(A \to \neg B) \to \neg A}{\neg A}\ \text{MP}}\ \text{MP}$$

□

Exercise 2.45. Show that $A \wedge B \to C \vdash A \to (B \to C)$.

Exercise 2.46. Show that $\vdash (A \to B) \to (A \to ((B \to C) \to C))$.

Exercise 2.47. Show: if $A_1, A_2 \vdash B$, then $\vdash A_1 \wedge A_2 \to B$.

Exercise 2.48. Show that: If $\vdash (A_1 \wedge A_2) \wedge A_3 \to B$, then $A_1, A_2, A_3 \vdash B$.

Exercise 2.49. Prove or refute without making use of the completeness theorem: If $\vdash A \to C$ and $\vdash B \to C$, then $A \vee B \vdash C$. You may make use of the logical axiom $(A \to C) \to ((B \to C) \to (A \vee B \to C))$.

Exercise 2.50. Using ∨-elimination, show that $A \vee B$, $B \to C \vdash A \vee C$.

Exercise 2.51. Use ¬-introduction to show: if $A \vdash B$, then $\neg B \vdash \neg A$.

Exercise 2.52. Using ¬-introduction and exercise 2.51, show that $\vdash A \vee \neg A$.

Exercise 2.53. Using ∨-elimination, ¬-introduction and weak negation elimination, show that $\neg A, \neg B \vdash \neg(A \vee B)$.

Exercise 2.54. Use ¬-introduction to show: if $A \vdash \neg A$, then $\vdash \neg A$.

Exercise 2.55. Prove or refute (by means of a counterexample): for all formulas A, B, if $\vdash A \vee B$, then $\vdash A$ or $\vdash B$. Carefully specify your arguments.

Exercise 2.56. Prove or refute (by means of a counterexample): for all formulas A, if not $\vdash A$, then $\vdash \neg A$. Carefully specify your arguments and do not use the completeness theorem.

Exercise 2.57. Prove or refute, carefully specifying your arguments and not making use of the completeness theorem:
a) If $\vdash A \to B$, then $A \vdash B$.   b) If $\vdash \neg A$, then not $\vdash A$.

Exercise 2.58. Show that $\neg A \vee B \vdash A \to B$. Next show: a) $A \to B, \neg(\neg A \vee B) \vdash \neg\neg A$ and b) $A \to B, \neg(\neg A \vee B) \vdash \neg A$. Conclude from a) and b) by ¬-introduction that $A \to B \vdash \neg\neg(\neg A \vee B)$ and hence $A \to B \vdash \neg A \vee B$.

Exercise 2.59 (Completeness). In this exercise we shall prove the *completeness theorem* for classical propositional logic along the lines of L. Kalmár, 1934-5.

Consider the truth table for a formula $E(A,B)$ built from the formulas A and B. To each entry (or line) of this truth table a corresponding deducibility relationship holds, as indicated below:

	A	B	$E(A,B)$	
u_1	1	1	$u_1(E)$	$A, B \vdash E_1^*$
u_2	1	0	$u_2(E)$	$A, \neg B \vdash E_2^*$
u_3	0	1	$u_3(E)$	$\neg A, B \vdash E_3^*$
u_4	0	0	$u_4(E)$	$\neg A, \neg B \vdash E_4^*$

where $E_i^* = E$ if $u_i(E) = 1$ and $E_i^* = \neg E$ if $u_i(E) = 0$ ($i = 1, 2, 3, 4$).

a) Establish the first two deducibility relationships for $E = A \wedge B$ and the last two for $E = A \vee B$.

b) Using the result mentioned above prove the completeness theorem for classical propositional logic: if $\models E$, then $\vdash E$.

2.7.2 Natural Deduction*

Hilbert's proof system, presented in Section 2.6, has several axiom schemas and only one rule, Modus Ponens. In his *Untersuchungen über das logische Schliessen* G. Gentzen [9] introduced a different, but equivalent, proof system which has several rules, but no axioms. This proof system is called Gentzen's system of *Natural Deduction*. Logical proofs in this system are very similar to the informal proofs in daily reasoning, which makes the search for a logical proof in this system much easier than in a Hilbert-type proof system. Before the rules are presented some of them will be discussed and the notation explained.

$\rightarrow$-*Introduction*: Suppose B is derived from the assumption A (and perhaps other assumptions as well); notation:
$$\begin{array}{c} A \\ \vdots \\ B \end{array}$$
Then one can derive $A \rightarrow B$, cancelling the assumption A; notation:

$$\begin{array}{c} [A]^i \\ \vdots \\ B \\ \hline A \rightarrow B \end{array} \quad i \quad \text{where } i \text{ is a natural number.}$$

Note that this rule corresponds to the deduction theorem (Theorem 2.24).

$\neg$-*Introduction*: Suppose a contradiction (B and $\neg B$) is derived from one or more assumption formulas among which is A. Notation:

2.7 Syntax: Meta-logical Results

$$A$$
$$\vdots$$
$$B \quad \neg B$$

Then one can obtain a deduction of $\neg A$ from the assumptions without A. Notation:

$$[A]^i$$
$$\vdots$$
$$\frac{B \quad \neg B}{\neg A} i$$

$\vee$-*Elimination* : Suppose one has a deduction of C from the assumption A and another deduction of the same formula C from the assumption B, where in both cases other assumptions may be present. Then one can obtain a deduction of C from the assumption $A \vee B$, cancelling the assumptions A and B. Notation:

$$[A]^i \quad [B]^i$$
$$\vdots \quad \vdots$$
$$\frac{A \vee B \quad C \quad C}{C} i$$

Having explained how to read the more complicated rules of natural deduction, below all Gentzen rules for *natural deduction* are presented.

GENTZEN'S INTRODUCTION RULES	GENTZEN'S ELIMINATION RULES

&I $\quad \dfrac{A \quad B}{A \wedge B}$

&E $\quad \dfrac{A \wedge B}{A} \qquad \dfrac{A \wedge B}{B}$

$\vee I$ $\quad \dfrac{A}{A \vee B} \qquad \dfrac{B}{A \vee B}$

$\vee E$ $\quad \dfrac{A \vee B \quad C \quad C}{C}$ with $[A], [B]$ discharged

$\rightarrow I$ $\quad \dfrac{\begin{array}{c}[A]\\ \vdots \\ B\end{array}}{A \rightarrow B}$

$\rightarrow E$ $\quad \dfrac{A \quad A \rightarrow B}{B}$

$\neg I$ $\quad \dfrac{\begin{array}{c}[A]\\ \vdots \\ B \quad \neg B\end{array}}{\neg A}$

$w \neg E$ $\quad \dfrac{A \quad \neg A}{B}$ (w = weak)

$d \neg E$ $\quad \dfrac{\neg \neg A}{A}$ (d = double)

The reader should note the analogy with the Introduction and Elimination rules in Theorem 2.25, but he should also see the difference. For instance, $A \vdash A \vee B$ says that $A \vee B$ can be obtained from A and the logical axioms by applying the rule Modus Ponens a finite number of times, while $\dfrac{A}{A \vee B}$ itself is a rule of inference in the natural deduction system, as Modus Ponens is a rule of inference in the axiomatic system of Section 2.6. In other words, $A \vdash A \vee B$ says that $\dfrac{A}{A \vee B}$ is a *derived rule* of inference in the axiomatic system of Section 2.6.

Example 2.18. Below are some examples of deductions in Gentzen's system of Natural Deduction.

(i) $(A \to B) \to ((B \to C) \to (A \to C))$

$$\begin{array}{c}
\dfrac{{}^1[A] \quad [A \to B]^3}{B} \to E \\
\dfrac{\qquad\qquad\qquad [B \to C]^2}{C} \to E \\
(1)\ \dfrac{C}{A \to C} \to I \\
(2)\ \dfrac{}{(B \to C) \to (A \to C)} \to I \\
(3)\ \dfrac{}{(A \to B) \to ((B \to C) \to (A \to C))} \to I
\end{array}$$

The reader should note the analogy with the way in which we intuitively verify that $(A \to B) \to ((B \to C) \to (A \to C))$ is true.
To show: $(A \to B) \to ((B \to C) \to (A \to C))$.
So suppose $A \to B$; then to show $(B \to C) \to (A \to C)$.
So suppose $B \to C$; then to show $A \to C$.
So suppose A; then to show C.
Now from A and $A \to B$ it follows that B. And from B and $B \to C$ it follows that C. So C follows from A, $B \to C$ and $A \to B$. Hence $A \to C$ follows from $B \to C$ and $A \to B$. Therefore $(B \to C) \to (A \to C)$ follows from $A \to B$. Consequently, $(A \to B) \to ((B \to C) \to (A \to C))$.

(ii) $\neg\neg A \to A$

$$\begin{array}{c}
(1)\ \dfrac{[\neg\neg A]^1}{A}\ d\ \neg E \\
\dfrac{}{\neg\neg A \to A} \to I
\end{array}$$

(iii) $A \to \neg\neg A$

$$\begin{array}{c}
(1)\ \dfrac{{}^2[A] \quad [\neg A]^1}{\neg\neg A}\ \neg I \\
(2)\ \dfrac{}{A \to \neg\neg A} \to I
\end{array}$$

2.7 Syntax: Meta-logical Results

(iv) In the deduction of $A \vee \neg A$ below, the reader should again note the analogy with the way in which we intuitively show that $A \vee \neg A$ is true. Suppose that $\neg(A \vee \neg A)$. Then, since $A \vee \neg A$ follows from A, $\neg A$. But also, since $A \vee \neg A$ follows from $\neg A$, $\neg(\neg A)$. So from $\neg(A \vee \neg A)$ it follows that both $\neg A$ and $\neg(\neg A)$. Therefore, by $\neg$-introduction, $\neg\neg(A \vee \neg A)$ and hence, by double $\neg$-elimination, $A \vee \neg A$.

$$(1) \frac{\dfrac{{}^3[\neg(A \vee \neg A)] \quad \dfrac{[A]^1}{A \vee \neg A}\vee I}{\neg A}\neg I}{} \quad (2) \frac{\dfrac{{}^3[\neg(A \vee \neg A)] \quad \dfrac{[\neg A]^2}{A \vee \neg A}\vee I}{\neg\neg A}\neg I}{}$$

$$(3) \frac{\neg\neg(A \vee \neg A)}{A \vee \neg A} d\neg E \quad \neg I$$

Definition 2.12 (Deducibility in natural deduction). a) Let Γ be a (possibly infinite) set of formulas. B is *deducible from* Γ in Gentzen's system of Natural Deduction := B can be obtained by one or more (but finitely many) applications of Gentzen's rules of natural deduction from uncancelled assumptions that belong to the set Γ. **Notation**: $\Gamma \vdash_{ND} B$.

b) In case Γ is empty, we say that B is *provable* in Gentzen's system of natural deduction. **Notation**: $\vdash_{ND} B$.

Example 2.19. In Example 2.18 we have seen:
$A \to B \vdash_{ND} (B \to C) \to (A \to C)$ $\quad \vdash_{ND} (A \to B) \to ((B \to C) \to (A \to C))$
$\neg\neg A \vdash_{ND} A$ $\quad \vdash_{ND} \neg\neg A \to A$
$A \vdash_{ND} \neg\neg A$ $\quad \vdash_{ND} A \to \neg\neg A$
$\quad \vdash_{ND} A \vee \neg A$

Once having shown Theorem 2.25 (introduction and elimination rules), one easily sees that Gentzen's system of natural deduction is equivalent to the axiomatic (Hilbert-type) system of Section 2.6.

Theorem 2.26. $\Gamma \vdash B$ *iff* $\Gamma \vdash_{ND} B$.

Proof. i) Suppose $\Gamma \vdash B$. One easily checks that all the axioms of (classical) propositional logic are provable in Gentzen's system of natural deduction. Modus Ponens *MP* is precisely Gentzen's rule $\to E$. It follows that $\Gamma \vdash_{ND} B$.

ii) Suppose $\Gamma \vdash_{ND} B$. a) If B is an element of Γ, then $\Gamma \vdash B$.

b) Theorem 2.25 shows that all steps made in Gentzen's rules of natural deduction are also available for the notion of (Hilbert-type) deducibility of Section 2.6. More precisely, Gentzen's rule $\vee E$, for instance, says that if $\Delta, A \vdash_{ND} C$ and $\Delta, B \vdash_{ND} C$, then $\Delta, A \vee B \vdash_{ND} C$ for any set Δ of formulas. Now suppose (by induction hypothesis) that $\Delta, A \vdash C$ and $\Delta, B \vdash C$; then by $\vee$-elimination in Theorem 2.25, $\Delta, A \vee B \vdash C$. By a) and b) it follows (by induction on the length of a given ND-deduction of B from Γ in Gentzen's system of natural deduction) that $\Gamma \vdash B$. □

Exercise 2.60. Show that: i) $\neg(A \wedge B) \vdash_{ND} \neg A \vee \neg B$, ii) $\neg A \vee \neg B \vdash_{ND} \neg(A \wedge B)$. Keep in mind the way in which we would intuitively verify that the conclusion follows from the premisses.

Exercise 2.61. i) Show that $A \vdash B \to A$ and follow the proof of Theorem 2.26, part i), to convert the given deduction of $B \to A$ from A in Hilbert's system into a deduction of $B \to A$ from A in Gentzen's system of natural deduction.
ii) Show that $A \to B \vdash_{ND} \neg B \to \neg A$ and follow the proof of Theorem 2.26, part ii) to show that $A \to B \vdash \neg B \to \neg A$.

2.8 Tableaux

In this section we will introduce another notion of provability and of deducibility, which is based on the work of E. Beth [2] and of G. Gentzen [9], and equivalent to the corresponding notions defined in Section 2.6. The advantage of Beth's and Gentzen's notions is that the search for a deduction of B from $A_1, \ldots, A_n$ becomes a mechanical matter and is not achieved by the method of trial and error, as is (sometimes) the case for the historically older notions of Section 2.6, which are essentially based on the work of G. Frege [7] (1848-1925) and B. Russell [25] (1872-1970). This advantage is obtained by reducing the number of axiom-schemes to one, essentially $A \to A$, and by replacing the axioms by T and F rules, two for each connective. The presentation chosen here is close to the one of R. Smullyan [23] and was introduced by M. Fitting [6].

Definition 2.13 (Signed formula). A *signed formula* is any expression of the form $T(A)$ or $F(A)$, where A is a formula.

In the case of classical logic, the intended meanings of $T(A)$ and $F(A)$, in Beth's semantic tableaux rules, are as follows: $T(A)$: A is true, $F(A)$: A is false. (The intended meanings of $T(A)$ and $F(A)$ for modal and intuitionistic logic are different.)

If it is clear from the context what is meant, we will simply write TA instead of $T(A)$ and FA instead of $F(A)$. For instance, instead of $T(B \wedge C)$ we will mostly write $T B \wedge C$.

Definition 2.14 (Sequent). A *sequent* S is any finite set of signed formulas.

For example, $\{T P_1 \to P_2, F \neg P_1 \wedge P_2, F \neg P_2 \vee (P_2 \to P_1)\}$ is a sequent. In Gentzen's approach the intended meaning of a sequent $\{TB_1, \ldots, TB_m, FC_1, \ldots, FC_n\}$ is as follows: if B_1 and ... and B_m, then C_1 or ... or C_n.

Below we present the T- and F- *tableaux rules for classical propositional logic*; next we will explain how to read them, either as *semantic tableaux rules* in the sense of Beth or as *Gentzen-type rules*. In what follows, S will always denote a sequent.

2.8 Tableaux

$$T\land \quad \frac{S, T\,B\land C}{S, TB, TC} \qquad\qquad F\land \quad \frac{S, F\,B\land C}{S, FB \mid S, FC}$$

$$T\lor \quad \frac{S, T\,B\lor C}{S, TB \mid S, TC} \qquad\qquad F\lor \quad \frac{S, F\,B\lor C}{S, FB, FC}$$

$$T\to \quad \frac{S, T\,B\to C}{S, FB \mid S, TC} \qquad\qquad F\to \quad \frac{S, F\,B\to C}{S, TB, FC}$$

$$T\neg \quad \frac{S, T\,\neg B}{S, FB} \qquad\qquad F\neg \quad \frac{S, F\,\neg B}{S, TB}$$

Notation: S, TA stands for $S \cup \{TA\}$, i.e., the set containing all signed formulas in S and in addition TA; and S, FA similarly stands for $S \cup \{FA\}$. Instead of $\{TB_1, \ldots, TB_m, FC_1, \ldots, FC_n\}$ we often simply write $TB_1, \ldots, TB_m, FC_1, \ldots, FC_n$. For example, by $\{TD, FE\}, TA$ we mean $\{TD, FE, TA\}$, but we will usually write TD, FE, TA.

Since $S, T\,B\land C$ stands for $S\cup\{T\,B\land C\}$, and since this latter set is equal to $S\cup\{T\,B\land C, T\,B\land C\}$, the following rule

$$\frac{S, T\,B\land C}{S, T\,B\land C, TB, TC}$$

is a derived rule. So, in any application of any rule the T-signed or the F-signed formula to which the rule is applied may be repeated in the lower half of the rule.

Beth's semantic tableaux rules The rules given above can be read in two ways.

First, *read downwards*, as *semantic tableaux rules* in the sense of E. Beth, *interpreting the signed formulas* rather than the sequents. For example, in the case of rule $T\to$: if $B\to C$ is true ($T\,B\to C$), then there are two possibilities, B is false (FB) *or* C is true (TC). And in the case of rule $F\to$: if $B\to C$ is false ($F\,B\to C$), then B is true (TB) and C is false (FC).

This way of reading the rules is derived from E. Beth's [2] method of semantic tableaux. A formula B is called *tableau-deducible from* given formulas $A_1, \ldots, A_n$ if it turns out to be impossible that $A_1, \ldots, A_n$ are all 1 and B is 0; more precisely, if all sequents which result from application of the rules to the supposition $TA_1, \ldots, TA_n$, FB ($A_1, \ldots, A_n$ are all 1 and B is 0) and to which no further rules can be applied, turn out to be contradictory, i.e., for all such sequents there is an atomic formula P such that both TP (P is true) and FP (P is false) occur in it (see Def. 2.16 and 2.18).

Note that we essentially have used this idea in exercise 2.11 to verify that, for instance, $\models (P\to Q)\to(\neg Q\to\neg P)$ or, equivalently, $(P\to Q)\models(\neg Q\to\neg P)$, by showing that it is impossible that in some line of the truth table $(P\to Q)$ is 1 and $(\neg Q\to\neg P)$ is 0. In the left column of Example 2.20 we apply the tableaux rules to $T\,(P\to Q), F\,(\neg Q\to\neg P)$ and in the right column of Example 2.20 we give the interpretation of the left column in the sense of E. Beth.

Example 2.20.

$T\ (P \to Q),\ F(\neg Q \to \neg P)$	Suppose in some line of its truth table $(P \to Q)$ is 1 and $\neg Q \to \neg P$ is 0.
$TP \to Q,\ T\neg Q,\ F\neg P$	Then $P \to Q$ is 1, $\neg Q$ is 1 and $\neg P$ is 0.
$TP \to Q,\ FQ,\ F\neg P$	So, $P \to Q$ is 1, Q is 0 and $\neg P$ is 0 in that line.
$TP \to Q,\ FQ,\ TP$	So, $P \to Q$ is 1, Q is 0 and P is 1 in that line.
$FP,\ FQ,\ TP \mid TQ,\ FQ,\ TP$	So, P is 0, Q is 0 and P is 1, or, Q is 1, Q is 0 and P is 1 in that same line. And both are impossible.

Informally, we say that the left column in Example 2.20 is a *tableau* $\mathcal{T}$ with initial branch $\mathcal{B}_0 = \{T\ (P \to Q),\ F(\neg Q \to \neg P)\}$. This tableau $\mathcal{T}$ consists of two tableau branches $\mathcal{B}_{31}$ and $\mathcal{B}_{32}$, with $\mathcal{B}_{31} = \{T\ (P \to Q),\ F(\neg Q \to \neg P),\ T\neg Q,\ F\neg P,\ FQ,\ TP,\ FP\}$, containing all signed formulas in the left half of the tableau and $\mathcal{B}_{32} = \{T\ (P \to Q),\ F(\neg Q \to \neg P),\ T\neg Q,\ F\neg P,\ FQ,\ TP,\ TQ\}$, containing all signed formulas in the right half of the tableau. The branch $\mathcal{B}_{31}$ is *closed* because it contains TP and FP, and the branch $\mathcal{B}_{32}$ is *closed* because it contains TQ and FQ. Both branches are *completed*, i.e., for each signed formula in the branch the corresponding T- or F-rule has been applied.

Definition 2.15 ((Tableau) Branch). (a) A *tableau branch* is a set of signed formulas. A branch is *closed* if it contains signed formulas TA and FA for some formula A. A branch that is not closed is called *open*.
(b) Let $\mathcal{B}$ be a branch and TA, resp. FA, a signed formula occurring in $\mathcal{B}$. TA, resp. FA, is *fulfilled* in $\mathcal{B}$ if (i) A is atomic, or (ii) $\mathcal{B}$ contains the bottom formulas in the application of the corresponding rule to A, and in case of the rules $T\lor$, $F\land$ and $T\to$, $\mathcal{B}$ contains one of the bottom formulas in the application of these rules.
(c) A branch $\mathcal{B}$ is *completed* if $\mathcal{B}$ is closed or every signed formula in $\mathcal{B}$ is fulfilled in $\mathcal{B}$.

More formally, in Example 2.20 we call $\mathcal{B}_0 = \{T\ (P \to Q),\ F(\neg Q \to \neg P)\}$ the initial branch and $\mathcal{T}_0 = \{\mathcal{B}_0\}$ a tableau (with initial branch $\mathcal{B}_0$).

Let $\mathcal{B}_1 = \{T\ (P \to Q),\ F(\neg Q \to \neg P),\ T\neg Q,\ F\neg P\}$. Then $\mathcal{T}_1 = \{\mathcal{B}_1\}$ is called a *one-step expansion* of $\mathcal{T}_0$, because there is a signed formula in $\mathcal{B}_0$, to wit $F(\neg Q \to \neg P)$, such that $\mathcal{B}_1 = \mathcal{B}_0 \cup \{T\neg Q,\ F\neg P\}$.

Let $\mathcal{B}_2 = \{T\ (P \to Q),\ F(\neg Q \to \neg P),\ T\neg Q,\ F\neg P,\ FQ\}$. Then $\mathcal{T}_2 = \{\mathcal{B}_2\}$ is again a *one-step expansion* of $\mathcal{T}_1$.

Let $\mathcal{B}_3 = \{T\ (P \to Q),\ F(\neg Q \to \neg P),\ T\neg Q,\ F\neg P,\ FQ,\ TP\}$. Then $\mathcal{T}_3 = \{\mathcal{B}_3\}$ is a *one-step expansion* of $\mathcal{T}_2$.

Finally, let $\mathcal{B}_{31} = \{T\ (P \to Q),\ F(\neg Q \to \neg P),\ T\neg Q,\ F\neg P,\ FQ,\ TP,\ FP\}$ and $\mathcal{B}_{32} = \{T\ (P \to Q),\ F(\neg Q \to \neg P),\ T\neg Q,\ F\neg P,\ FQ,\ TP,\ TQ\}$. Then $\mathcal{T}_4 = \{\mathcal{B}_{31},\ \mathcal{B}_{32}\}$ is called a *one-step expansion* of $\mathcal{T}_3$, because there is a signed formula in $\mathcal{B}_3$, to wit $T\ (P \to Q)$, such that $\mathcal{B}_{31} = \mathcal{B}_3 \cup \{FP\}$ and $\mathcal{B}_{32} = \mathcal{B}_3 \cup \{TQ\}$.

$\mathcal{T}_0,\ \mathcal{T}_1,\ \mathcal{T}_2,\ \mathcal{T}_3$ and $\mathcal{T}_4$ are all tableaux with initial branch $\mathcal{B}_0$.

The branches $\mathcal{B}_0,\ \mathcal{B}_1,\ \mathcal{B}_2$ and $\mathcal{B}_3$ are not closed and not completed. But the branches $\mathcal{B}_{31}$ and $\mathcal{B}_{32}$ are completed and both are also closed.

2.8 Tableaux

We shall call, for instance, $\mathscr{T}_3 = \{\mathscr{B}_3\}$ a *tableau* with initial branch or sequent $\mathscr{B}_0$, because there is a sequence $\mathscr{T}_0, \mathscr{T}_1, \ldots, \mathscr{T}_3$ such that $\mathscr{T}_0 = \{\mathscr{B}_0\}$ and each $\mathscr{T}_{i+1}$ is a one-step expansion of $\mathscr{T}_i$ ($0 \leq i < 3$). This tableau $\mathscr{T}_3$ (with initial branch $\mathscr{B}_0$) is not yet completed, because its only branch $\mathscr{B}_3$ is not completed: the $T \to$ rule has not yet been applied to $T(P \to Q)$. And $\mathscr{T}_3 = \{\mathscr{B}_3\}$ is open, because it contains an open branch, to wit $\mathscr{B}_3$ itself. The tableau $\mathscr{T}_4 = \{\mathscr{B}_{31}, \mathscr{B}_{32}\}$, however, is completed, because each of its branches is completed and also closed, because all its branches are closed.

Definition 2.16 (Tableau). (a) A set of branches $\mathscr{T}$ is a *tableau* with initial branch $\mathscr{B}_0$ if there is a sequence $\mathscr{T}_0, \mathscr{T}_1, \ldots, \mathscr{T}_n$ such that $\mathscr{T}_0 = \{\mathscr{B}_0\}$, each $\mathscr{T}_{i+1}$ is a one-step expansion of $\mathscr{T}_i$ ($0 \leq i < n$) and $\mathscr{T} = \mathscr{T}_n$.
(b) We say that a finite $\mathscr{B}$ has tableau $\mathscr{T}$ if $\mathscr{T}$ is a tableau with initial branch $\mathscr{B}$.
(c) A tableau $\mathscr{T}$ is *open* if some branch $\mathscr{B}$ in it is open, otherwise $\mathscr{T}$ is *closed*.
(d) A tableau is *completed* if each of its branches is completed, i.e., no application of a tableau rule can change the tableau.

Example 2.21.
We make a tableau starting with $T(P \to Q), F(P \wedge Q)$:

$$T(P \to Q), F(P \wedge Q)$$
$$FP, F(P \wedge Q) \mid TQ, F(P \wedge Q)$$
$$FP, FP \mid FP, FQ \mid TQ, FP \mid TQ, FQ$$

Let $\mathscr{B}_1$ be the leftmost branch, consisting of the formulas $T(P \to Q), F(P \wedge Q), FP$ and FP, i.e., $\mathscr{B}_1 = \{T(P \to Q), F(P \wedge Q), FP, FP\}$. Let $\mathscr{B}_2$ be the second branch from the left, so $\mathscr{B}_2 = \{T(P \to Q), F(P \wedge Q), FP, FQ\}$. Let $\mathscr{B}_3$ be the third branch from the left, so $\mathscr{B}_3 = \{T(P \to Q), F(P \wedge Q), TQ, FP\}$. Finally, let $\mathscr{B}_4$ be the rightmost branch, i.e., $\mathscr{B}_4 = \{T(P \to Q), F(P \wedge Q), TQ, FQ\}$.

Then $\mathscr{T} = \{\mathscr{B}_1, \mathscr{B}_2, \mathscr{B}_3, \mathscr{B}_4\}$ is a tableau with $\mathscr{B}_0 = \{T(P \to Q), F(P \wedge Q)\}$ as initial branch. Branch $\mathscr{B}_4$ is completed and closed, because it contains TQ and FQ. The branches $\mathscr{B}_1, \mathscr{B}_2, \mathscr{B}_3$ are completed and open. Hence, the tableau $\mathscr{T} = \{\mathscr{B}_1, \mathscr{B}_2, \mathscr{B}_3, \mathscr{B}_4\}$ is completed, because all of its branches are completed and the tableau $\mathscr{T}$ is open, since at least one of its branches is open.

From the formulation of the tableaux rules, we see immediately that our tableaux have the so-called *subformula property*: each formula in any sequent of a tableau is a subformula of some formula occurring in the preceding sequents. For that reason, any tableau (in classical propositional logic) is necessarily a *finite* sequence of sequents. For instance, all formulas in the tableau in Example 2.20 are subformulas of $P \to Q$ and/or $\neg Q \to \neg P$.

From the examples in Section 2.6 it is clear that a Hilbert-type proof system does not have the subformula property. For instance, we have given a deduction of $A \to C$ from $A \to B$ and $B \to C$; in this deduction we have used the formula $A \to (B \to C)$ and even more complex ones, which are subformulas of neither the premisses nor the conclusion. Modus Ponens is responsible for this: E may be deduced from D and $D \to E$; but $D \to E$ is not a subformula of E and D is not necessarily one.

Definition 2.17 (Tableau-deduction). (a) A (logical) *tableau-deduction* of B from $A_1, \ldots, A_n$ (in propositional logic) is a tableau $\mathscr{T}$ with $\mathscr{B}_0 = \{TA_1, \ldots, TA_n, FB\}$ as initial branch, such that all branches of $\mathscr{T}$ are closed.

In case $n = 0$, i.e., there are no premises $A_1, \ldots, A_n$, this definition reduces to:
(b) A (logical) *tableau-proof* of B (in classical propositional logic) is a tableau $\mathscr{T}$ with $\mathscr{B}_0 = \{FB\}$ as initial sequent, such that all branches of $\mathscr{T}$ are closed.

Example 2.22. (a) The following is a tableau-deduction of $\neg P \vee \neg Q$ from $\neg(P \wedge Q)$.

$$T \neg(P \wedge Q), \; F \neg P \vee \neg Q$$
$$F \, P \wedge Q, \; F \neg P \vee \neg Q$$
$$F \, P \wedge Q, \; F \neg P, \; F \neg Q$$
$$F \, P \wedge Q, \; TP, \; F \neg Q$$
$$F \, P \wedge Q, \; TP, \; TQ$$
$$FP, \, TP, \, TQ \; | \; FQ, \, TP, \, TQ$$

(b) The following is a tableau-proof of $((P \to Q) \to P) \to P$, i.e., Peirce's law.

$$F \, ((P \to Q) \to P) \to P$$
$$T \, (P \to Q) \to P, \; FP$$
$$F \, P \to Q, \; FP \; | \; TP, \, FP$$
$$TP, \, FQ, \, FP \, |$$

Definition 2.18 (Tableau-deducible). (a) B is *tableau-deducible from* $A_1, \ldots, A_n$ (in classical propositional logic) if there exists a tableau-deduction of B from $A_1, \ldots, A_n$.
Notation: $A_1, \ldots, A_n \vdash' B$. By $A_1, \ldots, A_n \nvdash' B$ we mean: not $A_1, \ldots, A_n \vdash' B$.
(b) B is *tableau-provable* (in classical propositional logic) if there exists a tableau-proof of B. **Notation**: $\vdash' B$.
(c) For Γ a (possibly infinite) set of formulas, B is *tableau-deducible from* Γ if there exists a finite list $A_1, \ldots, A_n$ of formulas in Γ such that $A_1, \ldots, A_n \vdash' B$.
Notation: $\Gamma \vdash' B$.

Example 2.23. (a) $\neg(P \wedge Q) \vdash' \neg P \vee \neg Q$, because in Example 2.22 (a) we have given a tableau-deduction of $\neg P \vee \neg Q$ from $\neg(P \wedge Q)$. One also easily checks that, equivalently, $\vdash' \neg(P \wedge Q) \to \neg P \vee \neg Q$.
(b) $\vdash' ((P \to Q) \to P) \to P$, because in Example 2.22 (b) we have given a tableau-proof of $((P \to Q) \to P) \to P$. One also easily checks that, equivalently, $(P \to Q) \to P \vdash' P$.

Note that by our definitions $A \vdash' B$ is trivially equivalent to $\vdash' A \to B$ (because a tableau starting with $F \, A \to B$ continues with TA, FB), while the corresponding result for $\vdash$ (Theorem 2.23 and 2.24) was not trivial at all.

It is important to note that the T- and F-rules and hence the notions of 'tableau-provable' and 'tableau-deducible from' are purely *syntactic*, i.e., they only refer to the forms of the formulas: for instance, rule $T\wedge$ tells us that any time we see an expression of the form $T \, B \wedge C$ we must write down the expressions TB and TC immediately below it; and a formula B is tableau-provable if starting with FB we end up with sequents which all contain both TP and FP for some atomic formula P.

2.8 Tableaux

Whether a formula B is tableau-provable or not only depends on the *form* of B, and precisely this justifies our use of the expression 'B is tableau-provable'.

So, we had good semantic reasons to choose the rules and the notions of 'tableau-provable' and 'tableau-deducible from' as they are, but once having these rules and these notions, we can forget the intuitive (semantic) motivation behind them and like a computer or machine/robot play with them in a purely syntactic way, i.e., apply the rules of the game, forgetting about their underlying ideas.

Gentzen-type rules A second way to read the T- and F- tableaux rules is to read them *upwards*, as *Gentzen-type rules*, interpreting the sequents rather than the signed formulas. Remember that a sequent $\{TA_1,\ldots,TA_n,FB_1,\ldots,FB_k\}$ is read as: if A_1 and ... and A_n, then B_1 or ... or B_k.

For example, taking $S = \{TD, FE\}$, rule $T \to$ becomes

$$TD, FE, T\,B \to C$$
$$TD, FE, FB \mid TD, FE, TC$$

and is read upwards as follows:

if	(*) D implies E or B	(TD, FE, FB),
and	(**) D and C imply E	(TD, FE, TC),
then	D and $B \to C$ imply E	$(TD, FE, T\,B \to C)$.

That rule $T \to$, read in this way, is intuitively correct is easily seen as follows: suppose (*), (**), D and $B \to C$; then by (*), E or B; if B, then by $B \to C$ also C; and hence by (**) E.

And again taking $S = \{TD, FE\}$, rule $F \to$ becomes

$$TD, FE, F\,B \to C$$
$$TD, FE, TB, FC$$

and is read upwards as follows:

if	(*) D and B imply E or C	(TD, FE, TB, FC),
then	D implies E or $B \to C$	$(TD, FE, F\,B \to C)$.

That rule $F \to$, read in this way, is intuitively correct is seen as follows: suppose (*) and D; if $\neg B$, then $B \to C$ and hence E or $B \to C$; and if B, then D and B, and hence by (*), E or C; so, also E or $B \to C$.

This way of reading the rules is derived from G. Gentzen's system in [9]. Gentzen thought his rules reflected (the elementary steps in) the actual reasoning of human beings. With this reading the notion of tableau-provability is explained (see Def. 2.18) in terms of reducing a formula according to the rules to axioms essentially of the type $P \to P$. More precisely, a formula B is tableau-provable if $\{FB\}$ (to be read as: $\to B$ or B) can be obtained by applying the rules to sequents of the form $\{\ldots, TP, FP, \ldots\}$ (to be read as: if ... and P, then P or ...), which can be conceived of as axioms.

Decidability Evidently, it is easy to decide whether a given sequence of symbols is a formula (of propositional logic). It is also easy to decide whether a given sequence of formulas is a (Hilbert-type) deduction (see Section 2.6) of a given formula B from given premises $A_1, \ldots, A_n$. And similarly, it is easy to decide whether a given tableau is a tableau-deduction of a given formula B from given premises $A_1, \ldots, A_n$.

But the question whether, given any formulas $A_1, \ldots A_n$ and B, there *exists* a Hilbert-type deduction of B from $A_1, \ldots A_n$, is not so easy to decide: one may search for such a deduction without finding one and this may be due to the fact that one is not smart enough – in which case one may continue trying to find one –, but also due to the fact that there is no such deduction – in which case one better stops searching. The deeper reason behind this is that Hilbert-type deductions do not have the subformula property: if one searches for a deduction of B from given premises, one may try any formula D, not necessarily a subformula of the given formulas, in order to apply Modus Ponens to D and $D \to B$.

Interestingly, for any propositional formulas $A_1, \ldots, A_n, B$, the question whether B is a valid (or logical) consequence of $A_1, \ldots, A_n$ is *decidable*, i.e., there is a decision procedure (algorithm, mechanical test) which yields in finitely many steps an answer 'yes' or 'no': make the truth table of the formulas in question and check whether B is 1 in all lines where the premises $A_1, \ldots, A_n$ are all 1.

Similarly, for any propositional formulas $A_1, \ldots, A_n, B$, the question whether there *exists* a tableau-deduction of B from given premises $A_1, \ldots, A_n$ is *decidable*, since there is a decision procedure which yields in finitely many steps an answer 'yes' or 'no': given $A_1, \ldots, A_n$ and B, start a tableau with $\{TA_1, \ldots, TA_n, FB\}$ as initial sequent and apply all possible tableau rules as frequently as possible; because of the subformula property, after finitely many steps the tableau will be finished; if all tableau branches are closed, then one has a tableau-deduction of B from $A_1, \ldots, A_n$, and if some completed tableau branch is open, one can from any open completed tableau branch read off a line in the truth table in which $A_1, \ldots, A_n$ are all 1 and B is 0, hence showing that $A_1, \ldots, A_n \not\models B$. We shall prove this (completeness) result in Section 2.9, but will illustrate this result now with an example.

Example 2.24. We wonder whether from $P \to Q$ and $\neg P$ one may deduce $\neg Q$. So, we start a tableau with $\{TP \to Q, T\neg P, F\neg Q\}$:

$$TP \to Q, T\neg P, F\neg Q$$
$$TP \to Q, FP, F\neg Q$$
$$TP \to Q, FP, TQ$$
$$FP, FP, TQ \mid TQ, FP, TQ$$

For instance, the left tableau branch is completed but open, i.e. not closed. From it one may immediately read off a counterexample, i.e., a line in the truth table in which the premises $P \to Q$ and $\neg P$ are 1 and $\neg Q$ is 0: corresponding with the occurrence of FP in the left completed tableau branch give P the value 0 and corresponding with the occurrence of TQ in the left completed tableau branch give Q the value 1.

P	Q	$P \to Q$	$\neg P$	$\neg Q$
0	1	1	1	0

This shows that $P \to Q, \neg P \not\models \neg Q$.

2.8 Tableaux

Once we have shown in Sect. 2.9 that the three notions $A_1,\ldots,A_n \models B$, $A_1,\ldots,A_n \vdash B$, and $A_1,\ldots,A_n \vdash' B$, although intensionally quite different, are equivalent, we have also a decision procedure for the question whether, given formulas $A_1,\ldots,A_n$, B, there *exists* a (Hilbert-type) deduction of B from $A_1,\ldots,A_n$. The significance of this latter result is that the Hilbert-type system of Section 2.6, which does not have the subformula property, is equivalent to the tableaux system of this section, which does have the subformula property. (This result is essentially based on the work of G. Gentzen, 1934-5.)

In order to show that our notions of tableau-deducibility (Def. 2.18) and (Hilbert-type) deducibility (Def. 2.11) are equivalent, we first prove the following.

Theorem 2.27. *(i) If B is tableau-deducible from $A_1,\ldots,A_n$, i.e., $A_1,\ldots,A_n \vdash' B$, then B is deducible from $A_1,\ldots,A_n$, i.e., $A_1,\ldots,A_n \vdash B$. In particular, for $n = 0$: (ii) If $\vdash' B$, then $\vdash B$.*

Proof. Suppose $A_1,\ldots,A_n \vdash' B$, i.e., B is tableau-deducible from $A_1,\ldots,A_n$. It suffices to show:

for every sequent $S = \{TD_1,\ldots,TD_k, FE_1,\ldots,FE_m\}$ in a tableau-deduction of B from $A_1,\ldots,A_n$ it holds that $D_1,\ldots,D_k \vdash E_1 \vee \ldots \vee E_m$. (*)

Consequently, because $\{TA_1,\ldots,TA_n, FB\}$ is the first (upper) sequent in any given tableau-deduction of B from $A_1,\ldots,A_n$, we have that $A_1,\ldots,A_n \vdash B$.

The proof of (*) is tedious, but has a simple plan: the statement is true for the closed sequents in a tableau-deduction, and the statement remains true if we go up in the tableau-deduction via the T and F rules.

Basic step: Any closed sequent in a tableau-deduction of B from $A_1,\ldots,A_n$ is of the form $\{TD_1,\ldots,TD_k, TP, FP, FE_1,\ldots,FE_m\}$. So, we have to show that $D_1,\ldots,D_k, P \vdash P \vee E_1 \vee \ldots \vee E_m$. And this is straightforward: $D_1,\ldots,D_k, P \vdash P$ and $P \vdash P \vee E_1 \vee \ldots \vee E_m$.

Induction step: We have to show that for all rules the following is the case: if (*) holds for all lower sequent(s) in the rule (induction hypothesis), then (*) holds for the upper sequent in the rule. For convenience, we will suppose that $S = \{TD, FE\}$ in all rules.

Rule $T\wedge$: $TD, FE, T\,B\wedge C$
                   TD, FE, TB, TC

Suppose $D, B, C \vdash E$ (induction hypothesis). To show: $D, B\wedge C \vdash E$. This follows immediately, because $B\wedge C \vdash B$ and $B\wedge C \vdash C$.

Rule $F\wedge$: $TD, FE, F\,B\wedge C$
                   $TD, FE, FB \mid TD, FE, FC$

Suppose $D \vdash E \vee B$ and $D \vdash E \vee C$ (induction hypothesis). To show: $D \vdash E \vee (B \wedge C)$. It suffices to show that $E \vee B, E \vee C \vdash E \vee (B \wedge C)$. Now it is clear that $B, E \vdash E \vee (B \wedge C)$ and $B, C \vdash E \vee (B \wedge C)$. Hence, by $\vee$-elimination, $B, E \vee C \vdash E \vee (B \wedge C)$. But also $E, E \vee C \vdash E \vee (B \wedge C)$. Hence, again by $\vee$-elimination, $E \vee B, E \vee C \vdash E \vee (B \wedge C)$.

Rule T∨: $\qquad TD, FE, T\,B\vee C$
$\qquad\qquad\qquad TD, FE, TB \mid TD, FE, TC$
Suppose $D, B \vdash E$ and $D, C \vdash E$ (induction hypothesis). To show: $D, B\vee C \vdash E$. This follows from the induction hypothesis by $\vee$-elimination.

Rule F∨: $\qquad TD, FE, F\,B\vee C$
$\qquad\qquad\qquad TD, FE, FB, FC$
Suppose $D \vdash (E\vee B)\vee C$ (induction hypothesis). To show: $D \vdash E\vee (B\vee C)$. It suffices to show that $(E\vee B)\vee C \vdash E\vee (B\vee C)$.
It is clear that $E \vdash E\vee (B\vee C)$ and also $B \vdash E\vee (B\vee C)$. Hence, by $\vee$-elimination, $E\vee B \vdash E\vee (B\vee C)$. Since also $C \vdash E\vee (B\vee C)$, again by $\vee$-elimination, $(E\vee B)\vee C \vdash E\vee (B\vee C)$.

Rule T→: $\qquad TD, FE, T\,B\to C$
$\qquad\qquad\qquad TD, FE, FB \mid TD, FE, TC$
Suppose $D \vdash E\vee B$ and $D, C \vdash E$ (induction hypothesis). To show: $D, B\to C \vdash E$.
By Exercise 2.50 $E\vee B, B\to C \vdash E\vee C$; hence, by the first induction hypothesis, $D, B\to C \vdash E\vee C$. $\qquad\qquad\qquad\qquad\qquad\qquad\qquad\qquad\qquad\qquad\qquad$ (1)
From the second induction hypothesis, by the deduction theorem, $D \vdash C \to E$. (2)
By Exercise 2.50 $E\vee C, C\to E \vdash E\vee E$; hence, from (1) and (2): $D, B\to C \vdash E\vee E$. But by $\vee$-elimination $E\vee E \vdash E$. Hence $D, B\to C \vdash E$.

Rule F→: $\qquad TD, FE, F\,B\to C$
$\qquad\qquad\qquad TD, FE, TB, FC$
Suppose $D, B \vdash E\vee C$ (induction hypothesis). To show: $D \vdash E\vee (B\to C)$.
From weak negation elimination, applying the deduction theorem, it follows that $\neg B \vdash B\to C$; hence $D, \neg B \vdash B\to C$. Hence $D, \neg B \vdash E\vee (B\to C)$. $\qquad$ (1)
By Exercise 2.50 $E\vee C, C\to (B\to C) \vdash E\vee (B\to C)$. So, since $C\to (B\to C)$ is an axiom, it follows that $E\vee C \vdash E\vee (B\to C)$. So, by the induction hypothesis, $D, B \vdash E\vee (B\to C)$. $\qquad\qquad\qquad\qquad\qquad\qquad\qquad\qquad\qquad\qquad\qquad$ (2)
From (1) and (2), by $\vee$-elimination $D, B\vee\neg B \vdash E\vee (B\to C)$. But, by Exercise 2.52, $\vdash B\vee\neg B$. Hence, $D \vdash E\vee (B\to C)$.

Rule T¬: $\qquad TD, FE, T\,\neg B$
$\qquad\qquad\qquad TD, FE, FB$
Suppose $D \vdash E\vee B$ (induction hypothesis). To show: $D, \neg B \vdash E$. In order to do this, it suffices to prove that $E\vee B, \neg B \vdash E$.
By Exercise 2.53 $\neg B, \neg E \vdash \neg(E\vee B)$ and hence also $E\vee B, \neg B, \neg E \vdash \neg(E\vee B)$. But also $E\vee B, \neg B, \neg E \vdash E\vee B$. Hence, by $\neg$-introduction $E\vee B, \neg B \vdash \neg\neg E$. So, by double negation elimination $E\vee B, \neg B \vdash E$.

Rule F¬: $\qquad TD, FE, F\,\neg B$
$\qquad\qquad\qquad TD, FE, TB$
Suppose $D, B \vdash E$ (induction hypothesis). To show: $D \vdash E\vee \neg B$.
From the induction hypothesis, $D, B \vdash E\vee\neg B$. $\qquad\qquad\qquad\qquad\qquad\qquad$ (1)
From $\neg B \vdash E\vee\neg B$ it follows that $D, \neg B \vdash E\vee\neg B$. $\qquad\qquad\qquad\qquad$ (2)
From (1) and (2) it follows by $\vee$-elimination that $D, B\vee\neg B \vdash E\vee\neg B$. By Exercise 2.52 $\vdash B\vee\neg B$ and hence $D \vdash E\vee\neg B$. $\qquad\qquad\qquad\qquad\qquad\qquad$ □

2.8 Tableaux

With the help of tableaux we may give a constructive proof of the *interpolation theorem*.

Theorem 2.28 (Interpolation theorem for propositional logic). *Suppose $A \vdash' B$, $\nvdash' \neg A$ and $\nvdash' B$. Then there is a formula C such that every atomic formula that occurs in C also occurs in both A and B (so, C is in the joint vocabulary of A and B) and $A \vdash' C$ and $C \vdash' B$.*

Example 2.25. $(P \vee \neg Q) \wedge R \vdash' (Q \rightarrow P) \vee S$. Then for $C = P \vee \neg Q$, we have $(P \vee \neg Q) \wedge R \vdash' C$ and $C \vdash' (Q \rightarrow P) \vee S$.

Proof. Let A and B as mentioned in the interpolation theorem. Because $A \vdash' B$, any completed tableau starting with the initial sequent $\{TA, FB\}$ is closed, i.e., all its branches are closed. (*)

Since $\nvdash' \neg A$ we know that any completed tableau starting with $F \neg A$ (or, equivalently, TA) has at least one open (completed) branch $\mathscr{B}$. And since $\nvdash' B$, we know there any completed tableau starting with the initial sequent $\{FB\}$ has at least one open branch. Let $\mathscr{T}_A$ be a completed tableau starting with TA and $\mathscr{T}_B$ a completed tableau sarting with FB. We may assume that a tableau is closed if and only if it is atomically closed, i.e., every branch contains for some *atomic* formula P both TP and FP. For any open branch $\mathscr{B}$ in $\mathscr{T}_A$, we define the sets $\mathscr{B}^1$ and $\mathscr{B}^0$: $\mathscr{B}^1 = \{P \mid TP$ occurs in $\mathscr{B}$ and FP occurs in some open branch of $\mathscr{T}_B\}$ and $\mathscr{B}^0 = \{\neg P \mid FP$ occurs in $\mathscr{B}$ and TP occurs in some open branch of $\mathscr{T}_B\}$.

By (*) the union of $\mathscr{B}^0$ and $\mathscr{B}^1$ is non empty and so the following sentence is well-defined: $C(\mathscr{B}) :=$ the conjunction of all formulas in $\mathscr{B}^1 \cup \mathscr{B}^0$. Finally, the sentence C is defined as the disjunction of all formulas $C(\mathscr{B})$, where $\mathscr{B}$ is an open branch in the given tableau $\mathscr{T}_A$ starting with TA. Clearly, C is in the joint vocabulary of A and B. After some thinking it becomes clear that $A \vdash' C$ and $C \vdash' B$. □

Let us illustrate the proof for Example 2.25, where $A = \neg(Q \wedge \neg P) \wedge R$ and $B = (Q \rightarrow P) \vee S$. Let $\mathscr{T}_A$ be the following completed tableau starting with $F \neg(\neg(Q \wedge \neg P) \wedge R)$:

$$F \neg(\neg(Q \wedge \neg P) \wedge R)$$
$$T \neg(Q \wedge \neg P) \wedge R$$
$$T \neg(Q \wedge \neg P), TR$$
$$F \ Q \wedge \neg P, TR$$
$$F \ \neg P, TR \mid FQ, TR$$
$$TP, TR \mid FQ, TR$$

Both the left branch $\mathscr{B}_L$ and the right branch $\mathscr{B}_R$ of this tableau are open. Now, by definition, $\mathscr{B}_L^1 = \{P\}$, since there is an open branch starting with $F(Q \rightarrow P) \vee S$ that contains FP:

$$F \ (Q \rightarrow P) \vee S$$
$$F \ (Q \rightarrow P), FS$$
$$TQ, FP, FS$$

Note that $\mathscr{B}_L^0$ is empty. So, by definition, $C(\mathscr{B}_L) = P$.

By definition, $\mathscr{B}_R^1$ is empty and $\mathscr{B}_R^0 = \{\neg Q\}$, since there is an open branch starting with $F(Q \to P) \vee S$ that contains TQ. So, by definition, $C(\mathscr{B}_R) = \neg Q$. Finally, $C = C(\mathscr{B}_L) \vee C(\mathscr{B}_R) = P \vee \neg Q$.

Exercise 2.62. (a) Show, by using $\neg$-introduction, that $A \to B \vdash \neg(A \wedge \neg B)$.
(b) Show that $A \to B \vdash' \neg(A \wedge \neg B)$.
(c) Show that $A \to B \models \neg(A \wedge \neg B)$ by verifying that it is impossible that $A \to B$ is 1 and $\neg(A \wedge \neg B)$ is 0 in some line of the truth table. Note the analogy in (b) and (c).

Exercise 2.63. (a) Show, by using the deduction theorem three times, that $\vdash (A \to B) \to ((B \to C) \to (A \to C))$.
(b) Show that $\vdash' (A \to B) \to ((B \to C) \to (A \to C))$.
(c) Show that $\models (A \to B) \to ((B \to C) \to (A \to C))$ by verifying that it is impossible that this formula is 0 in some line of its truth table. Note the analogy in (b) and (c).

Exercise 2.64. Prove the following statements:
(a) $A \to B, \neg A \to B \vdash' B$
(b) $\neg B \to \neg A \vdash' A \to B$
(c) $\neg(A \wedge B) \vdash' \neg A \vee \neg B$
(d) $\neg(A \wedge \neg B) \vdash' A \to B$
(e) $A \to B \vdash' \neg A \vee B$
(f) $A \to B \vee C \vdash' (A \to B) \vee (A \to C)$

Exercise 2.65. a) Translate the following argument in the language of propositional logic.
    If it rains [R], then John goes for a walk [W].
    If it does not rain, then John makes a bicycle tour [B].
    John does not make a bicycle tour.
    Therefore: John goes for a walk.
b) Construct a tableau-deduction of the putative conclusion from the premises or a counterexample (i.e., a line in the truth table in which all premises are 1 and the putative conclusion is 0) from a failed attempt to do so.

Exercise 2.66. a) Translate the following argument in the language of propositional logic.
    If it rains [R], then John does not go for a walk.
    If John goes for a walk [W], then he is happy [H].
    It does not rain.
    Therefore: John is happy.
b) Construct a tableau-deduction of the putative conclusion from the premises or a counterexample (i.e., a line in the truth table in which all premises are 1 and the putative conclusion is 0) from a failed attempt to do so.

Exercise 2.67. (a) Verify that the (logical) axioms for (classical) propositional calculus of Section 2.6 are tableau-provable.
(b) Check that it is not a simple matter to prove: if $\vdash' A$ and $\vdash' A \to B$, then $\vdash' B$. Hence, the converse of Theorem 2.27, if $\vdash A$, then $\vdash' A$, to be shown in Section 2.9, is not a trivial result. However, one easily shows that $A, A \to B \vdash' B$ does hold.

Exercise 2.68. Show right from the definitions that
(a) if $\vdash' A$ or $\vdash' B$, then $\vdash' A \vee B$;
(b) if $\vdash' A \wedge B$, then $\vdash' A$ and $\vdash' B$.

2.9 Completeness of classical propositional logic

Exercise 2.69. (a) Show that $\neg P, (P \to Q) \to P \vdash P$ by using weak negation elimination and the deduction theorem.
(b) Show that $P \vee \neg P, (P \to Q) \to P \vdash P$ by using (a) and $\vee$-elimination.
(c) Show that $\vdash ((P \to Q) \to P) \to P$ (Peirce's law) by using (b), Exercise 2.52. and the deduction theorem.
Compare the complexity of the proof of $\vdash ((P \to Q) \to P) \to P$ with the simplicity of the proof of $\vdash' ((P \to Q) \to P) \to P$. Note also that, although in Peirce's law implication is the only connective, we needed weak negation- and $\vee$-elimination in order to show that Peirce's law is (logically) provable (see Exercise 2.44).

2.9 Completeness of classical propositional logic

So far we have established the following results; for convenience, we use the Greek letter Γ to indicate a (possibly infinite) collection of formulas.
Theorem 2.27: if $\Gamma \vdash' B$, then $\Gamma \vdash B$.
Theorem 2.20: if $\Gamma \vdash B$, then $\Gamma \models B$ (soundness). In this section we shall prove *completeness*, i.e., every valid consequence of given premises Γ can be (logically) deduced from Γ: if $\Gamma \models B$, then $\Gamma \vdash' B$.

This shows that the three notions $\Gamma \vdash' B$ (B is tableau-deducible from Γ), $\Gamma \vdash B$ (B is deducible from Γ) and $\Gamma \models B$ (B is a valid consequence of Γ) are equivalent.

The intuitive 'B is a logical consequence of the premises in Γ' (without reference to the structure of the atomic formulas in B and Γ) has been made mathematically precise in three different ways: $\Gamma \vdash' B$, $\Gamma \vdash B$ and $\Gamma \models B$. Since these three mathematical notions, although intensionally quite different, turn out to be equivalent, we may say (after the results we are about to prove) that *we indeed have captured in a mathematically definite sense the intuitive notion of 'B is a logical conclusion from Γ'*. (See also the discussion following Theorem 2.21.)

In proving the completeness of classical propositional logic, a procedure of searching for a tableau-deduction of B from given premises $A_1, \ldots, A_n$ is presented, which will end after finitely many steps and then either gives such a deduction or shows that such a deduction cannot exist. This algorithm thus yields a decision procedure for the (classical) propositional logic. This shall provide us an opportunity to dwell upon automated theorem proving.

Given formulas B and $A_1, \ldots, A_n$, the tableaux rules suggest a *procedure of searching for a tableau-deduction* of B from $A_1, \ldots, A_n$:
start with $TA_1, \ldots, TA_n, FB$ and apply all the appropriate rules in some definite fixed order, the choice of ordering being unimportant (at least, if we do not care about efficiency); in an application of rule $T \to$ to, for example, $S, T P \to Q$ we make two branches, one with S, FP and the other with S, TQ and similarly for applications of the rules $F\wedge$ and $T\vee$.

Example 2.26. 1) The tableau starting with $F(P \to Q) \to (Q \to \neg P)$ is composed of the following two branches:

$$F\ (P \to Q) \to (Q \to \neg P) \quad \text{and } F(P \to Q) \to (Q \to \neg P)$$
$$T\ P \to Q,\ F\ Q \to \neg P \qquad\qquad T\ P \to Q,\ F\ Q \to \neg P$$
$$FP,\ F\ Q \to \neg P \qquad\qquad\qquad TQ,\ F\ Q \to \neg P$$
$$FP,\ TQ,\ F\ \neg P \qquad\qquad\qquad TQ,\ TQ,\ F\ \neg P$$
$$FP,\ TQ,\ TP \qquad\qquad\qquad\quad TQ,\ TQ,\ TP$$

The first branch for $(P \to Q) \to (Q \to \neg P)$ is closed; the second one is completed and open. Note that if we assign the value 1 to both P and Q, corresponding with the fact that both TP and TQ occur in the open branch, the formula $(P \to Q) \to (Q \to \neg P)$ is assigned the value 0, corresponding with the fact that $F\ (P \to Q) \to (Q \to \neg P)$ occurs in the open branch. We shall see in Lemma 2.2 that this is not accidental.

2) The tableau starting with $T\ P \to Q,\ F\ \neg Q \to \neg P$ is composed of the following two branches:

$$T\ P \to Q,\ F\ \neg Q \to \neg P \text{ and } TP \to Q,\ F\ \neg Q \to \neg P$$
$$FP,\ F\ \neg Q \to \neg P \qquad\qquad TQ,\ F\ \neg Q \to \neg P$$
$$FP,\ T\ \neg Q,\ F\ \neg P \qquad\qquad TQ,\ T\ \neg Q,\ F\ \neg P$$
$$FP,\ FQ,\ F\ \neg P \qquad\qquad\quad TQ,\ FQ,\ F\ \neg P$$
$$FP,\ FQ,\ TP \qquad\qquad\qquad TQ,\ FQ,\ TP$$

Both branches starting with $T\ P \to Q,\ F\ \neg Q \to \neg P$ are closed. Note that the two branches together yield a tableau-deduction of $\neg Q \to \neg P$ from $P \to Q$, just as a tableau-proof of $(P \to Q) \to (\neg Q \to \neg P)$. The correctness of this statement is not accidental either and follows immediately from the definition of a tableau-deduction and the structure of our procedure of searching for a tableau-deduction; see Lemma 2.3.

Definition 2.19. Let τ be a completed tableau branch which is open. Then i_τ is the interpretation defined by $i_\tau(P) = 1$ if TP occurs in τ, $i_\tau(P) = 0$ if TP does not occur in τ.

Lemma 2.2. *Let τ be a completed tableau branch which is open. Then for each formula E: a) if TE occurs in τ, then $i_\tau(E) = 1$, and
b) if FE occurs in τ, then $i_\tau(E) = 0$.*

Proof. The proof is by induction on the construction of E. Let τ be a completed tableau branch which is open.
Basic step. If $E = P$ (atomic formula) and TP occurs in τ, then by definition $i_\tau(P) = 1$. If $E = P$ and FP occurs in τ, then - since τ is open - TP does not occur in τ and hence by definition $i_\tau(P) = 0$.
Induction step. Suppose that a) and b) have been shown for C and D (induction hypothesis). We want to prove a) and b) for $C \wedge D, C \vee D, C \to D$ and $\neg C$.

If $E = C \wedge D$ and $T\ C \wedge D$ occurs in τ, then - because τ is completed - both TC and TD occur in τ. Hence, by the induction hypothesis, $i_\tau(C) = 1$ and $i_\tau(D) = 1$. So, $i_\tau(C \wedge D) = 1$.
If $E = C \wedge D$ and $F\ C \wedge D$ occurs in τ, then - because τ is completed - FC occurs in τ or FD occurs in τ. Hence, by the induction hypothesis, $i_\tau(C) = 0$ or $i_\tau(D) = 0$. So, $i_\tau(C \wedge D) = 0$.

2.9 Completeness of classical propositional logic

The other cases, $E = C \vee D$, $E = C \to D$ and $E = \neg C$, are treated similarly. □

Lemma 2.3. *If all branches in a tableau with initial sequent $\{TA_1, \ldots, TA_n, FB\}$ are closed, then $A_1, \ldots, A_n \vdash' B$.*

Proof. This follows from the definition of a tableau with $\{TA_1, \ldots, TA_n, FB\}$ as initial sequent and from the observation that there are only finitely many different branches in such a tableau. □

Lemma 2.2 and 2.3 together yield the completeness theorem.

Theorem 2.29 (completeness of classical propositional logic).
a) *If $A_1, \ldots, A_n \models B$, then $A_1, \ldots, A \vdash' B$. In particular, if $n = 0$:*
b) *If $\models B$, then $\vdash' B$.*

Proof. Suppose $A_1, \ldots, A_n \models B$. Apply the procedure of searching for a tableau-deduction of B from $A_1, \ldots, A_n$. If there were a completed tableau branch τ starting with $TA_1, \ldots, TA_n, FB$ which is open, then by Lemma 2.2, because $TA_1, \ldots, TA_n$ and FB occur in such a τ, $i_\tau(A_1) = \ldots = i_\tau(A_n) = 1$ and $i_\tau(B) = 0$. This would contradict that $A_1, \ldots, A_n \models B$. Hence, all tableau branches starting with $TA_1, \ldots, TA_n$, FB are closed. So, by Lemma 2.3, $A_1, \ldots, A_n \vdash' B$. □

Remark 2.2. Our procedure of searching for a tableau-deduction of B from given premises $A_1, \ldots, A_n$ will end after finitely many steps and then either give a tableau-deduction of B from $A_1, \ldots, A_n$, indicating that $A_1, \ldots, A_n \vdash' B$, or an interpretation i such that $i(A_1) = \ldots = i(A_n) = 1$ and $i(B) = 0$, indicating that $A_1, \ldots, A_n \not\models B$.

Corollary 2.4 (Decidability of classical propositional logic). *Classical propositional logic is decidable, i.e., we have an effective method (algorithm) to decide, given any finite set of formulas $B, A_1, \ldots, A_n$, whether B is tableau-deducible from $A_1, \ldots, A_n$ or not.*

Note that in Section 2.3 we have already given an effective method (algorithm) to decide whether or not B is a valid consequence of $A_1, \ldots, A_n$ for any finite set of formulas $A_1, \ldots, A_n$.

The tableaux system for classical propositional calculus can easily be modified and/or completed to a tableaux system for intuitionistic logic and for many intensional (modal) logics. In all cases the completeness proof given above can be adapted to a completeness proof for the logic in question. This type of proof has an advantage over some other completeness proofs in that it is *constructive*.

Automated theorem proving In the case of the classical propositional calculus an effective method has been given above to decide, given any finite set of formulas $B, A_1, \ldots, A_n$, whether B is tableau-deducible from $A_1, \ldots, A_n$ or not. This algorithm can be formulated in an appropriate programming language such as Prolog (see, for instance, Kogel-Ophelders [17]) and then a computer, when provided with formulas $B, A_1, \ldots, A_n$, is able to compute whether B is a theorem on the basis of the hypotheses $A_1, \ldots, A_n$ or not.

So, a computer, provided with the appropriate software, is able to simulate reasoning and in that case one may say that it disposes of *Artificial Intelligence*. By adding to such a computer-program a number of data, $A_1, \ldots, A_n$, concerning a small and well-described subject, the so-called *knowledge base*, the computer is able to draw conclusions from those data. If $A_1, \ldots, A_n$ represent someone's expertise, one speaks of an *expert system*. And if the knowledge base consists of Euclid's axioms for geometry or Peano's axioms for number theory or of axioms for some other part of mathematics, one speaks of *automated theorem proving*.

So the basic ideas underlying expert-systems and automated theorem proving are very simple. However, in practice there may be a lot of complications. Without being exhaustive let us mention some of them.

1. The language of propositional logic may be too restrictive. For instance, in Chapter 1 we have already seen that the argument

> All men are mortal.
> Socrates is a man.
> Therefore, Socrates is mortal.

cannot be adequately formulated in the propositional language. For that reason the propositional language will be extended to the predicate language in Chapter 4.

2. However, if one adapts the construction of a completed tableau with initial branch $\{TA_1, \ldots, TA_n, FB\}$ to the case that $B, A_1, \ldots, A_n$ are formulas of the predicate language, this construction no longer yields a decision: if no logical deduction exists, the tableau construction may continue forever, without ever knowing that this construction will come to an end; so, in this case the tableau construction may not stop. For more details see Subsection 4.4.2.

3. Even in the case of the propositional language, the time and space needed to search for a logical deduction of B from $A_1, \ldots, A_n$ may grow very fast in the event n is big or $B, A_1, \ldots, A_n$ are (very) complex; see Subsection 2.3.1.

4. If the knowledge base consists of Peano's axioms for number theory (see Section 5), this knowledge base contains the axiom schema of induction, and hence infinitely many axioms. Searching for a logical deduction of a given formula B from infinitely many axioms requires a *strategy*, without which such a search is hopeless.

5. If the knowledge base consists of someone's expertise, it may contain uncertain and/or incomplete information. For instance, it may be likely, but uncertain, that there is oil in the ground. An expert-system may have to deal with uncertain knowledge and then its conclusions will have a certain degree of probability, which has to be computed. This is a far from trivial matter. Also the information in the knowledge base may be incomplete in order to be able to draw a certain conclusion.

6. Building an expert-system is more than just providing an inference mechanism: the system should also be able to explain *how* the conclusion was established or *why* the conclusion cannot be drawn.

2.10 Paradoxes; Historical and Philosophical Remarks

2.10.1 Paradoxes

Paradoxes have been important for making progress in science and philosophy. In what follows a number of statements of the type $B \rightleftarrows \neg B$ are presented. Because statements of this type cannot possibly be true, in other words are inconsistent, these results are known as *paradoxes*. The reader easily checks the following theorem:

Theorem 2.30. *a) For each formula B,* $\models \neg(B \rightleftarrows \neg B)$.
b) If $A_1, \ldots, A_n \models B \rightleftarrows \neg B$, *then* $\models \neg(A_1 \wedge \ldots \wedge A_n)$.

So, if for some formula B, $B \rightleftarrows \neg B$ is a valid consequence of hypotheses $A_1, \ldots, A_n$, then at least one of the hypotheses must be false. In practice, the problem frequently is that we are not aware of the hypotheses we are using in deriving a paradox.

In his paper *'Paradox'*, W.V. Quine [21] distinguishes three types of paradox: antinomies, veridical and falsidical paradoxes. Below we shall discuss these three types and consider examples of each of them.

Antinomies There is the old *paradox of the liar*: A man says that he is lying. If he speaks the truth, he is lying. And if he is lying, he speaks the truth. Hence, he speaks the truth if and only if he does not.

A more recent version of this paradox is the one of A. Tarski [24] in his *'Truth and Proof'*. Consider the following sentence.

s: The underlined sentence is false

Here *s* is just an abbreviation for: the underlined sentence is false. But what is the object the name 'the underlined sentence' refers to? Up till now there is no underlined sentence. By underlining sentence *s*, we achieve that sentence *s* says of itself that it is false, just as the man in the paradox says of himself that he is lying.

s: <u>The underlined sentence is false</u>

When one refers to an object, one usually uses a name for that object. One and the same object may have different names. For instance, 'Harrie de Swart' and 'the author of this book' are two different names for the same person. Usually, when referring to a sentence or, more generally, a linguistic object, one may form its name by putting the sentence in question between quotation marks. But another name for that same sentence may be formed by underlining the sentence in question, after which 'the underlined sentence' is another name for the same sentence. So, having underlined sentence *s*, *s* has (at least) the following two names: '*s*'; the underlined sentence. Consequently, by replacing one name by another one:

(1) '*s*' is false if and only if the underlined sentence is false.

On the other hand we have the *principle of adequacy*: for each sentence *p*, '*p*' is true if and only if *p*; where '*p*' is again a name for the sentence *p*. For example, 'snow is white' is true if and only if snow is white. Now using this principle of adequacy, we find

's' is true if and only if s, i.e.,

(2) 's' is true if and only if the underlined sentence is false.

(1) and (2) together yield: 's' is false if and only if 's' is true.

The paradox of the liar, in one form or another, is a special kind of paradox, an *antinomy*: an absurd statement, that cannot be true, with a correct argument, and whose premises are not in themselves absurd. However, if $B \rightleftarrows \neg B$ is a valid consequence of premises $A_1, \ldots, A_n$, we know we have to revise our premises. It is typical of an antinomy that we are very surprised that such a revision is necessary, because the premises accepted seem more than plausible and seem completely in accordance with our intuition. In order to be able to 'solve' an antinomy, a major revision in our way of thinking is necessary. Because everything we do in the derivation of an antinomy seems so natural and evident, we are generally not very conscious of what precisely our premises are.

Through all ages the antinomies have caused concern to philosophers. According to a foolish tradition preserved by Diogenes Laertius, Diodorus Cronus (ca. 300 B.C.) committed suicide because he was not immediately able to solve the logical puzzle posed by the paradox of the liar. (See W & M Kneale [16], p. 113.)

In his paper *Truth and Proof*, A. Tarski [24] argues that the paradox of the liar forces us to give up our silent assumption that object language and meta language do not have to be distinguished. But when we say that a sentence 's' is true, we are saying something *about* sentence s. If s belongs to a language L_0, the sentence ' 's' is true' is a statement *about* a sentence of L_0 and hence a statement in the metalanguage L_1 of L_0. If we take care to distinguish predicates $true_0$, $true_1$, $false_0$, $false_1$, and so on, for the truth/falsity predicates in the different languages, the paradox of the liar disappears:

Again, let s be an abbreviation for: the underlined sentence is $true_0$. Next, let us underline this sentence:

s: The underlined sentence is $false_0$

Then, again replacing one name by another one:

(1a) 's' is $false_1$ if and only if the underlined sentence is $false_1$.

And by the principle of adequacy

(2a) 's' is $true_1$ if and only if the underlined sentence is $false_0$.

And now (1a) and (2a) are no longer contradictory!

If we wish to avoid contradictions, we must insist that what we ordinarily call English is in reality an infinite sequence $L_0, L_1, L_2, \ldots$ of languages, in which L_{n+1} is a metalanguage in relation to L_n.

Another way to escape the antinomy of the liar is by introducing a technical restriction on the class of sentences regarded as possessing a truth value. According to Ryle [22], sentences of the form 'the such-and-such sentence is false' should not be regarded as having a truth value unless it is possible to attach a 'namely-rider'. For instance, in 'the first thing that Plato said to Aristotle is true' we can insert a clause,

2.10 Paradoxes; Historical and Philosophical Remarks

'the first thing that Plato said to Aristotle, namely ..., is true', which may alter its meaning, but does not alter its truth-value. But in the paradoxical 'the underlined sentence is false', if we try to insert such a clause, 'the underlined sentence, namely 'the underlined sentence is false', is false' we get a new description (indirect) of a sentence which must again be supplied with a namely-rider. As this process never ends, the original sentence has no truth value, whereas in the Plato example, we get down to something of the form '...' is true, where the quoted part does not involve the notions of truth and falsehood.

The paradox of the liar is an antinomy at the level of sentences. At the level of subjects and singular descriptions there is the *antinomy of Berry*, to be discussed in Chapter 4. And at the level of predicates there is the *antinomy of Russell*, better known as *Russell's paradox*, which will be discussed in Chapter 3.

Besides antinomies, like those of the liar, of Berry and of Russell, W.V. Quine also distinguishes other, less serious, paradoxes: veridical and falsidical paradoxes.

Veridical paradoxes A *veridical* or *truth-telling paradox* is a paradoxical statement that on reflection turns out to yield a somewhat astonishing, but true, proposition.

Example 2.27. 1. Frederic has reached the age of twenty-one without having more than five birthdays.
2. The **barber paradox**: In a certain village there is a barber who shaves precisely those men in the village who do not shave themselves. Question: does the barber shave himself? Each man in the village is shaved by the barber if and only if he does not shave himself. Hence, in particular, the barber shaves himself if and only if he does not shave himself.

Both paradoxes are alike in the sense that at first sight they seem to prove absurdities by decisive arguments. The Frederic-paradox is a truth-telling paradox if we conceive the statement as the abstract truth that one can be $4n$ ($n = 0, 1, 2, \ldots$) years old at one's n^{th} birthday, namely if one has been born on February 29. The barber-paradox contains a reductio ad absurdum: from the, not explicitly mentioned, premiss that such a barber exists, we derive an absurdity of the form $B \rightleftarrows \neg B$. Hence the assumption is false and no village can have a barber who shaves all and only those men in the village who do not shave themselves.

The difference between an antinomy and a veridical paradox is that in the latter case we are only slightly astonished that we have to give up one of the premisses like the existence of a village-barber as described above, while in the case of an antinomy we are forced to give up very fundamental ideas and a major revision in our way of thinking is needed.

Falsidical paradoxes A *falsidical paradox* is a paradoxical statement that really is false, the argument backing it up containing some impossible hidden assumption or involving a fallacy. Typical examples of falsidical paradoxes are:

Example 2.28. 1. The comic mis-proof that $2 = 1$: Let $x = 1$. Then $x^2 = x$. Hence $x^2 - 1 = x - 1$. Dividing both sides by $x - 1$, we conclude that $x + 1 = 1$. Hence, because $x = 1$, $2 = 1$.

2. Three men agree to share a hotel room overnight, splitting the charge of $ 30 three ways, with each man paying $ 10. After they have gone to their room, the clerk realizes he should only have charged them $ 25 and sends the bellboy up with $ 5 to be returned to them. The bellboy, realizing how hard it will be to make change, pockets $ 2 and returns $ 1 to each man. Thus the men have each paid $ 9, for a total of $ 27 and the bellboy has $ 2, for a total of $ 29. One dollar of the original thirty is missing.

3. Zeno's paradox of Archilles and the Tortoise.

```
     A
     | - - - - - - - - - | o o o o o o o |. . . . |
     0                   T
                         | - - - - - - - | o o o o |. . . . |
                         1              1.1      1.11     1.111
```

Suppose A(rchilles) and the T(ortoise) start to run at the same time and A runs 10 times as fast as T does. Suppose also that in the starting position A is in position 0, one mile behind T, which hence is in position 1. While A runs from 0 to the starting position of T, T covers a distance of 0.1 mile since its velocity has been supposed to be $\frac{1}{10}$ of that of A. And while A runs from position 1 to position 1.1, T covers a distance of 0.01 mile, thus arriving at position 1.11. And while A runs from position 1.1 to position 1.11, T runs from position 1.11 to position 1.111. And so on. Consequently, A will never pass T.

In a falsidical paradox there is always a fallacy or some impossible hidden assumption in the argument and in addition the statement must look absurd and be false.

In the 'proof' of $2 = 1$ we divided by $x - 1$, which is 0 because x was supposed to be 1. In the hotel paradox the number 2 is added wrongly to 27: 2 should be subtracted from 27 in order to determine the price, 25 dollars, of the hotel room.

In the case of Archilles and the Tortoise the impossible hidden assumption is that the infinite process of Archilles running to the position where the tortoise was a moment ago, lasts infinitely long. In fact, however, if Archilles needs 0.1 hour for one mile, the infinite process will last only $0.1 + 0.01 + 0.001 + \ldots = 0.111\ldots = \frac{1}{9}$ hour, which is less than 0.12 hours. Within this time Archilles and the Tortoise will arrive at the same position and Archilles will pass the Tortoise. The process of Archilles passing the Tortoise may be thought of as consisting of infinitely many steps, but this infinite process is actually completed in $\frac{1}{9}$ hour (6 minutes and 40 seconds).

Only the antinomies cause a crisis of thought. Only an antinomy produces a self-contradiction via accepted means of reasoning. Only an antinomy requires that some tacitly accepted and trusted patterns of reasoning be made explicit and henceforth be avoided or revised.

The falsidical paradox of Zeno must have been a real antinomy in his day. It was thought as evident that a process consisting of infinitely many steps would last infinitely long. It is only because of the mathematical achievements of the 18th and 19th century that we know that some infinite sums, for example, $0.1 + 0.01 + 0.001 + \ldots = 0.111\ldots = \frac{1}{9}$ and $\frac{1}{2} + \frac{1}{4} + \frac{1}{8} + \frac{1}{16} + \ldots = 1$, are finite, while others, for

2.10 Paradoxes; Historical and Philosophical Remarks   101

example, $\frac{1}{2} + \frac{1}{3} + \frac{1}{4} + \frac{1}{5} + \ldots$, are not. What is an antinomy for the one is a falsidical paradox for the other, given a lapse of a couple of thousands of years.

In the case of the paradox of Archilles and the Tortoise one should realize that points of space and time do not occur in our perception, but are mathematical idealizations. Points of space and time belong to the language of mathematics, not to the language of our perception. If we talk about Archilles passing the infinitely many points (positions) the Tortoise was a moment ago, we are speaking in terms of our mathematical model and not in terms of what we perceive.

Exercise 2.70. Is the following paradox an antinomy, a veridical or a falsidical one? A judge tells a condemned prisoner that he will be hanged either on Monday, Tuesday, Wednesday, Thursday or Friday of the next week, but that the day of the hanging will come as a surprise: he will not know until the last moment that he is going to be hanged on that day. The prisoner reasons that if the first four days go by without the hanging, he will *know* on Friday, that he is due to be hanged that day. So it cannot be on Friday that he will be hanged. But now with Friday eliminated, if the first three days go by without the hanging, he will know on Thursday that he is due to be hanged that day, and it would not be a surprise. So it cannot be Thursday. In the same way he rules out Wednesday, Tuesday and Monday, and convinces himself that he cannot be hanged at all. But he is very surprised on Wednesday when the executioner arrives at his cell. (See also Exercise 6.12 and its solution.)

Exercise 2.71. Is the following paradox an antinomy, a veridical or a falsidical one? A crocodile seizes a baby, and tells the mother that he will return it if the next thing she says to him is the truth, but will eat it if the next thing she says is false. The mother says 'you will eat the baby'. The crocodile will eat the baby if and only if he will let it go.

Exercise 2.72. (From S.C. Kleene [15], p. 40) The following riddle also turns upon the paradox of the liar. A traveller has fallen among cannibals. They offer him the opportunity to make a statement, attaching the conditions that if his statement be true, he will be boiled, and if it be false, he will be roasted. What statement should he make? (A form of this riddle occurs in Cervantes' "Don Quixote" (1605), II, 51.)

Exercise 2.73. (From S.C. Kleene [15], p. 37, 38) Every municipality in Holland must have a mayor, and no two may have the same mayor. Sometimes the mayor is a non-resident of the municipality. Suppose a law is passed setting aside a special area S exclusively for such non-resident mayors, and compelling all non-resident mayors to reside there. Suppose further that there are so many non-resident mayors that S has to be constituted a municipality. Where shall the mayor of S reside? (Mannoury, cf. van Dantzig [5])

Exercise 2.74. (From S.C. Kleene [15], p. 38) Suppose the Librarian of Congress compiles, for inclusion in the Library of Congress, a bibliography of all those bibliographies in the Library of Congress which do not list themselves. (Gonseth 1933) Should that bibliography list itself?

Exercise 2.75. From *Attic Nights* by Aulus Gellius, Book V, x:

> Among fallacious arguments the one which the Greeks call ἀντιστρέφον seems to be by far the most fallacious. Some of our own philosophers have rather appropriately termed such arguments *reciproca*, or 'convertible'. The fallacy arises from the fact that the argument that is presented may be turned in the opposite direction and used against the one who has offered it, and is equally strong for both sides of the question. An example is the well-known argument which Protagoras, the keenest of all sophists, is said to have used against his pupil Euathlus.
>
> For a dispute arose between them and an altercation as to the fee which had been agreed upon, as follows: Euathlus, a wealthy young man, was desirous of instruction in oratory and the pleading of causes. He became a pupil of Protagoras and promised to pay him a large sum of money, as much as Protagoras had demanded. He paid half of the amount at once, before beginning his lessons, and agreed to pay the remaining half on the day when he first pleaded before jurors and won his case. Afterwards, when he had been for some little time a pupil and follower of Protagoras, and had in fact made considerable progress in the study of oratory, he nevertheless did not undertake any cases. And when the time was already getting long, and he seemed to be acting thus in order not to pay the rest of the fee, Protagoras formed what seemed to him at the time a wily scheme; he determined to demand his pay according to the contract, and brought suit against Euathlus.
>
> And when they had appeared before the jurors to bring forward and to contest the case, Protagoras began as follows: 'Let me tell you, most foolish of youths, that in either event you will have to pay what I am demanding, whether judgement be pronounced for or against you. For if the case goes against you, the money will be due me in accordance with the verdict, because I have won; but if the decision be in your favour, the money will be due me according to our contract, since you will have won a case.'
>
> To this Euathlus replied: 'I might have met this sophism of yours, tricky as it is, by not pleading my own cause but employing another as my advocate. But I take greater satisfaction in a victory in which I defeat you, not only in the suit, but also in this argument of yours. So let me tell you in turn, wisest of masters, that in either event I shall not have to pay what you demand, whether judgement be pronounced for or against me. For if the jurors decide in my favour, according to their verdict nothing will be due you, because I have won; but if they give judgement against me, by the terms of our contract I shall owe you nothing, because I have not won a case.'
>
> Then the jurors, thinking that the plea on both sides was uncertain and insoluble, for fear that their decision, for whichever side it was rendered, might annul itself, left the matter undecided and postponed the case to a distant day. Thus a celebrated master of oratory was refuted by his youthful pupil with his own argument, and his cleverly devised sophism failed. [From the English translation by John C. Rolfe of *The Attic Nights of Aulus Gellius*, Book V, section X. Reprinted, Cambridge, Mass., 1967. The Loeb Classical Library, 195, pp. 404-409.]

2.10.2 Historical and Philosophical Remarks

Stoic Logic Aristotle is generally seen as the founding father of logic. Only at the beginning of the 20th century it became clear, among others by the work of the Polish logician Łukasiewicz, that in fact the Stoics (± 300 B.C.) developed a kind of propositional logic, while the logic of Aristotle is a small part of what we now call predicate logic, to be studied in Chapter 4. A typical inference-schema of the

2.10 Paradoxes; Historical and Philosophical Remarks

Stoics runs as follows:

> If the first, then the second.
> The first.
> Therefore, the second.

As a concrete example of this type of inference, they were accustomed to give:

> If it is day, then it is light.
> It is day.
> Therefore, it is light.

A typical Aristotelian syllogism is: If all things with the predicate (property) P also satisfy the predicate Q, and all things with the predicate Q also satisfy the predicate R, then all things with the predicate P also satisfy the predicate R. A concrete instance of this would be: If all birds are animals and all animals are mortal, then also all birds are mortal.

As pointed out by Łukasiewicz, the Stoics were discussing the truth conditions for implication. The truth-functional account, as in our truth table for $\to$, is first known to have been proposed by Philo of Megara ca. 300 B.C. in opposition to the view of his teacher Diodorus Cronus. We know of this through the writings of Sextus Empiricus some 500 years later, the earlier documents having been lost. According to Sextus,

> Philo says that a sound conditional is one that does not begin with a truth and end with a falsehood. ... But Diodorus says it is one that neither could nor can begin with a truth and end with a falsehood. [Kneale, [16], p. 128]

There can be no doubt that what Sextus refers to is precisely the truth-functional connective that we have symbolized by the $\to$, for he says elsewhere,

> So according to him there are three ways in which a conditional may be true, and one in which it may be false. For a conditional is true when it begins with a truth and ends with a truth, like 'if it is day, it is light'; and true also when it begins with a falsehood and ends with a falsehood, like 'If the earth flies, the earth has wings'; and similarly a conditional which begins with a falsehood and ends with a truth is itself true, like 'If the earth flies, the earth exists'. A conditional is false only when it begins with a truth and ends with a falsehood, like 'If it is day, it is night'. [Kneale [16], p. 130]

So Sextus reports Philo as attributing truth values to conditionals just as in our truth table for $\to$, except for the order in which he lists the cases. Diodorus probably had in mind what later was called *strict implication*; see Chapter 6.

One of the Stoic principles noted by Łukasiewicz is as follows: an argument is valid if and only if the conditional proposition having the conjunction of the premisses as antecedent and the conclusion as consequent is logically true. The similarity of this principle to our Theorem 2.4 is obvious.

According to the Stoics, there were five basic types of undemonstrated, i.e., self evident, argument:

1. If the first, then the second; but the first. Therefore, the second.
2. If the first, then the second; but not the second. Therefore not the first.
3. Not both the first and the second; but the first. Therefore not the second.
4. Either the first or the second; but the first. Therefore not the second.
5. Either the first or the second; but not the second. Therefore the first.

These arguments are basic, it was maintained, in the sense that every valid argument can be reduced to them. Sextus Empiricus gives us two very clear examples of the analysis of an argument into its component basic arguments:

6. If the first, then if the first then the second; but the first. Therefore the second. (Composition of two type 1 undemonstrated arguments.)

7. If the first and the second, then the third; but not the third; on the other hand the first. Therefore not the second. (Composition of a type 2 and a type 3 undemonstrated argument.)

One of the theorems attributed to Chrysippus is:

8. Either the first or the second or the third; but not the first; and not the second. Therefore the third. (Composition of two type 5 undemonstrated arguments.)

Chrysippus himself is reported to have said that even dogs make use of this sort of argument. For when a dog is chasing some animal and comes to the junction of three roads, if he sniffs first at the two roads down which the animal did not run, he will rush off down the third road without stopping to smell. [See B. Mates [19], pp. 67-82 and W. & M. Kneale [16], pp. 158-176.]

Consequentiae In the Middle Ages several treatises on *consequentiae* were written. One of the more interesting ones is *In Universam Logicam Quaestiones*, formerly attributed to John Duns the Scot (1266-1308), but later to a Pseudo-Scot (? John of Cornwall). As we learn from Kneale [16], pp. 278-280, the Pseudo-Scot distinguishes various kinds of *consequentiae*.

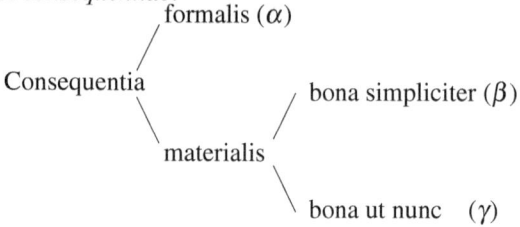

$$\text{Consequentia} \begin{cases} \text{formalis } (\alpha) \\ \text{materialis} \begin{cases} \text{bona simpliciter } (\beta) \\ \text{bona ut nunc } (\gamma) \end{cases} \end{cases}$$

Examples:

(α) Socrates currit et Socrates est albus, igitur album currit.
  Socrates walks and Socrates is white, so something white walks.
(β) Homo currit, igitur animal currit.
  A man walks, therefore a living being walks.
(γ) Socrates currit, igitur album currit.
  Socrates walks, therefore something white walks.

Consequentiae formales are inferences made exclusively on the basis of the forms of the expressions involved. In *consequentiae materiales* the meaning of the premisses and the conclusion also has to be taken into account. But *consequentiae materiales* can always be reduced to consequentiae formales by making explicit the silently assumed premises. For instance, 'Socrates currit, igitur album currit' (Socrates walks, so something white walks) can be reduced to 'Socrates currit et Socrates est albus, igitur album currit' (Socrates walks and Socrates is white, so something white walks). The *Consequentiae materiales bona simpliciter* are those

2.10 Paradoxes; Historical and Philosophical Remarks

inferences in which the silently assumed premisses are necessary, like, for instance, 'omnis homo est animal' (every man is a living being). When the silently assumed premisses are contingent (not necessary), like, for instance, 'Socrates est albus' (Socrates is white), the Pseudo-Scot speaks of *consequentiae materiales bona ut nunc*.

Because of their amusing character, we present below two theorems and their proofs, as given by the Pseudo-Scot.

1. *Ad quamlibet propositionem implicantem contradictionem de forma sequitur quaelibet alia propositio in consequentia formali* (From a proposition which implies a formal contradiction, any proposition follows as a 'consequentia formalis').
2. *Ad quamlibet propositionem impossibilem sequitur quaelibet alia propositio non consequentia formali sed consequentia materiali bona simpliciter* (From a proposition which is impossible, any proposition follows not as a 'consequentia formalis' but as a 'consequentia materialis bona simpliciter').

Kneale [16], pp. 281-282, gives the following reconstruction of the proof of 1.

$$\frac{\frac{\text{Socrates exists and}}{\text{Socrates does not exist}} \quad \frac{\text{Socrates exists and Socrates does not exist}}{\text{Socrates exists}}}{\frac{\text{Socrates does not exist} \quad \text{Socrates exists or a man is an ass}}{\text{a man is an ass}}}$$

And the Pseudo-Scot gives the following two proofs of 2:

1. Using 1., the *consequentia* 'A man is an ass and a man is not an ass, therefore you are in Rome' is formally valid. Since it is impossible that a man is an ass, it is necessary that a man is not an ass. And the Pseudo-Scot concludes that the *consequentia materialis* 'A man is an ass, therefore you are at Rome' is bona simpliciter, being reducible to a formally valid *consequentia* by addition of a necessarily true premise.
2. Supposing that 'A man is not an ass' is necessarily true, the Pseudo-Scot also gives the following derivation.

$$\frac{\frac{\text{A man is an ass}}{\text{A man is an ass or you are at Rome}} \quad \text{A man is not an ass}}{\text{you are at Rome}}$$

Suggested reading on Medieval Logic: W. & M. Kneale, *The Development of Logic*; L.M. de Rijk, *Logica Modernorum*; P. Boehner, *Medieval Logic*; E. Moody, *Truth and Consequence in Medieval Logic*.

Frege's Begriffsschrift (1879) Although an algebra of logic was initiated by Boole in 1847 and De Morgan in that same year, the propositional logic properly appeared with Frege's *Begriffsschrift* in 1879, and in Russell's work, especially in the *Principia Mathematica* by Whitehead and Russell, 1910-13.

> The imprecision and ambiguity of ordinary language led Frege (1848-1925) to look for a more appropriate tool; he devised a new mode of expression, a language that deals with the 'conceptual content' and that he came to call 'Begriffsschrift'. This ideography is a 'formula language', that is, a *lingua characterica*, a language written with special symbols, 'for pure thought', that is, free from rhetorical embellishments, ... [Heijenoort [12], p. 1]

In the preface to his Begriffsschrift, Frege makes the following remarks about his work (the following translations are by J. van Heijenoort [12], p. 6-7).

> (p. X) Its first purpose, therefore, is to provide us with the most reliable test of the validity of a chain of inferences and to point out every presupposition that tries to sneak in unnoticed, so that its origin can be investigated. That is why I decided to forgo expressing anything that is without significance for the inferential sequence. In § 3 I called what alone mattered to me the *conceptual content* [begrifflichen Inhalt].

> (p.XI) I believe that I can best make the relation of my ideography to ordinary language [Sprache des Lebens] clear if I compare it to that which the microscope has to the eye. Because of the range of its possible uses and the versatility with which it can adapt to the most diverse circumstances, the eye is far superior to the microscope. Considered as an optical instrument, to be sure, it exhibits many imperfections, which ordinarily remain unnoticed only on account of its intimate connection with our mental life. But, as soon as scientific goals demand great sharpness of resolution, the eye proves to be insufficient. The microscope, on the other hand, is perfectly suited to precisely such goals, but that is just why it is useless for all others.

> (p.XII) If it is one of the tasks of philosophy to break the domination of the word over the human spirit by laying bare the misconceptions that through the use of language often almost unavoidably arise concerning the relations between concepts and by freeing thought from that with which only the means of expression of ordinary language, constituted as they are, saddle it, then my ideography, further developed for these purposes, can become a useful tool for the philosopher. To be sure, it too will fail to reproduce ideas in a pure form, and this is probably inevitable when ideas are represented by concrete means; but, on the one hand, we can restrict the discrepancies to those that are unavoidable and harmless, and, on the other, the fact that they are of a completely different kind from those peculiar to ordinary language already affords protection against the specific influence that a particular means of expression might exercise. [J. van Heijenoort [12], p. 6-7]

The notation that Frege introduces in his Begriffsschrift has not survived. It presents difficulties in printing and takes up a large amount of space. But, as Frege himself says, 'the comfort of the typesetter is certainly not the summum bonum, and the notation undoubtedly allows one to perceive the structure of a formula at a glance and to perform substitutions with ease.'

In § 5 of his Begriffsschrift Frege introduces the notation for our $B \to A$. Our $C \to (B \to A)$ is represented by Frege as:

while Frege represents our $(C \to B) \to A$ by:

In section 7 of his Begriffsschrift Frege represents our $\neg A$ by:

Frege presents the propositional calculus in a version that uses the conditional and negation as primitive connectives. Frege renders our $A \vee B$ by $\neg B \to A$, i.e.,

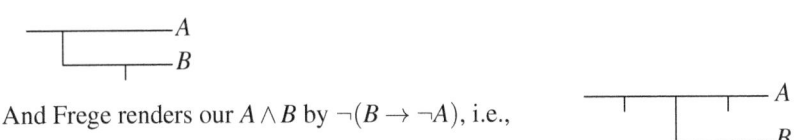

2.10 Paradoxes; Historical and Philosophical Remarks

And Frege renders our $A \wedge B$ by $\neg(B \to \neg A)$, i.e.,

The distinction between 'and' and 'but' is of the kind that is not expressed in the present ideography. [G. Frege, *Begriffsschrift*, § 7.]

Conversational implicature P. Grice in the 1967 William James Lectures (published in 1989 in [10]) works out a theory in *pragmatics* which he calls the theory of *conversational implicature*. Generally speaking, in conversation we usually obey or try to obey rules something like the following:

QUANTITY: Be informative
QUALITY: Tell the truth
RELATION: Be relevant
MODE: Avoid obscurity, prolixity, etc.

If the fact that *A* has been said, plus the assumption that the speaker is observing the above rules, plus other reasonable assumptions about the speaker's purposes and intentions in the context, logically entails that *B*, then we can say *A conversationally implicates B*.

It is possible for *A* to conversationally implicate many things which are in no way part of the *meaning* of *A*. For example, if X says 'I'm out of gas' and Y says 'there's a gas station around the corner', Y's remark conversationally implicates that the station in question is open, since the information that the station is there would be *irrelevant* to X's predicament otherwise. If X says 'Your hat is either upstairs in the back bedroom or down in the hall closet', this remark conversationally implicates 'I don't know which', since if X did know which, this remark would not be the most *informative* one he could provide.

Grice shows how philosophers have sometimes mistaken conversational implicatures for elements of meaning. For instance, Strawson sometimes claims not-knowing-which must be part of the *meaning* of 'or' (and therefore the traditional treatment of disjunction in logic is misleading or false). Grice claims this is mistaking the conversational implicature cited above for an aspect of meaning.

Sometimes it is possible to *cancel* a conversational implicature by adding something to one's remark. For example, in the gas station case, 'I'm not sure whether it's open' and in the hat case, 'I know, but I'm not saying which' (one might say this if locating the hat was part of some sort of parlor game). The possibility of cancellation shows that the conversational implicatures definitely are not part of the *meaning* of the utterance.

Conditionals In the examples below the conditional in (1) is in the *indicative* mood, while the conditional in (2) is a *subjunctive* one.
(1) If Oswald did not kill Kennedy, someone else did.
(2) If Oswald had not killed Kennedy, someone else would have.
(These examples are from E. Adams, *Subjunctive and Indicative Conditionals*, Foundations of Language 6: 89-94, 1970.)

(1) is true: someone killed Kennedy; but (2) is probably false. Therefore, different analyses are needed for indicative and for subjunctive conditionals.

A *counterfactual* conditional is an expression of the form 'if A were the case, then B would be the case', where A is supposed to be false. Not all subjunctive conditionals are counterfactual. Consider the argument, 'The murderer used an ice pick. But if the butler had done it, he wouldn't have used an ice pick. So the murderer must have been someone else.'. If this subjunctive conditional were a counterfactual, then the speaker would be presupposing that the conclusion of his argument is true. (This example is from R.C. Stalnaker, *Indicative Conditionals*, in W.L. Harper, e.a., *IFS*.)

In Chapter 6 we shall discuss counterfactuals and subjunctive conditionals in general. In this section we will restrict our attention from now on to indicative conditionals.

In Section 2.4 we have considered the so-called *paradoxes of material implication*: the following two inferences for material implication $\to$ are valid, whereas the corresponding English versions seem invalid.

$$\frac{\neg A}{A \to B} \qquad \frac{\text{There is no oil in my coffee}}{\text{If there is oil in my coffee, then I like it}}$$

$$\frac{B}{A \to B} \qquad \frac{\text{I'll ski tomorrow}}{\text{If I break my leg today, then I'll ski tomorrow}}$$

So, the truth-functional reading of 'if ..., then ...', in which $A \to B$ is equivalent to $\neg A \vee B$, seems to conflict with judgments we ordinarily make. The paradoxical character of these inferences disappears if one realizes that:
1. the material implication $A \to B$ has the same truth-table as $\neg A \vee B$;
2. speaking the truth is only one of the conversation rules one is expected to obey in daily discourse; one is also expected to be as relevant and informative as possible.
Now, if one has at one's disposal the information $\neg A$ (or B, respectively) and at the same time provides the information $A \to B$, i.e., $\neg A \vee B$, then one is speaking the truth, but a truth calculated to mislead, since the premiss $\neg A$ (or B, respectively) is so much simpler and more informative than the conclusion $A \to B$. If one knows the premiss $\neg A$ (or B, respectively), the conversation rules force us to assert this premiss instead of $A \to B$. Quoting R. Jeffrey:

> Thus defenders of the truth-functional reading of everyday conditionals point out that the disjunction $\neg A \vee B$ shares with the conditional 'if A, then B' the feature that normally it is not to be asserted by someone who is in a position to deny A or to assert B. ...
>
> Normally, then, conditionals will be asserted only by speakers who think the antecedent false or the consequent true, but do not know which. Such speakers will think they know of some connection between the components, by virtue of which they are sure (enough for the purposes at hand) that the first is false or the second is true. [R. Jeffrey [13], pp. 77-78]

Summarizing in a slogan:
        indicative conditional = material implication + conversation rules.

So H.P. Grice uses principles of conversation to explain facts about the use of conditionals that seem to conflict with the truth-functional analysis of the ordinary in-

2.10 Paradoxes; Historical and Philosophical Remarks

dicative conditional. In his paper 'Indicative Conditionals' (in W.L. Harper, e.a. (eds.), *IFS*), R.C. Stalnaker follows another strategy, rejecting the material conditional analysis. And in his book 'Causal Necessity', Brian Skyrms claims that the indicative conditional cannot be construed as the material implication '$\rightarrow$' plus conversational implicature. The dispute between advocates of the truth-functional account of conditionals and the advocates of other, more complex but seemingly more adequate accounts is as old as logic itself.

Frege, Russell, Hilbert In his *Begriffsschrift* (page 2) of 1879 Gottlob Frege distinguishes the notations $-A$ for 'the proposition that A' and $\vdash A$ for 'it is a fact that A'. Frege calls A in $-A$ and in $\vdash A$ 'der Inhalt' (the *content*) and '$\vdash A$' 'ein Urteil' (a *judgment*). In Chapter II of his book Frege gives the first axiomatic formulation of classical propositional (and predicate) logic, namely, the following system $\mathscr{P}_F$, presented below in our own notation.

$A \rightarrow (B \rightarrow A)$ (Begriffsschrift, p. 26, form. 1)
$(C \rightarrow (B \rightarrow A)) \rightarrow ((C \rightarrow B) \rightarrow (C \rightarrow A))$ (Begriffsschrift, p. 26, form. 2)
$(D \rightarrow (B \rightarrow A)) \rightarrow (B \rightarrow (D \rightarrow A))$ (Begriffsschrift, p. 35, form. 8)
$(B \rightarrow A) \rightarrow (\neg A \rightarrow \neg B)$ (Begriffsschrift, p. 43, form. 27)
$\neg\neg A \rightarrow A$ (Begriffsschrift, p. 44, form. 31)
$A \rightarrow \neg\neg A$ (Begriffsschrift, p. 47, form. 41)

together with Modus Ponens.

It is probably correct to say that Frege's work only became well-known through Russell. The following formulation $\mathscr{P}_R$ of classical propositional logic was used by Whitehead and Russell in *Principia Mathematica* in 1910 (see part I, page 13).

$A \vee A \rightarrow A$
$B \rightarrow A \vee B$
$A \vee B \rightarrow B \vee A$
$A \vee (B \vee C) \rightarrow B \vee (A \vee C)$
$(B \rightarrow C) \rightarrow (A \vee B \rightarrow A \vee C)$

together with Modus Ponens.

The following formulation $\mathscr{P}$ of propositional logic has implication and negation as primitive connectives and Modus Ponens as its only rule:

$A \rightarrow (B \rightarrow A)$
$(A \rightarrow (B \rightarrow C)) \rightarrow ((A \rightarrow B) \rightarrow (A \rightarrow C))$
$(\neg A \rightarrow \neg B) \rightarrow (B \rightarrow A)$

Defining $A \wedge B := \neg(A \rightarrow \neg B)$ and $A \vee B := (A \rightarrow B) \rightarrow B$, the axioms for $\wedge$ and $\vee$ in Section 2.6 become formulas containing no connectives other than $\rightarrow$ and $\neg$ and are deducible (using *MP*) from the three axiom schemes given above. So, by expressing $\wedge$ and $\vee$ in terms of $\rightarrow$ and $\neg$, formulations such as $\mathscr{P}$ are obtained, in which the number of axioms is small.

In their *Grundlagen der Mathematik* (1934) D. Hilbert (1862-1943) and P. Bernays (1888-1977) presented the following axiom system $\mathscr{P}_H$ for the classical propositional calculus. This system contains axioms for each of the connectives $\rightarrow$, $\wedge$, $\vee$ and $\neg$.

$$\left.\begin{array}{l} A \to (B \to A) \\ (A \to (B \to C)) \to ((A \to B) \to (A \to C)) \end{array}\right\} \to$$

$$\left.\begin{array}{l} A \wedge B \to A \\ A \wedge B \to B \\ A \to (B \to A \wedge B) \end{array}\right\} \wedge$$

$$\left.\begin{array}{l} A \to A \vee B \\ B \to A \vee B \\ (A \to C) \to ((B \to C) \to (A \vee B \to C)) \end{array}\right\} \vee$$

$$(\neg A \to \neg B) \to (B \to A) \qquad \neg$$

Formulations of *intuitionistic propositional logic* can be obtained by replacing the negation axiom of $\mathscr{P}_H$ by suitable different axioms, for instance, by $(A \to \neg A) \to \neg A$ and $\neg A \to (A \to B)$; see Chapter 8.

For more historical details the reader is referred to section 29 of A. Church [4]. *Introduction to Mathematical Logic*.

Scientific Explanation, Inductive Logic Some, but not all, scientific explanations are deductive arguments the premises of which consist of general laws and particular facts. A trivial example is the following explanation.

If someone drops his pencil, it falls to the ground. (L_1)
I drop my pencil. (C_1)
Therefore, my pencil falls to the ground. (E)

L_1 is a *general law*, i.e., a universal statement expressing that each time some condition P is satisfied, then without exception some condition Q will occur. C_1 is a *particular fact*. And E is the *explanandum*, the statement which has to be explained.

Explanations of this kind are called *deductive-nomological explanations*. (The Greek word 'nomos' means 'law'.) Their general form is

$$\left.\begin{array}{l} L_1, L_2, \ldots, L_r \quad \text{(universal laws)} \\ \underline{C_1, C_2, \ldots, C_k \quad \text{(particular facts)}} \\ E \end{array}\right\} \begin{array}{l} \text{Explanans} \\ \\ \text{Explanandum} \end{array}$$

In a deductive-nomological explanation the explanandum follows logically or deductively from the explanans.

Probabilistic explanations are different in that i) the laws are in terms of relative frequencies, and ii) the explanandum does not logically follow from the explanans, but can only be expected with a certain degree of probability, called *inductive* or *logical probability*. The following is an example of a probabilistic explanation.

Example 2.29. The statistical probability of catching the measles, when exposed to them, is $\frac{3}{4}$. The statistical probability of catching pneumonia, when exposed to it, is $\frac{1}{7}$. Jim was exposed to the measles and to pneumonia. Therefore, the inductive or logical probability that Jim catches both the measles and pneumonia is $\frac{3}{4} \times \frac{1}{7} = \frac{3}{28}$.

The main question in *inductive logic* is how to determine the inductive probability for the explanandum, given the statistical probabilities in the explanans. This

2.10 Paradoxes; Historical and Philosophical Remarks   111

problem is in part still unsettled. Note that inductive or logical probability is a relation between statements, while statistical probability is a relation between (kinds of) events.

References for further reading: 1. Hempel, R., *Philosophy of Natural Science*; 2. Carnap, R., *Logical foundations of probability*; 3. Carnap, R. and Jeffrey, R., *Studies in inductive logic and probability*; 4. Jeffrey, R., *The logic of decision*; 5. Swinburne, R., *An introduction to confirmation theory*.

Syntax – Semantics The *syntax* of a language is concerned only with the *form* of the expressions, while the *semantics* is concerned with their *meaning*.

So, the rules according to which the well-formed expressions of a language are formed and the rules belonging to a logical proof system, such as Modus Ponens, belong to the syntax of the language in question. These rules can be manipulated mechanically; a machine can be instructed to apply the rule Modus Ponens and to write down a B once it sees both A and $A \to B$, while the machine does not know the meanings of A, B and $\to$. The notions of (logical) proof and deduction, as well as the notions of (logical) provability and deducibility, clearly belong to the syntax: they are only concerned with the form of the formulas involved.

On the other hand, truth tables belong to the semantics, because they say how the truth value (meaning) of a composite proposition is related to the truth values (meanings) of the components from which it is built. The notions of validity and valid consequence also belong to the semantics: they are concerned with the meaning of the formulas in question.

Leibniz (1646-1716) We will here pay attention to only a few aspects of Leibniz. For more information the reader is referred to Kneale [16] and to Mates [20], Chapter 12. What follows in this subsection is based on these works.

One of Leibniz' ideals was to develop a *lingua philosophica* or *characteristica universalis*, an artificial language that in its structure would mirror the structure of thought and that would not be affected with ambiguity and vagueness like ordinary language. His idea was that in such a language the linguistic expressions would be pictures, at it were, of the thoughts they represent, such that signs of complex thoughts are always built up in a unique way out of the signs for their composing parts. Leibniz believed that such a language would greatly facilitate thinking and communication and that it would permit the development of mechanical rules for deciding all questions of consistency or consequence. The language, when it is perfected, should be such that 'men of good will desiring to settle a controversy on any subject whatsoever will take their pens in their hands and say *Calculemus* (let us calculate)'. If we restrict ourselves to propositional logic, Leibniz' ideal has been realized: classical propositional logic is decidable; see Section 2.9. However, A. Church and A. Turing proved in 1936 that extending the propositional language with the quantifiers 'for all' ($\forall$) and 'for some' ($\exists$), the resulting predicate logic is undecidable, i.e., there is no mechanical method to test logical consequence (in predicate logic), let alone philosophical truth.

Leibniz also developed a theory of identity, basing it on *Leibniz' Law*: eadem sunt quorum unum potest substitui alteri salva veritate – those things are the same if one may be substituted for the other with preservation of truth. Leibniz' Law is also called the *substitutivity of identity* or the *principle of extensionality* and it is frequently formulated as follows.

$$a = b \to (\ldots a \ldots \rightleftarrows \ldots b \ldots)$$

where $\ldots a \ldots$ is a context containing occurrences of the name a, and $\ldots b \ldots$ is the same context in which one or more occurrences of a has been replaced by b; if a is b, then what holds for a holds for b and vice versa. In the propositional calculus we have a similar principle of the *substitutivity of material equivalents*:

$$(A \rightleftarrows B) \to (\ldots A \ldots \rightleftarrows \ldots B \ldots).$$

Leibniz made a distinction between *truths of reason* and *truths of fact*. The truths of reason are those which could not possibly be false, i.e., – in modern terminology – which are *necessarily true* . Examples of such truths are: no bachelor is married, $2 + 2 = 4$, living creatures cannot survive fire, and so on. Truths of fact are called contingent truths nowadays; for example, unicorns do not exist, Amsterdam is the capital of the Netherlands, and so on. Leibniz spoke of the truths of reason as *true in all possible worlds*. He imagined that there are many *possible worlds* and that our actual world is one of them. '$2 + 2 = 4$' is true not only in this world, but also in any other world. 'Amsterdam is the capital of the Netherlands' is true in this world, but we can think of another world in which this proposition is false. In 1963, S. Kripke extended the notion of possible world with an *accessibility relation* between possible worlds, which enabled him to give adequate semantics for the different modal logics, as we will see in Chapter 6. The idea is that some worlds are accessible from the given world, and some are not. For instance, one could postulate (and one usually does) that worlds with different mathematical laws are not accessible from the present world.

2.11 Solutions

Solution 2.1. i) $P_1 \wedge P_2 \to \neg P_3$; ii) $P_1 \wedge (P_2 \to \neg P_3)$; iii) $P_1 \vee (P_2 \to P_3)$; iv) $(P_2 \vee P_1) \to P_3$; v) $P_1 \to (P_2 \to \neg P_3)$

Solution 2.2. i) If it is the case that if John works hard then he goes to school, then John is not wise. ii) John does not work hard or John is wise. iii) It is not the case that John works hard or that John is wise; in other words, John does not work hard and John is not wise. iv) John does not go to school and John is wise. v) It is not the case that both John goes to school and John is wise; in other words, John does not go to school or John is not wise.

Solution 2.3. 1. P_1 or $\forall x[P(x)]$; 2. P_2 or $\forall x[\neg P(x)]$; 3. $\neg P_1$ or $\neg \forall x[P(x)]$.

Solution 2.4. Only the expressions P_1, $\neg P_8$, $P_1 \wedge \neg P_8$, and $(P_1 \wedge P_2) \to \neg P_3$ are formulas of propositional logic. All other expressions contain symbols which do not occur in the alphabet of propositional logic.

2.11 Solutions

Solution 2.5. Let Φ be the property defined by $\Phi(n) := 1+2+\ldots n = \frac{1}{2}n(n+1)$.
1. 0 has the property Φ, since $0 = \frac{1}{2}0(0+1)$.
2. Suppose n has the property Φ, i.e., $1+2+\ldots n = \frac{1}{2}n(n+1)$ (induction hypothesis). Then we have to show that $n+1$ also has the property Φ, i.e., $1+2+\ldots n + (n+1) = \frac{1}{2}(n+1)((n+1)+1)$.
Proof: According to the induction hypothesis, $1+2+\ldots n+(n+1) = \frac{1}{2}n(n+1) + (n+1) = (\frac{1}{2}n+1)(n+1) = \frac{1}{2}(n+1)(n+2)$.

Solution 2.6. Atomic formulas have no or zero parentheses, so as many left parentheses as right parentheses.
Assume that A and B have as many left parentheses as right parentheses (induction hypothesis). Then evidently the formulas $(A \rightleftarrows B)$, $(A \rightarrow B)$, $(A \wedge B)$, $(A \vee B)$ and $(\neg A)$ also have as many left parentheses as right parentheses.

Solution 2.7. We restrict ourselves to showing that $\neg(A \wedge B)$ has the same truth table as $\neg A \vee \neg B$. Although the formulas A and B may have been composed of many atomic formulas $P_1, \ldots, P_n$ and hence their truth tables may consist of many lines, 2^n, in the end there are at most 4 possible different combinations of 1 (true) and 0 (false) for A and B. Hence, it suffices to restrict ourselves to these maximally 4 possible different combinations:

A	B	$A \wedge B$	$\neg(A \wedge B)$	$\neg A$	$\neg B$	$\neg A \vee \neg B$
1	1	1	0	0	0	0
1	0	0	1	0	1	1
0	1	0	1	1	0	1
0	0	0	1	1	1	1

Solution 2.8. Below are the truth tables for the formulas from exercise 2.1 and 2.2.

			— 2.1 —					— 2.2 —				
P_1	P_2	P_3	i	ii	iii	iv	v	i	ii	iii	iv	v
1	1	1	0	0	1	1	0	0	1	0	0	0
1	1	0	1	1	1	0	1	1	0	0	0	1
1	0	1	1	1	1	1	1	1	1	0	1	1
1	0	0	1	1	1	0	1	1	0	0	0	1
0	1	1	1	0	1	1	1	0	1	0	0	0
0	1	0	1	0	0	0	1	1	1	1	0	1
0	0	1	1	0	1	1	1	0	1	0	1	1
0	0	0	1	0	1	1	1	1	1	1	0	1

Solution 2.9. $A \vee \neg A$ has the value 1 and $A \wedge \neg A$ has the value 0 in all lines of the truth table. Hence, in each line of the truth table
a) $(A \vee \neg A) \rightarrow B$ is 0 iff B is 0,
b) $(A \vee \neg A) \wedge B$ is 0 iff B is 0, and
c) $(A \wedge \neg A) \vee B$ is 0 iff B is 0.
Therefore $(A \vee \neg A) \rightarrow B$, $(A \vee \neg A) \wedge B$ and $(A \wedge \neg A) \vee B$ have the same truth table as B.

Solution 2.10.

A	B	$A \to B$	$(A \to B) \to B$	$B \to A$	$(B \to A) \to A$	$A \lor B$
1	1	$1 \to 1 = 1$	$1 \to 1 = 1$	1	$1 \to 1 = 1$	1
1	0	$1 \to 0 = 0$	$0 \to 0 = 1$	1	$1 \to 1 = 1$	1
0	1	$0 \to 1 = 1$	$1 \to 1 = 1$	0	$0 \to 0 = 1$	1
0	0	$0 \to 0 = 1$	$1 \to 0 = 0$	1	$1 \to 0 = 0$	0

Alternatively, one might argue as follows: $(A \to B) \to B$ is 0 iff ($A \to B$ is 1 and B is 0) iff (A is 0 and B is 0) iff $A \lor B$ is 0. Similarly for $(B \to A) \to A$.

Solution 2.11. We restrict ourselves to c) $(P \to Q) \to (\neg Q \to \neg P)$. Suppose in some line of its truth table this formula has the value 0. Then in that line $P \to Q$ is 1 and $\neg Q \to \neg P$ is 0. Hence, $P \to Q$ is 1, $\neg Q$ is 1 and $\neg P$ is 0 in that same line. So, $P \to Q$ is 1, Q is 0 and P is 1 in that same line. Then, either P is 0, Q is 0 and P is 1 in that line, or Q is 1, Q is 0 and P is 1 in that line. Both are impossible, so the original formula cannot have the value 0 in some line of its truth table.

Solution 2.12. a) In order that $P \lor Q \to P \land Q$ is 0 in some line, $P \lor Q$ must be 1 and $P \land Q$ 0 in that same line. So, at least one of P, Q must be 0. By taking the value of the other formula 1, one achieves that $P \lor Q$ is 1, while $P \land Q$ is 0:

P	Q	$P \lor Q$	$P \land Q$
1	0	1	0
0	1	1	0

b) is treated similarly.

Solution 2.13. 1 B, 2 A, 3 B, 4 C, 5 B, 6 C, 7 C, 8 C, 9 C, 10 C.

Solution 2.14. Each formula A built by means of connectives from only one atomic formula P must have one of the following four truth tables.

P	A
1	1 1 0 0
0	1 0 1 0

These four truth tables are the tables of $P \to P$, P, $\neg P$ and $P \land \neg P$, respectively.

Solution 2.15. Straightforward

Solution 2.16. * Let G be a group. If G can be ordered, then clearly every subgroup of G, generated by finitely many elements of G, can be ordered. Conversely, suppose every such subgroup of G can be ordered. $(*)$
Now, consider the propositional language built from atomic formulas $P_{a,b}$, where a,b are elements of G. Let Γ be the following set of formulas in this language.
$P_{a,a}$ for every element a in G.
$P_{a,b} \lor P_{b,a}$ for all a,b in G.
$P_{a,b} \to \neg P_{b,a}$ for all a,b in G with $a \neq b$.
$P_{a,b} \land P_{b,c} \to P_{a,c}$ for all a,b,c in G.
$P_{a,b} \to P_{ac,bc} \land P_{ca,cb}$ for all a,b,c in G.
Proposition 1: Every finite subset of Γ has a model.

2.11 Solutions

Proof: Let Γ' be a finite subset of Γ. In Γ' there occur only finitely many elements of G. Let G' be the subgroup of G, generated by these finitely many elements. By the hypothesis $(*)$, G' can be ordered by some relation $\leq$. Now, let $u(P_{a,b}) = 1$ if $a \leq b$, and $u(P_{a,b}) = 0$ if $a > b$. Then u is a model of Γ'.

By the compactness theorem it follows from Proposition 1 that Γ has a model, say v. Now, let $a \leq b := v(P_{a,b}) = 1$. Since v is a model of Γ, $\leq$ is an ordering of G.

Solution 2.17. * If a graph on V is k-chromatic, then clearly every finite subgraph of it is k-chromatic. Conversely, suppose R is a graph on V such that every finite sub-graph of R is k-chromatic. $\qquad(*)$

Now, consider the propositional language built from atomic formulas $P_{i,x}$, where $i \in \{1,\ldots,k\}$ and $x \in V$. And let Γ be the following set of formulas.

$P_{i,x} \to \neg P_{j,x}$ for all $i, j \leq k$ with $i \neq j$ and all $x \in V$.
$P_{1,x} \vee \ldots \vee P_{k,x}$ for all $x \in V$.
$P_{i,x} \to \neg P_{i,y}$ for all $i \leq k$ and all $x, y \in V$ such that xRy.

Proposition 1: Every finite subset of Γ has a model.

Proof: Let Γ' be a finite subset of Γ. In Γ' there occur only finitely many elements of V. Let R' be the sub-graph of R obtained by restricting R to the set V' of these finitely many elements. By hypothesis $(*)$, R' is k-chromatic, i.e., there is a partition of V' into k disjoint sets $W_1, \ldots, W_k$, such that two elements of V' connected by R' do not belong to the same W_i. Now, let $u(P_{i,x}) = 1$ if $x \in W_i$, and $u(P_{i,x}) = 0$ if $x \notin W_i$. Then u is a model of Γ'.

By the compactness theorem it follows from proposition 1 that Γ has a model, say v. Now, let $V_i := \{x \in V \mid v(P_{i,x}) = 1\}$ for $i = 1, \ldots, k$. Then $V_1, \ldots, V_k$ is a partition of V such that two elements of V, connected by R, do not belong to the same V_i. In other words, R is k-chromatic.

Solution 2.18. * Let B and G be sets. $R \subseteq B \times G$, such that (i) for all $x \in B$, $R_{\{x\}}$ is finite, and (ii) for every finite subset $B' \subseteq B$, $R_{B'}$ has at least as many elements as B'. Consider a propositional language with as atomic formulas all expressions $H_{x,y}$ with $x \in B$ and $y \in G$. Let Γ contain the following formulas:

$H_{x,y_1} \vee \ldots \vee H_{x,y_n}$ for any $x \in B$, where $R_{\{x\}} = \{y_1, \ldots, y_n\}$.
$\neg(H_{x,y_1} \wedge H_{x,y_2})$ for any $x \in B$, $y_1, y_2 \in G$ with $y_1 \neq y_2$.
$\neg(H_{x_1,y} \wedge H_{x_2,y})$ for any $x_1, x_2 \in B$, $y \in G$ with $x_1 \neq x_2$.

If u is a model of Γ, then $f : B \to G$, defined by $f(x) = y$ if $u(H_{x,y}) = 1$, is an injection from B to G.

In order to show that Γ has a model, by the compactness theorem it suffices to show that each finite subset Γ' of Γ has a model. So, let Γ' be a finite subset of Γ. Let $B' := \{x \in B \mid H_{x,y} \text{ occurs in } \Gamma' \text{ for some } y \in G\}$, and $G' := \{y \in G \mid H_{x,y} \text{ occurs in } \Gamma' \text{ for some } x \in B\}$. Since B' and G' are finite, there is an injection $f' : B' \to G'$, such that if $f'(x) = y$, then $R(x,y)$. Define u' as follows: $u'(H_{x,y}) = 1$ iff $f'(x) = y$. Then u' is a model of Γ'.

Solution 2.19.

P_1	P_2	P_3	$P_1 \to P_2$	$\neg P_2 \vee P_3$	$P_1 \to P_3$	$P_3 \to P_1$
1	1	1	1	1	1	1
1	1	0	1	0	0	1
1	0	1	0	1	1	1
1	0	0	0	1	0	1
0	1	1	1	1	1	0
0	1	0	1	0	1	1
0	0	1	1	1	1	0
0	0	0	1	1	1	1

Let P_1 stand for: the government raises taxes for its citizens; P_2 for: the unemployment grows; and P_3 for: the income of the state decreases. Then the argument has the following structure: $P_1 \to P_2, \neg P_2 \vee P_3 \models P_1 \to P_3$. Notice that $\neg P_2 \vee P_3$ has the same truth table as $P_2 \to P_3$. One easily checks that in each line of the truth table starting with P_1, P_2, P_3 in which both premises are 1, also the conclusion is 1.

There are four lines in which all premises are true: line 1, 5, 7 and 8. In each of these lines the conclusion $P_1 \to P_3$ is 1 too. Therefore, $P_1 \to P_2, \neg P_2 \vee P_3 \models P_1 \to P_3$.

Solution 2.20. Let P_1 stand for: Europe may form a monetary union; P_2 for: Europe is a political union; and P_3 for: all European countries are member of the union. Then the argument has the following structure: $P_1 \to P_2, \neg P_2 \vee P_3 \models P_3 \to P_1$, which is false, because there is at least one line in the truth table in which all premises are 1, while the putative conclusion $P_3 \to P_1$ is 0; see lines 5 and 7 in the table of solution 2.19. Therefore, $P_1 \to P_2, \neg P_2 \vee P_3 \not\models P_3 \to P_1$.

Solution 2.21. c) There is no line in the truth table in which both A and $\neg A$ are 1, so there is no line in the truth table in which both A and $\neg A$ are 1 and B is 0, i.e., $A, \neg A \models B$.

Solution 2.22. Let W stand for: John wins the lottery; J for: John makes a journey; and S for: John succeeds for logic. Then the structure of the argument is the following one: $\neg W \vee J, \neg J \to \neg S, W \vee S \models J$. Notice that the first premiss has the same truth table as $W \to J$ and that the second premiss has the same truth table as $S \to J$. Hence, the structure of the argument is equivalent to $W \to J, S \to J, W \vee S \models J$, which clearly is valid. Checking the truth table will confirm this.

Solution 2.23. Let T stand for: Turkey joins the EU; L for: the EU becomes larger; and S for: the EU becomes stronger. Then the argument has the following structure: $T \to L, \neg(S \wedge \neg L) \models \neg T \vee S$. Notice that $\neg(S \wedge \neg L)$ has the same truth table as $S \to L$ and that the conclusion $\neg T \vee S$ has the same truth table as $T \to S$. Hence the structure of the argument is equivalent to $T \to L, S \to L \models T \to S$, which clearly does not hold: if T and L are 1 and S is 0, then the premises are both 1, while the conclusion is 0. Making a truth table will confirm this.

Solution 2.24. 1) Assume $\models A \rightleftarrows (A \to B)$. To show: $\models A$ and $\models B$. So, suppose A were 0 in some line of its truth table. Then $A \rightleftarrows (A \to B)$ would be $0 \rightleftarrows (0 \to 0/1) = (0 \rightleftarrows 1) = 0$ in that line, contradicting the assumption. Therefore, $\models A$. In a similar

2.11 Solutions

way $\models B$ can be shown.

2) Assume $A \models \neg A$. To show: $\models \neg A$. So, suppose $\neg A$ were 0 in some line of its truth table, i.e. A were 1 in that line. Then, by assumption, also $\neg A$ would be 1 in that same line. Contradiction. Therefore, $\models \neg A$.

3) Assume $A \rightarrow B \models A$. To show: $\models A$. So, suppose A were 0 in some line of its truth table. Then $A \rightarrow B$ would be 1 in that same line and hence, by assumption, A would be 1 in that same line. Contradiction. Therefore, $\models A$.

Solution 2.25. a) Counterexample: let $A = P$ (atomic) and $B = Q$ (atomic). Then not $\models P \rightarrow Q$, but not $\models P$ and not $\models \neg Q$.
b) Proof: $\neg(A \rightarrow B)$ has the same truth table as $A \wedge \neg B$. So, if $\models \neg(A \rightarrow B)$, then $\models A \wedge \neg B$. Hence, by Theorem 2.14, $\models A$ and $\models \neg B$.
c) Counterexample: let $A = P$ (atomic) and $B = Q$ (atomic). Then not $\models P \wedge Q$, but not $\models \neg P$ and not $\models \neg Q$.
d) Counterexample: let $A = P$ (atomic) and $B = \neg P$. Then $\models \neg(P \wedge \neg P)$, but not $\models \neg P$ and not $\models \neg\neg P$. Notice that $A = P$ and $B = Q$ with P, Q atomic, is not a counterexample, because $\models \neg(P \wedge Q)$ does not hold.
e) Counterexample: $A = P$ (atomic) and $B = Q$ (atomic). Then not $\models P \vee Q$, but not $\models \neg P$ and not $\models \neg Q$.
f) Proof: $\neg(A \vee B)$ has the same truth table as $\neg A \wedge \neg B$. So, if $\models \neg(A \vee B)$, then $\models \neg A \wedge \neg B$. Hence, by Theorem 2.14, $\models \neg A$ and $\models \neg B$.

Solution 2.26. (a1) and (a2) For $i = 1, \ldots, n$, $A_1, \ldots, A_i, \ldots, A_n \models A_i$, since for every line in the truth table, if all of $A_1, \ldots, A_i, \ldots, A_n$ are 1, then also A_i is 1.
(b1) Assume $A_1, A_2, A_3 \models B_1$ and $A_1, A_2, A_3 \models B_2$ and $B_1, B_2 \models C$, i.e., for every line in the truth table, if all of A_1, A_2, A_3 are 1, then also B_1 is 1 and B_2 is 1; and for every line in the truth table, if all of B_1, B_2 are 1, then also C is 1. Therefore, for every line in the truth table, if all of A_1, A_2, A_3 are 1, then also C is 1, i.e., $A_1, A_2, A_3 \models C$.
(b2) Similarly.

Solution 2.27. 1) Assume $A \models B$ and $A \models \neg B$ and suppose that in some line of the truth table $\neg A$ is 0, i.e., A is 1. Then, because of $A \models B$, B is 1 in that line and, because of $A \models \neg B$, $\neg B$ is 1 (and hence B is 0) in that line of the truth table. Contradiction. So, there is no line in which A is 1. Therefore $\models \neg A$.
2) Assume $A \models C$ and $B \models C$ and, in order to show that $A \vee B \models C$, suppose $A \vee B$ is 1 in some line of the truth table. Then A is 1 or B is 1 in that line. In the first case it follows from $A \models C$ and in the second case it follows from $B \models C$ that C is 1 in that line.

Solution 2.28. (a) Right. There is no line in the truth table in which $A \rightarrow B \vee C$ is 1 and $(A \rightarrow B) \vee (A \rightarrow C)$ is 0.
(b) Wrong. Counterexample: for P, Q atomic, $\models (P \rightarrow Q) \vee (P \rightarrow \neg Q)$, but not $\models P \rightarrow Q$ and not $\models P \rightarrow \neg Q$. (See also Theorem 2.13 (b))
(c) Assume $A \models B$ (1). To show: $B \rightarrow C \models A \rightarrow C$. So, suppose $B \rightarrow C$ is 1 in some line of the truth table (2). Then we have to show that also $A \rightarrow C$ is 1 in that line. So, suppose A is 1 in that same line (3). Then, because of (1), B is 1 in that line and hence, because of (2), C is 1 in that line, which had to be proved.

Solution 2.29. Assume $T \wedge A \wedge B \models P$. To show: $\neg P \models \neg T \vee \neg A \vee \neg B$. So, suppose $\neg P$ is 1 in some line of the truth table. Then P is 0 in that line and hence, by assumption, $T \wedge A \wedge B$ is 0 in that line. Then $\neg(T \wedge A \wedge B)$ is 1 and hence $\neg T \vee \neg A \vee \neg B$ is 1 in the given line. Therefore, $\neg P \models \neg T \vee \neg A \vee \neg B$.

Solution 2.30. Proof of a): Assume $A \models B$. To show: $\neg B \models \neg A$. So, suppose $\neg B$ is 1 in an arbitrary line of the truth table. Then B is 0 in that line and hence, by assumption, A is 0 in that line. Therefore $\neg A$ is 1 in that line, which had to be shown.

Proof of b): Assume $A \models B$ and $A, B \models C$. To show $A \models C$. So, suppose A is 1 in an arbitrary line of the truth table. Then, because of $A \models B$, A and B are 1 in that line and hence, by $A, B \models C$, C is 1 in that line, which had to be shown.

Proof of c): Assume $A \vee B \models A \wedge B$. And suppose A and B have different values in some line of the truth table ($1-0$ or $0-1$ respectively). Then $A \vee B$ is 1 in that line, while $A \wedge B$ is 0 in that line, contradicting $A \vee B \models A \wedge B$. Therefore A and B have the same truth table.

An alternative proof: Suppose $A \vee B \models A \wedge B$. This means that the formulas A and B are such that in the standard truth table for $A \vee B$ and for $A \wedge B$ line 2 (A is 1, B is 0) and line 3 (A is 0 and B is 1) do not occur. So, only line 1 (A is 1 and B is 1) and line 4 (A is 0 and B is 0) may occur. Hence, A and B have the same truth table.

Solution 2.31.

B	J	S	Brown's testimony $\neg J \wedge S$	Jones' testimony $\neg B \rightarrow \neg S$	Smith's testimony $S \wedge (\neg B \vee \neg J)$
1	1	1	0	1	0
1	1	0	0	1	0
1	0	1	1	1	1
1	0	0	0	1	0
0	1	1	0	0	1
0	**1**	**0**	**0**	**1**	**0**
0	0	1	1	0	1
0	0	0	0	1	0

a) Yes, for the three testimonies are all true in the third line of the truth table.
b) $\neg J \wedge S \models S \wedge (\neg B \vee \neg J)$, i.e., Smith's testimony follows from that of Brown.
c) The assumption that everybody is innocent means in terms of the truth tables that the first line applies. Since in this line Brown's and Smith's testimonies are false, Brown and Smith commit perjury in this case.
d) There is only one line (namely the third one) in which everyone's testimony is true. In this line B and S are 1 and J is 0. So, in this case Brown and Smith are innocent and Jones is guilty.
e) Line 6 in the truth table is the only line in which the innocent tells the truth and the guilty tells lies. From line 6 we read off that in this case Brown and Smith are guilty and tell lies and that Jones is innocent and tells the truth.

Solution 2.32. Let P, Q, R be the statement 'Pro wins', 'Quick wins, 'the Runners win', respectively.

2.11 Solutions

P Q R	Trainer of Pro $R \to \neg Q$	Trainer of Quick $Q \vee R$	Trainer of Runners R
1 1 1	0	1	1
1 1 0	1	1	0
1 0 1	1	1	1
1 0 0	**1**	**0**	**0**
0 1 1	0	1	1
0 1 0	1	1	0
0 0 1	1	1	1
0 0 0	1	0	0

a) The assumption that everyone's statement is true means in terms of the truth tables that the third or seventh line applies. Assuming there is at most one winner, the third line does not apply. So, the Runners win.

b) If only the trainer of the winning club makes a true statement, Pro wins the tournament, as can be seen from the fourth line.

Solution 2.33. (a) $\neg P \wedge Q \wedge R$ (see the outline of the proof of Theorem 2.16).
(b) $(P \wedge Q \wedge R) \vee (\neg P \wedge Q \wedge R) \vee (\neg P \wedge \neg Q \wedge R)$.
(c) $P \wedge \neg P$.
(d) $\neg((P \wedge Q \wedge R) \vee (\neg P \wedge Q \wedge \neg R))$. Note: the table of $(P \wedge Q \wedge R) \vee (\neg P \wedge Q \wedge \neg R)$ corresponds with the negation of column (d).

Solution 2.34. $\neg A$ has the same truth table as $\neg A \vee \neg A$ and hence as $A \downarrow A$.
$A \vee B$ has the same truth table as $\neg(\neg A) \vee \neg(\neg B)$, hence as $\neg A \downarrow \neg B$ and therefore as $(A \downarrow A) \downarrow (B \downarrow B)$.
$A \wedge B$ has the same truth table as $\neg(\neg A \vee \neg B)$, hence as $\neg(A \downarrow B)$ and therefore as $(A \downarrow B) \downarrow (A \downarrow B)$.

Solution 2.35. i) $\wedge$ can be expressed in terms of $\vee$ and $\neg$, for $A \wedge B$ has the same truth table as $\neg(\neg A \vee \neg B)$; similarly, $\vee$ can be expressed in terms of $\wedge$ and $\neg$, for $A \vee B$ has the same truth table as $\neg(\neg A \wedge \neg B)$.
ii) $\{\to, \neg\}$ is complete, for according to Theorem 2.16 $\{\wedge, \vee, \neg\}$ is complete and both $\wedge$ and $\vee$ can be expressed in terms of $\to$ and $\neg$: $A \wedge B$ has the same truth table as $\neg(A \to \neg B)$ and $A \vee B$ has the same truth table as $(A \to B) \to B$.
$\{\to, \neg\}$ is independent, for $\to$ cannot be expressed in terms of $\neg$; more precisely, $A \to B$ does not have the same truth table as A, $\neg A$, $\neg\neg A$, B, $\neg B$ or $\neg\neg B$; and $\neg$ cannot be expressed in terms of $\to$; for suppose A is 1, then $\neg A$ is 0 and one can show that any formula, built from A and $\to$ only, is 1 if A is 1.
iii) In a similar way one shows that $\{\wedge, \neg\}$ and $\{\vee, \neg\}$ are both complete and independent.

Solution 2.36. Suppose $|$ is a binary connective such that every truthfunctional connective of (1 or) 2 arguments can be expressed in it.   (*)
Then, in particular, there must be a formula A built from P and $|$ only, such that $\neg P$ has the same truth table as A (α). Now, if $1 \mid 1 = 1$, one can show that any formula, built from P and $|$ only, will have the value 1 if P is 1 (β). However, $\neg 1 = 0$ (γ). From

(α), (β) and (γ) it follows that $1 \mid 1 = 0$. In a similar way one shows that $0 \mid 0 = 1$. Consequently, the connective "|" must have one of the following four truth tables.

P	Q					
1	1	0	0	0	0
1	0	0	0	1	1
0	1	0	1	0	1
0	0	1	1	1	1

We will show next that the values of $1 \mid 0$ and $0 \mid 1$ should be the same, so that only the first and the fourth column remain and $\mid$ must be either $\uparrow$ or $\downarrow$.

If $1 \mid 0 \neq 0 \mid 1$, then one can show that any formula, built from P, Q and $\mid$ only, will get a different truth value if we interchange the P and the Q in it, giving P and Q the values 1 and 0 respectively (a). Under the assumption (*) there must be a formula B built from P, Q and $\mid$ only, such that $P \wedge Q$ has the same truth table as B (b). However, $1 \wedge 0 = 0 \wedge 1$ (c). From (a), (b) and (c) it follows that $1 \mid 0 = 0 \mid 1$.

Solution 2.37. i) $(P \to (Q \to P)) \wedge (P \to Q \vee P)$ has the same truth table as $(\neg P \vee (\neg Q \vee P)) \wedge (\neg P \vee (Q \vee P))$.

ii) The following formulas have the same truth table:

$(P \to \neg(Q \to P)) \wedge (P \to Q \wedge P)$ $(\neg P \vee \neg(\neg Q \vee P)) \wedge (\neg P \vee (Q \wedge P))$
$(\neg P \vee (\neg \neg Q \wedge \neg P)) \wedge (\neg P \vee (Q \wedge P))$ $(\neg P \vee (Q \wedge \neg P)) \wedge (\neg P \vee (Q \wedge P))$
$(\neg P \vee Q) \wedge (\neg P \vee \neg P) \wedge (\neg P \vee Q) \wedge (\neg P \vee P)$ $(\neg P \vee Q) \wedge \neg P$

iii) $(P \to \neg(Q \to P)) \vee (P \to Q \wedge P)$ has the same truth table as:
$((\neg P \vee Q) \wedge (\neg P \vee \neg P)) \vee ((\neg P \vee Q) \wedge (\neg P \vee P))$
$((\neg P \vee Q) \wedge \neg P) \vee (\neg P \vee Q)$
$((\neg P \vee Q) \vee (\neg P \vee Q)) \wedge ((\neg P \vee Q) \vee \neg P)$
$(\neg P \vee Q)$.

Solution 2.38. a) $R \to \neg K$, $K \vdash \neg R$. The following list of formulas is a deduction of $\neg R$ from the premisses $R \to \neg K$ and K:
1. K premiss
2. $K \to (R \to K)$ axiom 1
3. $R \to K$ MP applied to 1 and 2.
4. $(R \to K) \to ((R \to \neg K) \to \neg R)$ axiom 7
5. $(R \to \neg K) \to \neg R$ MP applied to 3 and 4.
6. $R \to \neg K$ premiss
7. $\neg R$ MP applied to 5 and 6.

b) Suppose $\neg K \to R$, $K \vdash \neg R$. Then by the soundness theorem $\neg K \to R$, $K \models \neg R$. Making the truth table shows that this is false. Therefore, $\neg K \to R$, $K \not\vdash \neg R$.

Solution 2.39. The following schemas are deductions of the last formula in the schema from the formulas mentioned as premisses.

(a) $\dfrac{\text{premiss} \quad \text{premiss}}{B} \text{MP}$
     $A \quad A \to B$

(b) $\text{premiss} \quad \dfrac{\text{premiss} \quad A \overset{3}{\to} (B \to A \wedge B)}{\dfrac{B \to A \wedge B}{A \wedge B} \text{MP}} \text{MP}$

2.11 Solutions

(c) $\dfrac{\text{premiss} \quad 4a}{A \wedge B \quad A \wedge B \to A}$ MP
$\qquad\qquad\qquad A$

(d) $\dfrac{\text{premiss} \quad 4b}{A \wedge B \quad A \wedge B \to B}$ MP
$\qquad\qquad\qquad B$

(e) $\dfrac{\text{premiss} \quad 5a}{A \quad A \to A \vee B}$ MP
$\qquad\qquad\quad A \vee B$

(f) $\dfrac{\text{premiss} \quad 5b}{B \quad B \to A \vee B}$ MP
$\qquad\qquad\quad A \vee B$

(g) $\dfrac{\text{premiss} \quad 8}{\neg\neg A \quad \neg\neg A \to A}$ MP
$\qquad\qquad\quad A$

(h) $\dfrac{\text{premiss}\ B \to A \quad \dfrac{\text{premiss}\ A \to B \quad 9\ (A \to B) \to ((B \to A) \to (A \rightleftarrows B))}{(B \to A) \to (A \rightleftarrows B)} \text{MP}}{A \rightleftarrows B}$ MP

(i) $\dfrac{\text{premiss} \quad 10a}{A \rightleftarrows B \quad (A \rightleftarrows B) \to (A \to B)}$ MP
$\qquad\qquad\qquad A \to B$

(j) $\dfrac{\text{premiss} \quad 10b}{A \rightleftarrows B \quad (A \rightleftarrows B) \to (B \to A)}$ MP
$\qquad\qquad\qquad B \to A$

Solution 2.40. The following list of formulas is a deduction of B from A and $\neg A$:

1.	A	premiss
2.	$A \to (\neg B \to A)$	axiom 1
3.	$\neg B \to A$	from 1 and 2 by MP
4.	$\neg A$	premiss
5.	$\neg A \to (\neg B \to \neg A)$	axiom 1
6.	$\neg B \to \neg A$	from 4 and 5 by MP
7.	$(\neg B \to A) \to ((\neg B \to \neg A) \to \neg\neg B)$	axiom 7
8.	$(\neg B \to \neg A) \to \neg\neg B$	from 3 and 7 by MP
9.	$\neg\neg B$	from 6 and 8 by MP
10.	$\neg\neg B \to B$	axiom 8
11.	B	from 9 and 10 by MP.

Solution 2.41. (a) $P \vee Q \not\models P \wedge Q$, since there is a line in the truth table in which $P \vee Q$ is 1 and $P \wedge Q$ is 0. According to the soundness theorem: if $P \vee Q \vdash P \wedge Q$, then $P \vee Q \models P \wedge Q$. Therefore, $P \vee Q \not\vdash P \wedge Q$.
(b), (c) and (d) are shown in a similar way.

Solution 2.42. $P \to Q$, $P \vdash R \vee Q$. The following list of formulas is a deduction of $R \vee Q$ from P and $P \to Q$.
1. P premiss
2. $P \to Q$ premiss
3. Q MP applied to 1 and 2.
4. $Q \to R \vee Q$ axiom
5. $R \vee Q$ MP applied to 3 and 4.

Solution 2.43. $S \to H$, $\neg I \to \neg S \vdash^? I \to H$. Suppose this were true. Then because of the soundness theorem $S \to H$, $\neg I \to \neg S \models I \to H$. One easily checks from the truth table that this is not the case. Therefore, $S \to H$, $\neg I \to \neg S \not\vdash I \to H$.

Solution 2.44. We leave the proof of (i) and (ii) to the reader. (iii) If $u(A) = 0$ and $u(A \to B) = 0$, then only the first line of the table applies; so $u(B) = 0$.
(iv) In the sixth line of the table $u(A) = 1$ and $u(B) = 2$. Hence, $u(((A \to B) \to A) \to A) = ((1 \to 2) \to 1) \to 1 = (2 \to 1) \to 1 = 0 \to 1 = 1$. If Peirce's law were generated by the production method consisting of the two logical axioms for $\to$ only, then because of (i), (ii) and (iii) $((A \to B) \to A) \to A$ would have the value 0 in every line of the table.

Solution 2.45. We show that $A \wedge B \to C$, A, $B \vdash C$. Then by two applications of the deduction theorem it follows that $A \wedge B \to C \vdash A \to (B \to C)$.

$$\cfrac{\cfrac{\text{premiss}}{B} \quad \cfrac{\text{premiss} \overset{3}{A} \quad A \overset{3}{\to} (B \to A \wedge B)}{B \to A \wedge B}}{\cfrac{A \wedge B \quad A \wedge B \to C}{C}} \text{premiss}$$

Solution 2.46. We show that $A \to B$, A, $B \to C \vdash C$. Then by three applications of the deduction theorem it follows that $\vdash (A \to B) \to (A \to ((B \to C) \to C))$.

$$\cfrac{\cfrac{\text{premiss}}{A} \quad \cfrac{\text{premiss}}{A \to B}}{\cfrac{B \quad B \to C}{C}} \text{premiss}$$

Solution 2.47. Suppose $A_1, A_2 \vdash B$, i.e., there exists a deduction of B from A_1, A_2. We show that $A_1 \wedge A_2 \vdash B$. Then by one application of the deduction theorem it follows that $\vdash A_1 \wedge A_2 \to B$.

$$\cfrac{\text{premiss}}{A_1 \wedge A_2} \quad \overset{4a}{A_1 \wedge A_2 \to A_1} \quad \cfrac{\text{premiss}}{A_1 \wedge A_2} \quad \overset{4b}{A_1 \wedge A_2 \to A_2}$$

$$\begin{cases} A_1 & A_2 \\ \text{given deduction} & \\ \text{of } B \text{ from } A_1, A_2 & B \end{cases}$$

Solution 2.48. Suppose $\vdash (A_1 \wedge A_2) \wedge A_3 \to B$. Let (α) be a (logical) proof of $(A_1 \wedge A_2) \wedge A_3 \to B$. Then the following schema is a deduction of B from A_1, A_2, A_3. Note that we first deduce $(A_1 \wedge A_2) \wedge A_3$ from A_1, A_2, A_3 and next use $\vdash (A_1 \wedge A_2) \wedge A_3 \to B$ in order to deduce B.

$$\cfrac{\cfrac{\text{premiss}}{A_2} \quad \cfrac{\text{premiss } \overset{3}{A_1} \quad A_1 \overset{3}{\to} (A_2 \to A_1 \wedge A_2)}{A_2 \to A_1 \wedge A_2}}{\cfrac{\text{premiss}}{A_3} \quad \cfrac{A_1 \wedge A_2 \quad A_1 \wedge A_2 \overset{3}{\to} (A_3 \to A_1 \wedge A_2 \wedge A_3)}{A_3 \to A_1 \wedge A_2 \wedge A_3}}{\cfrac{(\alpha) \quad A_1 \wedge A_2 \wedge A_3}{B}}}$$

2.11 Solutions

Solution 2.49. Proof: Suppose $\vdash A \to C$ and $\vdash B \to C$. The following list of formulas is a deduction of C from $A \lor B$:
1. $A \to C$ deducible
2. $(A \to C) \to ((B \to C) \to (A \lor B \to C))$ axiom
3. $(B \to C) \to (A \lor B \to C)$ MP applied to 1 and 2.
4. $B \to C$ deducible
5. $A \lor B \to C$ MP applied to 3 and 4.
6. $A \lor B$ premiss
7. C MP applied to 5 and 6.

Solution 2.50. $A, B \to C \vdash A$ and $A \vdash A \lor C$; hence, $A, B \to C \vdash A \lor C$. (1)
$B, B \to C \vdash C$ and $C \vdash A \lor C$; hence, $B, B \to C \vdash A \lor C$. (2)
From (1) and (2), by $\lor$-elimination, $A \lor B, B \to C \vdash A \lor C$.

Solution 2.51. Suppose $A \vdash B$. Then, by Corollary 2.3, $A, \neg B \vdash B$.
But also $A, \neg B \vdash \neg B$. Hence, by $\neg$-introduction, $\neg B \vdash \neg A$.

Solution 2.52. $A \vdash A \lor \neg A$. Hence, by Exercise 2.51, $\neg(A \lor \neg A) \vdash \neg A$. (a)
$\neg A \vdash A \lor \neg A$. Hence, by Exercise 2.51, $\neg(A \lor \neg A) \vdash \neg(\neg A)$. (b)
From (a) and (b), by $\neg$-introduction, $\vdash \neg\neg(A \lor \neg A)$. Hence, by double negation elimination, $\vdash A \lor \neg A$.

Solution 2.53. By weak negation elimination $\quad \neg A, \neg B, A \vdash C$ (1)
and $\quad \neg A, \neg B, B \vdash C$. (2)
From (1) and (2), by $\lor$-elimination, $\neg A, \neg B, A \lor B \vdash C$. (I)
By weak negation elimination $\quad \neg A, \neg B, A \vdash \neg C$ (a)
and $\quad \neg A, \neg B, B \vdash \neg C$. (b)
From (a) and (b), by $\lor$-elimination, $\neg A, \neg B, A \lor B \vdash \neg C$. (II)
From (I) and (II), by $\neg$-introduction, $\neg A, \neg B \vdash \neg(A \lor B)$.

Solution 2.54. Suppose $A \vdash \neg A$. Because of $A \vdash A$, by $\neg$-introduction, $\vdash \neg A$.

Solution 2.55. Counterexample: $A = P$ (it rains) en $B = \neg P$ (it does not rain). $\models P \lor \neg P$ (it is always true that it rains or does not rain). Hence, because of the *completeness theorem*, $\vdash P \lor \neg P$.
But $\not\vdash P$. For suppose $\vdash P$; then, because of the *soundness theorem*, $\models P$ (it is always true that it rains), which is false. Therefore $\not\vdash P$.
Similarly, $\not\vdash \neg P$. For suppose $\vdash \neg P$; then, because of the *soundness theorem*, $\models \neg P$ (it is always true that it does not rain; it never rains), which is false. Therefore, $\not\vdash \neg P$.

Solution 2.56. Counterexample: $A = P$ (it rains). From the truth table we know that $\not\models P$ (it is not always true that it rains). So, because of the *soundness theorem* $\not\vdash P$.
However, $\not\vdash \neg P$. For suppose $\vdash \neg P$; then because of the *soundness theorem* $\models \neg P$ (it is always true that it does not rain; it never rains), which is false. Therefore, $\not\vdash \neg P$.

Solution 2.57. a) Proof: Suppose $\vdash A \to B$. The following list of formulas is a deduction of B from A:

A premiss
$A \to B$ deducible
B MP applied to 1 and 2.
b) Proof: Suppose $\vdash \neg A$. Then, because of the *soundness theorem*, $\models \neg A$.   (*)
We want to show that not $\vdash A$. So, suppose that $\vdash A$; then, because of the *soundness theorem*, $\models A$; but this contradicts (*). Therefore, not $\vdash A$.

Solution 2.58. We have seen in Exercise 2.40 that $A, \neg A \vdash B$. Hence, by the deduction theorem $\neg A \vdash A \to B$ (1). Also, by applying axiom 1, $B \to (A \to B)$, we know that $B \vdash A \to B$ (2). From (1) and (2) by $\vee$ elimination: $\neg A \vee B \vdash A \to B$.
a) $A \to B, \neg(\neg A \vee B), \neg A \vdash \neg A \vee B$ and $A \to B, \neg(\neg A \vee B), \neg A \vdash \neg(\neg A \vee B)$. Hence, by $\neg$-introduction, $A \to B, \neg(\neg A \vee B) \vdash \neg\neg A$.
b) By $\neg$-introduction, $A \to B, \neg(\neg A \vee B) \vdash \neg A$.

Solution 2.59. a) $A, B \vdash A \wedge B$, by using the axiom $A \to (B \to A \wedge B)$.
Proof of $A, \neg B \vdash \neg(A \wedge B)$: $A, \neg B, A \wedge B \vdash \neg B$ and $A, \neg B, A \wedge B \vdash B$ Hence, by reductio ad absurdum ($\neg$-introduction), $A, \neg B \vdash \neg(A \wedge B)$.
$\neg A, B \vdash A \vee B$ because $B \vdash A \vee B$.
$\neg A, \neg B \vdash \neg(A \vee B)$; see Exercise 2.53.
b) Suppose $\models E$. Then $E_1^* = E_2^* = E_3^* = E_4^* = E$.
Therefore $A, B \vdash E$ and $A, \neg B \vdash E$; hence by $\vee$-elimination: $A, B \vee \neg B \vdash E$.
Also $\neg A, B \vdash E$ and $\neg A, \neg B \vdash E$; hence by $\vee$-elimination: $\neg A, B \vee \neg B \vdash E$.
By Exercise 2.52, $\vdash B \vee \neg B$ and consequently, $A \vdash E$ and $\neg A \vdash E$.
Hence, by $\vee$-elimination: $A \vee \neg A \vdash E$ and therefore, $\vdash E$.

Solution 2.60. i)

$$
\begin{array}{c}
\dfrac{\neg A \vee \neg B \qquad \dfrac{{}^3[\neg A] \quad \dfrac{{}^1[A \wedge B]}{A}\wedge E}{\neg(A \wedge B)}\neg I \qquad \dfrac{{}^3[\neg B] \quad \dfrac{{}^2[A \wedge B]}{B}\wedge E}{\neg(A \wedge B)}\neg I}{\neg(A \wedge B)}\vee E
\end{array}
$$

(1), (2), (3)

ii)

$$
\dfrac{{}^3[\neg(\neg A \vee \neg B)] \quad \dfrac{[\neg A]^1}{\neg A \vee \neg B}\vee I}{\dfrac{\neg\neg A}{A}d\neg E}\neg I
$$

$$
\dfrac{{}^3[\neg(\neg A \vee \neg B)] \quad \dfrac{[\neg B]^2}{\neg A \vee \neg B}\vee I}{\dfrac{\neg\neg B}{B}d\neg E}\neg I
$$

$$
\dfrac{\neg(A \wedge B) \qquad \dfrac{A \wedge B}{}\wedge}{\dfrac{\neg\neg(\neg A \vee \neg B)}{\neg A \vee \neg B}d\neg E}\neg I
$$

(1), (2), (3)

2.11 Solutions

Solution 2.61. i)

$$MP \frac{A \quad A \xrightarrow{axiom} (B \to A)}{B \to A} \qquad \frac{A \quad (1) \frac{\frac{[A]^1}{B \to A} \to I}{A \to (B \to A)} \to I}{B \to A} \to E$$

ii)
$$\frac{{}^1[A] \quad A \to B}{\underset{(2)}{\overset{(1)}{}} \frac{\frac{B \quad [\neg B]^2}{\neg A} \neg I}{\neg B \to \neg A} \to I}$$

This deduction starts as follows:

To this corresponds: (1) $A, A \to B, \neg B \vdash B$, and
(2) $A, A \to B, \neg B \vdash \neg B$.

The deduction continues as follows:
$$\frac{{}^1[A] \quad A \to B}{B \quad \neg B}$$

$$(1) \frac{{}^1[A] \quad A \to B}{\frac{B \quad \neg B}{\neg A}}$$

To this corresponds $A \to B, \neg B \vdash \neg A$, which follows from (1) and (2) by Theorem 2.25, $\neg$-introduction. And from $A \to B, \neg B \vdash \neg A$ it follows by Theorem 2.25, $\to$-introduction, that $A \to B \vdash \neg B \to \neg A$.

Solution 2.62. (a) $A \to B, A \wedge \neg B \vdash B$, and $A \to B, A \wedge \neg B \vdash \neg B$. Hence, by $\neg$-introduction, $A \to B \vdash \neg(A \wedge \neg B)$.
(b) The following schema is a tableau-deduction of $\neg(A \wedge \neg B)$ from $A \to B$:

$T A \to B, F \neg(A \wedge \neg B)$
$T A \to B, T A \wedge \neg B$
$T A \to B, TA, T \neg B$
$T A \to B, TA, FB$
$FA, TA, FB \mid TB, TA, FB$

(c) Suppose $A \to B$ is 1 and $\neg(A \wedge \neg B)$ is 0. Then $A \wedge \neg B$ is 1. So, A is 1 and $\neg B$ is 1. Hence, $A \to B$ is 1, A is 1 and B is 0. Then (A is 0, A is 1 and B is 0) or (B is 1, A is 1 and B is 0). Contradiction. Therefore, $A \to B \models \neg(A \wedge \neg B)$.

Solution 2.63. (a) $A \to B, B \to C, A \vdash C$. Hence, by using the deduction theorem three times, $\vdash (A \to B) \to ((B \to C) \to (A \to C))$.
(b)
$F (A \to B) \to ((B \to C) \to (A \to C))$
$T A \to B, F (B \to C) \to (A \to C)$
$T A \to B, T B \to C, F A \to C$
$T A \to B, T B \to C, TA, FC$
$FA, T B \to C, TA, FC \mid TB, T B \to C, TA, FC$
$TB, FB, TA, FC \mid TB, TC, TA, FC$

(c) Suppose $(A \to B) \to ((B \to C) \to (A \to C))$ is 0. Then $A \to B$ is 1, $B \to C$ is 1, A is 1 and C is 0. So, (A is 0 and 1) or (B, $B \to C$ and A are 1 and C is 0). In the latter case, B is 1 and 0 or C is 1 and 0. Contradiction.

Solution 2.64. (a) $T\ A \to B,\ T\ \neg A \to B,\ FB$
           $FA,\ T\ \neg A \to B,\ FB\ |\ TB,\ T\ \neg A \to B,\ FB$
    $FA,\ F\neg A,\ FB\ |\ FA,\ TB,\ FB\ |\ TB,\ T\ \neg A \to B,\ FB$
    $FA,\ TA,\ FB$ Note that all three tableau branches are closed.
(f) $T\ A \to B \vee C,\ F\ (A \to B) \vee (A \to C)$
         $T\ A \to B \vee C,\ F\ A \to B,\ F\ A \to C$
         $T\ A \to B \vee C,\ TA,\ FB,\ TA,\ FC$
    $FA,\ TA,\ FB,\ TA,\ FC\ |\ T\ B \vee C,\ TA,\ FB,\ TA,\ FC$
         $TB,\ TA,\ FB,\ TA,\ FC\ |\ TC,\ TA,\ FB,\ TA,\ FC$
Note that all three tableau branches are closed.

Solution 2.65. a) $R \to W, \neg R \to B, \neg B \models^? W$.
b) $T\ R \to W,\ T\ \neg R \to B,\ T\neg B,\ FW$
            $T\ R \to W,\ T\ \neg R \to B,\ FB,\ FW$
     $FR,\ T\ \neg R \to B,\ FB,\ FW\ |\ TW,\ T\ \neg R \to B,\ FB,\ FW$
 $FR,\ F\ \neg R,\ FB,\ FW\ |\ FR,\ TB,\ FB,\ FW$
 $FR,\ TR,\ FB,\ FW$
Note that all tableau branches are closed and hence: $R \to W, \neg R \to B, \neg B \vdash' W$.

Solution 2.66. a) $R \to \neg W, W \to H, \neg R \models^? H$.
b) $T\ R \to \neg W,\ T\ W \to H,\ T\neg R,\ FH$
            $T\ R \to \neg W,\ T\ W \to H,\ FR,\ FH$
        $T\ R \to \neg W,\ FW,\ FR,\ FH\ |\ T\ R \to \neg W,\ TH,\ FR,\ FH$
    $FR,\ FW,\ FR,\ FH\ |\ T\neg W,\ FW,\ FR,\ FH$
            $FW,\ FW,\ FR,\ FH$
Note that the two most left tableau branches are completed but open, i.e., not closed, while the third tableau branch is closed. From any open and completed tableau branch one read off a counterexample: give R, W and H value 0, corresponding with the occurrence of FR, FW, FH in the completed open tableau branch.

R	W	H ‖ $R \to \neg W$	$W \to H$	$\neg R$ ‖ H
0	0	0 ‖ 1	1	1 ‖ 0

Therefore: $R \to \neg W, W \to H, \neg R \not\models H$.

Solution 2.67. (a) The following schema is a tableau-proof of $A \to (B \to A)$:
        $F\ A \to (B \to A)$
        $TA,\ F\ B \to A$
        $TA,\ TB,\ FA$
The other axioms are treated similarly.
(b) $A, A \to B \vdash' B$, for the following schema is a tableau-deduction of B from $A, A \to B$: $TA,\ T\ A \to B,\ FB$
        $TA,\ FA,\ FB\ |\ TA,\ TB,\ FB$.
On the other hand, suppose $\vdash' A$ and $\vdash' A \to B$. Then there is a tableau-proof starting with FA and there is a tableau-proof starting with

2.11 Solutions

$FA \to B$
TA, FB.
In order to show that $\vdash' B$ one has to construct a tableau-proof starting with FB.

Solution 2.68. A tableau-proof of $A \lor B$ should start with: $F\ A \lor B$
FA, FB
So, if there is a tableau-proof starting with FA or there is a tableau-proof starting with FB, then $\vdash' A \lor B$.
(b) A tableau-proof of $A \land B$ starts with: $F\ A \land B$
$FA \mid FB$
The left part is a tableau-proof of A and the right part is a tableau-proof of B.

Solution 2.69. (a) $\neg P, P \vdash Q$ (weak negation elimination). Hence, by the deduction theorem, $\neg P \vdash P \to Q$. Therefore, $\neg P, (P \to Q) \to P \vdash P$.
(b) $P, (P \to Q) \to P \vdash P$. So, by (a) and $\lor$-elimination, $P \lor \neg P, (P \to Q) \to P \vdash P$.
(c) By Exercise 2.52, $\vdash P \lor \neg P$. Therefore, from (b), $(P \to Q) \to P \vdash P$. So, by the deduction theorem, $\vdash ((P \to Q) \to P) \to P$.

Solution 2.70. The prisoner should reason as follows: If I wake up on Friday morning, what can I conclude. One of two things. Either they will hang me today, or else the judge was lying when he said I would hang one day this week. Suppose I somehow knew that the judge's statement that I would hang one day this week was true. Then I would know that I was to die today, and I would then know that his statement about not knowing the day of my death was false. But since I do not know that his first statement is true, I have no idea what is going to happen. Shortly before noon, they come to get him. 'Now I know', says the prisoner. 'Both statements were true'.

Let A stand for 'the prisoner will be hanged on Monday, Tuesday, Wednesday or Thursday' and B for 'the prisoner will be hanged on Friday' and let $\Box B$ stand for 'one knows B', then it is shown in Exercise 6.12 that $A \lor B, \Box \neg A \not\vdash \Box B$, while $\Box(A \lor B), \Box \neg A \vdash \Box B$ does hold. See also W.V. Quine, *On a supposed antinomy*, in *The Ways of Paradox*, and F. Norwood, *The prisoner's card game*, in *The Mathematical Intelligencer*, Vol. 4, Number 3, 1982.

Solution 2.71. This paradox is veridical if we conceive it as making clear that the promise A of the crocodile is inconsistent, more precisely $A \models B \rightleftarrows \neg B$, where B stands for 'the crocodile will eat the baby'.

Solution 2.72. Let A be the statement made by the traveller. Then the condition of the cannibals may be expressed by $(A \to B \land \neg R) \land (\neg A \to \neg B \land R)$, where B stands for 'the traveller will be boiled', and R for 'the traveller will be roasted'. A should be such that the truth table of the condition has only 0's and hence A should be of the form (a), (b), (c) or (d).

B	R	$(A \to B \land \neg R) \land (\neg A \to \neg B \land R)$	A	(a)	(b)	(c)	(d)
1	1	0	0/1	0	0	1	1
1	0	0	0	0	0	0	0
0	1	0	1	1	1	1	1
0	0	0	0/1	0	1	0	1

So, the traveller should make one of the following four statements: (a) $\neg B \wedge R$, (b) $\neg B$, (c) R, (d) $B \to R$, which has the same truth table as $\neg B \vee R$.

Solution 2.73. Similar to the barber paradox. (See Exercise 2.27.)

Solution 2.74. Similar to the barber paradox. (See Exercise 2.27.)

Solution 2.75. Let 'W' stand for 'Euathlus wins the case' and 'P' for 'Euathlus has to pay'. Then according to the contract, $W \to P$ (1) and $\neg W \to \neg P$ (2), in other words, $W \rightleftarrows P$. But according to the verdict, $W \to \neg P$ (3) and $\neg W \to P$ (4), in other words, $W \rightleftarrows \neg P$. Note that $W \rightleftarrows P$ and $W \rightleftarrows \neg P$ are inconsistent. In his argument Protagoras uses both (4) and (1), while Euathlus uses both (3) and (2) in his argument.

References

1. Austin, A.Keith, An elementary approach to NP-completeness; in *The American Mathematical Monthly*, vol. 90, 1983, pp. 389-399.
2. Beth, E.W., *The Foundations of Mathematics*. North-Holland Publ. Co., Amsterdam, 1959.
3. Bruijn, N.G. de, A survey of the project Automath. In: J.P. Seldin and R.J. Hindley, *Essays on Combinatory Logic, Lambda Calculus and Formalism*. Academic Press, 1980.
4. Church, A., *Introduction to Mathematical Logic*. Princeton University Press, 1956.
5. Dantzig, D. van, Siginifics and its relation to semiotics. *Library of the Tenth International Congress of Philosophy* (Amsterdam, August 11-18, 1948), Vol. 2, Philosophical Essays. Veen, Amsterdam, 1948, pp. 176-189.
6. Fitting, M., *Proof methods for modal and intuitionistic logics*. Springer. 1983.
7. Frege, G., *Begriffsschrift*. Halle, 1879.
8. Frege, G., *Begriffsschrift und andere Aufsätze*. I Angelelli (ed.), Olms, Hildesheim, 1964.
9. Gentzen, G., Untersuchungen über das logische Schliessen. *Mathematische Zeitschrift*, Vol. 39, 1934-1935, 176-210; 405-431.
10. Grice, P., *Studies in the Way of Words*. Harvard University Press, 1989.
11. Harding, S.C., Can theories be refuted? *Essays on the Duhem-Quine Thesis*. Reidel Publishing Co., Dordrecht, 1976.
12. Heijenoort, J. van, *From Frege to Gödel*. A source book in mathematical logic 1879-1931. Harvard University Press, Cambridge, Mass. 1967.
13. Jeffrey, R., *Formal Logic, its scope and limits*. McGraw-Hill, New York, 1967, 1981.
14. Kleene, S.C., *Mathematical Logic*. John Wiley and Sons Inc., New York, 1967.
15. Kleene, S.C., *Introduction to Metamathematics*. North Holland, 1962.
16. Kneale, W. and M., *The Development of Logic*. Clarendon Press, Oxford, 1962.
17. Kogel, E.A. de, Ophelders, W.M.J., A Tableaux-based Automated Theorem Prover. Appendix A in H.C.M. de Swart, *LOGIC*, Volume II, *Logic and Computer Science*, Verlag Peter Lang, Frankfurt am Main, 1994.
18. Kreisel, G. and J. Krivine, *Elements of Mathematical Logic*. North-Holland, Amsterdam, 1967.
19. Mates, B., *Stoic Logic*. University of California Press, 1953, 1973.
20. Mates, B., *Elementary Logic*. Oxford University Press, London, 1965, 1972.
21. Quine, W.V., Paradox. *Scientific American*, April 1962. Reprinted in Quine, W.V., *The Ways of Paradox and other Essays*. New York, 1966.
22. Ryle, G., Heterologicality. *Analysis*, vol. 11 (1950-51).
23. Smullyan, R.M., *First-Order Logic*. Springer Verlag, Berlin, 1968.
24. Tarski, A., Truth and Proof. *Scientific American*, June 1969, pp. 63-77.
25. Whitehead, A.N., and Russell, B., *Principia Mathematica*. Vol. 1, 1910 (2nd ed. 1925), Vol. 2, 1912 (2nd ed. 1927), Vol. 3, 1913 (2nd ed. 1927). Cambridge University Press, England.

Chapter 3
Sets: finite and infinite

H.C.M. (Harrie) de Swart

Abstract Sets occur abundantly in mathematics and in daily life. But what is a set? Cantor (1845-1918) defined a set as a collection of all objects which have a certain property in common. Russell showed in 1902 that this assumption yields a contradiction, known as Russell's paradox, and hence is untenable. In 1908 Zermelo (1871-1953) weakened Cantor's postulate considerably and consequently had to add a number of additional axioms. We present the set theory of Zermelo-Fraenkel. Next we discuss relations and functions. We use the Hilbert hotel with as many rooms as there are natural numbers to illustrate a number of astonishing properties of sets which are equally large as the set $\mathbb{N}$ of the natural numbers. We shall discover that there are many sets which in a very precise sense are much larger than $\mathbb{N}$. We shall even see that for any set V, finite or infinite, there is a larger set $P(V)$, called the powerset of V. Amazingly, although all sets we experience in the world are finite, we are still able to imagine infinite sets like $\mathbb{N}$ and to see amazing properties of them. This reminds us of the statement by cardinal Cusanus (1400-1453) that in our pursuit of grasping the divine truths we may expect the strongest support of mathematics. Finally we point out that Kant was right that mathematical (true) propositions are not analytic, but synthetic, and that Russell and Frege's logicism, stating that all of mathematics may be reduced to logic, is wrong. What may be true is that mathematics can be reduced to logic plus set theory.

3.1 Russell's Paradox

We all know lots of sets. Here are a few examples: the set of all citizens of the Netherlands, the set of all players in a soccer team, the set of all triangles in a plane.

Another example is the set of the natural numbers 1, 2 and 3. This set is denoted by $\{1, 2, 3\}$. Then $3 \in \{1, 2, 3\}$ denotes: 3 is an element of the set $\{1, 2, 3\}$; and $7 \notin \{1, 2, 3\}$ denotes: $\neg(7 \in \{1, 2, 3\})$, i.e., 7 is not an element of the set $\{1, 2, 3\}$.

The numbers 0, 1, 2, 3, ... are called *natural numbers*. We may consider the infinite set of all natural numbers. This set is denoted by $\mathbb{N}$, in other words $\mathbb{N} = \{0, 1, 2, \ldots\}$. For example, $3 \in \mathbb{N}$ and $1024 \in \mathbb{N}$, but $-3 \notin \mathbb{N}$, $\frac{2}{3} \notin \mathbb{N}$ and $\sqrt{2} \notin \mathbb{N}$.

It turns out that many, if not all, notions from mathematics can be represented by sets. For instance, we shall see that the natural numbers $0, 1, 2, \ldots$ may be represented by sets. That means that set theory may be conceived as a foundation of mathematics, as a unifying theory in which all mathematics may be represented. So, from now on we shall assume that sets are our universe of discourse.

Cantor's naive comprehension principle But what is a set? G. Cantor (1845 - 1918) answered this question as follows: a set is by definition the collection of all objects which have a certain property A. This principle is now known as the *naive comprehension principle*: Let $A(x)$ express that (set) x has the property A. Then $\{x \mid A(x)\}$ is the set of all (sets) x which have the property A, i.e.,

for all (sets) y, $y \in \{x \mid A(x)\}$ iff $A(y)$.

For instance, let $A(x)$ stand for: x is a natural number. Then Cantor's naive comprehension principle tells us that $\{x \mid x \text{ is a natural number}\}$ is a set, which we may denote by $\mathbb{N}$.

However, in 1902 Bertrand Russell showed in a letter to Frege (see Heijenoort [6], p. 124) that the naive comprehension principle leads to a contradiction. The argument is extremely simple: apply the naive comprehension principle to the property $A(x)$: $x \notin x$. According to Cantor's principle, $\{x \mid x \notin x\}$ is a set V such that for all (sets) y, $y \in V$ iff $y \notin y$. In particular, taking for y the set V itself we get

$$V \in V \text{ iff } V \notin V.$$

Contradiction.

The argument above is known as *Russell's paradox*. Russell's argument shows that set theory with the naive comprehension principle is *inconsistent*. This was quite a shock to the community at the time, because set theory was (and still is) considered to be a foundation for all of mathematics.

One way to escape the paradox was indicated by Zermelo on the grounds of the following observation: the set involved in the derivation of the paradox turns out to be very large – the set of all sets not being an element of themselves. Zermelo noted that the full force of the naive comprehension principle was hardly ever used; one mostly uses it to create subsets of a given set. So, instead of the naive comprehension principle Zermelo put forward his *Aussonderungs Axiom* or separation axiom:

Separation Axiom: if V is a set and $A(x)$ a property, then also $\{x \in V \mid A(x)\}$ is a set, consisting of all elements **in** V which have the property A, i.e., such that for all (sets) y:

$$y \in \{x \in V \mid A(x)\} \text{ iff } y \in V \text{ and } A(y)$$

The separation axiom says that within a given set V we can collect all elements of V, which have a certain property A, into a subset $\{x \in V \mid A(x)\}$ of V. Cantor allowed

3.1 Russell's Paradox

this principle not only for a given set V, but also for the universe of all sets. And Russell showed that to be contradictory.

If we abandon the naive comprehension principle and adopt the separation axiom instead, we can no longer accept the proof of Russell's paradox. However, we may use the idea of Russell's proof to obtain, with the help of the separation axiom, a positive result. From the separation axiom it follows:

Theorem 3.1. *For any set V there is a set W, namely $W = \{x \in V \mid x \notin x\}$, such that $W \notin V$.*

Proof. Let V be a given set. According to the separation axiom, $W = \{x \in V \mid x \notin x\}$ is a set such that for all sets y, $y \in W$ iff $y \in V$ and $y \notin y$. In particular, since W itself is a set, we get

$$W \in W \text{ iff } W \in V \text{ and } W \notin W.$$

Now suppose $W \in V$; then $W \in W$ iff $W \notin W$. Contradiction. Therefore, $W \notin V$.

Making use of truth-tables (see Chapter 2) one may illustrate this proof as follows. The propositions $W \in W$ and $W \in V$ can be either true (1) or false (0), giving four possible combinations:

$W \in W$	$W \in V$	$W \notin W$	$W \in V \wedge W \notin W$	$W \in W \rightleftarrows W \in V \wedge W \notin W$
1	1	0	0	0
1	0	0	0	0
0	1	1	1	0
0	0	1	0	1

From the Separation Axiom it follows that $W \in W \rightleftarrows W \in V \wedge W \notin W$ is a true (1) proposition. Hence, we are in the 4th line of the truth table. And we can read off from that line that both $W \in W$ and $W \in V$ are false (0). In particular, $W \notin V$. □

From the Separation Axiom it follows that no set may contain all sets, in other words, the universe (or totality) of all sets is not a set.

Corollary 3.1. *The universe (or totality) of all sets is not a set.*

Proof. Suppose the universe of all sets were a set U. Then by definition of U, for all sets W, $W \in U$ (1). But if U were a set, it follows from Theorem 3.1 that there is a set W, namely $W = \{x \in U \mid x \notin x\}$, such that $W \notin U$ (2).
(1) and (2) are contradictory. Hence, the universe of all sets is not a set. □

Russell obtained his paradox from the naive comprehension principle by considering the 'set' $\{x \mid x \notin x\}$. By considering the set $\{x \in V \mid x \notin x\}$, given any set V, we did not obtain a paradox, but the positive and interesting results formulated in Theorem 3.1 and Corollary 3.1 instead.

Another way to escape Russell's paradox is to blame the contradiction on the expression $x \notin x$: $x \notin x$ produced a contradiction, so we must suppress $x \in x$. Russell, in his *theory of types*, has chosen this approach: assign type to variables (sets) and allow expressions such as $x \in y$ only if the type of x is one less than the type of y. So, the expression $x \in x$ is then grammatically not correct.

Since the separation axiom yields only new sets, given any set V in advance, we have to postulate the existence of at least one set, in order to be able to build other sets. E. Zermelo (1871-1953) laid down his system of axioms for sets in 1908. The extension of Fraenkel dates from 1922. Below we present the axioms ZF of Zermelo and Fraenkel. The axioms may be formulated in natural language, but they may also be formulated in the language of predicate logic, letting the variables range over sets and using only two binary predicate symbols: $\in$ (is element of) and $=$ (is equal to).

3.2 Axioms of Zermelo-Fraenkel for Sets

Empty set axiom: There exists a set without elements. In other words, there is a set x such that for all sets y, $y \notin x$.
Formulated in the predicate language just mentioned: $\exists x \forall y [\neg (y \in x)]$

There are many examples of empty sets in daily life: the set of living persons older than 150 years; the set of all persons with blue hair, the set of all natural numbers which are both even and odd, etc. Notice that the existence of the empty set also would follow from the naive comprehension principle: $\{x \mid x \neq x\}$, assuming that each thing is equal to itself.

Sets are, just like triangles and numbers, legitimate mathematical objects. So it makes perfectly good sense to ask whether two sets are identical or not. If two sets x and y are identical (equal), we write $x = y$, if not, $x \neq y$. Identical sets have exactly the same properties; so, if $x = y$, then every element of x is also an element of y and vice versa. One may wonder if, conversely, sets with exactly the same elements are identical. Consider, for example, the set V of all even numbers greater than zero and the set W of all sums of pairs of odd numbers. There is some reason to distinguish V and W: they are given in different ways. On the other hand, we feel (and mathematical practice confirms this) that definitions do not matter so much, it is rather content that counts. So, we make the explicit choice to consider sets as merely being determined by their elements. Hence, 'having the same elements' means 'being equal'.

Axiom of extensionality: Two sets are equal if and only if they have the same elements. As observed above, the 'only if' holds trivially.
Formulated in our predicate language: $x = y \rightleftarrows \forall z [z \in x \rightleftarrows z \in y]$.

The axiom of extensionality has among others the following consequences:

$$\{3, 4, 5\} = \{4, 3, 5\} \qquad \{2, 3\} \neq \{3, 4\}$$
$$\{3, 3, 7\} = \{3, 7\} \qquad \{0, 1\} \neq \{1, 2\}$$
$$\{2, 3\} = \{2, 3, 3\} \qquad \{2, \{3, 4\}\} \neq \{\{2, 3\}, 4\}$$

3.2 Axioms of Zermelo-Fraenkel for Sets

Notice that the only elements of $\{2,\{3,4\}\}$ are: 2 and $\{3,4\}$, while the only elements of $\{\{2,3\},4\}$ are: $\{2,3\}$ and 4. For instance, $2 \in \{2,\{3,4\}\}$, but $2 \notin \{\{2,3\},4\}$; and $\{2,3\} \in \{\{2,3\},4\}$, but $\{2,3\} \notin \{2,\{3,4\}\}$.

Since, by the extensionality axiom, a set is completely determined by its elements, there may be at most one empty set: if there were two sets without elements, they would have the same elements ($0 \rightleftarrows 0 = 1$) and hence, by the axiom of extensionality, be equal. The empty set axiom says that there is at least one empty set. By the axiom of extensionality there is at most one empty set. Hence, there is exactly one empty set. **Notation**: $\emptyset$.
By definition: $\forall y[y \notin \emptyset]$.

Given two sets V and W, we want to be able to construct a set whose elements are exactly V and W themselves. The existence of such a set would also follow from the naive comprehension principle: $\{x \mid x = V \text{ or } x = W\}$. So, we postulate:

Pairing Axiom: Given any sets v and w, there exists a set y, whose elements are exactly v and w.
Formulated in our predicate language: $\forall v \forall w \exists y \forall z[z \in y \rightleftarrows z = v \lor z = w]$.

Again, by the extensionality axiom, given sets v and w, the set whose existence is required by the pairing axiom is unique and is called the *unordered pair* $\{v,w\}$ of v and w. Because $\{v,w\}$ and $\{w,v\}$ have the same elements, they are equal.
So, for all (sets) z, $z \in \{v,w\}$ iff $z = v$ or $z = w$.

$\{v\} := \{v,v\}$ is the *singleton* of v. If v is a set, then so is $\{v\}$, because of the pairing axiom and the definition of $\{v\}$.

Now, with only a few axioms, the existence of infinitely many sets follows:

$$\emptyset, \{\emptyset\}, \{\{\emptyset\}\}, \{\{\{\emptyset\}\}\}, \ldots$$

$\emptyset$ (we repeat) is a set without elements. $\{\emptyset\}$, on the other hand, is a set with one element, namely $\emptyset$. Hence, $\emptyset \neq \{\emptyset\}$.
$\{\{\emptyset\}\}$ is the set with $\{\emptyset\}$ as its only element, while $\{\emptyset\}$ has $\emptyset$ as its only element. Hence, $\{\{\emptyset\}\} \neq \{\emptyset\}$, because $\emptyset \notin \{\{\emptyset\}\}$.
The Pairing Axiom also entails the existence of $\{\emptyset, \{\emptyset\}\}$, which is the set with $\emptyset$ and $\{\emptyset\}$ as its only elements.

Given two sets V and W we want to be able to construct the *union* $V \cup W$ of V and W such that for all z, $z \in V \cup W$ iff $z \in V \lor z \in W$. Its existence would follow from the naive comprehension principle: $\{x \mid x \in V \text{ or } x \in W\}$. Notice that in general, $V \cup W$ is a larger set than each of V and W separately.

Union axiom If v and w are sets, then there exists a set y such that for all (sets) z, $z \in y$ iff $z \in v$ or $z \in w$.
Formulated in our predicate language: $\forall v \forall w \exists y \forall z[z \in y \rightleftarrows z \in v \lor z \in w]$

Again, by the extensionality axiom, given sets V and W, the set required by the union axiom is unique and is called the *union* of V and W. **Notation**: $V \cup W$.
So, for all (sets) z,

$$z \in V \cup W \rightleftarrows z \in V \vee z \in W.$$

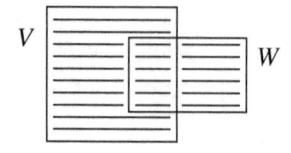

Example 3.1. $\{1,2\} \cup \{5,6\} = \{1,2,5,6\}$, $\{1,2\} \cup \{2\} = \{1,2\}$,
$\{1,2\} \cup \{2,6\} = \{1,2,6\}$, $\{1,2\} \cup \emptyset = \{1,2\}$.
$\{1,2\} \cup \{1,2\} = \{1,2\}$.

The union axiom allows us to construct the union of any two given sets v and w or, put differently, to form the union of all elements of the set $x = \{v, w\}$. A more general version of the union axiom, put forward by Zermelo, was the following.

Sumset Axiom: For every set x there exists a set y, whose elements are exactly the objects occurring in at least one element of x.
Formulated in our predicate language: $\forall x \exists y \forall z [z \in y \rightleftarrows \exists v[v \in x \wedge z \in v]]$.

Again, the extensionality axiom guarantees the uniqueness of the set y, given x. This unique set is called the *sum-set* of x. **Notation**: $\bigcup x$ or $\bigcup \{y \mid y \in x\}$.
Notice that $v \cup w = \bigcup \{v, w\}$.

Now we are able to define the natural numbers in terms of sets as follows.

Definition 3.1 (Successor function). $0 := \emptyset$.
The *successor function* S is defined by $S(n) = n \cup \{n\}$, also denoted by $n+1$.

Example 3.2. $0 := \emptyset$
$1 := 0 \cup \{0\}$. So, $1 = \{0\} = \{\emptyset\}$.
$2 := 1 \cup \{1\}$. So, $2 = \{0\} \cup \{1\} = \{0,1\} = \{\emptyset, \{\emptyset\}\}$.
$3 := 2 \cup \{2\}$. So, $3 = \{0,1\} \cup \{2\} = \{0,1,2\} = \{\emptyset, \{\emptyset\}, \{\emptyset, \{\emptyset\}\}\}$.
In general, for any natural number n, $n+1 := n \cup \{n\}$.

One easily checks by induction that for any natural n, defined in this way, $n = \{0, \ldots, n-1\}$ and that the sets $0, 1, 2, 3, \ldots$ are distinct pairwise. So, we have identified each natural number n with a certain standard set consisting of n elements. This definition of natural numbers in terms of sets justifies the use of natural numbers in the examples at the beginning of this section.

With very few axioms we have generated up till now infinitely many sets, but all of them are finite. But we also want to be able to deal with the infinite set of all natural numbers, which is so important in mathematics and its many applications. The existence of this set would follow easily from the naive comprehension principle: $\{x \mid x \text{ is a natural number}\}$. Since this naive comprehension principle had to be

3.2 Axioms of Zermelo-Fraenkel for Sets

replaced by the much weaker separation axiom we have to postulate the existence of at least one infinite set.

Axiom of Infinity: There is at least one set y that contains 0, i.e., $\emptyset$, and is such that for every $x \in y$ it also contains Sx, i.e., $x \cup \{x\}$.
Formulated in our predicate language: $\exists y[0 \in y \wedge \forall x[x \in y \rightarrow Sx \in y]]$

The set y whose existence is required by the axiom of infinity has clearly infinitely many members: 0, 1, 2, 3, ... But there might be many of such sets containing in addition other things. So, we take the smallest such set which contains 0 and with every number n its successor $Sn = n+1$ and denote it by $\mathbb{N}$. So, $0 \in \mathbb{N}$, $1 \in \mathbb{N}$, $2 \in \mathbb{N}$, etc. Notice that $\mathbb{N}$ has infinitely many members, but $\{\mathbb{N}\}$ has only one element: $\mathbb{N}$.

In order to be able to construct for instance the set of all even natural numbers, i.e., $\mathbb{N}_{even} = \{n \in \mathbb{N} \mid n \text{ is even}\}$, we need the separation axiom.

Separation Axiom: If x is a set and $A(z)$ a property, then also $\{z \in x \mid A(z)\}$ is a set, consisting of all elements **in** x which have the property A, i.e., such that for all z:

$$z \in \{z \in x \mid A(z)\} \text{ iff } z \in x \text{ and } A(z)$$

Formulated in our logical predicate language: $\forall x \exists y \forall z[z \in y \rightleftarrows z \in x \wedge A(z)]$ for any formula A in our logical predicate language.

The separation axiom says that within a given set x we can collect all elements of x, which have a given property A, into a subset $\{z \in x \mid A(z)\}$ of x. Notice that the separation axiom is in fact an axiom schema: it yields an axiom for any formula A. By the axiom of extensionality, given a set x and a property A, the set y, whose existence is demanded by the separation axiom, is uniquely determined and shall be denoted by $\{z \in x \mid A(z)\}$.

Given the separation axiom and the axiom of infinity, the existence of the empty set follows immediately: $\emptyset = \{z \in \mathbb{N} \mid z \neq z\}$, if we assume that for all z, $z = z$. Also, given the separation axiom, we may introduce some important set theoretical operations: *intersection* and *relative complement*.

Corollary 3.2 (Intersection). *Given any sets V and W, also the intersection $V \cap W := \{z \in V \mid z \in W\}$ of V and W is a set, such that for all z*

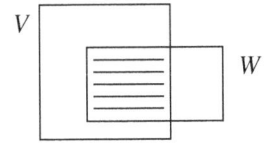

$$z \in V \cap W \rightleftarrows z \in V \wedge z \in W.$$

We may generalize the intersection as follows. If x is a non-empty set, say $v \in x$, then $\bigcap x := \{z \in v \mid \forall y[y \in x \rightarrow z \in y]\}$. Notice that $V \cap W = \bigcap \{V, W\}$.

Corollary 3.3 (relative complement). *Given any sets V and W, also the relative complement, $V - W := \{z \in V \mid z \notin W\}$ of W with respect to V, is a set, such that*

$$z \in V - W \rightleftarrows z \in V \wedge z \notin W.$$

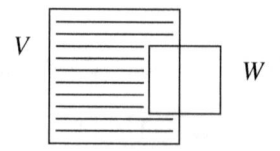

Notice that $V \cap W$ and $V - W$ are in general smaller sets than V, while $V \cup W$ in general is a larger set than V. The existence of $V \cap W$ and $V - W$ follows from the separation axiom, while the existence of $V \cup W$ requires the union axiom.

Example 3.3.

$\{1,2\} \cup \{2,3\} = \{1,2,3\}$ $\{1,2\} \cup \emptyset = \{1,2\}$ $\{1,2\} \cup \mathbb{N} = \mathbb{N}$
$\{1,2,3\} \cap \{2,3,4\} = \{2,3\}$ $\{1,2\} \cap \emptyset = \emptyset$ $\{2,3\} \cap \mathbb{N} = \{2,3\}$
$\{1,2,3\} - \{2,3,4\} = \{1\}$ $\{1,2,3\} - \emptyset = \{1,2,3\}$ $\{1,2,3\} - \mathbb{N} = \emptyset$

The reader may easily verify the following statements:
1. $\cap$ and $\cup$ are *idempotent*, i.e., $V \cap V = V$, respectively $V \cup V = V$, for any set V.
2. $\cap$ and $\cup$ are *commutative*, i.e., $V \cap W = W \cap V$, respectively $V \cup W = W \cup V$, for any sets V and W.
3. $\cap$ and $\cup$ are *associative*, i.e., $U \cap (V \cap W) = (U \cap V) \cap W$, respectively $U \cup (V \cup W) = (U \cup V) \cup W$, for any sets U, V, W.
4. $V \cap \emptyset = \emptyset$ and $V \cup \emptyset = V$ for any set V.

Theorem 3.2 (absorption laws). *For all sets V and W,*
$V \cap (V \cup W) = V$ and $V \cup (V \cap W) = V$.

Proof. By the axiom of extensionality we have to show that the two sets in question have the same elements, i.e., for all z, $z \in V \cap (V \cup W)$ iff $z \in V$ and $z \in V \cup (V \cap W)$ iff $z \in V$. This is straightforward. □

Theorem 3.3 (distributive laws). *For all sets U, V and W,*
$U \cap (V \cup W) = (U \cap V) \cup (U \cap W)$ and $U \cup (V \cap W) = (U \cup V) \cap (U \cup W)$.

Proof. By the axiom of extensionality we have to show that for all z, $z \in U \cap (V \cup W)$ iff $z \in (U \cap V) \cup (U \cap W)$, in other words, $z \in U \wedge (z \in V \vee z \in W)$ iff $(z \in U \wedge z \in V) \vee (z \in U \wedge z \in W)$. This is straightforward and also follows from the distributive laws of propositional logic in Theorem 2.10. □

When it is clear from the context that the complement of a set W is taken relative to a given universe U, $U - W$ is simply called the *complement* of W and denoted by W^c.

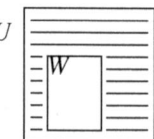

Theorem 3.4. *Let V^c and W^c be the complement of V, respectively W, relative to a given universe U. $(V \cup W)^c = V^c \cap W^c$ and $(V \cap W)^c = V^c \cup W^c$.*

3.2 Axioms of Zermelo-Fraenkel for Sets

Proof. We leave the proof to the reader as Exercise 3.3. □

In order to be able to formulate the powerset axiom we first have to introduce the notion of *subset*.

Definition 3.2 (Subset). W is a *subset* of $V :=$ every element of W is also an element of V, i.e., for every x, if $x \in W$, then also $x \in V$. **Notation**: $W \subseteq V$.

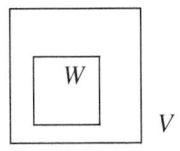

Notice that W is not a subset of V iff not all elements of W are elements of V, in other words, iff there is some $x \in W$ such that $x \notin V$. **Notation**: $\neg(W \subseteq V)$ or $W \not\subseteq V$.

Example 3.4.
$\{2,3\} \subseteq \{1,2,3,4\}$ $\{2,3\} \subseteq \{2,3\}$ $\emptyset \subseteq \{2,3\}$ $\{2,3\} \subseteq \mathbb{N}$
$\{2,3\} \not\subseteq \{3,4,5\}$ $\{1,\{2\}\} \not\subseteq \{1,2\}$ $\{1,2\} \not\subseteq \{1,\{2\}\}$ $\mathbb{N} \not\subseteq \{\mathbb{N}\}$

Definition 3.3 (Proper subset). W is a *proper subset* of $V := W \subseteq V$ and not $W = V$. **Notation**: $W \subset V$.

Example 3.5. $\{2,3\} \subset \{2,3,4\}$ and $\{2,3\} \subset \mathbb{N}$.

Warning: It is important not to confuse $\in$ and $\subseteq$:
$\{2\} \in \{\{2\},3\}$, but $\{2\} \not\subseteq \{\{2\},3\}$, the latter because $2 \in \{2\}$, but $2 \notin \{\{2\},3\}$.
$\{2,3\} \subseteq \{1,2,3\}$, but $\{2,3\} \notin \{1,2,3\}$.

Theorem 3.5. *For any set V, $\emptyset \subseteq V$ and $V \subseteq V$.*

Proof. Suppose that for some V, $\emptyset \not\subseteq V$, i.e., there would be an element $x \in \emptyset$ such that $x \notin V$. Because $\emptyset$ has no elements, this is impossible. Therefore, $\emptyset \subseteq V$. And because every element of V is an element of V, it follows that $V \subseteq V$. □

Example 3.6. $\emptyset \subseteq \emptyset$, but $\emptyset \notin \emptyset$.
$\emptyset \subseteq \{\emptyset\}$, and by definition of $\{\emptyset\}$ also $\emptyset \in \{\emptyset\}$.
$\emptyset \subseteq \{\{\emptyset\}\}$, but $\emptyset \notin \{\{\emptyset\}\}$, since the only element of $\{\{\emptyset\}\}$ is $\{\emptyset\}$.
$\{\emptyset\} \subseteq \{\emptyset\}$, but $\{\emptyset\} \notin \{\emptyset\}$, since the only element of $\{\emptyset\}$ is $\emptyset$.
$\{\emptyset\} \not\subseteq \{\{\emptyset\}\}$, because $\emptyset \in \{\emptyset\}$ while $\emptyset \notin \{\{\emptyset\}\}$, but $\{\emptyset\} \in \{\{\emptyset\}\}$.

Next we will determine for a few small finite sets all their subsets and the set of all their subsets. Let us start with $\emptyset$. The only subset of $\emptyset$ is $\emptyset$ itself. So, the set $P(\emptyset)$ of all subsets of $\emptyset$ is $\{\emptyset\}$.

The only subsets of the set $\{u\}$ are $\emptyset$ with zero elements and $\{u\}$ itself with one element. So, the set $P(\{u\})$ of all subsets of $\{u\}$ is $\{\emptyset, \{u\}\}$.

The subsets of $\{u,v\}$ can have 0, 1 or 2 elements and are, respectively, $\emptyset$ with zero elements, $\{u\}$ and $\{v\}$ with one element, and $\{u,v\}$ itself with two elements. So, the set $P(\{u,v\})$ of all subsets of $\{u,v\}$ is $\{\emptyset, \{u\}, \{v\}, \{u,v\}\}$. Notice that there are twice as many subsets of $\{u,v\}$ as there are subsets of $\{u\}$: all subsets of $\{u\}$, i.e., $\emptyset$ and $\{u\}$, are also a subset of $\{u,v\}$ and the other subsets of $\{u,v\}$ are obtained by adding the element v to the subsets of $\{u\}$.

The subsets of $\{u,v,w\}$ can have 0, 1, 2 or 3 elements and are, respectively, $\emptyset$ with zero elements, $\{u\}, \{v\}$ and $\{w\}$ with one element, $\{u,v\}, \{u,w\}$ and $\{v,w\}$ with two elements, and finally $\{u,v,w\}$ itself with three elements. So, the set $P(\{u,v,w\})$ of all subsets of $\{u,v,w\}$ is $\{\emptyset, \{u\}, \{v\}, \{w\}, \{u,v\}, \{u,w\}, \{v,w\}, \{u,v,w\}\}$. Notice that there are twice as many subsets of $\{u,v,w\}$ as there are subsets of $\{u,v\}$: all subsets of $\{u,v\}$, i.e., $\emptyset, \{u\}, \{v\}$ and $\{u,v\}$, are also a subset of $\{u,v,w\}$ and the other subsets of $\{u,v,w\}$ are obtained by adding the element w to the subsets of $\{u,v\}$.

This brings us to the following observation: each time that one adds one element w to a given finite set V, one obtains twice as many subsets: all the subsets of V plus all subsets of V with the new element w added. From this insight results the following theorem:

Theorem 3.6. *For each natural number n, if V is a finite set with n elements, then V has 2^n subsets.*

Proof. By mathematical induction. For $n = 0$: a set V with 0 elements is the empty set $\emptyset$, and this set has $2^0 = 1$ subset, namely $\emptyset$. Suppose the statement is true for $n = k$, i.e. any set with k elements has 2^k subsets (induction hypothesis). Then a set with $k+1$ elements has twice as many subsets, i.e., $2 \cdot 2^k = 2^{k+1}$ subsets. $\square$

For instance, if V has 10 elements, V has $2^{10} = 1024$ subsets. And if V has 20 elements, V has $2^{20} = 2^{10} \cdot 2^{10} = 1024 \cdot 1024$ subsets, that is more than one million!

Since sets of subsets occur abundantly in mathematics and since the existence of many of these sets does not follow from the set theoretic axioms introduced up till now, we postulate the following powerset axiom:

Powerset axiom: If V is a set, then also $P(V) = \{X \mid X \subseteq V\}$ is a set. We call $P(V)$ the *powerset* of V.

Formulated in our logical predicate language: $\forall v \exists y \forall x [x \in y \rightleftarrows x \subseteq v]$.

So, the elements of $P(V)$ are the subsets of V, i.e.,

$$X \in P(V) \text{ iff } X \subseteq V.$$

The name *powerset* refers to the fact that if V has n ($n \in \mathbb{N}$) elements, then by Theorem 3.6, $P(V)$ has 2^n elements.

This powerset axiom may look innocent, but is it? We have already seen that if V is a relatively small finite set, then $P(V)$ may become a relatively large set. And what will happen when we apply the P-operator to an infinite set, like $\mathbb{N}$? According to the powerset axiom, not only $P(\mathbb{N})$ is another set, but also $P(P(\mathbb{N})), P(P(P(\mathbb{N})))$, etc. are new sets. As we shall see later on in Section 3.6, these sets become so large that one may ask the question whether we are still able to construct these sets. In fact, the powerset axiom is the only set theoretic axiom which is not by everyone accepted in its full strength, in particular not by the intuitionists; see Chapter 8.

3.2 Axioms of Zermelo-Fraenkel for Sets

Up till now we have postulated the following axioms for set theory: empty set axiom, axiom of extensionality, pairing axiom, union axiom, sumset axiom, axiom of infinity, separation axiom, and powerset axiom. The set theory ZF of Zermelo-Fraenkel contains two more axioms: the axiom of replacement, which is the only contribution of Fraenkel, and the axiom of regularity (or foundation). We only mention these axioms here and refer to exercise 3.8 and to van Dalen, Doets, de Swart [3].

Axiom of Replacement: If for every x in V there is exactly one y such that $\Phi(x,y)$, then there exists a set W which contains precisely the elements y for which there is an $x \in V$ with the property $\Phi(x,y)$. In other words, the image of a set V under an operation (functional property Φ) is again a set.

Axiom of Regularity: Every non-empty set is disjoint from at least one of its elements.

The latter axiom guarantees that for any set x, $x \notin x$ and that there is no sequence $v_1, \ldots, v_n$ of sets such that $v_1 \in v_2$, $v_2 \in v_3$, $\ldots$, $v_{n-1} \in v_n$ and $v_n \in v_1$ (Exercise 3.8).

There are several set theoretical principles which are consistent with, but independent of the axioms of Zermelo-Fraenkel. The axioms of choice and the continuum hypothesis (see Section 3.6) are not treated here because of their more dubious status. See van Dalen, Doets, de Swart, [3] for an elaborate discussion.

Exercise 3.1. Which of the following propositions are true and which are false?

$\mathbb{N} \in \mathbb{N}$	$\{2,3\} \subseteq \{\mathbb{N}\}$	$\emptyset \in \emptyset$	$\{\emptyset\} \in \emptyset$
$\mathbb{N} \in \{\mathbb{N}\}$	$\{2\} \subseteq \{\mathbb{N}\}$	$\emptyset \subseteq \emptyset$	$\{\emptyset\} \subseteq \emptyset$
$\mathbb{N} \subseteq \mathbb{N}$	$\{2\} \subseteq \mathbb{N}$	$\emptyset \in \{\emptyset\}$	$\{\emptyset\} \subseteq \{\emptyset\}$
$\mathbb{N} \in \{\{\mathbb{N}\}\}$	$2 \in \{1, \{2\}, 3\}$	$\emptyset \subseteq \{\emptyset\}$	$\emptyset \subseteq \{\emptyset, \{\emptyset\}\}$
$\mathbb{N} \subseteq \{\mathbb{N}\}$	$\{2\} \in \{1, \{2\}, 3\}$	$\emptyset \in \{\{\emptyset\}\}$	$\emptyset \in \{\emptyset, \{\emptyset\}\}$
$\{1,2\} \in \mathbb{N}$	$\{1, \{2\}\} \subseteq \{1, \{2,3\}\}$	$\emptyset \subseteq \{\{\emptyset\}\}$	$\{\emptyset\} \subseteq \{\emptyset, \{\emptyset\}\}$
$\{1,2\} \subseteq \mathbb{N}$	$\{1, \{2\}\} \subseteq \{1, \{2\}, 3\}$	$\{\emptyset\} \in \{\{\emptyset\}\}$	$\{\emptyset\} \in \{\emptyset, \{\emptyset\}\}$
$\{1,2\} \in \{\mathbb{N}\}$	$\{-2, 2\} \subseteq \mathbb{N}$	$\{\emptyset\} \subseteq \{\{\emptyset\}\}$	$\emptyset \subseteq \{\{\emptyset, \{\emptyset\}\}\}$

Exercise 3.2. Prove or refute: a) $W \subseteq V$ iff $V \cap W = W$; b) $W \subseteq V$ iff $V \cup W = V$.

Exercise 3.3. Prove or refute: for all sets U, V and W,
a) $U - (V \cup W) = (U - V) \cap (U - W)$; b) $U - (V \cap W) = (U - V) \cup (U - W)$.

Exercise 3.4. Prove or refute: for all sets U, V and W,
a) if $U \in V$ and $V \in W$, then $U \in W$; b) if $U \subseteq V$ and $V \subseteq W$, then $U \subseteq W$.

Exercise 3.5. Determine $P(\emptyset)$, $P(P(\emptyset))$ and $P(P(P(\emptyset)))$.

Exercise 3.6. Prove:
(a) If $W \subseteq V$, then $P(W) \subseteq P(V)$; (b) If $P(W) \subseteq P(V)$, then $W \subseteq V$.
(c) If $P(W) = P(V)$, then $W = V$; (d) If $P(W) \in P(V)$, then $W \in V$.

Exercise 3.7. Prove or refute:
a) for all sets W, V, if $P(W) \in PP(V)$, then $W \in P(V)$.
b) for all sets W, V, if $W \in P(V)$, then $P(W) \in PP(V)$.
c) for all sets W, V, if $P(W) \subseteq PP(V)$, then $W \subseteq P(V)$.
d) for all sets W, V, if $W \subseteq P(V)$, then $P(W) \subseteq PP(V)$.

Exercise 3.8. Show that from the axiom of regularity it follows that i) for any set x, $x \notin x$, and ii) there is no sequence $v_1, \ldots, v_n$ such that $v_1 \in v_2, v_2 \in v_3, \ldots, v_{n-1} \in v_n$ and $v_n \in v_1$.

3.3 Historical and Philosophical Remarks

3.3.1 Mathematics and Theology

In Corollary 3.1 we have seen that from the separation axiom it follows that the universe of all sets itself is not a set. This reminds us of Cardinal Cusanus (1400-1453), who in his *De docta ignorantia* [2] says that in the pursuit of grasping the divine truths we may expect the strongest support from mathematics. Although he illustrated this statement with other examples, it seems fair to say that he might have used Corollary 3.1 as an illustration: the universe of all earthly things (God?) is itself not an earthly thing.

Also the insights about infinite sets to be discovered in Sections 3.5 and 3.6 may be considered as illustrations of his statement. Although we never experience infinite sets in daily life, we are still able to imagine them and even to gain insights into their amazing properties.

3.3.2 Ontology of mathematics

Since the integers, the rational and the real numbers can be defined in terms of sets and natural numbers, it follows that these numbers can ultimately be defined in set-theoretical terms (see van Dalen, Doets, de Swart, [3]). Through practical experience mathematicians have found that most well-known concepts, such as the notion of number, function, triangle, and so on, can be defined in set-theoretical terms. This has led to the slogan 'Everything is a set', meaning that all objects from mathematical practice turn out to be representable in terms of sets. Consequently, every mathematical proposition can be reduced to a proposition about sets. It turns out that most, if not all, mathematical theorems – after translation in terms of sets – can be deduced logically from the axioms of set theory.

Set-theoretical Axioms

|

logical reasoning

|

mathematical theorems

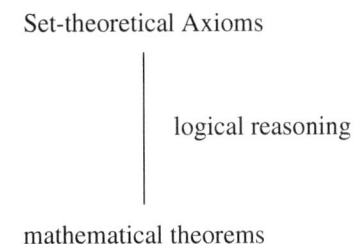

So one might say that the axioms of *ZF* (Zermelo-Fraenkel) determine the *ontology of mathematics*: all mathematical objects are conceived as sets and the axioms of Zermelo-Fraenkel postulate the existence of certain sets, leaving room for extension with possibly more axioms, and they specify what the characteristic properties of these mathematical objects (sets) are. In this sense the axioms of *ZF* can be considered to be a foundation for (the greater part of) mathematics.

The axioms of Zermelo-Fraenkel (*ZF*) may be described informally. But we have also seen that the set theory of Zermelo-Faenkel may be *formalized* by:
1. first introducing the predicate language with only two binary predicate symbols = and ∈ with 'is equal to', respectively 'is element of' as intended interpretation, such that all statements about sets may be expressed in this language;
2. and next by specifying the axioms of *ZF* in this language, such that statements about sets (mathematical objects) may be logically deduced from these axioms.

3.3.3 Analytic-Synthetic

In his *Critique of Pure Reason* (1781) Immanuel Kant [7] makes a distinction between analytic and synthetic judgments. Kant calls a judgment *analytic* if its predicate is contained (though covertly) in the subject, in other words, the predicate adds nothing to the conception of the subject. Kant gives 'All bodies are extended' (Alle Körper sind ausgedehnt) as an example of an analytic judgment; I need not go beyond the conception of *body* in order to find extension connected with it. If a judgment is not analytic, Kant calls it *synthetic*; a synthetic judgment adds to our conception of the subject a predicate which was not contained in it, and which no analysis could ever have discovered therein. Kant mentions 'All bodies are heavy' (Alle Körper sind schwer) as an example of a synthetic judgment.

Also in his *Critique of Pure Reason* Kant makes a distinction between *a priori* knowledge and *a posteriori* knowledge. A priori knowledge is knowledge existing altogether independent of experience, while a posteriori knowledge is empirical knowledge, which has its sources in experience.

Sometimes one speaks of *logically necessary* truths instead of analytic truths and of *logically contingent* truths instead of synthetic truths, to be distinguished from physically necessary truths (truths which physically could not be otherwise, true in all physically possible worlds). The distinction between necessary and contingent truth is a *metaphysical* one, to be distinguished from the *epistemological* distinction between *a priori* and *a posteriori* truths.. Although these – the metaphysical and the epistemological – are certainly different distinctions, it was controversial whether they coincide in extension, that is, whether all and only necessary truths are *a priori* and all and only contingent truths are *a posteriori*.

In his *Critique of Pure Reason* Kant stresses that *mathematical judgments are both a priori and synthetic*. 'Proper mathematical propositions are always *judgments a priori*, and not empirical, because they carry along with them the conception of necessity, which cannot be given by experience.' Why are mathematical judgments synthetic? Kant considers the proposition $7+5 = 12$ as an example. 'The conception of twelve is by no means obtained by merely cogitating the union of seven and five; and we may analyse our conception of such a possible sum as long as we will, still we shall never discover in it the notion of twelve.' We must go beyond this conception of $7+5$ and have recourse to an intuition which corresponds to counting using our fingers: first take seven fingers, next five fingers extra, and then by starting to count right from the beginning we arrive at the number twelve.

```
7:     1  1  1  1  1  1  1
5:                          1  1  1  1  1
7+5:   1  1  1  1  1  1  1  1  1  1  1  1
       1  2  3  4  5  6  7  8  9  10 11 12
```

'Arithmetical propositions are therefore always synthetic, of which we may become more clearly convinced by trying large numbers.' Geometrical propositions are also synthetic. As an example Kant gives 'A straight line between two points is the shortest', and explains 'For my conception of *straight* contains no notion of *quantity*, but is merely *qualitative*. The conception of the *shortest* is therefore wholly an addition, and by no analysis can it be extracted from our conception of a straight line.'

In more modern terminology, following roughly a 'Fregean' account of analyticity, one would define a proposition A to be *analytic* iff either
(i) A is an instance of a logically valid formula; e.g., 'No unmarried man is married' has the logical form $\neg\exists x[\neg P(x) \wedge P(x)]$, which is a valid formula, or
(ii) A is reducible to an instance of a logically valid formula by substitution of synonyms for synonyms; e.g., 'No bachelor is married'.

In his *Two dogmas of empiricism* W.V. Quine [8] is sceptical of the analytic-synthetic distinction. Quine argues as follows. In order to define the notion of analyticity we used the notion of synonymy in clause (ii) above. However, if one tries to explain this latter notion, one has to take recourse to other notions which directly or indirectly will have to be explained in terms of analyticity.

3.3.4 Logicism

Logicism dates from about 1900, its most important representatives being G. Frege in his *Grundgesetze der Arithmetik* I, II (1893, 1903) and B. Russell in his *Principia Mathematica* (1903), together with A.N. Whitehead. The program of the logicists

3.3 Historical and Philosophical Remarks

was to reduce mathematics to logic. What do they mean by this? In his Grundgesetze der Arithmetik Frege defines the natural numbers in terms of sets as follows: 1 := the class of all sets having one element, 2 := the class of all sets having two elements, and so on. Next Frege shows that all kinds of properties of natural numbers can be logically deduced from a *naive comprehension principle*: if $A(x)$ is a property of an object x, then there exists a set $\{x \mid A(x)\}$ which contains precisely all objects x which have property A. (See Section 3.1.)

Logicism tried to introduce mathematical notions by means of explicit definitions; mathematical truths would then be logical consequences of these definitions. Mathematical propositions would then be reducible to logical propositions and hence mathematical truths would be analytic, contrary to what Kant said.

The greatest achievement of Logicism is that it succeeded in reducing great parts of mathematics to one single (formal) system, namely, set theory. The logicists believed that by doing this they reduced all of mathematics to logic without making use of any non-logical assumptions, hence showing that mathematical truths are analytic. However, what they actually did was reduce mathematics to logic PLUS set theory. And the axioms of set theory have a non-logical status! The axioms of set theory are – in Kant's terminology – synthetic, and surely not analytic. In his later years Frege came to realize that the axioms of set theory (see Section 3.2) are not a part of logic and gave up Logicism, which he had founded himself. The interested reader is referred to K. Gödel [4], *Russell's mathematical logic*.

Another way to see that a mathematical truth like $7 + 5 = 12$ is synthetic is to realize that $7 + 5 = 12$ is not a logically valid formula; it is true under the intended interpretation, but not true under all possible interpretations. $7 + 5 = 12$ can be logically deduced from the axioms of Peano for (formal) number theory (see Chapter 5), but it cannot be proved by the axioms and rules of formal logic alone.

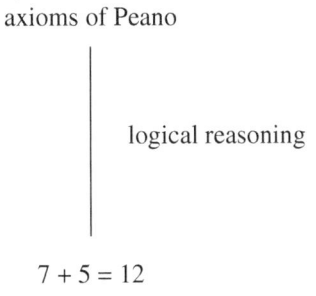

axioms of Peano

logical reasoning

$7 + 5 = 12$

Again, Peano's axioms are true under the intended interpretation, but are not (logically) valid and hence they do not belong to logic.

3.4 Relations, Functions and Orderings*

3.4.1 Ordered pairs and Cartesian product

In the plane the pairs $(4,2)$ and $(2,4)$ indicate different points.

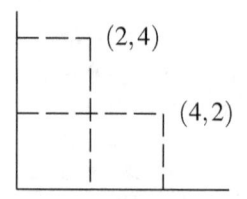

The order of the numbers 2 and 4 is of importance here, in the same way that the order of letters is of importance in constructing words: 'pin' and 'nip' contain the same letters, but in a different order. A pair of objects, say v and w, in which their order is relevant, is called the *ordered pair* of v and w, written (v,w). Sometimes the notation $<v,w>$ is used. This is different from the ordinary (unordered) pair $\{v,w\}$, which is the same as $\{w,v\}$. Ordered pairs have the characteristic property

$$(v,w) = (x,y) \text{ iff } v = x \text{ and } w = y. \tag{*}$$

Unordered pairs do not have this property, since $\{v,w\} = \{w,v\}$ even for $v \neq w$.

We can introduce the notion of ordered pair as a primitive notion (i.e., undefined) and introduce the above-mentioned property (*) as an axiom. However, it is a wise rule not to introduce more primitive notions than necessary ('Ockham's razor') and hence we shall define a set, which behaves as an ordered pair, i.e., which satisfies the desired property (∗).

Definition 3.4 (Ordered pair). $(v,w) := \{\{v\},\{v,w\}\}$.

This is not the only definition which will work: see Exercise 3.9. We must now show that this definition satisfies (*).

Theorem 3.7. $(v,w) = (x,y)$ *iff* $v = x$ *and* $w = y$.

Proof. The implication from right to left is trivial. So suppose $(v,w) = (x,y)$, i.e., $\{\{v\},\{v,w\}\} = \{\{x\},\{x,y\}\}$. If two sets are equal, then they have the same elements. Hence, $\{v\} = \{x\}$ and $\{v,w\} = \{x,y\}$ or $\{v\} = \{x,y\}$ and $\{v,w\} = \{x\}$. In the first case it follows that $v = x$ and $w = y$. In the second case we can conclude: $v = x = y$ and $v = w = x$; so, also in this case, $v = x$ and $w = y$. □

The following theorem holds for Definition 3.4 of ordered pairs.

Theorem 3.8. *If* $v \in V$ *and* $w \in W$, *then* $(v,w) \in PP(V \cup W)$.

Proof. Suppose $v \in V$ and $w \in W$. Then:
 (i) $v \in V \cup W$, so $\{v\} \subseteq V \cup W$, in other words, $\{v\} \in P(V \cup W)$, and
 (ii) $w \in V \cup W$, so $\{v,w\} \subseteq V \cup W$, in other words, $\{v,w\} \in P(V \cup W)$.
 From (i) and (ii) it follows that $\{\{v\},\{v,w\}\} \subseteq P(V \cup W)$, in other words, $\{\{v\},\{v,w\}\} \in PP(V \cup W)$. □

3.4 Relations, Functions and Orderings*

We can generalize the notion of ordered pair to the notion of ordered *n-tuple*:

Definition 3.5 (Ordered *n*-tuple). For $n \in \mathbb{N}$, $n \geq 1$:
$$(v) := v,$$
$$(v_1, \ldots, v_n, v_{n+1}) := ((v_1, \ldots, v_n), v_{n+1}).$$

By means of mathematical induction one easily verifies that the object $(v_1, \ldots, v_n)$, ($n \in \mathbb{N}$, $n \geq 1$), defined above, indeed behaves as an ordered *n*-tuple.

Theorem 3.9. $(x_1, \ldots, x_n) = (y_1, \ldots, y_n)$ iff $x_1 = y_1$ and ... and $x_n = y_n$.

Proof. For $n = 1$, $(x_1) = x_1$ and $(y_1) = y_1$, so the proposition holds for $n = 1$.
Now suppose (induction hypothesis) that the proposition holds for n, i.e., $(x_1, \ldots, x_n) = (y_1, \ldots, y_n)$ iff $x_1 = y_1$ and ... and $x_n = y_n$. Next suppose that $(x_1, \ldots, x_n, x_{n+1}) = (y_1, \ldots, y_n, y_{n+1})$, i.e., $((x_1, \ldots, x_n), x_{n+1}) = ((y_1, \ldots, y_n), y_{n+1})$. Then by Theorem 3.7, $(x_1, \ldots, x_n) = (y_1, \ldots, y_n)$ and $x_{n+1} = y_{n+1}$. Hence, by the induction hypothesis, $x_1 = y_1$ and ... and $x_n = y_n$ and $x_{n+1} = y_{n+1}$. □

The *Cartesian product* $V \times W$ of two sets V and W is by definition the set of all ordered pairs (v, w) with $v \in V$ and $w \in W$.

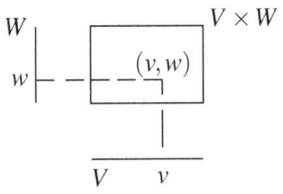

Definition 3.6 (Cartesian Product). $V \times W := \{x \mid$ there is some $v \in V$ and there is some $w \in W$ such that $x = (v, w)\}$, in other words, $V \times W := \{(v, w) \mid v \in V \wedge w \in W\}$.

Example 3.7.
$\{2, 3\} \times \{4\} = \{(2, 4), (3, 4)\}$, $\{2, 3\} \times \{4, 5\} = \{(2, 4), (3, 4), (2, 5), (3, 5)\}$,
$\{1\} \times \{4, 5\} = \{(1, 4), (1, 5)\}$, $\mathbb{R} \times \mathbb{R} = \{(x, y) \mid x \in \mathbb{R} \wedge y \in \mathbb{R}\}$.

So, $\mathbb{R} \times \mathbb{R}$ corresponds to the set of all points in the Euclidean plane:

'There is some $v \in V$ and there is some $w \in W$ such that $x = (v, w)$' can be formulated in our logical symbolism as follows: $\exists v \in V \; \exists w \in W \; [\, x = (v, w) \,]$.
So, $V \times W = \{x \mid \exists v \in V \; \exists w \in W \; [\, x = (v, w) \,]\}$.

From Definition 3.6 and Theorem 3.8 we immediately conclude:

Corollary 3.4. $V \times W = \{x \in PP(V \cup W) \mid \exists v \in V \; \exists w \in W \; [\, x = (v, w) \,]\}$, *or simply*
$$V \times W = \{(v, w) \in PP(V \cup W) \mid v \in V \wedge w \in W\}.$$

From Corollary 3.4, the Axiom of Union, the Powerset Axiom and the Separation Axiom it follows that: if V and W are sets, then so is $V \times W$.

$\{2\} \times \{4\} = \{(2,4)\}$, but $\{4\} \times \{2\} = \{(4,2)\}$. So, it is not true that for all sets V and W, $V \times W = W \times V$; in other words, the operation $\times$ is not commutative. The operation $\times$ is not associative either (see Exercise 3.11).

Instead of $V \times V$ we usually write V^2.

Example 3.8. $\{3,4\}^2 = \{3,4\} \times \{3,4\} = \{(3,3),(3,4),(4,3),(4,4)\}$.

More generally, we define V^n ($n \in \mathbb{N}$, $n \geq 1$) inductively by:

Definition 3.7. $V^1 := V$, and $V^{n+1} := V^n \times V$.

Example: $\{3,4\}^3 = \{3,4\}^2 \times \{3,4\} = \{((3,3),3),((3,3),4),((3,4),3),((3,4),4),((4,3),3),((4,3),4),((4,4),3),((4,4),4)\}$.

More generally, we define the Cartesian product with finitely many factors:

Definition 3.8. $X_{i=1}^1 V_i = V_1$ and $X_{i=1}^{n+1} V_i = (X_{i=1}^n V_i) \times V_{n+1}$.

Example 3.9. Let $V_1 = \{1,2\}$, $V_2 = \{3,4\}$ and $V_3 = \{7,8,9\}$. Then $X_{i=1}^3 V_i = (V_1 \times V_2) \times V_3 = (\{1,2\} \times \{3,4\}) \times \{7,8,9\}$.

3.4.2 Relations

We start with a few examples of *binary* relations R between the elements of a set V and the elements of a set W (or: between V and W). Instead of xRy – to be read as: x is in relation R to y – one also writes $R(x,y)$.

Example 3.10.
1. $V = M(\text{en})$ $W = W(\text{omen})$ $xRy := x$ is a son of y
2. $V = \mathbb{N}$ $W = \mathbb{N}$ $xRy := y = x+1$
3. $V = \mathbb{N}$ $W = \mathbb{R}$ $xRy := y = \sqrt{x}$
4. $V = \mathbb{N}^2$ $W = \mathbb{N}^2$ $(m,n)R(p,q) := m-n = p-q$
5. $V = \mathbb{N} \times (\mathbb{Z} - \{0\})$ $W = V$ $(m,n)R(p,q) := \frac{m}{n} = \frac{p}{q}$
6. $V = \mathbb{N}$ $W = P(\mathbb{N})$ $xRy := x \in y$.

Below are some examples of a *ternary* relation R between the elements of a set V, the elements of a set W and the elements of a set U:
1. $V = M(\text{en})$, $W = W(\text{omen})$, $U = P(\text{eople})$; $R(x,y,z) := x$ and y are parents of z.
2. $V = W = U = \mathbb{N}$; $R(x,y,z) := x+y = z$.

For reasons of efficiency, we will at this point discuss only binary relations.

The adagium 'everything is a set' also applies to relations. A relation R between sets V and W can be represented by the set $\{(v,w) \in V \times W \mid vRw\}$. For instance, the relations in Example 3.10, 1 and 2 can be represented by the sets:

3.4 Relations, Functions and Orderings*

1. $\{(x,y) \in M \times W \mid x \text{ is a son of } y\}$
2. $\{(x,y) \in \mathbb{N} \times \mathbb{N} \mid y = x+1\}$

So, we may represent the mathematical notion of 'relation' by a set: each binary relation R between the elements of a set V and those of a set W determines a subset of $V \times W$; and, conversely, each subset of $V \times W$ determines a binary relation between the elements of V and those of W. Hence, the following definition makes sense.

Definition 3.9 (Relation). R is a (binary) *relation* between V and $W := R \subseteq V \times W$.
Notation: $xRy := (x,y) \in R$. One sometimes uses $R(x,y)$ instead of xRy.

For $R \subseteq V \times W$ we define the *domain* and the *range* of R: The domain of R is the set of all elements x in V which are related to at least one element y in W; the range of R is the set of all elements y in W which are related to at least one element x in V.

Definition 3.10 (Domain and Range).
$\text{Dom}(R) := \{x \in V \mid \exists y \in W[\, xRy\,]\}$ *domain of* R
$\text{Ran}(R) := \{y \in W \mid \exists x \in V[\, xRy\,]\}$ *range of* R

For the relations in Example 3.10 Dom(R) and Ran(R) are respectively:

	Dom(R)	Ran(R)
1.	the set of all men	the set of all mothers with at least one son
2.	$\mathbb{N}$	$\mathbb{N} - \{0\}$
3.	$\mathbb{N}$	$\{y \in \mathbb{R} \mid \exists x \in \mathbb{N}\,[\,y = \sqrt{x}\,]\}$
4.	$\mathbb{N}^2$	$\mathbb{N}^2$
5.	$\mathbb{N} \times (\mathbb{Z} - \{0\})$	$\mathbb{N} \times (\mathbb{Z} - \{0\})$
6.	$\mathbb{N}$	$P(\mathbb{N}) - \{\emptyset\}$

If $R \subseteq V \times V$, then R is simply a relation on V. Example 3.10, 2 gives a relation on $\mathbb{N}$, Example 3.10, 4 a relation on $\mathbb{N}^2$ and Example 3.10, 5 a relation on $\mathbb{N} \times (\mathbb{Z} - \{0\})$.

Since a relation R between (the elements of) V and (the elements of) W may be represented by the set $\{(x,y) \in V \times W \mid xRy\}$, the set theoretic operations of intersection, union, and complement also apply to relations: $R \cap S$, $R \cup S$ and $\overline{R}$.

Similarly, the set theoretic predicates of inclusion and equality apply to relations R and S: $R \subseteq S$ and $R = S$.

Below we define two special operations on relations: the *converse* $\check{R}$, also called the *transposition* R^{T}, of R, and the composition $R;S$ of two relations R and S.

Definition 3.11 (Converse relation). Let R be a relation between V and W. Then the *converse* relation $\check{R}$ of R is the relation between W and V, defined by $w\check{R}v := vRw$. In set-theoretic terms, $\check{R} := \{(w,v) \in W \times V \mid (v,w) \in R\}$.

For the relations in Example 3.10, 1 - 4, the converse relations are respectively:
1. $\{(y,x) \in W \times M \mid y \text{ is the mother of } x\}$,
2. $\{(y,x) \in \mathbb{N} \times \mathbb{N} \mid x = y - 1\}$,
3. $\{(y,x) \in \mathbb{R} \times \mathbb{N} \mid x = y^2\}$,
4. $\{(p,q),(m,n) \in \mathbb{N}^2 \times \mathbb{N}^2 \mid p - q = m - n\}$.

Note that in example 4, $\check{R} = R$.

Let R be a relation between sets U and V and S a relation between sets V and W. Then the composition $R;S$ of R and S is the relation between U and W defined by $x(R;S)z :=$ there is some $y \in V$ such that xRy and ySz. In set theoretic terms:

Definition 3.12 (Composition). Let $R \subseteq U \times V$ and $S \subseteq V \times W$. Then $R;S := \{(x,z) \in U \times W \mid \exists y \in V \ [\ (x,y) \in R \wedge (y,z) \in S\ ]\}$ is called the *composition* of R and S. Instead of $R;S$ one also writes $R \circ S$ and (in case R and S are functions) also $S \circ R$.

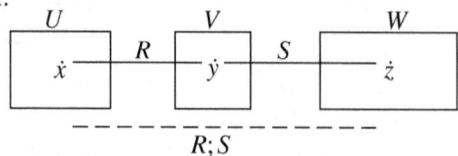

$$\begin{array}{ccc} U & V & W \\ \boxed{\dot{x}} \xrightarrow{R} & \boxed{\dot{y}} \xrightarrow{S} & \boxed{\dot{z}} \end{array}$$

$$R;S$$

Example 3.11. 1. Let R be the relation of Example 3.10, 2, $R \subseteq \mathbb{N} \times \mathbb{N}$, defined by $xRy := y = x+1$, and let S be the relation of Example 3.10, 3, $S \subseteq \mathbb{N} \times \mathbb{R}$, defined by $ySz := z = \sqrt{y}$. Then
$R;S = \{(x,z) \in \mathbb{N} \times \mathbb{R} \mid \exists y \in \mathbb{N} \ [\ (x,y) \in R \wedge (y,z) \in S\ ]\ \}$
$ = \{(x,z) \in \mathbb{N} \times \mathbb{R} \mid \exists y \in \mathbb{N} \ [\ y = x+1 \wedge z = \sqrt{y}\ ]\ \}$
$ = \{(x,z) \in \mathbb{N} \times \mathbb{R} \mid z = \sqrt{x+1}\ \}$.
In other words, $x(R;S)z := z = \sqrt{x+!}$.
2. Let M be the set of all Men and $R \subseteq M \times M$ with $xRy := y$ is the father of x. Then
$R;R = \{(x,z) \in M \times M \mid \exists y \in M \ [\ (x,y) \in R \wedge (y,z) \in R]\ \}$
$ = \{(x,z) \in M \times M \mid \exists y \in M \ [\ y$ is the father of x and z is the father of $y\ ]\ \}$
$ = \{(x,z) \in M \times M \mid z$ is the grandfather of $x\ \}$.
In other words: $x(R;R)z := z$ is the grandfather of x.

Finally, we define some special relations: the empty relation O, the universal relation L and the identity relation I.

Definition 3.13. Let V and W be any sets. Then:
$\mathsf{L} := \{(x,y) \mid x \in V \wedge y \in W\}$ is the *universal* relation between V and W. So, $x\mathsf{L}y$ for any $x \in V$ and for any $y \in W$.
$\mathsf{O} := \emptyset$ is the *empty* relation between V and W. So, not $x\mathsf{O}y$, for any $x \in V$ and for any $y \in W$.
$\mathsf{I} := \{(x,x) \mid x \in V\}$ is the *identity* relation on V (or the *diagonal* of $V \times V$). So, $x\mathsf{I}x$ for any $x \in V$.

Notice that in fact we have for any two sets V and W a universal, an empty and an identity relation.

Also notice that in case V and W are finite sets, a relation R between V and W may be represented by a Boolean matrix. For instance, let R be the relation between $V = \{1,2,3\}$ and $W = \{1,2,3,4,5,6\}$ defined by $xRy := y = 2 \cdot x$. Then R may be represented by the following Boolean matrix:

	1	2	3	4	5	6
1		■				
2				■		
3						■

3.4 Relations, Functions and Orderings*

A Boolean matrix interpretation of relations is well suited for many purposes and also used as one of the graphical representations of relations within *RelView*, a software tool for the evaluation of relation-algebraic expressions. The RelView system is an interactive tool for computer-supported manipulation of relations represented as Boolean matrices or directed graphs.

3.4.3 Equivalence Relations

$25 \neq 13$ and $13 \neq 1$, but 25 o'clock = 13 o'clock = 1 o'clock.
$26 \neq 14$ and $14 \neq 2$, but 26 o'clock = 14 o'clock = 2 o'clock.
and so on.

In reading off the clock we call two natural numbers equal if their difference is a multiple of twelve. Therefore, we consider the following relation R on the set $\mathbb{N}$ of the natural numbers: $nRm := n - m$ is a multiple of twelve.

In symbols: $nRm := \exists k \in \mathbb{Z} \, [\, n - m = 12 \cdot k \,]$.

Definition 3.14 (Equivalence relation). A relation R on a set V is an *equivalence relation* on $V := R$ is reflexive, symmetric and transitive, where
R is *reflexive* := for all $x \in V$, xRx;
R is *symmetric* := for all $x, y \in V$, if xRy, then yRx;
R is *transitive* := for all $x, y, z \in V$, if xRy and yRz, then xRz.

Example 3.12. 1. The relation R on the set $\mathbb{N}$, defined by $nRm := n - m$ is a multiple of twelve, is an equivalence relation on $\mathbb{N}$.
2. The relation = on $\mathbb{N}$ is an equivalence relation.
3. The relation R on the set $\mathbb{N}^2$, defined by $(m,n)R(p,q) := m+q = n+p$ (or $m-n = p-q$), is an equivalence relation on $\mathbb{N}^2$.
4. The relation R on the set $\mathbb{N} \times (\mathbb{Z} - \{0\})$, defined by $(m,n)R(p,q) := m \cdot q = n \cdot p$ (or $\frac{m}{n} = \frac{p}{q}$), is an equivalence relation on $\mathbb{N} \times (\mathbb{Z} - \{0\})$.
5. The relation *is parallel to or is equal to* on the set of all straight lines in the Euclidean plane is an equivalence relation.

Definition 3.15 (Equivalence class). Let R be an equivalence relation on a set V. The *equivalence class* $[v]_R$, also called *v modulo R*, of an element v of V with respect to R is by definition the subset of V, consisting of all those elements w in V for which vRw. Instead of $[v]_R$ one sometimes writes v/R.

$$[v]_R := \{w \in V \mid vRw\}$$

v is called a *representative* of the class $[v]_R$. Note that if R is an equivalence relation on V, then for all $v, w \in V$, vRw iff $[v]_R = [w]_R$.

Example 3.13. We now give the equivalence classes $[v]_R$ for the equivalence relation R on $\mathbb{N}$ from Example 3.12, 1, where $nRm := n - m$ is a multiple of 12.

$[0]_R = \{0, 12, 24, 36, \ldots\}$, $[12]_R = [0]_R$, $[24]_R = [0]_R$.
$[1]_R = \{1, 13, 25, 37, \ldots\}$, $[13]_R = [1]_R$, $[25]_R = [1]_R$.
$[2]_R = \{2, 14, 26, 38, \ldots\}$, $[14]_R = [2]_R$, $[26]_R = [2]_R$.
$\vdots$

$[11]_R = \{11, 23, 35, 47, \ldots\}$, $[23]_R = [11]_R$, $[35]_R = [11]_R$.
Thus, it would be more appropriate to indicate the numerals on the clock by $[1]_R, [2]_R, \ldots, [11]_R, [12]_R$ instead of $1, 2, \ldots, 11, 12$.

One may show that the integers and the rational numbers can be defined in terms of the natural numbers, making use of the equivalence relations R from Example 3.12, 3 and 4 respectively. So, roughly speaking, one may say that the natural numbers form the basis of all mathematics. For instance, $-1 := [(1,2)]_R$ with $(m,n)R(p,q) := m+q = n+p$ (or $m-n = p-q$) and $\frac{2}{3} := [(2,3)]_R$ with $(m,n)R(p,q) := m \cdot q = n \cdot p$ (or $\frac{m}{n} = \frac{p}{q}$). See van Dalen, Doets, de Swart, [3].

Definition 3.16 (Quotient set). Let R be an equivalence relation on V. The *quotient set* V/R or V *modulo* R is the set of all equivalence classes $[v]_R$ with $v \in V$. In other words: $V/R := \{[v]_R \mid v \in V\}$.

As an example let us consider the quotient set from Example 3.13 above, where R is the equivalence relation on $\mathbb{N}$ defined by $nRm := n - m$ is a multiple of twelve.

$$\mathbb{N}/R = \{[1]_R, [2]_R, \ldots, [11]_R, [12]_R\}.$$

$\mathbb{N}/R$ has twelve elements, corresponding to the twelve numerals on the clock. The twelve different elements of $\mathbb{N}/R$ are pairwise disjoint, i.e., $[n]_R \cap [m]_R = \emptyset$ for $n \neq m$ and $1 \leq n, m \leq 12$, and together they form the whole set $\mathbb{N}$, more precisely,

$$[1]_R \cup [2]_R \cup \ldots \cup [11]_R \cup [12]_R = \mathbb{N}.$$

Therefore we call $\mathbb{N}/R$ a *partition* of $\mathbb{N}$:

$$\mathbb{N} \begin{cases} [1]_R = \{1, 13, 25, 37, \ldots\} \\ \quad \vdots \\ [11]_R = \{11, 23, 35, 47, \ldots\} \\ [12]_R = \{0, 12, 24, 36, \ldots\} \end{cases}$$

Definition 3.17 (Partition). A collection U consisting of subsets of V is a *partition* of $V :=$ 1) $V =$ the union of all elements of U, and 2) the different elements of U are pairwise disjoint.

Clearly, every partition U consisting of subsets of V defines an equivalence relation R: xRy iff x and y belong to the same element of U. Conversely,

Theorem 3.10. *If R is an equivalence relation on V, then V/R is a partition of V.*

Proof. We have to show: 1) $V =$ the union of all elements in V/R, and 2) the different elements of V/R are pairwise disjoint.
1) Let $v \in V$. Then $v \in [v]_R$. Conversely, if $w \in [v]_R$, then $w \in V$.

3.4 Relations, Functions and Orderings*

2) Suppose $[v]_R \neq [w]_R$. Then not vRw. (1)
Now suppose $[v]_R \cap [w]_R \neq \emptyset$. Then for some $u \in V$, $u \in [v]_R$ and $u \in [w]_R$. But then vRu and uRw, and consequently – since R is an equivalence relation – vRw. This is a contradiction of (1). Therefore, $[v]_R \cap [w]_R = \emptyset$ if $[v]_R \neq [w]_R$. □

3.4.4 Functions

Let V and W be sets. 'f is a (total) *function* or mapping from V to W' means intuitively: f assigns to each $v \in V$ a uniquely determined $w \in W$. *Notation*: $f : V \to W$.

For each $v \in V$, the uniquely determined $w \in W$, which is assigned by f to v, is called the *image* (under f) of v. *Notation*: $w = f(v)$.

An example from daily life is the function f from the set M of all men to the set W of all women, which assigns to every person x his or her mother $f(x)$.

Example 3.14. Examples of functions $f : V \to W$:

1. $V = \{1, 2, 3\}$, $W = \{4, 5, 6\}$, $f(1) = 4$
    $f(2) = 4$
    $f(3) = 6$

2. $V = \{1, 2, 3\}$, $W = \{4, 5, 6\}$, $f(1) = 4$
    $f(2) = 5$
    $f(3) = 6$

3. $V = \{1, 2, 3\}$, $W = \{4, 5\}$, $f(1) = 4$
    $f(2) = 4$
    $f(3) = 5$

4. $V = \{1, 2, 3\}$, $W = \{4, 5, 6\}$, $f(1) = 5$
    $f(2) = 4$
    $f(3) = 6$

5. $V = \mathbb{N}$, $W = \mathbb{N}$, $\begin{cases} f(n) = 0 \text{ if } n \text{ is even,} \\ f(n) = 1 \text{ if } n \text{ is odd.} \end{cases}$

6. $V = \mathbb{N}$, $W = P(\mathbb{N})$, $f(n) = \{n\}$.

7. $V = \mathbb{N}^2$, $W = \mathbb{Z}$, $f((n, m)) = n - m$.

8. $V = \mathbb{R}_+$ with $\mathbb{R}_+ := \{x \in \mathbb{R} \mid x > 0\}$, $W = \mathbb{R}$, $f(x) = \log(x)$.

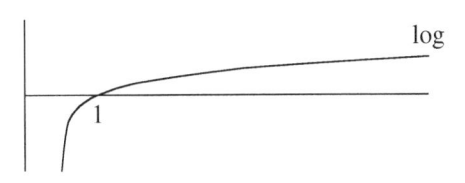

If $f : V \to W$, then f determines a set of ordered pairs, namely, $\{(v,w) \in V \times W \mid w = f(v)\}$. This set, known as the *graph* of f, has the property that for each v in V there is a unique element w in W such that (v,w) is in the set (namely $w = f(v)$). Conversely, each subset of $V \times W$ with this special property will determine a function $f : V \to W$.

The graphs of the functions from Example 3.14 are respectively:
1. $\{(1,4),(2,4),(3,6)\}$,
2. $\{(1,4),(2,5),(3,6)\}$,
3. $\{(1,4),(2,4),(3,5)\}$,
4. $\{(1,5),(2,4),(3,6)\}$,
5. $\{(n,m) \in \mathbb{N}^2 \mid (n \text{ is even} \wedge m = 0) \vee (n \text{ is odd} \wedge m = 1)\}$,
6. $\{(n,y) \in \mathbb{N} \times P(\mathbb{N}) \mid y = \{n\}\}$,
7. $\{((n,m),y) \in \mathbb{N}^2 \times \mathbb{Z} \mid y = n - m\}$,
8. $\{(x,y) \in \mathbb{R}_+ \times \mathbb{R} \mid y = \log(x)\}$.

Any function can thus be represented by its graph. In fact, it is common in set theory to identify a function with its graph and thus reduce the notion of function to the notion of set. This is what we will do.

Definition 3.18 (Function). f is a (total) *function* from V to $W := f$ is a relation between V and W, such that for each $v \in V$ there is a unique $w \in W$ such that $(v,w) \in f$. **Notation**: $f : V \to W$.

Because a function $f : V \to W$ is by definition a relation, Definition 3.10 defines the domain $\text{Dom}(f)$ and the range $\text{Ran}(f)$ of f. It is evident that for $f : V \to W$, $\text{Dom}(f) = V$ and $\text{Ran}(f) = \{w \in W \mid \exists v \in V \, [\, w = f(v)\,]\}$. For instance, for the function f in Example 3.14, 1, $\text{Ran}(f) = \{4,6\}$; and in Example 3.14, 2, $\text{Ran}(f) = \{4,5,6\}$.

We shall maintain the notation introduced earlier, that we write $f(v)$ for the unique $w \in W$ such that $(v,w) \in f$. Thus we have, for all $v \in V$, $w \in W$: $w = f(v)$ if and only if $(v,w) \in f$. From time to time we will write $v \mapsto f(v)$ for $(v,f(v)) \in f$.

Sometimes it is convenient to have at one's disposal also the notion of *partial function*. Intuitively, a partial function f from V to W assigns to some (not necessarily all) $v \in V$ a uniquely determined $w \in W$.

Definition 3.19 (Partial function). f is a *partial function* from V to $W := f$ is a relation between V and W, such that for all $v \in V$ and $w, w' \in W$, if $(v,w) \in f$ and $(v,w') \in f$, then $w = w'$.

If f is a partial function from V to W, then $\text{Dom}(f) := \{v \in V \mid \text{there is a } w \in W \text{ such that } (v,w) \in f\}$. If f is a (total) function from V to W, then $\text{Dom}(f) = V$.

Definition 3.20. If $f : V \to W$ and $V' \subseteq V$, then $f(V') := \{f(v) \mid v \in V'\}$.
If $f : V \to W$ and $W' \subseteq W$, then $f^{-1}(W') := \{v \in V \mid f(v) \in W'\}$.

The notation $f(V')$ may be ambiguous, because a subset of V may at the same time be an element of V.

Remark: Let W be any set. Then $\emptyset \subseteq \emptyset \times W$. Further, because $\emptyset$ has no elements, it follows that for each $v \in \emptyset$ there is a unique $w \in W$ such that $(v,w) \in \emptyset$. Hence, by Definition 3.18, $\emptyset$ is a function from $\emptyset$ to W, in other words $\emptyset : \emptyset \to W$. Since $\emptyset$ is the only relation with $\text{Dom}(\emptyset) = \emptyset$, $\emptyset$ is also the only function from $\emptyset$ to W.

If $f : V \to W$, then $f \subseteq V \times W$ and hence, $f \in P(V \times W)$.

3.4 Relations, Functions and Orderings*

Definition 3.21 (Set of all functions $f : V \to W$).
$W^V :=$ the set of all functions $f : V \to W$, i.e., $W^V := \{f \in P(V \times W) \mid f : V \to W\}$.
So, if V and W are sets, then by the separation axiom W^V is a set too.

Example 3.15. The set $\{1,2,3\}^{\{5,6\}}$ has $3^2 = 9$ elements $f_1, \ldots, f_9$, the functions $f_1, \ldots, f_9$ being defined by the following scheme:

	f_1	f_2	f_3	f_4	f_5	f_6	f_7	f_8	f_9
5	1	1	1	2	2	2	3	3	3
6	1	2	3	1	2	3	1	2	3

i.e., $\begin{cases} f_1(5) = 1, f_2(5) = 1, \ldots, f_9(5) = 3, \\ f_1(6) = 1, f_2(6) = 2, \ldots, f_9(6) = 3. \end{cases}$

The reader should check for him or her self that $\{5,6\}^{\{1,2,3\}}$ has $2^3 = 8$ elements.

Theorem 3.11. *If W is a set with m elements and V is a set with n elements ($m, n \in \mathbb{N}$), then W^V has m^n elements.*

So, if W is a set with 10 elements and V has 6 elements, then there are, by this theorem, 10^6, i.e., one million, functions $f : V \to W$.

Proof. Throughout the following argument, let $m \in \mathbb{N}$ be fixed, and let W be a fixed set with m elements. Let $\Phi(n) :=$ if V is any set with n elements, then W^V has m^n elements. Then Theorem 3.11 says: for every $n \in \mathbb{N}$, $\Phi(n)$.
 By induction it suffices to show: $\Phi(0)$ and for all $k \in \mathbb{N}$, $\Phi(k) \to \Phi(k+1)$.
Induction basis $\Phi(0)$: if V has 0 elements, i.e., $V = \emptyset$, then $\emptyset$ is the only function from V to W; hence, $W^V = \{\emptyset\}$; so $W^\emptyset$ has $m^0 = 1$ element.
Induction step $\Phi(k) \to \Phi(k+1)$: Suppose $\Phi(k)$, i.e., if V is any set with k elements, then W^V has m^k elements. We must now show that $\Phi(k+1)$ holds. So let $\{v_1, \ldots, v_k, v_{k+1}\}$ be a set with $k+1$ elements. By the induction hypothesis $\Phi(k)$ there are m^k different functions from $\{v_1, \ldots, v_k\}$ to W.

$$\begin{array}{c|cccc} & f_1 & f_2 & \cdots & f_{m^k} \\ v_1 & * & * & & * \\ v_2 & * & * & & * \\ \vdots & & & & \\ v_k & * & * & & * \\ v_{k+1} & & & & \end{array}$$

For each i, $1 \leq i \leq m^k$, there are now m different possible choices for $f_i(v_{k+1})$. Thus, there are $m \cdot m^k = m^{k+1}$ different functions from $\{v_1, \ldots, v_k, v_{k+1}\}$ to W. □

In mathematics (especially analysis) one frequently uses sequences of objects. We can now give an exact formulation of the notion of sequence.

Definition 3.22 (Sequence). An (infinite) *sequence* of elements of V is a function f from $\mathbb{N}$ to V. **Notation**: $f(0), f(1), f(2), \ldots$.
A (finite) *sequence* of elements of V is a function f from $\{0, \ldots, n\}$ to V, for some $n \in \mathbb{N}$. **Notation**: $f(0), \ldots, f(n)$.

The functions $f : V \to W$ in Example 3.14, 2, 4, 6 and 8 have the property that they assign distinct elements of W to distinct elements of V; in other words: for all $v, v' \in V$, if $v \neq v'$, then $f(v) \neq f(v')$, or (equivalently): for all $v, v' \in V$, if $f(v) = f(v')$, then $v = v'$. We call such functions *injective* (one-to-one). Notice that the other functions in Example 3.14 do not have this property.

Definition 3.23 (Injection). $f : V \to W$ is *injective* or an *injection* := for all $v, v' \in V$, if $v \neq v'$, then $f(v) \neq f(v')$. In logical notation: $\forall x \in V \; \forall x' \in V \; [\, x \neq x' \to f(x) \neq f(x') \,]$. **Notation**: Intuitively, the existence of an injection $f : V \to W$ means that the set V cannot be larger than W; therefore we write $f : V \leq_1 W$ to indicate that $f : V \to W$ is injective.

The functions $f : V \to W$ in Example 3.14, 2, 3, 4, 7 and 8 have the property that each element $w \in W$ is the image (under f) of an element $v \in V$. We call such functions *surjective* (onto). Note that the other functions in Example 3.14 do not have this property.

Definition 3.24 (Surjection). $f : V \to W$ is *surjective* or a *surjection* := for every $w \in W$ there is a $v \in V$ such that $w = f(v)$. In logical notation: $\forall y \in W \; \exists x \in V \; [\, y = f(x) \,]$. In other words, $f : V \to W$ is surjective if and only if $\text{Ran}(f) = W$.

The functions in Example 3.14, 1 and 5 are neither injective nor surjective. Those in Example 3.14, 2, 4 and 8 have both properties. We call such functions *bijective*.

Definition 3.25 (Bijection). $f : V \to W$ is *bijective* or a *bijection* := f is both injective and surjective. **Notation**: Intuitively, the existence of a bijection $f : V \to W$ means that the sets V and W are equally large; therefore one writes $f : V =_1 W$ to indicate that $f : V \to W$ is bijective.

A bijection $f : V \to W$ gives a **one-one correspondence** *between the elements of V and the elements of W: for each $v \in V$ there is exactly one (f is a function) $w \in W$ such that $w = f(v)$ and for each $w \in W$ there is at least one (f is surjective) and precisely one (f is injective) $v \in V$ such that $w = f(v)$.*

Definition 3.26 (Canonical function). Let R be an equivalence relation on V. The canonical function $f : V \to V/R$ is defined by $f(x) := [x]_R$. It is of course surjective, but in general not injective.

Definition 3.27 (Characteristic function). Let $U \subseteq V$. The *characteristic function* $K_U : V \to \{0, 1\}$ *of U* is defined by $K_U(v) = \begin{cases} 1 & \text{if } v \in U, \\ 0 & \text{if } v \notin U. \end{cases}$

In the special case that $U \subseteq \mathbb{N}$, the characteristic function $K_U : \mathbb{N} \to \{0, 1\}$ of U may be represented by the infinite sequence $K_U(0), K_U(1), K_U(2), K_U(3), \ldots$ of 0's and 1's (see Definition 3.22). For instance, let $U = \{0, 2, 4, 6, \ldots\}$, then $K_U = 1\,0\,1\,0\,1\,0\,1\,\ldots$.

Since we have defined a function $f : V \to W$ as a set $\{(v, w) \in V \times W \mid w = f(v)\}$ of ordered pairs, the equality relation between functions is thereby determined. Let

3.4 Relations, Functions and Orderings*

$f : V \to W$ and $g : V \to W$. Then, by the axiom of extensionality: $f = g$ iff f and g have the same elements, i.e., for all $v \in V$ and for all $w \in W$, $(v,w) \in f$ iff $(v,w) \in g$. In other words, $f = g :=$ for all $v \in V$ and for all $w \in W$, $w = f(v)$ iff $w = g(v)$. So, for $f, g : V \to W$, $f = g$ iff for all $v \in V$, $f(v) = g(v)$.

In logical notation: $f = g := \forall x \in V[f(x) = g(x)]$.

Theorem 3.12. *The function $K : P(V) \to \{0,1\}^V$, defined by $K(U) := K_U$ (i.e., K assigns to each subset U of V the characteristic function K_U of U) is a bijection.*

Proof. We first show that K is injective. So, suppose $U_1 \neq U_2$, i.e., there is some $v \in V$ such that ($v \in U_1$ and $v \notin U_2$) or ($v \in U_2$ and $v \notin U_1$). Then ($K_{U_1}(v) = 1$ and $K_{U_2}(v) = 0$) or ($K_{U_2}(v) = 1$ and $K_{U_1}(v) = 0$). So, there is a $v \in V$ such that $K_{U_1}(v) \neq K_{U_2}(v)$, and hence $K_{U_1} \neq K_{U_2}$.

Next we show that K is surjective. Suppose $f \in \{0,1\}^V$. Let $U_f := \{v \in V \mid f(v) = 1\}$. Then for all $v \in V$, $K_{U_f}(v) = 1$ iff $v \in U_f$, i.e, for all $v \in V$, $K_{U_f}(v) = 1$ iff $f(v) = 1$. Hence, for all $v \in V$, $K_{U_f}(v) = f(v)$. Therefore, $f = K_{U_f}$. □

Let $f : U \to V$ and $g : V \to W$. Since f and g are (special) relations, the composition $f;g$ of f and g has been defined according to Definition 3.12.

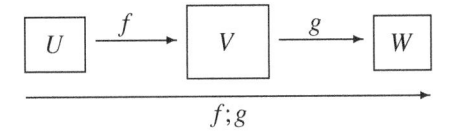

$$U \xrightarrow{f} V \xrightarrow{g} W$$
$$f;g$$

Applying $f;g$ to an element $x \in U$, we first apply f to x and next g to $f(x)$, resulting in $g(f(x))$. So, in the case of the composition of functions $f : U \to V$ and $g : V \to W$ it is attractive to write $g \circ f$ instead of $f;g$, where $(g \circ f)(x) := g(f(x))$.

Definition 3.28 (Composition of functions). Let $f : U \to V$ and $g : V \to W$. Then the *composition* $g \circ f : U \to W$ of f and g is defined by $(g \circ f)(x) = g(f(x))$.

Example 3.16. Let $f : \mathbb{N} \to \mathbb{Z}$ be defined by $f(n) := -n$. Let $g : \mathbb{Z} \to \mathbb{Q}$ be defined by $g(m) := \frac{1}{2}m$. Then $g \circ f : \mathbb{N} \to \mathbb{Q}$ is defined by $(g \circ f)(n) = -\frac{1}{2}n$.

If $f : V \to W$ is a bijection, then there is – because f is surjective – for each $w \in W$ at least one $v \in V$ such that $w = f(v)$, and – because f is injective – there is for each $w \in W$ at most one $w \in V$ such that $w = f(v)$. Hence, if $f : V \to W$ is a bijection, then for each $w \in W$ there is precisely one $v \in V$ such that $w = f(v)$.

Definition 3.29 (Inverse function). Let $f : V \to W$ be a bijection. Then the *inverse function* $f^{-1} : W \to V$ is defined by $f^{-1}(w) :=$ the unique element v in V such that $w = f(v)$.

Note that the inverse function f^{-1} of a bijection f equals the converse $\check{f}$ of f (see Definition 3.11). If $f : V \to W$ is a bijection, then $f^{-1} \circ f : V \to V$ is the identity function on V and $f \circ f^{-1} : W \to W$ is the identity function on W.

Example 3.17. Let $\mathbb{N}_{even}$ be the set of all even natural numbers and define $f : \mathbb{N} \to \mathbb{N}_{even}$ by $f(n) := 2n$. Then $f : \mathbb{N} \to \mathbb{N}_{even}$ is a bijection and $f^{-1} : \mathbb{N}_{even} \to \mathbb{N}$ is defined by $f^{-1}(m) := \frac{1}{2}m$.

Let $\mathbb{R}_+$ be the set of all real numbers greater than 0 and define $f : \mathbb{R}_+ \to \mathbb{R}$ by $f(x) := \log(x)$ (see Example 3.14, 8). Then $f : \mathbb{R}_+ \to \mathbb{R}$ is a bijection and $f^{-1} : \mathbb{R} \to \mathbb{R}_+$ is defined by $f^{-1}(x) := e^x$.

Definition 3.30. Let $f : V \to W$ and $V_0 \subseteq V$. Then the *restriction* $f\lceil V_0 : V_0 \to W$ is defined by $(f\lceil V_0)(x) := f(x)$.

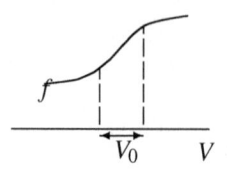

Example 3.18. Let $f : \mathbb{R} \to \mathbb{R}$ be defined by $f(x) := \sin \pi x$. Then $f\lceil \mathbb{Z} : \mathbb{Z} \to \mathbb{R}$ is defined by $(f\lceil \mathbb{Z})(m) = \sin \pi m = 0$ (for $m \in \mathbb{Z}$).

3.4.5 Orderings

We start with giving six examples of an ordering relation R on a given set V.

Example 3.19.
1. $V = P(\{v,w\}) = \{\emptyset, \{v\}, \{w\}, \{v,w\}\}$ with $xRy := x \subseteq y$.
2. $V = \{1, 2, 3, 4, 6, 8, 12, 24\}$ with $xRy := x$ is a divisor of y.
3. V is the set M of all men with $xRy := x$ is at least as old (in years) as y.
4. $V = \mathbb{Z}$ with $xRy := x \leq y$.
5. $V = \mathbb{N}$ with $xRy := x \leq y$.
6. $V = \mathbb{N} \times \mathbb{N}$ and $(n,m)R(x,y) := n \leq x$ or $(n = x$ and $m \leq y)$.
   $(0,0), (0,1), (0,2), \ldots, (1,0), (1,1), (1,2), \ldots, (2,0), \ldots$

The ordering in example 6 is similar to the well-known ordering of words in a dictionary. Therefore we call this ordering the *lexicographic ordering* on $\mathbb{N} \times \mathbb{N}$.

Definition 3.31 (Partial ordering).
A relation R on a set V is a *partial ordering* on $V :=$
1. R is *reflexive*, i.e., for all $x \in V$, xRx, and
2. R is *anti-symmetric*, i.e., for all $x, y \in V$, if xRy and yRx, then $x = y$, and
3. R is *transitive*, i.e., for all $x, y, z \in V$, if xRy and yRz, then xRz.

3.4 Relations, Functions and Orderings*

The reader should check that all relations in Example 3.19 are a partial ordering on the given set V. Instead of 'R is a partial ordering on V' one sometimes says: V is a set, partially ordered by R, or: R partially orders V, or: (V, R) is a partially ordered set. If it is clear from the context what partial ordering relation is involved, we may write: V is a partially ordered set.

The relations 1 and 2 in Example 3.19 do not have the property that any two elements are comparable via R: for instance, for $v \neq w$, $\{v\} \not\subseteq \{w\}$ and $\{w\} \not\subseteq \{v\}$. The other relations in Example 3.19 do have the property that for all $x, y \in V$, xRy or yRx (or both). In the case that R expresses the (weak) preference of an agent (voter) or a society over the elements of a set V of alternatives or candidates, reading xRy as 'the agent judges x is at least as good as y', 'xRy and yRx' expresses that the agent is indifferent between x and y. Anti-symmetry then expresses that indifference between two distinct elements of V does not occur and transitivity expresses that the preference of the agent is rational.

Definition 3.32 (Complete relation). A relation R on a set V is *complete* := for all $x, y \in V$, xRy or yRx. In other words, any two elements in V are related via R.

Notice that a complete relation on V is by definition reflexive: taking $x = y$, $(xRy$ or $yRx)$ implies xRx.

Definition 3.33 (Weak ordering). A relation R on a set V is a *weak ordering* on V := R is complete and transitive.

The relations in Example 3.19, 3, 4, 5 and 6 are a weak ordering on the given set V. Notice that the third relation is not anti-symmetric: two different men may have the same age; however, the fourth, fifth and sixth are anti-symmetric.

Definition 3.34 (Linear ordering). R is a *linear* or *total* ordering or simply an *ordering* on V := R is weak ordering on V that in addition is anti-symmetric, i.e.,
1. R is complete: for all $x, y \in V$, xRy or yRx; and hence, in particular, xRx;
2. R is anti-symmetric: for all $x, y \in V$, if xRy and yRx, then $x = y$.
3. R is transitive: for all $x, y, z \in V$, if xRy and yRz, then xRz.

Relation 3 in Example 3.19 is not a linear ordering; the relations 4, 5 and 6 in Example 3.19 are linear orderings on the given sets. Whenever we refer to a subset W of a partially or totally ordered set (V, R), we will usually think of this subset W as being partially, resp. totally ordered by the restriction of R to W, i.e., $R \cap (W \times W)$.

Let R be a weak (preference) ordering on a set V of alternatives, reading xRy as: the agent (voter, judge) weakly prefers x to y, in other words: the agent judges that x is at least as good as y. Then we can express 'the agent strictly prefers x to y' by: xRy and not yRx, which we abbreviate by xPy.

Definition 3.35 (Strict associated ordering). Let R be an ordering on V. The *strict associated ordering* P of R on V is defined by $xPy := xRy$ and not yRx.

Theorem 3.13. *Let R be a (total or linear) ordering on V. Let $xPy := xRy$ and not yRx. Then P satisfies the following properties: 1. for all $x \in V$, not xPx;*

2. P is asymmetric, i.e, for all $x, y \in V$, if xPy, then not yPx;
3. P is transitive; and
4. P is connected, i.e., for all $x, y \in V$, xPy or $x = y$ or yPx.

Proof. Let R be a (total or linear) ordering on V and let $xPy := xRy$ and not yRx.
1. From this definition follows immediately that not xPx.
2. Suppose xPy, i.e., xRy and not yRx. Then certainly not yPx.
3. Suppose xPy and yPz, i.e,, xRy and yRz and hence, by transitivity of R, xRz. Also, not yRx and not zRy. In order to show xPz, we still have to show that not zRx. So, suppose zRx. Then by xRy and the transitivity of R, zRy. Contradiction.
4. It suffices to show: if $x \neq y$, then xPy or yPx. So suppose $x \neq y$. Then, because R is anti-symmetric: not xRy or not yRx (1). Because R is complete: xRy or yRx (2). From (1) and (2) follows: (not xRy and yRx) or (not yRx and xRy), i.e., yPx or xPy. □

The ordered set $(\mathbb{N}, \leq)$ has the property that each non-empty subset of $\mathbb{N}$ has a least (with respect to $\leq$) element. The ordered sets $(\mathbb{Z}, \leq)$ and $(\mathbb{Q}, \leq)$ do not have this property.

Definition 3.36 (Well-ordering). A relation R on a set V is a *well-ordering* on $V :=$
1. R is an (total) ordering on V, and
2. each non-empty subset of V has a least element (with respect to R), i.e., an element $x \in V$ such that for all $y \in V$, xRy.

So, the set $(\mathbb{N}, \leq)$ is well-ordered, but the sets $(\mathbb{Z}, \leq)$ and $(\mathbb{Q}, \leq)$ are not.

3.4.6 Structures and Isomorphisms

Frequently one is not interested in how the elements of a given set have been constructed, only in how they behave under certain given relations (and operations) on the set. For instance, given a certain set V of people, one may be interested only in how the people in the set behave under the relation 'is father of', or under the relation 'is older than', or under the relation 'is stronger than'; and sometimes one is interested in more than one relation on the same set. This brings us to the notion of structure.

Definition 3.37 (Structure). $\langle V, R_0, \ldots, R_k \rangle$ is a (relational) *structure* := V is a set and $R_0, \ldots, R_k$ are relations on V.

Remark: A more general notion of structure is obtained by considering sets together with certain relations and operations on them; see, for instance, [3].

Example 3.20. Examples of (relational) structures:
1. $\langle$ {Charles, John, Peter}, is older than $\rangle$;
2. $\langle$ {Charles, John, Peter}, is older than, is stronger than $\rangle$;
3. $\langle \mathbb{N}, < \rangle$, where $m < n := m$ is less than n;
4. $\langle \mathbb{N}, <, | \rangle$, where $m \mid n := m$ is a divisor of n;
5. $\langle \mathbb{N}_{even}, < \rangle$, where $\mathbb{N}_{even}$ is the set of all even natural numbers;
6. $\langle \mathbb{N}_{even}, <, | \rangle$.

3.4 Relations, Functions and Orderings*   159

Now, let us suppose that John is older than Charles and that Charles is older than Peter. Then there is no difference, as far as order properties are concerned, between the set {Charles, John, Peter} together with the ordering relation 'is older than' and the set {1, 2, 3} together with the ordering relation <. In both cases we get the same picture or structure:

$$
\begin{array}{ccc}
\text{John} & & 1 \\
\text{Charles} & & 2 \\
\text{Peter} & & 3
\end{array}
$$

where the vertical line denotes in the left picture the relation 'is older than' and in the right picture the relation 'is less than'. For that reason we call the two structures ⟨ {Charles, John, Peter}, is older than ⟩ and ⟨ {1, 2, 3}, < ⟩ *isomorphic*.

Definition 3.38 (Isomorphism). Let $\langle V, R_0, \ldots, R_k \rangle$ and $\langle W, S_0, \ldots, S_k \rangle$ be two (relational) structures such that for each $i = 0, \ldots, k$, R_i and S_i have the same number, say n_i, of arguments; for convenience, suppose $n_i = 2$ for all i. Let $f : V \to W$.
f is an *isomorphism* from $\langle V, R_0, \ldots, R_k \rangle$ to $\langle W, S_0, \ldots, S_k \rangle :=$
1. f is a bijection from V to W, and
2. for all $i = 0, \ldots, k$ and for all $v, w \in V$, $R_i(v, w)$ iff $S_i(f(v), f(w))$.

Example 3.21 (Isomorphisms).

1) $f : \{\text{John, Charles, Peter}\} \to \{1, 2, 3\}$, defined by $f(\text{John}) = 1$, $f(\text{Charles}) = 2$, $f(\text{Peter}) = 3$, is an isomorphism from $\langle \{\text{John, Charles, Peter}\}, \text{is older than} \rangle$ to $\langle \{1, 2, 3\}, < \rangle$, under the supposition that John is older than Charles and that Charles is older than Peter.
2) $f : \mathbb{N} \to \mathbb{N}_{even}$, defined by $f(n) = 2n$, is an isomorphism from $\langle \mathbb{N}, < \rangle$ to $\langle \mathbb{N}_{even}, < \rangle$ and likewise an isomorphism from $\langle \mathbb{N}, <, | \rangle$ to $\langle \mathbb{N}_{even}, <, | \rangle$, where $|$ is the divisibility-relation.

$$
\begin{array}{l}
\mathbb{N} \quad : 0\ 1\ 2\ 3\ 4\ \ldots \\
f \downarrow \\
\mathbb{N}_{even} : 0\ 2\ 4\ 6\ 8\ \ldots
\end{array}
$$

3) Let us suppose that John is the father of Charles and that Charles is the father of Peter. Then the function f, defined in 1), is NOT an isomorphism from $\langle \{\text{John, Charles, Peter}\}, \text{is father of} \rangle$ to $\langle \{1,2,3\}, < \rangle$, since $1 < 3$, i.e., $f(\text{John}) < f(\text{Peter})$, but not (John is the father of Peter).
4) $f : \mathbb{N} \to \mathbb{Z}$, defined by $f(2n) = n$ and $f(2n-1) = -n$, is a bijection from $\mathbb{N}$ to $\mathbb{Z}$, but it is not an isomorphism from $\langle \mathbb{N}, < \rangle$ to $\langle \mathbb{Z}, < \rangle$, since $0 < 1$, but not $f(0) < f(1)$.

$$
\begin{array}{l}
\mathbb{N} : 0\quad 1\quad 2\quad 3\quad 4\ \ldots \\
f \downarrow \\
\mathbb{Z} : 0\ -1\ 1\ -2\ 2\ \ldots
\end{array}
$$

Definition 3.39 (Isomorphic). $\langle V, R_0, \ldots, R_k \rangle$ is *isomorphic* to $\langle W, S_0, \ldots, S_k \rangle :=$ there is at least one isomorphism f from $\langle V, R_0, \ldots, R_k \rangle$ to $\langle W, S_0, \ldots, S_k \rangle$.

Example 3.22 (Isomorphic).

1) Supposing that John and Peter are equally strong and that Charles is stronger than John and Peter, $\langle\{\text{Charles, John, Peter}\}$, is stronger than $\rangle$ is isomorphic to $\langle\{2,4,6\},|\,\rangle$.

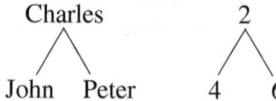

```
       Charles              2
        /   \             /   \
     John  Peter         4     6
```

In the left picture the line denotes the relation 'is stronger than' and in the right picture it denotes the relation 'is divisor of'. However, $\langle\{\text{Charles, John, Peter}\}$, is stronger than $\rangle$ is (under the same supposition as above) NOT isomorphic to $\langle\{1,2,3\},<\rangle$.

```
  . 1
  | 2
  . 3
```

2) $f: \mathbb{N} \to \mathbb{N}_{even}$, defined by $f(0) = 2, f(1) = 0, f(n) = 2n$ for $n \geq 2$, is not an isomorphism from $\langle \mathbb{N}, <\rangle$ to $\langle \mathbb{N}_{even}, <\rangle$, although f is a bijection from $\mathbb{N}$ to $\mathbb{N}_{even}$, because $0 < 1$, but not $2 = f(0) < f(1) = 0$.

$$\mathbb{N} \quad : 0\ 1\ 2\ 3\ 4\ \ldots$$
$$f \downarrow$$
$$\mathbb{N}_{even} : 2\ 0\ 4\ 6\ 8\ \ldots$$

Nevertheless, $\langle \mathbb{N}, <\rangle$ is isomorphic to $\langle \mathbb{N}_{even}, <\rangle$, since there is an isomorphism from $\langle \mathbb{N}, <\rangle$ to $\langle \mathbb{N}_{even}, <\rangle$, namely $f: \mathbb{N} \to \mathbb{N}_{even}$ defined by $f(n) = 2n$ for all $n \in \mathbb{N}$.

$$\mathbb{N} \quad : 0\ 1\ 2\ 3\ 4\ \ldots$$
$$f \downarrow$$
$$\mathbb{N}_{even} : 0\ 2\ 4\ 6\ 8\ \ldots$$

Exercise 3.9. We provide alternative notions of ordered pair:
a) $(v,w) := \{\{v,\emptyset\}, \{w,\{\emptyset\}\}\}$, and b) $(v,w) := \{\{v,\emptyset\}, \{w\}\}$.
Prove that for these definitions it holds that $(v,w) = (x,y)$ iff $v = x$ and $w = y$.

Exercise 3.10. Prove that the operation $\times$ (Cartesian Product) is distributive with respect to union and intersection, i.e., $U \times (V \cup W) = (U \times V) \cup (U \times W)$, and $U \times (V \cap W) = (U \times V) \cap (U \times W)$.

Exercise 3.11. Give an example to show that the operation $\times$ (Cartesian Product) is not associative, i.e., that not for all sets U, V and W, $U \times (V \times W) = (U \times V) \times W$.

Exercise 3.12. Let $R = \{(0,1),(0,3),(0,4),(2,1),(1,2),(4,7)\}$. Compute Dom($R$), Ran($R$) and $\check{R}$. Is R a function? Let $S = \{(1,4),(3,2),(5,0)\}$. Compute $R;S$ and $S;R$.

Exercise 3.13. a) Let U be a partition of V and define for $v,w \in V$, $vSw :=$ there is a set W in U such that both $v,w \in W$. Show that S is an equivalence relation on V.
b) Let R be an equivalence relation on V. Then $V/R = \{[v]_R \mid v \in V\}$ is the partition of V belonging to R (see Theorem 3.10). Let S be the equivalence relation belonging to V/R according to a). Prove that R and S are identical.

3.4 Relations, Functions and Orderings*

Exercise 3.14. Check whether each of the following relations on $\mathbb{Z}$ is an equivalence relation or not.
a) $R = \{(x,y) \in \mathbb{Z}^2 \mid x+y < 3\}$ b) $R = \{(x,y) \in \mathbb{Z}^2 \mid x \text{ is a divisor of } y\}$
c) $R = \{(x,y) \in \mathbb{Z}^2 \mid x+y \text{ is even }\}$ d) $R = \{(x,y) \in \mathbb{Z}^2 \mid x = y \text{ or } x = -y\}$

Exercise 3.15. Prove that in each of the following cases $\{V_r \mid r \in \mathbb{R}\}$ is a partition of $\mathbb{R} \times \mathbb{R}$. Describe geometrically the members of this partition. Find the equivalence relations corresponding to the partitions (see Exercise 3.13).
a) $V_r = \{(x,y) \in \mathbb{R}^2 \mid y = x+r\}$, b) $V_r = \{(x,y) \in \mathbb{R}^2 \mid x^2+y^2 = r\}$.
Hint: $y = x+r$ is the equation of a line and $x^2+y^2 = r$ is the equation of a circle.

Exercise 3.16. For each $n \in \mathbb{Z}$ let $V_n = \{m \in \mathbb{Z} \mid \exists q \in \mathbb{Z} [\, m = n+5q\,]\}$. Prove that $\{V_n \mid n \in \mathbb{Z}\}$ is a partition of $\mathbb{Z}$.

Exercise 3.17. Give an example of a relation, which is transitive and symmetric, but not reflexive.

Exercise 3.18. Spot the flaw in the following argument: Let R be transitive and symmetric. Then xRy and yRz implies xRz for all x, y and z. Also $xRy \to yRx$ holds for all x and y. Now take any x and y such that xRy; then, by the preceding lines, xRx. Hence R is reflexive.

Exercise 3.19. Draw diagrams for the following partially ordered sets:
a) The set of all subsets of a set with 3 elements, partially ordered by $\subseteq$.
b) The set of natural numbers $1, \ldots, 25$, partially ordered by divisibility.

Exercise 3.20. Determine which of the following sets are relations, functions, injections, surjections or bijections from $\{1,2,3,4\}$ to $\{1,2,3,4\}$:
a) $R_1 = \{(3,1),(4,2),(4,3),(2,3)\}$, b) $R_2 = \{(2,3),(1,2),(3,2),(4,3)\}$,
c) $R_3 = \{(2,1),(1,2),(4,3),(3,4)\}$, d) $R_1;R_2$ and e) $\check{R}_3$.

Exercise 3.21. Let $f : U \to V$ and $g : V \to W$. Prove: a) if $g \circ f$ is injective, then f is injective; and b) if $g \circ f$ is surjective, then g is surjective.
Let $f^* : \mathbb{N} \to \mathbb{N}$ be defined by $f^*(n) = n+1$ and let $g^* : \mathbb{N} \to \mathbb{N}$ be defined by $g^*(0) = 0$ and $g^*(n+1) = n$. Prove, using f^* and g^*, that not for all f and g:
c) if $g \circ f$ is injective, then g is injective; d) if $g \circ f$ is surjective, then f is surjective;
e) if $g \circ f$ is bijective, then f or g is bijective.

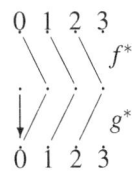

$$\begin{array}{cccc} 0 & 1 & 2 & 3 \\ & & & \end{array}$$
f^*
g^*
$$\begin{array}{cccc} 0 & 1 & 2 & 3 \end{array}$$

Exercise 3.22. Let $f : V \to W$. Prove: $\check{f}$ is a function from W to V iff f is bijective.

Exercise 3.23. $f : U \to V$ and $g : V \to W$. Prove: a) If f and g are injective, then $g \circ f$ is injective; b) If f and g are surjective, then $g \circ f$ is surjective; c) If f and g are bijective, then $g \circ f$ is bijective.

Exercise 3.24. Prove that $f : \mathbb{N} \times \mathbb{N} \to \mathbb{N}$, defined by $f(n,m) := 2^m(2n+1) - 1$, is injective.

Exercise 3.25. Prove: a) $\langle \mathbb{N}, < \rangle$ is not isomorphic to $\langle \mathbb{Z}, < \rangle$, i.e., there is no isomorphism from $\langle \mathbb{N}, < \rangle$ to $\langle \mathbb{Z}, < \rangle$; and b) $\langle \mathbb{Z}, < \rangle$ is not isomorphic to $\langle \mathbb{Q}, < \rangle$.

Exercise 3.26. Prove: $\langle \{2,4,6,12\}, / \rangle$ is isomorphic to $\langle P(\{1,2\}), \subseteq \rangle$.

3.5 The Hilbert Hotel; Denumerable Sets

All sets we experience in daily life are finite. That is why we think that a proper part is smaller than its whole. For instance, $\{2, 3\}$ is a smaller set than $\{1, 2, 3\}$. We shall see that this law for finite sets does not hold anymore for infinite sets.

The numbers $0, 1, 2, 3, \ldots$ are called *natural numbers*. $\mathbb{N} = \{0, 1, 2, \ldots\}$ is the set of all natural numbers. So, for example, $3 \in \mathbb{N}$, $5 \in \mathbb{N}$ and $1024 \in \mathbb{N}$, while $-3 \notin \mathbb{N}$, $\frac{2}{3} \notin \mathbb{N}$ and $\sqrt{2} \notin \mathbb{N}$. The numbers $\ldots, -3, -2, -1, 0, 1, 2, 3, \ldots$ are called *integers*. $\mathbb{Z} = \mathbb{N} \cup \{-1, -2, -3, \ldots\}$ is the set of all integers. Note that each natural number is an integer, but not conversely. Examples: $2 \in \mathbb{Z}$, $-2 \in \mathbb{Z}$, $0 \in \mathbb{Z}$, $3 \in \mathbb{Z}$, $\frac{2}{3} \notin \mathbb{Z}$ and $\sqrt{2} \notin \mathbb{Z}$.

Numbers of the form $\frac{p}{q}$, where $p \in \mathbb{Z}$, $q \in \mathbb{N}$, $q \neq 0$ (and p and q relatively prime) are called *rational numbers*. $\mathbb{Q}^+ = \mathbb{N} \cup \{\frac{1}{1}, \frac{1}{2}, \frac{1}{3}, \ldots\} \cup \{\frac{2}{1}, \frac{2}{2}, \frac{2}{3}, \ldots\} \cup \{\frac{3}{1}, \frac{3}{2}, \frac{3}{3}, \ldots\} \cup \ldots$ is the set of all positive rational numbers. Examples: $\frac{1}{4} \in \mathbb{Q}^+$, $2 \in \mathbb{Q}^+$, $0 \in \mathbb{Q}^+$, $\frac{3}{5} \in \mathbb{Q}^+$, $\sqrt{2} \notin \mathbb{Q}^+$, $\pi \notin \mathbb{Q}^+$. By $\mathbb{Q}$ we mean the set of all positive and negative rational numbers. Note that all integers and hence also all natural numbers are rational.

There are many, many, numbers which are not rational. Already the Greeks knew that $\sqrt{2}$ cannot be written as a quotient of the form $\frac{p}{q}$. The same holds for many other numbers, such as $\sqrt{5}$, $\log 2$, π and Euler's constant e. By $\mathbb{R}$ we mean the set of all *real numbers*. This set contains all natural numbers, all integers and all rational numbers, but also all limits of convergent sequences of rational numbers, such as $\sqrt{2}$, $\log 2$, π and e. For a precise definition of real numbers in terms of sets, see van Dalen, Doets, de Swart [3], section 12.

In this section we shall see that the set $\mathbb{N}$ of all natural numbers is as large as the set $\mathbb{Z}$ of all integers and also as large as the set $\mathbb{Q}$ of all rational numbers. In Section 3.6 we shall see that the set $\mathbb{R}$ of all real numbers is larger than each of the sets $\mathbb{N}$, $\mathbb{Z}$ and $\mathbb{Q}$, which are equally large.

From a classical or platonistic point of view the sets $\mathbb{N}$, $\mathbb{Z}$, $\mathbb{Q}$ and $\mathbb{R}$ are *actually infinite*, i.e., the Creator has created these sets, just like the planets, as a completed totality, prior to and independently of any human process of generation and as though they can be spread out completely for our inspection. Mathematicians are like astronomers who try to discover properties of the objects which have been created in its full totality by the Creator.

Since for an intuitionist like L.E.J. Brouwer (1891-1966) mathematical objects are my own mental constructions, from an intuitionistic point of view the infinite is

3.5 The Hilbert Hotel; Denumerable Sets

treated only as *potential* or *becoming* or *constructive*, i.e., the set $\mathbb{N}$ of the natural numbers is identified with the construction process for its elements: start with 0 and add 1 to each natural number which has already been constructed before. And it was one of the main achievements of Brouwer to solve the problem how we can talk constructively about the non-denumerable set $\mathbb{R}$ of the real numbers; see Chapter 8.

What does it mean that 'set V has just as many elements as set W'? The proper formulation of this question makes use of the notion of *one-one correspondence* or *matching*. For example, the set {Plato, Augustine, Wittgenstein} has just as many elements as the set {chair 1, chair 2, chair 3}, simply because we can match these sets in a suitable way:

$$\begin{array}{ccc} \text{Plato} & \text{------} & \text{chair 1} \\ \text{Augustine} & \text{------} & \text{chair 2} \\ \text{Wittgenstein} & \text{------} & \text{chair 3} \end{array}$$

From an intuitive point of view a one-one correspondence between two sets V and W is a prescription or function f that associates with every element v in V exactly one element $f(v)$ in W in such a way that conversely for every element w in W there is exactly one v in V with $w = f(v)$. More technically, a one-one correspondence between V and W is a bijective function from V to W; see Section 3.4.

Early scientists were rather puzzled by the effects of the matching-concept. In 1638 Galileo noticed that we can match the set of squares of the positive integers and the set of positive integers itself:

$$\begin{array}{cccccc} 1 & 2 & 3 & 4 & \ldots & n & \ldots \\ \downarrow & \downarrow & \downarrow & \downarrow & & \downarrow & \\ 1 & 4 & 9 & 16 & \ldots & n^2 & \ldots \end{array}$$

This was considered paradoxical, in view of Euclid's proposition that 'the whole is greater than its part' (circa 300 B.C.). However, if one thinks of billiard-balls, being labeled 1, 2, 3, 4, ... on one occasion and the same balls being labeled 1, 4, 9, 16, ... on another occasion, it becomes quite obvious that the sets in question can be matched and hence have as many elements.

This is essentially Gödel's defense that the following definition is *the* natural one for comparing sets in magnitude, also in the case of infinite sets; see Gödel [5], *What is Cantor's continuum problem?*.

Definition 3.40 (Equipollent). V is *equally great as* or *equipollent to* W (V and W are of the *same cardinality*) iff there exists a one-one correspondence between (the elements of) V and (the elements of) W. **Notation**: $V =_1 W$.

One easily may verify the following:

Theorem 3.14. *For all sets U, V and W,*
i) $V =_1 V$; *ii) if* $V =_1 W$, *then* $W =_1 V$; *iii) if* $U =_1 V$ *and* $V =_1 W$, *then* $U =_1 W$.

Proof. i) The identity function which associates with every element $x \in V$ this same x, is a one-one correspondence (bijection) between V and V. ii) Let $f : V \to W$ be a one-one correspondence (bijection) between V and W, then the inverse function

$f^{-1}: W \to V$ (see Definition 3.29) is a one-one correspondence between W and V.
iii) Let $f: U \to V$ be a one-one correspondence between U and V and $g: V \to W$ a one-one correspondence between V and W. Then the composition $g \circ f: U \to W$ (see Definition 3.28) is a one-one correspondence between U and W. □

Definition 3.41 (Finite). a) V is *finite* iff there is some natural number $n \in \mathbb{N}$ such that $V =_1 \{x \in \mathbb{N} \mid x < n\}$. b) V is *infinite* iff V is not finite.

Example 3.23. $\{Plato, Augustine, Wittgenstein\} =_1 \{1,2,3\}$; not $\{1,2,3\} =_1 \{1,2\}$.

Example 3.24. $\mathbb{N} =_1 P$, where P is the set of prime numbers. In book IX of Euclid's 'Elements' (300 B.C.) it is shown that there are infinitely many prime numbers. Euclid proceeds by constructing for each finite set of primes a prime which does not belong to it. Using the fact that there are infinitely many primes, we can find a bijection from $\mathbb{N}$ to P by running through $\mathbb{N}$ and checking whether each number is a prime. This is basically the method known as the sieve of Eratosthenes.

Theorem 3.15. $\mathbb{N} =_1 \mathbb{N}_{even}$, where $\mathbb{N}_{even} := \{x \in \mathbb{N} \mid x \text{ is even}\}$.

Proof. The correspondence or function f that associates with each natural number n in $\mathbb{N}$ the even natural number $f(n) = 2n$ in $\mathbb{N}_{even}$ is one-one: f associates with each natural number n in $\mathbb{N}$ exactly one even natural number in $\mathbb{N}_{even}$, namely $f(n) = 2n$, in such a way that conversely for every even natural number $m = 2n$ in $\mathbb{N}_{even}$ there is exactly one natural number n in $\mathbb{N}$, namely $n = \frac{m}{2}$, such that $f(n) = m$.

$\mathbb{N}$:	0	1	2	3	4	...	n	...
	↓	↓	↓	↓	↓		↓	
$\mathbb{N}_{even}$	0	2	4	6	8	...	$2n$	...

□

Hence, the proposition 'the whole is greater than its part' (Euclid) turns out to be false for infinite sets: $\mathbb{N}_{even}$ is a proper subset of $\mathbb{N}$, but $\mathbb{N}_{even}$ is still equipollent to $\mathbb{N}$. However, it is easy to see that 'a proper part is smaller than the whole' is true for finite sets.

$\mathbb{N} =_1 V$ means that there is a one-one correspondence v between the sets $\mathbb{N}$ and V:

$\mathbb{N}$:	0	1	2	3	4	5	...
V:	$v(0)$	$v(1)$	$v(2)$	$v(3)$	$v(4)$	$v(5)$	...

If V is equipollent to $\mathbb{N}$, we say that V is *denumerable*: $V = \{v(0), v(1), v(2), ...\}$.

Definition 3.42 (Denumerable; Enumerable; Countable).
V is *denumerable* $:= \mathbb{N} =_1 V$.
V is *enumerable* or *countable* $:= V$ is finite or denumerable.
A one-one correspondence v between $\{0, 1, ..., n\}$ or $\mathbb{N}$ respectively and V is called an *enumeration* of V.

Remark 3.1. The usage of the terminology is not firmly established. Instead of 'denumerable' some authors use *countably infinite*.

3.5 The Hilbert Hotel; Denumerable Sets

Suppose somewhere in heaven is a hotel, called the *Hilbert hotel*, after the German mathematician and philosopher David Hilbert (1862 – 1943), with as many rooms as there are natural numbers. We also suppose that in every room there is exactly one guest: $g_0, g_1, g_2, g_3, \ldots$.

room:	0	1	2	3	4	5	...
	g_0	g_1	g_2	g_3	g_4	g_5	

So, the Hilbert hotel is full in the sense that there is a one-one correspondence g between the set of room numbers $\{0,1,2,\ldots\}$ and the set $\{g_0,g_1,g_2,\ldots\}$ of guests.

At a certain day two new guests, g_{-1} and g_{-2}, arrive at the reception and both ask for a private room; neither the two new guests nor the existing guests want to share a room with somebody else. The receptionist, who studied mathematics and philosophy, had to think a little while, but found an easy solution: let all the existing guests move two rooms; then the first two rooms are becoming free and can be given to the two new guests. The result is the following room assignment:

room:	0	1	2	3	4	5	...
	g_{-2}	g_{-1}	g_0	g_1	g_2	g_3	

We see that the two sets $\{g_0,g_1,g_2,\ldots\}$ and $\{g_{-2},g_{-1},g_0,g_1,g_2,\ldots\}$ are equally large: the number of rooms did not change. We also see that $\mathbb{N} = \{0,1,2,\ldots\}$ is as large as $\{-2,-1\} \cup \mathbb{N}$, in other words: there is a one-one correspondence f between these two sets: $f(0) = -2$, $f(1) = -1$ and $f(n+2) = n$ for $n \geq 0$.

Theorem 3.16. a) $\mathbb{N} =_1 \{-2,-1\} \cup \mathbb{N}$, in other words, $\{-2,-1\} \cup \mathbb{N}$ is denumerable. b) More generally: if W is a finite set $\{w_0, \ldots, w_{k-1}\}$, $k \geq 1$, and V is a denumerable set, then $W \cup V$ is denumerable.

Proof. a) The function f from $\mathbb{N}$ to $\{-2,-1\} \cup \mathbb{N}$, defined by $f(0) = -2$, $f(1) = -1$, and $f(n+2) = n$ for $n \geq 0$, is a one-one correspondence between these two sets: f assigns to every $n \in \mathbb{N}$ exactly one element in $\{-2,-1\} \cup \mathbb{N}$, such that conversely for every element m in $\{-2,-1\} \cup \mathbb{N}$ there is exactly one $n \in \mathbb{N}$, namely $n = m+2$, with $m = f(n)$.
b) Suppose $W = \{w_0, \ldots, w_{k-1}\}$, $k \geq 1$, and V is denumerable, i.e., there is a one-one correspondence v between $\mathbb{N}$ and V. Hence, $V = \{v_0,v_1,v_2,\ldots\}$. Then the function f from $\mathbb{N}$ to $W \cup V$, defined by $f(0) = w_0, \ldots, f(k-1) = w_{k-1}$, and for each $n \in \mathbb{N}$, $f(n+k) = v_n$, is a one-one correspondence between $\mathbb{N}$ and $W \cup V$. □

So far, so good! But at a certain day all denumerably many guests, except g_0, of the Hilbert hotel want to invite a personal friend and to give him or her a private room; again nobody is willing to share his room with somebody else. Say guest g_i wants to invite g_{-i} for each $i \geq 1$. This situation is pictured in the following schema:

room:	0	1	2	3	4	5	...
	g_0	g_1	g_2	g_3	g_4	g_5	

$g_{-1}\quad g_{-2}\quad g_{-3}\quad g_{-4}\quad g_{-5}\quad ...$

The receptionist looks concerned; she could host finitely many new guests, but now she is asked to host countably many new guests, each wanting a separate room. But ... after some thinking she found a solution: let all old guests move to the room with number twice the old room number; by doing that all rooms with an odd number become empty and the new guests can be hosted in these odd numbered rooms. So, the new room assignment looks as follows:

room:	0	1	2	3	4	5	...
	g_0	g_{-1}	g_1	g_{-2}	g_2	g_{-3}	

The receptionist is proud, the guests are happy and the Hilbert hotel is doing good business. Guest g_0 can stay in room number 0, guest g_1 moves to room number 2, guest g_2 moves to room number 4, guest g_3 moves to room number 6, etc. By doing this the rooms 1, 3, 5, ... with an odd number become available and the new guests $g_{-1}, g_{-2}, g_{-3}, ...$ can occupy these rooms.

We see that the set $\mathbb{N}$ is as large as the set $\{-1, -2, -3, ...\} \cup \mathbb{N}$, this is $\mathbb{Z}$, while at the same time $\mathbb{N}$ is a proper subset of $\mathbb{Z}$.

Theorem 3.17. *a)* $\mathbb{N} =_1 \mathbb{N} \cup \{-1, -2, -3, ...\} = \mathbb{Z}$; *i.e.,* $\mathbb{Z}$ *is denumerable.*
b) More generally: if V and W are denumerable, then also $V \cup W$ is denumerable.

Proof. a) With the even natural numbers 0, 2, 4, ... in $\mathbb{N}$ we can associate respectively the numbers 0, 1, 2, ... in $\mathbb{Z}$ and with the odd natural numbers 1, 3, 5, ... in $\mathbb{N}$ we can associate respectively the numbers $-1, -2, -3, ...$ in $\mathbb{Z}$. More precisely, the function f from $\mathbb{N}$ to $\mathbb{Z}$, defined by $f(2n) = n$ and $f(2n-1) = -n$ is a one-one correspondence between $\mathbb{N}$ and $\mathbb{Z}$.
b) Suppose V and W are denumerable, i.e., there are one-one correspondences v and w between $\mathbb{N}$ and V, respectively W. Hence, $V = \{v(0), v(1), v(2), ...\}$ and $W = \{w(0), w(1), w(2), ...\}$. Then $v(0), w(0), v(1), w(1), v(2), w(2), ...$ is an enumeration of $V \cup W$. More precisely, the function f from $\mathbb{N}$ to $V \cup W$, defined by $f(2n) = v(n)$ and for $n \geq 1$, $f(2n-1) = w(n-1)$ is a one-one correspondence between $\mathbb{N}$ and $V \cup W$. □

So far the Hilbert hotel had overcome all difficulties. The real problem started only the next day, when each of the denumerably many guests announced that he or she wants to accommodate denumerably many friends. Each guest g_i, $i \in \mathbb{N}$, wants to invite denumerably many friends $g_{i1}, g_{i2}, g_{i3}, ...$ How should the receptionist provide everybody with a private room?

3.5 The Hilbert Hotel; Denumerable Sets

room:	0	1	2	3	4	5	...
	g_0	g_1	g_2	g_3	g_4	g_5	

g_{01} g_{11} g_{21} g_{31} g_{41} g_{51} ...

g_{02} g_{12} g_{22} g_{32} g_{42} g_{52} ...

g_{03} g_{13} g_{23} g_{33} g_{43} g_{53} ...

$\vdots$ $\vdots$ $\vdots$ $\vdots$ $\vdots$ $\vdots$ ...

Although the first thought of the receptionist was that this may be impossible, after thinking approximately fifteen minutes she found a solution. For convenience, she identifies g_0 with g_{00}, g_1 with g_{10}, g_2 with g_{20}, etc. Let $V_0 = \{g_{00}, g_{01}, g_{02}, \ldots\}$, $V_1 = \{g_{10}, g_{11}, g_{12}, \ldots\}$, $V_2 = \{g_{20}, g_{21}, g_{22}, \ldots\}$, etc. Then the diagram below at the left hand side shows all the guests who have to be accommodated in a private room, i.e., $V_0 \cup V_1 \cup V_2 \cup \ldots$. Making a systematic 'walk' through the schema of guests, as indicated in the diagram below at the right hand side, gives an enumeration of all the guests in $V_0 \cup V_1 \cup V_2 \cup \ldots$. The receptionist assigns to guest g_{ij}, the jth friend of g_i, room number $\frac{1}{2}(i+j)(i+j+1) + j$. So, guest $g_0 = g_{00}$ gets room 0, guest $g_1 = g_{10}$ gets room 1, guest g_{01} gets room 2, guest $g_2 = g_{20}$ gets room 3, guest g_{11} gets room 4, guest g_{02} gets room 5, guest $g_3 = g_{30}$ gets room 6, guest g_{21} gets room 7, guest g_{12} gets room 8, guest g_{03} gets room 9, guest $g_4 = g_{40}$ gets room 10, etc.

More precisely, the function f from $V_0 \cup V_1 \cup V_2 \cup \ldots$ to $\mathbb{N}$, defined by $f(g_{ij}) = \frac{1}{2}(i+j)(i+j+1) + j$, is a one-one correspondence between the two sets in question: f assigns to every g_{ij} exactly one natural (room) number $f(g_{ij})$, such that conversely for every natural number n in $\mathbb{N}$ there is exactly one guest g_{ij} with $n = f(g_{ij})$.

V_0	V_1	V_2	V_3	V_4	...					
g_{00}	g_{10}	g_{20}	g_{30}	g_{40}	...	0	1	3	6	10 ...
g_{01}	g_{11}	g_{21}	g_{31}	g_{41}	...	2	4	7	11	...
g_{02}	g_{12}	g_{22}	g_{32}	g_{42}	...	5	8	12		...
g_{03}	g_{13}	g_{23}	g_{33}	g_{43}	...	9	13			...
$\vdots$	$\vdots$	$\vdots$	$\vdots$	$\vdots$		$\vdots$	$\vdots$	$\vdots$	$\vdots$	$\vdots$

Theorem 3.18. *a) The union of denumerably many denumerable sets $V_0, V_1, V_2, \ldots$ is denumerable.*
b) The set $\mathbb{Q}^+$ of all rational numbers greater or equal than 0 is denumerable.

Proof. a) Let $V_0 = \{v_{00}, v_{01}, v_{02}, \ldots\}$, $V_1 = \{v_{10}, v_{11}, v_{12}, \ldots\}$, $V_2 = \{v_{20}, v_{21}, v_{22}, \ldots\}$, etc. be denumerably many denumerable sets. Then the function f from $V_0 \cup V_1 \cup V_2 \cup$

... to $\mathbb{N}$, defined by $f(v_{ij}) = \frac{1}{2}(i+j)(i+j+1)+j$, is a one-one correspondence between the two sets in question: f assigns to every v_{ij} exactly one natural number $f(v_{ij}) \in \mathbb{N}$, in such a way that conversely for every natural number $n \in \mathbb{N}$ there is exactly one $v_{ij} \in V_0 \cup V_1 \cup V_2 \cup \ldots$ with $f(v_{ij}) = n$.

b) Identifying g_{ij} with the rational number $\frac{i}{j}$, leaving out g_{i0} for all $i \in \mathbb{N}$ and taking away all double occurrences of the same rational number, such as $\frac{1}{2} = \frac{2}{4} = \frac{3}{6} = \ldots$, we obtain an enumeration of all rational numbers $\frac{i}{j} \geq 0$ with $i,j \in \mathbb{N}$ and $j > 0$. □

Corollary 3.5. $\mathbb{Q}$ *is denumerable.*

Proof. $\mathbb{Q} = \mathbb{Q}^+ \cup \mathbb{Q}^-$, where $\mathbb{Q}^- = \{x \in \mathbb{Q} \mid x < 0\}$. According to Theorem 3.18 $\mathbb{Q}^+$ is denumerable. In the same way one may prove that $\mathbb{Q}^-$ is denumerable. And by Theorem 3.17 the union of two denumerable sets is again denumerable. □

Exercise 3.27. a) Prove that a) $\mathbb{Z} =_1 \mathbb{N}_{even}$; b) $\mathbb{N}_{even} =_1 \mathbb{N}_{odd}$, where $\mathbb{N}_{even} = \{x \in \mathbb{N} \mid x \text{ is even}\}$ and $\mathbb{N}_{odd} = \{x \in \mathbb{N} \mid x \text{ is odd}\}$.

Exercise 3.28. a) Prove that the set $\{0,1\}^*$ of all finite sequences of 0's and 1's is denumerable.
b) Let Σ be an alphabet, i.e., a finite set of symbols. And let Σ^* be the set of all words over Σ, i.e., Σ^* is the set of all finite sequences of elements of Σ. Prove that Σ^* is denumerable. Hint: Note that a) is a special case of b) by taking $\Sigma = \{0,1\}$.
c) Conclude that the set of all expressions in English is denumerable.

Exercise 3.29. Let V be a enumerable set. Prove that the set V^* of all finite sequences of elements of V is denumerable.

3.6 Non-enumerable Sets

In Section 3.5 we have seen that the infinite sets $\mathbb{N}$, $\mathbb{Z}$ and $\mathbb{Q}$ have the same cardinality, i.e., are equally large, although clearly $\mathbb{N}$ is a proper subset of $\mathbb{Z}$ and $\mathbb{Z}$ is a proper subset of $\mathbb{Q}$. One might be inclined to think that all infinite sets are equally large. Nothing is less true! We shall see in this section that there are many sets which are larger than the sets $\mathbb{N}$, $\mathbb{Z}$ and $\mathbb{Q}$. But first we have to explain what we mean by 'being larger than'. The natural definition of $V <_1 W$, V is smaller than W, is:
1. V may be embedded into W, i.e., there is a function f from V to W such that for all $x,y \in V$, if $x \neq y$, then $f(x) \neq f(y)$ (f is injective), but
2. there is no one-one correspondence between the (elements of) V and W. More precisely, for every function f from V to W there will be at least one element $w \in W$ such that there is no $v \in V$ with $f(v) = w$. In other words, there is no surjection $f : V \to W$ and hence there cannot be a bijection $f : V \to W$.

A first example of a non-enumerable set is the set $\{0,1\}^{\mathbb{N}}$, i.e., the set of all functions $f : \mathbb{N} \to \{0,1\}$. Since a function $f : \mathbb{N} \to \{0,1\}$ may be identified with the infinite sequence $f(0), f(1), f(2), \ldots$ of zero's and one's, the set $\{0,1\}^{\mathbb{N}}$ is also called the set of all infinite sequences of zero's and one's.

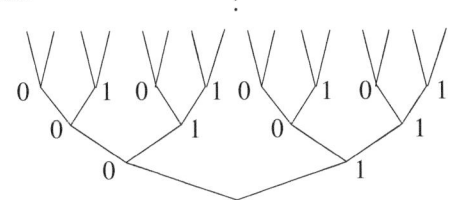

3.6 Non-enumerable Sets

It is easy to see that $\mathbb{N}$ can be embedded into $\{0,1\}^{\mathbb{N}}$: let $F : \mathbb{N} \to \{0,1\}^{\mathbb{N}}$ be defined by $F(n)$ = the infinite sequence of zero's and one's with $F(n)(i) = 0$ for $i \neq n$ and $F(n)(n) = 1$, i.e., $F(n)$ is the sequence of zero's and one's with a 1 only at the nth place. Evidently, an infinite sequence with two or more one's does not belong to the range of F. In Theorem 3.19 we shall prove that any function $F : \mathbb{N} \to \{0,1\}^{\mathbb{N}}$ will 'forget' some elements of $\{0,1\}^{\mathbb{N}}$, more precisely, that for any such function F there is an infinite sequence s of zero's and one's such that $s \neq F(i)$ for all $i \in \mathbb{N}$.

Definition 3.43 (Smaller than). V is smaller than $W :=$
a) There exists an embedding from V into W, i.e., there exists a function $f : V \to W$ such that for all $x, y \in V$, if $x \neq y$, then $f(x) \neq f(y)$ (f is injective),
b) but there is no surjection, and hence no bijection or one-one correspondence, $f : V \to W$. **Notation**: $V <_1 W$.

Theorem 3.19. $\mathbb{N} <_1 \{0,1\}^{\mathbb{N}}$.

Proof. a) There is an injection $F : \mathbb{N} \to \{0,1\}^{\mathbb{N}}$; for instance, the function F with $F(n)(i) = 0$ for all $i \neq n$ and $F(n)(n) = 1$.
b) We show that each $F : \mathbb{N} \to \{0,1\}^{\mathbb{N}}$ is not surjective, in other words, that for each such function F there is an infinite sequence s in $\{0,1\}^{\mathbb{N}}$ such that $s \neq F(i)$ for all $i \in \mathbb{N}$. So, let $F : \mathbb{N} \to \{0,1\}^{\mathbb{N}}$. Then for all $i \in \mathbb{N}$, $F(i)$ is an infinite sequence of zero's and one's. The sequences $F(i)$ may be represented in, for instance, the following diagram:

	0	1	2	3	...
$F(0) =$	0	1	0	0	...
$F(1) =$	1	0	1	0	...
$F(2) =$	0	0	1	1	...
$F(3) =$	1	1	0	0	...
$\vdots$					
$s =$	1	1	0	1	...

Construct the infinite sequence s by interchanging the zero's and one's at the diagonal $F(0)(0)$, $F(1)(1)$, $F(2)(2)$, $F(3)(3)$,..., i.e., define $s(i) := 1 - F(i)(i)$. Then for all $i \in \mathbb{N}$, s differs from $F(i)$ at place i, in other words $s(i) \neq F(i)(i)$. So, $s \neq F(i)$ for all $i \in \mathbb{N}$ and therefore $F : \mathbb{N} \to \{0,1\}^{\mathbb{N}}$ is not surjective. □

Remark 3.2. The method used in the proof of Theorem 3.19 is called *diagonalisation* or the *diagonal method* of Cantor.

Next we will show that $P(\mathbb{N}) =_1 \{0,1\}^{\mathbb{N}}$, from which it follows by Theorem 3.19 that $\mathbb{N} <_1 P(N)$.

Theorem 3.20. *For any set V, $P(V) =_1 \{0,1\}^V$.*

Proof. The function $K : P(V) \to \{0,1\}^V$, defined by $K(U) = K_U$, where K_U is the characteristic function of U, is a bijection from $P(V)$ to $\{0,1\}^V$. First we show that K is injective. So, suppose $U_1 \neq U_2$, say $v \in U_1$ and $v \notin U_2$. Then $K_{U_1}(v) = 1$ and $K_{U_2}(v) = 0$ and therefore $K(U_1) \neq K(U_2)$. To show that K is surjective, suppose $f \in \{0,1\}^V$. Taking $U := \{v \in V \mid f(v) = 1\}$, it follows that $f = K(U)$, since $f(v) = 1$ iff $v \in U$, i.e., iff $K_U(v) = 1$. □

Now that we have established that $P(\{1,\ldots,n\}) =_1 \{0,1\}^{\{1,\ldots,n\}}$ we can use Theorem 3.11 to determine the number of elements of $P(\{1,\ldots,n\})$, namely 2^n, the number of elements of $\{0,1\}^{\{1,\ldots,n\}}$. So, we know that for finite sets V the power set of V is much larger than the set V itself. A similar proposition, Theorem 3.21, holds for infinite sets, only we cannot expect to prove it by just counting. Cantor provided us with a revolutionary technique for this purpose: *diagonalisation*.

From Theorem 3.19 and Theorem 3.20 follows Cantor's theorem:

Corollary 3.6 (Cantor's Theorem). $\mathbb{N} <_1 P(\mathbb{N})$. *So, there are more subsets of $\mathbb{N}$ than there are natural numbers.*

Proof. By Theorem 3.19 there is an injection $f : \mathbb{N} \to \{0,1\}^{\mathbb{N}}$. By Theorem 3.20 there is a bijection $g : P(\mathbb{N}) \to \{0,1\}^{\mathbb{N}}$. Then $g^{-1} \circ f : \mathbb{N} \to P(\mathbb{N})$ is an injection. Suppose there were a surjection $f : \mathbb{N} \to P(\mathbb{N})$. Then $g \circ f : \mathbb{N} \to \{0,1\}^{\mathbb{N}}$ would be a surjection (see Exercise 3.23), which contradicts Theorem 3.19. □

More generally, we shall prove that any set V is smaller than its powerset $P(V)$. It is easy to see that any set V can be embedded into its powerset $P(V)$: with every element $v \in V$ corresponds the set $\{v\} \in P(V)$, more precisely, the function f from V to $P(V)$, defined by $f(v) = \{v\}$, assigns to different elements in V different elements of $P(V)$. Clearly, this function is not a one-one correspondence between V and $P(V)$: for instance, if W is a subset of V with two or more elements, then there is no $v \in V$ such that $f(v) = W$. Even stronger, below we shall show that there cannot exist a one-one correspondence between V and $P(V)$. Consequently, for any set V, the powerset $P(V)$ of V is larger than V itself. Note that we already verified this for finite sets, see Theorem 3.6.

Theorem 3.21. *For any set V, $V <_1 P(V)$.*

Proof. Clearly, the function f from V to $P(V)$, defined by $f(v) = \{v\}$, assigns to different elements of V different elements of $P(V)$. So, f embeds V into $P(V)$. Next we have to show that there cannot exist a one-one correspondence between V and $P(V)$. So, suppose g is any function from V to $P(V)$. Then we have to show that there is a set $W \in P(V)$, i.e., $W \subseteq V$, such that for no $v \in V$, $W = g(v)$. Take $W = \{v \in V \mid v \notin g(v)\}$. Then indeed, there is no $v \in V$ such that $W = g(v)$.

3.6 Non-enumerable Sets   171

For suppose for some $v_0 \in V$, $W = \{v \in V \mid v \notin g(v)\} = g(v_0)$. Then for all x, $x \in W$ iff $x \in g(v_0)$, i.e., for all $x \in V$, $x \notin g(x)$ iff $x \in g(v_0)$. In particular, taking $x = v_0$, $v_0 \notin g(v_0)$ iff $v_0 \in g(v_0)$. Contradiction. □

The preceding theorem is an eye-opener: it says in particular that

$$\mathbb{N} <_1 P(\mathbb{N}) <_1 P(P(\mathbb{N})) <_1 P(P(P(\mathbb{N}))) <_1 \ldots.$$

So, there are many degrees of infinity: the degree of infinity of $\mathbb{N}$ is smaller than the one of $P(\mathbb{N})$, which in its turn is smaller than the one of $P(P(\mathbb{N}))$, etc.

Definition 3.44 (Interval). For $a, b \in \mathbb{R}$ let $[a,b] := \{x \in \mathbb{R} \mid a \leq x \leq b\}$; $(a,b) := \{x \in \mathbb{R} \mid a < x < b\}$; $[a,b) := \{x \in \mathbb{R} \mid a \leq x < b\}$; and $(a,b] := \{x \in \mathbb{R} \mid a < x \leq b\}$. $[a,b]$ is called the *closed* (at both sides) interval between a and b, while (a,b) is called the *open* (at both sides) interval between a and b.

Next we will prove that $\mathbb{N}$ is not only smaller than $P(\mathbb{N})$, but also smaller than $[0, 1]$, the set of all real numbers between 0 and 1. We will present a direct proof here for historical reasons. The proof below is Poincaré's proof. The first direct proof was presented by Cantor.

Theorem 3.22. $\mathbb{N} <_1 [0, 1]$. *So, there are more real numbers between 0 and 1 than there are natural numbers.*

Proof. It is easy to construct an embedding from $\mathbb{N}$ into $[0, 1]$. For instance, $f : \mathbb{N} \to [0, 1]$, defined by $f(0) = 0$ and $f(n) = \frac{1}{n}$ for $n \geq 1$, is an injection. Next we have to show that there cannot exist a surjection $g : \mathbb{N} \to [0, 1]$. To do so, we shall prove that for any function $g : \mathbb{N} \to [0, 1]$ we can construct a real number b between 0 and 1 such that $b \neq g(n)$ for any $n \in \mathbb{N}$. So, let $g : \mathbb{N} \to [0, 1]$. Given this g, we can construct a chain $S_0, S_1, S_2, \ldots$ of segments (in $\mathbb{Q}$), where each segment is contained in the preceding one and the length of the segments is decreasing to 0, such that for every $n \in \mathbb{N}$, $g(n)$ is not an element of S_n.

Note that $[0, 1] = [0, \frac{1}{3}] \cup [\frac{1}{3}, \frac{2}{3}] \cup [\frac{2}{3}, 1]$. At least one of those three subsets does not contain $g(0)$, say S_0.

Suppose $S_0, \ldots, S_n$ have already been defined, such that
1. for all i, $0 \leq i \leq n$, $g(i)$ is not an element of S_i,
2. for all i, $0 \leq i < n$, $S_{i+1} \subseteq S_i$, and
3. for all i, $0 \leq i \leq n$, the length of S_i equals 3^{-i-1}.

Let $S_n = [p_n, q_n]$. Now S_n is the union of

$$[p_n, \tfrac{2p_n+q_n}{3}],\ [\tfrac{2p_n+q_n}{3}, \tfrac{p_n+2q_n}{3}] \text{ and } [\tfrac{p_n+2q_n}{3}, q_n].$$

At least one of those three subsets of S_n does not contain $g(n+1)$, say S_{n+1}. This chain of segments $S_0, S_1, S_2, \ldots$ determines a real number b (which in general will not be a rational number), such that for every $n \in \mathbb{N}$, b occurs in S_n, and hence, $b \in [0, 1]$. Now for every $n \in \mathbb{N}$, $g(n)$ does not occur in S_n, while b does occur in S_n. Hence, for every $n \in \mathbb{N}$, $b \neq g(n)$. □

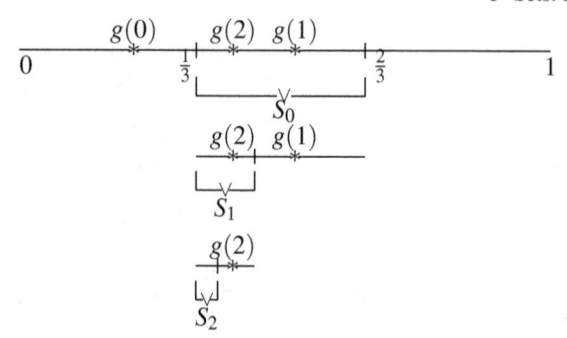

Theorem 3.22 tells us that [0, 1] is not enumerable, more precisely, for each enumeration $g : \mathbb{N} \to [0, 1]$ of elements of [0, 1], a real number b (between 0 and 1) can be constructed such that b does not occur in that enumeration, i.e., for all $n \in \mathbb{N}$, $b \neq g(n)$. On the other hand, we can define only countably many individual real numbers (between 0 and 1). This restriction is inherent to our language. Next we are going to show that [0, 1] is equipollent to $\mathbb{R}$ and consequently, by Theorem 3.22, that $\mathbb{N} <_1 \mathbb{R}$. In order to do so, we first show:

Theorem 3.23. $[0, 1] =_1 (0, 1)$.

Proof. Consider the following denumerable subset of [0, 1]: $\{1, 0, \frac{1}{2}, \frac{1}{3}, \frac{1}{4}, \ldots\}$. Now let $f : [0, 1] \to (0, 1)$ be defined as follows:

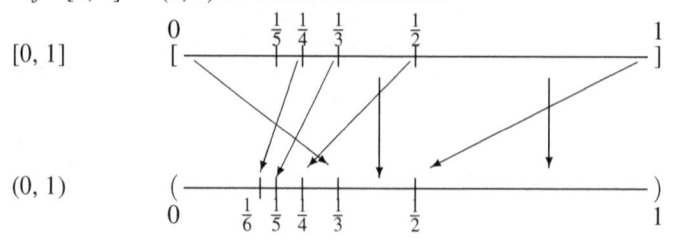

$f(1) = \frac{1}{2}$, $f(\frac{1}{n}) = \frac{1}{n+2}$ if $n \geq 2$,
$f(0) = \frac{1}{3}$, $f(x) = x$ if $x \notin \{1, 0, \frac{1}{2}, \frac{1}{3}, \frac{1}{4}, \ldots\}$.

Clearly, f is a bijection from [0, 1] to (0, 1). Therefore $[0, 1] =_1 (0, 1)$. $\square$

In the proof of Theorem 3.23 we have used the fact that the uncountable sets [0, 1] and (0, 1) have an denumerable subset, $\{1, 0, \frac{1}{2}, \frac{1}{3}, \frac{1}{4}, \ldots\}$ and $\{\frac{1}{2}, \frac{1}{3}, \frac{1}{4}, \ldots\}$ respectively. More generally, one can show:

Theorem 3.24. *If V contains a denumerable subset, then there is a proper subset of V, which is equipollent to V. Hence, Euclid's axiom 'the whole is greater than its proper part' does not hold for such sets V.*

Proof. Let $\{x_0, x_1, x_2, \ldots\}$ be a denumerable subset of V. Then $V - \{x_0\}$ is a proper subset of V, which is equipollent to V. For the function $g : V \to V - \{x_0\}$, defined by $g(x) = x$ if $x \notin \{x_0, x_1, x_2, \ldots\}$, and $g(x_i) = x_{i+1}$ for all $i \in \mathbb{N}$, is a bijection from V to $V - \{x_0\}$. $\square$

3.6 Non-enumerable Sets

By using an argument similar to the proof of Theorem 3.23 (see Exercise 3.30) one can show:

Theorem 3.25. *For $a,b \in \mathbb{R}$, $[a,b] =_1 (a,b] =_1 [a,b) =_1 (a,b)$.*

Amazingly, the length of an interval of real numbers does not change the cardinality (number of elements) of the interval. Compare an interval of real numbers with an elastic. By stretching the elastic out, its length becomes larger, but the number of points in the elastic does not change.

Theorem 3.26. *For $a,b,c,d \in \mathbb{R}$ with $a < b$ and $c < d$, $(a,b) =_1 (c,d)$.*

Proof. First we translate a and c to 0.
$f_1 : (a,b) \to (0, b-a)$, defined by $f_1(x) = x - a$, is a bijection from (a,b) to $(0, b-a)$.
$f_2 : (c,d) \to (0, d-c)$, defined by $f_2(x) = x - c$, is a bijection from (c,d) to $(0, d-c)$.
Next we stretch (or shrink) $(0, b-a)$: $f_3 : (0, b-a) \to (0, d-c)$, defined by $f_3(x) = \frac{d-c}{b-a}x$, is a bijection from $(0, b-a)$ to $(0, d-c)$. Then $f_2^{-1} \circ f_3 \circ f_1 : (a,b) \to (c,d)$ is a bijection from (a,b) to (c,d). □

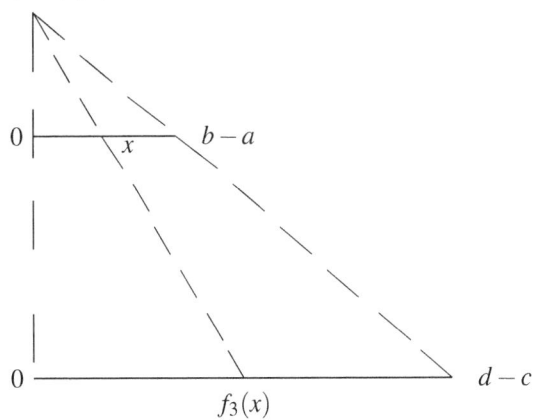

Next we show that any interval (of finite length) of real numbers is equipollent to the set $\mathbb{R}$ of all real numbers.

Theorem 3.27. $(-1, 1) =_1 \mathbb{R}$.

Proof. $f : (-1, 1) \to \mathbb{R}$, defined by $f(x) = tg(\frac{\pi}{2}x)$, is a bijection from $(-1, 1)$ to $\mathbb{R}$. □

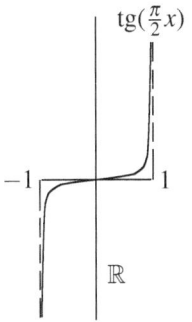

$(-1, 1)$ is again a proper subset of $\mathbb{R}$, which is equipollent to $\mathbb{R}$. So, there are as many real numbers between -1 and 1 as there are real numbers on a straight line.

By Theorem 3.23, $[0, 1] =_1 (0, 1)$, by Theorem 3.26, $(0, 1) =_1 (-1, 1)$ and by Theorem 3.27, $(-1, 1) =_1 \mathbb{R}$. Hence, $[0, 1] =_1 \mathbb{R}$. Since, according to Theorem 3.22, $\mathbb{N} <_1 [0, 1]$, it follows that:

Theorem 3.28. $\mathbb{N} <_1 \mathbb{R}$

By Theorem 3.19 and Theorem 3.22 we know that $\{0, 1\}^{\mathbb{N}}$, i.e., the set of all infinite sequences of zero's and one's, and $[0, 1]$, i.e., the set of all real numbers between 0 and 1, each are larger than $\mathbb{N}$. But how do the cardinalities of these two sets compare; in other words, is one larger than the other or are they equipollent?

It is known that each real number in $(0, 1]$ has a unique non-terminating decimal extension. For instance, $1 = 0.999\ldots$, $\frac{1}{3} = 0.333\ldots$ and $0.5 = 0.4999\ldots$. Hence, $(0, 1] =_1 \{0, 1, \ldots, 9\}^{\mathbb{N}}$. One can also show that $\{0, 1, \ldots, 9\}^{\mathbb{N}} =_1 \{0, 1\}^{\mathbb{N}}$ (see [3], section 18). Hence, it follows that:

Theorem 3.29. $\mathbb{R} =_1 (0, 1] =_1 \{0, 1, \ldots, 9\}^{\mathbb{N}} =_1 \{0, 1\}^{\mathbb{N}} =_1 P(\mathbb{N})$.

Traditionally $\mathbb{R}$ is called the *continuum*. The term is, however, also metaphorically used for $\{0, 1\}^{\mathbb{N}}$, which is usually written as $2^{\mathbb{N}}$.

Summarizing: The sets in each column below are equipollent and are strictly smaller than any set in a column to the right of it.

$\mathbb{N}$	$\{0, 1\}^{\mathbb{N}}$	$PP(\mathbb{N})$	$PPP(\mathbb{N})$	$\ldots$
$\mathbb{Z}$	$P(\mathbb{N})$	$P(\mathbb{R})$	$PP(\mathbb{R})$	
$\mathbb{Q}$	(a, b)			
	$\mathbb{R}$			

One may say that there are infinitely many degrees of infinity. As far as our limited experience goes, it turns out that (leaving aside larger sets, such as $PP(\mathbb{N}), PPP(\mathbb{N})$) most familiar infinite sets are either denumerable or equipollent to the continuum. A natural question to ask is whether there are sets which are larger than $\mathbb{N}$ and smaller than $\mathbb{R}$.

Cantor conjectured in 1878 that each infinite subset of $\mathbb{R}$ is either denumerable or equipollent to the continuum. This conjecture is known as the *Continuum Hypothesis (CH)*. A precise formulation reads:

Cantor's Continuum Hypothesis: there is no set $V \subseteq \mathbb{R}$ such that $\mathbb{N} <_1 V <_1 \mathbb{R}$.

So far the continuum hypothesis has withstood all attempts to settle it. From the work of Gödel (1938) and Cohen (1963) we know that the Continuum Hypothesis is consistent with, but at the same time independent of, the basic axioms of set theory (such as given by Zermelo and Fraenkel). The matter of its truth or falsity in the intended universe of set theory however remains unsettled. Gödel, in his paper *What is Cantor's Continuum Problem?* in [5] has analysed the evidence, which turns out to be rather in favour of a rejection.

Exercise 3.30. Prove that $(0, 1] =_1 [0, 1]$, $(0, 1] =_1 (0, 1)$ and $(0, 1] =_1 [0, 1)$.

Exercise 3.31. Let Σ be an alphabet, i.e., a finite set of symbols. L is a language over $\Sigma := L \subseteq \Sigma^*$, where Σ^* is the set of all finite sequences of elements of Σ. Prove that the set of languages over Σ is uncountable.

Exercise 3.32. Prove that $[0,1] =_1 [0,2]$ and that $(0,1) =_1 (0,3)$.

Exercise 3.33. V is *Dedekind infinite* := there is an injective function with domain V and whose range is a proper subset of V. Prove: V is infinite iff V is Dedekind infinite. Hint: If V is infinite, then by the axiom of choice V has an denumerable subsct. Next use Theorem 3.24. For the axiom of choice see van Dalen, e.a. [3].

3.7 Solutions

Solution 3.1.

$\mathbb{N} \notin \mathbb{N}$	$\{2,3\} \not\subseteq \{\mathbb{N}\}$	$\emptyset \notin \emptyset$	$\{\emptyset\} \notin \emptyset$
$\mathbb{N} \in \{\mathbb{N}\}$	$\{2\} \not\subseteq \{\mathbb{N}\}$	$\emptyset \subseteq \emptyset$	$\{\emptyset\} \not\subseteq \emptyset$
$\mathbb{N} \subseteq \mathbb{N}$	$\{2\} \subseteq \mathbb{N}$	$\emptyset \in \{\emptyset\}$	$\{\emptyset\} \subseteq \{\emptyset\}$
$\mathbb{N} \notin \{\{\mathbb{N}\}\}$	$2 \notin \{1, \{2\}, 3\}$	$\emptyset \subseteq \{\emptyset\}$	$\emptyset \subseteq \{\emptyset, \{\emptyset\}\}$
$\mathbb{N} \not\subseteq \{\mathbb{N}\}$	$\{2\} \in \{1, \{2\}, 3\}$	$\emptyset \notin \{\{\emptyset\}\}$	$\emptyset \in \{\emptyset, \{\emptyset\}\}$
$\{1,2\} \notin \mathbb{N}$	$\{1,\{2\}\} \not\subseteq \{1,\{2,3\}\}$	$\emptyset \subseteq \{\{\emptyset\}\}$	$\{\emptyset\} \subseteq \{\emptyset, \{\emptyset\}\}$
$\{1,2\} \subseteq \mathbb{N}$	$\{1,\{2\}\} \subseteq \{1,\{2\},3\}$	$\{\emptyset\} \in \{\{\emptyset\}\}$	$\{\emptyset\} \in \{\emptyset, \{\emptyset\}\}$
$\{1,2\} \notin \{\mathbb{N}\}$	$\{-2,2\} \not\subseteq \mathbb{N}$	$\{\emptyset\} \not\subseteq \{\{\emptyset\}\}$	$\emptyset \subseteq \{\{\emptyset, \{\emptyset\}\}\}$

Solution 3.2. a) $W \subseteq V$ iff $V \cap W = W$. Proof: We have to show that
(i) if $W \subseteq V$, then $V \cap W = W$, and conversely, (ii) if $V \cap W = W$, then $W \subseteq V$.
Proof of (i): Suppose $W \subseteq V$. In order to show that $V \cap W = W$ it suffices – by the axiom of extensionality – to show that $V \cap W$ and W have the same elements. Clearly, each element of $V \cap W$ is also an element of W. Conversely, that each element of W also is an element of $V \cap W$ follows from the assumption that $W \subseteq V$.
Proof of (ii): Suppose $V \cap W = W$. To show: $W \subseteq V$. So, let $x \in W$. Then it follows from $V \cap W = W$ that $x \in V \cap W$. Hence, $x \in V$.
b) $W \subseteq V$ iff $V \cup W = V$ is shown in a similar way.

Solution 3.3. a) To show: $U - (V \cup W) = (U - V) \cap (U - W)$.
Proof: $x \in U - (V \cup W) \rightleftarrows x \in U \land \neg (x \in V \cup W)$
$\rightleftarrows x \in U \land \neg (x \in V \lor x \in W)$
$\rightleftarrows x \in U \land (x \notin V \land x \notin W)$
$\rightleftarrows (x \in U \land x \notin V) \land (x \in U \land x \notin W)$
$\rightleftarrows x \in (U - V) \land x \in (U - W)$
$\rightleftarrows x \in (U - V) \cap (U - W)$.
b) $U - (V \cap W) = (U - V) \cup (U - W)$ is shown in a similar way.

Solution 3.4. a) Let $U = \emptyset$, $V = \{\emptyset\}$ and $W = \{\{\emptyset\}\}$. Then $U \in V$ and $V \in W$, but $U \notin W$. b) Proof: Suppose that $U \subseteq V$ and $V \subseteq W$, i.e., $\forall x[x \in U \to x \in V]$ and $\forall x[x \in V \to x \in W]$. Then it follows that $\forall x[x \in U \to x \in W]$, i.e., $U \subseteq W$.

Solution 3.5. $\emptyset$ has only one subset: $\emptyset$. So, $P(\emptyset) = \{\emptyset\}$.
$P(\emptyset) = \{\emptyset\}$ has $2^1 = 2$ subsets: $\emptyset$ and $\{\emptyset\}$. So, $P(P(\emptyset)) = \{\emptyset, \{\emptyset\}\}$.
$P(P(\emptyset)) = \{\emptyset, \{\emptyset\}\}$ has $2^2 = 4$ subsets: $\emptyset$, $\{\emptyset\}$, $\{\{\emptyset\}\}$ and $\{\emptyset, \{\emptyset\}\}$.
So, $P(P(P(\emptyset))) = \{\emptyset, \{\emptyset\}, \{\{\emptyset\}\}, \{\emptyset, \{\emptyset\}\}\}$.

Solution 3.6. (a) Suppose that $W \subseteq V$. Then $\forall x[x \subseteq W \rightarrow x \subseteq V]$, in other words, $\forall x[x \in P(W) \rightarrow x \in P(V)]$ and this means precisely that $P(W) \subseteq P(V)$.
(b) Suppose $P(W) \subseteq P(V)$, i.e., $\forall x[x \in P(W) \rightarrow x \in P(V)]$, in other words, $\forall x[x \subseteq W \rightarrow x \subseteq V]$. Now we know $W \subseteq W$. Hence also $W \subseteq V$.
(c) Suppose $P(W) = P(V)$. Then $P(W) \subseteq P(V)$ and $P(V) \subseteq P(W)$. Hence, applying (b) twice, $W \subseteq V$ and $V \subseteq W$. Hence $W = V$.
(d) Suppose $P(W) \in P(V)$, i.e., $P(W) \subseteq V$. Now $W \in P(W)$, and so $W \in V$.
Warning: The converse of (d), if $W \in V$, then $P(W) \in P(V)$, does not hold. Counterexample: Let $W := \{\emptyset\}$ and $V := \{\{\emptyset\}\}$. Then $P(W) = \{\emptyset, \{\emptyset\}\}$ and $P(V) = \{\emptyset, \{\{\emptyset\}\}\}$. So $P(W) \notin P(V)$, while $W \in V$.

Solution 3.7. a) Proof: Suppose $P(W) \in PP(V)$. This is equivalent to $P(W) \subseteq P(V)$. Since $W \in P(W)$, it follows that $W \in P(V)$.
b) Proof: Suppose $W \in P(V)$, i.e., $W \subseteq V$. Then $\forall x[x \subseteq W \rightarrow x \subseteq V]$, i.e., $\forall x[x \in P(W) \rightarrow x \in P(V)]$, i.e., $P(W) \subseteq P(V)$, or equivalently, $P(W) \in P(P(V))$.
c) Proof: Suppose $P(W) \subseteq PP(V)$, i.e., $\forall x[x \in P(W) \rightarrow x \in PP(V)]$. $W \subseteq W$, so $W \in P(W)$; therefore, $W \in PP(V)$; in other words, $W \subseteq P(V)$.
d) Proof: Suppose $W \subseteq P(V)$. Then $\forall x[x \subseteq W \rightarrow x \subseteq P(V)]$, i.e., $\forall x[x \in P(W) \rightarrow x \in P(P(V))]$, or, equivalently, $P(W) \subseteq P(P(V))$.

Solution 3.8. i) $\{v\} \neq \emptyset$. So, by the regularity axiom, there is some $z \in \{v\}$ such that $z \cap \{v\} = \emptyset$, i.e., $v \cap \{v\} = \emptyset$. Now suppose $v \in v$. Then $v \in v$ and $v \in \{v\}$; so, $v \cap \{v\} \neq \emptyset$. Contradiction. Therefore, by the regularity axiom it follows that $v \notin v$.
ii) $\{v_1, \ldots, v_n\} \neq \emptyset$. So, by the regularity axiom, there is some $z \in \{v_1, \ldots, v_n\}$ such that $z \cap \{v_1, \ldots, v_n\} = \emptyset$. Now suppose $v_1 \in v_2 \wedge v_2 \in v_3 \wedge \ldots v_{n-1} \in v_n \wedge v_n \in v_1$. Then there is no $z \in \{v_1, \ldots, v_n\}$ such that $z \cap \{v_1, \ldots, v_n\} = \emptyset$. Contradiction.

Solution 3.9. a) From right to left is trivial. From left to right: Suppose $(v, w) = (x, y)$, i.e., $\{\{v, \emptyset\}, \{w, \{\emptyset\}\}\} = \{\{x, \emptyset\}, \{y, \{\emptyset\}\}\}$. So, these two sets have the same elements; hence, (i) $\{v, \emptyset\} = \{x, \emptyset\}$ and $\{w, \{\emptyset\}\} = \{y, \{\emptyset\}\}$, or (ii) $\{v, \emptyset\} = \{y, \{\emptyset\}\}$ and $\{w, \{\emptyset\}\} = \{x, \emptyset\}$. In case (i) $v = x$ and $w = y$. In case (ii) it follows from $\emptyset \neq \{\emptyset\}$ that $v = \{\emptyset\}$ and $y = \emptyset$; $w = \emptyset$ and $x = \{\emptyset\}$. Hence, $v = x$ and $w = y$.
b) From right to left is trivial. So, suppose $(v, w) = (x, y)$, i.e., $\{\{v, \emptyset\}, \{w\}\} = \{\{x, \emptyset\}, \{y\}\}$. So, these two sets have the same elements. Hence, (i) $\{v, \emptyset\} = \{x, \emptyset\}$ and $\{w\} = \{y\}$, or (ii) $\{v, \emptyset\} = \{y\}$ and $\{w\} = \{x, \emptyset\}$. In case (i), $v = x$ and $w = y$. In case (ii), $v = y = \emptyset$ and $w = x = \emptyset$; so, again $v = x$ and $w = y$.

Solution 3.10.
$(u, v) \in U \times (V \cup W)$ iff $u \in U$ and $v \in V \cup W$
$\qquad u \in U$ and $(v \in V$ or $v \in W)$
$\qquad (u \in U$ and $v \in V)$ or $(u \in U$ and $v \in W)$
$\qquad (u, v) \in U \times V$ or $(u, v) \in U \times W$
$\qquad (u, v) \in (U \times V) \cup (U \times W)$.

3.7 Solutions

Solution 3.11. Counterexample: Let $U = \{1\}$, $V = \{2\}$, $W = \{3\}$. Then $U \times (V \times W) = \{1\} \times \{(2,3)\} = \{(1,(2,3))\}$, which is different from $(U \times V) \times W = \{(1,2)\} \times \{3\} = \{((1,2),3)\}$, since $(1,(2,3)) \neq ((1,2),3)$.

Solution 3.12. Dom(R) = $\{0,1,2,4\}$, Ran(R) = $\{1,2,3,4,7\}$. R is not a function, because $0 \in$ Dom(R) and there is more than one $z \in$ Ran(R) such that $(0,z) \in R$. $\check{R} = \{(1,0),(3,0),(4,0),(1,2),(2,1),(7,4)\}$. $R;S = \{(0,4),(0,2),(2,4)\}$. $S;R = \{(1,7),(3,1),(5,1),(5,3),(5,4)\}$.

Solution 3.13. a) Let U be a partition of V. To prove: S is reflexive, symmetric and transitive. (1) S is reflexive. Suppose $v \in V$. Then there is precisely one set $W \in U$ such that $v \in W$; hence, vSv. (2) S is symmetric. Suppose $v,w \in V$ and vSw, i.e., there is a set W in U such that both v and w are elements of W. Then also w and v are elements of W; hence, wSv. (3) S is transitive. Suppose $u,v,w \in V$ and uSv and vSw. Then for some W_1 in U both $u \in W_1$ and $v \in W_1$. Also for some W_2 in U both $v \in W_2$ and $w \in W_2$. Since U is a partition of V and $v \in W_1 \cap W_2$, it follows that $W_1 = W_2$. So, $u \in W_1$ and $w \in W_1$; therefore uSw.
b) vSw is defined as follows: there is a set $[u]_R$ in V/R such that $v,w \in [u]_R$, i.e., vRu and wRu. To prove: vSw iff vRw. From left to right: suppose vSw, i.e., vRu and wRu for some $u \in V$. Then vRu and uRw. Hence, vRw. From right to left: Suppose vRw. Then vRv and wRv. Hence, there is a set $[u]_R$ in V/R, namely $[v]_R$, such that $v \in [u]_R$ and $w \in [u]_R$, i.e., vSw.

Solution 3.14. a) R is neither reflexive nor transitive. b) R is not symmetric. c) R is an equivalence relation. d) R is an equivalence relation.

Solution 3.15. To prove: (1) $\cup \{V_r \mid r \in \mathbb{R}\} = \mathbb{R} \times \mathbb{R}$. (2) The elements of $\{V_r \mid r \in \mathbb{R}\}$ are pairwise disjoint. a) Proof for $V_r = \{(x,y) \in \mathbb{R}^2 \mid y = x + r\}$: (1) For any $x, y \in \mathbb{R}$ take $r := y - x$. Then $(x,y) \in V_r$. (2) Suppose $r \neq r'$. Then, clearly, $V_r \cap V_{r'} = \emptyset$. Geometrically, V_r as defined above is a straight line cutting the y-axis in r and the x-axis in $-r$. The equivalence relation R is defined by $(x_1,y_1)R(x_2,y_2) :=$ for some $r \in \mathbb{R}$, $(x_1,y_1) \in V_r$ and $(x_2,y_2) \in V_r$. Hence, $(x_1,y_1)R(x_2,y_2)$ iff $y_1 - x_1 = y_2 - x_2$.
b) The proof that $\{V_r \mid r \in \mathbb{R}\}$ with $V_r := \{(x,y) \in \mathbb{R}^2 \mid r = x^2 + y^2\}$ is a partition of $\mathbb{R} \times \mathbb{R}$ is analogous to the proof given in a). In this case V_r is a circle with centre $(0,0)$ and radius r. The equivalence relation R is defined by $(x_1,y_1)R(x_2,y_2) := x_1^2 + y_1^2 = x_2^2 + y_2^2$.

Solution 3.16. To prove: (1) $\cup\{V_n \mid n \in \mathbb{Z}\} = \mathbb{Z}$. (2) The different elements of $\{V_n \mid n \in \mathbb{Z}\}$ are pairwise disjoint. Proof: (1) Take any $m \in \mathbb{Z}$. Then there is an $n \in \mathbb{Z}$ such that $m = n + 5 \cdot q$ for some $q \in \mathbb{Z}$. So, $m \in V_n$.
(2) $V_0 = \{\ldots, -10, -5, 0, 5, 10, \ldots\}$, $V_5 = V_0$,
$V_1 = \{\ldots, -9, -4, 1, 6, 11, \ldots\}$, $V_6 = V_1$,
$V_2 = \{\ldots, -8, -3, 2, 7, 12, \ldots\}$, $V_7 = V_2$,
$V_3 = \{\ldots, -7, -2, 3, 8, 13, \ldots\}$, $V_8 = V_3$,
$V_4 = \{\ldots, -6, -1, 4, 9, 14, \ldots\}$, $V_9 = V_4$, etc.

Solution 3.17. For any set V consider the empty relation $R_\emptyset$ on V, i.e., for all $x,y \in V$, not $xR_\emptyset y$. Clearly, $R_\emptyset$ is not reflexive, but $\forall x,y \in V[xR_\emptyset y \rightarrow yR_\emptyset x]$ is logically true, since $xR_\emptyset y$ is false; in a similar way one sees that $R_\emptyset$ is transitive.

Solution 3.18. The argument presupposes there is at least one pair (x,y) such that xRy. This argument is not valid if $R = \emptyset$.

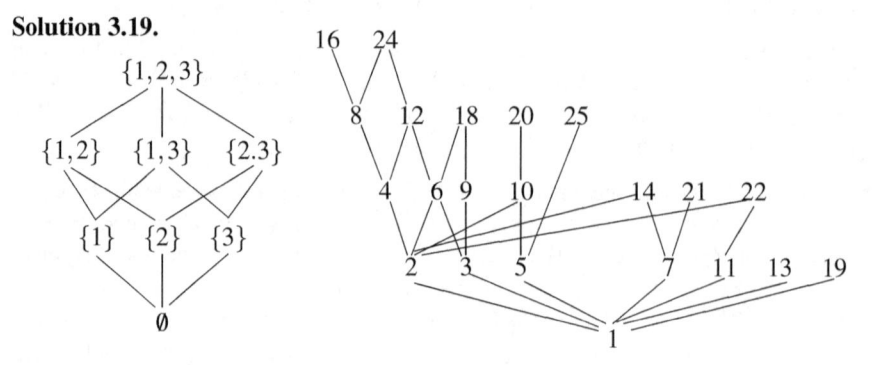

Solution 3.19.

{1,2,3}

{1,2} {1,3} {2,3}

{1} {2} {3}

$\emptyset$

16 24

8 12 18 20 25

4 6 9 10 14 21 22

2 3 5 7 11 13 19

1

Solution 3.20. a) R_1 is a relation between $\{1,2,3,4\}$ and $\{1,2,3,4\}$. b) R_2 is a function from $\{1,2,3,4\}$ to $\{1,2,3,4\}$. c) R_3 is a bijection from $\{1,2,3,4\}$ to $\{1,2,3,4\}$. d) $R_1;R_2 = \{2,2),(3,2),(4,2),(4,3)\}$ is a relation between $\{1,2,3,4\}$ and $\{1,2,3,4\}$. e) $\check{R}_3$ is a bijection from $\{1,2,3,4\}$ to $\{1,2,3,4\}$.

Solution 3.21. a) Proof: Suppose $g \circ f : U \to W$ is injective, $x \neq x'$ and $f(x) = f(x')$. Then, because $g : V \to W$ is a function, $g(f(x)) = g(f(x'))$. But $g \circ f$ is injective. So, we have a contradiction.
b) Proof: Suppose $g \circ f : U \to W$ is a surjection. Then for every $w \in W$ there is $u \in U$ such that $w = g(f(u))$. Hence, for every $w \in W$ there is $v \in V$, namely $v = f(u)$, such that $w = g(v)$. In other words: $g : V \to W$ is surjective.
c) Counterexample: $g^* \circ f^* : \mathbb{N} \to \mathbb{N}$ is an injection; but $g^*(0) = 0$ and $g^*(1) = 0$; hence, $g^* : \mathbb{N} \to \mathbb{N}$ is not an injection.
d) Counterexample: $g^* \circ f^* : \mathbb{N} \to \mathbb{N}$ is a surjection, but there is no $n \in \mathbb{N}$ such that $0 = f^*(n)$.
e) Counterexample: $g^* \circ f^* : \mathbb{N} \to \mathbb{N}$ is a bijection, but $f^* : \mathbb{N} \to \mathbb{N}$ is not a surjection and $g^* : \mathbb{N} \to \mathbb{N}$ is not an injection.

Solution 3.22. Let $f : V \to W$. $\check{f} : W \to V :=$ for all $w \in W$ there is precisely one $v \in V$ such that $\check{f}(w) = v$, or equivalently, $f(v) = w$. Hence, $\check{f} : W \to V$ iff $f : V \to W$ is a bijection.

Solution 3.23. a) Proof: Suppose $f : U \to V$ and $g : V \to W$ are injective. Then for any $x, x' \in U$, if $x \neq x'$, then $f(x) \neq f(x')$ and $g(f(x)) \neq g(f(x'))$.
b) Proof: Suppose $f : U \to V$ and $g : V \to W$ are surjective. Then for every $w \in W$ there is $v \in V$ such that $w = g(v)$. Also for every $v \in V$ there is $u \in U$ such that $v = f(u)$. Hence, for every $w \in W$ there is $u \in U$ such that $w = g(f(u))$.
c) This follows immediately from a) and b).

Solution 3.24. $f : \mathbb{N} \times \mathbb{N} \to \mathbb{N}$, defined by $f(n,m) = 2^m(2n+1) - 1$, is injective. Proof: suppose that $2^m(2n+1) - 1 = 2^{m'}(2n'+1) - 1$. Then, supposing that $m \geq m'$, $2^{m-m'} = \frac{2n'+1}{2n+1}$. But $2^{m-m'}$ is even, except when $m = m'$; and an odd number

3.7 Solutions

divided by an odd number is again an odd number. So, $2^{m-m'} = 1$ and $m = m'$. Consequently, also $n = n'$.

Solution 3.25. a) Suppose f were an isomorphism from $\langle \mathbb{N}, < \rangle$ to $\langle \mathbb{Z}, < \rangle$. Let $f(0) = z$ with $z \in \mathbb{Z}$. Since f is an isomorphism, for all $k \in \mathbb{N}$, $z \le f(k)$. So, f is not surjective, since the elements of $\mathbb{Z}$ smaller than z are not in the range of f.
b) Suppose f were an isomorphism from $\langle \mathbb{Z}, < \rangle$ to $\langle \mathbb{Q}, < \rangle$. Let $f(0) = q_1$ and $f(1) = q_2$ with $q_1, q_2 \in \mathbb{Q}$. Then between q_1 and q_2 there is a rational number q with $q_1 < q < q_2$. But there is no integer i in $\mathbb{Z}$ between 0 and 1 such that $f(i) = q$. Hence, f is not surjective.

Solution 3.26. Let $f : \{2,4,6,12\} \to P(\{1,2\})$ be defined as follows: $f(2) = \emptyset$, $f(4) = \{1\}$, $f(6) = \{2\}$ and $f(12) = \{1,2\}$. Then f is a bijection and for all $n, m \in \{2,4,6,12\}$, n/m iff $f(n) \subseteq f(m)$.

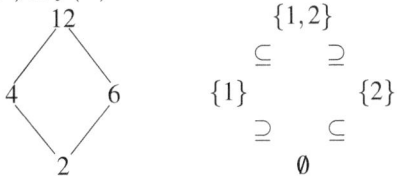

$$\begin{array}{cc}
12 & \{1,2\} \\
\diagup \diagdown & \subseteq \quad \supseteq \\
4 \quad 6 & \{1\} \quad \{2\} \\
\diagdown \diagup & \supseteq \quad \subseteq \\
2 & \emptyset
\end{array}$$

Solution 3.27. a) The function f from $\mathbb{Z}$ to $\mathbb{N}_{even}$, defined by $f(n) = 4n$ and $f(-n) = 4n - 2$ for any $n \in \mathbb{N}$, is a one-one correspondence between $\mathbb{Z}$ and $\mathbb{N}_{even}$:

$$\begin{array}{lcccccccc}
\mathbb{Z}: & 0 & -1 & 1 & -2 & 2 & -3 & 3 & \ldots \\
 & | & | & | & | & | & | & | & \\
\mathbb{N}: & 0 & 1 & 2 & 3 & 4 & 5 & 6 & \ldots \\
 & | & | & | & | & | & | & | & \\
\mathbb{N}_{even}: & 0 & 2 & 4 & 6 & 8 & 10 & 12 & \ldots
\end{array}$$

b) The function f from $\mathbb{N}_{even}$ to $\mathbb{N}_{odd}$, defined by $f(2n) = 2n + 1$ for all $n \in \mathbb{N}$, is a one-one correspondence between the two sets in question:

$$\begin{array}{lccccccc}
\mathbb{N}_{even}: & 0 & 2 & 4 & 6 & 8 & 10 & 12 & \ldots \\
 & | & | & | & | & | & | & | & \\
\mathbb{N}_{odd}: & 1 & 3 & 5 & 7 & 9 & 11 & 13 & \ldots
\end{array}$$

Solution 3.28. a) Let $\{0,1\}^n$ be the set of all finite sequences of 0's and 1's of length n ($n \in \mathbb{N}$). For each $n \in \mathbb{N}$, $\{0,1\}^n$ has 2^n elements. Now $\{0,1\}^*$ is the union of all sets $\{0,1\}^n$ with $n \in \mathbb{N}$. Hence, $\{0,1\}^*$ is the union of denumerably many finite sets and hence denumerable. b) Let Σ_n ($n \in \mathbb{N}$) be the set of all words over Σ of length n. Let k be the number of symbols (characters) in Σ. Then Σ_n has k^n elements. Now Σ^* is the union of all Σ_n with $n \in \mathbb{N}$. Hence, Σ^* is the union of denumerably many finite sets and hence denumerable.

Solution 3.29. Suppose V is enumerable. Let V_n ($n \in \mathbb{N}$) be the set of all finite sequences of elements of V of length n. For each $n \in \mathbb{N}$, V_n is enumerable. Now V^*, the set of all finite sequences of elements of V, is the union of all V_n with $n \in \mathbb{N}$. Hence, V^* is the union of denumerably many enumerable sets and hence, by Theorem 3.18, V^* is denumerable.

Solution 3.30. (i) $f : (0, 1] \to [0, 1]$, defined by $f(1) = 0$, $f(\frac{1}{n}) = \frac{1}{n-1}$ for $n \in \mathbb{N}$, $n \geq 2$, $f(x) = x$ if $x \notin \{1, \frac{1}{2}, \frac{1}{3}, \ldots\}$, is a bijection.
(ii) $f : (0, 1] \to (0, 1)$, defined by $f(\frac{1}{n}) = \frac{1}{n+1}$ for $n \in \mathbb{N}$, $n \geq 1$, $f(x) = x$ if $x \notin \{1, \frac{1}{2}, \frac{1}{3}, \ldots\}$, is a bijection.
(iii) $f : (0, 1] \to [0, 1)$, defined by $f(1) = 0$, $f(x) = x$ if $x \neq 1$, is a bijection.

Solution 3.31. By Exercise 3.28, Σ^* is denumerable. Hence, by Theorem 3.21, $P(\Sigma^*)$ is uncountable. And $P(\Sigma^*)$ is precisely the set of all languages over Σ, since L is a language over Σ iff $L \in P(\Sigma^*)$.

Solution 3.32. a) $f : [0, 1] \to [0, 2]$, defined by $f(x) = 2x$, is a bijection.
b) $f : (0, 1) \to (0, 3)$, defined by $f(x) = 3x$, is a bijection.

Solution 3.33. Suppose V is infinite. Then by the axiom of choice V has a denumerable subset $\{x_0, x_1, x_2, \ldots\}$. By Theorem 3.24, $g : V \to V - \{x_0\}$, defined by $g(x_i) = x_{i+1}$ and $g(x) = x$ if $x \notin \{x_0, x_1, x_2, \ldots\}$, is a bijection with domain V and range $V - \{x_0\}$. Conversely, suppose V is Dedekind infinite, i.e., there is an injective function with domain V and whose range is a proper subset of V. Then V cannot be finite. Therefore, V is infinite.

References

1. Arrow, K., E.Maskin, *Social Choice and Individual Values*. Yale University Press, 1951, 2012.
2. Cusanus, N. (Nikolaus von Kues), *De docta ignorantia*. In: *Philosophisch-Theologische Schriften* Band I, II, III. Herder Verlag, Wien, 1964, 1966, 1967.
3. Dalen, D. van, H.C. Doets, H.C.M. de Swart, *Sets: Naive, Axiomatic and Applied*. Pergamon Press, Oxford, 1978.
4. Gödel, K., *Russell's mathematical logic*. In: P.A. Schilpp (ed.), *The philosophy of Bertrand Russell* (Tudor, N.Y., 1944), and in: P. Benacerraf and H. Putnam, *Philosophy of Mathematics, Selected readings* (1964, 1983).
5. Gödel, K., What is Cantor's continuum problem? *American Mathematical Monthly*, vol. 54, 1947, pp. 515-525.
6. Heijenoort, J. van, *From Frege to Gödel*. A source book in mathematical logic 1879-1931. Harvard University Press, Cambridge, Mass., 1967.
7. Kant, I., *The Critique of Pure Reason*. William Benton, Publisher, Encyclopaedia Brittanica, 1952, 1978, in particular pp. 14-18 (Great Books of the Western World 42).
8. Quine, W.V., *Two Dogmas of Empiricism*. In: W.V. Quine, *From a Logical Point of View*. Harvard University Press, Cambridge, Mass., 1953, 1961, 1980.

Chapter 4
Predicate Logic

H.C.M. (Harrie) de Swart

Abstract In this chapter we extend the language of propositional logic to the one of predicate logic, in which we also can analyse arguments containing subjects and predicates, such as in, for example: All men are mortal; therefore: Socrates is mortal; and in: Socrates is a philosopher; therefore: someone is a philosopher. These simple arguments cannot be adequately dealt with in propositional logic. The semantic notions of logical consequence and logical validity and the syntactic notions of (logical) deducibility and provability are adapted to the language of predicate logic, and again it turns out that these two notions are extensionally equivalent (soundness and completeness).

4.1 Predicate Language

There are many arguments which cannot be analyzed adequately in propositional logic. An example is the following argument:

John is ill
Therefore: someone is ill.

If we translate the premiss and the conclusion into a propositional language, two atomic propositional formulas P_1 and P_2 respectively result. However, P_2 is not a valid consequence of P_1, while the argument above certainly is correct. The point is that P_1 and P_2 are two different atomic formulas not expressing the internal 'subject-predicate structure' of the premiss and the conclusion in the argument above. And it is the similarity in the internal structure of the premiss and the conclusion which is responsible for the correctness of the argument above.

So, we have to enrich the propositional language with symbols to indicate subjects, such as 'John' and 'someone' and symbols to indicate predicates, such as 'is ill'. In propositional logic, treated in Chapter 2, one can only analyze those arguments the correctness of which depends on the meaning of the propositional operations 'if …, then …', 'and', 'or' and 'not'. In *predicate logic*, also called *predicate calculus*, one can also analyze arguments the correctness of which depends on the 'subject-predicate structure' of the sentences involved.

With the help of a number of examples we introduce quantifiers, individual variables, constants and terms. Then we pay attention to the translation of English sentences into formulas of predicate logic and consider both intended and non-intended interpretations of these formulas. The scope of a quantifier and free and bound occurrences of a variable in a formula A are defined. A precise definition of the language of predicate logic is given, starting with an alphabet from which formulas can be built by means of connectives and quantifiers.

4.1.1 Quantifiers, Individual Variables and Constants

Below we give a number of examples of atomic propositions, grouping together those which have a similar internal (subject-predicate) structure.

1. Each of the numbers 2, 4, and 6 is even.
   All natural numbers are positive.
   All natural numbers are negative.
   All men are mortal.

The atomic propositions of group 1 all are of the following form:

> all objects (of a certain kind) have the property P;
> in other words: for each object x, x has the property P.

Notation: $\forall x[P(x)]$.

Here $P(a)$ stands for: a has the property P. In $P(a)$, 'a' is called an *individual variable* (or object variable) to emphasize that a ranges over the domain of individuals (or objects). The variable a indicates an open place, which may be filled by the name of a concrete individual, for instance, 'Socrates'. P(Socrates) then means: Socrates has the property P. $P(x)$ results from $P(a)$ by replacing a by x.

$\forall x$ is read as: for each object x. The symbol $\forall$ is called a *universal quantifier*. (The latin 'quantum' means 'how much'.) One might also use Ax (for All x) or $\bigwedge x$ instead of $\forall x$; the first one because it does not need any special symbol, the second one because of its analogy with $\wedge$ (and). For instance, 'each of the numbers 2, 4 and 6 is even' is equivalent to '2 is even and 4 is even and 6 is even'. However, in the case of an infinite domain as in 'all natural numbers are positive', for instance, the universal quantifier can be represented only by infinitely many conjunctions (0 is positive and 1 is positive and 2 is positive and ...). Such an expression with infinitely many conjunctions is not a formula, since formulas are by definition finite expressions. Therefore, we need quantifiers.

Instead of the variable x one may also use another variable y: $\forall x[P(x)]$ and $\forall y[P(y)]$ have the same meaning! They both mean: all objects (of a certain kind) have the property P.

2. At least one of the numbers 2, 3 and 4 is even.
   There is some natural number x such that $x > 0$.
   Some men are immortal.

The atomic propositions of group 2 are all of the following form:

some objects (of a certain kind) have the property P;
in other words: there is at least one object x such that x has the property P.

Notation: $\exists x[P(x)]$.

$\exists x$ is read as: there is at least one object x such that .... The symbol $\exists$ is called an *existential quantifier*. One might also use Ex (there Exists an x such that) or $\bigvee x$ instead of $\exists x$; the first one again because it does not need any special symbol, the second one because of its analogy with $\vee$ (or). For instance, 'There is some natural number x such that x is even' is analogous to '0 is even or 1 is even or 2 is even or ...'. Again, $\exists y[P(y)]$ and $\exists x[P(x)]$ have exactly the same meaning.

The predicate in an atomic proposition may be built from simpler predicates by means of 'if and only if (iff)', 'if ..., then ...', 'and', 'or' and 'not'. For instance, using $\rightleftarrows$ for 'iff', $\rightarrow$ for 'if ..., then ...', $\wedge$ for 'and', $\vee$ for 'or' and $\neg$ for 'not':
'For each number x, x is even iff x^2 is even' is of the form $\forall x[P(x) \rightleftarrows Q(x)]$.
'All animals having four legs are cows' is of the form $\forall x[P(x) \rightarrow Q(x)]$.
'Some natural numbers are positive and even" is of the form $\exists x[P(x) \wedge Q(x)]$.
'All natural numbers are positive or negative" is of the form $\forall x[P(x) \vee Q(x)]$.
'There is some natural number x such that not $x > 0$' is of the form $\exists x[\neg P(x)]$.

In an atomic proposition more than one quantifier may occur, as is the case in the following examples:
'All natural numbers are equal', or equivalently, 'for every natural number x and for every natural number y, $x = y$' is of the form $\forall x \forall y[R(x,y)]$.
'There are different natural numbers', or equivalently, 'there is a natural number x and there is a natural number y such that $x \neq y$' is of the form $\exists x \exists y[R(x,y)]$.

Here '$R(a,b)$' stands for: a is in the relation R to b. In $R(a,b)$, 'a' and 'b' are individual variables indicating open places which may be filled by the names of concrete individuals, for instance, by 'Janet' and 'Peter' respectively. 'R(Janet, Peter)' then means: Janet is in the relation R to Peter. '$R(x,y)$' results from '$R(a,b)$' by replacing a and b by x and y respectively.

In 'John loves Jane' we call 'John' the *subject* and ' - loves Jane' or 'a loves Jane' the *predicate* of the sentence. In logic we use the expression *predicate* in a more general way than in grammar. In grammar 'a loves Jane' is a *predicate*, but not 'John loves b' or 'a loves b'. In grammar, we call 'John' the *subject* and 'Jane' the *object* of the proposition 'John loves Jane'. In mathematics and in logic, but not in grammar, 'John loves b' and 'a loves b' are also called *predicates*, with one and two arguments respectively; and both 'John' and 'Jane' are called *subjects* of the proposition 'John loves Jane'. Notice that 'a loves Jane' assigns a proposition to each value of a; 'John loves b' assigns a proposition to each value of b and 'a loves b' assigns a proposition to each pair of values of a and b.

'*a* loves Jane' is a predicate with one argument, also called a *property*; and so is 'John loves *b*'. But '*a* loves *b*' is a predicate with two arguments, also called a *(binary) relation*. 'a_1 and a_2 are the parents of *b*' is an example of a 3-ary predicate, also called a *ternary relation*.

3. Every person has a mother; or equivalently: for every person x there is some person y such that x has y as mother.
   For every natural number there is a greater one; or equivalently: for every natural number x there is some natural number y such that $x < y$.

The atomic propositions of group 3 are all of the form:

for every object x there is an object y (possibly depending on x) such that x is in the relation R to y.

Notation: $\forall x \exists y [R(x,y)]$.

4. Someone is the mother of all persons; or equivalently: there is some person y such that for all persons x, x has y as mother.
   There is a greatest natural number; or equivalently: there is some natural number y such that for all natural numbers x, $x < y$.
   There is a least natural number; or equivalently: there is some natural number y such that for all natural numbers x, $y \leq x$.

The atomic propositions of group 4 all are of the form:

there is some object y (independent of any x) such that for all objects x (including y itself), x is in the relation R to y.

Notation: $\exists y \forall x [R(x,y)]$.

From the examples in group 3 and 4 it should become obvious that the reading of $\forall x \exists y [R(x,y)]$ is quite different from the reading of $\exists y \forall x [R(x,y)]$. So, the order of the quantifiers is very important. The following example may clarify the difference: It is true that for every natural number x there is a natural number y such that $x^2 = y$, which is of the form $\forall x \exists y [R(x,y)]$, but it is not true that there is a natural number y such that for *every* natural number x, $x^2 = y$, which is of the form $\exists y \forall x [R(x,y)]$.

In an atomic proposition the names of concrete individuals may occur, as is the case in the following examples.

5. Socrates is a man.
   Socrates is mortal.
   3 is odd.
   4 is even.

The atomic propositions of group 5 are all of the form:

c has the property P.

Notation: $P(c)$.

4.1 Predicate Language

The letter 'c' is used as the name for some concrete object. Different objects within the same context should be indicated by different names, for instance, $c_1, c_2, \ldots$. We call 'c_1', 'c_2', ... *individual constants*: throughout some context every occurrence of each of them is the name for the same object.

'All natural numbers are greater than or equal to zero' and 'everyone loves Janet' both are of the form $\forall x[R(x,c)]$, where $R(a,c)$ is to be read as: a is in the relation R to c. The symbol 'a' is an *individual variable* and the symbol 'c' is an *individual constant* .

From the atomic propositions considered above one can build composite propositions by means of the propositional operations studied in Chapter 2 on propositional logic. For instance, 'if all natural numbers are even, then all natural numbers are odd' is a composite proposition of the form $\forall x[P(x)] \to \forall x[Q(x)]$, not to be confused with the atomic proposition 'for each natural number x, if x is even, then x is odd', which is of the form $\forall x[P(x) \to Q(x)]$.

Note the difference between:

a) $\forall x[P(x)] \to \forall x[Q(x)]$: if *every* object x has the property P, then also *every* object x has the property Q.
b) $\forall x[P(x) \to Q(x)]$: for each (individual) object x, if x has the property P, then x also has the property Q.

In a) the implication $\to$ is between the two sentences $\forall x[P(x)]$ and $\forall x[Q(x)]$ to form a new sentence $\forall x[P(x)] \to \forall x[Q(x)]$. In b) the implication $\to$ is between the two predicates $P(x)$ and $Q(x)$ to form a new predicate $P(x) \to Q(x)$ and the formula in b) says that every object x has this property $P(x) \to Q(x)$. The formulas in a) and b) have quite different meanings! For instance, 'if all natural numbers are even (which is false), then all natural numbers are odd (which is also false)' is an instance of the formula in a) and is true ($0 \to 0 = 1$), while 'for each natural number x, if x is even, then x is odd' is an instance of the formula in b) and is false.

Similarly, 'if there is an even natural number, then there is a natural number not equal to itself' is a composite proposition of the form $\exists x[P(x)] \to \exists x[Q(x)]$ and false ($1 \to 0 = 0$), not to be confused with the atomic proposition 'there is some natural number x such that *if* x is even, *then* $x \neq x$', which is of the form $\exists x[P(x) \to Q(x)]$ and true, because 'if 3 is even, then $3 \neq 3$' is true ($0 \to 0 = 1$).

4.1.2 Translating English into Predicate Logic, Intended and Non-intended Interpretation

The English sentences 'John is ill' and 'Someone is ill' have the same noun phrase (NP) - verb phrase (VP) syntactic structure, while their translations into the predicate language do not have the same (logical) structure:

$$\text{John is ill} \quad I(j),$$
$$\text{Someone is ill} \quad \exists x[I(x)].$$

This makes automated translation of English into symbolic logic a non-trivial matter. The following six English sentences also have the same *NP-VP* structure, while their translations into predicate logic have quite different (logical) structures.

English sentences	*Usual translation into logic*
(1) John walks	$W(j)$
(2) Every student walks	$\forall x[S(x) \to W(x)]$
(3) Some student walks	$\exists x[S(x) \land W(x)]$
(4) No student walks	$\neg \exists x[S(x) \land W(x)]$
(5) Somebody walks	$\exists x[W(x)]$
(6) Nobody walks	$\neg \exists x[W(x)]$ or $\forall x[\neg W(x)]$

We have translated the sentences (1) - (6) into a formal predicate language the alphabet of which consists of the following symbols with the corresponding intended interpretation:

Symbols	*Intended interpretation*
$x, y, \ldots$	persons
j	John
$W; S$	is walking; being a student
$\rightleftarrows, \to, \land, \lor, \neg$	
$\forall, \exists$	
$[,], (,)$	

The translations of the sentences (1) - (6) are called *formulas* of this formal language. The interpretation of the connectives and the quantifiers has been fixed once and for all in Section 2.2 of Chapter 2 on propositional logic and in Subsection 4.1.1 at the beginning of this section; for this reason these symbols are called *logical symbols*. But the interpretation of the other symbols can be varied and therefore the symbols 'j', 'W' and 'S' are called *non-logical symbols*. Consider, for instance, the following non-intended interpretation:

Symbols	*Example of a non-intended interpretation*
$x, y, \ldots$	natural numbers
j	0
$W; S$	is even; is odd

Under this (non-intended) interpretation the meanings of the formulas above are as follows:

formula	*Non-intended interpretation, as specified above*
$W(j)$	0 is even;
$\forall x[S(x) \to W(x)]$	Every odd natural number is even;
$\exists x[S(x) \land W(x)]$	Some natural number is both odd and even;
$\neg \exists x[S(x) \land W(x)]$	No natural number is both odd and even;
$\exists x[W(x)]$	Some natural number is even;
$\neg \exists x[W(x)]$	No natural number is even.

The translation of the *correct* argument

If every student walks and John is a student, then John walks

4.1 Predicate Language

into propositional logic would be an *invalid* formula of the form $P \land Q \to R$, and hence such a translation is inadequate. However, the translation of this sentence into the predicate language specified above is

$$\forall x[S(x) \to W(x)] \land S(j) \to W(j). \qquad (*)$$

Now the reader can easily convince himself that this formula yields a true proposition for each possible interpretation (intended or non-intended): for every domain D, for every unary predicate S^* and W^* over D and for every element j^* in D, if all elements of D with the property S^* have the property W^* and j^* has the property S^*, then j^* also has the property W^*. For instance: if every Soccer player Wins the lottery and John is a Soccer player, then John Wins the lottery; and: if every Son of my father is Wealthy and John is a Son of my father, then John is Wealthy. For this reason the formula $(*)$ is called *valid*. *The validity of $(*)$ is guaranteed by the fixed meaning of the logical symbols $\forall$, $\to$ and $\land$ in this formula.*

Other examples of valid formulas of the formal language under consideration are: $\forall x[S(x) \to S(x)]$, $\forall x[\neg(W(x) \land \neg W(x))]$, and
$\forall x[S(x) \to W(x)] \land \forall x[S(x)] \to \forall x[W(x)]$.
We will study valid formulas more closely in Section 4.2.

Suppose we want to translate sentences about addition and multiplication of natural numbers into a logical language. Examples of such sentences are:
(1) for any natural number n, $n + 0 = n$,
(2) for any natural number n, $n \times 0 = 0$,
(3) there is no natural number n such that $n \times n = 2$.
Of course, we might translate these sentences into atomic propositional formulas P_1, P_2 and P_3 of propositional logic, respectively. This suffices, if we want to conclude, for instance, that the sentence ((1) or (2)) logically follows from sentence (1), because $P_1 \lor P_2$ is a valid consequence of P_1. However, if we want to conclude from sentence (1) that $2 + 0 = 2$, our translation into propositional formulas is not adequate. The proposition $2 + 0 = 2$ should be rendered by a different atomic formula Q and we know from Chapter 2 that Q is not a valid consequence of P_1; on the other hand, the proposition $2 + 0 = 2$ does follow from proposition (1). Therefore, a translation into the language of predicate logic, exhibiting the subject-predicate structure of the sentences involved, is needed.

We may take a predicate language with the following non-logical symbols, having the corresponding intended interpretation:

Non-logical symbols	Intended-interpretation
$0, 1, 2, \ldots$	zero, one, two, $\ldots$
$\equiv$	is equal to ($=$)
A	$A(a,b,c)$: a plus b equals c (Addition)
M	$M(a,b,c)$: a times b equals c (Multiplication)

The symbols $0, 1, 2, \ldots$ are individual constants. the symbol $\equiv$ is a binary predicate symbol, i.e., with two arguments, and the symbols A (Addition) and M (Multiplication) are ternary predicate symbols, i.e., with three arguments. The translations of sentences (1), (2) and (3) are now respectively: $\forall x[A(x,0,x)]$, $\forall x[M(x,0,0)]$,

$\neg\exists x[M(x,x,2)]$ and the translation of $2 + 0 = 2$ now becomes $A(2,0,2)$, which is a valid consequence of $\forall x[A(x,0,x)]$.

Of course, once having built these formulas one can forget about their origin and consider *non-intended interpretations*, like the following one.

x, y	: persons
0, 1, 2, ...	: John, Mary, Janet, ..., respectively
$a \equiv b$	: a loves b
$A(a,b,c)$	: a and b are the parents of c
$M(a,b,c)$	: a and b are the grandparents of c.

Needless to say that under this non intended interpretation the formula $A(2,0,2)$ yields a false proposition: Janet and John are the parents of Janet.

4.1.3 Scope, Bound and Free Variables

In $\forall x[A(x)]$ and in $\exists x[A(x)]$ we call $A(x)$ the *scope* of the quantifier $\forall x$. For example, in the expression

$$\exists x[R(a,x) \to S(x,a,b)] \to R(a,b)$$

the scope of the $\exists x$ is the part $R(a,x) \to S(x,a,b)$.

In the expression

$$\forall x \exists y[R(x,y) \to \exists z[S(y,z)]] \to \forall x[\neg R(x,a)]$$

the scope of the first occurrence of the $\forall x$ is the part $\exists y[R(x,y) \to \exists z[S(y,z)]]$, the scope of $\exists y$ is the part $R(x,y) \to \exists z[S(y,z)]$, the scope of $\exists z$ is the part $S(y,z)$ and the scope of the second occurrence of the $\forall x$ is $\neg R(x,a)$.

Similarly, in $\neg A$, $A \rightleftarrows B$, $A \to B$, $A \land B$ and $A \lor B$ we call the expression A or pair of expressions A, B the *scope* of the propositional connective in question.

Definition 4.1 (Bound/Free occurrence of a variable in a formula). An occurrence of a variable x in an expression A is said to be *bound* (or as a *bound variable*), if the occurrence is in a quantifier $\forall x$ or $\exists x$ or in the scope of a quantifier $\forall x$ or $\exists x$ (with the same x); otherwise, *free* (or as a *free variable*).

It has turned out that it is convenient to use different letters for free and bound variables:

$a_1, a_2, a_3, \ldots$ for free occurrences only and
$x_1, x_2, x_3, \ldots$ for bound occurrences only.

Example 4.1. In 'a_2 is the mother of a_1' and in '$a_2 > a_1$' both occurrences of a_1 and a_2 are free.
In $\exists x_2[x_2$ is the mother of $a_1]$ (a_1 has a mother) and in $\forall x_2[x_2$ is the mother of $a_1]$ (everyone is a mother of a_1), the occurrence of a_1 is free and both occurrences of x_2 are bound.

4.1 Predicate Language

In $\forall x_1 \exists x_2 [x_2$ is the mother of $x_1]$ (everyone has a mother) and in $\exists x_1 \forall x_2 [x_2$ is the mother of $x_1]$ (someone has everyone as mother) both occurrences of x_1 and both occurrences of x_2 are bound.
The occurrences of the variables a_1 and a_2 are free in $\exists x_1 \forall x_2 [R(x_1, a_1) \wedge R(x_2, a_2)]$, while both occurrences of the variables x_1 and x_2 are bound in this formula.

A variable a which occurs as a free variable (briefly, occurs free) in A is called a *free variable of A*, and A is then said to contain a as a *free variable* (briefly, to contain a free); and likewise for bound variables.

4.1.4 Alphabet and Formulas

In Subsection 4.1.2 we introduced two different formal predicate languages: one for expressing that certain students walk (John was one of them) and one for expressing certain properties of natural numbers. In the exercises at the end of this section several other predicate languages are introduced. All predicate languages have the individual variables, the connectives and quantifiers in common, they differ only in the choice of the individual constants and predicate symbols, which depends on the context. We do not want to study any particular one of these languages, but we want to study these languages in general, so that any of our results is applicable to each particular language we want to consider.

So, in order to retain flexibility for the applications, we shall assume throughout this chapter that we are dealing with one or another object language in which there is a class of *individual constants*

$$c_1, c_2, c_3, \ldots$$

and a class of *predicate symbols*

$$P_1, P_2, P_3, \ldots$$

where each P_i is supposed to be a different n_i-*place* predicate symbol, i.e., taking n_i arguments ($n_i = 0, 1, 2, \ldots$). By including the possibility that $n_i = 0$, we allow $P_1, P_2, \ldots$ to express atomic *propositions*. Consequently, the predicate calculus extends the propositional calculus. That is, any propositional language can be conceived of as a predicate language: instead of the atomic formula P_i, one can take a 0-ary predicate symbol P_i (with $n_i = 0$).

In Chapter 3 we introduced a formal predicate language for set theory and in Chapter 5 we shall introduce another predicate language for arithmetic, in which we can express properties of natural numbers. In *The Proper Treatment of Quantification in Ordinary English*, R. Montague presented a formal language in which a suitably restricted and regulated part of English or some other natural language can be expressed.

Thus, throughout this chapter our logical predicate language shall consist of the following symbols:

Definition 4.2 (Alphabet of predicate logic).

Symbols	Name
$a_1, a_2, a_3, \ldots$	free individual variables
$x_1, x_2, x_3, \ldots$	bound individual variables
$c_1, c_2, c_3, \ldots$	individual constants
$P_1, P_2, P_3, \ldots$	predicate symbols (each P_i is n_i-ary)
$\rightleftarrows, \rightarrow, \wedge, \vee, \neg$	connectives
$\forall, \exists$	quantifiers
(,), [,]	parentheses

Since the logical predicate language is the object of our study in this chapter, we shall call it the *object-language*. We shall study this language using English as metalanguage, i.e., as the language we use to talk *about* (formulas of) the object language.

In order to prevent writing subscripts and because $\forall x_1[P(x_1)]$ has the same meaning as $\forall x_3[P(x_3)]$, we agree to use x, y, z as (meta)variables over $x_1, x_2, x_3, \ldots$ and simply write $\forall x[P(x)]$ instead of $\forall x_1[P(x_1)], \forall x_2[P(x_2)], \ldots$. The use of the letter x in the expression 'if $A(a)$ is a formula and x is a bound variable, then $\forall x[A(x)]$ is a formula' is similar to the use of the letter n in the expression 'if n is a natural number, then also $n+1$ is natural number'. The letter n itself is not a natural number, but may be replaced by any natural number $0, 1, 2, \ldots$ in the expression just mentioned. Similarly, the letter x itself is not a variable, but may be replaced by any variable $x_1, x_2, x_3, \ldots$. So, strictly speaking, the expression $\forall x[P(x)]$ itself is not a formula, but replacing x by x_1 (or $x_2, x_3, \ldots$) and P by P_1 yields a formula $\forall x_1[P_1(x_1)]$, which does belong to the object language.

In a similar way we agree to use the symbols a and b as names for free individual variables $a_1, a_2, \ldots$ in the object language; the symbols c and d as names for individual constants $c_1, c_2, \ldots$ in the object language; and the symbols P, Q, R and S as names for predicate symbols $P_1, P_2, \ldots$ in the object language. Strictly speaking, the symbols $a, b, x, y, z, c, d, P, Q, R$ and S themselves do not belong to the logical predicate language!

Definition 4.3 (Basic Term). A *basic term* is a free individual variable or an individual constant. Later in Definition 4.17 the notion of term will be generalized, allowing it to contain also function symbols.

Definition 4.4 (Atomic formulas). If P is an n-ary predicate symbol and $t_1, \ldots, t_n$ are *terms*, then $P(t_1, \ldots, t_n)$ is an *atomic formula*.

Example 4.2. Supposing that S (being a Student) and W (Walking) are unary predicate symbols, that M (having as Mother) is a binary predicate symbol, that a and b stand for any free individual variable $a_1, a_2, \ldots$, and that c and d stand for any individual constant $c_1, c_2, \ldots$, the following expressions are atomic formulas of predicate logic: $S(a), W(a), S(c), W(c); M(a,b), M(a,c)$ (cobi is the Mother of a), $M(c,a)$ (a is the Mother of cobi), $M(c,b), M(c,d)$ (cobi has dora as Mother).

The expression $P(t_1, \ldots, t_n)$ itself is not an atomic formula, but a meta-expression representing any atomic formulas. In particular, the expression $P(a)$ itself is not

4.1 Predicate Language

an atomic formula, but $P(a_1), P(a_2), P(a_3), \ldots$ are atomic formulas, if P is a unary predicate symbol in the alphabet of our predicate language, which is the object of our study.

Definition 4.5 (Formulas).

a) Each atomic formula is a formula.
b) If A and B are any formulas (either atomic formulas, or composite formulas already constructed), then $(A \rightleftarrows B)$, $(A \rightarrow B)$, $(A \wedge B)$, $(A \vee B)$ and $(\neg A)$ are (composite) formulas.
c) If $A(a)$ is any formula in which the free variable a occurs, and x is any bound variable not occurring in $A(a)$, then $\forall x[A(x)]$ and $\exists x[A(x)]$ are (composite) formulas, where $A(x)$ results from $A(a)$ by replacing every occurrence of a in $A(a)$ by x.
d) The only formulas are those given by a), b) and c).

Example 4.3. Supposing that S (being a Student) and W (Walking) are unary predicate symbols in our predicate language, by clause b), $S(a) \rightarrow W(a)$ is a formula of our logical predicate language, and by clause c) $\forall x[S(x) \rightarrow W(x)]$ is a formula of our predicate language. Supposing that $M(a,b)$ (a has b as Mother) is a binary predicate symbol in our predicate language, by clause c) $\exists y[M(a,y)]$ (a has a Mother) is a formula of our predicate language, and again by clause c), also $\forall x \exists y[M(x,y)]$ (everyone has a Mother) is a formula of our predicate language. And by applying clause b) again, $\forall x[S(x) \rightarrow W(x)] \wedge \forall x \exists y[M(x,y)]$ is also a formula of our predicate language.

Strictly speaking, assuming that M is a binary predicate symbol of our predicate language, $\exists y[M(a,y)]$ itself is not a formula of our predicate language, but, for instance, $\exists x_2[M(a_1,x_2)]$ is, expressing that 'a_1 has a mother'. And strictly speaking, $\forall x \exists y[M(x,y)]$ itself is not a formula of our predicate language, but $\forall x_1 \exists x_2[M(x_1,x_2)]$ is, expressing that 'everyone has a mother'.

We are using the symbols A, B, C, ..., A_1, A_2, A_3, ..., from the beginning of the Roman alphabet to stand for any formulas, not necessarily atomic. Such distinct letters as A, B, C, ... need not represent distinct formulas in contrast to the symbols P, Q, R, S, ... which represent distinct predicate symbols.

Assuming that $A(a)$ is a formula, the expression $\forall x[A(x)]$ itself is, strictly speaking, not a formula, since the letter x is a meta-variable representing any bound variable; but $\forall x[A(x)]$ becomes a formula when the letter x is replaced by any bound variable x_1 or x_2 or x_3 or ....

For instance, supposing again that S (is a Student) is a unary predicate symbol and that M (has as Mother) is a binary predicate symbol of our predicate language, $S(a_1)$ and $M(a_1,a_2)$, are atomic formulas of our predicate language and $\forall x_1[S(x_1)]$ (everyone is a Student) and $\exists x_2[M(a_1,x_2)]$ (a_1 has a mother) are composite formulas of our predicate language. Using, e.g., x_3 instead, we get different formulas $\forall x_3[S(x_3)]$ and $\exists x_3[M(a_1,x_3)]$ which have the same meaning as $\forall x_1[S(x_1)]$ and $\exists x_2[M(a_1,x_2)]$, respectively. This is why the meta-variable x is necessary in clause c) in Definition 4.5; had we written x_1 instead, we would be allowing only $\forall x_1[S(x_1)]$ (but not $\forall x_2[S(x_2)]$, $\forall x_3[S(x_3)]$, etc.) as a formula.

The quantifiers act as unary operators in building formulas, and with our other unary operator ¬ are ranked last under the convention for omitting parentheses. Thus, $\forall x A(x) \to B$ means $\forall x[A(x)] \to B$, not $\forall x[A(x) \to B]$.

Definition 4.6 (Closed Formula). A formula A is called *closed* if it contains no free occurrences of variables; otherwise, *open*. A closed formula is also called a *sentence*.

Example 4.4. Supposing that M is a binary predicate symbol of our predicate language, $M(a_1, a_2)$ (a_1 has a_2 as Mother) and $\exists x_2[M(a_1, x_2)]$ (a_1 has a Mother) are open formulas, while $M(c_1, c_2)$ (c_1 has c_2 as Mother), $\exists x_2[M(c_1, x_2)]$ (c_1 has a Mother) and $\forall x_1 \exists x_2[M(x_1, x_2)]$ (everyone has a Mother) are closed formulas.

Since formulas are built up from atomic formulas by successive applications of connectives and quantifiers to formulas already generated before, the following Theorem, called the *induction principle* (for predicate formulas), follows immediately from the definition of formulas. (See also Theorem 2.2.)

Theorem 4.1 (Induction principle for formulas). *Let Φ be a property of formulas, such that a) all atomic formulas have the property Φ,*
b) if A and B have the property Φ, then also $(A \rightleftarrows B)$, $(A \to B)$, $(A \wedge B)$, $(A \vee B)$ and $(\neg A)$ have the property Φ, and
c) if $A(a)$ has the property Φ, x does not occur in $A(a)$ and $A(x)$ results from $A(a)$ by replacing all occurrences of a in $A(a)$ by x, then also $\forall x[A(x)]$ and $\exists x[A(x)]$ have the property Φ.
Then all formulas have the property Φ.

For an application of this induction principle see the proof of Theorem 4.18.

Exercise 4.1. Let $G(a)$ stand for 'a is a girl' and $P(a)$ for 'a is pretty'.

a) Translate each of the following sentences into logical symbolism in an adequate way: (1) Every girl is pretty. (2) Some girl is pretty.
b) Explain why $\forall x[G(x) \wedge P(x)]$ is not a correct representation of the meaning of sentence (1) and why $\exists x[G(x) \to P(x)]$ is not a correct representation of the meaning of sentence (2).

Exercise 4.2. Let M be a binary predicate (relation) symbol with intended interpretation 'is married to', and c and d individual constants with Cod, respectively Diana, as intended interpretation. Translate the following sentences into formulas of predicate logic: 1. Cod is not married to Diana; 2. For all persons x and y, if x is married to y, then y is married to x; 3. Diana is married; 4. There is at least one person who is not married.

Exercise 4.3. Let $A(a, b)$ stand for 'a admires b'. Translate the following two sentences into logical symbolism.
(1) Everyone has someone whom he admires.
(2) There is someone whom everyone admires.
Note that 'everyone admires someone' is ambiguous and can have each of the two readings above.

4.1 Predicate Language

Exercise 4.4. Let $L(x,y)$ stand for 'x loves y'. Translate the following sentences into logical symbolism.
(1) All persons love each other.   (2) Some persons love each other.
(3) Every person loves someone.   (4) Someone is loved by everyone.
(5) Everyone is loved by someone.   (6) There is a person who loves everyone.

Exercise 4.5. Let $D(a)$ stand for 'a is a Dutchman', $C(a)$ for 'a is a kind of cheese', $W(a)$ for 'a is a kind of wine', $L(a,b)$ for 'a likes b', c for Chip, and d for Donald. Translate the following sentences into logical symbolism in an adequate way.

1. Donald likes all kinds of cheese.
2. Some Dutchmen like all kinds of cheese.
3. Donald likes some kinds of cheese.
4. All Dutchmen like at least one kind of cheese.
5. There is a kind of cheese which is liked by any Dutchman.
6. Chip doesn't like any kind of cheese.
7. All Dutchmen don't like any kind of cheese.
8. All Dutchmen like some kind of cheese and some kind of wine.
9. All Dutchmen who like some kind of cheese, also like some kind of wine.
10. If all Dutchmen like some kind of cheese, then all Dutchmen like some kind of wine.

Exercise 4.6. Consider the predicate language with the following non-logical symbols: the binary predicate symbol $\equiv$ and the individual constants c_1, c_2.

1 Translate the sentences below into this language in an adequate way.
    i) The Morning Star is the same as the Evening Star.
    ii) Every star identical to the Morning Star, is the same as the Evening Star.
2 For the formulas found in 1, consider the non-intended interpretation:
    $\forall x$: for all numbers x, ...; $\exists x$: there is some number x such that ....
    $\equiv$: is equal to (=); c_1: 3, c_2: 4.
    Are the readings of the formulas found in 1 i) and ii) under this interpretation true or false propositions?
3 Similar question as in 2, but now for the non-intended interpretation:
    $\forall x$: for all persons x ...; $\exists x$: there is some person x such that ....
    $\equiv$: was older than; c_1: Reagan, c_2: Nixon.

Exercise 4.7. Let $P(a)$ stand for 'a has the property P', and $a \equiv b$ for 'a equals b'. Translate each of the following sentences into logical symbolism, using the binary predicate symbol $\equiv$ for equality.

1. There is *at least one* x which has the property P.
2. There is *at most one* x which has the property P.
3. There is *exactly one* x which has the property P.
4. There are *at least two* objects which have the property P.
5. There are *at most two* objects which have the property P.
6. There are *exactly two* objects which have the property P.

$\exists!x[A(x)]$ is adopted as an abbreviation for the formula expressing *'there is exactly one x such that $P(x)$'* or *'there exists a unique x such that $P(x)$'*.

Exercise 4.8. Translate the following sentences containing the indefinite article 'a' into the language of predicate logic, using the unary predicate symbols C, A, M and W for 'being a Child', 'needs Affection', 'being a Man' and 'to Whistle', respectively: a) A child needs affection. b) A man was whistling. Notice that the indefinite article 'a' or 'an' sometimes has the force of 'all', sometimes of 'some'.

Exercise 4.9. Translate the following sentences containing the word 'any' into the language of predicate logic, using the unary predicate symbols M, O, B and S for 'being Mortal', 'being Older than 150 years', 'celebrating one's Birthday' and 'being Stupid' respectively, and using the propositional formula P for 'there is a party'.
a) For any x, x is mortal.
b) Not for any x, x is older than 150 years.
c) If anyone celebrates his or her birthday, then there is a party.
d) If John was stupid, then anyone is stupid.
Notice that the meaning of 'any' depends on the context. When an any-expression stands by itself, as in sentence a), 'any' has the same logical force as 'all'. But when an any-expression D is put into either of the context $\neg D$, as in sentence b), or $D \rightarrow E$, as in sentence c), the meaning of 'any' normally alters from 'all' to 'some'.

Exercise 4.10. Give an interpretation such that $\forall x[P(x) \rightarrow Q(x)]$ yields a true proposition, while $\exists x[P(x) \wedge Q(x)]$ yields a false proposition under this interpretation. This shows that from $\forall x[P(x) \rightarrow Q(x)]$ one may not conclude that $\exists x[P(x) \wedge Q(x)]$, although one may conclude from it that $\exists x[P(x) \rightarrow Q(x)]$.

Exercise 4.11. a) Give an interpretation such that $\forall x[P(x)] \rightarrow \forall x[Q(x)]$ yields a true proposition, while $\forall x[P(x) \rightarrow Q(x)]$ yields a false proposition under this interpretation. So, from $\forall x[P(x)] \rightarrow \forall x[Q(x)]$ one may not conclude that $\forall x[P(x) \rightarrow Q(x)]$.
b) Show in a similar way that from $\exists x[P(x) \rightarrow Q(x)]$ one may not conclude that $\exists x[P(x)] \rightarrow \exists x[Q(x)]$.

4.2 Semantics: Tarski's Truth Definition; Logical (Valid) Consequence

Let A be an atomic formula containing (free occurrences of) variables, a one-place predicate symbol P or a 2-place predicate (or relation) symbol R and individual constants c and d. For instance, $A = P(c)$, $A = P(a)$, $A = R(c,d)$ or $A = R(a,c)$. In order to give a meaning to A, we have to give an interpretation of the symbols occurring in A. Such an *interpretation M* has to specify:

1. a *domain* or *universe of discourse* D; for instance, the set of all men or the set $\mathbb{N}$ of all natural numbers.
2. a unary predicate P^* or a binary predicate R^*, respectively, over the given domain, determining the meaning of the predicate symbols P and R; for instance, assuming the domain is $\mathbb{N}$, $P^*(a)$: a is even, $R^*(a,b)$: $a > b$.

4.2 Semantics: Tarski's Truth Definition; Logical (Valid) Consequence

3. elements c^* and d^* in the given domain, determining the meaning of the individual constants c and d.

So, let $M = \langle \mathbb{N}; P^*, R^*; c^*, d^* \rangle$ be the interpretation with domain $\mathbb{N}$, $P^*(a)$: a is even, $R^*(a,b)$: $a > b$; $c^* = 2$ and $d^* = 3$. Then under interpretation M the formula $P(c)$ yields the proposition $P^*(c^*)$, i.e., 2 is even, which happens to have the truth value 1. Therefore, we say that M is a *model* for the formula $P(c)$, i.e., $P(c)$ yields under interpretation M a true proposition. Notation: $M \models P(c)$.

And under interpretation M the formula $R(c,d)$ yields the proposition $R^*(c^*,d^*)$, i.e., $2 > 3$, which happens to have the truth value 0. Therefore, we say that M is not a *model* for the formula $R(c,d)$, i.e., $R(c,d)$ yields under interpretation M a false proposition. Notation: $M \not\models R(c,d)$.

An interpretation M for a formula A does specify the domain and the meanings of the predicate symbols and individual constants in A, but it does not specify the meaning of the variables that occur free in A. Given an interpretation M for formula A with domain D, a *valuation* v shall give a value in the given domain to the variables occurring free in A. So, let $M = \langle \mathbb{N}; P^*, R^*; c^*, d^* \rangle$ be the interpretation given above for the formula $P(a)$ or $R(a,c)$ respectively, and let v be the valuation which assigns to the free variable a the value 4, $v(a) = 4$, then under interpretation M and valuation v the formula $P(a)$ yields the proposition $P^*(4)$, i.e., 4 is even, which happens to have the truth value 1. Therefore, we say that interpretation M and valuation v make the formula $P(a)$ true. Notation: $M \models P(a)[v]$ or $M \models P(a)[4]$.

Under the interpretation M just given and valuation v with $v(a) = 4$, the formula $R(a,c)$ yields the proposition $R^*(4,c^*)$, i.e., $4 > 2$, which happens to have the truth value 1. So, interpretation M and valuation v make also the formula $R(a,c)$ true. Notation: $M \models R(a,c)[v]$ or $M \models R(a,c)[4]$.

So, an interpretation M for a formula A together with a valuation v assigns to A a truth value 1 or 0. In the first case we write $M \models A[v]$ and in the second case we write $M \not\models A[v]$.

If A is composed from atomic formulas by means of connectives, the truth tables tell us the truth value of A under a given interpretation and valuation. For instance, if $M = \langle \mathbb{N}; \text{is even}, > ; 2 \rangle$, then $M \models P(a) \wedge R(a,c)[4]$, since '4 is even and $4 > 2$' has truth value $1 \wedge 1 = 1$. But $M \not\models P(a) \wedge R(a,c)[3]$, since '3 is even and $3 > 2$' has truth value $0 \wedge 1 = 0$. And $M \models P(a) \rightarrow R(a,c)[1]$, since 'if 1 is even, then $1 > 2$' has truth value $0 \rightarrow 0 = 1$.

Next, consider the formula $\forall x[P(x)]$.

If we let the individual variable x range over the set of all men and if we interpret the predicate symbol P as 'is mortal', then the atomic proposition 'all men are mortal' results and this proposition has truth value 1. So, for $M = \langle Men; \text{is mortal} \rangle$, M is a model for $\forall x[P(x)]$; notation: $M \models \forall x[P(x)]$. However, if we let the variable x range over the set of all natural numbers and if we interpret the predicate symbol P as 'is even', then the proposition 'all natural numbers are even' results and this proposition has truth value 0; so, for $M = \langle \mathbb{N}; \text{is even} \rangle$, M is not a model for $\forall x[P(x)]$; notation: $M \not\models \forall x[P(x)]$.

So depending on the interpretation of the individual variable x and the predicate symbol P, a true or false atomic proposition results from the formula $\forall x[P(x)]$:

	$\forall x[P(x)]$
$M = \langle Men; P^* \rangle$ with $P^*(x)$: x is mortal	1
$M = \langle \mathbb{N}; P^* \rangle$ with $P^*(x)$: x is even	0

In the following table for the two formulas $\forall x[P(x)]$ and $\exists x[Q(x)]$ we indicate on the left-hand side an interpretation and on the right-hand side the truth or falsity of the corresponding (atomic) proposition.

	$\forall x[P(x)]$	$\exists x[Q(x)]$
$\mathbb{N}$; $P^*(x)$: $x = x$, $Q^*(x)$: x is even	1	1
Men; $P^*(x)$: x is mortal, $Q^*(x)$: x is immortal	1	0
$\mathbb{N}$; $P^*(x)$: x is even, $Q^*(x)$: x is odd	0	1
$Pets$; $P^*(x)$: x is a dog, $Q^*(x)$: x is immortal	0	0

Above, we have given two interpretations of the symbols x and P, under which $\forall x[P(x)]$ yields a true proposition ('every natural number is equal to itself' and 'all men are mortal', respectively); and two interpretations under which $\forall x[P(x)]$ yields a false proposition ('all natural numbers are even' and 'all pets are dogs', respectively). So, $\forall x[P(x)]$, although not under all interpretations true, is true under at least one interpretation. For that reason we say that $\forall x[P(x)]$ is *satisfiable*.

'Not all men have black hair' is equivalent to 'there is some man who does not have black hair'. More generally, we see that $\neg \forall x[P(x)]$ (not all objects have the property P) has the same meaning as $\exists x[\neg P(x)]$ (there is some object which does not have the property P), no matter how we interpret the symbols x and P. Hence, we say that $\neg \forall x[P(x)] \rightleftarrows \exists x[\neg P(x)]$ is a *valid* or *always true* formula. So, we shall call a formula A *valid* or *always true* if A yields a true proposition under each possible interpretation of the individual and predicate-symbols which occur in A. Notation: $\models A$. Examples of valid formulas are:

1. $\models \neg \forall x[P(x)] \rightleftarrows \exists x[\neg P(x)]$
2. $\models \neg \exists x[P(x)] \rightleftarrows \forall x[\neg P(x)]$
3. $\models \forall x[P(x)] \rightleftarrows \neg \exists x[\neg P(x)]$
4. $\models \exists x[P(x)] \rightleftarrows \neg \forall x[\neg P(x)]$

In order to see the validity of the formula $A \rightleftarrows A$, we do not have to consider the internal structure of the formula A. However, in order to see the validity of the formula $\neg \forall x[P(x)] \rightleftarrows \exists x[\neg P(x)]$, which is a formula of the form $\neg A \rightleftarrows B$, we do have to consider the internal structure of the subformulas A and B from which this formula has been built. $\neg A \rightleftarrows B$ is not for all formulas A and B valid, but it is valid when A is $\forall x[P(x)]$ and B is $\exists x[\neg P(x)]$.

And we shall call B a *valid* or *logical consequence* of given premisses $A_1, \ldots, A_n$ if every interpretation M and valuation v which make all of the premisses $A_1, \ldots, A_n$ true also make B true. Notation: $A_1, \ldots, A_n \models B$.

For instance, $P(a) \models \exists x[P(x)]$ and $\forall x[P(x) \to Q(x)], P(a) \models Q(a)$.

4.2 Semantics: Tarski's Truth Definition; Logical (Valid) Consequence 197

After this introduction we shall give a precise definition of the notion of $M \models A$, which is *Tarski's truth definition* (1933), and of the notions of (logical) validity and valid (or logical) consequence.

Definition 4.7 (Interpretation). Let A be a formula, containing predicate symbols $P_1, \ldots, P_k$ and individual constants $c_1, \ldots c_l$. An *interpretation* or *structure* for A is a tuple $M = \langle D; P_1^*, \ldots, P_k^*; c_1^*, \ldots, c_l^* \rangle$, where

1. D is a non-empty set, called the *domain* or *universe of discourse*. All individual variables occurring bound in A are interpreted as ranging over this domain D. For instance, D is the finite set of all men or the infinite set $\mathbb{N}$ of all natural numbers. The requirement that the domain is non-empty is to guarantee that the following formula will be valid: $\forall x[P(x)] \rightarrow \exists x[P(x)]$.
2. For each n_i-ary predicate symbol P_i in A, P_i^* is a n_i-ary predicate over D. For instance, if P is a unary and R is a binary predicate symbol in A, and $D = \mathbb{N}$, then $P^*(n)$ might be 'n is even' and $R^*(n,m)$ might be '$n > m$'.
3. For each individual constant c_j in A, c_j^* is a concrete element of D. For instance, if c is an individual constant in A and $D = \mathbb{N}$, then c^* might be 2.

Note that the interpretation of the quantifiers and of the connectives in a formula A has been fixed once and for all in Section 4.1 and in the truth tables for the connectives (see Section 2.2). We are only free to vary the interpretation of the individual variables, the predicate symbols and the individual constants in A.

Given a formula A and an interpretation M for A with domain D, in order to give a meaning to A we still have to interpret the individual variables occurring free in A as elements of D.

Definition 4.8 (Valuation). Let A be a formula and M an interpretation for A with domain D. A *valuation* v for A assigns to each variable occurring free in A an element $v(a)$ in D.

Example 4.5. Let $A = P(a) \wedge R(a,c)$. Then $M = \langle \mathbb{N}; P^*, R^*; c^* \rangle$ with $P^*(a) :=$ 'a is even', $R^*(a,b) :=$ '$a > b$' and $c^* = 2$, is an interpretation for A; and v with $v(a) = 4$ is a valuation for A.

Next we shall give *Tarski's truth definition* (1933), which is not a definition of truth, but which defines the notion of $M \models A[v]$, i.e., 'interpretation M and valuation v make A true', or 'under interpretation M and valuation v formula A yields a proposition with truth value 1'.

Definition 4.9 (Tarski's truth definition, 1933). Let A be a formula containing predicate symbols $P_1, \ldots, P_k$ and individual constants $c_1, \ldots, c_l$.
Let $M = \langle D; P_1^*, \ldots, P_k^*; c_1^*, \ldots, c_l^* \rangle$ be an interpretation for A and let v be a valuation for the variables occurring free in A.
    We define $M \models A[v]$ by induction on the build-up of A:

- A is atomic, say $A = P_i(a_1, \ldots, a_k, c_1, \ldots, c_l)$.

    $M \models P_i(a_1, \ldots, a_k, c_1, \ldots, c_l)[v]$ iff $P_i^*(v(a_1), \ldots, v(a_k), c_1^*, \ldots, c_l^*)$.

For instance, if R is binary predicate symbol, $R^*(a,b) := $ '$a > b$', $c^* = 2$ and $v(a) = 4$, then $M \models R(a,c)$ $[v]$ iff $4 > 2$. If $v(a) = 4$, then instead of $M \models R(a,c)$ $[v]$ we shall also write $M \models R(a,c)$ $[4]$.

Notice that if A contains only the free variables $a_1, \ldots, a_k$, then only the values $v(a_1), \ldots, v(a_k)$ matter in the definition of $M \models A[v]$. In particular, if A contains no free occurrences of variables, then the valuation v in '$M \models A[v]$' does not matter. These properties are preserved throughout the definition of $M \models A[v]$. Instead of 'not $M \models A[v]$' we shall write: $M \not\models A[v]$. In such a case M is called a *countermodel for A* or a *counterexample to A*.

- $A = B \rightleftarrows C, A = B \to C, A = B \wedge C, A = B \vee C, A = \neg B$:

  1. $M \models B \rightleftarrows C$ $[v]$ iff ($M \models B[v]$ and $M \models C[v]$) or ($M \not\models B[v]$ and $M \not\models C[v]$).
  2. $M \models B \to C$ $[v]$ iff $M \not\models B[v]$ or $M \models C[v]$.
  3. $M \models B \wedge C$ $[v]$ iff $M \models B[v]$ and $M \models C[v]$.
  4. $M \models B \vee C$ $[v]$ iff $M \models B[v]$ or $M \models C[v]$.
  5. $M \models \neg B$ $[v]$ iff $M \not\models B[v]$.

This definition just follows the truth tables for the connectives given in Section 2.2. This may be easily seen if one realizes that a pair (M, v) consisting of an interpretation M and a valuation v assigns to every formula A a truth value 1 or 0. So, a pair (M, v) corresponds with a line in the truth table and one might write $(M,v)(A) = 1$ iff $M \models A[v]$ and $(M,v)(A) = 0$ iff not $M \models A[v]$. Then, for instance, clause 2 reads as follows: $(M,v)(B \to C) = 1$ iff $(M,v)(B) = 0$ or $(M,v)(C) = 1$.

- $A = \forall x[P(x)]$ or $A = \exists x[Q(x)]$

  In case $A = \forall x[P(x)]$ does not contain any free occurrences of variables, $M \models \forall x[P(x)]$ iff for every element d in the domain D of M, $M \models P(a)[d]$.

For instance, let $M = \langle \mathbb{N}; \geq 0 \rangle$, then $M \models \forall x[P(x)]$ since for every natural number d in $\mathbb{N}$, $M \models P(a)[d]$, i.e., for every natural number d, $d \geq 0$.

More generally, allowing $A = \forall x[P(x)]$ to contain also free occurrences of variables, $M \models \forall x[P(x)]$ $[v]$ iff for every d in the domain D of M, $M \models P(a)[d/v]$, where a is a (new) variable not occurring in $\forall x[P(x)]$ and d/v is the same valuation as v, except that d/v assigns to a the value d.

  In case $A = \exists x[Q(x)]$ does not contain any free variables, $M \models \exists x[Q(x)]$ iff there is at least one element d in the domain D of M, such that $M \models Q(a)[d]$.

For instance, let $M = \langle \mathbb{N}; \text{is even} \rangle$, then $M \models \exists x[Q(x)]$ since there is at least one natural number d in $\mathbb{N}$ such that $M \models Q(a)[d]$, i.e., there is a natural number d such that d is even.

More generally, allowing $A = \exists x[Q(x)]$ to contain also free occurrences of variables, $M \models \exists x[Q(x)]$ $[v]$ iff there is an element d in the domain D of M such that $M \models Q(a)[d/v]$, where a is a (new) variable not occurring in $\exists x[Q(x)]$ and d/v is the same valuation as v, except that d/v assigns to a the value d.

This finishes the definition of $M \models A[v]$. Notice that if A contains no free occurrences of variables, the valuation v does not play a role. Now Tarski's notion of $M \models A[v]$ (A

4.2 Semantics: Tarski's Truth Definition; Logical (Valid) Consequence

yields a true proposition under interpretation M and valuation v) has been defined, it is straightforward to define satisfiability and validity of a formula A.

Definition 4.10 (Satisfiable). Let A be a formula. A is *satisfiable* := there is an interpretation M for A and a valuation v such that $M \models A[v]$.

Example 4.6. $\forall x[P(x)]$ is satisfiable, since $M = \langle \mathbb{N}; \geq 0\rangle$ makes $\forall x[P(x)]$ true. However, $\forall x[P(x)] \wedge \exists x[\neg P(x)]$ is not satisfiable.

Definition 4.11 (Model). Let A be a formula and let M be an interpretation for A with domain D. M is a *model* of A := for all valuations v assigning elements of D to the variables occurring free in A, $M \models A[v]$. **Notation:** $M \models A$.
Instead of 'M is a model of A', one also says: M *makes A true* or A *is true in M*.
M is called a *countermodel* or *counterexample* for A if M is not a model for A, i.e., not $M \models A$. **Notation:** $M \not\models A$.

Example 4.7. Let $M = \langle \mathbb{N}; =\rangle$. Then $M \models a \equiv a$, since for all $n \in \mathbb{N}$, $n = n$.
Let $M = \langle \mathbb{N}; \geq; 0\rangle$. Then $M \models R(a,c)$, since for all natural numbers n in $\mathbb{N}$, $M \models R(a,c)[n]$, i.e., for all natural numbers n, $n \geq 0$. However, for $M = \langle \mathbb{N}; \geq; 2\rangle$ we have $M \not\models R(a,c)$, since there is a valuation v with $v(a) = 1$ such that $M \not\models R(a,c)[v]$, i.e., it is not the case that $1 \geq 2$.

Definition 4.12 (Closure). Let $A = A(a_1,\ldots,a_k)$ be a formula having $a_1,\ldots,a_k$ as the only free variables and not containing the bound variables $z_1,\ldots,z_k$. Then the *universal closure* of A is by definition the closed formula $\forall z_1 \ldots \forall z_k[A(z_1,\ldots,z_k)]$, where $A(z_1,\ldots,z_k)$ results from $A(a_1,\ldots,a_k)$ by replacing every occurrence of $a_1,\ldots,a_k$ by $z_1,\ldots,z_k$, respectively. **Notation:** $Cl(A)$.

Theorem 4.2. $M \models A$ iff $M \models Cl(A)$.

Proof. Evident from the definitions. For instance, for $M = \langle \mathbb{N}; \geq; 0\rangle$, $M \models R(a,c)$ iff $M \models \forall z[R(z,c)]$. □

Since every interpretation M (for a formula A) is a model of some formula B, one often uses the word *model* instead of 'interpretation' or 'structure'. The notion of $M \models A$ is the main notion of *model theory*. However, in *logic* one is not interested in the truth of formulas in individual interpretations M, but in the truth of formulas in *all* interpretations M (of the appropriate kind), in other words, in the validity of formulas.

Definition 4.13 (Validity). A is *valid* or *always true* := for all interpretations M for A, $M \models A$. **Notation:** $\models A$.

Example 4.8. $\models \forall x[R(x,c) \vee \neg R(x,c)]$; $\quad \models \forall x[P(x) \rightarrow P(x)]$;
$\models \forall x[P(x) \rightarrow Q(x)] \wedge P(c) \rightarrow Q(c)$; $\quad \models \neg \forall x[P(x)] \rightleftarrows \exists x[\neg P(x)]$.

Theorem 4.3. 1) $\models \forall x[P(x)] \rightleftarrows \forall y[P(y)]$ and 2) $\models \exists x[P(x)] \rightleftarrows \exists y[P(y)]$.

Proof. 1) Let $M = \langle D; P^* \rangle$ be an interpretation. Then $M \models \forall x[P(x)]$ iff $M \models \forall y[P(y)]$, because under interpretation M both formulas express the same proposition: all elements in D have the property P^*. So, every structure $\langle D; P^* \rangle$ is a model of $\forall x[P(x)] \rightleftarrows \forall y[P(y)]$. 2) is shown in a similar way. □

$\forall x \forall y[R(x,y)]$ and $\forall y \forall x[R(x,y)]$ express the same proposition: all objects are in the relation R with each other. Similarly, $\exists x \exists y[R(x,y)]$ and $\exists y \exists x[R(x,y)]$ express the same proposition: there are objects which are in the relation R to each other. Therefore:

Theorem 4.4.
$\models \forall x \forall y[R(x,y)] \rightleftarrows \forall y \forall x[R(x,y)]$ and $\models \exists x \exists y[R(x,y)] \rightleftarrows \exists y \exists x[R(x,y)]$.

Adapting the definition of 'valid consequence' for propositional logic to predicate logic, we say that B is a *valid (or logical) consequence* of $A_1, \ldots, A_n$, iff every interpretation which makes $A_1, \ldots, A_n$ simultaneously true also makes B true. For instance, $Q(c)$ is a logical consequence of $\forall x[P(x) \rightarrow Q(x)]$ and $P(c)$:

$$\forall x[P(x) \rightarrow Q(x)], P(c) \models Q(c)$$

since every interpretation which makes both $\forall x[P(x) \rightarrow Q(x)]$ and $P(c)$ true also makes $Q(c)$ true; in particular, for $M = \langle \text{Persons; is a man, is mortal; Caspar} \rangle$ we have: if all men are mortal and Caspar is a man, then Caspar is mortal.

Definition 4.14 (Valid (or logical) consequence). B is a *valid (or logical) consequence of* $A_1, \ldots, A_n$:= for every interpretation M and for all valuations v, if $M \models A_1[v]$ and ... and $M \models A_n[v]$, then $M \models B[v]$. **Notation**: $A_1, \ldots, A_n \models B$.

Example 4.9.

1. $\forall x[P(x) \rightarrow Q(x)], \exists x[R(x) \wedge \neg Q(x)] \models \exists x[R(x) \wedge \neg P(x)]$. This statement corresponds to Aristotle's syllogism 'Baroco' (see Subsection 4.7.4). For instance, the following argument is of this form:
   All logicians are philosophers.
   There are men who are not philosophers.
   Hence, there are men who are not logicians.
2. $\forall x[P(x) \rightarrow \neg Q(x)], \exists x[R(x) \wedge Q(x)] \models \exists x[R(x) \wedge \neg P(x)]$. This statement corresponds to Aristotle's syllogism 'Festino' (see Subsection 4.7.4).
3. $P(a), P(a) \rightarrow Q(a) \models Q(a)$

From the definition of $A_1, \ldots, A_n \models B$ it follows immediately that $A_1, \ldots, A_n \not\models B$ (B is not a logical consequence of $A_1, \ldots, A_n$) iff there is an interpretation M and a valuation v which make all of $A_1, \ldots, A_n$ true ($M \models A_1 \wedge \ldots \wedge A_n [v]$), but which make B false ($M \not\models B[v]$). Notice that if the formulas $A_1, \ldots, A_n$ and B are all closed, i.e., contain no free occurrences of variables, then the valuation v does not play any role.

Example 4.10. $\neg \forall x[P(x)] \not\models \forall x[\neg P(x)]$, since $M = \langle \mathbb{N}; P^* \rangle$, with $P^*(x) := x$ is even, makes $\neg \forall x[P(x)]$ true ('not all natural numbers are even' has truth value 1), while M makes $\forall x[\neg P(x)]$ false ('all natural numbers are not even' has truth value 0). In Exercise 4.10 we have shown that $\forall x[P(x) \rightarrow Q(x)] \not\models \exists x[P(x) \wedge Q(x)]$ and in Exercise 4.11 we have seen that $\forall x[P(x)] \rightarrow \forall x[Q(x)] \not\models \forall x[P(x) \rightarrow Q(x)]$.

4.2 Semantics: Tarski's Truth Definition; Logical (Valid) Consequence

The following theorem generalizes Theorem 2.4 for propositional logic to predicate logic.

Theorem 4.5.
a) $A \models B$ if and only if (iff) $\models A \to B$.
More generally,
b) $A_1, A_2 \models B$ if and only if (iff) $A_1 \models A_2 \to B$
                   if and only if (iff) $\models A_1 \to (A_2 \to B)$
                   if and only if (iff) $\models A_1 \wedge A_2 \to B$.
Even more generally,
c) $A_1, \ldots, A_n \models B$ if and only if (iff) $A_1, \ldots, A_{n-1} \models A_n \to B$
                      if and only if (iff) $\models (A_1 \wedge \ldots \wedge A_n) \to B$.

Proof. We shall prove the first statement of b). $A_1, A_2 \models B :=$ for every interpretation M and for every valuation v, if $M \models A_1[v]$ and $M \models A_2[v]$, then $M \models B[v]$.   (1)
$A_1 \models A_2 \to B :=$ for every interpretation M and for every valuation v, if $M \models A_1[v]$, then $M \models A_2 \to B[v]$ (2)
It is easy to see that (1) and (2) mean exactly the same, because $M \models A_2 \to B[v]$ means: if $M \models A_2[v]$, then $M \models B[v]$. □

Notice that $P(a) \not\models \forall x[P(x)]$, because from 'Antoine has property P' we cannot conclude that 'everyone has property P'. More precisely, let $M = \langle \mathbb{N}; \text{is even} \rangle$ and let $v(a) = 2$. Then $M \models P(a)[2]$, but $M \not\models \forall x[P(x)]$. However, the following does hold: if $M \models P(a)$, then $M \models \forall x[P(x)]$. For $M \models P(a)$ means: for every valuation v, $M \models P(a)[v]$, which means the same as: $M \models \forall x[P(x)]$ (see Theorem 4.2).

Corresponding to two possible treatments of the free individual variables in mathematical practice (see below), there are two different notions of 'valid consequence', the one defined in Def. 4.14 and the other to be defined in Def. 4.15 below.
$a^2 - 2a - 3 = 0$ is a *conditional equation*, since it expresses a condition on a. From this condition we should not infer that $2^2 - 2 \cdot 2 - 3 = 0$; however, from $a^2 - 2a - 3 = 0$ we can infer that $(a-3)(a+1) = 0$ and hence that $a = 3$ or $a = -1$. We may say that in these inferences the variable a is *held constant*, since it stands for the same number throughout the deductions. This inference can be written thus:
$a^2 - 2a - 3 = 0 \to a = 3 \vee a = -1$ or, equivalently, as
$\forall x[x^2 - 2x - 3 = 0 \to x = 3 \vee x = -1]$.   (1)
This inference corresponds with our definition of $A \models B$.

However, from $a + b = b + a$ one may conclude that $2 + 3 = 3 + 2$. In the inferences from $a + b = b + a$, the variables a and b are *general* or *allowed to vary*. Using only bound variables, the result of this inference can be written thus:
$\forall x \forall y[x + y = y + x] \to 2 + 3 = 3 + 2$.   (2)
This inference corresponds with our definition of $A \models^2 B$, as given in Def. 4.15 below.

Note that in (1) parentheses close *after* the $\to$, in (2) *before* the $\to$. Whether we choose to use interpretation (1) or (2) depends on the role the assumptions have in each case we want to infer consequences from assumptions.

Definition 4.15. B is a *valid consequence of $A_1, \ldots, A_n$ with all free variables general* := for every structure M, if $M \models A_1$ and ... and $M \models A_n$, then $M \models B$.
Notation: $A_1, \ldots, A_n \models^2 B$. So, $A_1, \ldots, A_n \models^2 B$ iff $Cl(A_1), \ldots, Cl(A_n) \models Cl(B)$, where $Cl(B)$ is the universal closure of B.

Theorem 4.6. *If $A \models B$, then $A \models^2 B$, but in general not conversely.*

Proof. Suppose $A \models B$, i.e., for every interpretation M and for every valuation v, if $M \models A[v]$, then $M \models B[v]$. (*)
To show: $M \models^2 B$. So, suppose that $M \models A$, i.e., for every valuation v, $M \models A[v]$. Then it follows from (*) that for every valuation v, $M \models B[v]$, i.e., $M \models B$.

To establish that in general the converse does not hold, note that $P(a) \models^2 \forall x[P(x)]$, i.e., $\forall x[P(x)] \models \forall x[P(x)]$, but $P(a) \not\models \forall x[P(x)]$, since for $M = \langle \mathbb{N}; \text{is even} \rangle$, $M \models P(a)[2]$ (2 is even), while $M \not\models \forall x[P(x)]$ (not all natural numbers are even). □

Many-sorted and higher-order predicate logic In order to avoid misunderstanding, it should be noted that also for formulas containing two or more quantifiers, like, for instance, $\forall x \exists y[R(x,y)]$, an interpretation contains only *one* (non-empty) domain or set for the bound individual variables of the formula, such that *all* individual variables x, y, etc., are to be interpreted as elements of that one domain. So, in $\forall x \exists y[R(x,y)]$, for instance, we are not allowed to let x range over the set of all Men and y range over the set of all Women; the variables x and y have to be interpreted as elements of the same set, for instance, the set of all persons. The expression 'for every man x there is some woman y such that $R(x,y)$' should be translated into our symbolism by a formula of the form $\forall x[M(x) \to \exists y[W(y) \wedge R(x,y)]]$, where M and W are unary predicate symbols for 'is a man' and 'is a woman' respectively.

The predicate logic we have presented thus far is *one-sorted*, i.e., the language contains only one sort of variables which have to be interpreted as elements of one and the same domain. One might also develop a *two-sorted* predicate logic having two sorts of variables, where the variables of the one sort should be interpreted as elements of a domain D_1 and the variables of the other sort as elements of a domain D_2. This corresponds more closely to mathematical practice, where frequently different sorts of variables are used; for instance, m, n, p, ... ranging over natural numbers and x, y, z, ... ranging over real numbers. The development of two-sorted predicate logic is similar to that of one-sorted predicate logic. The same holds for predicate logic with more than two sorts of variables.

The predicate calculus we have presented thus far is also *first-order*, i.e., one can only quantify over individuals and not over properties of individuals, nor over properties of properties of individuals, and so on. (For instance, 'being a colour' is a property of the property 'being red' of individuals.) In *second-order* logic, not only quantification over individual variables, $\forall x$, $\exists y$, ..., but also quantification over predicate variables is allowed: $\forall P$, $\exists Q$, ... This increases the expressive power of the language considerably. By iteration one can obtain *higher-order* predicate logic.

Exercise 4.12. Let $\mathbb{N}$ be the set of natural numbers and $M = \langle \mathbb{N}, P^*, Q^*, R^* \rangle$ with P^*: is even, Q^*: is odd, R^*: is less than ($<$). Which of the following statements are right

and which are wrong?
$M \models P(a)[2]$ $M \models \exists x[P(x)]$ $M \models \exists x[P(x)] \to \exists x[Q(x)]$ $M \models \forall x \exists y[R(x,y)]$
$M \models P(a)[5]$ $M \models \forall x[P(x)]$ $M \models \exists x[P(x)] \to \forall x[Q(x)]$ $M \models \exists y \forall x[R(x,y)]$
$M \models Q(a)[5]$ $M \models \exists x[Q(x)]$ $M \models \forall x[P(x)] \to \exists x[Q(x)]$ $M \models \exists x[P(x) \to Q(x)]$
$M \models Q(a)[2]$ $M \models \forall x[Q(x)]$ $M \models \forall x[P(x)] \to \forall x[Q(x)]$ $M \models \forall x[P(x) \to Q(x)]$

Exercise 4.13. Which of the following alternatives applies to the following below: (i) not satisfiable, (ii) satisfiable, but not valid, (iii) valid, and hence satisfiable?

1. $\exists x[P(x)] \to \forall x[P(x)]$
2. $\exists x[P(x)] \to \exists x[\neg P(x)]$
3. $\exists x[P(x)] \land \forall x[\neg P(x)]$
4. $\forall x[P(x)] \land \neg \exists x[P(x)]$
5. $\neg \forall x[P(x)] \to \forall x[\neg P(x)]$
6. $\forall x[\neg P(x)] \to \neg \forall x[P(x)]$
7. $\forall x \exists y[R(x,y)] \land \exists x \forall y[\neg R(x,y)]$
8. $\forall x \exists y[R(x,y)] \to \exists y \forall x[R(x,y)]$
9. $\exists x[P(x)] \land \exists x[Q(x)] \to \exists x[P(x) \land Q(x)]$
10. $\forall x[P(x) \lor Q(x)] \to \exists x[P(x)] \lor \forall x[Q(x)]$

Exercise 4.14 (Kleene [9]). Translate each of the following arguments into the language of predicate logic and establish whether the conclusion logically follows from the premises. If so, give a proof; if not, give a counterexample.

1. Each politician is a showman. Some showmen are insincere. Therefore, some politicians are insincere.
2. No professors are ignorant. All ignorant people are vain. Therefore, no professors are vain.
3. Only birds have feathers. No mammal is a bird. Therefore, each mammal is featherless.
4. Some masons are not strong. All carpenters are strong. Therefore, some carpenters are not masons.
5. Some plumbers are smart. There are no smart persons who are not careful. Therefore, some plumbers are careful.

Exercise 4.15 (Kleene [9]). The same question as in Exercise 4.14.

1. No animals are immortal. All cats are animals. Therefore, some cats are not immortal.
2. If anyone can solve this problem, some philosopher can solve it. Cabot is a philosopher and cannot solve the problem. Therefore, the problem cannot be solved.
3. Any mathematician can solve this problem if anyone can. Cabot is a mathematician and cannot solve the problem. Therefore, the problem cannot be solved.
4. Some healthy people are fat. No unhealthy people are strong. Therefore, some fat people are not strong.
5. Some students are studious. No student is unqualified. Therefore, some unqualified students are not studious.

Exercise 4.16. Prove or refute: $\forall x[P(x) \to Q(x)] \models \exists x[P(x) \land Q(x)]$.

Exercise 4.17. Let $R(a,b)$ stand for 'a is greater than b'. i) Translate the following sentences into the language of predicate logic using the binary predicate symbol R:

(a) For every natural number there is a greater one.
(b) There is no natural number which is greater than all natural numbers.
ii) Let A and B be the translations of (a), (b) respectively. Show that not $A \models B$.
iii) Intuitively, (b) seems to follow from (a). Why does not this contradict $A \not\models B$?
iv) Show that $\forall x \exists y [R(y,x)], \forall x \forall y [R(y,x) \to \neg R(x,y)] \models \neg \exists x \forall y [R(x,y)]$.

Exercise 4.18. Translate the following sentence into the language of predicate logic and show that the resulting formula is always true (valid). Take as domain the set of all men in a certain village and interpret $S(x,y)$ as 'x shaves y': there is no man (in the village) such that he shaves precisely those men (in the village) who do not shave themselves.

Exercise 4.19. Check that the following formulas are valid.
a. $\forall x \exists y [P(x) \to P(y)]$;
b. $\forall y \exists x [P(x) \to P(y)]$;
c. $\exists x \forall y [P(x) \to P(y)]$;
d. $\exists y \forall x [P(x) \to P(y)]$.

Exercise 4.20. Which of the following formulas are valid? Give either a proof or a counterexample.
a. $\forall x \exists y [R(x,y)] \to \exists x \forall y [R(x,y)]$;
b. $\exists x \forall y [R(x,y)] \to \forall x \exists y [R(x,y)]$;
c. $\forall x \exists y [R(x,y)] \to \forall x \exists y [R(y,x)]$;
d. $\exists x \forall y [R(x,y)] \to \exists y \forall x [R(x,y)]$;
e. $\exists x \forall y [R(x,y)] \rightleftarrows \exists y \forall x [R(y,x)]$;
f. $\forall x \exists y [R(x,y)] \rightleftarrows \forall y \exists x [R(y,x)]$.

4.3 Basic Results about Validity and Logical Consequence

Definition 4.16. $A \dashv\vdash B := A \models B$ and $B \models A$, i.e., for every interpretation M and valuation v, $M \models A\,[v]$ iff $M \models B\,[v]$. This is equivalent to $\models A \rightleftarrows B$.

In what follows it is important to realize that for $M = \langle D; P^* \rangle$ and d an element of D, $M \models P(a)[d]$ is equivalent to saying that the proposition $P^*(d)$ - d has the property P^* - has truth value 1. For instance, for $M = \langle \mathbb{N}; \text{is odd} \rangle$, $M \models P(a)[3]$ because the proposition '3 is odd' has truth value 1, while $M \not\models P(a)[2]$, because the proposition '2 is odd' has truth value 0.

4.3.1 Quantifiers and Connectives

We start with looking at combinations of the quantifiers $\forall$ and $\exists$, respectively, with negation $\neg$:

Theorem 4.7 (Quantifiers and Negation).
1) $\neg \forall x [P(x)] \dashv\vdash \exists x [\neg P(x)]$;
2) $\neg \exists x [P(x)] \dashv\vdash \forall x [\neg P(x)]$.
3) $\neg \forall x [P(x)] \not\models \forall x [\neg P(x)]$, although conversely, $\forall x [\neg P(x)] \models \neg \forall x [P(x)]$.
4) $\neg \exists x [P(x)] \models \exists x [\neg P(x)]$, but conversely, $\exists x [\neg P(x)] \not\models \neg \exists x [P(x)]$.

4.3 Basic Results about Validity and Logical Consequence 205

Proof. 1) Let $M = \langle D; P^* \rangle$ be an interpretation. Then $M \models \neg \forall x[P(x)]$ (i.e., not all elements in D have the property P^*) iff $M \models \exists x[\neg P(x)]$ (i.e., some element in D does not have the property P^*). So each model M is a model of $\neg \forall x[P(x)] \rightleftarrows \exists x[\neg P(x)]$.
2) is shown in a way similar: there is no element in D which has the property P^* iff all elements in D do not have the property P^*.
3) $M = \langle \mathbb{N}; \text{is odd} \rangle$ is a counterexample, since $M \models \neg \forall x[P(x)]$: the proposition 'not all natural numbers are odd' has truth value 1; but $M \not\models \forall x[\neg P(x)]$: the proposition 'all natural numbers are not odd' has truth value 0.
Conversely, suppose $M = \langle D; P^* \rangle$ is an interpretation and suppose $M \models \forall x[\neg P(x)]$, i.e., all elements in D have the property not-P^*. Then surely not all elements in D have the property P^*, i.e., $M \models \neg \forall x[P(x)]$.
4) Let $M = \langle D; P^* \rangle$ be an interpretation and suppose $M \models \neg \exists x[P(x)]$, i.e., there is no element d in D which has the property P^*, in other words, all elements in D have the property not-P^*. So, since D is non-empty, there is an element in D with the property not-P^*, i.e., $M \models \exists x[\neg P(x)]$.
Conversely, $M = \langle \mathbb{N}; \text{is odd} \rangle$ is a counterexample, for $M \models \exists x[\neg P(x)]$: the proposition 'there is a natural number that is not odd' has truth value 1; but $M \not\models \neg \exists x[P(x)]$: the proposition 'there is no odd natural number' has truth value 0. $\square$

Given a propositional formula A, one might let the variable x range over the lines of the truth table of A, and interpret $P(x)$ as 'formula A is 1 at line x'. Under this interpretation the formula $\neg \forall x P(x)$ yields the proposition 'not in all lines of the truth table A is 1', i.e., $\not\models A$, while the formula $\forall x[\neg P(x)]$ yields the proposition 'in all lines of the truth table A is 0', i.e., $\models \neg A$. Under this interpretation $\neg \forall x[P(x)] \not\models \forall x[\neg P(x)]$ expresses that from $\not\models A$ one may in general not conclude that $\models \neg A$, as we have already seen in Theorem 2.12.

Because the meaning of the universal quantitier $\forall$ is similar to the meaning of the connective $\land$, the following theorem is evident:

Theorem 4.8 ($\forall$ **and** $\land$). $\forall x[P(x)] \land \forall x[Q(x)] \models\mid \forall x[P(x) \land Q(x)]$

However, one has to be careful when combining a universal quantifier $\forall$ with the connective $\lor$. Consider the following argument:

Every gnome has a conical cap or is a Quaker.
Therefore: all gnomes have a conical cap or all gnomes are Quakers.

Translating this argument into the language of predicate logic we find:

$$\forall x[P(x) \lor Q(x)] \not\models \forall x[P(x)] \lor \forall x[Q(x)]$$

The following interpretation (or model) is a counterexample: $M = \langle \mathbb{N}; P^*, Q^* \rangle$ with $P^*(x)$: x is even, and $Q^*(x)$: x is odd. Then $M \models \forall x[P(x) \lor Q(x)]$: the proposition 'every natural number is even or odd' has truth value 1. But $M \not\models \forall x[P(x)] \lor \forall x[Q(x)]$: the proposition 'all natural numbers are even or all natural numbers are odd' has truth value 0.

Theorem 4.9 ($\forall$ **and** $\lor$). *a)* $\forall x[P(x) \lor Q(x)] \not\models \forall x[P(x)] \lor \forall x[Q(x)]$.
But conversely, b) $\forall x[P(x)] \lor \forall x[Q(x)] \models \forall x[P(x) \lor Q(x)]$.

Proof. We start with an informal proof of b): Suppose all things have the property P or all things have the property Q. If an individual thing has the property P, then it also has the property $P \vee Q$; and similarly, if an individual thing has the property Q, then it also has the property $P \vee Q$. So, in both cases it follows that all things have the property $P \vee Q$, i.e., $\forall x[P(x) \vee Q(x)]$.

More precisely: Suppose $M = \langle D; P^*, Q^* \rangle$ is an interpretation and $M \models \forall x[P(x)] \vee \forall x[Q(x)]$, i.e., 1) for every thing d in the domain D of M, $M \models P(a)[d]$ or 2) for every thing d in the domain D of M, $M \models Q(a)[d]$. In case 1) it follows from 'if $M \models P(a)[d]$, then $M \models P(a) \vee Q(a)\,[d]$' that for all things d in the domain D of M, $M \models P(a) \vee Q(a)\,[d]$, in other words, $M \models \forall x[P(x) \vee Q(x)]$. In case 2) it follows in a similar way that $M \models \forall x[P(x) \vee Q(x)]$. □

Given propositional formulas A and B, one may let the variable x range over the lines in the truth tables of A, B, interpret $P(x)$ as 'A is 1 in line x' and $Q(x)$ as 'B is 1 in line x'. Under this interpretation $\forall x[P(x) \vee Q(x)]$ yields the proposition 'in all lines of the truth table, A is 1 or B is 1', i.e., $\models A \vee B$. But under this same interpretation $\forall x[P(x)] \vee \forall x[Q(x)]$ yields the proposition 'in all lines of the truth table A is 1 or in all lines of the truth table B is 1', i.e., $\models A$ or $\models B$. Under this interpretation $\forall x[P(x) \vee Q(x)] \not\models \forall x[P(x)] \vee \forall x[Q(x)]$ expresses that in general from $\models A \vee B$ one may not conclude that $\models A$ or $\models B$, as we have already seen in Theorem 2.13.

Because the meaning of the existential quantifier $\exists$ is similar to the meaning of the connective $\vee$, the following theorem is evident:

Theorem 4.10 ($\exists$ and $\vee$). $\exists x[P(x)] \vee \exists x[Q(x)] \models\!\!\dashv \exists x[P(x) \vee Q(x)]$.

However, one has to be careful when combining an existential quantifier with the connective $\wedge$. Consider the following argument:

> There is gnome who has a conical cap and there is a gnome who is a Quaker.
> Therefore: there is a gnome who has a conical cap and is a Quaker.

Translating this argument into the language of predicate logic we find:

$$\exists x[P(x)] \wedge \exists x[Q(x)] \not\models \exists x[P(x) \wedge Q(x)]$$

The following interpretation (or model) is a counterexample: $M = \langle \mathbb{N}; P^*, Q^* \rangle$ with $P^*(x)$: x is even, and $Q^*(x)$: x is odd. Then $M \models \exists x[P(x)] \wedge \exists x[Q(x)]$: the proposition 'there is an even natural number and there is an odd natural number' has truth value 1. But $M \not\models \exists x[P(x) \wedge Q(x)]$: the proposition 'there is natural number that is both even and odd' has truth value 0.

Theorem 4.11 ($\exists$ and $\wedge$). a) $\exists x[P(x)] \wedge \exists x[Q(x)] \not\models \exists x[P(x) \wedge Q(x)]$. But conversely, b) $\exists x[P(x) \wedge Q(x)] \models \exists x[P(x)] \wedge \exists x[Q(x)]$.

Proof. We start with an informal proof of b): Suppose $\exists x[P(x) \wedge Q(x)]$, i.e., there is a thing d such that d has both the property P and the property Q. Then this d has the property P, so $\exists x[P(x)]$; and this same d has the property Q, so $\exists x[Q(x)]$.

More precisely: Suppose $M = \langle D; P^*, Q^* \rangle$ is a model and $M \models \exists x[P(x) \wedge Q(x)]$, i.e., there is a thing d in the domain D of M such that $M \models P(a) \wedge Q(a)\,[d]$. Then $M \models P(a)[d]$, hence $M \models \exists x[P(x)]$; and $M \models Q(a)[d]$ and hence $M \models \exists x[Q(x)]$. □

4.3 Basic Results about Validity and Logical Consequence

Consider the following argument:

> If all gnomes have a conical cap, then all gnomes are Quakers.
> Therefore: every gnome with a conical cap is a Quaker.

Translating this argument into the language of predicate logic we find:

$$\forall x[P(x)] \to \forall x[Q(x)] \not\models \forall x[P(x) \to Q(x)]$$

The following interpretation (or model) is a counterexample: $M = \langle \mathbb{N}; P^*, Q^* \rangle$ with $P^*(x)$: x is even, and $Q^*(x)$: x is odd. Then $M \models \forall x[P(x)] \to \forall x[Q(x)]$: the proposition 'if all natural numbers are even, then all natural numbers are odd' has truth value $0 \to 0 = 1$. But $M \not\models \forall x[P(x) \to Q(x)]$: the proposition 'for every natural number n, if n is even, then n is odd' has truth value 0.

Theorem 4.12 ($\forall$ and $\to$). *a)* $\forall x[P(x)] \to \forall x[Q(x)] \not\models \forall x[P(x) \to Q(x)]$. *But conversely, b)* $\forall x[P(x) \to Q(x)] \models \forall x[P(x)] \to \forall x[Q(x)]$.

Proof. We start with an informal proof of b): Suppose $\forall x[P(x) \to Q(x)]$, i.e., every thing with the property P also has the property Q. Next suppose $\forall x[P(x)]$, i.e., every thing has the property P. Then clearly it follows that every thing has the property Q. More precisely: Suppose $M = \langle D; P^*, Q^* \rangle$ is a model and $M \models \forall x[P(x) \to Q(x)]$, i.e., for every thing d in the domain D of M, $M \models P(a) \to Q(a)$ $[d]$. Suppose next that $M \models \forall x[P(x)]$, i.e., for every thing d in the domain D of M, $M \models P(a)[d]$. Then it clearly follows that for every thing d in the domain D of M, $M \models Q(a)[d]$, in other words, $M \models \forall x[Q(x)]$. □

Given propositional formulas A and B, one may let the variable x range over the lines in the truth tables of A, B, interpret $P(x)$ as 'A is 1 in line x' and $Q(x)$ as 'B is 1 in line x'. Under this interpretation $\forall x[P(x)] \to \forall x[Q(x)]$ yields the proposition 'if A is 1 in all lines of the truth table, then B is 1 in all lines of the truth table', i.e., if $\models A$, then $\models B$. But under this same interpretation $\forall x[P(x) \to Q(x)]$ yields the proposition 'in all lines x of the truth table, if A is 1 at line x, then also B is 1 at line x', i.e., $\models A \to B$. Under this interpretation $\forall x[P(x)] \to \forall x[Q(x)] \not\models \forall x[P(x) \to Q(x)]$ expresses that in general from 'if $\models A$, then $\models B$' one may not conclude that $\models A \to B$, as we have already seen in Theorem 2.11.

Consider the following argument:

> There is a gnome such that if he has a conical cap, then he is a Quaker.
> There is gnome who has a conical cap.
> Therefore: there is a gnome who is a Quaker.

Translating this argument into the language of predicate logic we find:

$$\exists x[P(x) \to Q(x)], \exists x[P(x)] \not\models \exists x[Q(x)]$$

The following interpretation is a counterexample: $M = \langle \mathbb{N}; P^*, Q^* \rangle$ with $P^*(x)$: x is even, and $Q^*(x)$: $x \neq x$. Then $M \models \exists x[P(x) \to Q(x)]$, since $M \models P(a) \to Q(a)$ [3]: the proposition 'if 3 is even, then $3 \neq 3$' has truth value $0 \to 0 = 1$. Also $M \models \exists x[P(x)]$: the proposition 'there is an even natural number' has truth value 1. But

$M \not\models \exists x[Q(x)]$: the proposition 'there is a natural number which is not equal to itself' has truth value 0.

Theorem 4.13 ($\exists$ **and** $\rightarrow$). *a)* $\exists x[P(x) \rightarrow Q(x)], \exists x[P(x)] \not\models \exists x[Q(x)]$, *or, equivalently,* $\exists x[P(x) \rightarrow Q(x)] \not\models \exists x[P(x)] \rightarrow \exists x[Q(x)]$.
But conversely, b) $\exists x[P(x)] \rightarrow \exists x[Q(x)] \models \exists x[P(x) \rightarrow Q(x)]$.

Proof. We start with an informal proof of b): Suppose $\exists x[P(x)] \rightarrow \exists x[Q(x)]$ (*) and $\neg \exists x[P(x) \rightarrow Q(x)]$. Then $\forall x[\neg (P(x) \rightarrow Q(x))]$, i.e., $\forall x[P(x) \wedge \neg Q(x)]$, in other words, $\forall x[P(x)] \wedge \forall x[\neg Q(x)]$. Hence, surely, $\exists x[P(x)]$ and hence by (*) $\exists x[Q(x)]$. Contradiction with $\forall x[\neg Q(x)]$.
More precisely: Suppose $M = \langle D; P^*, Q^* \rangle$ is a model and $M \models \exists x[P(x)] \rightarrow \exists x[Q(x)]$. (*) Case 1) $M \models \exists x[Q(x)]$, i.e., there is some element d in the domain D of M such that $M \models Q(a)[d]$. Then also $M \models P(a) \rightarrow Q(a)$ $[d]$ and hence, $M \models \exists x[P(x) \rightarrow Q(x)]$. Case 2) $M \not\models \exists x[Q(x)]$. Then by (*), $M \not\models \exists x[P(x)]$, i.e., $M \models \forall x[\neg P(x)]$. But then $M \models \forall x[P(x) \rightarrow Q(x)]$, since for every d in the domain D of M, $M \models P(a) \rightarrow Q(a)$ $[d]$ ($0 \rightarrow 0 = 1$). Hence, surely, $M \models \exists x[P(x) \rightarrow Q(x)]$. □

4.3.2 Two different quantifiers

Consider the following argument:

Every gnome has a teacher. .
Therefore: some gnome is the teacher of all gnomes.

Translating this argument into the language of predicate logic, reading $R(x,y)$ as 'x has y as teacher', we find:

$$\forall x \exists y[R(x,y)] \not\models \exists y \forall x[R(x,y)]$$

$M = \langle \mathbb{N}; < \rangle$ is a counterexample. $M \models \forall x \exists y[R(x,y)]$: the proposition 'for every natural number x there is a larger natural number y' has truth value 1; but $M \not\models \exists y \forall x[R(x,y)]$: the proposition 'there is a natural number y such that all natural numbers x are smaller than y' has truth value 0.

Theorem 4.14 (Interchanging Quantifiers). *a)* $\forall x \exists y[R(x,y)] \not\models \exists y \forall x[R(x,y)]$.
But conversely, b) $\exists y \forall x[R(x,y)] \models \forall x \exists y[R(x,y)]$.

Proof. We start with an informal proof of b): Suppose $\exists y \forall x[R(x,y)]$, i.e., there is a thing d such that each thing x stands in the relation R to d. Then clearly, for every x there is a thing y, namely d, such that x is in relation R to y. For instance, suppose there is someone, say Michael Jackson, such that all persons admire this one person. Then clearly, everyone admires at least one person, namely Michael Jackson.
More precisely: Let $M = \langle D; R^* \rangle$ be a model and suppose $M \models \exists y \forall x[R(x,y)]$, i.e., there is some d in the domain D of M such that $M \models \forall x[R(x,b)]$ $[d]$. Then clearly, $M \models \forall x \exists y[R(x,y)]$ because for every x one may take $y = d$. □

4.3 Basic Results about Validity and Logical Consequence

The following theorem says that a negation in front of a sequence of quantifiers may be pushed inside, provided one changes a universal quantifier $\forall$ into an existential quantifier $\exists$ and an existential quantifier $\exists$ into a universal quantifier $\forall$.

Theorem 4.15 (Negation in front of a sequence of Quantifiers).
1. $\neg \forall x \exists y [R(x,y)] \models\!\!\dashv \exists x \forall y [\neg R(x,y)]$.
2. $\neg \exists x \forall y [R(x,y)] \models\!\!\dashv \forall x \exists y [\neg R(x,y)]$.

Proof. 1. Let $M = \langle D; R^* \rangle$ be an interpretation. Then
$M \models \neg \forall x \exists y [R(x,y)]$ iff (by Theorem 4.7, 1)
$M \models \exists x \neg \exists y [R(x,y)]$ iff (by Theorem 4.7, 2)
$M \models \exists x \forall y [\neg R(x,y)]$.
2. $M \models \neg \exists x \forall y [R(x,y)]$ iff $M \models \forall x \neg \forall y [R(x,y)]$ iff $M \models \forall x \exists y [\neg R(x,y)]$. □

Warning Note that 'not $\models A$ (A is not valid)' means that not every interpretation M for A is a model of A, in other words, there is at least one interpretation M that makes A false. In such a case one may in general not conclude that $\models \neg A$, since there may be other interpretations which make A true.

For instance, we have seen in Theorem 4.9 that there are interpretations M which make the formula $\forall x [P(x) \lor Q(x)] \to \forall x [P(x)] \lor \forall x [Q(x)]$ false, but the interpretation $M = \langle \mathbb{N}; \text{is even}, x = x \rangle$ makes $\forall x [P(x) \lor Q(x)] \to \forall x [P(x)] \lor \forall x [Q(x)]$ true. Summarizing, the formula $\forall x [P(x) \lor Q(x)] \to \forall x [P(x)] \lor \forall x [Q(x)]$ yields a false proposition for some interpretations and a true proposition for others. Hence, neither the formula itself nor its negation is valid.

4.3.3 About the axioms and rules for $\forall$ and $\exists$

Later in this chapter the formula $\forall x [A(x)] \to A(t)$, where t is a term, will be chosen as an logical axiom schema for $\forall$, and $A(t) \to \exists x [A(x)]$ as a logical axiom schema for $\exists$. In the next theorem we verify that these formulas are valid or always true.

Theorem 4.16 (Validity of the logical axioms for the Quantifiers).
Let t be a term, and let $A(t)$ result from $A(x)$ by substituting t for all occurrences of x in $A(x)$. Then 1. $\models \forall x [A(x)] \to A(t)$, and 2. $\models A(t) \to \exists x [A(x)]$.

Proof. The formula $\forall x [A(x)] \to A(t)$ expresses that if all objects (of a certain kind) have the property A^* and t^* is one these objects, then also t^* has the property A^*. More formally: Let M be a structure with domain D and let v be a valuation assigning values in D to the individual variables occurring free in $A(x)$. We have to show that $M \models \forall x [A(x)] \to A(t) \, [v]$. So, suppose $M \models \forall x [A(x)] \, [v]$, i.e., for all d in D, $M \models A(a) \, [d/v]$ where a is a free variable not occurring in $A(x)$ and d/v is the same valuation as v except that it assigns d to the variable a. (*)
Now let t^* be the element in D assigned by the valuation v to the term t, i.e., $v(t) = t^*$. Then because of (*), $M \models A(a) \, [t^*/v]$, which is equivalent to $M \models A(t) \, [v]$.
The proof of 2. is similar to the proof of 1. □

Note that if y is a bound variable, $\forall x[A(x)] \to A(y)$ is *not* a formula; and even if it were a formula, in general, not $\models \forall x[A(x)] \to A(y)$. For instance, if $A(x) = \exists y[P(x,y)]$, $\not\models \forall x \exists y[P(x,y)] \to \exists y[P(y,y)]$, for $\langle \mathbb{N}; < \rangle$ is a counterexample to this formula. This demonstrates the usefulness of having two kinds of symbols for free and bound (occurrences of) individual variables. If one uses the same symbols for both free and bound occurrences of individual variables, then $\forall x[A(x)] \to A(y)$ is only valid under the condition that y is *free for* x in $A(x)$, i.e., if any free occurrence of x in $A(x)$ is replaced by an occurrence of y, then the resulting occurrence of y in $A(y)$ should also be free.

In Section 4.4 we shall introduce the following deduction rules for $\forall$ and for $\exists$, assuming that C does not contain the free variable a:

$$\frac{C \to P(a)}{C \to \forall x[P(x)]} \qquad \frac{P(a) \to C}{\exists x[P(x)] \to C}$$

Theorem 4.17 says that these rules are sound in the sense that for any interpretation M, if M makes the premiss true, then M also makes the conclusion true. But the same theorem says that $C \to P(a) \not\models C \to \forall x[P(x)]$ and that $P(a) \to C \not\models \exists x[P(x)] \to C$. Note the difference with the rule Modus Ponens, where we do have $A, A \to B \models B$. In propositional logic we have seen in Theorem 2.11 that '$A \models B$' is a stronger statement than 'if $\models A$, then $\models B$'. This becomes particularly evident in predicate logic, as can be seen from Theorem 4.17 below. For instance, from Theorem 4.17, 1 it follows that 'if $\models P(a)$, then $\models \forall x[P(x)]$' is true, while the stronger statement $P(a) \models \forall x[P(x)]$ is false. Items 2 and 3 of this theorem state that the logical deduction rules for $\forall$ and for $\exists$, to be introduced in Section 4.4, are sound.

Theorem 4.17 (Soundness of the deduction rules for the Quantifiers).
Let $A(a)$ be a formula containing a free variable a and let C be a formula not containing the free variable a. Let M be an interpretation.
1. *If $M \models A(a)$, then $M \models \forall x[A(x)]$. But $A(a) \not\models \forall x[A(x)]$.*
2. *If $M \models C \to A(a)$, then $M \models C \to \forall x[A(x)]$. But $C \to A(a) \not\models C \to \forall x[A(x)]$.*
3. *If $M \models A(a) \to C$, then $M \models \exists x[A(x)] \to C$. But $A(a) \to C \not\models \exists x[A(x)] \to C$.*

Proof. 1) Suppose $M \models A(a)$, i.e., for every d in the domain D of M, $M \models A(a)[d]$. In other words, $M \models \forall x[A(x)]$. On the other hand, let $M = \langle \mathbb{N}; \text{is even}\rangle$. Then $M \models P(a)[2]$ (2 is even), but $M \not\models \forall x[P(x)]$ (not all natural numbers are even). Hence $P(a) \not\models \forall x[P(x)]$.
2) Suppose $M \models C \to A(a)$ and C does not contain the variable a. Then by 1) $M \models \forall x[C \to A(x)]$ and hence, because C does not contain a, $M \models C \to \forall x[A(x)]$. On the other hand, for $C = Q \vee \neg Q$, $C \to A(a)$ is equivalent to $A(a)$ and $C \to \forall x[A(x)]$ is equivalent to $\forall x[A(x)]$. Hence, for $C = Q \vee \neg Q$, $C \to A(a) \models C \to \forall x[A(x)]$ iff $A(a) \models \forall x[A(x)]$, which according to 1) does not hold. Another way to see that $C \to A(a) \not\models C \to \forall x[A(x)]$ does not hold is as follows: from 'if it is September 5, then a(ntoine) has his birthday' one may not conclude 'if it is September 5, then everyone has his birthday'.
3) Suppose $M \models A(a) \to C$ and C does not contain a. That is, for every element d in

4.3 Basic Results about Validity and Logical Consequence

the domain of M, $M \models A(a) \to C\,[d]$ (*). We have to show: $M \models \exists x[A(x)] \to C$. So, suppose $M \models \exists x[A(x)]$, i.e., for some d in the domain of M, $M \models A(a)\,[d]$. Hence, because of (*) and because C does not contain a, $M \models C$. On the other hand, from 'if a(ntoine) has his birthday, then it is September 5' one may not conclude that 'if someone has his birthday, then it is September 5'. □

The condition in Theorem 4.17 that C does not contain the free variable a is necessary. To see this, let C be $A(a)$. Then $M \models A(a) \to A(a)$, but in general $M \not\models A(a) \to \forall x[A(x)]$; for instance, the proposition 'if a(ntoine) has his birthday, then everyone has his birthday' is false.

Also $M \models A(a) \to A(a)$, but in general $M \not\models \exists x[A(x)] \to A(a)$; for instance, the proposition 'if there is an even number, then 3 is even' is false.

4.3.4 Predicate Logic with Function Symbols*

In mathematics, but also in natural language, one frequently uses functions. For instance, the binary function + that assigns to any pair of natural numbers n and m the natural number $n+m$; the unary function *the-mother-of* that assigns to any person a his or her mother: *the-mother-of (a)*. So, it is convenient to extend the predicate language with function symbols:

$$f_1, f_2, f_3, \ldots$$

where each f_i is supposed to be k_i-ary, i.e., taking k_i arguments. Individual constants are then special function symbols, namely, function symbols f_i taking 0 arguments, i.e., $k_i = 0$. An example of a predicate language containing function symbols for addition and multiplication of natural numbers is given in Chapter 5.

With no function symbols present, the only terms - denoting elements of the domain D of a given interpretation M - are free individual variables and individual constants. But with function symbols present, we have to extend the notion of *term*.

Definition 4.17 (Terms). *Terms* are defined (inductively) as follows:
1. Each free individual variable is a term.
2. Each individual constant is a term.
3. If f_i is a k_i-ary function symbol and $t_1, \ldots, t_{k_i}$ are terms, then $f_i(t_1, \ldots, t_{k_i})$ is a term. Note that clause 2 can be treated as a special case of clause 3, taking $k_i = 0$.

Formulas are defined as before (see Definition 4.5), but now allowing the $t_1, \ldots, t_n$ in Definition 4.4 of 'atomic formula' to be any terms, instead of simply any free individual variables or individual constants.

If we extend the predicate language with function symbols, we have to adapt the definition of an interpretation or structure (Definition 4.7) accordingly.

Definition 4.18 (Interpretation). An *interpretation M for the predicate logic with function symbols* is by definition a tuple $\langle D; P_1^*, P_2^*, \ldots\,; f_1^*, f_2^*, \ldots \rangle$, such that:
1. D is a non-empty set, called the *domain* of M.

2. For any n_i-ary predicate symbol P_i, P_i^* is a n_i-ary predicate over D.
3. For any k_i-ary function symbol f_i, f_i^* is a function that assigns to any k_i tuple of elements of D an element of D.

For instance, if $\oplus$ is a 2-ary function symbol in the predicate language and M is an interpretation with domain $\mathbb{N}$, then the interpretation $\oplus^*$ of $\oplus$ might be the function $+$ that assigns to any pair (or 2-tuple) n, m of natural numbers the natural number $n+m$. If f_i is an individual constant, i.e., $k_i = 0$, then f_i^* is an element of D.

The definitions of $M \models A$ (M is a model of A), $\models B$ (B is valid) and of $A_1, \ldots, A_n \models B$ (B is a valid or logical consequence of $A_1, \ldots, A_n$) are as before, taking into consideration that now all structures M are interpretations for the predicate logic with function symbols.

All results stated for the predicate logic without function symbols also hold for the predicate logic with function symbols, where terms may also contain function symbols in addition to individual variables and constants.

Example 4.11. Let f be a binary (i.e., 2-ary) function symbol, $\equiv$ a binary predicate symbol, and a and b free individual variables. Then $f(a,b)$ and $f(b,a)$ are terms. Let $M = \langle \mathbb{N}; =; + \rangle$ be the model with domain $\mathbb{N}$, interpreting $\equiv$ as = (equality) and f as + (addition). Then $M \models f(a,b) \equiv f(b,a)$, because for all natural numbers n, m, $n+m = m+n$. Also $M \models f(a,b) \equiv a$ [2,0], because $2 + 0 = 2$, but $M \not\models f(a,b) \equiv a$, because, for instance, $M \not\models f(a,b) \equiv a$ [2,1], since $2+1 \neq 2$.

4.3.5 Prenex Form*

Definition 4.19 (Prenex Formula). A formula A is in *prenex (normal) form* if A consists of a (possibly empty) string of quantifiers followed by a formula without quantifiers. We also say that A is a *prenex formula*.

A simple example is the formula $\forall x \forall y \exists z [P(x,y) \wedge Q(y,x) \rightarrow P(z,z)]$. By pulling out quantifiers, we can reduce every formula to a formula in prenex form.

Theorem 4.18 (Prenex Normal Form Theorem). *For every formula A there is a prenex formula B such that $\models A \rightleftarrows B$ (or, equivalently, $A \mathrel{\vdash\!\!\dashv} B$).*

Proof. The proof is by induction on the complexity of the formula A (Theorem 4.1). *Induction basis*: for an atomic formula $P(t_1, \ldots, t_n)$ the theorem is trivially true. *Induction step for the connectives*: suppose $A = B \rightarrow C$, $B \wedge C$, $B \vee C$ or $\neg B$, and B, C are equivalent to prenex formulas B^*, C^* respectively (induction hypothesis). Then $B^* = (Q_1 y_1) \ldots (Q_n y_n) B^1$ and $C^* = (Q'_1 z_1) \ldots (Q'_m z_m) C^1$, where Q_i, Q'_j are quantifiers and B^1, C^1 open. By Theorem 4.3 all bound variables can be chosen distinct. Now A is semantically equivalent to $B^* \rightarrow C^*$, $B^* \wedge C^*$, $B^* \vee C^*$ or $\neg B^*$ respectively. By means of the *prenex operations* (a), (b), (c) and (d) below, we can convert the latter formula into a formula in prenex form.

(a) (1) Replace a part $Qx[B] \to C$ by $Q'x[B \to C]$,
where $Q'x$ is $\forall x$ if Qx is $\exists x$ and $Q'x$ is $\exists x$ if Qx is $\forall x$.
(2) Replace a part $B \to Qx[C]$ by $Qx[B \to C]$.
(b) (1) Replace a part $Qx[B] \wedge C$ by $Qx[B \wedge C]$.
(2) Replace a part $B \wedge Qx[C]$ by $Qx[B \wedge C]$.
(c) (1) Replace a part $Qx[B] \vee C$ by $Qx[B \vee C]$.
(2) Replace a part $B \vee Qx[C]$ by $Qx[B \vee C]$.
(d) Replace a part $\neg Qx[B]$ by $Q'x[\neg B]$ where Q' is as in (a).

It remains to be shown that if E' results from E by a prenex operation, then $\models E \rightleftarrows E'$. But this is straightforward; see Exercise 4.26.

Induction step for the quantifiers: Suppose $A = \forall x[B(x)]$ or $A = \exists x[B(x)]$ and $B(a)$ is equivalent to a prenex formula $B^*(a)$ (induction hypothesis). By Theorem 4.3 we can choose the bound variables in $B^*(a)$ distinct from x. Then $\forall x[B^*(x)]$ and $\exists x[B^*(x)]$ are prenex formulas and $\models A \rightleftarrows \forall x[B^*(x)]$ or $\models A \rightleftarrows \exists x[B^*(x)]$ respectively. □

The prenex normal form theorem states that for every formula A there is a prenex formula B which is equivalent to A. Being prenex, B consists of a finite string of quantifiers followed by a formula C without quantifiers, i.e., $B = Q_1x_1\ldots Q_nx_n[C]$. According to Theorem 2.18, C is equivalent to a formula C' in conjunctive normal form. So, by combining the prenex normal form theorem and the conjunctive normal form theorem (Theorem 2.18), every formula A is equivalent to a formula of the form

$$Q_1x_1\ldots Q_nx_n\,[(L_{i_1} \vee \ldots \vee L_{i_k}) \wedge \ldots \wedge (L_{j_1} \vee \ldots \vee L_{j_l})],$$

where each L_n is a literal, i.e., an atomic formula or the negation of an atomic formula. Any *logic program* in the programming language PROLOG, to be treated in Section 9.1, will be a formula of this form with all the quantifiers universal.

In 1936 A. Church and A. Turing proved independently that there is no decision procedure for validity of formulas in predicate logic (see Section 4.5). Nevertheless, there is a decision procedure for formulas in prenex normal form in the prefix of which no existential quantifier precedes any universal quantifier. In the exercises of Section 4.5 some other classes of formulas are given for which a decision procedure is known. Most of these classes consist of formulas having a prenex normal form of a particular type. For more of these results see A. Church [3].

4.3.6 Skolemization, Clausal Form*

If M is a model with domain D and $M \models \forall x \exists y[P(x,y)]$, then there must be some function $f^* : D \to D$ such that for all $d \in D$, $M \models P(a_1,a_2)[d,f^*(d)]$. This suggests introducing a function symbol f in our language and replacing $\forall x \exists y[P(x,y)]$ by the formula $\forall x[P(x,f(x))]$.

Let A be a formula in prenex normal form, the *Skolem (normal) form* of A is obtained by eliminating all existential quantifiers in (the prefix of) A as follows: for any expression of the form $\forall x_1 \ldots \forall x_k \exists y[B(x_1,\ldots,x_k,y,\ldots)]$ a new k-ary function symbol f is introduced and the original expression is replaced by $\forall x_1 \ldots \forall x_k[B(x_1,\ldots,x_k,f(x_1,\ldots,x_k),\ldots)]$.

Thus, the Skolem normal forms of $\exists x[P(x)], \forall x \exists y[P(x,y)], \forall x \exists y \forall z[P(x,y,z)]$ and $\forall x \exists y \forall z \exists u[P(x,y,z,u)]$ are $P(c), \forall x[P(x,f(x))], \forall x \forall z[P(x,f(x),z)]$ and $\forall x \forall z[P(x,f(x),z,g(x,z))]$ respectively, where c is a new individual constant and f and g are new function symbols.

Let $Sk(A)$ denote the Skolem normal form of A. Clearly, if $M \models Sk(A)$, then $M \models A$. But not conversely: $\langle \mathbb{N};$ is even; $3 \rangle \models \exists x[P(x)]$, but $\langle \mathbb{N};$ is even; $3 \rangle \not\models P(c)$. And if i is the identity function on $\mathbb{N}$, then $\langle \mathbb{N}; <; i \rangle \models \forall x \exists y[P(x,y)]$, but $\langle \mathbb{N}; <; i \rangle \not\models \forall x[P(x,f(x))]$. However, it is easy to see the following

Theorem 4.19. *1. $Sk(A) \models A$; but not conversely.*
2. $Sk(A)$ is satisfiable iff A is satisfiable.

It follows that if $\models Sk(A)$, then also $\models A$. But the converse does not hold: $\models \forall x \exists y[P(x) \rightarrow P(y)]$, but $\not\models \forall x[P(x) \rightarrow P(f(x))]$.

Definition 4.20 (Clausal Form). Given any formula A (of first-order predicate logic), the *clausal form* $C(A)$ of A is obtained as follows:
1. construct the prenex (normal) form A' of A; $A' = Q_1 x_1 \ldots Q_n x_n[M]$, where $Q_i = \forall$ or $\exists$ and M is quantifier-free;
2. construct the Skolem (normal) form A^* of A'; $A^* = \forall x_1 \ldots \forall x_k[M^*]$, M^* quantifier-free, but containing $(n-k)$ additional function-symbols;
3. construct the conjunctive normal form $(L_1 \vee \ldots \vee L_{n_1}) \wedge \ldots \wedge (L_k \vee \ldots \vee L_{n_k})$ of M^* (see Theorem 2.18).

Example 4.12. Let $A = \exists x[P(x)] \rightarrow \exists x[Q(x)]$. Then $A' = \forall x \exists y[P(x) \rightarrow Q(y)]$, $A^* = \forall x[P(x) \rightarrow Q(f(x))]$ and $C(A) = \forall x[\neg P(x) \vee Q(f(x))]$.

Theorem 4.20. *For the clausal form $C(A)$ of A the following holds:*
1. $C(A) \models A$. Consequently, if $C(A)$ is valid, then A is valid. But not conversely: $\models \forall x \exists y[P(x) \rightarrow P(y)]$, but $\not\models \forall x[\neg P(x) \vee P(f(x))]$.
2. A is satisfiable iff $C(A)$ is satisfiable.
3. The 'complexity' of $C(A)$ is lower than that of A, in the sense that $C(A)$ contains only universal quantifiers and no existential quantifiers that occur in the prenex (normal) form of A.

Automated theorem provers for logic based on *resolution* operate as follows. Given any assumption formulas $A_1,\ldots,A_n$ and given any formula B, they construct $\neg B$ and the clausal forms $C(A_1),\ldots,C(A_n)$ and $C(\neg B)$ of $A_1,\ldots,A_n$ and $\neg B$ respectively. Next they check whether a contradiction can be derived from $C(A_1),\ldots,C(A_n)$ and $C(\neg B)$ by resolution (or otherwise, for instance, by the tableaux-method). If so, then $C(A_1),\ldots,C(A_n)$, $C(\neg B)$ are not simultaneously satisfiable; hence, $A_1,\ldots,A_n$, $\neg B$ are not simultaneously satisfiable and therefore $A_1,\ldots,A_n \models B$. If not, then $C(A_1),\ldots,C(A_n)$, $C(\neg B)$ are simultaneously satisfiable (completeness); hence, $A_1,\ldots,A_n$, $\neg B$ are simultaneously satisfiable and therefore $A_1,\ldots,A_n \not\models B$.

4.3 Basic Results about Validity and Logical Consequence

Theorem 4.21. *Any definite logic program (see also Chapter 2 and 9) is actually a formula in clausal form.*

Proof. The structure of any definite *logic program* is by definition the following:

$$\left.\begin{array}{c} P_1 :\text{-} Q_1,\ldots,Q_{n_1}. \\ \vdots \\ P_k :\text{-} Q_k,\ldots,Q_{n_k}. \end{array}\right\} \text{I}$$

where P_i and Q_j are atomic formulas.

I stands for:
$$\left.\begin{array}{c} (P_1 \leftarrow Q_1 \wedge \ldots \wedge Q_{n_1}) \wedge \\ \vdots \\ \wedge (P_k \leftarrow Q_k \wedge \ldots \wedge Q_{n_k}) \end{array}\right\} \text{II}$$

or, equivalently, for:

$$\left.\begin{array}{c} (P_1 \vee \neg Q_1 \vee \ldots \vee \neg Q_{n_1}) \wedge \\ \vdots \\ \wedge (P_k \vee \neg Q_k \vee \ldots \vee \neg Q_{n_k}). \end{array}\right\} \text{III}$$

Remembering that $Cl(A)$ denotes the universal closure of A, III is short for:

$$\left.\begin{array}{c} Cl(P_1 \vee \neg Q_1 \vee \ldots \vee \neg Q_{n_1}) \wedge \\ \vdots \\ \wedge Cl(P_k \vee \neg Q_k \vee \ldots \vee \neg Q_{n_k}). \end{array}\right\} \text{IV}$$

IV, and hence also I is equivalent to a formula $\forall x_1 \ldots \forall x_k[(P_1 \vee \neg Q_1 \vee \ldots \vee \neg Q_{n_1}) \wedge \ldots \wedge (P_k \vee \neg Q_k \vee \ldots \vee \neg Q_{n_k})]$, which is in clausal form.

Exercise 4.21. let P be a unary and Q a 0-ary predicate symbol. Which of the following statements are right? Give either a proof or a counterexample.
1. $\forall x[P(x)] \rightarrow Q \models \forall x[P(x) \rightarrow Q]$ and 2. $\forall x[P(x) \rightarrow Q] \models \forall x[P(x)] \rightarrow Q$.
3. $\exists x[P(x)] \rightarrow Q \models \exists x[P(x) \rightarrow Q]$ and 4. $\exists x[P(x) \rightarrow Q] \models \exists x[P(x)] \rightarrow Q$.

Exercise 4.22. Are the following formulas valid or invalid? Give either a proof or a counterexample.
1. $(\forall x[P(x)] \rightarrow \exists x[Q(x)]) \rightleftarrows \exists x[P(x) \rightarrow Q(x)]$;
2. $(\exists x[P(x)] \rightarrow \forall x[Q(x)]) \rightleftarrows \forall x[P(x) \rightarrow Q(x)]$.

Exercise 4.23. Which of the following statements are right? Give either a proof or a counterexample.
1. $\forall x[P(x) \rightarrow Q(x)] \models (\exists x[P(x)] \rightarrow \exists x[Q(x)])$; and conversely?.
2. $\exists x[P(x) \rightarrow Q(x)] \models (\forall x[P(x)] \rightarrow \forall x[Q(x)])$; and conversely?

Exercise 4.24. Prove or refute: $\forall x \exists y[P(x) \rightarrow Q(y)] \models \exists y \forall x[P(x) \rightarrow Q(y)]$.

Exercise 4.25. (H. Wang) Prove: $\models \exists x \exists y \forall z[(P(x,y) \rightarrow P(y,z) \wedge P(z,z)) \wedge (P(x,y) \wedge Q(x,y) \rightarrow Q(x,z) \wedge Q(z,z))]$.

Exercise 4.26. Prove that the formulas in the prenex operations (a), (b), (c) and (d) in the proof of Theorem 4.18 are semantically equivalent.

Exercise 4.27. Following the proof of the prenex normal form theorem (Theorem 4.18) convert each of the following formulas into a formula in prenex form.
1. $\forall x[P(x)] \to \exists x[Q(x)]$; 2. $\exists x[P(x)] \to \forall x[Q(x)]$;
3. $\exists x[P(x,a)] \to \exists x[Q(x) \vee \neg \exists y[R(y)]]$.

Exercise 4.28. Find two prenex normal forms for the formula $\exists x[P(x)] \to \exists x[Q(x)]$. (See also Exercise 4.24)

4.4 Syntax: Provability and Deducibility

In this section we shall generalize the notions of (logical) provability ($\vdash B$) and (logical) deducibility ($A_1, \ldots, A_n \vdash B$), as defined for propositional logic in Section 2.6, to predicate logic.

It turns out that also for (classical) predicate logic one can select a small, finite, number of valid formula schemata, henceforth called (logical) axiom schemata, and rules of inference such that i) precisely all valid formulas can be obtained by finitely many applications of the rules to instances of the axiom schemata and such that ii) for any premises $A_1, \ldots, A_n$, precisely all valid consequences of $A_1, \ldots, A_n$ can be obtained by finitely many applications of the rules to $A_1, \ldots, A_n$ and to instances of the axiom schemata.

We have selected the following *axiom schemata* and *rules of inference for (classical) predicate logic*:

The axiom schemata $1, \ldots, 10b$ for (classical) propositional logic (see Section 2.6), together with the rule of inference Modus Ponens. However, the formulas in these axiom schemata and in applications of the rule Modus Ponens are now understood to be formulas of predicate logic. For the sake of completeness we repeat the axiom schemata for propositional logic and the rule Modus Ponens below:

1. $A \to (B \to A)$; 2. $(A \to B) \to ((A \to (B \to C)) \to (A \to C))$
3. $A \to (B \to A \wedge B)$; 4a. $A \wedge B \to A$; 4b. $A \wedge B \to B$
5a. $A \to A \vee B$; 5b. $B \to A \vee B$; 6. $(A \to C) \to ((B \to C) \to (A \vee B \to C))$
7. $(A \to B) \to ((A \to \neg B) \to \neg A)$; 8. $\neg \neg A \to A$
9. $(A \to B) \to ((B \to A) \to (A \rightleftarrows B))$;
10a. $(A \rightleftarrows B) \to (A \to B)$; 10b. $(A \rightleftarrows B) \to (B \to A)$

We add one axiom schema for $\forall$ and one for $\exists$ (compare Theorem 4.16):

the $\forall$-*schema* $\forall x[A(x)] \to A(t)$, and the $\exists$-*schema* $A(t) \to \exists x[A(x)]$,

where t is a term, i.e., a free individual variable or an individual constant.

To the rule Modus Ponens (MP), $\dfrac{A \quad A \to B}{B}$, we also add one rule of inference for $\forall$ and one for $\exists$:

4.4 Syntax: Provability and Deducibility

the $\forall$-rule $\dfrac{C \to A(a)}{C \to \forall x[A(x)]}$ and the $\exists$-rule $\dfrac{A(a) \to C}{\exists x[A(x)] \to C}$,

when C does not contain a (compare Theorem 4.17).

Warning For predicate logic we have three rules of inference:

$$MP \ \dfrac{A \ \ A \to B}{B}, \quad \forall \ \dfrac{C \to A(a)}{C \to \forall x[A(x)]} \text{ and } \exists \ \dfrac{A(a) \to C}{\exists x[A(x)] \to C},$$

if C does not contain a. However, there is an important difference between these rules. We do have $A, A \to B \models B$ and hence also the weaker statement: if $M \models A$ and $M \models A \to B$, then $M \models B$. But – as we have seen in Theorem 4.17 – we do not have $C \to A(a) \models C \to \forall x[A(x)]$, although we do have the weaker statement: if $M \models C \to A(a)$, then $M \models C \to \forall x[A(x)]$. Similarly for the $\exists$-rule.

The definition of a *(logical) proof of B* is similar to the one for propositional logic, taking into account that we have two more axiom schemata and two more rules of inference.

Definition 4.21 (Proof; Provable). Let B be a formula. A (logical, Hilbert-type) *proof* of B is a finite list of formulas with B as last formula in the list, such that every formula in the list is either an axiom of predicate logic (i.e, an instance of an axiom schema) or obtained by application of one of the rules to formulas earlier in the list. B is (logically) *provable* := there exists a (logical, Hilbert-type) proof of B.
Notation: $\vdash B$

Example 4.13. $\vdash \forall x[P(x) \to P(x)]$. Below is a (logical) proof of $\forall x[P(x) \to P(x)]$:

Ex. 2.12 $\begin{cases} \text{axiom 1} \quad \text{axiom 2} \\ \overline{P(a) \to P(a)} \quad \overline{(P(a) \to P(a)) \to (\text{axiom} \to (P(a) \to P(a)))} \end{cases}$

$\dfrac{\text{axiom} \to (P(a) \to P(a))}{\text{axiom} \to \forall x[P(x) \to P(x)]} \forall$

$MP \ \dfrac{\text{axiom}}{\forall x[P(x) \to P(x)]}$

The definition of a (logical) *deduction of B from $A_1, \ldots, A_n$* in predicate logic is similar to the one for propositional logic. However, in order to prevent that, for instance, one could deduce from $C \to P(a)$ (if it is September 5, then ad has his birthday) that $C \to \forall x[P(x)]$ (if it is September 5, then everyone has his birthday), in such a deduction *all free variables of $A_1, \ldots, A_n$* should be *held constant*, i.e., the $\forall$-rule and the $\exists$-rule may not be applied with respect to a free variable a occurring in $A_1, \ldots, A_n$, except preceding the first occurrence of $A_1, \ldots, A_n$ in the deduction.

Definition 4.22 (Deduction; Deducible).
1. A (logical, Hilbert-type) *deduction* of B from $A_1, \ldots, A_n$ (in classical predicate logic) is a finite list $B_1, \ldots, B_k$ of formulas, such that
(a) $B = B_k$ is the last formula in the list, and

(b) each formula in the list is either one of $A_1,\ldots,A_n$, or an axiom of predicate logic (i.e., an instance of one of the axiom schemata), or is obtained by application of one of the rules to formulas preceding it in the list, such that
(c) *all free variables of* $A_1,\ldots,A_n$ *are held constant*, i.e., the $\forall$-rule and the $\exists$-rule are not applied with respect to a free variable a occurring in $A_1,\ldots,A_n$, except preceding the first occurrence of $A_1,\ldots,A_n$ in the deduction.

2. B is (logically) *deducible from* $A_1,\ldots,A_n$:= there exists a (logical, Hilbert-type) deduction of B from $A_1,\ldots,A_n$. **Notation**: $A_1,\ldots,A_n \vdash B$. The symbol $\vdash$ may be read 'yields'. $A_1,\ldots,A_n \not\vdash B$ abbreviates: not $A_1,\ldots,A_n \vdash B$.

3. For Γ a (possibly infinite) set of formulas, B is *deducible from* Γ := there is a finite list $A_1,\ldots,A_n$ of formulas in Γ such that $A_1,\ldots,A_n \vdash B$. **Notation**: $\Gamma \vdash B$.

Example 4.14. $\forall x[P(x) \to Q(x)], P(c) \vdash Q(c)$. The following schema is a deduction of $Q(c)$ from $\forall x[P(x) \to Q(x)]$ and $P(c)$.

$$\dfrac{\text{premiss } P(c) \quad \dfrac{\forall x[P(x) \to Q(x)] \quad \forall x[P(x) \to Q(x)] \to (P(c) \to Q(c))}{P(c) \to Q(c)} \text{ MP}}{Q(c)} \text{ MP}$$

Example 4.15. $\forall x[P(x)] \vdash \exists x[P(x)]$. The following schema is a deduction of $\exists x[P(x)]$ from $\forall x[P(x)]$.

$$\text{MP} \dfrac{\dfrac{\text{premiss } \forall x[P(x)] \quad \forall x[P(x)] \to P(t)}{P(t)} \quad \dfrac{\exists\text{-schema}}{P(t) \to \exists x[P(x)]}}{\exists x[P(x)]} \text{ MP}$$

Warning: Note that according to our definition the schema $\dfrac{C \to A(a)}{C \to \forall x[A(x)]}$, where the variable a does not occur in C, is not a deduction of $C \to \forall x[A(x)]$ from $C \to A(a)$ (holding all free variables constant), since in this schema the $\forall$-rule is applied with respect to a free variable a occurring in the premiss.

This remark by itself does *not* establish that there is no deduction of $C \to \forall x[A(x)]$ from $C \to A(a)$; it only says that the given schema is not such a deduction. In order to establish that no (other) schema can be a deduction of $C \to \forall x[A(x)]$ from $C \to A(a)$ (holding all free variables constant), we have to prove the generalized soundness theorem:

if $A_1,\ldots,A_n \vdash B$, then $A_1,\ldots,A_n \models B$.

Since we have seen in Theorem 4.17 that $C \to A(a) \not\models C \to \forall x[A(x)]$, it follows by this theorem that $C \to A(a) \not\vdash C \to \forall x[A(x)]$, i.e., there is no deduction of $C \to \forall x[A(x)]$ from $C \to A(a)$ (holding all free variables constant).

Sometimes the free variables in the premisses $A_1,\ldots,A_n$ are allowed to vary, such as, for instance, in 'if it rains, then a takes an umbrella', when one means 'for all x, if it rains, then x takes an umbrella' or 'if it rains, then any a takes an umbrella'.

4.4 Syntax: Provability and Deducibility

Definition 4.23. B is *deducible from* $A_1, \ldots, A_n$ *allowing all free variables to vary* or *with all free variables general* $:= Cl(A_1), \ldots, Cl(A_n) \vdash B$. **Notation**: $A_1, \ldots, A_n \vdash^2 B$.

Example 4.16. $C \to A(a) \vdash^2 C \to \forall x[A(x)]$, i.e., $\forall x[C \to A(x)] \vdash C \to \forall x[A(x)]$, since the following schema is a deduction of $C \to \forall x[A(x)]$ from $\forall x[C \to A(x)]$ (holding all free variables constant):

$$\frac{\overset{\text{premiss}}{\forall x[C \to A(x)]} \quad \overset{\forall\text{-schema}}{\forall x[C \to A(x)] \to (C \to A(a))}}{\dfrac{C \to A(a)}{C \to \forall x[A(x)]} \; \forall} \; MP$$

where we have chosen the free variable a such that a does not occur in the premiss $\forall x[C \to A(x)]$ and hence not in C.

In a similar way one shows that $A(a) \to C \nvdash \exists x[A(x)] \to C$ (from 'if ad has his birthday, then it is May 5' it does not follow that 'if someone has his birthday, then it is May 5) and $A(a) \nvdash \forall x[A(x)]$ (from 'ad is an Alcoholic', it does not follow that 'everyone is an Alcoholic'), while we do have that $A(a) \to C \vdash^2 \exists x[A(x)] \to C$ (from 'if any a takes an umbrella, then it is raining', it follows that 'if someone takes an umbrella, then it is raining') and $A(a) \vdash^2 \forall x[A(x)]$ (from 'any a has the property A', it follows that 'everything has the property A').

The soundness theorem says that the logical axioms and rules of (classical) predicate logic are sound, i.e., every formula B which may be deduced from given premisses $A_1, \ldots, A_n$ by means of the logical axioms and rules is a logical (or valid) consequence of the given premisses.

Theorem 4.22 (Soundness). *a) If $A_1, \ldots, A_n \vdash B$, then $A_1, \ldots, A_n \models B$.*
Hence, in particular, if $\vdash B$, then $\models B$.
b) If $A_1, \ldots, A_n \vdash^2 B$, then $A_1, \ldots, A_n \models^2 B$.

Proof. a) Suppose $A_1, \ldots, A_n \vdash B$, i.e., there is a finite schema of formulas of the following form:

$$\frac{\dfrac{A_1 \; \cdots \; A_n \quad \text{axioms}}{\underline{\quad\quad\quad\quad\quad}}}{B} \; MP, \forall \text{ or } \exists$$

where the $\forall$- and $\exists$-rule are not applied with respect to a free variable a occurring in $A_1, \ldots, A_n$, except preceding the first occurrence of $A_1, \ldots, A_n$ in the deduction. (α)
Let M be an interpretation with domain D and let v be a valuation in D of the free individual variables. We have to show: if for all i, $1 \le i \le n$, $M \models A_i[v]$, then $M \models B[v]$. So suppose that for all i, $1 \le i \le n$, $M \models A_i[v]$. (1)
By Theorem 2.7 and Theorem 4.16. all axioms of predicate logic are valid. (2)
For an application of Modus Ponens, note that if $M \models C[v]$ and $M \models C \to D[v]$, then $M \models D[v]$. (3)
For an application of the $\forall$- or $\exists$-rule, note that due to the condition (α) stated above, interpretation M and valuation d/v make the premiss true for every d in D and hence M and valuation v make the conclusion of the rule true. (4).

From (1), (2), (3) and (4) it follows that $M \models B[v]$.
(b) follows from (a). □

Without proof we mention that the *deduction theorem* for propositional logic (Theorem 2.24) also holds for (classical) predicate logic: if $\Gamma, A \vdash B$, then also $\Gamma \vdash A \to B$. The introduction and elimination rules for the propositional connectives (Theorem 2.25) may be extended with introduction and elimination rules for the quantifiers.

Theorem 4.23 (Introduction and elimination rules for the quantifiers). *Let a be a free variable, $A(a)$ a formula, t a term and $A(t)$ the result of substituting t for the occurrences of a in $A(a)$. Also let Γ be a list of (zero or more) formulas, and C a formula. Then the following rules hold.*

    INTRODUCTION   ELIMINATION

$\forall$ *If $\Gamma \vdash A(a)$, then $\Gamma \vdash \forall x[A(x)]$,* $\forall x[A(x)] \vdash A(t)$
     provided Γ does not contain a.

$\exists$ $A(t) \vdash \exists x[A(x)]$ *If $\Gamma, A(a) \vdash C$, then $\Gamma, \exists x[A(x)] \vdash C$,*
                                                             provided Γ and C do not contain a.

Proof. $\forall$-*introduction*: Let E be an axiom not containing a. Suppose $\Gamma \vdash A(a)$. Then also $\Gamma, E \vdash A(a)$. By the deduction theorem (for predicate logic), $\Gamma \vdash E \to A(a)$. Since by hypothesis Γ and E do not contain a, we can apply the $\forall$-rule; hence $\Gamma \vdash E \to \forall x[A(x)]$. So $\Gamma, E \vdash \forall x[A(x)]$ and since E is an axiom, $\Gamma \vdash \forall x[A(x)]$.
$\forall$-*elimination*: From $\forall x[A(x)]$, by using the $\forall$-axiom, $\forall x[A(x)] \to A(t)$ and Modus Ponens, we may deduce $A(t)$.
$\exists$-*introduction*: From $A(t)$, by using the $\exists$-axiom, $A(t) \to \exists x[A(x)]$, and Modus Ponens, we may deduce $\exists x[A(x)]$.
$\exists$-*elimination* : Suppose $\Gamma, A(a) \vdash C$. Then by the deduction theorem, $\Gamma \vdash A(a) \to C$. Since Γ and C do not contain a, by the $\exists$-rule, $\Gamma \vdash \exists x[A(x)] \to C$. And therefore $\Gamma, \exists x[A(x)] \vdash C$. □

Exercise 4.29.

$$\cfrac{\cfrac{\text{premiss}}{A(a)} \qquad \cfrac{\cfrac{\cfrac{\text{axiom}}{C} \qquad \cfrac{A(a) \overset{1}{\to} (C \to A(a))}{C \to A(a)} \, MP}{C \to \forall x[A(x)]} \, \forall}{\forall x[A(x)]} \, MP}{}$$

Is this schema, in which C is an axiom not containing the free variable a, a deduction of $\forall x[A(x)]$ from $A(a)$ (holding all free variables constant)?

Exercise 4.30. Is the following schema a deduction of $\exists y \forall x[P(y,x,a)]$ from $\exists z \forall x[P(b,x,z)]$?

$$\cfrac{\cfrac{\text{premiss}}{\exists z \forall x[P(b,x,z)]} \quad \cfrac{\cfrac{\exists\text{-schema}}{\forall x[P(b,x,a)] \to \exists y \forall x[P(y,x,a)]}}{\exists z \forall x[P(b,x,z)] \to \exists y \forall x[P(y,x,a)]} \, \exists}{\exists y \forall x[P(y,x,a)]} \, MP$$

4.4 Syntax: Provability and Deducibility

Exercise 4.31. Is the following schema a deduction of $\exists z \forall y [P(z,y,b)]$ from $\exists x \forall y [P(x,y,b)]$?

$$\cfrac{\cfrac{\text{premiss}}{\exists x \forall y [P(x,y,b)]} \quad \cfrac{\forall y [P(a,y,b)] \to \exists z \forall y [P(z,y,b)]}{\exists x \forall y [P(x,y,b)] \to \exists z \forall y [P(z,y,b)]} \exists}{\exists z \forall y [P(z,y,b)]} MP$$

with $\exists$-schema on the top right.

Exercise 4.32. Show that i) $\exists x[A(x)] \vdash \exists z[A(z)]$ and ii) $\forall x[A(x)] \vdash \forall z[A(z)]$.

4.4.1 Natural Deduction

Gentzen's system of *Natural Deduction* for classical predicate logic is obtained by adding to the natural deduction rules for the connectives (see Subsection 2.7.2) the following *introduction and elimination rules for the quantifiers*.

INTRODUCTION

$\forall I \; \cfrac{A(a)}{\forall x[A(x)]}$
if the free variable a does not occur in any of the premisses of $A(a)$.

$\exists I \; \cfrac{A(t)}{\exists x[A(x)]}$
where t is a term

ELIMINATION

$\forall E \; \cfrac{\forall x[A(x)]}{A(t)}$
where t is a term

$\exists E \; \cfrac{\exists x[A(x)] \quad \begin{matrix}[A(a)]\\ \vdots \\ C\end{matrix}}{C}$
if the variable a in the cancelled formula $A(a)$ does not occur in C or any of the premisses in the righthand derivation.

That the conditions accompanying the rules above are necessary follows immediately from Theorem 4.17. The definition of $\Gamma \vdash_{ND} B$ (B is deducible from Γ in *Gentzen's system of natural deduction*) for classical predicate logic is similar to the one for classical propositional logic (see Definition 2.12), taking into account that now we also have introduction and elimination rules for the quantifiers. And again one easily shows that $\Gamma \vdash B$ iff $\Gamma \vdash_{ND} B$ (compare Theorem 2.26), where the if-part now follows from Theorem 2.25 and Theorem 4.23. Below are some examples of deductions in Gentzen's system of natural deduction.

Example 4.17. $\vdash_{ND} \forall x[A(x) \land B(x)] \to \forall x[A(x)] \land \forall x[B(x)]$:

$$\forall E \frac{[\forall x[A(x) \land B(x)]]^1}{A(a) \land B(a)} \quad \forall E \frac{[\forall x[A(x) \land B(x)]]^1}{A(a) \land B(a)} \quad \forall E$$
$$\land E \frac{A(a) \land B(a)}{A(a)} \quad \frac{B(a)}{\forall x[B(x)]} \land E$$
$$\forall I \frac{\forall x[A(x)]}{\forall x[A(x)]} \quad \forall x[B(x)]$$
$$\land I \frac{\forall x[A(x)] \land \forall x[B(x)]}{\forall x[A(x) \land B(x)] \to \forall x[A(x)] \land \forall x[B(x)]} \quad (1)$$

Example 4.18. $\vdash_{ND} (\exists x[A(x)] \to B) \to \forall x[A(x) \to B]$:

$$\frac{[\exists x[A(x)] \to B]^2 \quad \dfrac{[A(a)]^1}{\exists x[A(x)]} \exists I}{(1) \dfrac{B}{A(a) \to B} \to I} \to E$$
$$(2) \dfrac{\dfrac{A(a) \to B}{\forall x[A(x) \to B]} \forall I}{(\exists x[A(x)] \to B) \to \forall x[A(x) \to B]} \to I$$

The reader should be aware again that the logical proofs by natural deduction are very close to our informal way of verifying the formula in question. For instance, how do we prove informally that $(\exists x[A(x)] \to B) \to \forall x[A(x) \to B]$? We suppose that $\exists x[A(x)] \to B$ (2). Then we have to show that $\forall x[A(x) \to B]$. So, we let a be an arbitrary individual and we show that $A(a) \to B$. So, suppose $A(a)$ (1). Then $\exists x[A(x)]$ and hence by (2), B. Therefore $A(a) \to B$ under the assumption of $\exists x[A(x)] \to B$. Since a was arbitrary, it follows that $\forall x[A(x) \to B]$ under the assumption of $\exists x[A(x)] \to B$. Therefore, $(\exists x[A(x)] \to B) \to \forall x[A(x) \to B]$.

Exercise 4.33. Show that $\forall x[A \to B(x)] \vdash_{ND} A \to \forall x[B(x)]$ and $A \to \forall x[B(x)] \vdash_{ND} \forall x[A \to B(x)]$, realizing that a formal proof by natural deduction is very close to an informal proof of the formula in question.

4.4.2 Tableaux

To make this subsection self contained, we repeat some definitions from Section 2.8.

Definition 4.24 (Signed Formula; Sequent). A *signed formula* is any expression of the form $T(A)$ or $F(A)$, where A is a formula. Informally, we may read $T(A)$ as 'A is true' and $F(A)$ as 'A is false'. We frequently write TA and FA instead of $T(A)$ and $F(A)$, respectively. A *sequent* S is any finite set of signed formulas.

Below are the tableaux rules for the propositional connectives.

4.4 Syntax: Provability and Deducibility

$T \wedge$	$S, T\, B \wedge C$		$F \wedge$	$S, F\, B \wedge C$
	S, TB, TC			$S, FB \mid S, FC$
$T \vee$	$S, T\, B \vee C$		$F \vee$	$S, F\, B \vee C$
	$S, TB \mid S, TC$			S, FB, FC
$T \rightarrow$	$S, T\, B \rightarrow C$		$F \rightarrow$	$S, F\, B \rightarrow C$
	$S, FB \mid S, TC$			S, TB, FC
$T \neg$	$S, T\, \neg B$		$F \neg$	$S, F\, \neg B$
	S, FB			S, TB

The *tableaux rules* for (classical) predicate logic are the following ones:
1. The T- and F-rules for $\rightarrow$, $\wedge$, $\vee$ and $\neg$ of (classical) propositional logic (see Section 2.8), but now for any formulas of predicate logic.
2. To these, we add T- and F-rules for the quantifiers $\forall$ and $\exists$:

$T\, \exists \quad S, T\, \exists x[A(x)]$
$ S, TA(a)$
a new: *a* does not occur in S, $T \exists x[A(x)]$

$F\, \exists \quad S, F\, \exists x[A(x)]$
$ S, F\, \exists x[A(x)], FA(t)$
t being any term

$T\, \forall \quad S, T\, \forall x[A(x)]$
$ S, T\, \forall x[A(x)], TA(t)$
t being any term

$F\, \forall \quad S, F\, \forall x[A(x)]$
$ S, FA(a)$
a new: *a* does not occur in S, $F\, \forall x[A(x)]$

The extra condition in the rules $T\, \exists$ and $F\, \forall$ which the free individual variable a has to satisfy can be explained as follows.

If we read the rules downwards as semantic tableaux rules in the sense of E. Beth, interpreting the signed formulas rather than the sequents, the condition on a in the rule $T\, \exists$ and $F\, \forall$ is intuitively clear:

$T\, \exists$: Suppose $T\, \exists x[A(x)]$, i.e., there is at least one object with the property A. This object is not necessarily one of the objects already mentioned before.

$F\, \forall$: Suppose $F\, \forall x[A(x)]$, i.e., not all objects have the property A, or equivalently, at least one object does not have the property A. And again this object is not necessarily one of the objects already mentioned before.

If we read the rules upwards as Gentzen-type rules, interpreting the sequents rather than the signed formulas, rule $T\, \exists$, for instance, taking $S = \{FC\}$, becomes $\frac{FC,\, T\, \exists x[A(x)]}{FC,\, TA(a)}$ and is read as: if $A(a) \rightarrow C$, then $\exists x[A(x)] \rightarrow C$; the condition on a in rule $T\, \exists$ now corresponds to the condition 'a does not occur in C' in Theorem 4.17.

The definitions of '(tableau) branch', 'tableau', '*tableau proof* of B', 'B is *tableau-provable*' (notation: $\vdash' B$), a '*tableau-deduction* of B from $A_1, \ldots, A_n$' and of 'B is *tableau-deducible* from $A_1, \ldots, A_n$' (notation: $A_1, \ldots, A_n \vdash' B$) are similar to the definitions for propositional logic (see Definition 2.17 and 2.18), allowing that now we have two more T-rules and two more F-rules for the quantifiers.

Definition 4.25 ((Tableau) Branch). (a) A *tableau branch* is a (possibly infinite) set of signed formulas. A branch is *closed* if it contains signed formulas TA and FA

for some formula A. A branch that is not closed is called *open*.
(b) Let $\mathscr{B}$ be a branch and TA, resp. FA, a signed formula occurring in $\mathscr{B}$. TA, resp. FA, is *fulfilled* in $\mathscr{B}$ if (i) A is atomic, or (ii) $\mathscr{B}$ contains the bottom formulas in the application of the corresponding rule to A, and in case of the rules $T\vee$, $F\wedge$ and $T\rightarrow$, $\mathscr{B}$ contains one of the bottom formulas in the application of these rules.
(c) A branch $\mathscr{B}$ is *completed* if $\mathscr{B}$ is closed or every signed formula in $\mathscr{B}$ is fulfilled in $\mathscr{B}$.

Let $\mathscr{B} = \{T\forall x[P(x)],\ldots\}$ be a tableau branch and let $\mathscr{B}_a = \{T\forall x[P(x)], P(a),\ldots\}$. Then the tableau $\mathscr{T}_a = \{\mathscr{B}_a\}$ is called a *one-step expansion* of tableau $\mathscr{T} = \{\mathscr{B}\}$. And if $\mathscr{B} = \{T(P\rightarrow Q),\ldots\}$ is a tableau branch, $\mathscr{B}_1 = \{T(P\rightarrow Q), FP,\ldots\}$ and $\mathscr{B}_2 = \{T(P\rightarrow Q), TQ,\ldots\}$, then the tableau $\mathscr{T}' = \{\mathscr{B}_1, \mathscr{B}_2\}$ is called a *one-step expansion* of tableau $\mathscr{T} = \{\mathscr{B}\}$.

Definition 4.26 (Tableau). (a) A set of branches $\mathscr{T}$ is a *tableau* with initial branch $\mathscr{B}_0$ if there is a sequence $\mathscr{T}_0, \mathscr{T}_1, \ldots, \mathscr{T}_n$ such that $\mathscr{T}_0 = \{\mathscr{B}_0\}$, each $\mathscr{T}_{i+1}$ is a one-step expansion of $\mathscr{T}_i$ ($0 \leq i < n$) and $\mathscr{T} = \mathscr{T}_n$.
(b) We say that a branch $\mathscr{B}$ has tableau $\mathscr{T}$ if $\mathscr{T}$ is a tableau with initial branch $\mathscr{B}$.
(c) A tableau $\mathscr{T}$ is *open* if some branch $\mathscr{B}$ in it is open, otherwise $\mathscr{T}$ is *closed*.
(d) A tableau is *completed* if each of its branches is completed, i.e., no application of a tableau rule can change the tableau.

Definition 4.27 (Tableau-deduction; Tableau-proof).
(a) A (logical) *tableau-deduction* of B from $A_1,\ldots,A_n$ (in classical predicate logic) is a tableau $\mathscr{T}$ with $\mathscr{B}_0 = \{TA_1,\ldots,TA_n, FB\}$ as initial branch, such that all branches of $\mathscr{T}$ are closed.

In case $n = 0$, i.e., there are no premisses $A_1,\ldots,A_n$, this definition reduces to:
(b) A (logical) *tableau-proof* of B (in classical predicate logic) is a tableau $\mathscr{T}$ with $\mathscr{B}_0 = \{FB\}$ as initial sequent, such that all branches of $\mathscr{T}$ are closed.

Definition 4.28 (Tableau-deducible; Tableau-provable).
(a) B is *tableau-deducible* from $A_1,\ldots,A_n$ (in classical predicate logic) if there exists a tableau-deduction of B from $A_1,\ldots,A_n$. **Notation**: $A_1,\ldots,A_n \vdash' B$.
By $A_1,\ldots,A_n \not\vdash' B$ we mean: not $A_1,\ldots,A_n \vdash' B$.
(b) B is *tableau-provable* (in classical predicate logic) if there exists a tableau-proof of B. **Notation**: $\vdash' B$.
(c) For Γ a (possibly infinite) set of formulas, B is *tableau-deducible from* Γ if there exists a finite list $A_1,\ldots,A_n$ of formulas in Γ such that $A_1,\ldots,A_n \vdash' B$.
Notation: $\Gamma \vdash' B$.

Since $S \cup \{T\ \exists x[A(x)]\} = S \cup \{T\ \exists x[A(x)],\ T\ \exists x[A(x)]\}$ and $S \cup \{F\ \forall x[A(x)]\} = S \cup \{F\ \forall x[A(x)],\ F\ \forall x[A(x)]\}$, we have the following two derived rules:

$S,\ T\ \exists x[A(x)]$	and	$S,\ F\ \forall x[A(x)]$
$S,\ T\ \exists x[A(x)], TA(a)$		$S,\ F\ \forall x[A(x)], FA(a)$
provided a is new;		provided a is new.

This enables us to apply the rules $T\ \exists$ and $F\ \forall$ as frequently as we want, each time with a new variable.

4.4 Syntax: Provability and Deducibility   225

Example 4.19. $\vdash' \forall x[\neg P(x)] \to \neg \forall x[P(x)]$, since there is a tableau-proof (in classical predicate logic) of $\forall x[\neg P(x)] \to \neg \forall x[P(x)]$, i.e., there is a closed tableau $\mathscr{T} = \{\mathscr{B}_5\}$ with $\mathscr{B}_0 = \{F\ \forall x[\neg P(x)] \to \neg \forall x[P(x)]\}$ as initial branch, where $\mathscr{B}_5$ consists of all signed formulas in the schema below:

$$F\ \forall x[\neg P(x)] \to \neg \forall x[P(x)]$$
$$T\ \forall x[\neg P(x)],\ F\ \neg \forall x[P(x)]$$
$$T\ \forall x[\neg P(x)],\ T\ \forall x[P(x)]$$
$$T\ \neg P(a_1),\ T\ \forall x[\neg P(x)],\ T\ \forall x[P(x)]$$
$$F\ P(a_1),\ T\ \forall x[\neg P(x)],\ T\ \forall x[P(x)]$$
$$F\ P(a_1),\ T\ \forall x[\neg P(x)],\ T\ \forall x[P(x)],\ TP(a_1)$$
closure

Example 4.20. not $\vdash' \neg \forall x[P(x)] \to \forall x[\neg P(x)]$. If we try to construct a tableau-proof of $\neg \forall x[P(x)] \to \forall x[\neg P(x)]$, we find the following open, i.e., not closed, and not completed branch consisting of the signed formulas in the schema below:

$$F\ \neg \forall x[P(x)] \to \forall x[\neg P(x)]$$
$$T\ \neg \forall x[P(x)],\ F\ \forall x[\neg P(x)]$$
$$F\ \forall x[P(x)],\ F\ \forall x[\neg P(x)]$$
$$FP(a_1),\ F\ \forall x[P(x)],\ F\ \forall x[\neg P(x)]$$
$$FP(a_1),\ F\ \forall x[P(x)],\ F\ \forall x[\neg P(x)],\ F\ \neg P(a_2)$$
$$FP(a_1),\ F\ \forall x[P(x)],\ F\ \forall x[\neg P(x)],\ TP(a_2)$$
$$FP(a_3),\ FP(a_1),\ F\ \forall x[P(x)],\ F\ \forall x[\neg P(x)],\ TP(a_2)$$
$$FP(a_3),\ FP(a_1),\ F\ \forall x[P(x)],\ F\ \forall x[\neg P(x)],\ TP(a_2),\ F\ \neg P(a_4)$$
$$FP(a_3),\ FP(a_1),\ F\ \forall x[P(x)],\ F\ \forall x[\neg P(x)],\ TP(a_2),\ TP(a_4)$$
and so on

It is clear now that, no matter how far we continue this construction, we will never find a tableau-proof of the formula in question. We may apply the rules in a different order yielding a different open tableau, but not one that is closed. On the contrary, from the resulting open tableau with initial branch $\{F\ \neg \forall x[P(x)] \to \forall x[\neg P(x)]\}$, which consists of only one infinitely long, open branch, we can immediately read off a counterexample to $\neg \forall x[P(x)] \to \forall x[\neg P(x)]$ with the set $\mathbb{N}$ of all natural numbers as domain: the natural numbers $1, 3, 5, \ldots$ do not have the property P, corresponding with the occurrences of $FP(a_1), FP(a_3), FP(a_5), \ldots$ in the open branch and the natural numbers $2, 4, 6, \ldots$ do have the property P, corresponding with the occurrences of $TP(a_2), TP(a_4), TP(a_6), \ldots$ in the (open) branch. So let P^* be the predicate 'is even' over the natural numbers. Then $\langle \mathbb{N}; P^* \rangle$ is a counterexample to $\neg \forall x[P(x)] \to \forall x[\neg P(x)]$, since $\langle \mathbb{N}; P^* \rangle \models \neg \forall x[P(x)]$ (the sentence 'not all natural numbers are even' has truth value 1), while $\langle \mathbb{N}; P^* \rangle \not\models \forall x[\neg P(x)]$ (the sentence 'all natural numbers are not even' has truth value 0).

Note that $\langle \mathbb{N}; P^* \rangle \models P(a_2)$ [2] (the sentence '2 is even' has truth value 1), $\langle \mathbb{N}; P^* \rangle \models P(a_4)$ [4] (the sentence '4 is even' has truth value 1), and so on, corresponding with the occurrence of $TP(a_2), TP(a_4), \ldots$ in the open branch above. But $\langle \mathbb{N}; P^* \rangle \not\models P(a_1)$ [1] (the sentence '1 is even' has truth value 0), $\langle \mathbb{N}; P^* \rangle \not\models P(a_3)$ [3] (the sentence '3 is even' has truth value 0), and so on, corresponding with the occurrences of $FP(a_1), FP(a_3), \ldots$ in the open branch above.

Example 4.21. $\exists y \forall x[P(x,y)] \vdash' \forall x \exists y[P(x,y)]$, since there is a tableau-deduction of $\forall x \exists y[P(x,y)]$ from $\exists y \forall x[P(x,y)]$, i.e., there is a closed tableau $\mathscr{T} = \{\mathscr{B}_4\}$ starting with the initial branch $\mathscr{B}_0 = \{T\ \exists y \forall x[P(x,y)],\ F\ \forall x \exists y[P(x,y)]\}$, where $\mathscr{B}_4$ is the set consisting of all signed formulas in the schema below.

$$T\ \exists y \forall x[P(x,y)],\ F\ \forall x \exists y[P(x,y)]$$
$$T\ \forall x[P(x,a_1)],\ T\ \exists y \forall x[P(x,y)],\ F\ \forall x \exists y[P(x,y)]$$
$$T\ \forall x[P(x,a_1)],\ T\ \exists y \forall x[P(x,y)],\ F\ \forall x \exists y[P(x,y)],\ F\ \exists y[P(a_2,y)]$$
$$TP(a_2,a_1),\ T\ \exists y \forall x[P(x,y)],\ F\ \forall x \exists y[P(x,y)],\ F\ \exists y[P(a_2,y)]$$
$$TP(a_2,a_1),\ T\ \exists y \forall x[P(x,y)],\ F\ \forall x \exists y[P(x,y)],\ FP(a_2,a_1)$$
closure

Example 4.22. not $\forall x \exists y[P(x,y)] \vdash' \exists y \forall x[P(x,y)]$. If we try to construct a tableau-deduction of $\exists y \forall x[P(x,y)]$ from $\forall x \exists y[P(x,y)]$, we find the following open and not completed branch consisting of the signed formulas in the schema below, using some obvious abbreviations:

$$T\forall x \exists y[P(x,y)],\ F\exists y \forall x[P(x,y)]$$
$$T\exists y[P(a_1,y)],\ T\forall x \exists y,\ F\exists y \forall x$$
$$T\exists y[P(a_1,y)],\ T\forall x \exists y,\ F\exists y \forall x,\ F\forall x[P(x,a_1)]$$
$$TP(a_1,a_2),\ T\forall x \exists y,\ F\exists y \forall x,\ F\forall x[P(x,a_1)]$$
$$TP(a_1,a_2),\ T\forall x \exists y,\ F\exists y \forall x,\ FP(a_3,a_1)$$
$$T\exists y[P(a_2,y)],\ TP(a_1,a_2),\ T\forall x \exists y,\ F\exists y \forall x,\ FP(a_3,a_1)$$
$$T\exists y[P(a_2,y)],\ TP(a_1,a_2),\ T\forall x \exists y,\ F\exists y \forall x,\ FP(a_3,a_1),\ F\forall x[P(x,a_2)]$$
$$TP(a_2,a_4),\ TP(a_1,a_2),\ T\forall x \exists y,\ F\exists y \forall x,\ FP(a_3,a_1),\ F\forall x[P(x,a_2)]$$
$$TP(a_2,a_4),\ TP(a_1,a_2),\ T\forall x \exists y,\ F\exists y \forall x,\ FP(a_3,a_1),\ FP(a_5,a_2)$$
and so on

It is clear that, no matter how far we continue our construction, we will never find a tableau-deduction of $\exists y \forall x[P(x,y)]$ from $\forall x \exists y[P(x,y)]$. (Application of the rules in a different order may result in a different tableau, but not one that is closed.) On the contrary, from the resulting open branch, which is infinitely long, we can immediately read off a counterexample with the set $\mathbb{N}$ of all natural numbers as its domain: for each $n = 1, 2, 3, \ldots$, $P^*(n, 2n)$, corresponding with the occurrences of $TP(a_1,a_2)$, $TP(a_2,a_4)$, $TP(a_3,a_6)$, ... in the open branch and for each $n = 1, 2, 3, \ldots$, not $P^*(2n+1, n)$, corresponding with the occurrences of $FP(a_3,a_1)$, $FP(a_5,a_2)$, $FP(a_7,a_3)$, ... in the open branch. So, let P^* be the binary predicate over the natural numbers, defined by $P^*(n,m) := m = 2n$. Then $\langle \mathbb{N};\ P^* \rangle \models \forall x \exists y[P(x,y)]$ (the sentence 'for each natural number n there is a natural number m such that $m = 2n$' has truth value 1), but $\langle \mathbb{N};\ P^* \rangle \not\models \exists y \forall x[P(x,y)]$ (the sentence 'there is a natural number m such that for all natural numbers n, $m = 2n$' has truth value 0). Note that $\langle \mathbb{N};\ P^* \rangle \models \exists y[P(a_1,y)][1]$,

$\langle \mathbb{N};\ P^* \rangle \models P(a_1,a_2)[1,2]$ (the sentence '$2 = 2 \cdot 1$' has truth value 1),
$\langle \mathbb{N};\ P^* \rangle \models \exists y[P(a_2,y)][2]$,
$\langle \mathbb{N};\ P^* \rangle \models P(a_2,a_4)[2,4]$ (the sentence '$4 = 2 \cdot 2$' has truth value 1),
and so on,

corresponding with the occurrences of $T\ \exists y[P(a_1,y)]$, $TP(a_1,a_2)$, $T\ \exists y[P(a_2,y)]$,

4.4 Syntax: Provability and Deducibility

$TP(a_2, a_4)$, ... respectively in the open branch in Example 4.22.
But $\langle \mathbb{N}; P^* \rangle \not\models \forall x[P(x, a_1)][1]$,
$\quad \langle \mathbb{N}; P^* \rangle \not\models P(a_3, a_1)[3, 1]$ (the sentence '$1 = 2 \cdot 3$' has truth value 0),
$\quad \langle \mathbb{N}; P^* \rangle \not\models \forall x[P(x, a_2)][2]$,
$\quad \langle \mathbb{N}; P^* \rangle \not\models P(a_5, a_2)[5, 2]$ (the sentence '$2 = 2 \cdot 5$' has truth value 0),
$\quad$ and so on,
corresponding with the occurrences of $F \ \forall x[P(x,a_1)]$, $FP(a_3,a_1)$, $F \ \forall x[P(x,a_2)]$, $FP(a_5,a_2)$, ... respectively in the open branch in Example 4.22.

Like in classical propositional logic, the three notions $A_1, \ldots, A_n \vdash' B$, $A_1, \ldots, A_n \vdash B$ and $A_1, \ldots, A_n \models B$ turn out to be equivalent. Remember that the first two of these notions are syntactic, while the latter one is a semantic notion.

Theorem 4.24. *If $A_1, \ldots, A_n \vdash' B$, then $A_1, \ldots, A_n \vdash B$.*

Proof. The proof is a generalization of the proof of the corresponding theorem 2.27 for classical propositional logic, now in addition using the introduction and elimination rules for the quantifiers in Theorem 4.23.

In Theorem 4.22 we have already shown that classical predicate logic is sound.
Theorem (Soundness): if $A_1, \ldots, A \vdash B$, then $A_1, \ldots, A_n \models B$.

So in order to show the equivalence of model theory and proof theory for classical predicate logic, it remains to be shown that the completeness theorem holds:
Theorem (Completeness): if $A_1, \ldots, A_n \models B$, then $A_1, \ldots, A_n \vdash' B$.
We shall do so in Section 4.5.

This is Gödel's *completeness theorem* (1930) for classical predicate logic. The soundness theorem says that we do *not* have *too many* axioms or rules, i.e., every formula which may be deduced from given premisses by our axioms and rules is a logical consequence of those premisses. The completeness theorem, on the other hand, says that we have *enough* axioms and rules (for classical predicate logic), i.e., every formula which is a logical consequence of given premisses can be deduced from those premisses by finitely many applications of our axioms and rules.

In Section 2.6 we already paid attention to the *philosophical meaning of the completeness theorem*. Having the notions of validity and provability for (classical) predicate logic at our disposal, we can add the following observations to our discussion in Section 2.6 (assuming acquaintance with the notions of enumerable set and non-enumerable set, which are treated in Chapter 3).

a) The notions of validity and satisfiability refer to the totality of all structures, which is non-enumerable, while the equivalent proof-theoretic notions of provability and irrefutability (B is *irrefutable* := $\neg B$ is unprovable) refer only to the enumerable infinity of logical proofs. In other words, in the definition of $\models B$ we have a (universal) quantification over non-enumerably many structures, while in the definition of $\vdash B$ there is an (existential) quantification over only enumerably many logical proofs. So in the completeness theorem a reduction from the non-enumerably to the enumerably infinite is achieved.

b) The proof of Gödel's completeness theorem is more complex than the proofs of other theorems considered thus far. However, from a careful analysis of this proof one can draw some further conclusions which are philosophically interesting. These conclusions are formulated in the Compactness Theorem and the Löwenheim-Skolem Theorem, also to be treated in Section 4.5.

Exercise 4.34. Construct either a tableau-proof of the following formulas or construct a counterexample from an open branch in the tableau.
1. $(\exists x[P(x)] \to \exists x[Q(x)]) \to \exists x[P(x) \to Q(x)]$,
2. $\exists x[P(x) \to Q(x)] \to (\exists x[P(x)] \to \exists x[Q(x)])$, and 3. $\exists x \forall y [P(x) \to P(y)]$.

4.5 Completeness, Compactness and Löwenheim-Skolem

Given formulas B and $A_1, \ldots, A_n$, the tableaux rules suggest a *procedure of searching for a tableau-deduction* of B from $A_1, \ldots, A_n$:
start with $TA_1, \ldots, TA_n, FB$ and apply all the appropriate rules in some definite fixed order, the choice of ordering being unimportant (at least, if we do not care about efficiency); in an application of rule $T \to$ to, for example, S, T $P \to Q$ we make two branches, one with S, FP and the other with S, TQ and similarly for applications of the rules $F \wedge$ and $T \vee$. Owing to the rules $F \exists$ and $T \forall$, the systematic search for such a tableau-deduction now does not necessarily come to an end in finitely many steps, because new variables can be introduced again and again, which may or may not cause closure. In the proof of the completeness theorem below we suppose for reasons of simplicity that there are no individual constants in the language. The case in which there are individual constants (and/or function symbols) in the language is treated similarly.

Example 4.23. We wonder whether $\exists x[P(x) \to Q(x)] \vdash' \exists x[P(x)] \to \exists x[Q(x)]$. Start making a tableau with initial branch $\{T\ \exists x[P(x) \to Q(x)], F\ \exists x[P(x)] \to \exists x[Q(x)]\}$:

$$T\ \exists x[P(x) \to Q(x)], F\ \exists x[P(x)] \to \exists x[Q(x)]$$
$$T\ P(a_1) \to Q(a_1), F\ \exists x[P(x)] \to \exists x[Q(x)]$$
$$T\ P(a_1) \to Q(a_1), T\ \exists x[P(x)], F\ \exists x[Q(x)]$$

Because of rule $T \to$ we continue with two branches:
$FP(a_1), T\exists x[P(x)], F\exists x[Q(x)]$ and $TQ(a_1), T\exists x[P(x)], F\exists x[Q(x)]$
$FP(a_1), T\exists x[P(x)], F\exists x[Q(x)], FQ(a_1)$ $TQ(a_1), T\exists x[P(x)], F\exists x[Q(x)], FQ(a_1)$

So, the left – not yet completed – branch is open, while the right branch is closed. For the left branch we may continue with:

$$FP(a_1),\ T\exists x[P(x)],\ F\exists x[Q(x)],\ FQ(a_1)$$
$$FP(a_1),\ TP(a_2),\ T\exists x[P(x)],\ F\exists x[Q(x)],\ FQ(a_1)$$
$$FP(a_1),\ TP(a_2),\ T\exists x[P(x)],\ F\exists x[Q(x)],\ FQ(a_1),\ FQ(a_2)$$

By *ad hoc* considerations we can see that the left-most branch will never close, no matter how many more tableaux rules we apply. From this open branch we

4.5 Completeness, Compactness and Löwenheim-Skolem

can construct a counterexample $M = \langle \mathbb{N}; P^*, Q^* \rangle$ with the natural numbers as domain: by definition, $M \not\models P(a)[1]$, corresponding to the occurrence of $FP(a_1)$ in the left branch; $M \models P(a)[n]$ iff $n > 1$, corresponding to the occurrence of $TP(a_2), TP(a_3), \ldots$ in the left branch; and $M \not\models Q(a)[n]$ for all $n \in \mathbb{N}$, corresponding to the occurrence of $FQ(a_1), FQ(a_2), \ldots$ in the left branch.

Then $M \models \exists x[P(x) \to Q(x)]$, corresponding to the occurrence of $T \exists x[P(x) \to Q(x)]$ in the left branch, since $M \models P(a) \to Q(a)[1]$; also $M \models \exists x[P(x)]$, corresponding to the occurrence of $T \exists x[P(x)]$ in the left branch, since $M \models P(a)[2]$, but $M \not\models \exists x[Q(x)]$, corresponding to the occurrence of $F \exists x[Q(x)]$ in the left branch, since there is no natural number n with $M \models Q(a)[n]$.

Like in propositional logic, any completed open tableau branch τ in a tableau with initial branch $\{TA_1, \ldots, TA_n, FB\}$ yields a model M_τ (see Definition 4.29) of $A_1, \ldots, A_n$ in which B does not hold, showing that $A_1, \ldots, A_n \not\models B$ (see Lemma 4.1) and if all branches in a tableau with initial branch $\{TA_1, \ldots, A_n, FB\}$ are closed, then $A_1, \ldots, A_n \vdash' B$ (see Lemma 4.2).

Definition 4.29 (Model M_τ). Let τ be a completed open tableau branch. Then $M_\tau = \langle D; P_1^*, P_2^*, \ldots \rangle$ is the model defined by (1) D is the set of all natural numbers i such that a_i occurs in τ, and (2) $P_i^*(n_1, \ldots, n_i)$ iff $T P_i(a_{n_1}, \ldots, a_{n_i})$ occurs in τ. From the construction of M_τ it follows that one can always take $D = \mathbb{N}$.

Lemma 4.1. *Let τ be a completed open tableau branch. Then for each formula $E(a_{n_1}, \ldots, a_{n_k})$:*
a) if $TE(a_{n_1}, \ldots, a_{n_k})$ occurs in τ, then $M_\tau \models E(a_{n_1}, \ldots, a_{n_k})[n_1, \ldots, n_k]$,
b) if $FE(a_{n_1}, \ldots, a_{n_k})$ occurs in τ, then $M_\tau \not\models E(a_{n_1}, \ldots, a_{n_k})[n_1, \ldots, n_k]$.

Proof. The proof is by induction on the construction of E and generalizes the proof of the completeness theorem for classical propositional logic in Chapter 2.
Basic step: $E = P(a_{n_1}, \ldots, a_{n_k})$ is atomic. a) If $T P(a_{n_1}, \ldots, a_{n_k})$ occurs in τ, then by Definition 4.29, $M_\tau \models P(a_{n_1}, \ldots, a_{n_k}) [n_1, \ldots, n_k]$. b) If $F P(a_{n_1}, \ldots, a_{n_k})$ occurs in τ, then – since τ is open – $T P(a_{n_1}, \ldots, a_{n_k})$ does not occur in τ and hence, by Definition 4.29, $M_\tau \not\models P(a_{n_1}, \ldots, a_{n_k}) [n_1, \ldots, n_k]$.
Induction step. Suppose that a) and b) have been shown for C and D (induction hypothesis). We want to prove a) and b) for $C \wedge D, C \vee D, C \to D$ and $\neg C$.

If $E = C \wedge D$ and $T C \wedge D$ occurs in τ, then - because τ is completed - both TC and TD occur in τ. Hence, by the induction hypothesis, $M_\tau \models C [n_1, \ldots, n_k]$ and $M_\tau \models D [n_1, \ldots, n_k]$. So, $M_\tau \models C \wedge D [n_1, \ldots, n_k]$. If $E = C \wedge D$ and $F C \wedge D$ occurs in τ, then - because τ is completed - FC occurs in τ or TD occur in τ. Hence, by the induction hypothesis, $M_\tau \not\models C [n_1, \ldots, n_k]$ or $M_\tau \not\models D [n_1, \ldots, n_k]$. So, $M_\tau \not\models C \wedge D [n_1, \ldots, n_k]$.

The cases $E = C \vee D, E = C \to D$ and $E = \neg C$ are treated similarly. Next suppose that a) and b) have been shown for $A(a_i, a_{n_1}, \ldots, a_{n_k})$ (induction hypothesis). We want to prove a) and b) for $\forall x[A(x, a_{n_1}, \ldots, a_{n_k})]$ and for $\exists x[A(x, a_{n_1}, \ldots, a_{n_k})]$.

If $E = \forall x[A(x, a_{n_1}, \ldots, a_{n_k})]$ and TE occurs in τ, then - because τ is completed - $T A(a_i, a_{n_1}, \ldots, a_{n_k})$ occurs in τ for every $i \in D$. Hence, by the induction hypothesis, $M_\tau \models A(a_i, a_{n_1}, \ldots, a_{n_k})[i, n_1, \ldots, n_k]$ for every $i \in D$. So, $M_\tau \models$

$\forall x[A(x,a_{n_1},\ldots,a_{n_k})][n_1,\ldots,n_k]$. If $E = \forall x[A(x,a_{n_1},\ldots,a_{n_k})]$ and FE occurs in τ, then - because τ is completed - $F\,A(a_i,a_{n_1},\ldots,a_{n_k})$ occurs in τ for some $i \in D$ with a_i new. Hence, by the induction hypothesis, $M_\tau \not\models A(a_i,a_{n_1},\ldots,a_{n_k})[i,n_1,\ldots,n_k]$ for some $i \in D$. So, $M_\tau \not\models \forall x[A(x,a_{n_1},\ldots,a_{n_k})][n_1,\ldots,n_k]$.

The case $E = \exists x[A(x,a_{n_1},\ldots,a_{n_k})]$ is treated similarly. $\square$

Lemma 4.2. *If all branches in a tableau with initial branch $\{TA_1,\ldots,TA_n,FB\}$ are closed, then $A_1,\ldots,A_n \vdash' B$.*

Proof. Suppose all branches in a tableau with initial sequent $\{TA_1,\ldots,TA_n,FB\}$ are closed. Then for some natural number k they all close in less than k steps. For if not, then (by König's lemma 1926; see Exercise 4.43) there would be an open infinite tableau branch starting with $\{TA_1,\ldots,TA_n,FB\}$. The finitely many (closed) tableau branches together yield a tableau-deduction of B from $A_1,\ldots,A_n$. $\square$

Lemma 4.1 and Lemma 4.2 together yield the completeness theorem for classical predicate logic.

Theorem 4.25 (Completeness of classical predicate logic).
a) If $A_1,\ldots,A_n \models B$, then $A_1,\ldots,A_n \vdash' B$. In particular, if $n = 0$:
b) If $\models B$, then $\vdash' B$.

Proof. a) Suppose $A_1,\ldots,A_n \models B$. Apply the procedure of searching for a tableau-deduction of B from $A_1,\ldots,A_n$. Let $\mathcal{T}$ be the resulting completed tableau. Let $a_{n_1},\ldots,a_{n_k}$ be the free variables occurring in $A_1,\ldots,A_n,B$. If there were an open tableau branch τ in $\mathcal{T}$, then by Lemma 4.1 for all $i = 1,\ldots,n$, $M_\tau \models A_i[n_1,\ldots,n_k]$ and $M_\tau \not\models B[n_1,\ldots,n_k]$, contradicting $A_1,\ldots,A_n \models B$. Therefore, all branches in tableau $\mathcal{T}$ with initial branch $\{TA_1,\ldots,TA_n,FB\}$ are not open, i.e., not not closed and hence closed. So, by Lemma 4.2, $A_1,\ldots,A_n \vdash' B$.
b) is a special case ($n = 0$) of a). $\square$

The proof of the completeness theorem for (classical) predicate logic given above is close to Gödel's original proof, 1930; see van Heijenoort [6] for Gödel's proof. Another interesting completeness proof has been given by Henkin in [7].

4.5.1 Undecidability

In contrast to the case of propositional logic, the construction of a completed tableau with initial branch $\{TA_1,\ldots,TA_n,FB\}$ will in general not end after finitely many steps, because the quantifier rules $T\,\forall$ and $F\,\exists$ may be applied again and again. Although for many concrete formulas B we can make a decision about their being valid (provable) by constructing a completed tableau with $\{FB\}$ as initial branch by *ad hoc* considerations about the growth of the tableau branches – as in the examples 4.20 and 4.22: $\neg\forall x[P(x)] \to \forall x[\neg P(x)]$, resp. $\forall x\exists y[P(x,y)] \to \exists y\forall x[P(x,y)]$ –, constructing a completed tableau with initial branch $\{FB\}$ does not give a uniform

4.5 Completeness, Compactness and Löwenheim-Skolem

decision procedure for validity (provability) in predicate logic. It is not the case that given *any* formula B, our construction of a completed tableau with initial branch $\{FB\}$ will tell us after a finite number of steps (which, given B, can be determined in advance) whether B is tableau-provable (and hence valid) or not.

A *positive test* for validity (provability) is a mechanical test such that for each formula B, B is valid (provable) iff the test applied to input B gives a positive answer in finitely many steps. And a *negative test* for validity is a mechanical test such that for each formula B, B is invalid (not provable) iff the test applied to input B gives a negative answer in finitely many steps. A *decision procedure* for validity is now simultaneously both a positive and a negative test. Conversely, if one has both a positive and a negative test (possibly different) for validity, then one can obtain a decision procedure by applying the steps of both tests alternately to the input formula B. $(*)$

Note that constructing a completed tableau with initial branch $\{FB\}$ clearly is a positive test for validity: if it is applied to a valid formula B, the construction will come to an end after finitely many steps and provide a tableau-deduction of B. But our procedure does not give a negative test for validity: if it is applied to a non-valid formula B, the procedure may run forever without presenting an answer, because the rules $T \forall$ and $F \exists$ may be applied again and again. In 1936 A. Church (see Kleene [9], Section 45) and A. Turing [15] proved independently that there is no decision procedure for validity (provability) in classical predicate logic.

Theorem 4.26 (Church-Turing: Predicate logic is undecidable). *There is no decision procedure for validity (provability) in (classical) predicate logic.*

This theorem not only says that constructing a completed tableau with initial branch $\{FB\}$ does not give a decision procedure for validity of an arbitrary formula B, but also that no other decision procedure can exist. In other words, classical predicate logic is *undecidable*.

From the Church-Turing Theorem and remark $(*)$ above it follows that there can be no negative test for validity, since constructing a completed tableau with initial branch $\{FB\}$ is a positive test for validity.

And since A is not satisfiable if and only if $\neg A$ is valid, it follows that there is a negative test for satisfiability, but no positive test for satisfiability.

In the exercises we will consider some particular classes of formulas B, for which one can determine a natural number N such that constructing a completed tableau with initial branch $\{FB\}$ with only N applications of the $T \forall$ and $F \exists$-rules, provides a decision procedure for formulas in the class; for each formula B in the class it yields in finitely many steps either a tableau-proof of B or the conclusion that no tableau-deduction of B exists. So, for formulas B in the given class it holds that if no tableau-proof of B is found after N applications of the $T\forall$ and $F\exists$-rules in the construction of a completed tableau with initial branch $\{FB\}$, then there is no tableau-proof of B at all.

4.5.2 Compactness and Löwenheim-Skolem Theorems

Definition 4.30 (Validity and Satisfiability in a given Domain). Let D be a non-empty domain and B a formula.
a) B is *valid in* $D := M \models B\,[v]$ for all models M with domain D and for each valuation v in D.
b) B is *satisfiable in* $D :=$ there is at least one model M with domain D and at least one valuation v in D such that $M \models B\,[v]$.
c) A class Γ of formulas is *simultaneously satisfiable in* $D :=$ there is a model M with domain D and a valuation v in D such that $M \models B\,[v]$ for all B in Γ.

On close inspection we have shown in the proof of the Completeness Theorem 4.25 much more than is stated in the formulation of this theorem itself.

Theorem 4.27 (Löwenheim, 1915). a) *If not* $\vdash' B$, *then B is not valid in an enumerable domain ($\mathbb{N}$ or a finite subset of $\mathbb{N}$).*
b) *Löwenheim's theorem: if a formula B is satisfiable in any non-empty domain, then it is satisfiable in an enumerable domain ($\mathbb{N}$ or a finite subset of $\mathbb{N}$).*
c) *If $A_1, \ldots, A_n$ are simultaneously satisfiable in any non-empty domain, then they are simultaneously satisfiable in an enumerable domain ($\mathbb{N}$ or a finite subset of $\mathbb{N}$).*

Proof. a) Suppose not $\vdash' B$. Then, by Lemma 4.2, not all branches in a completed tableau with initial branch $\{FB\}$ are closed. Hence, there is an open branch τ in such a tableau. Let $a_{n_1}, \ldots, a_{n_k}$ be the free variables occurring in B. Then, by Lemma 4.1, since FB occurs in τ, $M_\tau \not\models B[n_1, \ldots, n_k]$. And M_τ is a model with $\mathbb{N}$ or a finite subset of $\mathbb{N}$ as domain (see Definition 4.29).
b) Suppose B is satisfiable (in some non-empty domain). Then not $\vdash' \neg B$. Let $a_{n_1}, \ldots, a_{n_k}$ be the free variables occurring in $\neg B$. Then by the proof of a), $M_\tau \not\models \neg B[n_1, \ldots, n_k]$, i.e., $M_\tau \models B[n_1, \ldots, n_k]$. And the domain of M_τ is $\mathbb{N}$ or a finite subset of $\mathbb{N}$. c) Follows from b) taking $B = A_1 \wedge \ldots \wedge A_n$. □

Theorem 4.28 (Compactness; Skolem). *Let $A_0, A_1, A_2, \ldots$ be an infinite list of formulas.* a) *Compactness (Gödel 1930): If, for each natural number k, $A_0, \ldots, A_k$ are simultaneously satisfiable, then $A_0, A_1, A_2, \ldots$ are simultaneously satisfiable in an enumerable domain ($\mathbb{N}$ or a finite subset of $\mathbb{N}$).*
b) *Skolem's (1920) generalization of Löwenheim (1915): If $A_0, A_1, A_2, \ldots$ are simultaneously satisfiable, then $A_0, A_1, A_2, \ldots$ are simultaneously satisfiable in an enumerable domain ($\mathbb{N}$ or a finite subset of $\mathbb{N}$).*

Proof. b) follows immediately from a). To prove a), suppose that for each natural number k, $A_0, \ldots, A_k$ are simultaneously satisfiable in some non-empty domain D. Construct a tableau with initial branch $\{TA_0, TA_1, TA_2, \ldots, F(P \wedge \neg P)\}$ by admitting step by step more and more assumption formulas A_i in the tableau construction (see also Kleene [9], Section 50).

Now suppose that all branches in this tableau would close. Then (by König's Lemma, see Exercise 4.43) there is a natural number m such that they all close in less than m steps. Let $A_0, \ldots, A_k$ be all formulas A_i occurring in the finitely

4.5 Completeness, Compactness and Löwenheim-Skolem 233

many closed branches. Then $A_0, \ldots, A_k \vdash' P \wedge \neg P$, contradicting the hypothesis of (a). So, there must be at least one open branch τ in the tableau with initial branch $\{TA_0, TA_1, TA_2, \ldots, F(P \wedge \neg P)\}$. Then, by Lemma 4.1, M_τ is a model with domain $\mathbb{N}$ (or a finite subset of $\mathbb{N}$) such that for all $i = 0, 1, 2, \ldots$, $M_\tau \models A_i[v]$ for some valuation v in the domain of M_τ. □

Let Γ be a (possibly infinite) set of sentences (closed formulas). We say that M is a *model* of Γ if M is a model of all sentences in Γ. We can specialize Theorem 4.28 to the case that Γ is an infinite list $A_0, A_1, A_2, \ldots$ of *closed* formulas (sentences).

Corollary 4.1. *Let Γ be a (possibly infinite) set of sentences (closed formulas).*
a) Compactness theorem: If each finite subset of Γ has a model, then Γ has a model.
b) Downward Löwenheim-Skolem theorem: If Γ has a model, then Γ has an enumerable model.

It is important to realize that the downward Löwenheim-Skolem theorem is due to the fact that first-order predicate languages contain only enumerably many symbols.

In addition to the 'downward-' there is also an 'upward-' Löwenheim-Skolem theorem, saying that under certain conditions, if Γ has a model, then it has arbitrarily large models. The interested reader is referred to van Dalen [4].

Once one has proved the completeness theorem, 'if $\Gamma \models B$, then $\Gamma \vdash B$', also for an infinite set Γ of premisses, the compactness theorem is an immediate consequence of it. The argument goes as follows: By definition, $\Gamma \vdash B$ iff there is a finite subset Γ' of Γ such that $\Gamma' \vdash B$. Therefore, using soundness and completeness, $\Gamma \models B$ iff there is a finite subset Γ' of Γ such that $\Gamma' \models B$. Taking for $B = P \wedge \neg P$, we obtain by contraposition: $\Gamma \not\models P \wedge \neg P$ iff for each finite subset Γ' of Γ, $\Gamma' \not\models P \wedge \neg P$. Or equivalently, Γ has a model iff each finite subset Γ' of Γ has a model.

For historical details concerning the Löwenheim-Skolem Theorem the reader is referred to van Heijenoort [6].

On the Meaning of the Compactness and Löwenheim-Skolem Theorem
From Theorem 4.28 b) it follows immediately that there can be no class Γ of formulas of first-order predicate logic such that Γ is simultaneously satisfiable in D iff D has non-enumerably many elements. Therefore, the expression 'having non-enumerably many elements' cannot be formulated in a first-order predicate language; in other words, *'non-enumerable' is not a first-order property*.

The Löwenheim-Skolem Theorem points out that *the expressive power of first-order predicate languages is restricted*: 'being non-enumerable' cannot be formulated in first-order logic. On the other hand, 'having infinitely many elements' is a first-order property; see Exercise 4.35.

As is explained in Exercise 4.35, there are classes $\Gamma_1, \Gamma_2, \Gamma_3, \ldots$ of formulas such that for each $n \in \mathbb{N}$, Γ_n is simultaneously satisfiable in D iff D contains at least n elements. So, the expressions 'having at least one element', 'having at least two elements', and so on, can all be formulated in an appropriate first-order predicate language. However, below we shall prove that 'having finitely many elements' cannot be formulated in a first-order predicate language.

Theorem 4.29. *There is no class Γ of formulas such that Γ is simultaneously satisfiable in D iff D contains finitely many elements.* (∗)

Proof. Suppose there was a class Γ of formulas such that (∗) holds. Now consider $\Delta := \Gamma \cup \Gamma_1 \cup \Gamma_2 \cup \ldots$, where for each n, Γ_n is a class of formulas expressing that there are at least n elements (see above). Then each finite subset of Δ is simultaneously satisfiable. So, by Theorem 4.28, Δ is simultaneously satisfiable in $\mathbb{N}$ or in a finite subset of $\mathbb{N}$. But by virtue of the formulas in Δ, Δ cannot be satisfiable in a finite subset of $\mathbb{N}$. So,5 Δ is simultaneously satisfiable in $\mathbb{N}$. Therefore, Γ is simultaneously satisfiable in $\mathbb{N}$. Contradiction with (∗). □

Summarizing *The Löwenheim-Skolem Theorems point out that the expressive power of first-order predicate languages is restricted: 'finite' and 'non-enumerable' are not first-order properties.*

4.5.3 Second-order Logic

It is interesting to note that the notions of 'finite' and of 'non-enumerable', which are not first-order properties (see Subsection 4.5.2), can be formulated in *second-order logic*. In second-order logic one is allowed to quantify not only over individual variables, but also over function variables and predicate variables. A second-order formula is a formula that contains at least one occurrence of a function or predicate variable. Here are some examples of second-order formulas, using x, y, z as individual variables, u as a function variable and X as a (unary) predicate variable:

Example 4.24. $\exists u \forall x[u(x) = x]$: there exists an identity function;
$\forall x \forall y \exists X[X(x) \wedge X(y)]$: every two individuals share some property;
$a = b \rightleftarrows \forall X[X(a) \rightleftarrows X(b)]$: a and b are equal iff they have the same properties (Leibniz' Law).

Now let Inf be the second-order sentence

$$\exists z \exists u [\forall x[z \neq u(x)] \wedge \forall x \forall y[x \neq y \rightarrow u(x) \neq u(y)]].$$

Inf is true in an interpretation with domain D iff there is an injective function (u) with domain D whose range is a proper subset of D ($z \notin \mathrm{Ran}(u)$). So, Inf is true in an interpretation iff the domain is Dedekind infinite (see Exercise 3.33). Consequently, 'being finite' can be expressed by the second-order formula $\neg Inf$.

Let En be the second-order sentence

$$\exists z \exists u \forall X[X(z) \wedge \forall x[X(x) \rightarrow X(u(x))] \rightarrow \forall x[X(x)]].$$

En is true in an interpretation iff the domain of the interpretation is enumerable (see Exercise 4.38). Consequently, 'non-enumerable' can be expressed by the second-order formula $\neg En$.

$\neg En$ is true in an interpretation iff the domain of the interpretation is non-enumerable. Since such interpretations exist, $\neg En$ is satisfiable. But $\neg En$ is not

4.5 Completeness, Compactness and Löwenheim-Skolem   235

satisfiable in any enumerable domain. Therefore: *the Löwenheim-Skolem theorem fails for second-order logic*. The compactness theorem and many other properties of first-order logic also fail for second-order logic. See Chapter 5 and Boolos [2], Chapter 22.

4.5.4 Skolem's Paradox

Below we shall work out the astonishing and philosophically interesting consequence of the Löwenheim-Skolem theorem known as *Skolem's paradox*.

Let Γ be the set of (closed) axioms of some axiomatic set theory formulated in a first-order predicate language, for instance, $\Gamma = ZF$ (Zermelo-Fraenkel set theory; see Chapter 3). It is generally believed that such a Γ is consistent, in other words, that there is some model M which makes all axioms in Γ true. But then it follows from the Löwenheim-Skolem theorem that Γ has an enumerable model. There are only enumerably many 'sets' in this model.   (1)

On the other hand, we know that Cantor's theorem (Corollary 3.6) is deducible from Γ, saying that the set $P(\mathbb{N})$ of all subsets of $\mathbb{N}$ is not enumerable.   (2)

(1) and (2) together constitute what is called *Skolem's paradox* (1922-3).

Skolem's paradox is not an antinomy (or real paradox), but rather a veridical (or truth-telling) paradox (see Section 2.10). It tells us the (astonishing) truth that there is an *enumerable model* which makes all axioms in Γ true, although it follows from Γ that there are *non-enumerably many sets*. How is this possible? How can we explain this phenomenon?

The set of all subsets in the model is indeed enumerable and therefore there is a bijective mapping from it to the set of natural numbers. But this mapping is *not* in the model; so it does not make invalid the theorem of set theory which states that there is no bijective mapping from the set $P(\mathbb{N})$ of all subsets of $\mathbb{N}$ to $\mathbb{N}$. More precisely: Let Γ be the axioms of set theory formulated in a first order predicate language. Then

$$\Gamma \vdash \neg \exists x [x \text{ is a bijection from } P(\mathbb{N}) \text{ to } \mathbb{N}].$$

Now let M be a countable model of Γ. Then

$$M \models \neg \exists x [x \text{ is a bijection from } P(\mathbb{N}) \text{ to } \mathbb{N}],$$

i.e., there is no bijection (in the sense of M) *in M* from the set $P(\mathbb{N})^M$ in M to the set $\mathbb{N}^M$ in M. This does not exclude that there is a bijection *outside of M* (i.e., being not a set or object of the model M) from the set $P(\mathbb{N})^M$ in M to the set $\mathbb{N}^M$ in M.

Skolem's 'paradox' may be further clarified by the following two observations.

1. From the axioms of set theory it follows that there are non-enumerably many subsets of a given infinite set. But given some set, we can actually *define* only enumerably many subsets of it. So, it is not that surprising that there is an enumerable model of the axioms of set theory.

2. Skolem's 'paradox' is the result of an application of the Löwenheim-Skolem theorem which was a by-product of the completeness theorem. These theorems are the result of considerations *about* the formal system as a whole and are not deducible within the formal system itself. This explains the possibility that looking from the outside to a formal system for set theory, the collection of all subsets of a given infinite set may be enumerable, while at the same time this collection is non-enumerable within the formal system itself.

We finish this section with a quotation from van Heijenoort [6], pp. 290-291:

> For Skolem the discrepancy between an intuitive set-theoretic notion and its formal counterpart leads to the 'relativity' of set-theoretic notions. Thus, two sets are equivalent if there exists a one-to-one mapping of the first onto the second; but this mapping is itself a collection of ordered pairs of elements. If, in a formalized set theory, this collection exists as a set, the two given sets are equivalent in the theory; if it does not, the sets are not equivalent in the theory and, when one set is that of the natural numbers as defined in the theory, the other becomes 'nondenumerable'. The existence of such a 'relativity' is sometimes referred to as the Löwenheim-Skolem paradox. But, of course, it is not a paradox in the sense of an antinomy; it is a novel and unexpected feature of formal systems.

Exercise 4.35. a) Show that $P(a_1), \neg P(a_2)$ are simultaneously satisfiable in D iff D contains at least two elements.
b) Find a class Γ of formulas such that Γ is simultaneously satisfiable in D iff D contains at least three elements.
c) Show that $\forall x[\neg P(x,x)]$, $\forall x \forall y \forall z[P(x,y) \wedge P(y,z) \to P(x,z)]$, $\forall x \exists y[P(x,y)]$ are simultaneously satisfiable in D iff D contains at least denumerably many elements.
d) Show that it is impossible to find a class Γ of formulas such that Γ is satisfiable in D iff D has non-enumerably many elements.

Thus if one attempts to characterize a mathematical structure by means of a set of axioms formulated in first order predicate logic, one is in a certain sense doomed to failure if that structure involves a non-enumerable infinity of elements.

Exercise 4.36. Let Γ be a (possibly infinite) set of formulas. Γ is *consistent* := for no formula B, $\Gamma \vdash B$ and $\Gamma \vdash \neg B$. Note that Γ is consistent iff there is at least one formula C such that $\Gamma \nvdash C$. Supposing that Γ is a finite set of formulas, show that Γ is consistent iff Γ has a model with an enumerable domain. (Skolem, 1922; see van Heijenoort [6], p. 293.)

Exercise 4.37. Using Skolem's result (1922) that Γ is consistent iff Γ has a model with an enumerable domain (see Exercise 4.36), prove the completeness theorem: if $A_1, \ldots, A_n \models B$, then $A_1 \ldots, A_n \vdash B$. Skolem himself did not make this step from his result (1922) to the completeness theorem (K. Gödel, 1930) for philosophical reasons. The notions of validity and valid consequence contain a universal quantification over all non-enumerably many structures and for that reason the idea of formulating the completeness theorem did not even occur to Skolem.

Exercise 4.38. Let En be the second-order formula

$$\exists z \exists u \forall X[X(z) \wedge \forall x[X(x) \to X(u(x))] \to \forall x[X(x)]].$$

4.6 Predicate Logic with Equality

Prove: En is true in an interpretation with domain D iff D is enumerable.

Exercise 4.39. Prove that there is a decision procedure (to test tableau-provability) for the class of formulas having a prenex normal form such that, in the prefix, no existential quantifier precedes any universal quantifier.

Exercise 4.40. A *monadic formula* contains by definition only unary (monadic) predicate symbols. Prove that each monadic formula is equivalent to a truth-functional composition of formulas of the form $\forall x[B(x)]$ and $\exists x[B(x)]$, where B does not contain any quantifiers.

Exercise 4.41. Prove that there is a decision procedure (to test tableau-provability) for the class of monadic formulas (see Exercise 4.40).

Exercise 4.42. Prove that there is a decision procedure (to test tableau-provability) for the class of formulas having a prenex normal form $\exists x \forall y[M]$, where M is the matrix, containing no individual variables except x and y and containing only one binary predicate symbol P. Similarly, if M contains two binary predicate symbols.

Solutions of the decision problem for special classes of more complex formulas can be found in Church [3], Section 46.

Exercise 4.43. Let $\mathbb{N}^*$ be the set of all n-tuples $(k_1, \ldots, k_n)$ of natural numbers, $n \in \mathbb{N}$. Let $s = (k_1, \ldots, k_n) \in \mathbb{N}^*$ and $t = (l_1, \ldots, l_m) \in \mathbb{N}^*$. Then the *concatenation* of s and t, denoted by st, is the $(n+m)$-tuple $(k_1, \ldots, k_n, l_1, \ldots, l_m)$. And s is a *prefix* of t := there is a tuple $s' \in \mathbb{N}^*$ such that $t = ss'$.

Let T be a subset of $\mathbb{N}^*$. T is a *tree* := a) for each $t \in T$, every prefix of t is also in T, and b) for each $t \in T$, and for every $i \in \mathbb{N}$, if $t(i) \in T$, then for every $j \leq i, t(j)$ is also in T. The elements of a tree T are called *nodes*. For $s, t \in T$, t is an *immediate successor* of s := for some $i \in \mathbb{N}, t = s(i)$. A *path* in T is a finite or infinite sequence $s_0, s_1, \ldots$ of nodes in T, starting with the empty tuple $()$, i.e., $s_0 = ()$, and such that each node s_{i+1} is an immediate successor of the preceding node s_i.

König's Lemma: Let T be a tree such that each node in T has only finitely many immediate successors. If there are arbitrarily long finite paths in T, then there is an infinite path in T. Prove König's Lemma and show that the lemma need not hold for trees in which some node has infinitely many immediate successors.

4.6 Predicate Logic with Equality

The predicate logic with equality arises from predicate logic (without equality) by giving one of the binary predicate symbols, say $\equiv$, special treatment. That is, in the predicate logic with equality one allows only interpretations (structures) which interpret $\equiv$ as equality ($=$); no other interpretations of $\equiv$ are allowed.

Definition 4.31 (Interpretation). Let M be an interpretation. M is an *interpretation* or model *for the predicate logic with equality* := M interprets $\equiv$ as $=$ (equality).

Here $\equiv$ is a particular binary predicate symbol in the predicate language, while $=$ is the name of the equality relation, which is a mathematical object. For convenience, one frequently writes $=$ instead of the logical predicate symbol $\equiv$, in which case the sign $=$ is used both as a symbol in the predicate language and as a symbol denoting mathematical equality, which is the interpretation of the predicate symbol $\equiv$.

Definition 4.32 (Validity). $\models A$ (A is *valid in the predicate logic with equality*) := for all interpretations M for the predicate logic with equality, $M \models A$. '$A_1, \ldots, A_m \models B$' in the predicate logic with equality is defined similarly.

A Hilbert-type proof system for the predicate logic with equality is obtained by adding to the axiom schemata and rules of inference for the predicate logic without equality the following formulas as further axioms:

$\forall x[x \equiv x]$ $\qquad\qquad\forall x \forall y[x \equiv y \to (P(\ldots, x, \ldots) \to P(\ldots, y, \ldots))]$
$\forall x \forall y \forall z[x \equiv y \to (x \equiv z \to y \equiv z)]$ $\quad \forall x \forall y[x \equiv y \to f(\ldots, x, \ldots) \equiv f(\ldots, y, \ldots)]$

One easily sees that these axioms are valid in the predicate logic *with* equality. Of course, they are not valid (in the predicate logic without equality): taking $\mathbb{N}$ as domain and interpreting $\equiv$ as $<$ (is less than), a false proposition results from $\forall x[x \equiv x]$. The provability and deducibility results, in particular the deduction theorem, already established for the predicate logic (without equality) all hold also for the predicate logic with equality.

Theorem 4.30. *In the predicate logic with equality:*

$\vdash \forall x[x \equiv x]$ $\qquad\qquad\qquad\qquad r \equiv s \vdash t \equiv r \rightleftarrows t \equiv s$
$\vdash \forall x \forall y[x \equiv y \to y \equiv x]$ $\qquad\quad r \equiv s \vdash r \equiv t \rightleftarrows s \equiv t$
$\vdash \forall x \forall y \forall z[x \equiv y \land y \equiv z \to x \equiv z]$ $\quad r \equiv s \vdash P(\ldots, r, \ldots) \rightleftarrows P(\ldots, s, \ldots)$
(where r, s and t are terms) $\qquad\quad r \equiv s \vdash f(\ldots, r, \ldots) \equiv f(\ldots, s, \ldots)$

Proof.
1) $\forall x[x \equiv x]$ is an axiom and hence provable in the predicate logic *with equality*.
2) To show that $\vdash \forall x \forall y[x \equiv y \to y \equiv x]$ in the predicate logic with equality, suppose $a_1 \equiv a_2$. From the second equality axiom: $a_1 \equiv a_2 \to (a_1 \equiv a_1 \to a_2 \equiv a_1)$. By Modus Ponens: $a_1 \equiv a_1 \to a_2 \equiv a_1$. From the first equality axiom: $a_1 \equiv a_1$. Applying Modus Ponens: $a_2 \equiv a_1$. Therefore, $a_1 \equiv a_2 \to a_2 \equiv a_1$. Therefore, $\forall x \forall y[x \equiv y \to y \equiv x]$ is provable in the predicate logic with equality.
3) To show that $r \equiv s \vdash P(\ldots, r, \ldots) \rightleftarrows P(\ldots, s, \ldots)$ in the predicate logic *with equality*, assume $r \equiv s$ and assume $P(\ldots, r, \ldots)$. From the third equality axiom: $r \equiv s \to (P(\ldots, r, \ldots) \to P(\ldots, s, \ldots))$. Then by two applications of Modus Ponens, $P(\ldots, s, \ldots)$. Conversely, assume $P(\ldots, s, \ldots)$. We have already shown that $\vdash \forall x \forall y[x \equiv y \to y \equiv x]$. So $\vdash r \equiv s \to s \equiv r$. Assuming $r \equiv s$, by Modus Ponens $s \equiv r$. Then from the third equality axiom, $P(\ldots, r, \ldots)$.
4) The other cases are similar. $\qquad\square$

Another equivalent proof system for the predicate logic with equality is obtained by adding to the axiom schemata and rules of inference for predicate logic the axiom $\forall x[x \equiv x]$ and the axiom schema $\forall x \forall y[x \equiv y \to (A(x) \to A(y))]$.

4.6 Predicate Logic with Equality   239

Theorem 4.31. *Let M be a model (structure) for the predicate logic (without equality) with domain D, such that M satisfies the equality axioms. Let R be the interpretation in M of $\equiv$. Then R is an equivalence relation on D (see Subsection 3.4.3). And there is a model M′ such that*
1. the domain of M′ is the quotient set D/R (see Subsection 3.4.3),
2. M′ interprets the binary predicate symbol $\equiv$ as the equality relation $=$; hence M′ is a model for the predicate logic with equality, and
3. for any formula $A = A(a_1, \ldots, a_n)$, $M \models A\, [d_1, \ldots, d_n]$ iff $M' \models A\, [[d_1]_R, \ldots, [d_n]_R]$, where for $d \in D$, $[d]_R$ is the equivalence class of d with respect to R.

Proof. Let $M = \langle D; P_1^M, P_2^M, \ldots; f_1^M, f_2^M, \ldots \rangle$ be a structure (for the predicate logic without equality), which satisfies the equality axioms. Then M also satisfies the formulas in Theorem 4.30, since these formulas are deducible from the equality axioms in the predicate logic (without equality). Let R be $\equiv^M$, i.e., the interpretation in M of the binary predicate symbol $\equiv$; R is not necessarily the equality relation. Since M satisfies the first three formulas in Theorem 4.30, $\forall x[x \equiv x]$, $\forall x \forall y[x \equiv y \rightarrow y \equiv x]$, $\forall x \forall y \forall z[x \equiv y \land y \equiv z \rightarrow x \equiv z]$, it follows that the relation R on D is reflexive, symmetric and transitive. In other words, R is an equivalence relation on D. As explained in Subsection 3.4.3, any equivalence relation on D separates D into disjoint non-empty equivalence classes. For $d \in D$, let $[d]_R$ be the equivalence class of d with respect to R, i.e.,

$$[d]_R := \{d' \in D \mid R(d,d')\} = \{d' \in D \mid M \models a_1 \equiv a_2\, [d,d']\}.$$

Define the model M' as follows: a) the domain of M' is the quotient set D/R of all equivalence classes $[d]_R$ with $d \in D$;
b) for any n-ary predicate symbol P, $P^{M'}([d_1]_R, \ldots, [d_n]_R) := P^M(d_1, \ldots, d_n)$;
c) for any n-ary function symbol f, $f^{M'}([d_1]_R, \ldots, [d_n]_R) := [f^M(d_1, \ldots, d_n)]_R$.
By Theorem 4.30, M satisfies in particular: $r \equiv s \rightarrow (P(\ldots, r, \ldots) \rightleftarrows P(\ldots, s, \ldots))$ and $r \equiv s \rightarrow f(\ldots, r, \ldots) \equiv f(\ldots, s, \ldots)$. Consequently, the definitions of $P^{M'}$ and $f^{M'}$, given above, are correct, i.e., if $[d_1]_R = [e_1]_R$ and $\ldots$ $[d_n]_R = [e_n]_R$, then $P^M(d_1, \ldots, d_n)$ iff $P^M(e_1, \ldots, e_n)$ and $f^M(d_1, \ldots, d_n) R f^M(e_1, \ldots, e_n)$.

M' interprets the binary predicate symbol $\equiv$ by the equality relation $=$ on D/R. For $\equiv^{M'}([d_1]_R, [d_2]_R) := \equiv^M(d_1, d_2)$; but $\equiv^M(d_1, d_2)$ iff $R(d_1, d_2)$; hence $\equiv^M(d_1, d_2)$ iff $[d_1]_R = [d_2]_R$. Therefore $\equiv^{M'}$ is the equality relation on D/R. By straightforward induction it follows from the definition of M' that $M \models A\, [d_1, \ldots, d_n]$ iff $M' \models A\, [[d_1]_R, \ldots, [d_n]_R]$. □

It is straightforward to check that the *soundness theorem* holds for the predicate logic with equality: if $\Gamma \vdash B$ in the predicate logic with equality, then $\Gamma \models B$ in the predicate logic with equality. Making use of Theorem 4.31 one can easily see that from the completeness theorem, the compactness theorem and the Löwenheim-Skolem theorem for the predicate logic (without equality) similar theorems follow for the predicate logic with equality.

Theorem 4.32. *Let Γ be a possibly infinite set of closed formulas, and let B be a closed formula.*

a) **Completeness for the predicate logic with equality**: *if $\Gamma \models B$ in the predicate logic with equality, then $\Gamma \vdash' B$ in the predicate logic with equality.*
b) **Compactness for the predicate logic with equality**: *Γ has a model (in the predicate logic with equality) if and only if every finite subset of Γ has a model (in the predicate logic with equality).*
c) **Downward Löwenheim-Skolem for the predicate logic with equality**: *If Γ is simultaneously satisfiable in the predicate logic with equality, then Γ is simultaneously satisfiable in an enumerable domain in the predicate logic with equality.*

Proof. a) Let $\Gamma = \{A_1, \ldots, A_n\}$. The construction of a complete tableau $\mathcal{T}$ with initial branch $\{TA_1, \ldots, TA_n, FB\}$ in the predicate logic with equality starts with $\{TA_1, \ldots, TA_n, TE_1, \ldots, TE_m, FB\}$, where $E_1, \ldots, E_m$ are the equality axioms for the predicate and function symbols occurring in $A_1, \ldots, A_n, B$. If all branches in $\mathcal{T}$ close, then by Lemma 4.2, $A_1, \ldots, A_n, E_1, \ldots, E_m \vdash' B$, i.e., $A_1, \ldots, A_n \vdash' B$ in the predicate logic with equality. If τ is an open branch in $\mathcal{T}$, then by Lemma 4.1 the model M_τ (see Definition 4.29) makes all of $A_1, \ldots, A_n, E_1, \ldots, E_m$ true and B false. Then by Theorem 4.31 there is a model M'_τ for the predicate logic with equality which makes $A_1, \ldots, A_n$ true and B false. Therefore, if $A_1, \ldots, A_n \models B$ in the predicate logic with equality, there can be no open branch starting with $TA_1, \ldots, TA_n, TE_1, \ldots, TE_m, FB$; in other words, in that case all such branches will close and hence $A_1, \ldots, A_n \vdash' B$ in the predicate logic with equality. In case that Γ contains infinitely many sentences, the construction of a complete tableau has to be adapted, such that at each step one more assumption formula in Γ is taken into consideration.
b) Because only a finite number of sentences in Γ can be used in a formal deduction, it follows that $\Gamma \vdash B$ iff there is a finite subset Γ' of Γ such that $\Gamma' \vdash B$. From the soundness and completeness theorems it follows that $\Gamma \models B$ iff for some finite subset Γ' of Γ, $\Gamma' \models B$. Taking for B the formula $\exists x[x \not\equiv x]$ and noting that $\exists x[x \not\equiv x]$ is not true in any structure, the result follows by contraposition.
c) Let M be a model for the predicate logic with equality, which makes all formulas in Γ simultaneously true. Construct a complete tableau $\mathcal{T}$ with initial branch $\{TA_1, TA_2, \ldots, TE_1, \ldots, TE_m, F(P \wedge \neg P)\}$ where $\{A_1, A_2, \ldots\} = \Gamma$ and $E_1, \ldots, E_m$ are the equality axioms for the predicate and function symbols occurring in Γ. If all branches would close, then $\Gamma \vdash' P \wedge \neg P$ in the predicate logic with equality, and hence $M \models P \wedge \neg P$. Contradiction. Therefore, there is at least one open branch τ in $\mathcal{T}$, which by Lemma 4.1 yields a model M_τ that satisfies (simultaneously) all formulas in Γ and the equality axioms. Then by Theorem 4.31 there is a model M'_τ for the predicate logic with equality which simultaneously satisfies all formulas in Γ. Since the domain D of M_τ is enumerable and since the domain of M'_τ is D modulo R for some equivalence relation R, the domain of M'_τ is also enumerable. □

Warning For instance $M = \langle \{0\}; = \rangle$ is a model of the formula $\exists x \forall y [x \equiv y]$ in the predicate logic with equality. So, by the downward Löwenheim-Skolem theorem for the predicate logic with equality (Theorem 4.32) the formula $\exists x \forall y [x \equiv y]$ has a model with an enumerable domain. Notice that this domain cannot be $\mathbb{N}$ or any other denumerable domain, since in the predicate logic with equality $\exists x \forall y [x \equiv y]$

4.6 Predicate Logic with Equality

expresses that there is exactly one element in the domain. Of course, in the predicate logic without equality, $\exists x \forall y [x \equiv y]$ does have a model with $\mathbb{N}$ as domain, for instance, $M = \langle \mathbb{N}; \leq \rangle$ is a model of $\exists x \forall y [x \equiv y]$.

For applications of the Compactness theorem in mathematics see Exercises 4.44, 4.45 and 4.46.

Exercise 4.44. * The *elementary theory of fields*, designated by FL, has as non-logical symbols the constants 0, 1 and -1 and the binary function symbols $+$ and $\cdot$. The non-logical axioms of FL are:

$\forall x \forall y \forall z [(x+y)+z = x+(y+z)]$ $\quad \forall x [x+0 = x]$
$\forall x [x+(-1 \cdot x) = 0]$ $\quad \forall x \forall y [x+y = y+x]$
$\forall x \forall y \forall z [(x \cdot y) \cdot z = x \cdot (y \cdot z)]$ $\quad \forall x [x \cdot 1 = x]$
$\forall x [x \neq 0 \rightarrow \exists y [x \cdot y = 1]]$ $\quad \forall x \forall y [x \cdot y = y \cdot x]$
$\forall x \forall y \forall z [x \cdot (y+z) = (x \cdot y) + (x \cdot z)]$ $\quad 0 \neq 1$.

The models of FL are just the fields. Let A_n be the formula $1+1+\ldots+1 = 0$, where there are n occurrences of 1 on the left. By adding the non-logical axioms $\neg A_2, \neg A_3, \ldots, \neg A_{n-1}, A_n$, we get the elementary theory $FL(n)$ of fields of characteristic n ($n \geq 2$). To get the elementary theory $FL(0)$ of fields of characteristic 0, we add all of the $\neg A_n$ as non-logical axioms. Prove the following assertions:
1. If $FL(n)$ is consistent, then $n = 0$ or n is prime. Hint: use the mathematical fact that the characteristic of a field is 0 or a prime number.
2. If $FL(0) \models B$, then there is an n_0 such that for every $n \geq n_0$, $FL(n) \models B$. Hint: use the compactness theorem.
3. We cannot replace the infinite number of non-logical axioms we added to FL to get $FL(0)$ by a finite number. Hint: use 2.
4. There is no extension of FL whose models are just the finite fields.

Exercise 4.45. * **Theorem**: Let R be a partial ordering on V. Then there is a complete partial ordering R' on V such that $R \subseteq R'$. Prove this theorem for finite sets V using mathematical induction on the number of elements of V and, using the compactness theorem, prove this theorem also for infinite sets.

Exercise 4.46. * Let $B = \langle V, \sqcap, \sqcup, \neg, 0, 1 \rangle$ be a Boolean algebra. An element v in V is called an *atom* (of B) if $v \neq 0$ and for all $y \in V$, if $y \leq v$ (i.e., $y \sqcap v = y$), then $y = 0$ or $y = v$. B is *atomic* := for all x in V, if $x > 0$, then there is a y in V such that y is an atom of B and $y \leq x$. Let $AT_B := \{v \in V \mid v$ is an atom of $B\}$.
1. Prove that every atomic Boolean algebra B is isomorphic to a subalgebra of a set-algebra $\langle P(W), \cap, \cup, C_W, \emptyset, W \rangle$. Hint: consider $f : V \rightarrow P(AT_B)$, defined by $f(w) := \{v \in V \mid v$ is an atom of B and $v \leq w\}$.
2. Using the compactness theorem, prove that every Boolean algebra can be embedded in an atomic Boolean algebra. Hint: use the mathematical fact that the smallest Boolean algebra, generated by finitely many elements, is finite and hence atomic.

4.7 About the Relation of Logic with other Disciplines

4.7.1 Logic and Philosophy of Language

4.7.1.1 Definite Descriptions

Both Russell (1872-1970) and Wittgenstein (1889-1951), for different sets of reasons, rejected Frege's [5] distinction between sense (Sinn) and reference (Bedeutung) (see Chapter 7). Frege's analysis of a sentence like 'The king of France is bald' would be that this sentence lacks a truth value (reference, Bedeutung), because the subject expression has no reference, but that the lack of a truth value does not render the sentence meaningless, since this sentence does have a sense (Sinn). Russell, having already rejected Frege's theory of sense and reference, explains how sentences like this one can be meaningful, while there is nothing for the proposition, expressed by the sentence, to be about. Russell claims in [14] that the sentence in question appears to be in subject-predicate form, but is not really so. Its grammatical form is misleading as to its logical form. Russell's analysis of 'The king of France is bald' is as follows:

$\exists x$ [x is king of France $\wedge$ x is bald $\wedge$ $\forall y$ [y is king of France $\rightarrow y = x$]], or equivalently, but shorter $\exists x$ [x is bald $\wedge$ $\forall y$ [y is king of France $\rightleftarrows y = x$]].

And since there is no king of France, this sentence is false.

Russell analyzed 'The king of France is bald' as no simple subject-predicate statement but as a far more complicated one, in which two different quantified variables occur. In Russell's theory, the deep structure of such statements is very different from what their surface grammar suggests. Russell does not give an explicit definition enabling one to replace a definite description by an equivalent one wherever it appears, but a *contextual definition*, which enables one to replace sentences containing definite descriptions by equivalent sentences not containing definite descriptions. Russell used the following 'iota'-notation:

$\iota x A(x)$ the unique x with property A, and
$C(\iota x A(x))$ the unique x with property A has property C
as shorthand for $\exists x [A(x) \wedge C(x) \wedge \forall y [A(y) \rightarrow y = x]]$.

Where the condition C is complex, the iota notation is ambiguous. Russell's simple example is well known:

$\neg B(\iota x F(x))$ The king of France is not bald.

Here the ambiguity of the iota notation corresponds to an ambiguity in the English, between these two:
1. $\neg(B(\iota x F(x)))$, i.e., $\neg \exists x [F(x) \wedge B(x) \wedge \forall y [F(y) \rightarrow y = x]]$: there is no object x such that x is king of France and x is bald and x is the only king of France. And this happens to be true.
2. $(\neg B)(\iota x F(x))$, i.e., $\exists x [F(x) \wedge (\neg B)(x) \wedge \forall y [F(y) \rightarrow y = x]]$: there is some object x such that x is king of France and x is not bald and x is the only king of France. And this happens to be false; so we have $\neg((\neg B)(\iota x F(x)))$.

4.7 About the Relation of Logic with other Disciplines 243

Note that this latter expression is not equivalent to $B(\iota x F(x))$, i.e., $\exists x[F(x) \wedge B(x) \wedge \forall y[F(y) \rightarrow y = x]]$ (the king of France is bald): $\neg((\neg B)(\iota x F(x)))$ is true, while $B(\iota x F(x))$ is false. In Russell's jargon, the definite description $\iota x F(x)$ has *narrow scope* in version 1 and *wide scope* in version 2.

A less confusing notation for definite descriptions would result by treating them as a kind of quantifier: $(Ix)(F(x), B(x))$ instead of $B(\iota x F(x))$. Then the sentence in version 1, $\neg(B(\iota x F(x)))$, would be rendered by $\neg(Ix)(F(x), B(x))$, and the sentence in version 2, $(\neg B)(\iota x F(x))$, by $(Ix)(F(x), \neg B(x))$. While it was somewhat strange to have both, $\neg(B(\iota x F(x)))$ and $\neg((\neg B)(\iota x F(x)))$ in the new notation this would become $\neg(Ix)(F(x), B(x))$ and $\neg(Ix)(F(x), \neg B(x))$, which looks similar to $\neg \forall x[A(x)]$ and $\neg \forall x[\neg A(x)]$. which does not look like a contradiction at all.

4.7.1.2 Analytic-Synthetic

Immanuel Kant in his *Critique of Pure Reason* [8] makes a distinction between analytic and synthetic judgments. Kant calls a judgment *analytic* if its predicate is contained (though covertly) in the subject, in other words, the predicate adds nothing to the conception of the subject. Kant gives 'All bodies are extended (Alle Körper sind ausgedehnt)' as an example of an analytic judgment; I need not go beyond the conception of *body* in order to find extension connected with it. If a judgment is not analytic, Kant calls it *synthetic*. So, a synthetic judgment adds to our conception of the subject a predicate which was not contained in it, and which no analysis could ever have discovered therein. Kant mentions 'All bodies are heavy (Alle Körper sind schwer)' as an example of a synthetic judgment.

Kant makes in [8] also a distinction between *a priori* knowledge and *a posteriori* knowledge. A priori knowledge is knowledge existing altogether independent of experience, while a posteriori knowledge is empirical knowledge, which has its sources in experience.

Sometimes one speaks of *logically necessary* truths instead of analytic truths and of *logically contingent* truths instead of synthetic truths, to be distinguished from physically necessary truths (truths which physically could not be otherwise, true in all physically possible worlds). The distinction between necessary and contingent truth is a *metaphysical* one, while the distinction between a priori and a posteriori truth is an *epistemic* one. Although these – the metaphysical and the epistemological – are certainly different distinctions, it is controversial whether they coincide in extension, that is, whether all and only necessary truths are *a priori* and all and only contingent truths are *a posteriori*.

In [8] Kant stresses that mathematical judgments are both a priori and synthetic. 'Proper mathematical propositions are always judgments *a priori*, and not empirical, because they carry along with them the conception of necessity, which cannot be given by experience'. Why are mathematical judgments synthetic? Kant considers the proposition $7 + 5 = 12$ as an example. 'The conception of twelve is by no means obtained by merely cogitating the union of seven and five; and we may analyse our conception of such a possible sum as long as we will, still we shall never discover in it the notion of twelve'. We must go beyond this conception of $7+5$ and have recourse to an intuition which corresponds to counting using our fingers: first take seven fingers, next five fingers extra, and then by starting to count right from the beginning we arrive at the number twelve.

```
 7:    1 1 1 1 1 1 1
 5:                  1 1 1 1 1
7+5:   1 1 1 1 1 1 1 1 1 1 1 1
       1 2 3 4 5 6 7 8 9 10 11 12
```

'Arithmetical propositions are therefore always synthetic, of which we may become more clearly convinced by trying large numbers'. Geometrical propositions are also synthetic. As an example Kant gives 'A straight line between two points is the shortest', and explains 'For my conception of *straight* contains no notion of *quantity*, but is merely *qualitative*. The conception of the *shortest* is therefore wholly an addition, and by no analysis can it be extracted from our conception of a straight line'.

In more modern terminology, following roughly a 'Fregean' account of analyticity, one would define a proposition A to be *analytic* iff either

(i) A is an instance of a logically valid formula; e.g., 'No unmarried man is married' has the logical form $\neg \exists x [\neg P(x) \wedge P(x)]$, which is a valid formula, or

(ii) A is reducible to an instance of a logically valid formula by substitution of synonyms for synonyms; e.g., 'No bachelor is married'.

W.V. Quine [13] is sceptical of the analytic/synthetic distinction. Quine argues as follows. In order to define the notion of analyticity we used the notion of synonymy in clause (ii) above. However, if one tries to explain this latter notion, one has to take recourse to other notions which directly or indirectly will have to be explained in terms of analyticity.

4.7.2 Logic and Philosophy of Science

It is an old problem to draw the line between scientifically meaningful and meaningless statements. Consider the following quotation, taken from Hume's *Enquiry Concerning Human Understanding*.

> When we run over libraries, persuaded of these principles, what havoc must we make? If we take in our hand any volume; of divinity or school metaphysics, for instance; let us ask, *Does it contain any abstract reasoning concerning quantity of number*? No. *Does it contain any experimental reasoning concerning matter of fact and existence*? No. Commit it then to the flames: for it can contain nothing but sophistry and illusion". (David Hume, 1711-1776)

As we learn from A.J. Ayer [1], the quotation above is a good formulation of the positivist's position. In the 1930's the adjective *logical* was added, resulting in the term *Logical Positivism*, which underscored the successes of modern logic and the expectation that the new logical discoveries would be very fruitful for philosophy. This logical positivism was typical of the *Vienna Circle*, a group of philosophers (among them Moritz Schlick, Rudolf Carnap and Otto Neurath), scientists and mathematicians (among them Karl Menger and Kurt Gödel). According to A.J. Ayer [1],

4.7 About the Relation of Logic with other Disciplines

Einstein, Russell and Wittgenstein had a clear kinship to the Vienna Circle and had a great influence upon it.

In order to draw a sharp distinction between scientifically meaningful statements and scientifically meaningless statements the *verification principle* was formulated: only those statements are scientifically meaningful which can be verified in principle; in other words, the meaning of a proposition is its method of verification. However, a proposition like 'all ravens are black', which has as logical form $\forall x[R(x) \rightarrow B(x)]$, cannot be verified due to the universal quantifier, $\forall$; at the same time we consider this proposition to be (scientifically) meaningful.

On the other hand, the proposition 'all ravens are black' can be conclusively falsified, since its negation 'not all ravens are black', being of the form $\neg \forall x[R(x) \rightarrow B(x)]$, is logically equivalent to 'some raven is not black', which has the logical form $\exists x[R(x) \wedge \neg B(x)]$, and hence can be verified. For this reason the *falsification principle* was formulated: only those statements are scientifically meaningful which can be falsified in principle. This principle seems to be more in conformity with scientific practice: hypotheses are set up and rejected as soon as experimental results force us to do so. However, Otto Neurath himself soon realized that a slightly more complex proposition, like 'all men are mortal', which has the logical form $\forall x \exists y[R(x,y)]$ (for every person there is a moment of time such that ...), can neither be verified (due to the universal quantifier, $\forall$) nor falsified, since its negation 'not all men are mortal', being of the form $\neg \forall x \exists y[R(x,y)]$, is equivalent to 'some men are immortal', which has the logical form $\exists x \forall y[\neg R(x,y)]$, and hence – again due to the universal quantifier – cannot be verified.

Falsification of $\forall x \exists y[R(x,y)]$ is equivalent to verification of $\neg \forall x \exists y[R(x,y)]$, i.e., verification of $\exists x \forall y[\neg R(x,y)]$, which is not possible in principle due to the universal quantifier. At the same time we want to consider a statement like 'all men are mortal' as (scientifically) meaningful. Therefore, we have to give up not only the verification principle, but also the falsification principle. This was already realized by Otto Neurath during his stay (1938-39) in the Netherlands (oral communication by Johan J. de Iongh).

Summarizing: statements of the form $\forall x \exists y[R(x,y)]$ cannot be verified due to the universal quantifier $\forall$ and cannot be falsified due to the existential quantifier $\exists$.

Instead of the verification or falsification principle, a weaker criterion was formulated, called the *confirmation principle*: a statement is scientifically meaningful if and only if it is to some degree possible to confirm or disconfirm it. One way to confirm (increase the degree of credibility of) universal generalizations like 'all ravens are black' is to find things that are both ravens and black, and one way to disconfirm this proposition is to find things that are ravens but not black. The problem with this confirmation principle is that 'all ravens are black', $\forall x[R(x) \rightarrow B(x)]$, is logically equivalent to 'all non-black things are non-ravens', $\forall x[\neg B(x) \rightarrow \neg R(x)]$, and according to the confirmation principle, the latter proposition is confirmed by observations of non-black non-ravens; thus observations of brown shoes, white chalk, etc., would confirm the proposition 'all ravens are black'. Various attempts have been made to give the verification principle, in this weaker form, a precise expression, but the results have not been altogether satisfactory. For instance, a solution might be found by replacing the material implication $\to$ in $\forall x[R(x) \to B(x)]$ by the counterfactual implication $\square\!\!\to$ (see Chapter 6), for $\forall x[A(x) \square\!\!\to B(x)]$ is not logically equivalent to $\forall x[\neg B(x) \square\!\!\to \neg A(x)]$.

4.7.3 Logic and Artificial Intelligence; Prolog

As already mentioned in Chapter 2, the language of logic can be used to represent knowledge. And the language of predicate logic is a richer tool than the propositional language. Suppose, for instance, that someone knows the following:

(1) john is a parent of bob (2) john is a parent of claudia
(3) john is male   (4) bob is male
(5) claudia is female   (6) x is brother of y if x and y have a parent in common and x is male.

Introducing a predicate language containing the individual constants j, b and c, the unary predicate symbols 'male' and 'female' and the binary predicate symbols 'parent' and 'brother', (1) to (6) can be represented by the following formulas:

(1a) parent(j,b). (2a) parent(j,c).
(3a) male(j).   (4a) male(b).
(5a) female(c).   (6a) brother(x,y) $\leftarrow$ parent(z,x) $\wedge$ parent(z,y) $\wedge$ male(x).

Note that (1) to (6) cannot be adequately formulated in a propositional language. In the programming language Prolog, to be treated in Section 9.1, these formulas are rendered as follows:

(1b) parent(j,b). (2b) parent(j,c).
(3b) male(j).   (4b) male(b).
(5b) female(c).   (6b) brother(X,Y) :- parent(Z,X), parent(Z,Y), male(X).

(1b) to (6b) constitute what is called a *logic program*; (1b) to (5b) are called a *fact* and (6b) is called a *rule* in the logic program.

In (6a) and (6b) all variables are understood to be quantified universally. So (6a) is short for

$$\forall x \forall y \forall z \ [\text{parent}(z,x) \wedge \text{parent}(z,y) \wedge \text{male}(x) \to \text{brother}(x,y)]$$
which is equivalent to
$$\forall x \forall y \ [\ \exists z \ [\ \text{parent}(z,x) \wedge \text{parent}(z,y)\ ] \wedge \text{male}(x) \to \text{brother}(x,y)].$$

(1b), ..., (6b), taken together, can be considered to form a *knowledge base* from which new knowledge can be obtained by logical reasoning. The programming language Prolog has a built-in inference mechanism. When provided with the database consisting of (1b), ..., (6b), Prolog will give the following answers to the following questions, respectively:

?-brother(b,c). Answer: yes (corresponding with the fact that 'brother(b,c)' is a valid consequence of the given database).

?-brother(c,b). Answer: no (corresponding with the fact that 'brother(c,b)' is not a valid consequence of the given database).

?-brother(X,c). (For which X, brother(X,c) ?) Answer: X = bob.

4.7.4 Aristotle's Organon

While Stoic Logic was primarily concerned with propositions, Aristotle's logic (see [10, 11, 12] was mainly concerned with predicate logic, at least with a (small) part of it. After Aristotle's death in 322 B.C. his students grouped together a number of his treatises on reasoning. This collection was called the *Organon*, or instrument of science. Its two best known contributions to logic are described below.

The doctrine of the square of opposition This doctrine occurs in one of the earlier works of the Organon, the *Peri Hermeneias* (On Exposition), also known under its Latin name, *De Interpretatione*. Because there is a *practical* interest in the winning of arguments, it is important to know what statements are opposed to each other and in what ways. However, the only statements considered are of the form 'P is Q' and 'P is not Q' with a universal or existential quantification. The doctrine can be summarized in the following figure, called the *square of opposition*. Neither the square of opposition itself nor the vowels A, E, I and O, by which the four types have been distinguished since the Middle Ages, occur in Aristotle's work.

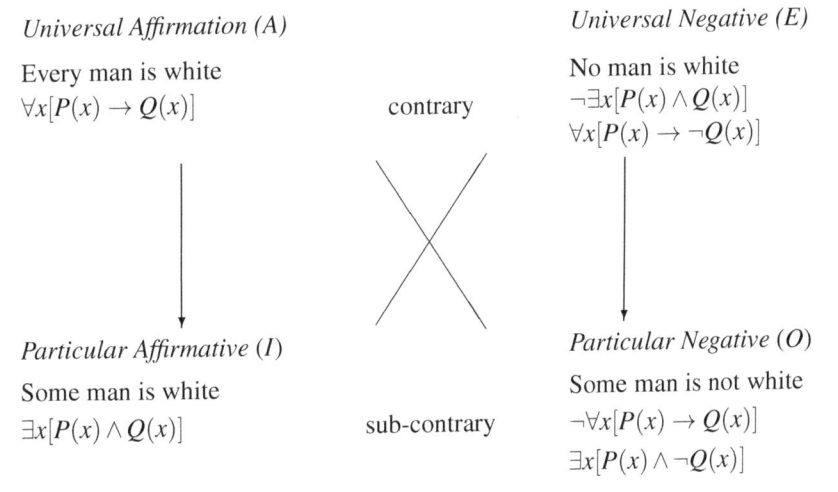

Universal Affirmation (A)

Every man is white
$\forall x[P(x) \rightarrow Q(x)]$

contrary

Universal Negative (E)

No man is white
$\neg \exists x[P(x) \wedge Q(x)]$
$\forall x[P(x) \rightarrow \neg Q(x)]$

Particular Affirmative (I)

Some man is white
$\exists x[P(x) \wedge Q(x)]$

sub-contrary

Particular Negative (O)

Some man is not white
$\neg \forall x[P(x) \rightarrow Q(x)]$
$\exists x[P(x) \wedge \neg Q(x)]$

Two statements are *contradictory* when they cannot both be true and cannot both be false. Two statements are *contrary* when they cannot both be true, but may both be false. Note that Aristotle here assumes implicitly that $\exists x[P(x)]$ is true.

Later logicians have said the two particular statements are *subaltern* to the universal statements under which they occur in the figure, and *sub-contrary* to each other. Again assuming that $\exists x[P(x)]$ is true, subcontraries cannot both be false, although they may both be true. Aristotle also assumes that each universal statement entails its subaltern, which again means that Aristotle is assuming implicitly the truth of $\exists x[P(x)]$.

Syllogisms In the *Prior Analytics*, one of the later works of the Organon, there is a *theoretical* interest in valid reasoning. However, Aristotle was only concerned with arguments of a particular form, called syllogisms. A *syllogism* is an argument consisting of two premisses and one conclusion, where the two premisses relate the terms of the conclusion to a third term, called the middle. For instance,

$$\models \underset{(A)}{\forall x[P(x) \to Q(x)]} \land \underset{(A)}{\forall x[Q(x) \to R(x)]} \to \underset{(A)}{\forall x[P(x) \to R(x)]}$$

corresponds to Aristotle's syllogism *b A r b A r A*. In this example, Q is the middle term, since it relates the terms P and R of the conclusion. Below is another example:

$$\models \underset{(E)}{\forall x[P(x) \to \neg Q(x)]} \land \underset{(I)}{\exists x[R(x) \land P(x)]} \to \underset{(O)}{\exists x[R(x) \land \neg Q(x)]}$$

corresponds to Aristotle's syllogism *f E r I O*. In this example, P is the middle term.

4.8 Solutions

Solution 4.1. a) (1) $\forall x[G(x) \to P(x)]$; (2) $\exists x[G(x) \land P(x)]$.
b) $\forall x[G(x) \land P(x)]$ says among other things that $\forall x[G(x)]$ (every individual is a girl), which is not implied by 'every girl is pretty'. 'Some girl is pretty', rendered by (2), implies that there is at least one girl ($\exists x[G(x)]$), who in addition is pretty. However, this is not implied by $\exists x[G(x) \to P(x)]$, which says that there is some individual x such that *if x is a girl, then x is pretty*.

Solution 4.2. 1. $\neg M(c,d)$; 2. $\forall x \forall y[M(x,y) \to M(y,x)]$; 3. $\exists x[M(x,d)]$; 4. $\exists x \forall y[\neg M(x,y)]$.

Solution 4.3. (1) $\forall x \exists y[A(x,y)]$; (2) $\exists y \forall x[A(x,y)]$.

Solution 4.4.
(1) $\forall x \forall y[L(x,y)]$: for all objects x and y, x is in the relation L to y.
(2) $\exists x \exists y[L(x,y)]$: there are objects x and y such that x is in the relation L to y.
(3) $\forall x \exists y[L(x,y)]$: for every object x there is at least one object y (possibly depending on x) such that x is in the relation L to y.
(4) $\exists y \forall x[L(x,y)]$: there is an object y such that for all x, x is in the relation L to y.
(5) $\forall y \exists x[L(x,y)]$: for every object y there is an object x (possibly depending on y) such that x is in the relation L to y. Interchanging x and y, this formula is equivalent to $\forall x \exists y[L(y,x)]$.
(6) $\exists x \forall y[L(x,y)]$: there is an object x such that for all objects y, x is in relation L to y. Interchanging x and y, this formula is equivalent to $\exists y \forall x[L(y,x)]$.

4.8 Solutions

Solution 4.5. 1. $\forall x[C(x) \to L(d,x)]$ 2. $\exists y[D(y) \land \forall x[C(x) \to L(y,x)]]$
3. $\exists x[C(x) \land L(d,x)]$ 4. $\forall y[D(y) \to \exists x[C(x) \land L(y,x)]]$
5. $\exists x[C(x) \land \forall y[D(y) \to L(y,x)]]$
6. $\neg \exists x[C(x) \land L(c,x)]$, or, equivalently, $\forall x[C(x) \to \neg L(c,x)]$
7. $\forall y[D(y) \to \neg \exists x[C(x) \land L(y,x)]]$ or $\forall y \forall x[D(y) \land C(x) \to \neg L(y,x)]$
8. $\forall y[D(y) \to \exists x[C(x) \land L(y,x)] \land \exists x[W(x) \land L(y,x)]]$
9. $\forall y[D(y) \land \exists x[C(x) \land L(y,x)] \to \exists x[W(x) \land L(y,x)]]$
10. $\forall y[D(y) \to \exists x[C(x) \land L(y,x)]] \to \forall y[D(y) \to \exists x[W(x) \land L(y,x)]]$

Solution 4.6. 1. i) $c_1 \equiv c_2$; ii) $\forall x[x \equiv c_1 \to x \equiv c_2]$.
2. i) '3 = 4' is false; ii) 'all numbers equal to 3 are equal to 4' is false.
3. i) 'Reagan was older than Nixon' is true; ii) 'all persons older than Reagan are older than Nixon' is true.

Solution 4.7.
1. $\exists x[P(x)]$, or equivalently, $\exists x \forall y[y \equiv x \to P(y)]$
2. $\exists x \forall y[P(y) \to y \equiv x]$
3. $\exists x \forall y[P(y) \rightleftarrows y \equiv x]$
4. $\exists x \exists y[\neg(x \equiv y) \land P(x) \land P(y)]$, or equivalently, $\exists x \exists y[x \not\equiv y \land \forall z[z \equiv x \lor z \equiv y \to P(z)]]$, or equivalently, $\exists x \exists y \forall z[x \not\equiv y \land (z \equiv x \lor z \equiv y \to P(z))]$.
5. $\exists x \exists y \forall z[x \not\equiv y \land (P(z) \to z \equiv x \lor z \equiv y)]$
6. $\exists x \exists y \forall z[x \not\equiv y \land (P(z) \rightleftarrows z \equiv x \lor z \equiv y)]$

Solution 4.8. a) $\forall x[C(x) \to A(x)]$; b) $\exists x[M(x) \land W(x)]$

Solution 4.9. a) $\forall x[M(x)]$.
b) $\neg \exists x[O(x)]$, or, equivalently, $\forall x[\neg O(x)]$.
c) $\exists x[B(x)] \to P$, or, equivalently, $\forall x[B(x) \to P]$.
d) $S(j) \to \forall x[S(x)]$ or, equivalently, $\forall x[S(j) \to S(x)]$.

Solution 4.10. 'For any natural number n, if $n \neq n$, then $n \neq n$' is a true sentence of the form $\forall x[P(x) \to Q(x)]$, but 'there is a natural number n such that $n \neq n$ and $n \neq n$' is a false sentence of the form $\exists x[P(x) \land Q(x)]$. If we assume that a domain is by definition non-empty, i.e., contains at least one element, then it follows from $\forall x[A(x)]$ that $\exists x[A(x)]$.

Solution 4.11.
a) 'If all natural numbers are even, then all natural numbers are odd' is a true ($0 \to 0 = 1$) sentence of the form $\forall x[P(x)] \to \forall x[Q(x)]$, but 'for each natural number n, if n is even, then n is odd' is a false sentence of the form $\forall x[P(x) \to Q(x)]$.
b) 'There is a natural number n such that *if n is even, then $n \neq n$*' is a true sentence of the form $\exists x[P(x) \to Q(x)]$, since, for instance, 'if 3 is even, then $3 \neq 3$' is true ($0 \to 0 = 1$). But 'if there is a natural number n such that n is even, then there is also a natural number n such that $n \neq n$' is a false ($1 \to 0 = 0$) sentence of the form $\exists x[P(x)] \to \exists x[Q(x)]$.

Solution 4.12. For $M = \langle \mathbb{N}, P^*, Q^*, R^* \rangle$ with P^*: is even, Q^*: is odd, R^*: is less than ($<$) we have:

$M \models P(a)[2]$ (2 is even) $\quad M \models \exists x[P(x)]$ (there is an even natural number)
$M \not\models P(a)[5]$ (not: 5 is even) $\quad M \not\models \forall x[P(x)]$ (not: all natural numbers are even)
$M \models Q(a)[5]$ (5 is odd) $\quad M \models \exists x[Q(x)]$ (there is an odd natural number)
$M \not\models Q(a)[2]$ (not: 2 is odd) $\quad M \not\models \forall x[Q(x)]$ (not: all natural numbers are odd)

$M \models \exists x[P(x)] \to \exists x[Q(x)]$ (true: $1 \to 1 = 1$)
$M \not\models \exists x[P(x)] \to \forall x[Q(x)]$ (false: $1 \to 0 = 0$)
$M \models \forall x[P(x)] \to \exists x[Q(x)]$ (true: $0 \to 1 = 1$)
$M \models \forall x[P(x)] \to \forall x[Q(x)]$ (true: $0 \to 0 = 1$)
$M \models \forall x \exists y[R(x,y)]$ (true: for each natural number x there is a natural number y such that $x < y$). But $M \not\models \exists y \forall x[R(x,y)]$ (false: there is a natural number y such that for each natural number x, $x < y$).
$M \models \exists x[P(x) \to Q(x)]$ (true: there is a natural number x such that *if* x is even, *then* x is odd; since 'if 3 is even, then 3 is odd' has truth value 1).
$M \not\models \forall x[P(x) \to Q(x)]$ (false: for all natural numbers x, *if* x is even, *then* x is odd; since 'if 2 is even, then 2 is odd' has truth value 0).

Solution 4.13. 1. $\exists x[P(x)] \to \forall x[P(x)]$ is satisfiable: $M = \langle \mathbb{N}; x = x \rangle$ is a model; but the formula is not valid: $M = \langle \mathbb{N}; \text{is odd} \rangle$ is a countermodel.
2. $\exists x[P(x)] \to \exists x[\neg P(x)]$ is satisfiable: $M = \langle \mathbb{N}; \text{is odd} \rangle$ is a model; but the formula is not valid: $M = \langle \mathbb{N}; x = x \rangle$ is a countermodel.
3. $\exists x[P(x)] \land \forall x[\neg P(x)]$ is not satisfiable, since by Theorem 4.7 $\forall x[\neg P(x)]$ means the same as $\neg \exists x[P(x)]$.
4. $\forall x[P(x)] \land \neg \exists x[P(x)]$ is not satisfiable, since by Theorem 4.7 $\neg \exists x[P(x)]$ means the same as $\forall x[\neg P(x)]$.
5. $\neg \forall x[P(x)] \to \forall x[\neg P(x)]$ is satisfiable: $M = \langle \mathbb{N}; \text{is negative} \rangle$ is a model; but the formula is not valid: $M = \langle \mathbb{N}; \text{is odd} \rangle$ is a countermodel.
6. $\forall x[\neg P(x)] \to \neg \forall x[P(x)]$ is valid and hence satisfiable.
7. $\forall x \exists y[R(x,y)] \land \exists x \forall y[\neg R(x,y)]$ is not satisfiable, since $\exists x \forall y[\neg R(x,y)]$ has the same meaning as $\neg \forall x \exists y[R(x,y)]$ (see Theorem 4.15).
8. $\forall x \exists y[R(x,y)] \to \exists y \forall x[R(x,y)]$ is satisfiable: $M = \langle \mathbb{N}; x \geq y \rangle$ is a model; but the formula is not valid: $M = \langle \mathbb{N}; x \leq y \rangle$ is a countermodel.
9. $\exists x[P(x)] \land \exists x[Q(x)] \to \exists x[P(x) \land Q(x)]$ is satisfiable: $M = \langle \mathbb{N}; \text{is even}, x = x \rangle$ is a model; but the formula is not valid: $M = \langle \mathbb{N}; \text{is even}, \text{is odd} \rangle$ is a countermodel.
10. $\forall x[P(x) \lor Q(x)] \to \exists x[P(x)] \lor \forall x[Q(x)]$ is valid and hence satisfiable: in case that $\neg \exists x[P(x)]$, it follows from $\forall x[P(x) \lor Q(x)]$ that $\forall x[Q(x)]$.

Solution 4.14. 1. Not $\forall x[P(x) \to S(x)], \exists x[S(x) \land I(x)] \models \exists x[P(x) \land I(x)]$. Counterexample: $M = \langle \mathbb{N}; P^*, S^*, I^* \rangle$ with $P^*(x)$: x is even, $S^*(x)$: $x = x$ and $I^*(x)$: x is odd.
2. Not $\neg \exists x[P(x) \land I(x)], \forall x[I(x) \to V(x)] \models \neg \exists x[P(x) \land V(x)]$. Counterexample: $M = \langle \mathbb{N}; P^*, V^*, I^* \rangle$ with $P^*(x)$: x is even, $V^*(x)$: x is even, and $I^*(x)$: $x \neq x$.
3. $\forall x[F(x) \to B(x)], \neg \exists x[M(x) \land B(x)] \models \forall x[M(x) \to \neg F(x)]$. Proof: Suppose the two premisses are true under an interpretation M. To show: $M \models \forall x[M(x) \to \neg F(x)]$. So, suppose $M \models M(a)[d]$ for an arbitrary element d in the domain of M. Then by the second premiss, $M \models \neg B(a)[d]$ and hence, by the first premiss, $M \models \neg F(a)[d]$.
4. Not $\exists x[M(x) \land \neg S(x)], \forall x[C(x) \to S(x)] \models \exists x[C(x) \land \neg M(x)]$. Counterexample:

$M = \langle \mathbb{N}; M^*, S^*, C^* \rangle$ with $M^*(x)$: x is even, $S^*(x)$: x is odd, and $C^*(x)$: $x \neq x$.
5. $\exists x[P(x) \wedge S(x)], \neg\exists x[S(x) \wedge \neg C(x)] \models \exists x[P(x) \wedge C(x)]$. Proof: the second premiss is equivalent to $\forall x[S(x) \rightarrow C(x)]$ and the first premiss says that some plumber is smart; so, by the second premiss, this plumber will also be careful.

Solution 4.15. 1. $\neg\exists x[A(x) \wedge I(x)], \forall x[C(x) \rightarrow A(x)] \not\models \exists x[C(x) \wedge \neg I(x)]$.
Counterexample: $M = \langle \mathbb{N}; C^*(x) : x \neq x, A^*(x) : x \text{ is even}, I^*(x) : x \text{ is odd} \rangle$.
2. $\exists x[S(x)] \rightarrow \exists x[P(x) \wedge S(x)], P(c) \wedge \neg S(c) \not\models \neg\exists x[S(x)]$.
Counterexample: $M = \langle \mathbb{N}; S^*(x) : x \text{ is even}, P^*(x) : x = x; c^* : 3 \rangle$.
3. $\forall x[M(x) \wedge \exists y[S(y)] \rightarrow S(x)], M(c) \wedge \neg S(c) \models \neg\exists x[S(x)]$.
Proof: Let $M = \langle D; M^*, S^*; c^* \rangle$ be a model of the two premisses. Then it is in particular a model of $M(c) \wedge \exists y[S(y)] \rightarrow S(c)$ (1) and of $M(c) \wedge \neg S(c)$ (2). Now suppose M were a model of $\exists x[S(x)]$. Then by (2) it would be a model of $M(c) \wedge \exists y[S(y)]$. So, by (1), M would be a model of $S(c)$, contradicting (2). Hence, M is a model of $\neg\exists x[S(x)]$.
4. $\exists x[H(x) \wedge F(x)], \neg\exists x[\neg H(x) \wedge S(x)] \not\models \exists x[F(x) \wedge \neg S(x)]$.
Counterexample: $M = \langle \mathbb{N}; H^*(x) : x = x, F^*(x) : x \text{ is even}, S^*(x) : x = x \rangle$.
5. $\exists x[S_1(x) \wedge S_2(x)], \neg\exists x[S_1(x) \wedge U(x)] \not\models \exists x[S_1(x) \wedge U(x) \wedge \neg S_2(x)]$.
Counterexample: $M = \langle \mathbb{N}; S_1^*(x) : x \text{ is even}, S_2^*(x) : x \text{ is even}, U^*(x) : x \neq x \rangle$.

Solution 4.16. $\forall x[P(x) \rightarrow Q(x)] \not\models \exists x[P(x) \wedge Q(x)]$. $M = \langle \mathbb{N}; P^*, Q^* \rangle$, with $P^*(x) = Q^*(x) := x \neq x$, is a counterexample: $M \models \forall x[P(x) \rightarrow Q(x)]$ (for every natural number x, if $x \neq x$, then $x \neq x$), but $M \not\models \exists x[P(x) \wedge Q(x)]$ (it is not the case that there is a natural number x such that $x \neq x$).

Solution 4.17. i) (a) $\forall x \exists y[R(y,x)]$; (b) $\neg\exists x \forall y[R(x,y)]$.
ii) $\forall x \exists y[R(y,x)] \not\models \neg\exists x \forall y[R(x,y)]$. Counterexample: Let $M = \langle \mathbb{N}; \leq \rangle$. Then $M \models \forall x \exists y[R(y,x)]$ (for every natural number n there is a natural number m such that $m \leq n$). But $M \not\models \neg\exists x \forall y[R(x,y)]$, since $M \models \exists x \forall y[R(x,y)]$ (there is a natural number n, namely 0, such that for all natural numbers m, $n \leq m$.
iii) Concluding (b) from (a) we use tacitly that if $m > n$, then not $n > m$. Suppose (a) and not (b). So, there is a natural number n greater than all natural numbers. From (a) it follows that there is a natural number m such that $m > n$ (1). However, by the choice of n, $n > m$ (2). However, (1) and (2) contradict $m > n \rightarrow \neg(n > m)$.
iv) To show: $\forall x \exists y[R(y,x)], \forall x \forall y[R(y,x) \rightarrow \neg R(x,y)] \models \neg\exists x \forall y[R(x,y)]$ So, suppose $M \models \forall x \exists y[R(y,x)]$, $M \models \forall x \forall y[R(y,x) \rightarrow \neg R(x,y)]$ and $M \models \exists x \forall y[R(x,y)]$. Then for some d_1 in the domain of M, $M \models \forall y[R(a_1,y)][d_1]$. Since $M \models \forall x \exists y[R(y,x)]$ it follows that $M \models \exists y[R(y,a_1)][d_1]$ and therefore for some d_2 in the domain of M, $M \models R(a_2,a_1)[d_2,d_1]$ (1). From $M \models \forall y[R(a_1,y)][d_1]$ it follows that also $M \models R(a_1,a_2)[d_1,d_2]$ (2). But (1) and (2) contradict that $M \models \forall x \forall y[R(y,x) \rightarrow \neg R(x,y)]$.

Solution 4.18. $\models \neg\exists y \forall x[S(y,x) \rightleftarrows \neg S(x,x)]$.
Proof: Suppose that $M \models \exists y \forall x[S(y,x) \rightleftarrows \neg S(x,x)]$. Then there is some element d in the domain D of M such that $M \models \forall x[S(a,x) \rightleftarrows \neg S(x,x)][d]$, where a is a new free variable. Hence, in particular, $M \models S(a,a) \rightleftarrows \neg S(a,a)[d]$, i.e., $M \models S(a,a)[d]$ iff $M \models \neg S(a,a)[d]$. Contradiction. So, for every interpretation M, $M \models \neg\exists y \forall x[S(y,x) \rightleftarrows \neg S(x,x)]$.

Solution 4.19. Let $M = \langle D; P^* \rangle$ be an interpretation.
a) $M \models \forall x[P(x) \to P(x)]$; hence, $M \models \forall x \exists y[P(x) \to P(y)]$.
b) $M \models \forall y[P(y) \to P(y)]$; hence, $M \models \forall y \exists x[P(x) \to P(y)]$.
c) To show: for any interpretation M, $M \models \exists x \forall y[P(x) \to P(y)]$. Case 1: $M \models \exists x[\neg P(x)]$, i.e., there is a d in the domain of M such that $M \models \neg P(a)\,[d]$. But then $M \models P(a) \to P(b)\,[d,d']$ for any valuation d' of the free variable b ($0 \to 0/1 = 1$). Therefore, $M \models \exists x \forall y[P(x) \to P(y)]$. Case 2: $M \models \neg \exists x[\neg P(x)]$, i.e., $M \models \forall x[P(x)]$, i.e., all objects in the domain of M have the property P^*. Then $M \models \forall x \forall y[P(x) \to P(y)]$ ($1 \to 1 = 1$). Hence, in particular, $M \models \exists x \forall y[P(x) \to P(y)]$.
d) To show: for any interpretation M, $M \models \exists y \forall x[P(x) \to P(y)]$. Case 1: $M \models \exists y[P(y)]$, i.e., there is some d in the domain of M such that $M \models P(b)[d]$. But then $M \models \forall x[P(x) \to P(b)]\,[d]$ ($0/1 \to 1 = 1$) and hence $M \models \exists y \forall x[P(x) \to P(y)]$. Case 2: $M \models \neg \exists y[P(y)]$, i.e., $M \models \forall y[\neg P(y)]$, that is, no element in the domain of M has the property P^*. But then $M \models \forall y \forall x[P(x) \to P(y)]$ ($0 \to 0 = 1$) and hence, in particular, $M \models \exists y \forall x[P(x) \to P(y)]$.

Solution 4.20. a) The formula $\forall x \exists y[R(x,y)] \to \exists x \forall y[R(x,y)]$ contains a transition from $\exists y$ to $\forall y$ and hence cannot be valid. Let $M = \langle \mathbb{N}; R^* \rangle$ with $R^*(d_1,d_2) := d_2$ is even (and $d_1 = d_1$). Then $M \models \forall x \exists y[R(x,y)]$ (there is some natural number which is even), but $M \not\models \exists x \forall y[R(x,y)]$ (it is not the case that all natural numbers are even).
b) The formula $\exists x \forall y[R(x,y)] \to \forall x \exists y[R(x,y)]$ contains a transition from $\exists x$ to $\forall x$ and hence cannot be valid. $M = \langle \mathbb{N}; R^* \rangle$, with $R^*(d_1,d_2) := d_1$ is even (and $d_2 = d_2$), is a counterexample.
c) Let $M = \langle \mathbb{N}; < \rangle$. Then $M \models \forall x \exists y[R(x,y)]$ (for every natural number x there is a greater one y), but $M \not\models \forall x \exists y[R(y,x)]$ (it is not the case that for every natural number x there is a smaller one y; there is no natural number less than 0).
d) The formula $\exists x \forall y[R(x,y)] \to \exists y \forall x[R(x,y)]$ contains again a transition from $\exists x$ to $\forall x$ and hence cannot be valid. See the counterexample in b).
e) and f) The right and left part of $\rightleftarrows$ express the same proposition; only the variables x and y have been interchanged.

Solution 4.21. 1. Let $M = \langle \mathbb{N};$ is even, $0 = 1 \rangle$. Then $M \models \forall x[P(x)] \to Q$: the proposition 'if all natural numbers are even, then $0 = 1$' has truth value $0 \to 0 = 1$. But $M \not\models \forall x[P(x) \to Q]$: it is not the case that for every natural number n, if n is even, then $0 = 1$ (for instance, 'if 2 is even, then $0 = 1$' has truth value $1 \to 0 = 0$). So, M is a counterexample.
2. Of course, $\forall x[P(x) \to Q] \models \exists x[P(x) \to Q]$. And $\exists x[P(x) \to Q] \not\models \forall x[P(x)] \to Q$. Hence, $\forall x[P(x) \to Q] \models \forall x[P(x)] \to Q$.
3. $\exists x[P(x)] \to Q \models \exists x[P(x) \to Q]$ follows from $\exists x[P(x)] \to Q \models \forall x[P(x) \to Q]$.
4. Let $M = \langle \mathbb{N};$ is even, $0 = 1 \rangle$. Then $M \models \exists x[P(x) \to Q]$: the proposition 'there is a natural number n such that if n is even, then $0 = 1$' has truth value 1 (for instance, 'if 3 is even, then $0 = 1$' has truth value $0 \to 0 = 1$). But $M \not\models \exists x[P(x)] \to Q$: the proposition 'if there is an even natural number, then $0 = 1$' has truth value $1 \to 0 = 0$.

Solution 4.22. 1. Suppose $M \models \forall x[P(x)] \to \exists x[Q(x)]$ and $M \not\models \exists x[P(x) \to Q(x)]$. Then $M \models \neg \exists x[P(x) \to Q(x)]$, i.e., $M \models \forall x[\neg(P(x) \to Q(x))]$. So, $M \models \forall x[P(x) \wedge$

4.8 Solutions 253

$\neg Q(x)]$; therefore, $M \models \forall x[P(x)]$ and $M \models \forall x[\neg Q(x)]$. Contradiction with $M \models \forall x[P(x)] \to \exists x[Q(x)]$. So, $\forall x[P(x)] \to \exists x[Q(x)]) \models \exists x[P(x) \to Q(x)]$.

Conversely, suppose $M \models \exists x[P(x) \to Q(x)]$ and $M \models \forall x[P(x)]$. Then for some element d in the domain of M, $M \models P(a) \to Q(a)$ $[d]$, where a does not occur in $\exists x[P(x) \to Q(x)]$. Since $M \models \forall x[P(x)]$, $M \models P(a)[d]$. So, $M \models Q(a)[d]$. Therefore, $M \models \exists x[Q(x)]$. This shows that also $\exists x[P(x) \to Q(x)] \models \forall x[P(x)] \to \exists x[Q(x)]$.

2. $\exists x[P(x)] \to \forall x[Q(x)]) \models \forall x[P(x) \to Q(x)]$, but $M = \langle \mathbb{N};$ is even, is even$\rangle$ is a counterexample to the converse formula, $\forall x[P(x) \to Q(x)] \to (\exists x[P(x)] \to \forall x[Q(x)])$.

Solution 4.23. 1. $\forall x[P(x) \to Q(x)] \models \exists x[P(x)] \to \exists x[Q(x)]$.
For suppose $M \models \forall x[P(x) \to Q(x)]$ (i) and $M \models \exists x[P(x)]$. Then for some element d in the domain of M, $M \models P(a)[d]$, where a is a new variable. From (i) it follows that $M \models Q(a)[d]$. Hence, $M \models \exists x[Q(x)]$.

However, conversely, $M = \langle \mathbb{N};$ is even, is odd$\rangle$ makes $\exists x[P(x)] \to \exists x[Q(x)]$ true: the proposition 'if there is an even natural number, then there is an odd natural number' has truth value $1 \to 1 = 1$. But $M \not\models \forall x[P(x) \to Q(x)]$: the proposition 'every even natural number is odd' has truth value 0.

2. $\exists x[P(x) \to Q(x)] \not\models \forall x[P(x)] \to \forall x[Q(x)]$. For $M = \langle \mathbb{N}; x = x,$ is even$\rangle$ is a counterexample. $M \models \exists x[P(x) \to Q(x)]$: the proposition 'there is some natural number n such that if $n = n$, then n is even' has truth value 1; for instance, 'if $2 = 2$, then 2 is even' has truth value $1 \to 1 = 1$. Also $M \models \forall x[P(x)]$. But $M \not\models \forall x[Q(x)]$.

However, conversely, we do have $\forall x[P(x)] \to \forall x[Q(x)] \models \exists x[P(x) \to Q(x)]$. For suppose $M \models \forall x[P(x)] \to \forall x[Q(x)]$ (i) and $M \not\models \exists x[P(x) \to Q(x)]$. Then $M \models \neg \exists x[P(x) \to Q(x)]$, i.e., $M \models \forall x[\neg(P(x) \to Q(x))]$. So, $M \models \forall x[P(x) \land \neg Q(x)]$. Therefore, $M \models \forall x[P(x)]$ and $M \models \forall x[\neg Q(x)]$. Contradiction with (i).

Solution 4.24. $\forall x \exists y[P(x) \to Q(y)] \models \exists y \forall x[P(x) \to Q(y)]$. For suppose $M \models \forall x \exists y[P(x) \to Q(y)]$ (i) and $M \models \neg \exists y \forall x[P(x) \to Q(y)]$. Then $M \models \forall y \exists x[P(x) \land \neg Q(y)]$. So, $M \models \exists x[P(x)]$ and $M \models \forall y[\neg Q(y)]$. Hence, for some element d in the domain of M, $M \models P(a)[d]$, where a is a new free variable. From (i) it follows that $M \models \exists y[P(a) \to Q(y)]$ $[d]$. So, $M \models \exists y[Q(y)]$. Contradiction with $M \models \forall y[\neg Q(y)]$.

Solution 4.25. Let W (Wang) be the formula in question and let M be a model.
Case 1: $M \models \exists x \exists y[\neg P(x,y)]$. Then $M \models W$.
Case 2: $M \not\models \exists x \exists y[\neg P(x,y)]$. Then $M \models \forall x \forall y[P(x,y)]$.
Subcase 2a): $M \models \exists x \exists y[\neg Q(x,y)]$. Then $M \models W$.
Subcase 2b): $M \not\models \exists x \exists y[\neg Q(x,y)]$. Then $M \models \forall x \forall y[Q(x,y)]$. Consequently, $M \models W$.

Solution 4.26. (a)(1) (i) To show: $\forall x[B(x)] \to C \models \exists x[B(x) \to C]$. Suppose $M \models \forall x[B(x)] \to C$ $[v]$ and $M \not\models \exists x[B(x) \to C]$ $[v]$. Then $M \models \neg \exists x[B(x) \to C]$ $[v]$, i.e., $M \models \forall x[\neg(B(x) \to C)]$ $[v]$. So, $M \models \forall x[B(x) \land \neg C]$ $[v]$; therefore, $M \models \forall x[B(x)]$ $[v]$ and $M \models \neg C$ $[v]$. Contradiction with $M \models \forall x[B(x)] \to C$ $[v]$.
Next we show: $\exists x[B(x) \to C] \models \forall x[B(x)] \to C$. So, suppose $M \models \exists x[B(x) \to C]$ $[v]$. Then for some element d in the domain D of M, $M \models B(a) \to C$ $[d/v]$, where a is new. Now suppose that $M \models \forall x[B(x)]$ $[v]$. Then $M \models B(a)$ $[d/v]$. So, $M \models C$ $[v]$.
(a)(1) (ii) To show: $\exists x[B(x)] \to C \models \forall x[B(x) \to C]$. Suppose $M \models \exists x[B(x)] \to C$ $[v]$

and $M \not\models \forall x[B(x) \to C]\ [v]$. Then $M \models \neg\forall x[B(x) \to C]\ [v]$, i.e., $M \models \exists x[\neg(B(x) \to C)]\ [v]$. So, $M \models \exists x[B(x) \wedge \neg C]\ [v]$; therefore, $M \models \exists x[B(x)]\ [v]$ and $M \models \neg C\ [v]$. Contradiction with $M \models \exists x[B(x)] \to C\ [v]$.

Next we show: $\forall x[B(x) \to C] \models \exists x[B(x)] \to C$. So, suppose $M \models \forall x[B(x) \to C]\ [v]$ and $M \models \exists x[B(x)]\ [v]$. Then for some element d in the domain D of M, $M \models B(a)\ [d/v]$, where a is new. Since $M \models \forall x[B(x) \to C]\ [v]$, it follows that $M \models B(a) \to C\ [d/v]$. So, $M \models C\ [v]$.

The validity of the other formulas is shown similarly.

Solution 4.27.
$$\models \forall x[P(x)] \to \exists x[Q(x)] \rightleftarrows \forall x[P(x)] \to \exists y[Q(y)]$$
$$\rightleftarrows \exists x[P(x) \to \exists y[Q(y)]]$$
$$\rightleftarrows \exists x \exists y[P(x) \to Q(y)].$$
$$\models \exists x[P(x)] \to \forall x[Q(x)] \rightleftarrows \exists x[P(x)] \to \forall y[Q(y)]$$
$$\rightleftarrows \forall x[P(x) \to \forall y[Q(y)]]$$
$$\rightleftarrows \forall x \forall y[P(x) \to Q(y)].$$
$$\models \exists x[P(x,a)] \to \exists x[Q(x) \vee \neg \exists y[R(y)]] \rightleftarrows \exists x[P(x,a)] \to \exists z[Q(z) \vee \forall y[\neg R(y)]]$$
$$\rightleftarrows \exists x[P(x,a)] \to \exists z \forall y[Q(z) \vee \neg R(y)]$$
$$\rightleftarrows \forall x[P(x,a) \to \exists z \forall y[Q(z) \vee \neg R(y)]]$$
$$\rightleftarrows \forall x \exists z[P(x,a) \to \forall y[Q(z) \vee \neg R(y)]]$$
$$\rightleftarrows \forall x \exists z \forall y[P(x,a) \to Q(z) \vee \neg R(y)].$$

Solution 4.28.
$$\models \exists x[P(x)] \to \exists x[Q(x)] \rightleftarrows \exists x[P(x)] \to \exists y[Q(y)]$$
$$\text{Theorem 4.18 (a)(1)} \rightleftarrows \forall x[P(x) \to \exists y[Q(y)]]$$
$$\text{Theorem 4.18 (a)(2)} \rightleftarrows \forall x \exists y[P(x) \to Q(y)].$$
$$\models \exists x[P(x)] \to \exists x[Q(x)] \rightleftarrows \exists x[P(x)] \to \exists y[Q(y)]$$
$$\text{Theorem 4.18 (a)(2)} \rightleftarrows \exists y[\exists x[P(x)] \to Q(y)]$$
$$\text{Theorem 4.18 (a)(1)} \rightleftarrows \exists y \forall x[P(x) \to Q(y)].$$

Solution 4.29. No, since the $\forall$-rule is applied with respect to the free variable a occurring in the premiss $A(a)$.

Solution 4.30. No, since the $\exists$-rule is applied to a formula of the form $A(a) \to C$ where C does contain a.

Solution 4.31. Yes.

Solution 4.32.

i)
$$\frac{\text{premiss}}{\exists x[A(x)]} \quad \frac{\exists\text{-schema}}{A(a) \to \exists z[A(z)]} \exists$$
$$\frac{\exists x[A(x)] \to \exists z[A(z)]}{\exists z[A(z)]} MP$$

ii)
$$\frac{\text{premiss}}{\forall x[A(x)]} \quad \frac{\forall\text{-schema}}{\forall x[A(x)] \to A(a)} \forall$$
$$\frac{\forall x[A(x)] \to \forall z[A(z)]}{\forall z[A(z)]} MP$$

4.8 Solutions

Solution 4.33.

$$\frac{\frac{\overbrace{\forall x[A \to B(x)]}^{\text{premiss}} \quad [A]^1}{A \to B(a)}}{\frac{B(a)}{\frac{\forall x[B(x)]}{A \to \forall x[B(x)]} \ (1)}}
\qquad
\frac{\frac{\overbrace{A \to \forall x[B(x)]}^{\text{premiss}} \quad [A]^1}{\forall x[B(x)]}}{\frac{\frac{B(a)}{A \to B(a)} \ (1)}{\forall x[A \to B(x)]}}$$

Solution 4.34. 1. A tableau-proof of $(\exists x[P(x)] \to \exists x[Q(x)]) \to \exists x[P(x) \to Q(x)]$:

$$F\ (\exists x[P(x)] \to \exists x[Q(x)]) \to \exists x[P(x) \to Q(x)]$$
$$T\ \exists x[P(x)] \to \exists x[Q(x)],\ F\ \exists x[P(x) \to Q(x)]$$
$$F\ \exists x[P(x)],\ F\ \exists x[P(x) \to Q(x)] \ |\ T\ \exists x[Q(x)],\ F\ \exists x[P(x) \to Q(x)]$$
$$FP(a),\ F\ \exists x[P(x) \to Q(x)] \ \ \ \ \ \ \ \ |\ TQ(a),\ F\ \exists x[P(x) \to Q(x)]$$
$$FP(a),\ F\ P(a) \to Q(a) \ \ \ \ \ \ \ \ \ \ \ \ \ \ \ |\ TQ(a),\ F\ P(a) \to Q(a)$$
$$FP(a), TP(a), FQ(a) \ \ \ \ \ \ \ \ \ \ \ \ \ \ \ \ \ |\ TQ(a), TP(a), FQ(a)$$

2. The schema below shows an open, not completed, branch in a tableau with initial branch $\{F\ \exists x[P(x) \to Q(x)] \to (\exists x[P(x)] \to \exists x[Q(x)])\}$:

$$F\ \exists x[P(x) \to Q(x)] \to (\exists x[P(x)] \to \exists x[Q(x)])$$
$$T\ \exists x[P(x) \to Q(x)],\ F\ \exists x[P(x)] \to \exists x[Q(x)]$$
$$T\ \exists x[P(x) \to Q(x)],\ T\ \exists x[P(x)],\ F\ \exists x[Q(x)]$$
$$T\exists x[\to],\ T\exists x,\ F\exists x,\ T\ P(a_1) \to Q(a_1)$$
$$T\exists x[\to],\ T\exists x,\ F\exists x,\ FP(a_1)$$
$$T\exists x[\to],\ T\exists x,\ F\exists x,\ FP(a_1), TP(a_2)$$
$$T\exists x[\to],\ T\exists x,\ F\exists x,\ FP(a_1), TP(a_2), FQ(a_1)$$
$$T\exists x[\to],\ T\exists x,\ F\exists x,\ FP(a_1), TP(a_2), FQ(a_1), FQ(a_2)$$
$$T\ P(a_3) \to Q(a_3),\ T\exists x,\ F\exists x,\ FP(a_1), TP(a_2), FQ(a_1), FQ(a_2)$$
$$FP(a_3),\ T\exists x,\ F\exists x,\ FP(a_1), TP(a_2), FQ(a_1), FQ(a_2)$$
$$FP(a_3),\ TP(a_4),\ F\exists x,\ FP(a_1), TP(a_2), FQ(a_1), FQ(a_2)$$
and so on,

where we have used some obvious abbreviations.

From this open branch we can read off a counterexample to the formula in question having as domain the set $\mathbb{N}$ of all natural numbers: the even natural numbers have the property P^*, corresponding to the occurrences of $TP(a_2)$, $TP(a_4)$, ..., the odd natural numbers do not have the property P^*, corresponding to the occurrences of $FP(a_1)$, $FP(a_3)$,... and all natural numbers have the property not-Q, corresponding to the occurrences of $FQ(a_1)$, $FQ(a_2)$, $FQ(a_3)$,...; take for P^* the predicate 'is even' and let $Q(a)$ be interpreted as $a \neq a$. Under this interpretation there results a true proposition from $\exists x[P(x) \to Q(x)]$, but a false proposition from $\exists x[P(x)] \to \exists x[Q(x)]$.

P.S. After application of rule $T\exists$ to $T\ \exists x[P(x) \to Q(x)]$ and to $T\ \exists x[P(x)]$ respectively, one may delete the occurrences of these signed formulas. If one does so, one finds a counterexample with a finite domain.

3. The schema below is a tableau-proof of $\exists x \forall y[P(x) \to P(y)]$:

$$F \; \exists x \forall y[P(x) \to P(y)]$$
$$F \; \exists x \forall y[P(x) \to P(y)], \; F \; \forall y[P(a_1) \to P(y)]$$
$$F \; \exists x \forall y[P(x) \to P(y)], \; F \; \forall y[P(a_1) \to P(y)], \; F \; P(a_1) \to P(a_2)$$
$$F \; \exists x \forall y, \; F \; \forall y[P(a_2) \to P(y)], \; F \; \forall y[P(a_1) \to P(y)], \; F \; P(a_1) \to P(a_2)$$
$$F \; \exists x \forall y, \; F \; P(a_2) \to P(a_3), \; F \; \forall y[P(a_1) \to P(y)], \; F \; P(a_1) \to P(a_2)$$
$$F \; \exists x \forall y, \; TP(a_2), \; FP(a_3), \; F \; \forall y[P(a_1) \to P(y)], \; TP(a_1), \; FP(a_2)$$

where we have used some obvious abbreviations.

Solution 4.35. a) trivial. b) $\{P(a_1), \neg P(a_2), \neg P(a_3), Q(a_2), \neg Q(a_3)\}$ is simultaneously satisfiable in D iff D contains at least three elements.

c) Suppose $M = \langle D; P^* \rangle$ is a model of the three formulas in question. Let d_1 be an element in D. Since $M \models \forall x \exists y[P(x,y)]$, there must be some element d_2 in D such that $M \models P(a_1,a_2)[d_1,d_2]$. Since $M \models \forall x[\neg P(x,x)]$, it follows that $d_2 \neq d_1$. From $M \models \forall x \exists y[P(x,y)]$ we conclude that $M \models P(a_2,a_3)[d_2,d_3]$ for some d_3 in D. Again $d_3 \neq d_2$, since $M \models \forall x[\neg P(x,x)]$. But also $d_3 \neq d_1$, since from $M \models \forall x \forall y \forall z[P(x,y) \land P(y,z) \to P(x,z)]$ it follows that $M \models P(a_1,a_3)[d_1,d_3]$ and $M \models \forall x[\neg P(x,x)]$. From $M \models \forall x \exists y[P(x,y)]$ it follows that there must be some element d_4 in D such that $M \models P(a_3,a_4)[d_3,d_4]$ and we can again show that $d_4 \neq d_3$, $d_4 \neq d_2$ and $d_4 \neq d_1$. By induction one shows that D contains at least denumerably many elements. Conversely, let $d_0, d_1, d_2, \ldots$ be denumerably many elements in D. Define $P^*(d_i, d_j)$ iff $i < j$. Then $\langle D; P^* \rangle$ is a model of the formulas in question.

d) By the Löwenheim-Skolem Theorem.

Solution 4.36. Let Γ be consistent and $A \in \Gamma$. Then $\Gamma \vdash A$; hence, $\Gamma \nvdash \neg A$. So, there is at least one formula C such that $\Gamma \nvdash C$. Conversely, suppose $\Gamma \nvdash C$ and for some formula B both $\Gamma \vdash B$ and $\Gamma \vdash \neg B$; then $\Gamma \vdash B \land \neg B$ and hence $\Gamma \vdash C$ for any formula C; contradiction.

Let $\Gamma = A_1, \ldots, A_n$ and suppose Γ is consistent. Then for some formula C, $\Gamma \nvdash C$, i.e., $A_1, \ldots A_n \nvdash C$. So, by the completeness theorem, $A_1, \ldots A_n \nvDash C$, i.e., there is a model M such that $M \models A_1 \land \ldots \land A_n \; [n_1, \ldots, n_k]$ and $M \nvDash C \; [n_1, \ldots, n_k]$, if $a_{n_1}, \ldots, a_{n_k}$ are the free variables in $A_1, \ldots, A_n, C$. So, Γ is satisfiable and hence, by Löwenheim's Theorem 4.27, Γ has an enumerable model. Conversely, if Γ has an enumerable model, then for no formula B, $\Gamma \vdash B$ and $\Gamma \vdash \neg B$.

Solution 4.37. Suppose $A_1, \ldots, A_n \models B$ and $A_1, \ldots, A_n \nvdash B$. Then $A_1, \ldots, A_n, \neg B \nvdash B$ and therefore $\Gamma = A_1, \ldots, A_n, \neg B$ is consistent. Using Skolem's result, formulated in Exercise 4.36, $A_1, \ldots, A_n, \neg B$ has an enumerable model, contradicting $A_1, \ldots, A_n \models B$. Therefore, $A_1, \ldots, A_n \vdash B$.

Solution 4.38. Let M be an interpretation for second-order logic with domain D. $M \models En$ iff there are $d \in D$ and $f : D \to D$ such that for all $V \subseteq D$, if 1. $d \in V$, and 2. for all $x \in D$, if $x \in V$, then $f(x) \in V$, then for all $x \in D$, $x \in V$.
Suppose $M \models En$. Take $V' := \{d, f(d), f(f(d)), \ldots\}$. Then V' satisfies 1. and 2. Therefore $D \subseteq V'$. Hence, D is enumerable. Conversely, suppose D is finite or $D = \{d_0, d_1, d_2, \ldots\}$. Then En is true in any interpretation with domain D.

4.8 Solutions

Solution 4.39. Let B be of the form $\forall x_1 \ldots \forall x_n \exists y_1 \ldots \exists y_m [A(x_1,\ldots,x_n, y_1,\ldots,y_m)]$ with A quantifier-free. For reasons of simplicity we suppose that $n = 2$ and $m = 1$. Then developing a completed tableau with initial branch $\{F\ \forall x_1 \forall x_2 \exists y[A(x_1,x_2,y)]\}$ goes as follows:

$F\ \forall x_1 \forall x_2 \exists y[A(x_1,x_2,y)]$
$\quad F\ \forall x_2 \exists y[A(a_1,x_2,y)]$ (a_1 new)
$\quad\quad F\ \exists y[A(a_1,a_2,y)]$ (a_2 new)
$FA(a_1,a_2,a_1),\ F\ \exists y[A(a_1,a_2,y)]$
$FA(a_1,a_2,a_1),\ FA(a_1,a_2,a_2),\ F\ \exists y[A(a_1,a_2,y)]$
$FA(a_1,a_2,a_1),\ FA(a_1,a_2,a_2),\ FA(a_1,a_2,a_3),\ F\ \exists y[A(a_1,a_2,y)]$
$\quad\quad\quad\quad\quad\vdots$ (propositional rules)

Further applications of rule $F\exists$ do not make sense. If $FA(a_1,a_2,a_1)$, $FA(a_1,a_2,a_2)$, $FA(a_1,a_2,a_3)$ with a_3 new does not provide a closed tableau (which is decidable), then there is no deduction of B.

Solution 4.40. In order to show that every monadic formula is equivalent to a truth-functional composition of formulas of the form $\forall x[B(x)]$ and $\exists x[B(x)]$, where B is quantifier-free, we proceed as follows. Let $A = Q_1 x_1 \ldots Q_n x_n [M]$ be a monadic formula in prenex normal form.
For instance, let $A_0 := \exists y \forall x [(\neg P(x) \vee Q(y)) \wedge (P(y) \vee \neg Q(x))]$.
STEP 1: a) If $Q_n = \forall$, replace M by its conjunctive normal form; if $Q_n = \exists$, replace M by its disjunctive normal form. (See Theorem 4.7.)
b) Replace $\forall x_n[C \wedge D]$ by $\forall x_n[C] \wedge \forall x_n[D]$; and replace $\exists x_n[C \vee D]$ by $\exists x_n[C] \vee \exists x_n[D]$ respectively. Applying step 1 b) to A_0 yields

$$A_0' := \exists y[\ \forall x[\neg P(x) \vee Q(y)] \wedge \forall x[P(y) \vee \neg Q(x)]\ ].$$

c) In the result of step 1b) replace expressions of the form $\forall x[E \vee F]$ by $\forall x[E] \vee F$, if x does not occur in F; and replace expressions of the form $\exists x[E \wedge F]$ by $\exists x[E] \wedge F$, if x does not occur in F. Applying step 1 c) to A_0' yields

$$A_1 := \exists y[\ (\forall x[\neg P(x)] \vee Q(y)) \wedge (P(y) \vee \forall x[\neg Q(x)])\ ].$$

d) Remove vacuous occurrences of quantifiers.
STEP $k+1$ ($k < n$): similar to step 1 with $n - k$ instead of n. Below we present the results of the different substeps of step 2 in the case of our example.
2a): $\exists y[\ (\forall x[\neg P(x)] \wedge P(y)) \vee (Q(y) \wedge P(y)) \vee$
$\quad\quad\quad\quad\quad\quad\quad\quad\quad (\forall x[\neg P(x)] \wedge \forall x[\neg Q(x)]) \vee (Q(y) \wedge \forall x[\neg Q(x)])]$
2b): $\exists y[\ \forall x[\neg P(x)] \wedge P(y)] \vee \exists y[Q(y) \wedge P(y)] \vee$
$\quad\quad\quad\quad\quad\quad\quad\quad\quad \exists y[\ \forall x[\neg P(x)] \wedge \forall x[\neg Q(x)]\ ] \vee \exists y[\ Q(y) \wedge \forall x[\neg Q(x)]\ ]$
2c) and d): $(\forall x[\neg P(x)] \wedge \exists y[P(y)]) \vee \exists y[Q(y) \wedge P(y)] \vee (\forall x[\neg P(x)] \wedge \forall x[\neg Q(x)]) \vee (\exists y[Q(y)] \wedge \forall x[\neg Q(x)])$.

Solution 4.41. Let A be a monadic formula. Let C be a truth-functional composition of formulas of the form $\forall x[B(x)]$ and $\exists x[B(x)]$, B quantifier-free, such that C is equivalent to A (see Exercise 4.40). Starting with FC and applying the propositional rules we find a sequent of the form

$$F\forall x[B_1(x)], T\exists x[B_2(x)], \ldots, T\forall x[B_3(x)], F\exists x[B_4(x)].$$

Next apply all $F\forall$- and $T\exists$-rules, yielding

$$FB_1(a_1), TB_2(a_2), \ldots, T\forall x[B_3(x)], F\exists x[B_4(x)] \ (a_1, a_2 \text{ new}).$$

For each $T\forall$- and $F\exists$-formula in this sequent finitely many applications of the corresponding rules suffice to find a closed tableau, if there is any. If $FB_1(a_1), TB_2(a_2),$ $\ldots, TB_3(a_1), TB_3(a_2), TB_3(a_3), FB_4(a_1), FB_4(a_2), FB_4(a_3), FB_4(a_4)$, where a_3 and a_4 are new, does not yield a closed tableau (which is decidable), then there is no closed tableau for the original formula.

Solution 4.42. Let B be a formula of the form $\exists x \forall y [M(x,y)]$, where M is quantifier-free, and suppose that the only predicate symbol appearing in M is a binary predicate symbol P. Our systematic search for a formal deduction of B starts as follows:

$$F \ \exists x \forall y[M(x,y)]$$
$$F \ \forall y[M(a_1,y)], F \ \exists x \forall y[M(x,y)]$$
$$F \ M(a_1,a_2), F \ \exists x \forall y[M(x,y)]$$
$$\vdots \text{ (propositional rules)}$$

The propositional rules applied to $F \ M(a_1,a_2)$ may give rise to signed atomic formulas of the form $P(a_1,a_1), P(a_1,a_2), P(a_2,a_1)$ and $P(a_2,a_2)$. Several branches may result, each containing the expression $F \ \exists x \forall y[M(x,y)]$. One more application of rule $F\exists$ yields at each branch:

$$F \ \forall y[M(a_2,y)], F \ \exists x \forall y[M(x,y)]$$
$$F \ M(a_2,a_3), \ F \ \exists x \forall y[M(x,y)]$$
$$\vdots \text{ (propositional rules)}$$

The propositional rules applied to $F \ M(a_2,a_3)$ may give rise to signed atomic formulas of the form $P(a_2,a_2), P(a_2,a_3), P(a_3,a_2)$ and $P(a_3,a_3)$. So, the only way closure can result from interaction of $F \ M(a_1,a_2)$ and $F \ M(a_2,a_3)$ is via $P(a_2,a_2)$. Applying rule $F\exists$ more than two times does not make sense: if not all branches are closed after two applications of rule $F\exists$, there is no deduction of $\exists x \forall y[M(x,y)]$. If M contains n binary predicate symbols, one has to allow 2^n applications of rule $F\exists$ in order that the construction of a completed tableau provides a decision procedure.

Solution 4.43. Suppose T is a tree such that each node in T has only finitely many immediate successors. For s a node in T, let $\Phi(s) :=$ there are arbitrarily long finite paths going through s. Let s_0 be the empty tuple ().
(1) $\Phi(s_0)$, by hypothesis.
(2) If $\Phi(s)$, then there is an immediate successor t of s such that $\Phi(t)$.
From (1) and (2) it follows that starting with (), we are thus always able to pick a next node with the property Φ, ad infinitum, yielding an infinite path in T.
Proof of (2): Let $s(0), \ldots, s(k)$ be the immediate successors of s. If the paths through $s(0), \ldots, s(k)$ were no longer than $l_0, \ldots, l_k$ respectively, then all paths through s would be no longer than $\max(l_0, \ldots, l_k)$. □

4.8 Solutions

In the following tree the empty node () has infinitely many successors and it has arbitrarily long finite paths, but there is no infinite path in it:

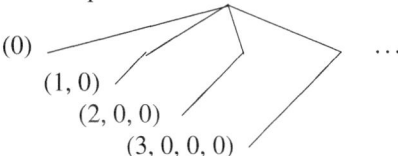

```
(0) ─────────────────────────  ...
     (1, 0)
       (2, 0, 0)
         (3, 0, 0, 0)
```

Solution 4.44. 1. Suppose $FL(n)$ is consistent. Then $FL(n)$ is satisfiable. Let M be a model of $FL(n)$. Then M is a field of characteristic n. Hence, $n = 0$ or n is prime.
2. If $FL(0) \models B$, then there is an n_0 such that for every $n \geq n_0$, $FL(n) \models B$. *Proof*: Suppose $FL(0) \models B$. Then it follows from the compactness theorem that there is a finite subset Γ' of $FL(0)$ such that $\Gamma' \models B$. Choose n_0 larger than all the n such that $\neg A_n$ occurs in Γ'.
3. We cannot replace the infinite number of non-logical axioms we added to FL to get $FL(0)$ by a finite number. *Proof*: Suppose we could. Let B be the conjunction of these non-logical axioms. B would be true in fields of characteristic 0 but in no other fields. $FL(0) \models B$. Choosing n_0 as in assertion 2, we would conclude that there are no fields of characteristic greater than n_0, which is absurd.
4. There is no extension Γ of FL whose models are just the finite fields. *Proof*: Suppose we had such an extension Γ. Let B_n be a formula which expresses that there are at least n individuals; for example, B_3 is $\exists x \exists y \exists z [x \neq y \wedge x \neq z \wedge y \neq z]$. Let Δ be obtained from Γ by adding all the B_n as non-logical axioms. Then Δ has no model. Then it follows from the compactness theorem that there is a finite subset Δ' of Δ which has no model. Choose n_0 larger than all the n such that B_n occurs in Δ' and choose a finite field M having more than n_0 elements. Then M is a model of Δ'. Contradiction.

Solution 4.45. 1. Let R be a partial ordering on V and V finite. Let n be the number of elements of V. If $n = 1$, then the proof is trivial. Suppose the induction hypothesis and let V have $n + 1$ elements. V has a minimal element, say v_0. Then $V - \{v_0\}$ is partially ordered by $R\lceil V - \{v_0\}$. By the induction hypothesis there is a complete partial ordering R_1 on $V - \{v_0\}$ such that $R \subseteq R_1$. Let $R' := R_1 \cup \{(v_0, v) \mid v \in V\}$. Then R' is a complete partial ordering on V such that $R \subseteq R'$.
2. Let R be a partial ordering on V and V infinite. Consider a language containing a binary predicate symbol $\leq$ and an individual constant c_v for each $v \in V$. Let Γ be the following set of sentences: If $v_1 R v_2$, then Γ contains $c_{v_1} \leq c_{v_2}$.
If $v_1 \neq v_2$, then Γ contains $\neg (c_{v_1} \equiv c_{v_2})$.
$\forall x [\, x \leq x\,], \forall x, y [\, x \leq y \wedge y \leq x \to x \equiv y\,], \forall x, y, z [\, x \leq y \wedge y \leq z \to x \leq z\,]$
$\forall x, y [\, x \leq y \vee y \leq x\,]$
If $M = \langle D; R \rangle$ is a model of Γ, then R yields a complete partial ordering on V. By the compactness theorem, it suffices to prove that every finite subset of Γ has a model. So, let Γ' be a finite subset of Γ and let $V' := \{v \in V \mid c_v \text{ occurs in } \Gamma'\}$. $R\lceil V'$ is a partial ordering on V' and V' is finite. Therefore, there is a complete partial ordering on V' which contains $R\lceil V'$. That is, Γ' has a model.

References

1. Ayer, A.J. (ed.), *Logical Positivism*. Editor's Introduction. The Free Press, New York, 1959.
2. Boolos, G., J. Burgess and R. Jeffrey, *Computability and Logic*. Cambridge University Press, 1974, 2010.
3. Church, A., *Introduction to Mathematical Logic*. Princeton University Press, 1956, 1996.
4. Dalen, D. van, *Logic and Structure*. Springer, Berlin, 1980, 2004.
5. Frege, G., *Uber Sinn und Bedeuting*, 1892. In Frege, G., *Philosophical Writings*. Translated by P. Geach and M. Black, 1970.
6. Heijenoort, J. van, *From Frege to Gödel*. A source book in mathematical logic 1879 - 1931. Harvard University Press, 1967, 2002.
7. Henkin, L., The completeness of the first-order functional calculus. *Journal of Symbolic Logic*, Vol. 14, 1949, pp. 159-166.
8. Kant, I., *The Critique of Pure Reason*. William Benton, Publisher, Encyclopaedia Brittanica, in particular pp. 14-18.
9. Kleene, S.C., *Mathematical Logic*. John Wiley and Sons, Inc. New York, 1967, 2002.
10. Kneale, W. & M., *The Development of Logic*. Clarendon Press, Oxford, 1962, 2008.
11. Łukasiewicz, j., *Aristotle's Syllogistic*. Clarendon Press, Oxford, 1957.
12. Łukasiewicz, j., *Elements of Mathematical Logic*. Pergamon Press, Oxford, 1963.
13. Quine, W.V., Two Dogmas of Empiricism. In W.V. Quine, *From a Logical Point of View*. Harvard University Press, 1953, 1999.
14. Russell, B., On Denoting. *Mind*, 1905, pp. 479-493. Reprinted in Lackey, D. (ed.), *Essays in Analysis by Bertrand Russell*.
15. Turing, A., On computable numbers, with an application to the Entscheidungsproblem, 1936-1937. Reprinted in Davis, M. The Undecidable, Raven Press, Hewlett, New York, 1965, 2004.

Chapter 5
Arithmetic: Gödel's Incompleteness Theorems

H.C.M. (Harrie) de Swart

Abstract We *formalize* elementary number theory, i.e., we introduce a formal language $\mathscr{L}$ for expressing properties of addition and multiplication of natural numbers, and a set $\mathscr{P}$ of non-logical axioms (of Peano) in order to be able to formally deduce those properties from $\mathscr{P}$.

Gödel's *first incompleteness theorem* says that not every formula in $\mathscr{L}$, which is true in the intended interpretation, can be deduced from $\mathscr{P}$; even worse, extending $\mathscr{P}$ consistently with further axioms does not remedy this incompleteness. Gödel's *second incompleteness theorem* follows from his first one and says that the consistency of $\mathscr{P}$ cannot be formally deduced from $\mathscr{P}$; similar results hold for consistent extensions of $\mathscr{P}$. A sketch of Gödel's incompleteness proofs is given.

It turns out that there are two non-isomorphic models of $\mathscr{P}$ (or of any consistent extension Γ of $\mathscr{P}$). However, if we also allow in our language quantifiers of the type $\forall X$, where X is a variable over properties of natural numbers (or subsets of $\mathbb{N}$), as is done in second-order logic, then there is one single formula $\mathscr{PA}$ such that any model of $\mathscr{PA}$ is isomorphic to the standard (or intended) interpretation.

5.1 Formalization of Elementary Number Theory

In *elementary number theory* or *arithmetic* one studies the properties of natural numbers with respect to addition and multiplication. In doing arithmetic one needs only a very restricted sub-language of English containing the following expressions:
1. The binary predicate or relation 'is equal to'.
2. The natural numbers: zero, one, two, three, and so on.
3. The functions of addition (plus) and multiplication (times).
4. Variables n, m for natural numbers. For instance, in: (n plus m) times (n plus m) equals (n times n) plus two times (n times m) plus (m times m).
5. The connectives 'if ..., then ...', 'and', 'or', 'not' and 'if and only if'. For instance, in: if n equals m, then (n times n) equals (m times m).
6. The quantifiers 'for all n, ...' and 'there is at least one n such that ...'. For in-

stance, in: for all natural numbers n, n plus zero equals n. And in: there is a natural number n such that (n times n) equals n.

Below we present a formal language $\mathscr{L}$ (Language for Arithmetic), rich enough to express properties of addition and multiplication of natural numbers. This language should contain non-logical symbols for:
1. the equality relation,
2. the individual natural numbers, and
3. the addition and multiplication functions.

Instead of introducing an individual constant c_n for each individual natural number n, we can take only one individual constant c_0 together with a unary function symbol s, to be interpreted as the successor function. Then $s(c_0)$ can play the role of c_1, $s(s(c_0))$ can play the role of c_2, and so on.

Definition 5.1 (Formal Language $\mathscr{L}$ for Arithmetic).
Alphabet of $\mathscr{L}$:

non-logical symbols:	$\equiv$	binary predicate symbol
	c_0	individual constant
	s	unary function symbol
	$\oplus, \otimes$	binary function symbols
logical symbols:	$a_1, a_2, a_3, \ldots$	free individual variables
	$x_1, x_2, x_3, \ldots$	bound individual variables
	$\rightleftarrows, \rightarrow, \wedge, \vee, \neg$	connectives
	$\forall, \exists$	quantifiers
	$(,),[,]$	parentheses

Definition 5.2 (Standard model of arithmetic).
$\mathscr{N} = \langle \mathbb{N}; =; 0; {}', +, \cdot \rangle$ is the *intended interpretation* of $\mathscr{L}$, i.e., $\mathscr{N}$ interprets the individual variables as natural numbers (i.e., as elements of $\mathbb{N}$), the symbol $\equiv$ as the equality relation $=$ between natural numbers, the symbol c_0 as the natural number 0, the symbol s as the successor function ${}' : \mathbb{N} \rightarrow \mathbb{N}$, defined by $n' = n + 1$, the symbol $\oplus$ as addition $+$ of natural numbers and the symbol $\otimes$ as multiplication $\cdot$ of natural numbers. The intended interpretation $\mathscr{N}$ of (the symbols in the formal language) $\mathscr{L}$ is also called the *standard model* for the formal language $\mathscr{L}$ or the *standard model of arithmetic*.

Warning: $\equiv$, c_0, s, $\oplus$ and $\otimes$ are just non-logical symbols in (the alphabet of) our object-language, which under different interpretations may get many different non-intended meanings: $\equiv$ might be interpreted as $<$ (less than), c_0 might be interpreted as 5, s might be interpreted as taking the square, $\oplus$ might be interpreted as exponentiation and so on. One should clearly distinguish between the symbols in our formal language, which under different interpretations may get many different meanings, and the intended interpretation of these symbols. c_0 is a symbol in the object-language and not (the name of) a natural number; 0 (zero), on the other hand, is the name of a natural number. Similarly, $\oplus$ is a function symbol, not a function; $+$ is (the name of) a function from $\mathbb{N}^2$ to $\mathbb{N}$, it is not a function symbol in the object-language. However, for reasons of easy notation, the following convention is adopted, mostly implicitly.

Convention: One uses $=$ instead of $\equiv$; 0 instead of c_0; $'$ instead of s; and $+$ and $\cdot$ instead of $\oplus$ and $\otimes$, respectively. So, the symbols $=, 0, ', +$ and $\cdot$ are used in two ways: 'par abus de language' as symbols in the formal language $\mathscr{L}$ for arithmetic with many possible interpretations and as the intended interpretation of the corresponding symbols in the language $\mathscr{L}$.

Under this convention the *alphabet of* $\mathscr{L}$ contains the following symbols.

Symbols	Name	Intended interpretation
$=$	binary predicate symbol	equality
0	individual constant	zero
$'$	unary function symbol	successor function
$+\,;\,\cdot$	binary function symbols	addition; multiplication
$a_1, a_2, \ldots$	free individual variables	natural numbers
$x_1, x_2, \ldots$	bound individual variables	natural numbers
$\rightleftarrows, \rightarrow, \wedge, \vee, \neg$	connectives	
$\forall, \exists$	quantifiers	
$(,),[,]$	parentheses	

Definition 5.3 (Terms of $\mathscr{L}$).
The *terms* of the language $\mathscr{L}$ for formal arithmetic are defined as follows:
1. Each free individual variable a is a term.
2. 0 is a term.
3. If r and s are terms, then $(r)'$, $(r+s)$ and $(r \cdot s)$ are also terms.

If no confusion is possible, parentheses are omitted as much as possible.

Example 5.1. Examples of terms of $\mathscr{L}$: $0, a_1, 0+a_1, (0+a_1)\cdot a_1, a_1 \cdot a_2, 0+a_1 \cdot a_1, 0'' \cdot a_1 + a_2 \cdot a_3$.

Since there is only one predicate symbol in the alphabet, the atomic formulas in the language $\mathscr{L}$ for formal number theory are of the form $=(r,s)$, where r and s are terms. Instead of $=(r,s)$ one usually writes $r=s$.

Definition 5.4 (Atomic formulas of $\mathscr{L}$).
If r and s are terms, then $r=s$ is an *atomic formula* of the language $\mathscr{L}$ for formal number theory.

From these atomic formulas complex formulas can be built in the usual way by means of connectives and quantifiers:

Definition 5.5 (Formulas of $\mathscr{L}$).
1. Every atomic formula of $\mathscr{L}$ is a formula of $\mathscr{L}$.
2. If A and B are formulas of $\mathscr{L}$, then also $(A \rightleftarrows B)$, $(A \rightarrow B)$, $(A \wedge B)$, $(A \vee B)$ and $(\neg A)$ are formulas of $\mathscr{L}$.
3. If $A(a)$ is a formula of $\mathscr{L}$ and x is a bound individual variable, then also $\forall x[A(x)]$ and $\exists x[A(x)]$ are formulas of $\mathscr{L}$, where $A(x)$ results from $A(a)$ by replacing one or more occurrences of a in $A(a)$ by x.

English sentences about addition and multiplication of natural numbers can be translated into formulas of the language $\mathscr{L}$ for formal number theory. Here are some examples:

(i) For all natural numbers n, m, (n plus m) times (n plus m) equals (n times n) plus two times (n times m) plus (m times m): $\forall x \forall y[(x+y)\cdot(x+y) = x\cdot x + 0''\cdot x\cdot y + y\cdot y]$.
(ii) For all natural numbers n, m, if n equals m, then n square equals m square: $\forall x \forall y[x = y \to x\cdot x = y\cdot y]$.
(iii) For all natural numbers n, n plus zero equals n: $\forall x[x+0 = x]$.
(iv) There is at least one natural number n such that n square equals n: $\exists x[x\cdot x = x]$.

Now consider the formula $\forall x[x+0 = x]$, or rather $\forall x[x \oplus c_0 \equiv x]$. This formula is true under the intended interpretation $\mathscr{N}$, in other words $\mathscr{N} \models \forall x[x+0 = x]$, but this formula is not under every interpretation true. For instance, let M be the structure $\langle \mathbb{Q}; >; 5, ', -, \cdot \rangle$, i.e., M has the set of rational numbers as domain, interprets $\equiv$ as 'is greater than $(>)$', c_0 as 5, and $\oplus$ as subtraction $(-)$. Under this interpretation $\forall x[x \oplus c_0 \equiv x]$ reads as follows: for all rational numbers x, $x - 5 > x$; and this happens to be false. Therefore, $M \not\models \forall x[x \oplus c_0 \equiv x]$. So, although $\forall x[x \oplus c_0 \equiv x]$ is true under the intended interpretation, it is not always true,, i.e., not under every interpretation true, in other words $\not\models \forall x[x \oplus c_0 \equiv x]$.

Of course, $\models \forall x[x \oplus c_0 \equiv x \vee \neg(x \oplus c_0 \equiv x)]$, i.e., $\forall x[x \oplus c_0 \equiv x \vee x \oplus c_0 \not\equiv x]$ is true in every interpretation. The validity of this formula rests upon the fixed meaning of the connectives and quantifiers, which for that reason are called *logical symbols*. The symbols $\equiv$, c_0, s, $\oplus$ and $\otimes$ are called *non-logical symbols*, because they do not belong to logic but come from mathematics; their meaning can vary depending on the context, in other words, they allow many different interpretations.

Since valid patterns of reasoning should be applicable universally, i.e., in any domain, mathematics, physics, economics or whatever, in logic we are interested in valid formulas, i.e., in formulas which are always true, in other words, which yield a true proposition in every interpretation of the non-logical symbols occurring in them. *But in elementary number theory (arithmetic) we are* of course *only interested in the intended interpretation*, and not in all possible interpretations.

Notice that $\mathscr{N} \models (a+1)\cdot(a+1) = a\cdot a + 2\cdot a + 1$, in other words, $\mathscr{N} \models \forall x[(x+1)\cdot(x+1) = x\cdot x + 2\cdot x + 1]$, because for every $n \in \mathbb{N}$, $\mathscr{N} \models (a+1)\cdot(a+1) = a\cdot a + 2\cdot a + 1$ $[a/n]$. But $\mathscr{N} \not\models (a+1)\cdot(a+1) = 4$, because, for instance, $\mathscr{N} \not\models (a+1)\cdot(a+1) = 4$ $[a/2]$, although $\mathscr{N} \models (a+1)\cdot(a+1) = 4$ $[a/1]$.

So far we have introduced a (first-order) formal language $\mathscr{L}$ for elementary number theory, in which propositions about addition and multiplication of natural numbers can be formulated. The next step is to select a number of arithmetic (non-logical) axioms, formulated in this language, in order to be able to deduce formally properties of natural numbers. To that purpose Guiseppe Peano formulated in 1891 the following set $\mathscr{P}$ of arithmetic axioms, named after him.

The Peano axioms are formulas in the formal language $\mathscr{L}$ for elementary number theory and these axioms are true in the intended interpretation. The induction axiom *schema* yields an induction axiom for any formula A in the language $\mathscr{L}$.

5.1 Formalization of Elementary Number Theory

Definition 5.6 (Axioms of Peano).

$\forall x \forall y \forall z [x = y \to (x = z \to y = z)]$ axiom for $=$

$\forall x \forall y [x' = y' \to x = y]$
$\forall x \forall y [x = y \to x' = y']$ axioms for $'$
$\forall x [\neg (x' = 0)]$

$\forall x [x + 0 = x]$ axioms for $+$
$\forall x \forall y [x + y' = (x + y)']$

$\forall x [x \cdot 0 = 0]$ axioms for $\cdot$
$\forall x \forall y [x \cdot y' = x \cdot y + x]$

$A(0) \wedge \forall x [A(x) \to A(x')] \to \forall x [A(x)]$ induction axiom *schema*

Now let $\mathscr{P}$ be the set of the axioms of Peano. One can verify that, for instance, $\mathscr{P} \vdash \forall x [x = x]$ and $\mathscr{P} \vdash \forall x \forall y [x + y = y + x]$ (see Exercise 5.1). And from experience we know that any formula which is true in the intended interpretation and *which one encounters in practice* can be formally deduced from $\mathscr{P}$. In fact, in [4], Sections 38-40, S.C. Kleene formally deduces a great number of such formulas from $\mathscr{P}$.

By the completeness theorem (for the predicate logic with equality) we know that for any formula A in $\mathscr{L}$, $\mathscr{P} \vdash A$ iff $\mathscr{P} \models A$, i.e., $\mathscr{P} \vdash A$ iff every interpretation that makes $\mathscr{P}$ true also makes A true, in other words, $\mathscr{P} \vdash A$ iff every model of $\mathscr{P}$ is also a model of A. In particular: if $\mathscr{P} \vdash A$, then A is true in the standard interpretation, in other words, if $\mathscr{P} \vdash A$, then $\mathscr{N} \models A$. But the question arises if the following holds:

$\mathscr{P} \vdash A$ iff A is true in the intended interpretation $\mathscr{N}$, i.e.,
$\mathscr{P} \vdash A$ iff $\mathscr{N} \models A$.

In Section 5.2 it will be made clear that this is not the case. Even worse, there is no consistent and axiomatizable extension Γ of $\mathscr{P}$ such that any formula A in $\mathscr{L}$ which is true in the intended interpretation can be formally deduced from Γ. This is Gödel's first *incompleteness theorem* (for formal number theory; 1931).

Summarizing: In this Section we have given a *formalization* of elementary number theory (arithmetic). That is:
1. We have introduced a formal language $\mathscr{L}$ for elementary number theory in which we can express properties of natural numbers with respect to addition and multiplication.
2. We have introduced an axiom system $\mathscr{P}$ for (formal) number theory in order to be able to deduce formally formulas from $\mathscr{P}$ which are true in the intended interpretation.

The result is called *formal number theory*, consisting of two components: the formal language $\mathscr{L}$ and the axioms $\mathscr{P}$ of Peano. For any formula A,
 if $\mathscr{P} \vdash A$, then A is true in the intended interpretation, i.e., $\mathscr{N} \models A$.
But according to Gödel's incompleteness theorem (1931), the converse,

if A is true in the intended interpretation, then $\mathscr{P} \vdash A$
is not for all formulas A in $\mathscr{L}$ true.

Therefore, Gödel's incompleteness theorem says that the proof power of $\mathscr{P}$ is restricted; more generally, that the proof power of any consistent and axiomatizable extension Γ of $\mathscr{P}$ is restricted.

Exercise 5.1. Prove: a) $\mathscr{P} \vdash \forall x[x = x]$; b) $\mathscr{P} \vdash \forall y \forall x[x' + y = (x+y)']$;
c) $\mathscr{P} \vdash \forall x \forall y[x + y = y + x]$.

5.2 Gödel's first Incompleteness Theorem

Definition 5.7 (Consistency). Let Γ be a set of formulas (in $\mathscr{L}$ or in any other language). Γ is *consistent* := there is no formula A such that $\Gamma \vdash A$ and $\Gamma \vdash \neg A$. (1)

Theorem 5.1. Γ *is consistent iff*
there is some formula A such that not $\Gamma \vdash A$ iff (2)
Γ *is satisfiable.* (3)

Proof. (1) implies (2), by the completeness theorem (2) implies (3), and (3) implies (1). □

Definition 5.8 (Axiomatizable). Let Γ be a set of formulas (in $\mathscr{L}$ or in any other language). Γ is *axiomatizable* := there is a subset Γ' of Γ such that:
1. Γ' is decidable, i.e., there is a decision method which decides for any formula A in the language whether A is in Γ' or A is not in Γ', and
2. for any formula A in the language, $\Gamma' \vdash A$ iff $\Gamma \vdash A$.
The elements of Γ' are called *axioms* for Γ.

The hope that any formula in $\mathscr{L}$ which is true under the intended interpretation, can be formally deduced from Peano's axioms, was dashed in 1931 by the incompleteness theorem of Kurt Gödel.

Theorem 5.2 (First Incompleteness Theorem for Arithmetic).
Let Γ be a consistent and axiomatizable extension of $\mathscr{P}$. Then there is a closed formula A_Γ (depending on Γ) in $\mathscr{L}$ such that
  1. A_Γ is true in the intended interpretation, i.e., $\mathscr{N} \models A_\Gamma$, but
  2. not $\Gamma \vdash A_\Gamma$, and
  3. not $\Gamma \vdash \neg A_\Gamma$.
2 and 3 together say that A_Γ is undecidable on the basis of Γ; i.e., the proof power of any consistent and axiomatizable extension Γ of $\mathscr{P}$ is restricted.

Of course, Gödel's incompleteness theorem does not hold if we take for Γ the set of all formulas in $\mathscr{L}$ which are true in the intended interpretation. But this set cannot be seen as an axiom system, more precisely, it is not axiomatizable.

Gödel's incompleteness theorem says that, given any consistent and axiomatizable extension Γ of $\mathscr{P}$, not every formula which is true in the intended interpretation

5.2 Gödel's first Incompleteness Theorem

can be formally deduced from Γ. Given any such Γ, the truth of A_Γ (in the standard model) can be seen semantically, but A_Γ cannot be formally deduced from Γ.

Since the set $\mathscr{P}$ of Peano's axioms satisfies the conditions in Theorem 5.2, Gödel's incompleteness theorem says in particular that there is a formula A_1 which is true in the intended interpretation, but which cannot be deduced from $\mathscr{P}$ (not $\mathscr{P} \vdash A_1$). Because A_1 is true in the intended interpretation, we might extend $\mathscr{P}$ with the formula A_1 to the set $\mathscr{P} \cup \{A_1\}$. But then, taking $\Gamma = \mathscr{P} \cup \{A_1\}$, Gödel's incompleteness theorem says that there is a formula A_2, depending on $\mathscr{P} \cup \{A_1\}$, such that A_2 is true in the intended interpretation and not $\mathscr{P}, A_1 \vdash A_2$. In a similar way we can find a formula A_3 such that A_3 is true in the intended interpretation and such that not $\mathscr{P}, A_1, A_2 \vdash A_3$, and so on.

Sketch of proof of Gödel's first incompleteness theorem

A detailed proof of Gödel's incompleteness theorem requires many pages. See, for instance, Kleene [4], Boolos, Burgess and Jeffrey [2], Smith [8], Nagel [5]. However, the heart of the proof can be explained in a few lines, if we postulate in addition that the formulas in Γ are true in $\mathscr{N}$, which only slightly strengthens the condition that Γ is consistent. The formula A_Γ in the language $\mathscr{L}$ for formal number theory, which is constructed given a set Γ satisfying the conditions of Theorem 5.2, means that A_Γ is not formally deducible from Γ; more precisely:

A_Γ is true (in the intended interpretation) if and only if not $\Gamma \vdash A_\Gamma$. (∗)

Hence, A_Γ is a sentence in $\mathscr{L}$ that says of itself that it is not deducible from Γ.

Now suppose A_Γ were false (in the intended interpretation). Then it follows from (∗) that $\Gamma \vdash A_\Gamma$. Because of the Soundness Theorem it follows that $\Gamma \models A_\Gamma$ and because Γ is supposed to be true (in the intended interpretation), it follows that A_Γ is true (in the intended interpretation). Contradiction. Therefore, A_Γ is not false, and hence true, in the intended interpretation. And hence it follows from (∗) that not $\Gamma \vdash A_\Gamma$.

Because A_Γ is true, $\neg A_\Gamma$ is false (in the intended interpretation). Now suppose $\Gamma \vdash \neg A_\Gamma$. Then by soundness, $\Gamma \models \neg A_\Gamma$. So, assuming that $\mathscr{N} \models \Gamma$, it would follow that $\mathscr{N} \models \neg A_\Gamma$, i.e., $\neg A_\Gamma$ is true in the intended interpretation. Contradiction. Therefore, not $\Gamma \vdash \neg A_\Gamma$. □

Corollary 5.1. *There exists a model of Peano's arithmetic $\mathscr{P}$ that is not the standard model $\mathscr{N}$.*

Proof. Since $\mathscr{P} \not\vdash A_\mathscr{P}$, we know by the completeness theorem for predicate logic that $\mathscr{P} \not\models A_\mathscr{P}$, i.e., there is a model M of $\mathscr{P}$ such that $M \not\models A_\mathscr{P}$. However, the standard model $\mathscr{N}$ is a model of $A_\mathscr{P}$, in other words, $A_\mathscr{P}$ is true in the intended interpretation. Therefore, M cannot be the standard model $\mathscr{N}$. □

5.2.1 Gödel-numbering

However, it still costs a lot of energy, given any Γ satisfying the conditions of Theorem 5.2, to construct a (closed) formula A_Γ in $\mathscr{L}$ satisfying the property $(*)$. The key idea is the *Gödel-numbering* of the symbols (letters) in the alphabet of $\mathscr{L}$, of the terms and of the formulas in the language $\mathscr{L}$ for formal number theory. Each symbol in the alphabet for formal number theory can be identified with a natural number, called the *Gödel-number* of that symbol. Different symbols are identified with different Gödel-numbers. For example, if we replace the free individual variables $a_1, a_2, \ldots$ by $a, (|, a), (|, (|, a)), \ldots$ respectively, then we can take the following correlation (identification) of natural numbers with the symbols in $\mathscr{L}$:

$\rightarrow$	$\wedge$	$\vee$	$\neg$	$\forall$	$\exists$	$=$	$+$	$\cdot$	$'$	0	a	$|$
3	5	7	9	11	13	15	17	19	21	23	25	27

Many other correlations are possible. There is nothing special about our particular Gödel-numbering.

A *Gödel-numbering* assigns to symbols, terms, formulas and deductions a natural number, called the *Gödel-number* of the expression, such that:

(i) it assigns different Gödel-numbers to different expressions;

(ii) the Gödel-number of any expression is effectively calculable;

(iii) one can effectively decide whether a natural number is the Gödel-number of some expression, and, if so, of what expression.

If A is an expression with Gödel-number n, we define $\ulcorner A \urcorner$ to be the expression $\bar{n}$, the *numeral* for n; $\bar{0} := 0$, $\bar{1} := 0'$, $\bar{2} := 0''$, $\ldots$; so, $\bar{n}$ is the term in $\mathscr{L}$ that corresponds to the natural number n.

Terms and formulas of $\mathscr{L}$ are finite sequences of symbols of (the alphabet of) $\mathscr{L}$ formed according to certain rules and hence they can be identified with finite sequences of natural numbers. And in its turn each finite sequence $k_1, \ldots, k_n$ of natural numbers can be identified with another natural number, for instance, with $p_1^{k_1} \cdots p_n^{k_n}$, where $p_1, \ldots, p_n$ are the first n prime numbers. Then the individual variable a_1, that is $(|, a)$, is identified with $2^{27} \cdot 3^{25}$ and the atomic formula $a_1 = 0$, that is $= (a_1, 0)$, is then identified with the natural number $2^{15} \cdot 3^{2^{27} \cdot 3^{25}} \cdot 5^{23}$.

Given a specific Gödel-numbering, if n is the Gödel-number of some formula, let $A_n(a)$ be the formula with Gödel-number n, so $\ulcorner A_n(a) \urcorner = \bar{n}$.

Now let Γ be a set of arithmetic axioms formulated in $\mathscr{L}$. Then a formal deduction of A from Γ is a finite sequence of formulas in $\mathscr{L}$, constructed according to certain rules, and hence can be identified with a finite sequence of natural numbers, and therefore with a natural number.

By correlating to different formal objects different natural numbers and by talking about the correlated natural numbers instead of the formal objects themselves, the meta-mathematical predicate '$A(a)$ is a formula, k is a natural number and b is a formal deduction of $A(\bar{k})$ from Γ' can be rendered by an arithmetical predicate $Ded_\Gamma(n, k, m)$ saying:

n is the Gödel-number of a formula, namely $A_n(a)$, and m is the Gödel-number of a formal deduction of $A_n(\bar{k})$ from Γ.

5.2 Gödel's first Incompleteness Theorem

So, using the Gödel-numbering, meta-mathematics becomes part of arithmetic. $\Gamma \vdash A_n(\bar{k})$ if and only if there is a natural number m such that $Ded_\Gamma(n,k,m)$.

Now consider the arithmetic predicate $Ded_\Gamma(n,n,m)$, which expresses: m is the Gödel-number of a formal deduction of $A_n(\bar{n})$ from Γ.

In section 52 of [4] S.C. Kleene proves that there is a formula $DED_\Gamma(a,a_1)$ of $\mathscr{L}$, such that for all natural numbers n,m:
(i) if $Ded_\Gamma(n,n,m)$ is true, then $\Gamma \vdash DED_\Gamma(\bar{n},\bar{m})$, and
(ii) if $Ded_\Gamma(n,n,m)$ is false, then $\Gamma \vdash \neg DED_\Gamma(\bar{n},\bar{m})$.

In order to prove (i) and (ii), one uses the supposition that Γ contains the axioms of Peano and that Γ is axiomatizable.

Next consider the formula $\neg \exists y[DED_\Gamma(a,y)]$, having a as the only free variable. This formula has a Gödel-number, say p, and hence equals $A_p(a)$ according to the notation introduced before.

Finally, consider the formula

$$A_\Gamma := A_p(\bar{p}) : \neg \exists y[DED_\Gamma(\bar{p},y)].$$

Then it holds that A_Γ is true in the intended interpretation if and only if there is no formal deduction of the formula $A_p(\bar{p})$ from Γ. But this latter formula $A_p(\bar{p})$ is A_Γ itself! Therefore:

A_Γ is true (in the intended interpretation) if and only if not $\Gamma \vdash A_\Gamma$ (∗)

So, using the Gödel-numbering, it is possible to construct a formula A_Γ of $\mathscr{L}$, which says of itself that it cannot be deduced from Γ.

Now it is easy to see that if Γ satisfies the conditions in Theorem 5.2, then not $\Gamma \vdash A_\Gamma$ and hence, by (∗), A_Γ is true (in the intended interpretation). For suppose Γ is consistent and $\Gamma \vdash A_\Gamma$. Let k be the Gödel-number of a formal deduction of A_Γ from Γ. Then $Ded_\Gamma(p,p,k)$ is true. So it follows from (i) that $\Gamma \vdash DED_\Gamma(\bar{p},\bar{k})$. Therefore $\Gamma \vdash \exists y[DED_\Gamma(\bar{p},y)]$. But we supposed that $\Gamma \vdash A_\Gamma$, i.e., $\Gamma \vdash \neg \exists y[DED_\Gamma(\bar{p},y)]$. Contradiction with the consistency of Γ. Therefore, if Γ is consistent, then not $\Gamma \vdash A_\Gamma$. And then according to (∗), A_Γ is true (in the intended interpretation).

This finishes our sketch of the proof of Gödel's first incompleteness theorem. For further details the reader is referred to section 42 and Chapter X of Kleene [4]. For a popular exposition of Gödel's work see Nagel and Newman [5], Hofstadter [3], and Smullyan [9].

Remark 5.1. The Liar's paradox results from considering a sentence A which says of itself that it is not true. By replacing 'A is not true' by 'A is not deducible from Γ', Gödel escapes a paradox and finds a deep philosophical insight instead.

Remark 5.2. In his proof of the incompleteness theorem, K. Gödel constructs – given any Γ satisfying the hypotheses of the theorem – a formula A_Γ, which in the intended interpretation says of itself that it is not deducible from Γ. By thinking *about* Γ and A_Γ, we then see that not $\Gamma \vdash A_\Gamma$ and hence that A_Γ is true (in the intended interpretation). The proof of Gödel's incompleteness theorem is – although very long and technically very smart – in essence very elementary. One can raise no objections against it which would not be at the same time objections against parts of traditional mathematics, which are generally considered to be unproblematic.

Remark 5.3. The formula A_Γ refers to itself, because it says about itself that it is not deducible from Γ. Such sentences are not of particular interest for mathematicians. However, Paris and Harrington [6] gave a strictly mathematical example of an incompleteness in first-order Peano arithmetic, which is mathematically simple and interesting and which does not require a numerical coding of logical notions.

Remark 5.4. From the definition of $\mathscr{P} \models A$ it follows immediately that for any formula A, if $\mathscr{P} \models A$, then A is true in the intended interpretation. $\hspace{1em}(\alpha)$
By Gödel's completeness theorem for the predicate logic, $\mathscr{P} \models A$ iff $\mathscr{P} \vdash A$. Therefore, by Gödel's incompleteness theorem for formal number theory, the converse of (α) does not hold, i.e., *not* for every formula A, if A is true in the intended interpretation, then $\mathscr{P} \models A$.

5.2.2 Provability predicate for $\mathscr{P}$

If A is a formula of the formal language $\mathscr{L}$ for arithmetic (see Section 5.1) with Gödel-number n, we define $\ulcorner A \urcorner$ to be the expression $\bar{n}$, the numeral for n; $\bar{1} = 0'$, $\bar{2} = 0''$, etc.

We shall assume, but not prove, the following FACT: By 'straightforwardly transcribing' in $\mathscr{L}$ the definition of *being deducible from* $\mathscr{P}$, where $\mathscr{P}$ is the set of Peano's axioms for arithmetic, making reference to Gödel-numbers instead of expressions, one can construct a formula $Prov(a)$ of $\mathscr{L}$, with the following properties:
(a) $Prov(a)$ expresses that a is the Gödel-number of a formula which is deducible from $\mathscr{P}$, and
(b) $Prov(a)$ is a *provability predicate* for $\mathscr{P}$, i.e.,
  (i) if $\mathscr{P} \vdash A$, then $\mathscr{P} \vdash Prov(\ulcorner A \urcorner)$;
  (ii) $\mathscr{P} \vdash Prov(\ulcorner B \to C \urcorner) \to (Prov(\ulcorner B \urcorner) \to Prov(\ulcorner C \urcorner))$;
  (iii) $\mathscr{P} \vdash Prov(\ulcorner A \urcorner) \to Prov(\ulcorner Prov(\ulcorner A \urcorner) \urcorner)$.
(c) In addition,
  (iv) if $\mathscr{P} \vdash Prov(\ulcorner A \urcorner)$, then $\mathscr{P} \vdash A$.

That $Prov(a)$ satisfies (i) may be seen as follows: Suppose $\mathscr{P} \vdash A$. Then there is a formal proof of A from $\mathscr{P}$. Let $\ulcorner A \urcorner$ be the Gödel number of A. Then the formula $Prov(\ulcorner A \urcorner)$ expresses that $\ulcorner A \urcorner$ is the Gödel number of a formula which is deducible from $\mathscr{P}$. Then $\mathscr{P} \vdash Prov(\ulcorner A \urcorner)$. (ii) A deduction of C can be obtained from deductions of B and of $B \to C$ by one more application of Modus Ponens. This argument can be formalized in $\mathscr{P}$. Showing that $Prov(a)$ satisfies (iii) is much harder: it involves showing that the argument that $Prov(a)$ satisfies (i) can be formalized in $\mathscr{P}$. To show (iv), suppose that $\mathscr{P} \vdash Prov(\ulcorner A \urcorner)$. Then $Prov(\ulcorner A \urcorner)$ is true in $\mathscr{N}$. Hence A is deducible from $\mathscr{P}$.

However, $Prov(a)$ does NOT meet the stronger condition $\mathscr{P} \vdash Prov(\ulcorner A \urcorner) \to A$. Löb's theorem says that if $\mathscr{P} \vdash Prov(\ulcorner A \urcorner) \to A$, then $\mathscr{P} \vdash A$.

For more details the reader is referred to Boolos and Jeffrey [1], Chapter 16, or to Boolos, Burgess and Jeffrey [2], Chapter 18.

5.3 Gödel's second Incompleteness Theorem

Theorem 5.3 (Second Incompleteness Theorem for Arithmetic). *Let Γ be a consistent and axiomatizable extension of $\mathscr{P}$. Let $Cons_\Gamma$ be a formula in $\mathscr{L}$, expressing the consistency of Γ. Then not $\Gamma \vdash Cons_\Gamma$.*

Gödel's second incompleteness theorem says that the consistency of Γ – provided that Γ satisfies the conditions mentioned above – cannot be proved by means which are available in Γ itself.

Since the standard model $\mathscr{N}$ is a model of the axioms $\mathscr{P}$ of Peano, we know that $\mathscr{P}$ is consistent. By Gödel's second theorem, the consistency proof for $\mathscr{P}$ just given cannot be formalized in $\mathscr{P}$ itself.

First we have to construct a formula $Cons_\Gamma$ in $\mathscr{L}$ expressing the consistency of Γ. Because $\mathscr{P} \subseteq \Gamma$, $\Gamma \vdash \neg(0 = 1)$. Consequently, Γ is consistent if and only if not $\Gamma \vdash 0 = 1$. Now let k be the Gödel-number of the formula $0 = 1$; therefore, $A_k(a)$ is the formula $0 = 1$ and $A_k(\bar{k})$ is the same formula, since a does not occur in $0 = 1$. The consistency of Γ can be expressed in $\mathscr{L}$ by the formula $\neg\exists y[DED_\Gamma(\bar{k},y)]$: there is no y such that y is the Gödel-number of a formal deduction of $A_k(\bar{k})$, i.e., $0 = 1$, from Γ. Let $Cons_\Gamma := \neg\exists y[DED_\Gamma(\bar{k},y)]$. Then $Cons_\Gamma$ is a formula in $\mathscr{L}$ expressing the consistency of Γ.

Proof (of Gödel's second theorem). Let Γ be an axiomatizable extension of $\mathscr{P}$. In Gödel's first incompleteness theorem we have shown informally:

(I) if Γ is consistent, then not $\Gamma \vdash A_\Gamma$, where A_Γ is the formula $A_p(\bar{p})$.

The statement that A_Γ is not deducible from Γ is expressed via the Gödel-numbering by $\neg\exists y[DED_\Gamma(\bar{p},y)]$, this is A_Γ itself. The statement that Γ is consistent is expressed by the formula $Cons_\Gamma$. Because the informal proof of (I) is so elementary, it can be completely formalized in $\mathscr{P}$ via the Gödel-numbering, and hence in Γ. Therefore,

(II) $\Gamma \vdash Cons_\Gamma \to A_\Gamma$.

Now suppose that $\Gamma \vdash Cons_\Gamma$. Then it follows from (II) that $\Gamma \vdash A_\Gamma$. Supposing that Γ is also consistent, this is in contradiction to Gödel's first incompleteness theorem. Therefore, if Γ is a consistent and axiomatizable extension of $\mathscr{P}$, then not $\Gamma \vdash Cons_\Gamma$. $\square$

5.3.1 Implications of Gödel's Incompleteness Theorems

In Chapter *X*, *Minds and Machines*, of his book *From Mathematics to Philosophy*, Hao Wang [10] discusses the implications of Gödel's incompleteness results with respect to the superiority of man over machine. In section 7 of this chapter Hao Wang presents as Gödel's opinion that the two most interesting rigorously proved results about minds and machines are:

> 1 The human mind is incapable of formulating (or mechanizing) all its mathematical intuitions. That is, if it has succeeded in formulating some of them, this very fact yields new intuitive knowledge, e.g., the consistency of this formalism. This fact may be called the 'incompletability' of mathematics. On the other hand, on the basis of what has been proved so far, it remains possible that there may exist (and even be empirically discoverable) a theorem-proving machine which in fact *is* equivalent to mathematical intuition, but cannot be *proved* to be so, nor even be proved to yield only *correct* theorems of finitary number theory.
>
> 2 The second result is the following disjunction: Either the human mind surpasses all machines (to be more precise: it can decide more number theoretical questions than any machine) or else there exist number theoretical questions undecidable for the human mind.
>
> Gödel thinks Hilbert was right in rejecting the second alternative. If it were true, it would mean that human reason is utterly irrational by asking questions it cannot answer, while asserting emphatically that only reason can answer them. Human reason would then be very imperfect . . . .

Wang also explains that Gödel considered the attempted proofs for the equivalence of mind and machines as fallacious. See also Searle [7].

5.4 Non-standard Models of Peano's Arithmetic

Let $\mathcal{N}$ be the intended interpretation or standard model of $\mathcal{L}$, the language for formal number theory, i.e., $\mathcal{N} := \langle \mathbb{N}; =; 0, ', +, \cdot \rangle$. Trivially, $\mathcal{N}$ is a model of $\mathcal{P}$, Peano's axioms. But $\mathcal{N}$ is not the only model of $\mathcal{P}$. Given $\mathcal{N}$, one can construct another model of $\mathcal{P}$ that is *isomorphic* but not identical to $\mathcal{N}$ by 'replacing' some element in the domain $\mathbb{N}$ of $\mathcal{N}$ by another object that is not in $\mathbb{N}$. We leave it to the reader to verify that the same sentences are true in isomorphic interpretations.

We now wonder whether any two models of $\mathcal{P}$ (or of some axiomatizable and consistent extension Γ of $\mathcal{P}$) are isomorphic. In that case, one would say that $\mathcal{P}$ (or Γ) characterizes its models 'up to isomorphism' and that it has 'essentially' only one model. The following theorem answers this question in the negative.

Theorem 5.4. *Let Γ be a consistent and axiomatizable extension of $\mathcal{P}$. Then there are two non-isomorphic models of Γ, both with enumerably infinite domains. (In other words, Γ is not aleph-null-categorical).*

Proof. Let Γ be a consistent and axiomatizable extension of $\mathcal{P}$. By Gödel's first incompleteness theorem, there is a sentence A_Γ such that A_Γ is true in $\mathcal{N}$, $\Gamma \not\vdash A_\Gamma$

5.4 Non-standard Models of Peano's Arithmetic

and $\Gamma \not\vdash \neg A_\Gamma$. By Gödel's completeness theorem (for predicate logic), it follows that $\Gamma \not\models A_\Gamma$ and $\Gamma \not\models \neg A_\Gamma$. Hence, there is a model M_1 of Γ such that $M_1 \models \neg A_\Gamma$ and there is a model M_2 of Γ such that $M_2 \models A_\Gamma$. By the Löwenheim-Skolem Theorem (for predicate logic), M_1 and M_2 may be assumed to have an enumerably infinite domain. Since $M_1 \models \neg A_\Gamma$ and $M_2 \models A_\Gamma$, M_1 and M_2 are non-isomorphic. □

Definition 5.9 (Non-standard model). Let M be an interpretation of the language $\mathscr{L}$ for formal number theory. M is a *non-standard model of arithmetic* := the same sentences are true in M as are true in $\mathscr{N}$, and M is not isomorphic to $\mathscr{N}$.

In Theorem 5.5 we prove the existence of non-standard models of arithmetic with enumerably infinite domains.

Theorem 5.5. *Let Δ be the set of all sentences of $\mathscr{L}$ that are true in $\mathscr{N}$. Then there is an interpretation M of $\mathscr{L}$ such that:*
1. M is a model of Δ,
2. M is not isomorphic to $\mathscr{N}$, and
3. M has an enumerably infinite domain.
1 and 2 say that M is a non-standard model of arithmetic. It follows that Δ is not aleph-null-categorical, i.e., it is not the case that any two models of Δ, which both have an enumerably infinite domain, are isomorphic.

Proof. Let Δ be the set of all sentences of $\mathscr{L}$ that are true in $\mathscr{N}$. Let $A_0, A_1, A_2, \ldots$ be an enumeration of all sentences in Δ. Now consider $\Delta' := \{A_0, a_1 \neq 0, A_1, a_1 \neq 0', A_2, a_1 \neq 0'', \ldots\}$. Then each finite subset of Δ' is simultaneously satisfiable. So, by the compactness theorem (for predicate logic), Δ' is simultaneously satisfiable in an enumerable domain. Say $M \models \Delta' \, [a_1^*]$, that is $M \models \Delta$ and $M \models a_1 \neq 0 \, [a_1^*]$, $M \models a_1 \neq 0' \, [a_1^*]$, $M \models a_1 \neq 0'' \, [a_1^*]$, and so on.

For any natural numbers m, n, if $m \neq n$, then $\bar{m} \neq \bar{n}$ is in Δ, where $\bar{1} := 0', \bar{2} := 0''$, etc. Since $M \models \Delta$, the domain of M is enumerably infinite.

The element a_1^* in the domain of M is not the denotation in M of $\bar{n}$ for any natural number n, while in any interpretation isomorphic to $\mathscr{N}$ every element in the domain is denoted by $\bar{n}$ for some natural number n. Hence, M is not isomorphic to $\mathscr{N}$. □

In Chapter 17 of [1], Boolos and Jeffrey investigate what non-standard models of arithmetic do look like.

5.4.1 Second-order Logic (continued)

In Subsection 4.5.3 on second-order logic we have already seen that the Löwenheim-Skolem theorem fails for second-order logic. In this subsection we will indicate other important differences between first- and second-order logic with respect to arithmetic.

First of all, in Theorem 5.5 we have seen that arithmetic (i.e., the set of sentences of $\mathscr{L}$ true in the standard model $\mathscr{N}$) has at least one model which is not isomorphic to $\mathcal{N}$. Below we will show that there is a single sentence, $\mathscr{PA}$, of second-order logic such that any model of $\mathscr{PA}$ is isomorphic to $\mathcal{N}$.

Let *Ind* be the second-order sentence

$$\forall X[\, X(0) \wedge \forall x[X(x) \to X(x')] \to \forall x[X(x)]\,].$$

When interpreted over $\mathcal{N}$, *Ind* formalizes the principle of mathematical induction. Therefore, *Ind* is true in $\mathcal{N}$, interpreting $\forall X$ as 'for all subsets of $\mathbb{N}$'. All of the enumerably many induction axioms of $\mathscr{P}$ (Peano's axioms) are logical consequences of the one second-order sentence *Ind*. Now let $\mathscr{PA}$ be the conjunction of *Ind* and the finitely many axioms of Peano which are not an induction axiom. *Ind* and hence $\mathscr{PA}$ are second-order sentences.

Theorem 5.6. *If* $M \models \mathscr{PA}$, *then M is isomorphic to* $\mathcal{N}$ *(the standard model).*

Proof. Let $M = \langle D; =; e, s, p, t \rangle$ be a model of $\mathscr{PA}$, where e, s, p and t are what M assigns to 0, $'$, $+$ and $\cdot$, respectively. Since M is a model of *Ind*, it follows that for any subset V of D

(†) if both e is in V and $s(d)$ is in V whenever d is in V (for all d in D), then $V = D$.

Define $h : \mathbb{N} \to D$ inductively by: $h(0) = e$, and $h(n') = s(h(n))$. In order to show that h is an isomorphism from $\mathcal{N}$ to M, we still have to prove:
a) h is a surjection from $\mathbb{N}$ to D,
b) h is an injection from $\mathbb{N}$ to D,
c) $h(m+n) = p(h(m), h(n))$, and
d) $h(m \cdot n) = t(h(m), h(n))$.

It is straightforward, but tedious, to prove b), c) and d), using the hypothesis of the theorem. We leave this as an exercise to the reader; or the reader may consult Chapter 18 of [1]. Here we restrict ourselves to the most crucial part of the proof, that is the proof of a). Note:
1) e is in the range of h,
2) if d is in the range of h, then $d = h(n)$ for some n, whence $h(n') = s(d)$, and so $s(d)$ is in the range of h.

It follows from (†) that the range of h equals D, i.e., h is a surjection. □

It is important to note that the proof above does not work for $\mathscr{P}$ instead of $\mathscr{PA}$, although the infinitely many induction axioms of $\mathscr{P}$ logically follow from *Ind*. The point is that 'd is in the range of h' cannot be expressed by any first-order formula A. There are more subsets of $\mathbb{N}$ than formulas in $\mathscr{L}$: there are only denumerably many formulas in $\mathscr{L}$, while there are uncountably many subsets of $\mathbb{N}$.

If $\mathscr{P} \models A$, then A is true in $\mathcal{N}$. But, by Gödel's first incompleteness theorem, the converse does not hold. However, any sentence A, which is true in $\mathcal{N}$, is a valid consequence of the second-order sentence $\mathscr{PA}$.

Corollary 5.2. *Suppose that A is a (first- or second-order) sentence of $\mathscr{L}$. Then* $\mathscr{PA} \models A$ *iff A is true in* $\mathcal{N}$.

5.5 Solutions

Proof. The 'only if' part is trivial. So, suppose A is true in $\mathcal{N}$. We want to show: $\mathscr{P}\mathscr{A} \models A$. So, let M be a model of $\mathscr{P}\mathscr{A}$. Then, by Theorem 5.6, M is isomorphic to $\mathcal{N}$. Since A is true in $\mathcal{N}$, it follows that A is true in M. □

A further Corollary of Theorem 5.6 is that the compactness theorem fails for second-order logic: there is an enumerable, unsatisfiable set of sentences (at least one of them is second-order), every finite subset of which is satisfiable.

Corollary 5.3. *Let* $\Gamma = \{\mathscr{P}\mathscr{A}, c \neq 0, c \neq 0', c \neq 0'', \ldots\}$, *where c is an individual constant. Then every finite subset of Γ is satisfiable, but Γ itself is not satisfiable.*

Proof. One easily sees that every finite subset of Γ is satisfiable. Now suppose Γ itself were satisfiable. Let M' be a model of Γ and let M be like M', but assigning nothing to c. Then M is a model of $\mathscr{P}\mathscr{A}$ and hence, by Theorem 5.6, M is isomorphic to $\mathcal{N}$. On the other hand, because all of $c \neq 0, c \neq 0', c \neq 0'', \ldots$ are true in M', M – having the same domain as M' – cannot be isomorphic to $\mathcal{N}$. Contradiction. Therefore Γ has no model. □

In Subsection 4.5.1 we have given an effective positive test for validity of first-order formulas. However, there is no effective positive test for validity of second-order sentences. The existence of such a test would imply that there is a decision procedure for truth in $\mathcal{N}$, which is not the case. For proofs of these results the reader is referred to Chapter 15 and 18 of [2].

5.5 Solutions

Solution 5.1. a) To show that $\mathscr{P} \vdash \forall x[x = x]$, we use the following abbreviations:
$A := \forall x \forall y \forall z[x = y \to (x = z \to y = z)]$
$B := \forall y \forall z[a_1 + 0 = y \to (a_1 + 0 = z \to y = z)]$
$C := \forall z[a_1 + 0 = a_1 \to (a_1 + 0 = z \to a_1 = z)]$
$D := a_1 + 0 = a_1 \to (a_1 + 0 = a_1 \to a_1 = a_1)$
Below we present a deduction of $\forall x[x = x]$ from Peano's axioms.
1. A; one of the axioms of Peano.
2. $A \to B$; one of the axioms of predicate logic.
3. B; Modus Ponens, 1, 2.
4. $B \to C$; one of the axioms of predicate logic.
5. C; Modus Ponens, 3, 4.
6. $C \to D$; one of the axioms of predicate logic.
7. D; Modus Ponens, 5, 6.
8. $\forall x[x + 0 = x]$; one of the axioms of Peano.
9. $\forall x[x + 0 = x] \to a_1 + 0 = a_1$; one of the axioms of predicate logic.
10. $a_1 + 0 = a_1$; Modus Ponens, 8, 9.
11. $a_1 + 0 = a_1 \to a_1 = a_1$; Modus Ponens, 7, 10.
12. $a_1 = a_1$; Modus Ponens, 10, 11.
13. $a_1 = a_1 \to (\text{axiom} \to a_1 = a_1)$; axiom schema 1.

14. axiom $\to a_1 = a_1$; Modus Ponens, 12, 13.
15. axiom $\to \forall x[x = x]$; $\forall$-rule, 14.
16. axiom.
17. $\forall x[x = x]$; Modus Ponens, 15, 16.

b) To show that $\mathscr{P} \vdash \forall y \forall x [x' + y = (x+y)']$. We use induction on y.
$y = 0$: $\forall x[x' + 0 = (x+0)']$; from the definition of $+$: $x' + 0 = x'$ and $x + 0 = x$.
Induction hypothesis: $\forall x[x' + y = (x+y)']$. To show: $\forall x[x' + y' = (x+y')']$. Proof:
$x' + y' := (x' + y)' =_{indhyp} ((x+y)')' := (x+y')'$.

c) To show that $\mathscr{P} \vdash \forall x \forall y [x + y = y + x]$. We use induction on x, using induction on y in the basis.
$x = 0$: To show $\forall y[0 + y = y + 0]$. We use induction on y:
  $y = 0$: $0 + 0 = 0 + 0$. Induction hypothesis: $0 + y = y + 0$.
  To show: $0 + y' = y' + 0$. Proof: $0 + y' := (0+y)' =_{indhyp} (y+0)' := y'$.
Induction hypothesis: $\forall y[x + y = y + x]$. To show: $\forall y[x' + y = y + x']$.
Proof: $y + x' := (y+x)' =_{indhyp} (x+y)'$ and according to b) $(x+y)' = x' + y$. □

References

1. Boolos, G. and R. Jeffrey, *Computability and Logic*. Cambridge University Press, London, 1974, 1980, 1982.
2. Boolos, G., J. Burgess and R. Jeffrey, *Computability and Logic*. Cambridge University Press, 5th edition, 2007.
3. Hofstadter, D.R., *Gödel, Escher, Bach: an eternal golden braid*. Basic Books, New York, 1979-1999.
4. Kleene, S.C., *Introduction to Metamathematics*. North-Holland, Amsterdam, 1952-2009.
5. Nagel, E. and J.R. Newman, *Gödel's Proof*. Routledge, London, 1971-2001.
6. Paris, J. and L. Harrington, A Mathematical Incompleteness in Peano Arithmetic. In: Barwise, J. (ed.), *Handbook of Mathematical Logic*, North-Holland, Amsterdam, 1977-1999.
7. Searle, J.R., Is the Brain's Mind a Computer Program? *Scientific American* 262 (1990) 20-25.
8. Smith, P., *An Introduction to Gödel's Theorems*. Cambridge Introductions to Philosophy, 2nd edition, 2013.
9. Smullyan, R.M, *Gödel's Incompleteness Theorems*. Oxford University Press, 1992.
10. Wang, H., *From Mathematics to Philosophy*. Routledge & Kegan Paul Ltd, 1974.

Chapter 6
Modal Logic

H.C.M. (Harrie) de Swart

Abstract Modal operators, like 'it is necessary that' or 'John knows that', express an attitude about the proposition to which they are applied. Modal logic studies the reasoning in modal contexts, extending classical logic in which only connectives and quantifiers are taken into account. There are many systems of modal logic, depending on the axioms one wants to accept for the modal operators. The semantics of the modal operators is in terms of possible worlds, where each possible world is supposed to satisfy classical logic. A proposition is necessarily true if it is true in every world accessible or imaginable from the given world. Also tableaux rules are available for the different systems of modal logic. Constructing a tableau-deduction in modal propositional logic of a formula from given premises, if it exists, is straightforward; and if it does not exist, one easily constructs a counterexample from a failed attempt to construct one. Epistemic logic is about the modal operator 'knowing that' and an interesting puzzle in this field is the one of the muddy children. The possible world semantics is useful to understand a number of phenomena in the philosophy of language: rigid designators and the 'de dicto - de re' distinction. Also strict implication and counterfactuals may be understood in terms of possible world semantics. In modal predicate logic we study the behavior of modal operators in combination with the quantifiers. We shall see that in order to make sense, modal contexts should be referentially transparent and at the same time extensionally opaque.

6.1 Modal Operators

Although *modal operators* seldom occur in scientific proza, they do occur in daily language: it is *possible* that Rotterdam is the capital of Holland; it is *impossible* that living creatures can survive fire; it is *necessary* that each object is equal to itself; John *knows* that Amsterdam is the capital of the Netherlands; Rhea *believes* that his wife is the best there is; it is *obligatory* to stop for a red traffic light; it is (not) *permitted* to have a gun; John will *always* love Janet.

A *modal operator* expresses an attitude about the proposition to which it is applied. One distinguishes *alethic* operators, such as 'it is necessary that' and 'it is possible that', *epistemic* operators, such as 'agent i knows that' and 'agent i believes that', *deontic* operators, such as 'it is obligatory that' and 'it is permitted that' and *tense* operators, such as 'it is and always will be true that'.

In modal logic one studies reasoning in modal texts, i.e., texts which contain modal operators; see, for instance, Exercise 6.1. One may distinguish:
- *modal propositional logic*: it studies the reasoning in texts containing not only the classical propositional connectives, denoted by $\rightleftarrows$, $\rightarrow$, $\wedge$, $\vee$ and $\neg$, but also the modal operators of necessity, denoted by $\Box$, and possibility, denoted by $\Diamond$; and
- *modal predicate logic*: it studies the reasoning in texts which in addition contain the quantifiers $\forall$ and $\exists$.

Frege's view in Section 4 of his *Begriffsschrift* [13] is that the notions of necessity and possibility belong to epistemology and involve a covert reference to human knowledge for which there is no place in pure logic.

C.I. Lewis' book *A Survey of Symbolic Logic* [25] from 1918 is generally considered to be the beginning of modern modal logic. Rejecting material implication as an adequate representation of 'if ..., then ...', C.I. Lewis put forward a logic of *strict implication*, in which the latter can be rendered in terms of necessity and material implication: $\Box(A \rightarrow B)$.

For a brief outline of the history of modal logic we refer the reader to the *Historical Introduction* of E.J. Lemmon [24], pp. 1-12. For Aristotle's modal logic and Megarian and Stoic Theories of Modality see W. & M. Kneale [20], pp. 81-96 and pp. 117-128 respectively.

In his *Reference and Modality*, W.V. Quine [32] argues that modal logic is problematic, because $\Box$, to be read as 'it is necessary that', and $\Diamond$, to be read as 'it is possible that', create a context for which Leibniz' Law does not seem to hold. His argument is as follows: let $a = 9$ and $b =$ the number of planets. Then $a = b$. But $\Box(9 > 7)$ is considered to be true, while $\Box$ (the number of planets > 7) is generally considered to be false; the number of planets might have been five, if it had pleased the Creator. So substitution of 'the number of planets' for '9' in $\Box(9 > 7)$ turns a truth into a falsehood, while the number of planets $= 9$. However, this argument is misleading, since the expression 9 refers to a natural number, while the expression 'the number of planets' is a function that assigns to every possible world a natural number. And the number 9 cannot be equal to the function 'the number of planets'. What is true is that 9 = the number of planets in this world, that $\Box(9 > 7)$ and that we hence also have to accept that $\Box$(the number of planets in this world > 7), which is not counter-intuitive at all. We shall come back to this issue in Subsection 6.6.2 and in Subsection 6.11.1.

Since the principle of extensionality (Leibniz' Law) at first sight does not seem to hold for contexts containing modal, epistemic or psychological operators, such contexts have come to be called non-extensional or *intensional*. See L. Linsky [28]. For a closer investigation of this issue see Subsection 6.11.1 on Modal Predicate Logic and Essentialism.

A model-theoretic description of modal logic in terms of *possible worlds* was developed, in particular by S.A. Kripke in his paper *Semantical Analysis of Modal Logic* [21]. The basic idea may be said to be to treat modal contexts as involving a reference to more than one possible world or possible state of affairs. □A holds in world w iff A holds in all worlds which are accessible from w and ◊A holds in world w iff there is some world accessible from w in which A holds.

Exercise 6.1. Translate the following argument in the language of modal propositional logic: If I want to succeed [S], then I should make many exercises [E]. If I want to make many exercises, then I should have a lot of free time [L]. It is impossible to have a lot of free time. Therefore, it is impossible to succeed.

6.2 Different systems of Modal Logic

Modal logic results from classical logic by adding one (or two) connectives to the language of classical logic:
□, to be read as: 'it is necessary that'; or as: 'it is obligatory that'; or as: 'agent i knows that', etc., and
◊, to be read as: 'it is possible that'; or as: 'it is permitted that', etc.

However, ◊$A \rightleftarrows \neg\Box\neg A$ is generally accepted as an axiom schema. Alternatively, one may define ◊A as $\neg\Box\neg A$: A is possible iff $\neg A$ is not necessary; and A is permitted iff $\neg A$ is not obligatory.

With □ and ◊ added as unary operators to the language of classical logic, □P, ◊P, □$P \rightarrow$ ◊P, □(□$P \rightarrow$ ◊P), □□P, □◊P, ◊□P, ◊◊P, and so on, become formulas of our extended language. Using ◊ we may translate the expression 'P is *contingent*' by ◊$P \land$ ◊$\neg P$; and the expression 'P is *compatible* with Q' as ◊$(P \land Q)$.

Since □ may have different (alethic, deontic, epistemic, tense) readings or interpretations, it comes as no surprise that there are many different axioms one may postulate for □. Even the meaning of the word 'necessary' may vary:
- logically necessary, like in: 'if I walk fast, then I walk fast' is logically necessary;
- physically necessary, like in: it is physically necessary that if I drop this pencil, then it falls to the ground;
- ethically necessary, like in: 'one should not kill' is ethically necessary.
However, in general the notion of necessity is not a very clear one: 'men are necessarily mortal' may mean 'all men are mortal' or 'from certain biological laws it follows that men are mortal' or 'from the history up till now it follows that men are mortal'; and the reader may discover other meanings as well.

Depending on the intended meaning of the modal operator □ one may accept or reject one or more axioms for □. For instance, □$A \rightarrow A$ seems plausible for the alethic interpretation of □: if A is (logically or physically) necessary, then A will be the case; but the same formula is not plausible for the deontic reading of □: from A is obligatory, it does not have to follow that A is actually the case. On the other hand, the formula □$A \rightarrow$ ◊A seems plausible for the deontic reading of □: if A is obligatory, then A is permitted.

By imposing different conditions on $\Box$, many modal logics result. Below we list some of the more important systems of modal logic.

The modal logic K (named after Kripke) results from classical propositional logic by adding to the axioms of (classical) propositional logic for $\rightarrow$, $\wedge$, $\vee$ and $\neg$ (see Section 2.6) and the rule Modus Ponens (from A and $A \rightarrow B$ deduce B) one axiom schema and one rule of inference for $\Box$:

axiom schema: $\Box(A \rightarrow B) \rightarrow (\Box A \rightarrow \Box B)$

rule: $\dfrac{\vdash A}{\vdash \Box A}$ i.e., if A is a theorem (of modal logic), then $\Box A$ is too.

The modal logic KT is obtained from K by adding the axiom schema

$$T: \Box A \rightarrow A.$$

The modal logic $S4 = KT4$ is obtained from KT by adding the axiom schema

$$4: \Box A \rightarrow \Box\Box A,$$

and the modal logic $S5 = KT4E$ is obtained from $KT4 = S4$ by adding the axiom schema

$$E: \Diamond A \rightarrow \Box \Diamond A.$$

Under the epistemic reading, the 4-axiom $\Box A \rightarrow \Box\Box A$ is called *positive introspection*: if I know A, then I know that I know A; and the E-axiom $\Diamond A \rightarrow \Box \Diamond A$ is called *negative introspection*: if I do not know $\neg A$, then I know that I do not know $\neg A$.

Definition 6.1. By $K-$ we shall mean any of the systems K, KT, $KT4 = S4$, or $KT4E = S5$.

Definition 6.2. The alphabet of the language of modal propositional logic consists of the following symbols:
$P_1, P_2, P_3, \ldots$, called propositional variables or atomic formulas;
the operators $\rightleftarrows$, $\rightarrow$, $\wedge$, $\vee$, $\neg$ and $\Box$; and the brackets (and).

Definition 6.3 (Formulas of modal propositional logic).
$P_1, P_2, P_3, \ldots$ are formulas of modal propositional logic;
If A and B are formulas of modal propositional logic, then also $(A \rightleftarrows B)$, $(A \rightarrow B)$, $(A \wedge B)$ and $(A \vee B)$ are formulas of modal propositional logic;
If A is a formula of modal propositional logic, then also $(\neg A)$ and $(\Box A)$ are formulas of modal propositional logic.

Definition 6.4. $\Diamond A := \neg \Box \neg A$.

Warning: $\Box \neg A$ or, equivalently, $\neg \Diamond A$ means '$\neg A$ is necessary' or, equivalently, A is impossible. Notice that in $\Box \neg A$ the negation concerns A. But $\neg \Box A$ or, equivalently, $\Diamond \neg A$ means 'A is not necessary' or, equivalently, $\neg A$ is possible. Notice that in $\neg \Box A$ the negation concerns $\Box$.

Convention We can minimize the need for parentheses by agreeing that we leave out the most outer parentheses in a formula and that in

6.2 Different systems of Modal Logic

$$\rightleftarrows, \rightarrow, \wedge, \vee, \neg, \Box$$

any connective has a higher rank than any connective to the right of it and a lower rank than any connective to the left of it.

According to this convention, $\Box A \wedge B \rightarrow C$ should be read as $((\Box A) \wedge B) \rightarrow C$, i.e., if A is necessary and (in addition) B, then C, because $\rightarrow$ has a higher rank than $\wedge$ and $\wedge$ has a higher rank than $\Box$. This formula is different from the formula $(\Box(A \wedge B)) \rightarrow C$, i.e., if $A \wedge B$ is necessary, then C, and also different from the formula $\Box((A \wedge B) \rightarrow C)$, i.e., it is necessary that if $A \wedge B$, then C. According to our convention, the formula $\Box \neg A \vee B$ should be read as $(\Box \neg A) \vee B$, because $\vee$ has a higher rank than $\neg$ and $\Box$, and not as $\Box((\neg A) \vee B)$, nor as $\Box(\neg(A \vee B))$, which mean quite something else.

Definition 6.5 (Deduction; deducible). Let $A_1, A_2, \ldots, A_n$ and B be formulas of modal propositional logic. A *deduction* of B from $A_1, A_2, \ldots, A_n$ in the modal propositional logic $K-$ is a finite sequence of formulas with B as last one, such that each formula in this sequence is either one of the formulas $A_1, A_2, \ldots, A_n$, or one of the logical axioms of $K-$, or is obtained by applying one of the rules to formula(s) earlier in the sequence.

B is *deducible from* $A_1, A_2, \ldots, A_n$ in $K-$ iff there exists a deduction of B from $A_1, A_2, \ldots, A_n$ in $K-$. **Notation**: $A_1, A_2, \ldots, A_n \vdash B$ in $K-$.

In case $n = 0$, i.e., there are no premisses $A_1, A_2, \ldots, A_n$, we say that B is *provable* in $K-$. **Notation**: $\vdash B$ in $K-$.

Example 6.1. $\vdash A \rightarrow \Diamond A$ in KT and also $\vdash \Box A \rightarrow \Diamond A$ in KT.

Proof. $\Box \neg A \rightarrow \neg A$ is an axiom of KT. Since $\Box \neg A \rightarrow \neg A \vdash A \rightarrow \neg \Box \neg A$ in classical propositional logic (contraposition), it follows that $\vdash A \rightarrow \Diamond A$ in KT. Both $\Box A \rightarrow A$ and $A \rightarrow \Diamond A$ are provable in KT and because $\Box A \rightarrow A$, $A \rightarrow \Diamond A \vdash \Box A \rightarrow \Diamond A$ in classical propositional logic, it follows that $\Box A \rightarrow \Diamond A$ is provable in KT. □

Exercise 6.2. Show that a) $A \rightarrow \Box \Diamond A$ and b) $\neg \Box A \rightarrow \Box \neg \Box A$ are provable in $S5$.

Exercise 6.3 (Cosmological argument for God's existence).
Let P stand for 'something exists' and Q for 'there is a perfect being (God exists)'. Show that: $\Diamond P$, $\Box(\Diamond P \rightarrow Q) \vdash \Box Q$ in $S5$. [From Hubbeling [19], Section 8; 'cosmological' because of the occurrence of $\Diamond P$]

Exercise 6.4 (Ontological proof of God's existence). Let Q stand for 'God exists'. Show that: $\Box(Q \rightarrow \Box Q)$, $\Diamond Q \vdash Q$ in $S5$. [This argument is Hartshorne's version of Anselm's ontological proof of God's existence (Anselm, Proslogion III); see Hubbeling [19], Section 8.]

Exercise 6.5. Find the mistake made in the following putative deduction in the modal logic $S5$ of Q (God exists) from $Q \rightarrow \Box Q$ and $\Diamond Q$.
1. $\Box Q \vee \neg \Box Q$
2. $\Box Q \vee \Box \neg \Box Q$ From 1 and exercise 6.2.
3. $\neg \Box Q \rightarrow \neg Q$ From the premiss $Q \rightarrow \Box Q$.
4. $\Box \neg \Box Q \rightarrow \Box \neg Q$ From 3 and the axioms and rule for $\Box$.
5. $\Box Q \vee \Box \neg Q$ From 2 and 4.
6. $\Box Q$ From 5 and the premiss $\Diamond Q$.
7. Q From 6 and $\Box Q \rightarrow Q$.

Exercise 6.6 (Ross's Paradox). Prove directly from the definitions:
i) $\vdash \Box A \to \Box(A \vee B)$ in K, and ii) $\vdash \Diamond A \to \Diamond(A \vee B)$ in K.
Notice that these theorems at first sight look counter-intuitive in the case of deontic logic, reading $\Box A$ as 'it is obligatory that A' or 'A ought to be the case'. See, however, the discussion of deontic logic in Section 6.3.

6.3 Possible World Semantics

Clearly, the truth of $\Box A$ depends on more than just the truth value of A. We say that $\Box A$ is true in the present world/situation w iff A is true in all worlds/situations w' which are accessibie/imaginable from w. And that $\Diamond A$ is true in world w iff there is a world w' accessible from w such that A is true in world w'. Consider, for instance, the following state of affairs: Jane is cleaning the street with water. So, in the present world/situation w_0, it does not rain ($\neg P$) and the street becomes wet (Q). In the present world/situation, Jane can imagine two other possible worlds, one (w_1) in which it does not rain ($\neg P$) and the street does not become wet ($\neg Q$) and another one (w_2) in which it does rain (P) and the street becomes wet (Q). We may model this state of affairs with the following (Kripke) model M:

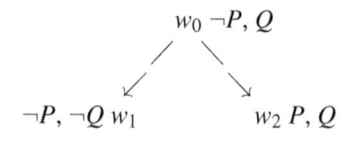

$$w_0 \ \neg P, Q$$
$$\swarrow \quad \searrow$$
$$\neg P, \neg Q \ w_1 \qquad w_2 \ P, Q$$

Given this state of affairs or Kripke model M, $\Box(P \to Q)$ (necessarily: if it rains, then the street becomes wet) is true in world w_0 because in every world Jane can imagine, i.e., in worlds w_0, w_1, w_2, it is true that if it rains, then the street becomes wet, in other words, in all three worlds, $\neg P$ is true or Q is true.

And $\Diamond P$ (it is possible that it rains) is true in world w_0, because Jane can imagine a world w', namely w_2, in which P is true.

We may describe this Kripke model M by the tuple $M = \langle \{w_0, w_1, w_2\}, R, \models \rangle$, where the accessibility relation R is defined by $w_0 R w_0$, $w_0 R w_1$ and $w_0 R w_2$, and where $\models$ is defined by $w_0 \not\models P$, $w_0 \models Q$, $w_1 \not\models P$, $w_1 \not\models Q$, $w_2 \models P$ and $w_2 \models Q$. Clearly, the picture contains all this information.

Of course, in world (situation) w_1 Jane may imagine two other possible worlds (situations): w_3, in which $\neg P$ and $\neg Q$ hold, and in addition the sun is shining (S), and w_4, in which P, Q and $\neg S$ are true. This state of affairs is then described by the following Kripke model M':

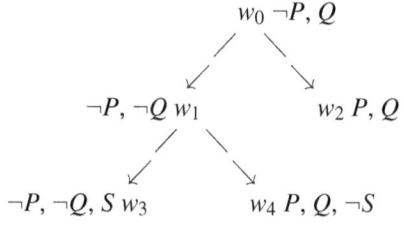

$$w_0 \ \neg P, Q$$
$$\swarrow \quad \searrow$$
$$\neg P, \neg Q \ w_1 \qquad w_2 \ P, Q$$
$$\swarrow \quad \searrow$$
$$\neg P, \neg Q, S \ w_3 \qquad w_4 \ P, Q, \neg S$$

6.3 Possible World Semantics

If in model M' it holds that world w_3 is accessible from world w_0, i.e., w_0Rw_3, then $\Diamond S$ (it is possible that the sun is shining) is true in world w_0 of model M'. However, if not w_0Rw_3, then $\Diamond S$ is not true in world w_0 of model M'.

This brings us to the general definition of a Kripke model.

Definition 6.6 (Kripke model). $M = \langle W, R, \models \rangle$ is a *Kripke model* iff

- W is a non-empty set, the elements of which are called *possible worlds*;
- R is a binary relation on W, called the *accessibility relation*; wRw' is to be read as: world w' is accessible from world w;
- $\models$ is a relation between the elements of W and the atomic formulas; $w \models P$ is to be read as: atomic formula P is true in world w.

In the case of deontic logic, wRw' is read as: w' is a (deontically) perfect alternative of w.

Definition 6.7 ($M, w \models A$)**.** Given a Kripke model $M = \langle W, R, \models \rangle$, we define $M, w \models A$ (to be read as: A is true (holds) in world w of model M) for arbitrary w in W and for arbitrary formulas A (of modal propositional logic) as follows:

- $M, w \models P := w \models P$ (P atomic).
- $M, w \models B \land C := M, w \models B$ and $M, w \models C$.
- $M, w \models B \lor C := M, w \models B$ or $M, w \models C$.
- $M, w \models B \rightarrow C :=$ not $M, w \models B$ or $M, w \models C$.
- $M, w \models \neg B :=$ not $M, w \models B$, also written as $M, w \not\models B$.
- $M, w \models \Box B :=$ for all w' in W, if wRw', then $M, w' \models B$.
- $M, w \models \Diamond B :=$ there is a world w' in W such that wRw' and $M, w' \models B$.

Note that the connectives $\land$, $\lor$, $\rightarrow$ and $\neg$ in each world w are treated as in classical logic; in other words, classical logic applies in each possible world, i.e., a Kripke model can be conceived as a collection of classical models, supplemented by an accessibility relation.

Definition 6.8 ($M \models A$)**.** Let $M = \langle W, R, \models \rangle$ be a Kripke model and A a formula. M is a Kripke *model* of A (or A is true in M) := for every world w in W, $M, w \models A$.
Notation: $M \models A$. 'Not $M \models A$' is also denoted by: $M \not\models A$.

It is easy to check that the axiom for K, i.e., $\Box(B \rightarrow C) \rightarrow (\Box B \rightarrow \Box C)$, is true in every world w of every Kripke model M, i.e., for all Kripke models M, $M \models \Box(B \rightarrow C) \rightarrow (\Box B \rightarrow \Box C)$. We shall say that $\Box(B \rightarrow C) \rightarrow (\Box B \rightarrow \Box C)$ is *valid*.

Proof. Suppose $M, w \models \Box(B \rightarrow C)$, i.e., for all w' in M, if wRw', then $M, w' \models B \rightarrow C$. (1)
Next, suppose $M, w \models \Box B$, i.e., for all w' in M, if wRw', then $M, w' \models B$. (2)
Then it follows from (1) and (2) that for all worlds w' in M, if wRw', then $M, w' \models C$, i.e., $M, w \models \Box C$. □

Instead of saying that $\Box(B \rightarrow C) \rightarrow (\Box B \rightarrow \Box C)$ is valid, we may also say that $\Box B \rightarrow \Box C$ is a *valid consequence* of $\Box(B \rightarrow C)$.

Definition 6.9 (Valid consequence; valid). B is a *valid consequence* of premises $A_1, \ldots, A_n$:= for all Kripke models M and for every world w in M, if $M, w \models A_1 \wedge \ldots \wedge A_n$, then $M, w \models B$. **Notation:** $A_1, \ldots, A_n \models B$.

In case $n = 0$, i.e., there are no premises, we say that B is *valid*, i.e., for all Kripke models M and for all worlds w in M, $M, w \models B$. **Notation:** $\models B$.

Notice that $A_1, \ldots, A_n \models B$ iff $\models A_1 \wedge \ldots \wedge A_n \to B$.

It is also easy to verify that the only rule for $\Box$ (if $\vdash A$, then $\vdash \Box A$) preserves validity: if $\models A$, then $\models \Box A$.

Proof. Suppose that $\models A$, i.e., for all Kripke models M and for every world w in M,
$$M, w \models A. \tag{1}$$
We have to show that for all M and for all w in M, $M, w \models \Box A$, i.e., for all w' in M, if wRw', then $M, w' \models A$. This follows trivially from (1). □

So, we have shown the following theorem:

Theorem 6.1.
1. $\models \Box(B \to C) \to (\Box B \to \Box C)$*; equivalently:* $\Box(B \to C) \models (\Box B \to \Box C)$.
2. if $\models A$*, then* $\models \Box A$.

The $\Box$-axiom for KT, $\Box A \to A$, is not in all worlds of all Kripke models true. The following Kripke model $M = \langle \{w_0, w_1\}, R, \models \rangle$ with $w_0 R w_1$, but not $w_0 R w_0$, is a counterexample:

$$w_0$$
$$\downarrow$$
$$w_1 \; P$$

$M, w_0 \models \Box P$, but $M, w_0 \not\models P$. In world w_0 of this Kripke model M, P (stopping for a red traffic light) is obligatory, meaning that P is true in all deontically perfect alternatives of w_0, but P does not have to be true in w_0.

It is easy to see that $\Box A \to A$ holds precisely in those Kripke models $M = \langle W, R, \models \rangle$ in which the accessibility relation R is *reflexive*, i.e., for all w in M, wRw. For if $M, w \models \Box A$, and R is reflexive, then clearly $M, w \models A$.

Deontic logic If one reads $\Box A$ as 'it ought to be the case that A' (or, equivalently, as 'A is obligatory') and $\Diamond A$ as 'it is permitted that A' (or, equivalently, as 'A is permissible'), one speaks of *deontic logic*. In that case wRw' is read as: w' is a deontically perfect alternative to w. Consequently, $w \models \Box A$ iff A is the case in all deontically perfect alternatives to w, and $w \models \Diamond A$ iff there is a deontically perfect alternative to w, in which A is true.

It is clear that in deontic logic $\Box A \to A$ and $A \to \Diamond A$ do not hold. This means that in general the accessibility relation R should not be reflexive. On the other hand, $\Box A \to \Diamond A$ should be valid in deontic logic. A necessary and sufficient condition on R in order to achieve this is that for each world w in a given Kripke model M there is a w' in M such that wRw'. This condition also rules out $\Box A \wedge \Box \neg A$ (something is obligatory and forbidden).

6.3 Possible World Semantics

However, certain theorems are not dependent upon any condition concerning R. Some of these theorems have been viewed with suspicion because of their paradoxical appearance as deontic principles. For example, A. Ross illustrated the oddity of $\Box A \to \Box(A \vee B)$ by substituting 'I mail a letter' for A and 'I burn the letter' for B. The result 'if I ought to mail a letter, then I ought to mail or burn it' is known as *Ross's paradox*. A similar substitution may reveal the strangeness of $\Diamond A \to \Diamond(A \vee B)$. (See Exercise 6.6.) However, although $\Box(A \vee B)$ is true if $\Box A$ is true, according to Grice's [16] conversation rules, discussed in Section 2.10.2, it is simply misleading to say $\Box(A \vee B)$, when one knows $\Box A$. For more information on deontic logic the reader is referred to Hilpinen [18].

Also the $\Box$-axiom for S4, $\Box A \to \Box\Box A$, is not in all worlds of all Kripke models true. The following Kripke model $M = \langle \{w_0, w_1, w_2\}, R, \models \rangle$ with $w_0 R w_1$, $w_1 R w_2$, but not $w_0 R w_2$, is a counterexample:

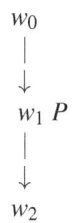

w_0
$\downarrow$
$w_1 \ P$
$\downarrow$
w_2

$M, w_0 \models \Box P$, because $M, w_1 \models P$. But $M, w_0 \not\models \Box\Box P$, because $w_0 R w_1$ and $M, w_1 \not\models \Box P$, the latter because $M, w_2 \not\models P$.

It is easy to see that $\Box A \to \Box\Box A$ holds precisely in those Kripke models $M = \langle W, R, \models \rangle$ in which the accessibility relation R is *transitive*, i.e., for all w, w', w'' in M, if wRw' and $w'Rw''$, then wRw''.

Proof. Let M be a Kripke model, w a world in M, and suppose $M, w \models \Box A$, i.e., for all w' in M, if wRw', then $M, w' \models A$. (1)
We have to show that $M, w \models \Box\Box A$, i.e., for all w' in M, if wRw', then $M, w' \models \Box A$. So, suppose that wRw'. (2)
We have to show that $M, w' \models \Box A$, i.e., for all w'' in M, if $w'Rw''$, then $M, w'' \models A$. So, suppose that $w'Rw''$. (3)
Assuming that R is transitive, it follows from (2) and (3) that wRw''. Hence, from (1): $M, w'' \models A$. □

Finally, also the $\Box$-axiom for S5, $\Diamond A \to \Box \Diamond A$, is not in all worlds of all Kripke models true. The following Kripke model $M = \langle \{w_0, w_1, w_2\}, R, \models \rangle$ with $w_0 R w_1$ and $w_0 R w_2$, but not $w_2 R w_1$, is a counterexample:

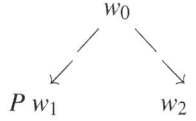

w_0
↙ ↘
$P \ w_1 \quad w_2$

$M, w_0 \models \Diamond P$, because $w_0 R w_1$ and $M, w_1 \models P$. But $M, w_0 \not\models \Box \Diamond P$, because $w_0 R w_2$ and $M, w_2 \not\models \Diamond P$.

It is not difficult to see that $\Diamond A \to \Box \Diamond A$ holds precisely in those Kripke models $M = \langle W, R, \models \rangle$ in which the accessibility relation R is transitive and *symmetric*, i.e., for all w and w' in M, if wRw', then also $w'Rw$.

Proof. Let M be a Kripke model, w a world in M, and suppose $M, w \models \Diamond A$, i.e., there is some world w_0 in M such that wRw_0 and $M, w_0 \models A$. (1)

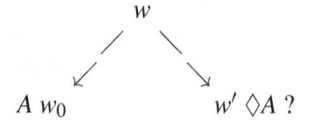

$$
\begin{array}{c}
w \\
\swarrow \quad \searrow \\
A\ w_0 \qquad\qquad w'\ \Diamond A\ ?
\end{array}
$$

We have to show that $M, w \models \Box \Diamond A$, i.e., for all w' in M, if wRw', then $M, w' \models \Diamond A$. So, suppose wRw'. (2)
We have to show that $M, w' \models \Diamond A$. Now, assuming that R is symmetric, it follows from (2) that also $w'Rw$. (3)
Assuming that R is transitive, it follows from (3) and (1) that $w'Rw_0$. And because $M, w_0 \models A$ (1), it follows that $M, w' \models \Diamond A$. □

We collect the preceding results in the following theorem.

Theorem 6.2.
For every Kripke model $M = \langle W, R, \models \rangle$, $M \models \Box(A \to B) \to (\Box A \to \Box B)$.
For every Kripke model $M = \langle W, R, \models \rangle$ with R reflexive, $M \models \Box A \to A$.
For every Kripke model $M = \langle W, R, \models \rangle$ with R transitive, $M \models \Box A \to \Box\Box A$.
For every Kripke model $M = \langle W, R, \models \rangle$ with R transitive and symmetric, $M \models \Diamond A \to \Box \Diamond A$.

Definition 6.10 (Kripke model for $K-$). Let $M = \langle W, R, \models \rangle$ be a Kripke model. M is a Kripke model for KT iff R is reflexive. M is a Kripke model for $KT4 = S4$ iff R is reflexive and transitive. M is a Kripke model for $KT4E = S5$ iff R is reflexive, transitive and symmetric.

Definition 6.11 (Valid consequence in $K-$). B is a *valid consequence* of premises $A_1, \ldots, A_n$ in $K- :=$ for all Kripke models M for $K-$ and for every world w in M, if $M, w \models A_1 \wedge \ldots \wedge A_n$, then $M, w \models B$. **Notation**: $A_1, \ldots, A_n \models B$ in $K-$.
In case $n = 0$, i.e., there are no premises, we say that B is *valid* in $K-$, i.e., for all Kripke models M for $K-$ and for every world w in M, $M, w \models B$.
Notation: $\models B$ in $K-$.

From Theorems 6.1 and 6.2 the following *soundness* theorem results, saying that any formula that may be logically deduced in $K-$ from given premises is a valid consequence in $K-$ of those premises:

Theorem 6.3 (Soundness of modal propositional logic).
If $A_1, \ldots, A_n \vdash B$ in $K-$, then $A_1, \ldots, A_n \models B$ in $K-$.

Proof. Suppose $A_1, \ldots, A_n \vdash B$ in K, i.e., there is a finite schema of formulas with B as last one, such that every formula A in this schema is either one of $A_1, \ldots, A_n$ or an

6.4 Epistemic logic

axiom of classical propositional logic or the $\Box$-axiom of K or obtained by the rule Modus Ponens to two preceding formulas C and $C \to D$ in the schema or obtained by application of the rule for $\Box$ to a preceding formula E in the schema such that $\vdash E$. We have to show that $A_1, \ldots, A_n \models B$ in K. So, let M be a Kripke model, w be a world in M and suppose $M, w \models A_1 \wedge \ldots \wedge A_n$. Notice that:

1. If A is an axiom of propositional logic or A is the $\Box$-axiom for K, then $M, w \models A$.
2. If $M, w \models C$ and $M, w \models C \to D$, then $M, w \models D$.
3. If $\models E$, then by Theorem 6.1 $\models \Box E$.

Hence, from 1, 2 and 3: $A_1, \ldots, A_n \models B$ in K.

The proofs for KT, $S4$ and $S5$ are similar. $\Box$

Exercise 6.7. Prove that $A \not\models \Box A$, although by Theorem 6.1: if $\models A$, then $\models \Box A$.

Exercise 6.8. Prove or refute: a) $\models \Box(A \wedge B) \rightleftarrows (\Box A \wedge \Box B)$; b) $\models \Box(A \vee B) \rightleftarrows (\Box A \vee \Box B)$; c) $\models \Diamond(A \vee B) \rightleftarrows (\Diamond A \vee \Diamond B)$.

6.4 Epistemic logic

In *epistemic logic* $\Box A$ is read as 'I know that A'. More generally, $\Box_i A$ is read as 'agent i knows that A', if one wants to consider more than one agent. Then $wR_i w'$ is read as: in world w agent i considers – on the ground of his knowledge – world w' as an (epistemic) alternative.

Because of the validity of $(\Box(A \to B) \wedge \Box A) \to \Box B$, epistemic logic is not concerned with actual occurrent knowledge, but with virtual or *implicit knowledge*. If a knower (or agent) knows A and $A \to B$, he or she also knows B, at least in principle, although one may not explicitly be aware of this.

In epistemic logic, one frequently uses K (Knowing) instead of the $\Box$-operator. For instance, K_A for 'A(lice) knows A' and K_B for 'B(ob) knows A'.

As an example with two agents, consider the following state of affairs: A(lice) works in an office without windows, it is raining (P), but as far as Alice knows also $\neg P$ might be the case. B(ob) works in an office with windows, has been informed that it will rain all day and considers it possible that an important letter will arrive today (Q). We may model this state of affairs by the following Kripke model $M = \langle \{w_0, w_1, w_2\}, R_A, R_B, \models \rangle$ with $w_0 R_A w_1$ and $w_0 R_B w_2$, R_A and R_B both reflexive, transitive and symmetric, $w_0 \models P$, but $w_1 \not\models P$, $w_2 \models P$ and $w_2 \models Q$.

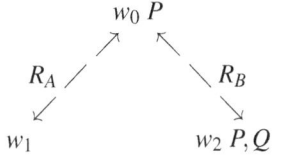

$$
\begin{array}{ccc}
 & w_0\ P & \\
R_A \swarrow & & \searrow R_B \\
w_1 & & w_2\ P, Q
\end{array}
$$

Clearly, $M, w_0 \not\models K_A P$ (in world w_0 of model M, Alice does not know P), because $w_0 R_A w_1$ and $w_1 \not\models P$ (Alice can imagine w_1 in which it does not rain). But $M, w_0 \models$

$K_B P$ (in world w_0 of model M, Bob knows P), because P holds in both worlds Bob can imagine: w_0 and w_2.

$M, w_0 \models K_A(K_B P \vee K_B \neg P)$ (Alice knows in world w_0 of model M that Bob knows if P holds), because $M, w_0 \models K_B P$ (Bob knows in world w_0 that P) and $M, w_1 \models K_B \neg P$ (because from world w_1 Bob can only imagine w_1).

$M, w_0 \models \neg K_B(\neg K_A P)$ (Bob does not know in w_0 that Alice does not know P), because $w_0 R_B w_2$ and $M, w_2 \models K_A P$ (from world w_0 Bob can imagine world w_2 and in w_2 Alice knows P, because the only world she can imagine from w_2 is w_2 itself).

As this example suggests, epistemic logic can be used for the formal description of the knowledge of 'agents' in distributed systems. A nice illustration is the *muddy children puzzle*. See also Exercise 6.9.

6.4.1 Muddy Children Puzzle; Reasoning about Knowledge

Imagine the following state of affairs. Two children are playing outside and their father asks them to come home. Both have mud on their foreheads, but they do not know themselves. Each child can see the other child, but not him- or herself; there are no mirrors. The father does not allow the children to talk to each other and says: at least one of you has mud on his forehead (*P*). If you *know* you have mud on your forehead, please step forward.

No child will step forward: each child sees the other child with mud on its forehead and considers it possible to be clean (without mud) himself. Notice that already before the statement of the father each child knows that *P*, but does not know that the other child knows *P*. After the statement of the father *P* has become common knowledge, in particular, now each child knows that the other child also knows *P*.

Since no child steps forward, the father repeats his request and asks again: if you *know* you have mud on your forehead, please step forward. Now both children step forward. Why? Because they can perfectly reason about knowledge: if there were only one child with mud, after the first statement/request of the father this child would know that he is the one with mud and step forward. Since no one stepped forward, there must be (at least) two children with mud.

We may model the state of affairs before the statement of the father by the following Kripke model M, where m_i stands for 'child i, $i = 1, 2$, has mud on his forehead', and R_i is the accessibility relation for child i.

$$\begin{array}{ccccc}
m_1, m_2 & w_1 \leftarrow\!-\!-\!-\!-\!- R_1 -\!-\!-\!-\!- \rightarrow w_2 & \neg m_1, m_2 \\
& \uparrow & & \uparrow & \\
& R_2 & & R_2 & \\
& \downarrow & & \downarrow & \\
m_1, \neg m_2 & w_3 \leftarrow\!-\!-\!-\!-\!- R_1 -\!-\!-\!-\!- \rightarrow w_4 & \neg m_1, \neg m_2
\end{array}$$

Before the statement of the father there are four possible worlds/situations, described by w_1, w_2, w_3 and w_4. For instance, in world w_1 child 1 sees that child 2 has mud on his forehead, but child 1 can imagine to have no mud himself, i.e.,

6.4 Epistemic logic

world w_2 is accessible from world w_1 for child 1: $w_1R_1w_2$. Conversely, from world w_2 child 1 can imagine world w_1: $w_2R_1w_1$. In a similar way, from world w_1 child 2 can easily imagine world w_3 and conversely: $w_1R_2w_3$ and $w_3R_2w_1$. The relations R_1 and R_2 are reflexive, transitive and symmetric.

Notice that $M,w_1 \models K_1m_2 \wedge K_2m_1$. In addition, for each world w in M, $M,w \models \neg K_1m_1$ and $M,w \models \neg K_2m_2$.

By the statement P of the father, world w_4 is eliminated and only three possible worlds are left, as described by the following Kripke model M':

$$
\begin{array}{ccc}
m_1,m_2 & w_1 \leftarrow\!-\!-\!-\!-\!-\,R_1\,-\!-\!-\!-\!-\!\rightarrow w_2 & \neg m_1,m_2 \\
 & \uparrow & \\
 & R_2 & \\
 & \downarrow & \\
m_1,\neg m_2 & w_3 &
\end{array}
$$

After the first statement P of the father, child 1 still does not know that he has mud on his forehead, because it sees child 2 with mud. This corresponds with $M',w_1 \models \neg K_1m_1$. Similarly, $M,w_1 \models \neg K_2m_2$.

If there would be only one child with mud, that is, if w_2 or w_3 would be the actual world, then, of course, after the first statement P of the father, the child with mud would *know* he has mud on his forehead, since he sees that the other child has no mud on his forehead. This corresponds with $M',w_2 \models K_2m_2$ and $M',w_3 \models K_1m_1$. So, if after the first statement/request of the father no child steps forward, each perfect logician will know that there must be at least two children with mud, in other words that world w_2 and w_3 do not occur and that only world w_1 is left. The new state of affairs is described by the Kripke model M'' containing only one possible world, i.e., w_1. And $M'',w_1 \models K_1m_1 \wedge K_2m_2$.

Exercise 6.9 (J.J.Ch. Meyer). Consider the following Kripke model M consisting of four possible worlds w_1, w_2, w_3, w_4, two agents A(lice) and B(ob) with reflexive and transitive accessibility relations R_A and R_B respectively, and suppose that R_A, R_B and $\models$ are defined as indicated in the following picture.

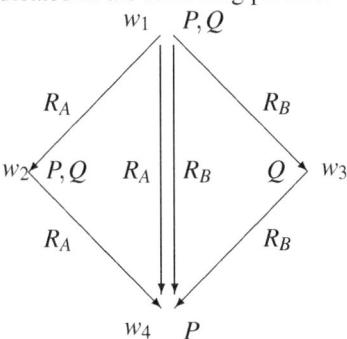

Check that
$M,w_1 \models Q$, $M,w_1 \models \neg K_AQ$, $M,w_1 \models \neg K_BK_AP$, $M,w_1 \models \neg K_B\neg K_AQ$,
$M,w_1 \models K_AP$, $M,w_1 \models \neg K_BQ$, $M,w_1 \models \neg K_A\neg K_BP$, $M,w_1 \models \neg K_A\neg K_BQ$,
$M,w_1 \models \neg K_BP$, $M,w_1 \models K_AK_AP$, $M,w_1 \models K_A\neg K_AQ$, $M,w_1 \models K_B\neg K_BQ$.

6.5 Tableaux for Modal Logics

A *tableaux system* for the modal logics K, KT and $S4$ is obtained by adding T and F rules for the modal operator $\Box$ to the T and F rules for the connectives $\rightarrow, \wedge, \vee, \neg$ of classical propositional logic, given in Section 2.8 and listed below. Now TA is read as: A is true in world w; and FA as: A is false in world w. We do not give the tableaux rules for $S5$ here, because they are complicated and hence somewhat artificial; the interested reader is referred to de Swart [36]. In the *tableaux rules* below, S is a *sequent*, i.e., a set of T- or F-signed formulas.

$$T\wedge \quad \frac{S,\, T\, B \wedge C}{S,\, TB,\, TC} \qquad\qquad F\wedge \quad \frac{S,\, F\, B \wedge C}{S,\, FB \mid S,\, FC}$$

$$T\vee \quad \frac{S,\, T\, B \vee C}{S,\, TB \mid S,\, TC} \qquad\qquad F\vee \quad \frac{S,\, F\, B \vee C}{S,\, FB,\, FC}$$

$$T\rightarrow \quad \frac{S,\, T\, B \rightarrow C}{S,\, FB \mid S,\, TC} \qquad\qquad F\rightarrow \quad \frac{S,\, F\, B \rightarrow C}{S,\, TB,\, FC}$$

$$T\neg \quad \frac{S,\, T\, \neg B}{S,\, FB} \qquad\qquad F\neg \quad \frac{S,\, F\, \neg B}{S,\, TB}$$

For K there is no $T\Box$ rule, but only a $F\Box$ rule: $\quad F\Box \dfrac{S,\, F\Box A}{S_\Box,\, FA}$

For KT (or KM or M or T) the $T\Box$ and $F\Box$ rules are:

$$T\Box \quad \frac{S,\, T\Box A}{S,\, T\Box A,\, TA} \qquad\qquad F\Box \quad \frac{S,\, F\Box A}{S_\Box,\, FA}$$

and for $S4$ these rules are:

$$T\Box \quad \frac{S,\, T\Box A}{S,\, T\Box A,\, TA} \qquad\qquad F\Box \quad \frac{S,\, F\Box A}{S_{T\Box},\, FA}$$

where $S_\Box := \{TB \mid T\Box B \in S\}$ and $S_{T\Box} := \{T\Box B \mid T\Box B \in S\}$, i.e., $S_\Box$ contains all expressions TB for which $T\Box B$ occurs in S and $S_{T\Box}$ is the set of all expressions $T\Box B$ which occur in S. We have drawn a line in the rules $F\Box$ in order to stress that in the transition from S to $S_\Box$ and $S_{T\Box}$, resp., some signed formulas may get lost.

The T- and F-rules for the propositional connectives follow the truth tables for these connectives. For instance, $B \rightarrow C$ is true in world w ($T\, B \rightarrow C$) iff B is false in w (FB) or C is true in w (TC); and $B \rightarrow C$ is false in w ($F\, B \rightarrow C$) iff B is true in w (TB) and C is false in w (FC). For obvious reasons the rules $T \rightarrow$, $T\vee$ and $F\wedge$ are called *split-rules*.

The intuitive motivation behind the T-rule for $\Box$ is this one: if $\Box A$ is true in a world w, then also A will be true in world w, at least if w is accessible from itself, i.e., when R is reflexive. So, this $T\Box$ rule will apply in KT and in $S4$, but not in K.

The intuitive motivation behind the F-rule for $\Box$ is the following one: if $\Box A$ is false in world w, then there must be a world w', accessible from w, in which A is false. Since F-signed formulas (which are aupposed to be false in w) do not have

6.5 Tableaux for Modal Logics

to be false in w', these formulas are not copied. In general, also T-signed formulas (which are supposed to be true in w) do not have to be true in w' and hence are not copied. There is one exception: If a T-signed formula $\Box B$ is true in w, then B will be true in w'; and even $\Box B$ will be true in w', if the accessibility relation R is transitive. So, we have different $F\Box$ rules for K and KT on the one hand, and for $S4$ on the other hand.

$A_1, \ldots, A_n \vdash' B$ (B is *tableau-deducible* from $A_1, \ldots, A_n$) in K, KT or $S4$, resp., is defined in a similar way as in Definition 2.18, the only difference being that there are two more rules for $\Box$.

Example 6.2. Let us verify that $\Box(A \to B) \vdash' \Box A \to \Box B$ in K. We construct a tableau starting with the premiss(es) T-signed and the putative conclusion F-signed; informally: we suppose the premises are true and the putative conclusion false. Next we apply the T and F rules for the different connectives and modal operator.

$$T\Box(A \to B),\ F\ (\Box A \to \Box B)$$
$$T\Box(A \to B),\ T\Box A,\ F\Box B$$
$$\overline{T(A \to B),\ TA,\ FB}$$
$$FA,\ TA,\ FB \mid TB,\ TA,\ FB$$

Since both 'branches' close, i.e., contain TC and FC for some formula C, this schema is by definition a tableau-deduction (in K) of $\Box A \to \Box B$ from $\Box(A \to B)$. Therefore, we have shown that $\Box(A \to B) \vdash' \Box A \to \Box B$ (in K), i.e., one can construct such a tableau-deduction. Informally: the supposition that the premises are true and the conclusion false turns out to be untenable.

Example 6.3. Let us verify that $\vdash' \Box A \to A$ in KT, but not in K:

$$F\ (\Box A \to A)$$
$$T\Box A,\ FA$$
$$TA,\ FA$$

The only 'branch' is closed, and hence $\vdash' \Box A \to A$ in KT.

Notice that this tableau-proof does not hold in K, because there is no $T\Box$ rule for K. If we make a tableau in K for $\Box A \to A$ we find:

$F\ (\Box A \to A)$
$T\Box A,\ FA \qquad w$

which does not close. In fact, we have constructed a Kripke counterexample $M = \langle \{w\}, R, \models \rangle$ in K, with, by definition, not wRw and $w \not\models A$, corresponding with the occurrence of FA in w. $M, w \models \Box A$, since there is no world accessible from w in which A is not true. But $M, w \not\models A$.

Example 6.4. Let us verify that $\vdash' \Box A \to \Box\Box A$ in $S4$, but not in KT:

$$F\ (\Box A \to \Box\Box A)$$
$$T\Box A,\ F\Box\Box A$$
$$T\Box A,\ TA,\ F\Box\Box A$$
$$\overline{T\Box A,\ F\Box A}$$
$$T\Box A,\ TA,\ F\Box A$$
$$\overline{T\Box A,\ FA}$$
$$T\Box A,\ TA,\ FA$$

The only 'branch' of this tableau is closed, and hence $\vdash' \Box A \to \Box\Box A$ in S4.

Notice that this tableau-proof does not hold in *KT*. A tableau starting with $F (\Box A \to \Box\Box A)$ in *KT* will look as follows and does not close:

$F (\Box A \to \Box\Box A)$
$T\Box A, F\Box\Box A$ $w_0\ A$
$T\Box A, TA, F\Box\Box A$ $\downarrow$
$\overline{TA, F\Box A}$ $w_1\ A$
                                           $\downarrow$
FA w_2

In fact, we have constructed a Kripke countermodel $M = \langle\{w_0, w_1, w_2\}, R, \models\rangle$ in *KT*, with $w_0 R w_1$, $w_1 R w_2$, but not $w_0 R w_2$, R reflexive, but not transitive, and by definition $w_0 \models A$, $w_1 \models A$, but $w_2 \not\models A$, corresponding with the occurrence of *TA* in w_0 and w_1 and the occurrence of *FA* in w_2. Then, corresponding with the occurrence of $T\Box A$ in w_0, $M, w_0 \models \Box A$, since $M, w_0 \models A$ and $M, w_1 \models A$, but, corresponding with the occurrence of $F\Box\Box A$ in w_0, $M, w_0 \not\models \Box\Box A$, since $M, w_1 \not\models \Box A$. Notice that if R were transitive, we would not have that $M, w_0 \models \Box A$.

Example 6.5. We shall try to construct a tableau proof of the S5-axiom $\Diamond A \to \Box \Diamond A$ in S4. So, we start with $F(\Diamond A \to \Box \Diamond A)$:

$$F(\Diamond A \to \Box \Diamond A)$$
$$T\Diamond A, F\Box\Diamond A$$
$$T\neg\Box\neg A, F\Box\neg\Box\neg A$$
$$F\Box\neg A, F\Box\neg\Box\neg A$$

At this point there are two possibilities to continue: we may proceed with $F\Box\neg A$ losing the second *F*-signed formula, or we may proceed with $F\Box\neg\Box\neg A$ losing the first *F*-signed formula. Either way, we do not get closure and hence we do not find a tableau proof in S4 of $\Diamond A \to \Box \Diamond A$:

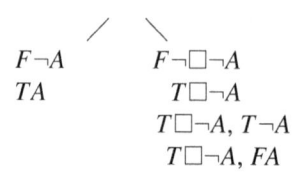

$$\begin{array}{cc} \swarrow & \searrow \\ F\neg A & F\neg\Box\neg A \\ TA & T\Box\neg A \\ & T\Box\neg A, T\neg A \\ & T\Box\neg A, FA \end{array}$$

We shall call the resulting tree the *search tree* for the conjecture $\vdash' \Diamond A \to \Box \Diamond A$ in S4. From this search tree one can immediately read off a Kripke counterexample $M = \langle\{w_0, w_1, w_2\}, R, \models\rangle$ in S4 for this formula, with, by definition, $w_0 R w_1$, $w_0 R w_2$, R reflexive and transitive, but not symmetric, and $w_1 \models A$, corresponding with the occurrence of *TA* in w_1:

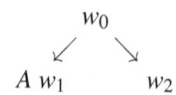

$$\begin{array}{c} w_0 \\ \swarrow \quad \searrow \\ A\ w_1 \qquad w_2 \end{array}$$

Then, corresponding with the occurrence of $T\Diamond A$ in w_0, $M, w_0 \models \Diamond A$, since $w_0 R w_1$ and $M, w_1 \models A$. But, corresponding with the occurrence of $F\Box\Diamond A$ in w_0, $M, w_0 \not\models$

$\Box\Diamond A$, since $w_0 R w_2$ and $M, w_2 \not\models \Diamond A$, corresponding with the occurrence of $F\Diamond A$ in w_2. Notice: if R were symmetric, we would have $M, w_2 \models \Diamond A$, because in that case symmetry would guarantee $w_2 R w_0$ and next transitivity would guarantee $w_2 R w_1$.

Example 6.6. The following tableau $\mathscr{T}$ with initial branch $\mathscr{B}_0 = \{T\Box(P \wedge Q),$ $F\, (\Box P \wedge (\Box Q \vee \Box R))\}$ is a tableau-deduction of $\Box P \wedge (\Box Q \vee \Box R)$ from $\Box(P \wedge Q)$ in K:

$$T\Box(P \wedge Q),\, F\,(\Box P \wedge (\Box Q \vee \Box R))$$
$$T\Box(P \wedge Q),\, F\,\Box P \quad | \quad T\Box(P \wedge Q),\, F(\Box Q \vee \Box R)$$
$$T\Box(P \wedge Q),\, F\,\Box P \quad | \quad T\Box(P \wedge Q),\, F\,\Box Q,\, F\,\Box R$$
$$\overline{T(P \wedge Q),\, FP \quad | \quad T(P \wedge Q),\, FQ}$$
$$TP,\, TQ,\, FP \quad | \quad TP,\, TQ,\, FQ$$

Notice that both branches are closed, i.e., contain for some formula C both TC and FC. Also notice that in the right branch, instead of applying the $F\Box$ rule to $F\Box Q$, we might also have applied the $F\Box$ rule to $F\Box R$, in which case the right branch would finish with TP, TQ, FR and hence would not close.

Let branch $\mathscr{B}_1 = \mathscr{B}_0 \cup \{F\Box P\}$ and branch $\mathscr{B}_2 = \mathscr{B}_0 \cup \{F(\Box Q \vee \Box R)\}$. Then tableau $\mathscr{T}_1 = \{\mathscr{B}_1, \mathscr{B}_2\}$ is called a *one-step expansion* in K of tableau $\mathscr{T}_0 = \{\mathscr{B}_0\}$.

Let branch $\mathscr{B}_{11} = \mathscr{B}_1$ and branch $\mathscr{B}_{21} = \mathscr{B}_2 \cup \{F\Box Q, F\Box R\}$. Then tableau $\mathscr{T}_2 = \{\mathscr{B}_{11}, \mathscr{B}_{21}\}$ is called a *one-step expansion* in K of tableau $\mathscr{T}_1$.

Let branch $\mathscr{B}_{111} = \mathscr{B}_{11}^* \cup \{T(P \wedge Q), FP\}$ and let $\mathscr{B}_{211} = \mathscr{B}_{21}^* \cup \{T(P \wedge Q), FQ\}$, where $\mathscr{B}^*$ indicates that the formulas in $\mathscr{B}$ do not count towards closure anymore. Then tableau $\mathscr{T}_3 = \{\mathscr{B}_{111}, \mathscr{B}_{211}\}$ is called a *one-step expansion* in K of $\mathscr{T}_2$.

Finally, let branch $\mathscr{B}_{1111} = \mathscr{B}_{111} \cup \{TP, TQ\}$ and $\mathscr{B}_{2111} = \mathscr{B}_{211} \cup \{TP, TQ\}$. Then tableau $\mathscr{T}_4 = \{\mathscr{B}_{1111}, \mathscr{B}_{2111}\}$ is called a *one-step expansion* in K of $\mathscr{T}_3$.

Definition 6.12 ((Tableau) Branch). (a) A *tableau branch* is a set of signed formulas. A branch is *closed* if it contains signed formulas TA and FA for some formula A. A branch that is not closed is called *open*.
(b) Let $\mathscr{B}$ be a branch and TA, resp. FA, a signed formula occurring in $\mathscr{B}$. TA, resp. FA, is *fulfilled* in $\mathscr{B}$ if (i) A is atomic, or (ii) $\mathscr{B}$ contains the bottom formulas in the application of the corresponding T or F rule to A, and in case of the rules $T\vee$, $F\wedge$ and $T\rightarrow$, $\mathscr{B}$ contains one of the bottom formulas in the application of these rules.
(c) A branch $\mathscr{B}$ is *completed* if $\mathscr{B}$ is closed or every signed formula in $\mathscr{B}$ is fulfilled in $\mathscr{B}$.

Definition 6.13 (Tableau). (a) A set $\mathscr{T}$ of branches is a *tableau* in $K-$ with initial branch $\mathscr{B}_0$ if there is a sequence $\mathscr{T}_0, \mathscr{T}_1, \ldots, \mathscr{T}_n$ such that $\mathscr{T}_0 = \{\mathscr{B}_0\}$, each $\mathscr{T}_{i+1}$ is a one-step expansion in $K-$ of $\mathscr{T}_i$ ($0 \leq i < n$) and $\mathscr{T} = \mathscr{T}_n$.
(b) We say that a finite $\mathscr{B}$ has tableau $\mathscr{T}$ if $\mathscr{T}$ is a tableau with initial branch $\mathscr{B}$.
(c) A tableau $\mathscr{T}$ in $K-$ is *open* if some branch $\mathscr{B}$ in it is open, otherwise $\mathscr{T}$ is *closed*.
(d) A tableau is *completed* if each of its branches is completed; informally, no application of a tableau rule can change the tableau.

Definition 6.14 (Tableau-deduction; Tableau-proof).
(a) A *tableau-deduction* of B from $A_1, \ldots, A_n$ in $K-$ is a tableau $\mathscr{T}$ in $K-$ with $\mathscr{B}_0 = \{TA_1, \ldots, TA_n, FB\}$ as initial branch, such that all branches of $\mathscr{T}$ are closed.

In case $n = 0$, i.e., there are no premises $A_1, \ldots, A_n$, this definition reduces to:
(b) A *tableau-proof* of B in $K-$ is a tableau $\mathscr{T}$ in $K-$ with $\mathscr{B}_0 = \{FB\}$ as initial sequent, such that all branches of $\mathscr{T}$ are closed.

Definition 6.15 (Tableau-deducible; Tableau-provable).
(a) B is *tableau-deducible from* $A_1, \ldots, A_n$ in $K- :=$ there exists a tableau-deduction of B from $A_1, \ldots, A_n$ in $K-$. **Notation**: $A_1, \ldots, A_n \vdash' B$ in $K-$.
(b) B is *tableau-provable* in $K- :=$ there exists a tableau-proof of B in $K-$.
Notation: $\vdash' B$ in $K-$. And for Γ a (possibly infinite) set of formulas,
(c) B is *tableau-deducible from* Γ in $K- :=$ there exists a finite list $A_1, \ldots, A_n$ of formulas in Γ such that $A_1, \ldots, A_n \vdash' B$ in $K-$. **Notation**: $\Gamma \vdash' B$ in $K-$.

Example 6.7. a) As seen in Example 6.2, $\Box(A \to B) \vdash' (\Box A \to \Box B)$ in K.
b) As seen in Example 6.3, $\Box A \vdash' A$ in KT, or, equivalently, $\vdash' \Box A \to A$ in KT.
c) As seen in Example 6.4, $\Box A \vdash' \Box\Box A$ in $S4$ or, equivalently, $\vdash' \Box A \to \Box\Box A$ in $S4$.

Example 6.8. We wonder whether $\Diamond\Box P \vdash' \Box\Diamond P$ in $S4$. We start a tableau with $T\Diamond\Box P$, $F\Box\Diamond P$ in $S4$:

$$T\Diamond\Box P,\ F\Box\Diamond P$$
$$F\Box\neg\Box P,\ F\Box\Diamond P$$

We may continue with $F\Box\neg\Box P$, losing $F\Box\Diamond P$ and we may continue with $F\Box\Diamond P$, losing $F\Box\neg\Box P$. If one of these two options would give closure, we would have found a tableau deduction of $\Box\Diamond P$ from $\Diamond\Box P$ in $S4$. However, it turns out that either way does not give closure:

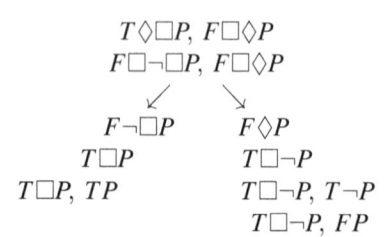

$$T\Diamond\Box P,\ F\Box\Diamond P$$
$$F\Box\neg\Box P,\ F\Box\Diamond P$$

```
        ↙           ↘
     F¬□P           F◇P
      T□P           T□¬P
   T□P, TP      T□¬P, T¬P
                  T□¬P, FP
```

We shall call the resulting tree the *search tree* for the conjecture $\Diamond\Box P \vdash' \Box\Diamond P$ in $S4$. From this search tree with both branches open we may immediately read off a Kripke counterexample $M = \langle \{w_0, w_1, w_2\}, R, \models \rangle$ in $S4$ with, by definition, $w_0 R w_1$, $w_0 R w_2$, R reflexive and transitive, $w_1 \models P$, corresponding with the occurrence of TP in w_1, and $w_2 \not\models P$, corresponding with the occurrence of FP in w_2:

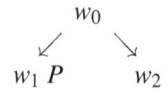

```
         w0
       ↙    ↘
    w1 P     w2
```

Clearly, $M, w_0 \models \Diamond\Box P$, since $M, w_1 \models \Box P$, but $M, w_0 \not\models \Box\Diamond P$, since $M, w_2 \not\models \Diamond P$.

6.5 Tableaux for Modal Logics   295

Example 6.9. We wonder whether $\Box\Diamond P \vdash' \Diamond\Box P$ in *S4*.
We start a tableau with $T\Box\Diamond P, F\Diamond\Box P$ in *S4*, i.e.,

$$T\Box\neg\Box\neg P, F\neg\Box\neg\Box P$$
$$T\Box\neg\Box\neg P, T\neg\Box\neg P, F\neg\Box\neg\Box P$$
$$T\Box\neg\Box\neg P, T\neg\Box\neg P, T\Box\neg\Box P$$
$$T\Box\neg\Box\neg P, F\Box\neg P, T\Box\neg\Box P$$
$$T\Box\neg\Box\neg P, F\Box\neg P, T\Box\neg\Box P, T\neg\Box P$$
$$T\Box\neg\Box\neg P, F\Box\neg P, T\Box\neg\Box P, F\Box P \quad (*)$$

At this stage we have applied the $T\Box$ rule as many times as possible and we now have two signed formulas of the form $F\Box$. If we apply the $F\Box$ rule to either one of them, we loose the other. So, there are two possibilities to go on; if one of them would give closure, we would have a tableau deduction of $\Diamond\Box P$ from $\Box\Diamond P$.

$$T\Box\neg\Box\neg P, F\neg P, T\Box\neg\Box P \qquad T\Box\neg\Box\neg P, T\Box\neg\Box P, FP$$
$$T\Box\neg\Box\neg P, TP, T\Box\neg\Box P$$

$T\Box\neg\Box\neg P$ will give $T\neg\Box\neg P$ and next $F\Box\neg P$ again, and $T\Box\neg\Box P$ will give $T\neg\Box P$ and next $F\Box P$. So, the tableau will continue with

$$F\Box\neg P, TP, F\Box P \qquad\qquad F\Box\neg P, F\Box P, FP.$$

So, we are essentially back at line (*) with $F\Box\neg P$ and $F\Box P$, from where the situation repeats itself. However, no branch will ever close and we read off the following Kripke counterexample *M* in *S4*:

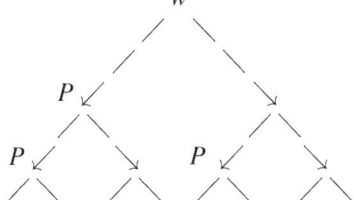

Clearly, $M, w \models \Box\Diamond P$, i.e., for every w' in M with wRw' there is a w'' in M such that $w'Rw''$ and $M, w'' \models P$; but $M, w \not\models \Diamond\Box P$, i.e., there is no w' in M with wRw' such that for all all w'' in M, if $w'Rw''$, then $M, w'' \models P$. Hence, $\Box\Diamond P \not\models \Diamond\Box P$.

The examples given above suggest a general procedure which, given a conjecture $A_1, \ldots, A_n \vdash' B$ in $K-$, will either construct a tableau-deduction of B from the premisses $A_1, \ldots, A_n$ in $K-$ or yield a Kripke counterexample in $K-$. We shall describe this procedure in more detail in Section 6.7 and prove that the three notions $A_1, \ldots, A_n \vdash B$ in $K-$, $A_1, \ldots, A_n \models B$ in $K-$, and $A_1, \ldots, A_n \vdash' B$ in $K-$, are equivalent.

Exercise 6.10. Translate the following argument in the language of modal propositional logic and either construct a tableau-deduction in K of the putative conclusion from the premisses or construct a Kripke counterexample in K.
It is not the case that: if John works hard [W], then he will necessarily succeed [S]. Therefore, it is possible that: if John works hard, then he will not succeed.

Exercise 6.11. Translate the following argument in the language of modal propositional logic and either construct a tableau-deduction in K of the putative conclusion from the premises or construct a Kripke counterexample in K.
It is possible that: if John fails [J], then he will give a party [P].
Therefore, if John fails, then it is possible that he will give a party.

Exercise 6.12. Prove that i) $K_i(A \vee B), K_i \neg A \vdash' K_i B$ in K,
but ii) $A \vee B, K_i \neg A \not\vdash' K_i B$ in K, neither in $S4$.
This explains the paradox in Exercise 2.70: let A stand for 'the prisoner will be hanged on Monday, Tuesday, Wednesday or Thursday' and let B stand for 'the prisoner will be hanged on Friday'. Then $A \vee B$ is the judge's statement that the prisoner would hang one day this week. Read $K_i E$ as 'prisoner i knows (on Friday morning) that E'. See also the answer to Exercise 2.70.

Exercise 6.13. Prove or refute in K: a) $\Box A \vee \neg \Box A$; b) $\Box A \vee \Box \neg A$.
Prove or refute in KT: c) $\Diamond A \vee \neg \Diamond A$; d) $\Diamond A \vee \Diamond \neg A$.

Exercise 6.14. Prove: $\vdash' \Box A \to \Box(A \vee B)$ in K and $\vdash' \Diamond A \to \Diamond(A \vee B)$ in K (cf. Exercise 6.6).

Exercise 6.15. Prove that K, KT and $S4$ have the *disjunction property*:
if $\vdash' \Box A \vee \Box B$, then $\vdash' \Box A$ or $\vdash' \Box B$.

Exercise 6.16. Prove or refute in KT: a) $\Diamond P \to \Diamond \Diamond P$; b) $\Diamond \Diamond P \to \Diamond P$.
Prove or refute in $S4$: c) $P \to \Box \Diamond P$; d) $(P \to Q) \to \neg \Diamond(P \wedge \neg Q)$.

6.6 Applications of Possible World Semantics

6.6.1 Direct Reference

There are at least two problems in the traditional theory of meaning:
1. In the traditional view, a proper name, like 'Jane', is identified with a description, such as 'the woman John is married to'. Now suppose that John is a bachelor. Then it would follow that Jane does not exist. This example makes clear that a person can be referred to by his or her name even if the description of the person in question does not apply to that person.
2. According to the traditional theory, a tiger, for instance, is identified with an object which has certain properties, among which the property of having sharp teeth. Consequently, the statement 'tigers have sharp teeth' is analytic; this seems to be counter-intuitive.

In the traditional theory, the conjunction of properties which a tiger is supposed to have is called the intension of the word 'tiger' and is supposed to be the *essence* of tiger. In the traditional theory as well, intension determines extension. Similarly, in the traditional view, the proper name 'Aristotle' is identified with a description

6.6 Applications of Possible World Semantics

such as 'the most well-known man who studied under Plato'. As a consequence, the proposition 'Aristotle studied under Plato' would be an analytic truth. This is again against our intuition.

Typical of the theory of direct reference is the position, held by Kripke, Donnellan and others, that proper names and nouns standing for natural kinds refer independently of identifying descriptions. In his paper [9], Donnellan distinguished between two kinds of use for definite descriptions – the *attributive use* and the *referential use*. In order to make this distinction clear, Donnellan considered the use of the definite description 'Smith's Murderer' in the following two cases.

> Suppose first that we come upon poor Smith foully murdered. From the brutal manner of the killing and the fact that Smith was the most lovable person in the world, we might exclaim 'Smith's murderer is insane'. I will assume, to make it a simpler case, that in a quite ordinary sense we do not know who murdered Smith. ... This, I shall say, is an *attributive* use of the definite description. [[9], 285-286]

So, in the case of the attributive use, the speaker wants to say something about whoever or whatever fits the description even if he does not know who or what that is. On the other hand,

> Suppose that Jones has been charged with Smith's murder and has been placed on trial. Imagine that there is a discussion of Jones' odd behavior at his trial. We might sum up our impression of his behavior by saying 'Smith's murderer is insane'. If someone asks to whom we are referring by using this description, the answer here is 'Jones'. This, I shall say, is a *referential use* of the definite description.

So, if the description 'Smith's murderer' is used referentially, the speaker is referring to Jones, even in the case that Jones turns out to be innocent. Note that in this case the description refers to Jones although it does not apply to Jones. To give another example, suppose someone asks me at a party who Mr. X is. I answer 'the man at the door with a glass of sherry in his hand'. Now suppose that the person referred to actually has a glass of white wine in his hand. Again the description may refer successfully without applying to the object referred to. These examples make clear that descriptions, when used referentially, do not always apply to the object they refer to. When using a description referentially, we have a definite object in mind whether or not it does fit the description.

According to the theory of *direct reference*, brought out by Keith Donnellan, Saul Kripke and others, proper names, like 'Aristotle', 'Thales' and 'Jane', and nouns standing for natural kinds, like 'gold', 'water' and 'tiger', have no intension (Sinn) in the traditional sense, but only have reference; and this reference is established by a *causal chain* rather than by an associated description. For example, the reference to the person called 'Aristotle' is determined by a causal chain as follows. The person in question is given a name in a 'baptism' with the referent present. Next this name is handed on from speaker to speaker. It is in this way that we use the name 'Aristotle' referring to the person in question. We do not have to have any description of Aristotle; the information 'Aristotle was a philosopher' may be completely new to the one who is using the name 'Aristotle'.

It is typical of the theory of direct reference that proper names, like 'Jane', refer to some definite object, even when the description we supply, such as 'the woman John is married to', does not apply to that object. This description may help us fix the reference, but it should not be taken to be the meaning of the name. And a similar view is held for nouns standing for natural kinds, like 'gold', 'water' and 'tiger'. The meaning of the word 'tiger' is its reference; identifying descriptions, such as 'a tawny-coloured animal with sharp teeth', only help us to fix the reference of this term.

Summarizing, according to the theory of direct reference, the meaning of a proper name or a natural kind term is its reference; the descriptions given in connection with these terms only help the hearer to pick out what the speaker has in mind.

6.6.2 Rigid Designators

In his paper *Naming and Necessity*, Kripke [22] in addition holds the view that a proper name, like 'Aristotle', is a *rigid designator*, i.e., it designates the very same object in all possible worlds in which this object exists. Thus, in the sentence 'Aristotle might have been a carpenter', the proper name 'Aristotle' refers to the same individual referred to in the sentence 'Aristotle was the philosopher who was a pupil of Plato and taught Alexander'. The definite description 'the most well-known man who studied under Plato', though it designates Aristotle in the actual world, may designate other individuals in other possible worlds; for it is possible that Aristotle did not study under Plato. Contrary to the traditional theory of meaning, according to the theory of direct reference, the statement 'Aristotle studied under Plato' is not necessarily true (and hence not analytic).

Now, if a and b are rigid designators and $a = b$ is true (in this world), then $a = b$ must be true in all worlds (accessible from this one) and hence $\Box(a = b)$ is true. So, it follows from the thesis that proper names are rigid designators that all true identity statements of the form $a = b$, where a and b are proper names, are necessarily true. In particular, it follows that 'Hesperus is Phosphorus (the morning star is the evening star)' and 'Tully is Cicero', if true (in this world) are necessarily true. On the other hand, we do not know a priori that Hesperus (the Morning Star) is Phosphorus (the Evening Star); this was discovered by empirical observation. Therefore, Kripke [23] claims in his paper *Identity and Necessity* that sentences like 'Hesperus is Phosphorus' and 'Tully is Cicero' if true (in this world) are *necessarily true* and at the same time are *a posteriori*.

Kripke extends his insights about proper names to nouns standing for natural kinds, such as 'gold', 'water' and 'tiger'. These nouns are rigid designators too, i.e., they refer to the same substance in all possible worlds in which this substance exists. Let us consider some interesting consequences of this point of view. 'Gold' being a rigid designator, the sentence 'gold is the element with atomic number 79', if true (in this world), will be true in all worlds (accessible from this one) and hence be necessarily true. Similarly, 'water' being a rigid designator, the sentence 'water has the chemical structure H$_2$O', if true (in this world), will be true in any world (accessible from this one) and hence be necessarily true. So both propositions, if true (in this world), are *necessarily true* and at the same time *a posteriori*. Kripke defines a sentence *A* to be *analytic* if it is both necessary and a priori. Consequently, sentences like 'Hesperus is Phosphorus', 'Tully is Cicero', 'gold is the element with atomic number 79' and 'water is H$_2$O' are NOT analytic, since they are a posteriori, although necessarily true, if true (in this world).

Let stick *S* denote the standard meter in Paris. Then, by definition, stick *S* is one meter long. Therefore, the epistemological status of the statement 'stick *S* is one meter long' is that this statement is an *a priori* truth. Conceiving 'one meter' as a rigid designator, indicating the same length in all possible circumstances (worlds), the metaphysical status of 'stick *S* is one meter long' will be that of a *contingent* statement, since the length of stick *S* can vary with the temperature, humidity and so on. So, assuming that 'one meter' is a rigid designator, the sentence 'stick *S* is one meter long' is both *a priori* and *contingent*, i.e., not necessarily true.

Similarly, the sentence 'water boils at 100 degrees Celcius' will be a priori and at the same time contingent, i.e., not necessarily true, if we conceive '100 degrees Celcius' as a rigid designator.

6.6.3 De dicto - de re distinction

If one wants to translate the sentence

It is possible that a Republican will win

into a logical formula, it becomes evident that this sentence is ambiguous. Using ◊ for 'it is possible that', the predicate symbol *R* for 'being a Republican' and the symbol *W* for 'will win', there are two different translations of the sentence in question:

$$(1)\ \exists x[R(x) \wedge \Diamond W(x)],\ \text{and}$$
$$(2)\ \Diamond \exists x[R(x) \wedge W(x)].$$

(1) says, literally, that there is some particular individual who actually is a Republican and who may possibly win.
(2) says, literally, that it is possible that some Republican or other will win.

(1) is called the *de re* or referential reading of the sentence above. Typical of the *de re* reading is that the possibility operator ◊ occurs within the scope of the (existential) quantifier.

(2) is called the *de dicto* or non-referential reading of the sentence above. Typical of the *de dicto* reading is that the (existential) quantifier occurs within the scope of the possibility operator ◊.

The example above demonstrates that sentences containing modalities such as 'possibly', 'necessarily', 'John believes that ...', etc., in combination with existential or universal quantifiers may give rise to ambiguities. Speaking in terms of possible worlds:

(1) says that in the given world there is a person who is a Republican (in the given world) and who will win in some world accessible from the given one;
(2) says that there is a world accessible from the given one in which there is a person who in that world is Republican and will win.

The proposition 'John finds a unicorn' can be properly translated as $\exists x[U(x) \wedge F(j,x)]$, where $U(a)$ stands for 'a is a unicorn', j stands for 'John' and $F(a,b)$ stands for 'a finds b'. But $\exists x[U(x) \wedge S(j,x)]$, where $S(a,b)$ stands for 'a seeks b' would be an improper translation of 'John seeks a unicorn', because the use of the existential quantifier commits us to an ontology in which unicorns do exist. Note that 'John finds a unicorn' and 'John seeks a unicorn' provide an extensional and an intensional context respectively (see Section 6.11).

In his paper [30], R. Montague develops a 'categorial' language in which 'John seeks a unicorn' can be properly translated.

6.6.4 Reasoning about Knowledge

Suppose three children, A(d), B(ob) and C(od), have played outside and two of them, say A and B, have mud on their forehead; they can see each other, but not themselves (there are no mirrors) and they do not communicate with each other. However, they are all perfect logicians! Let P be the proposition:

P: there is at least one child with mud on its forehead.

Notice that each child knows P, because A sees B, B sees A and C sees both A and B. But A does not know that B knows that P, because if A has no mud on its forehead, B sees nobody with mud. So, P is not *common knowledge*.

Now the father of the children announces P. By this announcement, P becomes *common knowledge*, in particular, everybody now knows that everybody knows P. For instance, A now knows that B knows P.

Next, the father asks each child (for the first time) to step forward if he *knows* to have mud on his forehead. What will happen? No child will step forward: A sees B with mud, B sees A with mud, and C sees both A and B with mud. So, no child has a reason to step forward.

Because after the first request no child steps forward, it becomes common knowledge that there must be at least two children with mud; if there were only one child with mud, this child would see no one else with mud and hence know he must be the one with mud. Consequently, if the father asks each child for the second time to step forward if he or she *knows* to have mud on the forehead, child A and B will step forward: A knows that there are at least two children with mud and only sees B with mud, and similarly for B.

Let m_A be the proposition 'A has mud on his forehead' and c_B the proposition 'B is clean'. By definition, $w_{m_A c_B m_C}$, abbreviated by w_{mcm} or even mcm, is the world in which m_A, c_B and m_C are true, i.e., $w_{mcm} \models m_A \wedge \neg m_B \wedge m_C$. We may model the

6.6 Applications of Possible World Semantics

initial situation described above - before the father has said anything - by a Kripke model $M = \langle W, R_A, R_B, R_C, \models \rangle$ with eight possible worlds and three accessibility relations R_A, R_B and R_C.

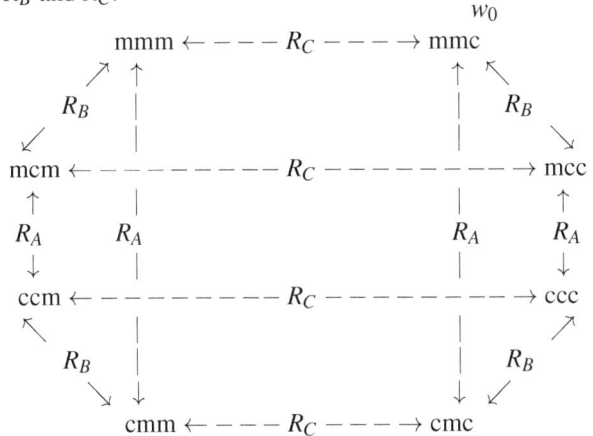

In our story, the actual world is $w_0 = w_{mmc}$. Because the children cannot see themselves, A, for instance, cannot distinguish between w_{mmm} and w_{cmm}. So, the accessibility relations R_A, R_B and R_C are reflexive and symmetric.

Notice that in world w_0 of this Kripke model M, A does not know that m_A, since A cannot distinguish between w_0 and w_{cmc}, in which m_A does not hold. In other words, $M, w_0 \not\models K_A m_A$, since $w_0 R_A w_{cmc}$, and $M, w_{cmc} \not\models m_A$. The proposition P, expressing that there is at least one child with mud, can now be rendered by $P = m_A \vee m_B \vee m_C$. In world w_0 of this Kripke model M, A does not know that B knows that P, because A cannot distinguish between w_0 and w_{cmc}, in which B does not know P, because B cannot distinguish between w_{cmc} and w_{ccc}. In other words, $M, w_0 \not\models K_A(K_B P)$, because $M, w_{cmc} \not\models K_B P$.

Once the father has announced the proposition P, each child eliminates the world w_{ccc}; the new situation is now modelled by the Kripke model M':

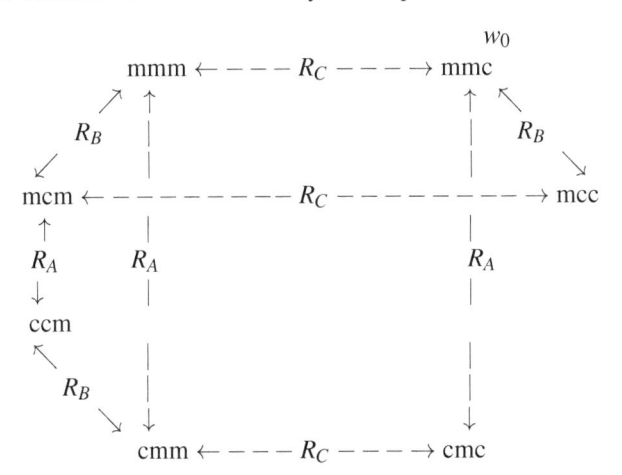

Notice: $M',w_0 \models K_A(K_B P)$, because $M',w_0 \models K_B P$ (B sees in w_0 that A has mud) and $M',w_{cmc} \models K_B P$ (B sees in w_{cmc} that A and C are clean).

In case that exactly one child, say A, has mud on his forehead, i.e., in world w_{mcc} of Kripke model M', we have $M',w_{mcc} \models K_A m_A$, because the only world accessible for A from w_{mcc} is w_{mcc}, in which m_A is true (A sees that B and C are clean). Similarly, $M',w_{cmc} \models K_B m_B$ and $M',w_{ccm} \models K_C m_C$. So, after announcing the proposition P, if there were only one child with mud, the child in question would know that he has mud on his forehead and would step forward. Once it becomes clear that no child knows that he has mud on his forehead, it follows that the three possible worlds w_{mcc}, w_{cmc} and w_{ccm} are cancelled and the only remaining possible worlds are depicted in the following Kripke model M'':

$$\begin{array}{c} w_0 \\ \text{mcm} \leftarrow R_B \rightarrow \text{mmm} \leftarrow R_C \rightarrow \text{mmc} \\ \uparrow \\ R_A \\ \downarrow \\ \text{cmm} \end{array}$$

Now, clearly, $M'',w_0 \models K_A m_A \wedge K_B m_B$, so A and B will step forward. Similarly, $M'',w_{mcm} \models K_A m_A \wedge K_C m_C$ and $M'',w_{cmm} \models K_B m_B \wedge K_C m_C$.

If no child would step forward after the second request of the father, it would follow that the worlds w_{mmc}, w_{mcm} and w_{cmm} are eliminated from model M'' and only world w_{mmm} would remain, resulting in the Kripke model M''', consisting of only one world w_{mmm}. And $M''',w_{mmm} \models K_A m_A \wedge K_B m_B \wedge K_C m_C$.

More generally, one may prove (see, for instance, Fagin, e.a. [10]):

Theorem 6.4. *If there are k, $k = 1,2,\ldots$, children with mud on the forehead, after announcing the proposition that there is at least one child with mud, the father has to state his request - to step forward once one knows that one has mud on the forehead - k times, before each child with mud knows that he has mud on his forehead. After i ($i < k$) rounds of questioning, it is common knowledge that at least $i+1$ children have mud on their foreheads.*

6.6.5 Common Knowledge

As seen in Subsection 6.6.4 common knowledge plays an important role in the muddy children puzzle. But common knowledge is also relevant for reaching agreement or for coordinating actions. We shall illustrate this by the *coordinated attack problem* informally as follows:

There are two hills with a valley in between. On the hills are two divisions of an army, each with its own general and in the valley is the enemy. If both divisions attack the enemy simultaneously they will surely win, but if only one division attacks, it will be defeated and have serious losses. So each general wants to be absolutely sure that both divisions attack at the same time. Say, general 1 wants to coordinate a simultaneous attack at dawn the next day and the generals are only able to communicate by means of a messenger (telephones are not available). The messenger, however, may get lost or may be captured by the enemy. How long will it take the generals to coordinate an attack?

> Suppose general 1 sends a messenger with the message P (*we attack at dawn tomorrow morning*) to general 2. Initially, we have that $K_1 P$ and $\neg K_2 P$, where K_i is the knowledge operator for general $i \in \{1,2\}$. Even if the message is in fact delivered, general 1 does not know that it was delivered: $\neg K_1(K_2 P)$; hence he cannot be sure that general 2 will attack simultaneously. So, given his state of knowledge, general 1 will not attack. General 2 knows this and does not want to take the risk of attacking alone; hence, he cannot attack on the basis of receiving the message of general 1. The only thing he can do is sending a messenger to general 1, acknowledging that he received the message and achieving that $K_1(K_2 P)$. However, even if general 1 receives this acknowledgment, he is in a similar position as general 2 was in when he received the original message. Now general 2 does not know that the acknowledgment was delivered: $\neg K_2(K_1(K_2 P))$. Because general 2 knows that without receiving the acknowledgment general 1 will not attack, general 2 cannot attack as long as he considers it possible that general 1 did not receive the acknowledgment. So, general 1 should send a message to general 2 in order to achieve that $K_2(K_1(K_2 P))$. However, the problem now is that $\neg K_1(K_2(K_1(K_2 P)))$, and so on. It turns out that no number of successful deliveries of acknowledgments can allow the generals to attack. Notice that, even if all the acknowledgments sent are received, *common knowledge* of P and hence coordination is not achieved, because of the uncertainty about what might have happened with the messengers.

Given a set $N = \{1,2\}$ of agents (persons, computers) and a formula A, we may define the the notions of 'everyone knows A' and 'A is common knowledge'.

Definition 6.16 (Common Knowledge). $EA := K_1 A \wedge K_2 A$ (everybody knows A); $E^0 A := A$ and for $k = 0, 1, \ldots$, $E^{k+1} A := E(E^k A)$. In particular, $E^1 A = E(E^0 A) = K_1 A \wedge K_2 A$ and $E^2 A = E(E^1 A) = K_1(K_1 A \wedge K_2 A) \wedge K_2(K_1 A \wedge K_2 A)$, which in S4 and S5 is equivalent to $K_1 A \wedge K_1(K_2 A) \wedge K_2(K_1 A) \wedge K_2 A$.
$CA := A \wedge EA \wedge E^2 A \wedge E^3 A \wedge \ldots$ (*A is common knowledge*).

Notice that strictly speaking CA is not a formula in our language, because it is an infinite conjunction. For the Kripke semantics and the syntaxis (axiom and rule) of common knowledge see Fagin, e.a. [10] and Meyer and van der Hoek [29].

6.7 Completeness of Modal Propositional Logic

Let $K-$ be any of the modal systems K, KT or $KT4 = S4$. We shall prove completeness of modal logic, i.e., that any valid consequence in $K-$ of given premises may be logically deduced by the tableaux rules of $K-$ from those premises:
if $A_1, \ldots, A_n \models B$ in $K-$, then $A_1, \ldots, A_n \vdash' B$ in $K-$ (Theorem 6.7). (1)
We shall also prove:
if $A_1, \ldots, A_n \vdash' B$ in $K-$, then $A_1, \ldots, A_n \vdash B$ in $K-$ (Theorem 6.9). (2)
In Theorem 6.3 we have already shown the soundness of modal logic:
if $A_1, \ldots, A_n \vdash B$ in $K-$, then $A_1, \ldots, A_n \models B$ in $K-$. (3)

From (1), (2) and (3) it follows that the three notions $A_1,\ldots,A_n \vdash B$ in $K-$, $A_1,\ldots,A_n \models B$ in $K-$, and $A_1,\ldots,A_n \vdash' B$ in $K-$ are equivalent.

In order to prove completeness of modal logic, we define a *procedure to construct a counterexample* to a given conjecture that $A_1,\ldots,A_n \vdash' B$ in $K-$ with the following property: if the procedure fails, i.e., does not yield a counterexample, we have in fact constructed a tableau-deduction of B from $A_1,\ldots,A_n$ in $K-$. The procedure makes use of the tableaux rules and produces 'trees' which we shall call *search trees*.

Definition 6.17 (Procedure to construct a counterexample). In order to construct a oounterexample to the conjecture that $A_1,\ldots,A_n \vdash' B$ in $K-$, we must construct a Kripke model M for $K-$ such that for some world w in M, $M, w \models A_1 \wedge \ldots \wedge A_n$, but $M, w \not\models B$.

Step 1: Start with $\{TA_1,\ldots,TA_n, FB\}$ and apply all tableaux rules for the propositional connectives and the $T\square$ rule in $K-$ as frequently as possible. However, in case one of the split-rules $T \rightarrow$, $T\vee$ and $F\wedge$ is applied, we make two search trees: one with the left split and one with the right split. Notice that for a tableau-deduction both search trees have to close.

For instance, consider the conjecture $\Diamond P \vdash' \square \Diamond P \wedge \Diamond \Diamond P$ in KT:

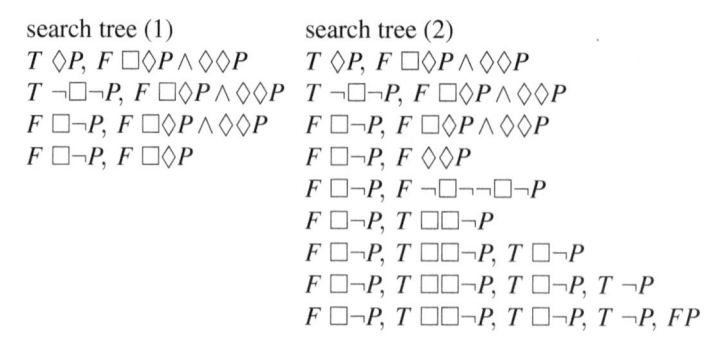

search tree (1)	search tree (2)
$T \Diamond P$, $F \square \Diamond P \wedge \Diamond \Diamond P$	$T \Diamond P$, $F \square \Diamond P \wedge \Diamond \Diamond P$
$T \neg \square \neg P$, $F \square \Diamond P \wedge \Diamond \Diamond P$	$T \neg \square \neg P$, $F \square \Diamond P \wedge \Diamond \Diamond P$
$F \square \neg P$, $F \square \Diamond P \wedge \Diamond \Diamond P$	$F \square \neg P$, $F \square \Diamond P \wedge \Diamond \Diamond P$
$F \square \neg P$, $F \square \Diamond P$	$F \square \neg P$, $F \Diamond \Diamond P$
	$F \square \neg P$, $F \neg \square \neg \neg \square \neg P$
	$F \square \neg P$, $T \square \square \neg P$
	$F \square \neg P$, $T \square \square \neg P$, $T \square \neg P$
	$F \square \neg P$, $T \square \square \neg P$, $T \square \neg P$, $T \neg P$
	$F \square \neg P$, $T \square \square \neg P$, $T \square \neg P$, $T \neg P$, FP

In the transition of the third line to the fourth line we apply the rule $F\wedge$ to $F \square \Diamond P \wedge \Diamond \Diamond P$, which causes a split. At that stage we make two search trees, one with the left split signed formula $F \square \Diamond P$ and one with the right split signed formula $F \Diamond \Diamond P$. One continues to apply all possible rules, except the $F\square$ rule, as frequently as possible.

At this stage we have partially constructed one, two (or more) search trees, each consisting of one node labeled with signed formulas. A labeled node w in which all tableaux rules except the $F\square$-rule have been applied as frequently as possible will be called *logically complete*. Intuitively, this means that one has fully described which formulas are true and which formulas are false in the present world w. Next we continue to expand each search tree by one or more applications of the $F\square$ rule.

Step 2 Each labeled node w in a search tree τ which is logically complete may contain one or more signed formulas of the form $F \square A$. For each of the signed formulas of the form $F \square A$ in a labeled node w we construct a new node w', declare w' accessible from w in the given search tree τ, i.e., $wR_\tau w'$, and label this node w' with the formulas $S_\square$, FA or $S_{T\square}$, FA which result from applying the rule $F\square$ to $S, F \square A$ in K, KT or $S4$, respectively. Notice that formulas that occur in labeled

6.7 Completeness of Modal Propositional Logic

node w may not occur anymore in node w' and that for closure it suffices that at least one of the successor nodes contains TA and FA for some formula A.

Next we apply step 1 again, but now starting with $S_\Box$, FA or $S_{T\Box}$, FA, depending on the system K, KT or $S4$, resulting in one or more logically complete nodes (worlds) w'. Step 1 and 2 are repeated as frequently as possible.

For search-tree (1) above one can apply the $F\Box$ rule to $F\ \Box\neg P$, losing the $F\ \Box\Diamond P$ signed formula, and we can apply the $F\Box$ rule to $F\ \Box\Diamond P$, losing the $F\ \Box\neg P$ signed formula. For a tableau-deduction only one of these two options has to yield closure. So, we have two options to go on with search tree (1):

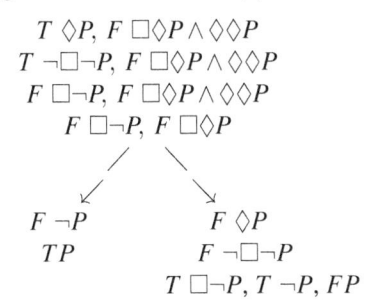

$$T\ \Diamond P,\ F\ \Box\Diamond P \wedge \Diamond\Diamond P$$
$$T\ \neg\Box\neg P,\ F\ \Box\Diamond P \wedge \Diamond\Diamond P$$
$$F\ \Box\neg P,\ F\ \Box\Diamond P \wedge \Diamond\Diamond P$$
$$F\ \Box\neg P,\ F\ \Box\Diamond P$$

$F\ \neg P$ $F\ \Diamond P$
TP $F\ \neg\Box\neg P$
         $T\ \Box\neg P,\ T\ \neg P,\ FP$

Whatever we do, we do not get closure. However, the nice thing is that we have constructed a search tree τ, starting with $T\ \Diamond P$, $F\ \Box\Diamond P \wedge \Diamond\Diamond P$, in this case consisting of three nodes labeled with signed formulas, which yields a Kripke counterexample $M = \langle \{w_0, w_1, w_2\}, R_\tau, \models \rangle$ to the conjecture that $\Diamond P \vdash' \Box\Diamond P \wedge \Diamond\Diamond P$ in KT:

$$w_0$$
$$P\ w_1 \quad\quad w_2$$

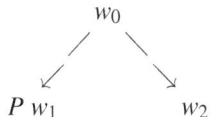

By definition, $w_0 R_\tau w_1$, $w_0 R_\tau w_2$, $w_1 \models P$, corresponding with the occurrence of TP in node w_1 and $w_2 \not\models P$, corresponding with the occurrence of FP in node w_2. One easily verifies that $M, w_0 \models \Diamond P$, because $M, w_1 \models P$, but $M, w_0 \not\models \Box\Diamond P$ and hence $M, w_0 \not\models \Box\Diamond P \wedge \Diamond\Diamond P$, because $M, w_2 \not\models \Diamond P$.

For search tree (2) there is only one formula of the form $F\ \Box A$ in the upper node. Application of Step 2 results in the following search tree in KT, consisting of two nodes:

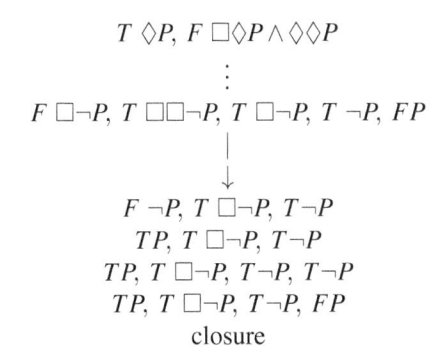

$$T\ \Diamond P,\ F\ \Box\Diamond P \wedge \Diamond\Diamond P$$
$$\vdots$$
$$F\ \Box\neg P,\ T\ \Box\Box\neg P,\ T\ \Box\neg P,\ T\ \neg P,\ FP$$
$$\downarrow$$
$$F\ \neg P,\ T\ \Box\neg P,\ T\neg P$$
$$TP,\ T\ \Box\neg P,\ T\neg P$$
$$TP,\ T\ \Box\neg P,\ T\neg P,\ T\neg P$$
$$TP,\ T\ \Box\neg P,\ T\neg P,\ FP$$
$$\text{closure}$$

However, because search tree (1) does not close, we have not found a tableau-deduction of $\Box\Diamond P \wedge \Diamond\Diamond P$ from $\Diamond P$ in KT. Instead, search tree (1) did not close and yielded a Kripke counterexample to the conjecture $\Diamond P \vdash' \Box\Diamond P \wedge \Diamond\Diamond P$ in KT. In our example, after executing step 1, 2 and 1 once more, the two search trees are finished and cannot be extended anymore.

Definition 6.18 (Search tree).
A search tree τ for the conjecture $A_1, \ldots, A_n \vdash' B$ in $K-$ is a set of nodes, labeled with signed formulas, with a relation R_τ between the nodes, such that:
0. The upper node contains $TA_1, \ldots, TA_n, FB$.
1. In case of K, $wR_\tau w' := w'$ is an immediate successor of w, i.e., w' results from the application of the $F\Box$ rule to a formula of the form $F \Box A$ in w.
In case of KT, $wR_\tau w' := w = w'$ or w' is an immediate successor of w.
In case of $KT4 = S4$, $wR_\tau w' := w = w'$ or w' is a (not necessarily immediate) successor of w.
2. For each node w in the search tree τ:
a) if $F\ C \to D$ occurs in w, then TC occurs in w **and** FD occurs in w;
b) if $T\ C \wedge D$ occurs in w, then TC occurs in w **and** TD occurs in w;
c) if $F\ C \vee D$ occurs in w, then FC occurs in w **and** FD occurs in w;
d) if $T\ \neg C$ occurs in w, then FC occurs in w;
e) if $F\ \neg C$ occurs in w, then TC occurs in w.
3. For each node w in the search tree τ:
a) if $T\ C \to D$ occurs in w, then FC occurs in w **or** TD occurs in w;
b) if $F\ C \wedge D$ occurs in w, then FC occurs in w **or** FD occurs in w;
c) if $T\ C \vee D$ occurs in w, then TC occurs in w **or** TD occurs in w.
4. For each node w in the search tree τ:
a) if $T\ \Box C$ occurs in w, then for all w' in τ with $wR_\tau w'$, TC occurs in w';
b) if $F\ \Box C$ occurs in w, then for some w' in τ with $wR_\tau w'$, FC occurs in w'.

Definition 6.19 (Closed/open search tree).
A search tree τ is *closed* if it contains at least one node labeled with TA and FA for some formula A. Otherwise, the search tree is called *open*.

Theorem 6.5. *Let τ be an open search tree for the conjecture $A_1, , \ldots, A_n \vdash' B$ in $K-$ with upper node w_0. Let W_τ the set of nodes in τ and let R_τ be defined as in Definition 6.18. Define $w \models P := TP$ occurs in w. Then $M_\tau = \langle W_\tau, R_\tau, \models \rangle$ is a Kripke countermodel in $K-$ to the conjecture that $A_1, \ldots, A_n \vdash' B$. More precisely, $M_\tau, w_0 \models A_1 \wedge \ldots \wedge A_n$, but $M_\tau, w_0 \not\models B$.*

Proof. Let τ be an open search tree with w_0 as upper node, containing $TA_1, \ldots, TA_n$, FB. Let $M_\tau = \langle W_\tau, R_\tau, \models \rangle$ be the corresponding Kripke model, as defined in the theorem. We shall prove by induction:
1) If TA occurs in w, then $M_\tau, w \models A$.
2) If FA occurs in w, then $M_\tau, w \not\models A$.
Since $TA_1, \ldots, TA_n, FB$ occur in the top node w_0, it follows that $M_\tau, w_0 \models A_1 \wedge \ldots \wedge A_n$, but $M_\tau, w_0 \not\models B$. Therefore, $A_1, \ldots, A_n \not\models B$ in $K-$.
Induction basis Let $A = P$ be atomic. If TP occurs in w, then by definition $w \models P$,

6.7 Completeness of Modal Propositional Logic   307

i.e., $M_\tau, w \models P$. If FP occurs in w, then - since τ is open - TP does not occur in w and hence by definition $w \not\models P$, i.e., $M_\tau, w \not\models P$.

Induction step Suppose 1) and 2) hold for C and D (induction hypothesis). We shall prove that 1) and 2) hold for $C \to D, C \wedge D, C \vee D, \neg C$ and $\Box C$.

Let $A = C \to D$ and suppose $T\ C \to D$ occurs in w. Then according to Definition 6.18, 3 a), FC is in w or TD is in w. So, by the induction hypothesis, $M_\tau, w \not\models C$ or $M_\tau, w \models D$. Consequently, $M_\tau, w \models C \to D$.

Let $A = C \to D$ and suppose $F\ C \to D$ occurs in w. Then according to Definition 6.18, 2 a), TC is in w and FD is in w. So, by the induction hypothesis, $M_\tau, w \models C$ and $M_\tau, w \not\models D$. Consequently, $M_\tau, w \not\models C \to D$.

The cases that $A = C \wedge D, A = C \vee D$ and $A = \neg C$ are treated similarly.

Let $A = \Box C$ and suppose $T\ \Box C$ occurs in w. Then according to Definition 6.18, 4 a), for every node w' in τ with $wR_\tau w'$, TC occurs in w'. So, by the induction hypothesis, for all w' in τ, if $wR_\tau w'$, then $M_\tau, w' \models C$ and hence $M_\tau, w \models \Box C$.

Let $A = \Box C$ and suppose $F\ \Box C$ occurs in w. Then according to Definition 6.18, 4 b), there is a node w' in τ with $wR_\tau w'$ such that FC occurs in w'. So, by the induction hypothesis, $M_\tau, w' \not\models C$ and hence $M_\tau, w \not\models \Box C$.   □

Theorem 6.6. *If all search trees for the conjecture $A_1, \ldots, A_n \vdash' B$ in $K-$ are closed, i.e., contain closure in one of their branches, then $A_1, \ldots, A_n \vdash' B$ in $K-$.*

Proof. Suppose all search trees for the conjecture $A_1, \ldots, A_n \vdash' B$ in $K-$ are closed. Then it follows from the construction of the search trees that the closed branches together form a tableau-deduction of B from $A_1, \ldots, A_n$ in $K-$.   □

Example 6.10. We construct the search trees for the conjecture $\Diamond(P \wedge Q) \vdash' \Diamond P \wedge (\Diamond Q \vee \Box P)$ in K. Step 1 yields two partial search trees each consisting of one node:

$T \Diamond(P \wedge Q), F \Diamond P \wedge (\Diamond Q \vee \Box P)$ $T \Diamond(P \wedge Q), F \Diamond P \wedge (\Diamond Q \vee \Box P)$
$T \Diamond(P \wedge Q), F \Diamond P$ $T \Diamond(P \wedge Q), F \Diamond Q \vee \Box P$
$T \neg\Box\neg(P \wedge Q), F \neg\Box\neg P$ $T \neg\Box\neg(P \wedge Q), F \neg\Box\neg Q, F \Box P$
$F \Box\neg(P \wedge Q), T \Box\neg P$ $F \Box\neg(P \wedge Q), T \Box\neg Q, F \Box P$

Because there is no $T\Box$ rule for K, step 1 finishes here. The only rule which may be applied next is the rule $F\Box$ for K. Applying step 2 to the last sequents of step 1 we get:

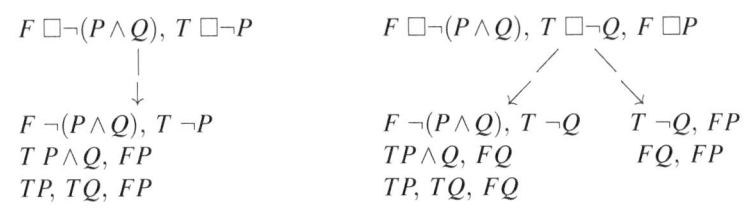

$F \Box\neg(P \wedge Q), T \Box\neg P$ $F \Box\neg(P \wedge Q), T \Box\neg Q, F \Box P$
$\qquad\qquad\downarrow$ $\qquad\swarrow \qquad\searrow$
$F \neg(P \wedge Q), T \neg P$ $F \neg(P \wedge Q), T \neg Q \qquad T \neg Q, FP$
$T P \wedge Q, FP$ $TP \wedge Q, FQ \qquad\qquad FQ, FP$
TP, TQ, FP TP, TQ, FQ

The leftmost search tree consists of one branch with two nodes, and is closed. The rightmost search tree consists of two branches and three nodes; its left branch is closed and its right branch is open. The two closed branches together form a tableau-deduction in K of $\Diamond P \wedge (\Diamond Q \vee \Box P)$ from $\Diamond(P \wedge Q)$.

Theorem 6.7 (Completeness).
If $A_1,\ldots,A_n \models B$ in $K-$, then $A_1,\ldots,A_n \vdash' B$ in $K-$.

Proof. Suppose $A_1,\ldots,A_n \models B$ in $K-$. Construct all search trees for the conjecture $A_1,\ldots,A_n \vdash' B$ in $K-$. If one of them is open, say τ, then by Theorem 6.5, $M_\tau, w_0 \models A_1 \wedge \ldots \wedge A_n$, while $M_\tau, w_0 \not\models B$. This contradicts the assumption $A_1,\ldots,A_n \models B$ in $K-$. Hence, there can be no open search tree for the conjecture $A_1,\ldots,A_n \vdash' B$ in $K-$. That is, all search trees for this conjecture are closed. So, by Theorem 6.6, $A_1,\ldots,A_n \vdash' B$ in $K-$. □

In the case of K, resp. KT, our procedure to construct a counterexample to the conjecture $A_1,\ldots,A_n \vdash' B$ will stop after finitely many steps and then either yield a Kripke counterexample or a tableau-deduction of B from $A_1,\ldots,A_n$ in K, resp. KT. In the case of $S4$, this procedure does not necessarily stop after finitely many steps (see Example 6.9), but nevertheless after finitely many steps it will become clear whether one has constructed a Kripke counterexample in $S4$ or a tableau-deduction of B from $A_1,\ldots,A_n$ in $S4$. Therefore, the modal propositional logics K, KT and $S4$ are *decidable*.

Theorem 6.8 (Decidability). *The modal propositional logics K, KT and $S4$ are decidable, i.e., there is a procedure to decide whether $A_1,\ldots,A_n \vdash' B$ in K, KT, resp. $S4$, in finitely many steps.*

In order to prove that the three notions of formal deducibility in $K-$, Kripke valid consequence in $K-$ and tableau-deducibility in $K-$ are equivalent we still have to show the following theorem.

Theorem 6.9. *If $A_1,\ldots,A_n \vdash' B$ in $K-$, then $A_1,\ldots,A_n \vdash B$ in $K-$.*

Proof. The proof is a generalization of the analogue for classical propositional logic; see Theorem 2.27. Suppose $A_1,\ldots,A_n \vdash' B$ in $K-$, i.e., B is tableau-deducible from $A_1,\ldots,A_n$ in $K-$. It suffices to show:
for every sequent $S = \{TD_1,\ldots,TD_k, FE_1,\ldots,FE_m\}$ in a tableau-deduction of B from $A_1,\ldots,A_n$ in $K-$ it holds that $D_1,\ldots,D_k \vdash E_1 \vee \ldots \vee E_m$ in $K-$. (*)
Consequently, because $\{TA_1,\ldots,TA_n, FB\}$ is the first (upper) sequent in any given tableau-deduction of B from $A_1,\ldots,A_n$ in $K-$, it follows that $A_1,\ldots,A_n \vdash B$ in $K-$.

The proof of (*) is tedious, but has a simple plan: the statement is true for the final sequents in a tableau-deduction in $K-$, and the statement remains true if we go up in the tableau-deduction in $K-$ via the T and F rules.
Basic step: Any final sequent in a tableau-deduction of B from $A_1,\ldots,A_n$ in $K-$ is of the form $\{TD_1,\ldots,TD_k, TP, FP, FE_1,\ldots,FE_m\}$. So, we have to show that $D_1,\ldots,D_k, P \vdash P \vee E_1 \vee \ldots \vee E_m$. And this is straightforward: $D_1,\ldots,D_k, P \vdash P$ and $P \vdash P \vee E_1 \vee \ldots \vee E_m$.
Induction step: We have to show that for all rules of $K-$ the following is the case: if (*) holds for all lower sequents in the rule (induction hypothesis), then (*) holds for the upper sequent in the rule.

In the proof of Theorem 2.27 we have already shown the induction step for the T- and F-rules for the connectives. So, we may restrict ourselves to the T- and F-rules for $\Box$ in system $K-$.

6.8 Strict Implication

Induction step for rule $F\Box$ in K: For convenience, we will suppose that $S = \{T\Box C, TD, FE\}$. So, consider:
$$\frac{T\Box C,\ TD,\ FE,\ F\Box A}{TC,\ FA}$$
By the induction hypothesis, we have $C \vdash A$ in K. We have to show: $\Box C, D \vdash E \vee \Box A$ in K. This is straightforward: from $C \vdash A$ in K follows $\Box C \vdash \Box A$ in K and hence, $\Box C, D \vdash E \vee \Box A$ in K.

Induction step for rule $T\Box$ in KT: For convenience, we will suppose that $S = \{TD, FE\}$. So, consider:
$$\frac{T\Box C,\ TD,\ FE}{TC,\ TD,\ FE}$$
By the induction hypothesis, we have $C, D \vdash E$ in KT. We have to show: $\Box C, D \vdash E$ in KT. This is straightforward, because $\Box C \rightarrow C$ is an axiom of KT.

The other T- and F-rules for $\Box$ in $K-$ are treated similarly. □

Exercise 6.17. Construct a counterexample showing that the cosmological proof of God's existence in $S5$, given in Exercise 6.3, does not hold in $S4$:
$\Diamond P, \Box(\Diamond P \rightarrow Q) \not\vdash \Box Q$ in $S4$.

Exercise 6.18. Construct a counterexample showing that the ontological proof of God's existence in $S5$, given in Exercise 6.4, does not hold in $S4$:
$\Box(Q \rightarrow \Box Q), \Diamond Q \not\vdash Q$ in $S4$.

Exercise 6.19. Prove or refute: a) $\Box(S \rightarrow E), \Box(E \rightarrow L), \neg\Diamond L \vdash' \neg\Diamond S$ in K.
b) $S \rightarrow \Box E, E \rightarrow \Box L, \neg\Diamond L \vdash' \neg\Diamond S$ in $S4$ (confer Exercise 6.1).

6.8 Strict Implication

The material implication, $\rightarrow$, of classical propositional logic is characterized in terms of its truth table: $P \rightarrow Q$ is 0 (false) if and only if P is 1 (true) and Q is 0 (false). Through the ages objections have been raised against the 'only if': if P is 0, then $P \rightarrow Q$ is 1. Although there are many arguments in favor of the truth table of $P \rightarrow Q$, as we have seen in Section 2.2, also objections have been raised, in particular the so-called *paradoxes of material implication*:
a) $\neg A \models A \rightarrow B$: if A is false, then from A follows any proposition B;
b) $B \models A \rightarrow B$: if B is true, then B follows from any proposition A.
So, from 'I do not break my leg' it logically follows that 'if I break my leg, then I go for skying' and from 'I like my coffee' it logically follows that 'if there is oil in my coffee, then I like my coffee'; see Section 2.4. In the same section we have seen that P. Grice [16] explains these paradoxes by pointing out that one should take into account not only the truth conditions of the propositions asserted, but also the pragmatic principles governing discourse: $A \rightarrow B$ is normally not to be asserted by someone who is in the position to deny A or to assert B.

The dispute between advocates of the truth-functional account of conditionals, given in Section 2.2, and the advocates of other - more complex but seemingly more adequate - accounts is as old as logic itself. The truth-functional account is first known to have been proposed by Philo of Megara ca. 300 B.C. in opposition to the view of his teacher Diodorus Cronus. We know of this through the writings of Sextus Empiricus some 500 years later, the earlier documents having been lost; see Section 2.10.2. Sextus reports Philo as attributing truth values to conditionals just as in our truth table for $\to$. Diodorus probably had in mind what later was called *strict implication*.

Rejecting material implication as an adequate representation of 'if ..., then ...', in 1918 C.I. Lewis [25] put forward *strict implication*, $\mapsto$, which can be rendered in terms of necessity and material implication: $\Box(A \to B)$.

Definition 6.20. *Strict implication*, $\mapsto$, is defined by $A \mapsto B := \Box(A \to B)$.

It is easy to show that the versions for strict implication of the paradoxes of material implication do not hold. According to Exercise 6.20:
a) not $\neg A \vdash' A \mapsto B$ in S4; and b) not $B \vdash' A \mapsto B$ in S4.

However, the definition of strict implication leads to the so-called *paradoxes of strict implication*. According to Exercise 6.21:
a) $\Box \neg A \vdash' A \mapsto B$ in K: an impossible proposition A implies every proposition B.
b) $\Box B \vdash' A \mapsto B$ in K: a necessary proposition B is implied by every proposition A.
c) $Q \vdash' P \mapsto P$ in K and d) $\vdash' \neg Q \land Q \mapsto P$ in K.
The problem with these paradoxes is that for the provability of an inference from A to B, A should be *relevant* to B. See Section 6.10.

Exercise 6.20. Prove: not $\neg A \vdash' A \mapsto B$ in S4 and not $B \vdash' A \mapsto B$ in S4.

Exercise 6.21. Prove the following so-called *paradoxes of strict implication*:
a) $\Box \neg A \vdash' A \mapsto B$ in K; b) $\Box B \vdash' A \mapsto B$ in K;
c) $Q \vdash' P \mapsto P$ in K;   d) $\vdash' \neg Q \land Q \mapsto P$ in K.

6.9 Counterfactuals

Counterfactuals are expressions of the form $A \,\Box\!\!\to B$, to be read as 'if it were the case that A, then it would be the case that B', where A is supposed to be false. Unlike material, strict and relevant implication, the counterfactual
a) is not transitive, i.e., not $\dfrac{A \,\Box\!\!\to B \quad B \,\Box\!\!\to C}{A \,\Box\!\!\to C}$,
b) does not have the property of contraposition $\dfrac{A \,\Box\!\!\to B}{\neg B \,\Box\!\!\to \neg A}$, and
c) does not have the property of strengthening $\dfrac{A \,\Box\!\!\to B}{A \land C \,\Box\!\!\to B}$.

The following counterexamples are from D. Lewis [26]:
a) If J. Edgar Hoover had been born a Russian, then he would have been a communist. If he had been a communist, he would have been a traitor.
Therefore: If he had been born a Russian, he would have been a traitor.

6.9 Counterfactuals

b) If Boris had gone to the party, Olga would still have gone.
Therefore: If Olga had not gone, Boris would still not have gone.
Suppose that Boris wanted to go, but stayed away solely in order to avoid Olga, so the conclusion is false; but Olga would have gone all the more willingly if Boris had been there, so the premiss is true.
c) If I walked on the lawn, no harm at all would come of it. Therefore:
If I and everyone else walked on the lawn, no harm at all would come of it.

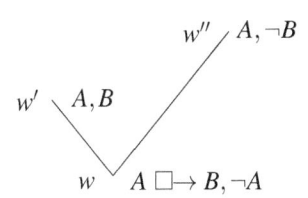

```
                    w″  A, ¬B
       w′  A, B
              ╲      ╱
               ╲    ╱
                w   A □→ B, ¬A
```

We say that $A \Box\!\!\rightarrow B$ is true in world w iff either A is impossible in w or there is an accessible $A \wedge B$-world w', which is *closer to w* than every $A \wedge \neg B$-world is (R. Stalnaker, D. Lewis, ± 1970), where a C-world is simply a world in which C is true.

Example 6.11. a) A young child to his father: If you would bring that big tree home (A), I would make matches from it (B). This proposition is true in the present world because the child considers the antecedent A to be impossible.
b) If you would jump out of the window at the 20th floor (A), you would get injured (B). This proposition is true in the present world w, because there is world w' in which $A \wedge B$ is true and which is closer to w than each world w'' in which $A \wedge \neg B$ is true.
c) If you would jump out of the window at the 20th floor, you would change into a bird. This proposition is not true in the present world w, because we cannot imagine a world w' in which $A \wedge B$ is true and which is closer to w than any world in which $A \wedge \neg B$ is true.

Given a Kripke model $M = \langle W, R, \models \rangle$, we assume that for each w in W there is a binary relation $<_w$ on W, where $w' <_w w''$ is to be read as: w' is *closer to w* than w''. Furthermore, we assume that R is reflexive, and
1. if wRw' and not wRw'', then $w' <_w w''$;
2. for all w, w' in W, if $w \neq w'$, then $w <_w w'$ and not $w' <_w w$.

Definition 6.21 ($M \models_w A \Box\!\!\rightarrow B$). Let $M = \langle W, R, \models, < \rangle$ be a Kripke model, where for each w in W, $<_w$ is a binary relation on W, satisfying the conditions just mentioned. $M \models_w A \Box\!\!\rightarrow B := M \models_w \neg \Diamond A$ or there is some world w' in W such that a) wRw' and $M \models_{w'} A \wedge B$, and b) for all w'' in W, if $M \models_{w''} A \wedge \neg B$, then $w' <_w w''$.

For an illustration we refer to Exercise 6.22.

Under the conditions just mentioned, counterfactuals with true antecedents reduce to material conditionals. More precisely, the following two inference-patterns are valid:

$$\text{(a)} \ \frac{A \wedge \neg B}{\neg(A \Box\!\!\rightarrow B)} \ \text{ and } \ \text{(b)} \ \frac{A \wedge B}{A \Box\!\!\rightarrow B};$$

that is, our truth conditions guarantee that whenever the premiss is true in a world of a given model M, then so is the conclusion; see Exercise 6.23.

The validity of the first inference-pattern (a) also guarantees the validity of the inference from a counterfactual to a material conditional and the validity of Modus Ponens for a counterfactual conditional: $\dfrac{A \square\!\!\rightarrow B}{A \rightarrow B}$ and $\dfrac{A \quad A \square\!\!\rightarrow B}{B}$. We also have the inference: $\dfrac{\square(A \rightarrow B)}{A \square\!\!\rightarrow B}$; see Exercise 6.24.

One can develop possible-world semantics for counterfactuals, a notion of validity ($\models A$) and a notion of provability ($\vdash A$) such that a counterfactual formula A is valid if and only if A is provable. See D. Lewis' paper [27], pp. 441-443, or his monograph [26]; de Swart [37] and Gent [14].

Let $A \diamond\!\!\rightarrow B$ stand for 'if A were the case, then B might be the case'. Then it is plausible to have $A \diamond\!\!\rightarrow B$ iff $\neg(A \square\!\!\rightarrow B)$. The reader can check for himself that, given plausible assumptions about comparative similarity of worlds where Bizet and Verdi would be compatriots, both
(1) if Bizet and Verdi were compatriots, then Bizet might be Italian, and
(2) If Bizet and Verdi were compatriots, then Bizet might not be Italian,
are true. For further reading see Harper e.a. [17].

Exercise 6.22. Let C stand for: Bizet and Verdi are compatriots. Let B_F, B_I and B_D stand for: Bizet is French, Italian, Dutch, respectively. And similarly, V_F, V_I and V_D for: Verdi is French, Italian, Dutch, respectively. Let w be the actual world, in which B_F and V_I hold, of the following Kripke model M. Verify that in the Stalnaker-Lewis analysis of counterfactuals:
a) $M, w \models C \square\!\!\rightarrow (B_F \wedge V_F) \vee (B_I \wedge V_I)$ b) $M, w \not\models C \square\!\!\rightarrow B_I$,
c) $M, w \not\models C \square\!\!\rightarrow V_F$ d) $M, w \models C \square\!\!\rightarrow \neg B_D \wedge \neg V_D$.

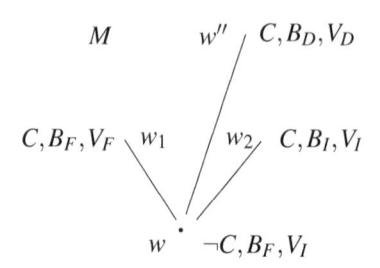

$$M \qquad w'' \quad C, B_D, V_D$$

$$C, B_F, V_F \quad w_1 \qquad w_2 \quad C, B_I, V_I$$

$$w \quad \neg C, B_F, V_I$$

Exercise 6.23. Let $M = \langle W, R, \models \rangle$ be a Kripke model with R reflexive and for each w in W, let $<_w$ be a binary relation on W satisfying: 1. if wRw' and not wRw'', then $w' <_w w''$ and 2. if $w \neq w'$, then $w <_w w'$ and not $w' <_w w$. Prove:
a) if $M, w \models A \wedge \neg B$, then $M, w \models \neg(A \square\!\!\rightarrow B)$, and
b) if $M, w \models A \wedge B$, then $M, w \models A \square\!\!\rightarrow B$.

Exercise 6.24. Under the conditions mentioned in Exercise 6.23 prove that:
if $M, w \models \square(A \rightarrow B)$, then $M, w \models A \square\!\!\rightarrow B$.

6.10 Weak and Relevant Implication; Entailment*

Exercise 6.25. Show that not $\models (P \mapsto Q) \vee (Q \mapsto P)$ in $S4$ and that also not $\models (P \square\!\!\rightarrow Q) \vee (Q \square\!\!\rightarrow P)$, while $\models (P \rightarrow Q) \vee (Q \rightarrow P)$.

6.10 Weak and Relevant Implication; Entailment*

In his paper *The weak theory of implication*, A. Church [8] succeeds in excluding the paradoxes of strict implication (see Exercise 6.21) without also excluding at the same time arguments which everyone regards as valid. In his paper A. Church presents essentially the following axiom schemes for what he calls *weak implication*, but what one might also call *relevant implication*, and which we denote by $\Rightarrow$:

1. $A \Rightarrow A$
2. $(A \Rightarrow B) \Rightarrow ((B \Rightarrow C) \Rightarrow (A \Rightarrow C))$
3. $(A \Rightarrow (B \Rightarrow C)) \Rightarrow ((A \Rightarrow B) \Rightarrow (A \Rightarrow C))$
4. $(A \Rightarrow (B \Rightarrow C)) \Rightarrow (B \Rightarrow (A \Rightarrow C))$

together with the rule Modus Ponens, $\dfrac{B \quad B \Rightarrow C}{C}$.

$\Rightarrow$ satisfies principles of relevance in the following mathematically definite sense:

$$A_1, \ldots, A_{n-1} \vdash^* A_n \Rightarrow B \text{ iff } A_1, \ldots, A_{n-1}, A_n \vdash^* B,$$

where $A_1, \ldots, A_{n-1}, A_n \vdash^* B$ (B is deducible from $A_1, \ldots, A_n$) means that B can be obtained by a finite number of applications of Modus Ponens to $A_1, \ldots, A_{n-1}, A_n$ and to instances of the axiom schemes 1, 2, 3, 4, *such that all of $A_1, \ldots, A_{n-1}, A_n$ actually are used in the deduction of B*; more precisely, such that B gets the relevance-index $\{1, \ldots, n-1, n\}$ if we assign to each A_i ($1 \leq i \leq n$) the index $\{i\}$ and to each consequence of an application of Modus Ponens the union of the indices of its premisses.

For instance, $A \Rightarrow B, B \Rightarrow C \vdash^* A \Rightarrow C$, for the following schema is a deduction (in the new sense) of $A \Rightarrow C$ from $A \Rightarrow B$ and $B \Rightarrow C$:

$$\dfrac{A \Rightarrow B_{\{1\}} \qquad (A \Rightarrow B) \Rightarrow ((B \Rightarrow C)) \Rightarrow (A \Rightarrow C))}{\dfrac{B \Rightarrow C_{\{2\}} \qquad ((B \Rightarrow C) \Rightarrow (A \Rightarrow C))_{\{1\}}}{(A \Rightarrow C)_{\{1,2\}}}}$$

However, it is not the case that $Q \vdash^* P \Rightarrow P$ (see Anderson & Belnap, [2]), while $Q \vdash P \rightarrow P$ does hold, since in '$A \vdash B$' it is not demanded that A actually is used in the deduction of B and $\vdash P \rightarrow P$ holds; see Section 2.6.

We define $M = \langle S, \emptyset, \cup, \models \rangle$ to be a *model* (for the logic of weak or relevant implication) if and only if

1. S is a collection of sets, closed under $\cup$; the elements of S are to be regarded as pieces of information;
2. $\emptyset$ is the empty set (regarded as the empty piece of information);

3. $a \cup b$ is the union of a and b (see Chapter 3);
4. $\models$ is a relation between elements of S and atomic formulas P; '$a \models P$' is to be read as: P is true on the basis of the information in a.

For M a model and A a formula built from atomic formulas by means of $\Rightarrow$ only, we define $M \models_a A$ (A is true on the basis of the information a of the model M) as follows:

$M \models_a P$ iff $a \models P$ (P atomic);
$M \models_a B \Rightarrow C$ iff for all b in S, not $M \models_b B$ or $M \models_{a \cup b} C$.

M is a model for A ($M \models^* A$) iff $M \models_\emptyset A$.
A is *valid* ($\models^* A$) iff for all models M, $M \models^* A$. And B is a *valid consequence* of $A_1, \ldots, A_n$ ($A_1, \ldots, A_n \models^* B$) iff $\models^* A_1 \Rightarrow (A_2 \Rightarrow \ldots (A_n \Rightarrow B) \ldots)$.

Exercise 6.26. Prove the (deduction) theorem: $A_1, \ldots, A_{n-1} \vdash^* A_n \Rightarrow B$ iff $A_1, \ldots, A_{n-1}, A_n \vdash^* B$. Hint: the proof of the 'if'-part of the (deduction) theorem proceeds by replacing in a given deduction of B from $A_1, \ldots, A_{n-1}, A_n$ each expression C_c with $n \in c$ by $(A_n \Rightarrow C)_{c-\{n\}}$. For Modus Ponens, $\dfrac{D_d \quad (D \Rightarrow E)_e}{E_{d \cup e}}$, four different cases arise, depending on whether $n \in d$ and/or $n \in e$; and the axioms have been chosen such that the resulting schema can easily be supplemented to a deduction of $A_n \Rightarrow B$ from $A_1, \ldots, A_{n-1}$.

Exercise 6.27. a) Prove the *Soundness Theorem*:
$$\text{if } A_1, \ldots, A_n \vdash^* B, \text{ then } A_1, \ldots, A_n \models^* B.$$
In [38] A. Urquhart also proves the converse of this statement, i.e., completeness.
b) Prove that $\not\models^* Q \Rightarrow (P \Rightarrow P)$, and hence $Q \not\vdash^* P \Rightarrow P$. In general, the relevant implication versions of the original paradoxes of strict implication do not hold.

Exercise 6.28. Prove that $\vdash^* A \Rightarrow ((A \Rightarrow A) \Rightarrow A)$. This says that if A is true, then it follows from $A \Rightarrow A$. But it seems reasonable to suppose that any logical consequence of $A \Rightarrow A$ should necessarily be true (see Anderson & Belnap [2], p. 23). We therefore consider *entailment*, $\twoheadrightarrow$, defined by $P \twoheadrightarrow Q := \Box(P \Rightarrow Q)$, which was essentially considered for the first time by W. Ackermann [1] in his *Begründung einer strengen Implikation*. In this paper W. Ackermann presents essentially the following axiomatic system for $\twoheadrightarrow$:
1. $A \twoheadrightarrow A$
2. $(A \twoheadrightarrow B) \twoheadrightarrow ((B \twoheadrightarrow C) \twoheadrightarrow (A \twoheadrightarrow C))$
3. $(A \twoheadrightarrow (B \twoheadrightarrow C)) \twoheadrightarrow ((A \twoheadrightarrow B) \twoheadrightarrow (A \twoheadrightarrow C))$
4. $(A \twoheadrightarrow B) \twoheadrightarrow (((A \twoheadrightarrow B) \twoheadrightarrow C) \twoheadrightarrow C)$

together with the rule Modus Ponens $\dfrac{P \quad P \twoheadrightarrow Q}{Q}$.

Entailment satisfies both principles of relevance and principles of necessity in certain mathematically definite senses: all valid entailments are necessarily valid and in all valid entailments the antecedent is relevant to the succedent.

6.11 Modal Predicate Logic

A *possible world semantics* for the modal predicate logics is obtained by demanding that a Kripke model M contains for every world w in M a domain or universe $U(w)$ such that if wRw', then $U(w)$ is a subset of $U(w')$.
$M, w \models \forall x[A(x)] :=$ for every individual d in $U(w)$, $M, w \models A(a)[d]$, and
$M, w \models \exists x[A(x)] :=$ there is some d in $U(w)$ such that $M, w \models A(a)[d]$.

'It is possible that unicorns exist' can be rendered by $\Diamond \exists x[P(x)]$, and is likely to be true. But 'there is an object which possibly is a unicorn', to be rendered by $\exists x[\Diamond P(x)]$, is generally held to be false.

In terms of possible worlds, the difference can be explained as follows, using $U(w)$ for the universe of world w:
$M, w \models \Diamond \exists x[P(x)] :=$ there is a world w' in M accessible from w (wRw') such that $M, w' \models \exists x[P(x)]$, i.e., there is a world w' in M accessible from w such that there is an individual d in the universe $U(w')$ of w' which is a unicorn in w'.
$M, w \models \exists x[\Diamond P(x)] :=$ there is an object d in the universe $U(w)$ of w such that $M, w \models \Diamond P(a)[d]$, i.e., there is an object d in the universe $U(w)$ of w such that there is a world w' in M accessible from w (wRw') in which d is a unicorn.

Supposing that if wRw', then the universe $U(w)$ of w is a subset of the universe $U(w')$ of w', we find that

$$\text{if } M, w \models \exists x[\Diamond P(x)], \text{ then } M, w \models \Diamond \exists x[P(x)],$$

but not conversely. Hence the following statements hold, but not conversely:
$\models \exists x[\Diamond \neg A(x)] \to \Diamond \exists x[\neg A(x)]$.
$\models \neg \Diamond \exists x[\neg A(x)] \to \neg \exists x[\Diamond \neg A(x)]$.
$\models \neg \neg \Box \neg \exists x[\neg A(x)] \to \neg \exists x[\neg \Box \neg \neg A(x)]$.
$\models \Box \forall x[A(x)] \to \forall x[\Box A(x)]$.

Again, the difference between $\Box \forall x[A(x)]$ and $\forall x[\Box A(x)]$ may be explained best in terms of possible world semantics:
$M, w \models \Box \forall x[A(x)] :=$ for every world w' in M with wRw' and for every object d in the universe $U(w')$ of w', $M, w' \models A(a)[d]$; but
$M, w \models \forall x[\Box A(x)] :=$ for every object d in the universe $U(w)$ of w and for every world w' in M with wRw', $M, w' \models A(a)[d]$.

A *Hilbert-type proof system* for the modal predicate logics K, KT, $S4$, and $S5$, is obtained by adding to the axioms and rules for the respective modal propositional logics the (classical) axioms and rules for the quantifiers:
$\forall$ axiom: $\forall x[A(x)] \to A(t)$ and $\exists$ axiom: $A(t) \to \exists x[A(x)]$ for any term t.
$\forall$ rule: from $C \to A(a)$ deduce $C \to \forall x[A(x)]$, provided a does not occur in C.
$\exists$ rule: from $A(a) \to C$ deduce $\exists x[A(x)] \to C$, provided a does not occur in C.

Let us show that $\vdash \Box \forall x[A(x)] \to \forall x[\Box A(x)]$ in K:
1. $\forall x[A(x)] \to A(a)$ by the axiom for $\forall$.
2. $\Box(\forall x[A(x)] \to A(a))$ from 1 by the rule for $\Box$.

3. $\Box \forall x[A(x)] \to \Box A(a)$ from 2 and the axiom for $\Box$, using Modus Ponens.
4. $\Box \forall x[A(x)] \to \forall x[\Box A(x)]$ from 3 by the rule for $\forall$.

A *tableaux proof system* for the modal predicate logics K, KT and $S4$ is obtained by adding the T- and F-rules for $\forall$ and $\exists$ to the tableaux rules for the connectives and $\Box$:

$$S, T\ \forall x[A(x)] \qquad\qquad S, F\ \forall x[A(x)]$$
$$S, T\ \forall x[A(x)], TA(t) \qquad S, FA(a)\ \text{with}\ a\ \text{new}$$

$$S, T\ \exists x[A(x)] \qquad\qquad S, F\ \exists x[A(x)]$$
$$S, T\ A(a)\ \text{with}\ a\ \text{new} \qquad S, F\ \exists x[A(x)], FA(t)$$

Soundness and completeness of the modal predicate logics with respect to the appropriate Kripke semantics can again be shown by generalizing the proofs for the propositional case in Section 6.7.

Although $\Box \forall x[A(x)] \to \forall x[\Box A(x)]$ is formally provable in K and hence Kripke-valid, the converse formula $\forall x[\Box A(x)] \to \Box \forall x[A(x)]$, called the *Barcan formula*, is not Kripke-valid. A Kripke counterexample in $S4$ can be obtained by trying to construct a tableau-proof of this formula; we do not succeed in finding such a proof, but instead we find an open search tree from which we can immediately read off a counterexample.

$$T\ \forall x[\Box A(x)], F\Box \forall x[A(x)]$$
$$T\Box A(a_1), T\ \forall x[\Box A(x)], F\Box \forall x[A(x)]$$
$$TA(a_1), T\Box A(a_1), T\ \forall x[\Box A(x)], F\Box \forall x[A(x)]$$

$$T\Box A(a_1), \qquad F\ \forall x[A(x)]$$
$$T\Box A(a_1), \qquad FA(a_2)$$
$$T\Box A(a_1), TA(a_1), \qquad FA(a_2)$$

M
$\{a_1\} \quad A(a_1)$

$\{a_1, a_2\} \quad A(a_1)$

Let $M = \langle \{w_1, w_2\}, R, \models \rangle$ be the Kripke model in $S4$ consisting of two worlds w_1, w_2 with $w_1 R w_2$, $U(w_1) = \{a_1\}$, $U(w_2) = \{a_1, a_2\}$, $w_1 \models A(a_1)$ and $w_2 \models A(a_1)$, but $w_2 \not\models A(a_2)$, corresponding with the occurrence of $TA(a_1)$ in w_1 and in w_2 and the occurrence of $FA(a_2)$ in w_2.

$M, w_1 \models \forall x[\Box A(x)] :=$ all objects in $U(w_1)$ have the property $\Box A$ in w_1. This is the case, because $M, w_1 \models A(a_1)$ and $M, w_2 \models A(a_1)$. But $M, w_1 \models \Box \forall x[A(x)] :=$ for all worlds w' in M accessible from w_1 each individual in $U(w')$ has the property A in w'. Since $w_1 R w_2$, a_2 in $U(w_2)$ and by definition $w_2 \not\models A(a_2)$, it follows that $M, w_1 \not\models \Box \forall x[A(x)]$.

Exercise 6.29. Show that $\vdash' \Box \forall x[A(x)] \to \forall x[\Box A(x)]$ in K.

6.11.1 Modal Predicate Logic and Essentialism

Leibniz' law says that those things are the same of which one may be substituted for the other with preservation of truth. In contemporary treatments of identity this law is presented as follows:

6.11 Modal Predicate Logic

$$(1) \models a = b \rightarrow (\ldots a \ldots \rightleftarrows \ldots b \ldots)$$

where $\ldots a \ldots$ is a context containing occurrences of the name a, and $\ldots b \ldots$ is the same context except that one or more occurrences of a have been replaced by b: if $a = b$, then what holds for a also holds for b and vice versa.

In the propositional calculus we have a similar principle, called the replacement theorem:

$$(2) \models (A \rightleftarrows B) \rightarrow (\ldots A \ldots \rightleftarrows \ldots B \ldots).$$

And the analogue of the replacement theorem for predicate logic is principle

$$(3) \models (P(a) \rightleftarrows Q(a)) \rightarrow (\ldots P(a) \ldots \rightleftarrows \ldots Q(a) \ldots).$$

Quine [33] and Føllesdal [12] have argued that in order to make sense of quantified modal logic, modal contexts should be *referentially transparent*, i.e., principle (1) should hold also for modal contexts, and at the same time they should be *extensionally opaque*, i.e., the principles (2) and (3) should NOT hold for modal contexts.

According to Quine, in order to be able to quantify into modal contexts, these contexts should be referentially transparent: $\exists x[\Box(x > 7)]$ holds because $\Box(9 > 7)$ is true; but 9 = the number of planets (in this world); so, $\Box$ (the number of planets (in this world) > 7) should hold.

Therefore, quantified modal logic only makes sense if we accept principle (1) also for modal contexts:

$$(1) \models a = b \rightarrow (\ldots a \ldots \rightleftarrows \ldots b \ldots).$$

Principle (1) says that whatever is asserted to be true of an object, must be true of it regardless of how it is referred to. In other words, modal contexts should be *referentially transparent*, i.e., if two singular terms refer to the same object, they are interchangeable with preservation of truth (also in modal contexts).

Principle (1) says in particular that $\models a = b \rightarrow (\Box(a = a) \rightleftarrows \Box(a = b))$. And since $\Box(a = a)$ is valid, it follows that

$$(1^*) : \models a = b \rightarrow \Box(a = b).$$

For instance, Hesperus and Phosphorus are two different names referring to the same object (the planet Venus), i.e., Hesperus = Phosphorus, and hence, $\Box$(Hesperus = Phosphorus).

Principle (1*) says that if a and b refer to the same object, say o, in this world, then they refer to the same object (but possibly different from o) in any world accessible from this one. Hence, if a is a rigid designator (i.e., refers to the same object in any world accessible from this one), then b is also a rigid designator. In fact, Kripke already argued that proper names and nouns for natural kinds are rigid designators; see Subsection 6.6.2.

On the other hand, if we accept one of the principles (2) or (3) also for modal contexts (i.e., contexts $\ldots A \ldots$ or $\ldots P(a) \ldots$ containing modalities), then it even follows that $\models B \rightleftarrows \Box B$ for any proposition B. In other words, the extension of (2) or (3) from classical propositional logic or predicate logic respectively to modal logic would collapse necessity into truth. The arguments are simple.

Suppose (2) would hold also for modal contexts. Then in particular $\models (A \rightleftarrows B) \rightarrow (\Box A \rightleftarrows \Box B)$. Taking for A the expression $a = a$, it follows that $\models B \rightarrow \Box B$, since both $a = a$ and $\Box(a = a)$ hold. Because we usually also assume the converse, $\models \Box B \rightarrow B$, it follows that $\models B \rightleftarrows \Box B$.

Suppose (3) would hold also for modal contexts. Next suppose B is true. Taking $P(a) := a = a$ and $Q(a) := a = a \wedge B$, $P(a) \rightleftarrows Q(a)$ is true and hence, by principle (3), $\Box(a = a) \rightleftarrows \Box(a = a \wedge B)$ is true. Since $\Box(a = a \wedge B)$ is equivalent to $\Box(a = a) \wedge \Box B$, it follows that $\Box B$ is true. So, we have shown that from principle (3) it follows that $\models B \rightarrow \Box B$ and therefore $\models B \rightleftarrows \Box B$.

Consequently, the principles (2) and (3) should NOT hold for modal contexts. In other words, modal contexts should be *extensionally opaque*; that is, – formulated negatively – general terms and sentences with the same extension (truth value in the case of sentences) must in general not be interchangeable with preservation of truth. Such interchangeability would amount to the collapse of modal distinctions. Formulated positively, extensional opacity means that some properties belong to things necessarily, while other properties belong to things only accidently.

So, in order to make sense of quantified modal logic, modal contexts should be *referentially transparent*, i.e., principle (1) should hold also for modal contexts, and at the same time they should be *extensionally opaque*, i.e., the principles (2) and (3) should NOT hold for modal contexts. From this it is immediately clear that a satisfactory semantics for the modalities must distinguish between expressions which refer (singular terms) and expressions which have extension (general terms and sentences, the extension of a sentence being its truth value). Therefore, a Fregean semantics, according to which all expressions are considered to be referring, cannot be appropriate for modal logic. However, as already has been pointed out by J.R. Searle, Frege's extension of the notion of reference to predicates and sentences is not very natural:

> ... an expression refers to an object only because it conveys something true of that object. But a predicate does not convey something true of a concept nor does a sentence convey something true of a truth value. [Searle [34], p. 3]

Summarizing, if we want quantified modal logic to make sense, we have to accept principle (1) also for modal contexts; in other words, modal contexts should be *referentially transparent*: whatever is true of an object is true of it regardless of how it is referred to (α). On the other hand, in order to avoid that necessity collapses into truth, we should not accept the principles (2) and (3) for modal contexts. In other words, modal contexts should be *extensionally opaque*: among the predicates true of an object, some are necessarily true of it, others only accidentally (β). And *essentialism* is just this combination of (α) and (β). See also Perrick [31].

6.12 The Modal Logic GL

The axioms of the modal logic GL (Gödel's Logic, also called the *Logic of Provability*) are the following:

6.12 The Modal Logic *GL*

the axioms of classical propositional logic;
$\Box(B \to C) \to (\Box B \to \Box C)$;
$\Box A \to \Box(\Box A)$; and
$\Box(\Box A \to A) \to \Box A$.

The two rules of *GL* are Modus Ponens and necessitation (from $\vdash A$ infer $\vdash \Box A$).

The axioms and rules of *GL* resemble facts about $Prov(a)$, in particular (i), (ii), (iii) and (iv) in Subsection 5.2.2:

(i) if $\mathscr{P} \vdash A$, then $\mathscr{P} \vdash Prov(\ulcorner A \urcorner)$;
(ii) $\mathscr{P} \vdash Prov(\ulcorner B \to C \urcorner) \to (Prov(\ulcorner B \urcorner) \to Prov(\ulcorner C \urcorner))$;
(iii) $\mathscr{P} \vdash Prov(\ulcorner A \urcorner) \to Prov(\ulcorner Prov(\ulcorner A \urcorner) \urcorner)$.
(iv) if $\mathscr{P} \vdash Prov(\ulcorner A \urcorner)$, then $\mathscr{P} \vdash A$.

However, $Prov(a)$ does NOT meet the stronger condition $\mathscr{P} \vdash Prov(\ulcorner A \urcorner) \to A$.

Theorem 6.10. $\vdash A$ *in GL iff for every Kripke model* $M = \langle W, R, \models \rangle$ *with W finite and non-empty, R transitive and irreflexive (i.e., for all $w \in W$, not wRw), $M \models A$.*

For a proof of this theorem the reader is referred to Boolos, Burgess and Jeffrey, [5], Chapter 27. Here we restrict ourselves to the remark that $\Box(B \to C) \to (\Box B \to \Box C)$ holds in any Kripke model and that $\Box A \to \Box\Box A$ holds in any Kripke model $M = \langle W, R, \models \rangle$ with R transitive. In Exercise 6.30 the reader is asked to prove that $\Box(\Box A \to A) \to \Box A$ holds in any finite Kripke model $M = \langle W, R, \models \rangle$ with R transitive and irreflexive. Note that $\Box A \to A$ does NOT hold in such Kripke models, which corresponds to the fact that NOT $\mathscr{P} \vdash Prov(\ulcorner A \urcorner) \to A$. The weaker statement (iv) 'if $\mathscr{P} \vdash Prov(\ulcorner A \urcorner)$, then $\mathscr{P} \vdash A$' does hold, which corresponds to the fact that if $\vdash \Box A$ in *GL*, then also $\vdash A$ in *GL*.

Definition 6.22. Let ϕ be a function that assigns to each atomic formula of modal propositional logic a sentence in the formal language $\mathscr{L}_A$ for arithmetic. For any formula A of modal propositional logic, the formula A^ϕ in $\mathscr{L}_A$ is inductively defined as follows: $P_i^\phi := \phi(P_i)$ for any atomic formula P_i, $i = 1, 2, \ldots$;
$(B \to C)^\phi := B^\phi \to C^\phi$;
$\bot^\phi := \bar{0} = \bar{1}$;
$(\Box B)^\phi := Prov(\ulcorner B^\phi \urcorner)$.
$\land$, $\lor$ and $\neg$ are treated similarly to $\to$.

The following theorems bring out an important connection between the formal system *GL* of modal logic and the formal system $\mathscr{P}$ for arithmetic.

Theorem 6.11 (Arithmetical Soundness).

$$\text{If } \vdash A \text{ in GL, then for all } \phi, \mathscr{P} \vdash A^\phi.$$

Proof. We restrict ourselves to the following observations:
If A is an axiom of propositional logic, then clearly $\mathscr{P} \vdash A^\phi$.
Let A be $\Box(B \to C) \to (\Box B \to \Box C)$. Then, by (ii) above, $\mathscr{P} \vdash A^\phi$.
Let A be $\Box B \to \Box\Box B$. Then, by (iii) above, $\mathscr{P} \vdash A^\phi$.
Corresponding to Modus Ponens: if $\mathscr{P} \vdash A^\phi$ and $\mathscr{P} \vdash (A \to B)^\phi$, then $\mathscr{P} \vdash B^\phi$.
Corresponding to the necessitation rule of *GL*: if $\mathscr{P} \vdash A^\phi$, then, by (i) above, also

$\mathscr{P} \vdash (\Box A)^\phi$, i.e., $\mathscr{P} \vdash Prov(\ulcorner A^\phi \urcorner)$.
It thus remains to show that $\mathscr{P} \vdash A^\phi$, where A is an axiom $\Box(\Box B \to B) \to \Box B$. For a proof of this the reader is referred to Boolos, Burgess and Jeffrey [5], Chapter 27.

Theorem 6.12 (Arithmetical completeness theorem).

If for all ϕ, $\mathscr{P} \vdash A^\phi$, then $\vdash A$ in GL.

This was proved by R. Solovay [35] and is also proved in Boolos [6], Chapter 12.

Exercise 6.30. Show that $\Box(\Box A \to A) \to \Box A$ holds in any Kripke model $M = \langle W, R, \models \rangle$ with W finite and non-empty, R transitive and irreflexive.

6.13 Solutions

Solution 6.1. Depending on what the speaker has in mind, at least two translations are possible: $\Box(S \to E), \Box(E \to L), \neg \Diamond L \vdash^? \neg \Diamond S$, and $S \to \Box E, E \to \Box L, \neg \Diamond L \vdash^? \neg \Diamond S$. The first argument is correct, the second incorrect; see Exercise 6.19.

Solution 6.2. a) In Example 6.1 we have seen that $\vdash A \to \Diamond A$ in KT and $\Diamond A \to \Box \Diamond A$ is an axiom of $S5$. By propositional logic: $A \to \Diamond A, \Diamond A \to \Box \Diamond A \vdash A \to \Box \Diamond A$. Therefore, $\vdash A \to \Box \Diamond A$ in $S5$. b) $\Diamond \neg A \to \Box \Diamond \neg A$ is an axiom of $S5$ and by propositional logic $\vdash \Diamond \neg A \rightleftarrows \neg \Box A$. Hence, $\vdash \neg \Box A \to \Box \neg \Box A$ in $S5$. This is called *negative introspection*: if I do not know A, then I know that I do not know A.

Solution 6.3. $\Diamond P, \Box(\Diamond P \to Q) \vdash \Box Q$ in $S5$:

prem	axiom $S5$	prem	axiom
$\Diamond P$	$\Diamond P \to \Box \Diamond P$	$\Box(\Diamond P \to Q)$	$\Box(\Diamond P \to Q) \to (\Box \Diamond P \to \Box Q)$

$\Box \Diamond P$ $\qquad\qquad\qquad\qquad\qquad\qquad\qquad$ $\Box \Diamond P \to \Box Q$

$\qquad\qquad\qquad\qquad\qquad\qquad\qquad\qquad\qquad$ $\Box Q$

Solution 6.4.
By propositional logic, $\qquad (Q \to \Box Q) \vdash (\neg \Box Q \to \neg Q)$.
So, by the $\Box$-axiom and MP, $\quad \Box(Q \to \Box Q) \vdash \Box(\neg \Box Q \to \neg Q)$ in K.
Again by the $\Box$-axiom and MP, $\Box(Q \to \Box Q) \vdash (\Box \neg \Box Q \to \Box \neg Q)$ in K. $\qquad$ (1)
According to Exercise 6.2: $\vdash \neg \Box Q \to \Box \neg \Box Q$ in $S5$. But $\vdash \neg A \to B$ iff $\vdash A \lor B$ by propositional logic; therefore: $\vdash \Box Q \lor \Box \neg \Box Q$ in $S5$. $\qquad$ (2)
From (1) and (2): $\Box(Q \to \Box Q) \vdash (\Box Q \lor \Box \neg Q)$ in $S5$.
Hence, by propositional logic, $\Box(Q \to \Box Q), \Diamond Q \vdash \Box Q$ in $S5$.

Solution 6.5. The mistake is made in the transition from 3. to 4.: from the premiss $Q \to \Box Q$ it follows that $\neg \Box Q \to \neg Q$; but we do not have $\vdash \neg \Box Q \to \neg Q$ and therefore we cannot apply the rule of necessitation, which would yield $\vdash \Box(\neg \Box Q \to \neg Q)$ and then, by the axiom for $\Box$ and MP, $\vdash \Box \neg \Box Q \to \Box \neg Q$.

6.13 Solutions

Solution 6.6. $A \to A \vee B$ is an axiom of propositional logic, so $\vdash A \to A \vee B$. Hence, by the $\Box$-rule, $\vdash \Box(A \to A \vee B)$ in K. Hence, by the $\Box$-axiom of K and Modus Ponens, $\vdash \Box A \to \Box(A \vee B)$ in K.
We want to show that $\vdash \Diamond A \to \Diamond(A \vee B)$ in K. By contraposition it suffices to show that $\vdash \Box \neg(A \vee B) \to \Box \neg A$. We know from propositional logic that $\vdash \neg(A \vee B) \to \neg A$. By the $\Box$-rule it follows: $\vdash \Box(\neg(A \vee B) \to \neg A)$ in K. Therefore, by the $\Box$-axiom and Modus Ponens, $\vdash \Box \neg(A \vee B) \to \Box \neg A$ in K.

Solution 6.7. Let $M = \langle \{w_0, w_1\}, R, \models \rangle$ be the Kripke model (for K) with $w_0 R w_1$, $w_0 \models P$ and not $w_1 \models P$. Then $M, w_0 \models P$, but not $M, w_0 \models \Box P$.

Solution 6.8. a) To show: for all Kripke models M and for all w in M, $M, w \models \Box(A \wedge B)$ iff $M, w \models \Box A \wedge \Box B$. This is true because for any w' in M with wRw', $M, w' \models A \wedge B$ iff $M, w' \models A$ and $M, w' \models B$.
b) However, $\not\models \Box(A \vee B) \rightleftarrows (\Box A \vee \Box B)$. $M = \langle \{w_0, w_1\}, R, \models \rangle$ with R reflexive, $w_0 R w_1$, $w_0 \models P$ and $w_1 \models Q$, is a counterexample: $M, w_0 \models \Box(P \vee Q)$, but $M, w_0 \not\models \Box P$ and $M, w_0 \not\models \Box Q$.
c) To show: for all Kripke models M and for all w in M, $M, w \models \Diamond(A \vee B)$ iff $M, w \models \Diamond A$ or $M, w \models \Diamond B$. This is true because: there is a w' with wRw' such that $M, w' \models A \vee B$ iff there is w' with wRw' such that $M, w' \models A$ or there is a w' with wRw' such that $M, w' \models B$.

Solution 6.9. By definition, $M, w_1 \models Q$.
$M, w_1 \models K_A P$, because w_1, w_2, w_4 are accessible from w_1 for Alice and $M, w_1 \models P$, $M, w_2 \models P$ and $M, w_4 \models P$.
$M, w_1 \models \neg K_B P$, because $w_1 R_B w_3$ and $M, w_3 \not\models P$.
$M, w_1 \models \neg K_A Q$ and $M, w_1 \models \neg K_B Q$ are shown in a similar way.
$M, w_1 \models K_A K_A P := M, w_1 \models K_A P$ and $M, w_2 \models K_A P$ and $M, w_4 \models K_A P$, which are all true.
$M, w_1 \models K_B K_A P := M, w_1 \models K_A P$ and $M, w_3 \models K_A P$ and $M, w_4 \models K_A P$. Because $w_3 R_A w_3$ and $M, w_3 \not\models P$, it follows that $M, w_3 \not\models K_A P$ and hence $M, w_1 \not\models K_B K_A P$, i.e., $M, w_1 \models \neg K_B K_A P$.
$M, w_1 \models K_A \neg K_B P := M, w_1 \models \neg K_B P$ and $M, w_2 \models \neg K_B P$ and $M, w_4 \models \neg K_B P$, because w_1, w_2, w_4 are accessible from w_1 for Alice. However, $M, w_2 \models K_B P$ and $M, w_4 \models K_B P$. Hence, $M, w_1 \not\models K_A \neg K_B P$, i.e., $M, w_1 \models \neg K_A \neg K_B P$.
The other cases are treated similarly.

Solution 6.10. $\neg(W \to \Box S) \vdash' \Diamond(W \to \neg S)$ in K, since the following tableau in K is closed:
$T \neg(W \to \Box S)$, $F \Diamond(W \to \neg S)$
$F (W \to \Box S)$, $T \Box \neg(W \to \neg S)$
TW, $F \Box S$, $T \Box \neg(W \to \neg S)$
$\overline{FS, T \neg(W \to \neg S)}$
$FS, F (W \to \neg S)$
$FS, TW, F \neg S$
FS, TW, TS

Solution 6.11. $\Diamond(J \to P) \not\vdash' J \to \Diamond P$ in K, since we can construct a counterexample:

$T \neg\Box\neg(J \to P), F (J \to \Diamond P)$
$F \Box\neg(J \to P), TJ, F\neg\Box\neg P$ $w_1\ J$
$\underline{F \Box\neg(J \to P), TJ, T\Box\neg P}$ |
    $F \neg(J \to P), T\neg P$ ↓
      $T(J \to P), FP$ w_2
        FJ, FP

$M = \langle \{w_1, w_2\}, R, \models \rangle$, with $w_1 R w_2$ and $w_1 \models J$, is a counterexample in K: $M, w_1 \models \Diamond(J \to P)$, since $M, w_2 \models J \to P$, and $M, w_1 \models J$, but $M, w_1 \not\models \Diamond P$.

Solution 6.12. The tableau in i) is a tableau-deduction of $K_i B$ from $K_i(A \vee B)$ and $K_i \neg A$ in K. The search tree in ii) yields a counterexample M in $S4$ against the conjecture $A \vee B, K_i \neg A \vdash' K_i B$ in $S4$.

i) $\underline{T\ K_i(A \vee B), T\ K_i\neg A, F\ K_i B}$ ii) $T(A \vee B), T\ K_i\neg A, F\ K_i B$
    $\underline{T\ (A \vee B), T\ \neg A, FB}$ $TB, T\ K_i\neg A, T\neg A, F\ K_i B$ $w_0\ B$
    $TA, FA, FB\ |\ TB, FA, FB$ $\underline{TB, T\ K_i\neg A, FA, F\ K_i B}$ |
                                                   $T\ K_i\neg A, FB$ ↓
                                                   $T\ K_i\neg A, T\neg A, FB$ |
                                                   $T\ K_i\neg A, FA, FB$ w_1

$M = \langle \{w_0, w_1\}, R, \models \rangle$, with $w_0 R w_1$ and $w_0 \models B$, is a countermodel in $S4$: $M, w_0 \models A \vee B$, $M, w_0 \models K_i \neg A$, but $M, w_0 \not\models K_i B$, since $M, w_1 \not\models B$.

Solution 6.13. $\Box A \vee \neg \Box A$ is tableau-provable in K, but $\Box A \vee \Box \neg A$ is not. $\Diamond A \vee \neg \Diamond A$ is tableau-provable in KT, and $\Diamond A \vee \Diamond \neg A$ too:

a) $F\ \Box A \vee \neg \Box A$ b) $F\ \Box A \vee \Box \neg A$ c) $F\ \Diamond A \vee \neg \Diamond A$ d) $F\ \Diamond A \vee \Diamond \neg A$
   $F\Box A, F\neg\Box A$ $F\Box A, F\Box\neg A$ $F\neg\Box\neg A, F\neg\neg\Box\neg A$ $F\neg\Box\neg A, F\neg\Box A$
   $\underline{F\Box A, T\Box A}$ ↙   ↘ $\underline{T\Box\neg A, F\Box\neg A}$ $\underline{T\Box\neg A, T\Box A}$
   FA, TA FA $F\neg A$ $T\neg A, F\neg A$ $T\neg A, TA$
   closure TA FA, TA FA, TA

The tableaux in a), c) and d) are closed, while the tableau in b) yields a Kripke counterexample $M = \langle \{w_0, w_1, w_2\}, R, \models \rangle$ in K with $w_0 R w_1$, $w_0 R w_2$, $w_1 \not\models A$ and $w_2 \models A$: $M, w_0 \not\models \Box A \vee \Box \neg A$.

Solution 6.14. Both tableaux below are closed and hence are a tableau-deduction of $\Box A \to \Box(A \vee B)$ and $\Diamond A \to \Diamond(A \vee B)$ in K, respectively.

$F\Box A \to \Box(A \vee B)$ $F\Diamond A \to \Diamond(A \vee B)$
$\underline{T\Box A, F\Box(A \vee B)}$ $T\neg\Box\neg A, F\neg\Box\neg(A \vee B)$
$TA, F A \vee B$ $\underline{F\Box\neg A, T\Box\neg(A \vee B)}$
TA, FA, FB $F\neg A, T\neg(A \vee B)$
closure $TA, F A \vee B$
                                         TA, FA, FB

Solution 6.15. Suppose $\vdash' \Box A \vee \Box B$ in K, KT or $S4$, i.e., there is a closed tableau starting with: $F\ \Box A \vee \Box B$
                       $F\Box A, F\Box B$

This tableau will continue with either FA or FB and one of these two will be closed.

6.13 Solutions 323

In the first case a tableau starting with $F\Box A$ will give closure and in the second case, a tableau starting with $F\Box B$ will give closure.

Solution 6.16.
a) tableau in KT for: b) search tree in KT for: countermodel M

$\quad F\;\Diamond P\to \Diamond\Diamond P$ $\quad F\;\Diamond\Diamond P\to \Diamond P$

$\quad T\;\neg\Box\neg P,\; F\;\neg\Box\Box\neg P$ $\quad T\;\neg\Box\Box\neg P,\; F\neg\Box\neg P$

$\quad F\;\Box\neg P,\; T\;\Box\Box\neg P,\ldots$ $\quad F\;\Box\Box\neg P,\; T\;\Box\neg P,\ldots$ w_0

$\quad \overline{F\;\neg P,\; T\Box\neg P}$ $\quad \overline{F\Box\neg P,\; T\neg P}$ $\downarrow$

$\quad F\;\neg P,\; T\Box\neg P,\; T\neg P$ $\quad F\Box\neg P,\; FP$ w_1

$\quad TP,\; T\Box\neg P,\; FP$ $\quad F\;\neg P$ $\downarrow$

$\quad$ closure $\quad TP$ no closure $w_2\; P$

The tableau in a) closes and hence is a tableau-proof of $\Diamond P\to\Diamond\Diamond P$ in KT. $M=\langle\{w_0,w_1,w_2\},R,\models\rangle$, with R reflexive, not transitive, w_0Rw_1, w_1Rw_2, not w_0Rw_2 and $w_2\models P$, is a counterexample in KT against $\Diamond\Diamond P\to\Diamond P$: $M,w_0\models\Diamond\Diamond P$, because $M,w_1\models\Diamond P$, since $M,w_2\models P$. But $M,w_0\not\models\Diamond P$, because $M,w_1\not\models P$.

c) search tree in $S4$ for: countermodel M' d) search tree in $S4$ for: M''

$\quad F\;P\to\Box\Diamond P$ $\quad w_0\; P$ $\quad T\;P\to Q,\; F\;\neg\Box\neg(P\wedge\neg Q)$ w_0

$\quad TP,\; F\;\Box\neg\Box\neg P$ $\quad |$ $\quad FP,\; F\;\Box\neg(P\wedge\neg Q)$ $|$

$\quad \overline{F\neg\Box\neg P}$ $\quad \downarrow$ $\quad \overline{F\;\neg(P\wedge\neg Q)}$ $\downarrow$

$\quad T\Box\neg P,\; T\neg P$ $\quad w_1$ $\quad T(P\wedge\neg Q)$ $w_1\; P$

$\quad T\Box\neg P,\; FP$ $\quad$ $\quad TP,\; T\;\neg Q,\; FQ$

$M'=\langle\{w_0,w_1\},R,\models\rangle$, with w_0Rw_1 and $w_0\models P$, is a countermodel in $S4$ for $P\to\Box\Diamond P$, because $M',w_0\models P$, but $M',w_0\not\models\Box\Diamond P$, since $M',w_1\not\models\Diamond P$.
$M''=\langle\{w_0,w_1\},R,\models\rangle$, with w_0Rw_1 and $w_1\models P$, is a countermodel in $S4$ for $(P\to Q)\to\neg\Diamond(P\wedge\neg Q)$, because $M'',w_0\models P\to Q$, but $M'',w_0\models\Diamond(P\wedge\neg Q)$, since $M'',w_1\models P\wedge\neg Q$.

Solution 6.17. The following search tree for the conjecture $\Diamond P,\Box(\Diamond P\to Q)\vdash'\Box Q$ in $S4$ does not close and hence yields a Kripke counterexample in $S4$:

$\quad T\Diamond P,\; T\;\Box(\Diamond P\to Q),\; F\Box Q$

$\quad T\;\neg\Box\neg P,\; T\;\Box(\Diamond P\to Q),\; T\;\Diamond P\to Q,\; F\Box Q$ $\quad Q\; w_1$

$\quad F\;\Box\neg P,\; T\;\Box(\Diamond P\to Q),\; TQ,\; F\Box Q$

$\quad\swarrow\qquad\searrow\qquad\qquad\qquad\qquad\swarrow\searrow$

$F\;\neg P,\; T\;\Box(\Diamond P\to Q)\qquad T\;\Box(\Diamond P\to Q),\; FQ$

$TP,\; T\;\Diamond P\to Q\qquad\quad T\;\Diamond P\to Q,\; FQ\qquad\quad P,Q,\; w_2\quad w_3$

$TP,\; TQ\qquad\qquad\qquad\quad F\Diamond P,\; FQ$

$\qquad\qquad\qquad\qquad\qquad\; T\Box\neg P,\; T\;\neg P,\; FQ$

$\qquad\qquad\qquad\qquad\qquad\; T\Box\neg P,\; FP,\; FQ$

$M=\langle\{w_1,w_2,w_3\},R,\models\rangle$, with w_1Rw_2, w_1Rw_3, R reflexive and transitive, $w_1\models Q$ and $w_2\models P\wedge Q$, is a counterexample in $S4$ for the conjecture in question: $M,w_1\models\Diamond P$, $M,w_1\models\Box(\Diamond P\to Q)$ because in every world in which $\Diamond P$ is true, Q is true too, but $M,w_1\not\models\Box Q$.

Solution 6.18. The following search tree for the conjecture $\Box(Q\to\Box Q),\Diamond Q\vdash' Q$ in $S4$ does not close and hence yields a counterexample in $S4$:

$T \Box(Q \to \Box Q), T \Diamond Q, FQ$
$T \Box(Q \to \Box Q), F \Box\neg Q, FQ$
$T \Box(Q \to \Box Q), T Q \to \Box Q, F \Box\neg Q, FQ \qquad w_1$
$T \Box(Q \to \Box Q), FQ, F \Box\neg Q, FQ$
$\qquad | \qquad\qquad\qquad\qquad\qquad\qquad \downarrow$
$T \Box(Q \to \Box Q), F\neg Q$
$T \Box(Q \to \Box Q), T Q \to \Box Q, F\neg Q \qquad w_2\ Q$
$T \Box(Q \to \Box Q), T \Box Q, F\neg Q$
$T \Box(Q \to \Box Q), T \Box Q, TQ$

$M = \langle \{w_1, w_2\}, R, \models \rangle$, with $w_1 R w_2$, R reflexive, and $w_2 \models Q$ is a Kripke counterexample in S4 for the conjecture in question: $M, w_1 \models \Box(Q \to \Box Q)$ because in every world in which Q is true, $\Box Q$ is true too, $M, w_1 \models \Diamond Q$, but $M, w_1 \not\models Q$.

Solution 6.19. a) We start a tableau in K with the T-signed premisses and the F-signed putative conclusion:

$T\Box(S \to E), T\Box(E \to L), T\neg\Diamond L, F\neg\Diamond S$
$T\Box(S \to E), T\Box(E \to L), F\neg\Box\neg L, T\neg\Box\neg S$
$T\Box(S \to E), T\Box(E \to L), T\Box\neg L, F\Box\neg S$
$\overline{T(S \to E), T(E \to L), T\neg L, F\neg S}$
$T(S \to E), T(E \to L), FL, TS$
$FS, T(E \to L), FL, TS \mid TE, T(E \to L), FL, TS$
$\qquad\qquad\qquad\qquad\quad \mid TE, FE, FL, TS \mid TE, TL, FL, TS$

Since all branches close: $\Box(S \to E), \Box(E \to L), \neg\Diamond L \vdash' \neg\Diamond S$ in K.
b) The following search tree for the conjecture $S \to \Box E, E \to \Box L, \neg\Diamond L \vdash' \neg\Diamond S$ in S4 does not close and hence yields a Kripke counterexample in S4:

$T (S \to \Box E), T(E \to \Box L), T\neg\Diamond L, F\neg\Diamond S$
$FS, FE, F\Diamond L, T\Diamond S$
$FS, FE, T \Box\neg L, F \Box\neg S \qquad w_1$
$FS, FE, T \Box\neg L, T\neg L, FL, F \Box\neg S$
$\qquad\qquad | \qquad\qquad\qquad\qquad \downarrow$
$T \Box\neg L, F\neg S$
$T \Box\neg L, T\neg L, FL, TS \qquad w_2\ S$

$M = \langle \{w_1, w_2\}, R, \models \rangle$, with $w_1 R w_2$, R reflexive, and $w_2 \models S$, is a Kripke counterexample in S4 for the conjecture in question: $M, w_1 \models S \to \Box E$ because $M, w_1 \not\models S$, $M, w_1 \models E \to \Box L$ because $M, w_1 \not\models E$, $M, w_1 \models \neg\Diamond L$, but $M, w_1 \models \Diamond S$.

Solution 6.20. The search trees for the conjectures $\neg A \vdash' \Box(A \to B)$ and $B \vdash' \Box(A \to B)$ in S4 do not close and hence yield counterexamples in S4:

$T \neg A, F \Box(A \to B) \quad w_1 \qquad\qquad TB, F \Box(A \to B) \quad w_1\ B$
$FA, F \Box(A \to B) \qquad\qquad\qquad\qquad | \qquad\qquad\qquad |$
$\qquad | \qquad\qquad\qquad\qquad\qquad\qquad | \qquad\qquad\qquad \downarrow$
$\qquad \downarrow \qquad\qquad\qquad\qquad\qquad\qquad | \qquad\qquad$
$FA \to B \quad w_2\ A \qquad\qquad\qquad FA \to B \quad w_2\ A$
$TA, FB \qquad\qquad\qquad\qquad\qquad TA, FB$

$M = \langle \{w_1, w_2\}, R, \models \rangle$, with $w_1 R w_2$, R reflexive, $w_1 \models B$ and $w_2 \models A$, is a Kripke

6.13 Solutions   325

counterexample in S4 for both conjectures: $M, w_1 \models \neg A$, $M, w_1 \models B$, but $M, w_1 \not\models \Box(A \to B)$, since $M, w_2 \models A$ and $M, w_2 \not\models B$.

Solution 6.21. The following tableaux in K are all closed and hence:
$\Box \neg A \vdash' \Box(A \to B)$, $\Box B \vdash' \Box(A \to B)$, $Q \vdash \Box(P \to P)$ and $\vdash' \Box(\neg Q \land Q \to P)$ in K:

$T \Box \neg A$, $F \Box(A \to B)$	$T \Box B$, $F \Box(A \to B)$	TQ, $F \Box(P \to P)$	$F \Box(\neg Q \land Q \to P)$
$T \neg A$, $F A \to B$	TB, $F A \to B$	$F P \to P$	$F \neg Q \land Q \to P$
FA, TA, FB	TB, TA, FB	TP, FP	$T \neg Q \land Q$, FP
			FQ, TQ, FP

Solution 6.22. a) $M, w \models C \Box\!\!\!\to (B_F \land V_F) \lor (B_I \land V_I)$: there is a world, namely w_1 (or w_2), such that $M, w_1 \models C \land ((B_F \land V_F) \lor (B_I \land V_I))$ and such that for any w'', if $M, w'' \models C \land \neg((B_F \land V_F) \lor (B_I \land V_I))$, then $w_1 <_w w''$.
b) $M, w \not\models C \Box\!\!\!\to B_I$ because there is no $C \land B_I$-world which is closer to w than any $C \land \neg B_I$-world; w_1 is as close to w as w_2.
c) and d) are treated similarly.

Solution 6.23. a) Suppose $M, w \models A \land \neg B$ (1) and $M, w \models A \Box\!\!\!\to B$. From (1) $M, w \models A$ and hence, since R is reflexive, $M, w \models \Diamond A$. So, from the definition of $M, w \models A \Box\!\!\!\to B$, it follows that there is some world w' in W such that (2) wRw' and $M, w' \models A \land B$, and (3) for all w'' in W, if $M, w'' \models A \land \neg B$, then $w' <_w w''$.
From (1) and (3) it follows that $w' <_w w$. And since $M, w \models \neg B$ and $M, w' \models B$ we know that $w \neq w'$ and therefore, by assumption, not $w' <_w w$. Contradiction. So, if $M, w \models A \land \neg B$, then $M, w \models \neg(A \Box\!\!\!\to B)$.
b) Suppose $M, w \models A \land B$. Since R is reflexive we have that wRw and $M, w \models A \land B$. So, in order to show that $M, w \models A \Box\!\!\!\to B$ it suffices to show that for all w'' in W, if $M, w'' \models A \land \neg B$, then $w <_w w''$. So, suppose $M, w'' \models A \land \neg B$. Now, $M, w \models B$ and $M, w'' \models \neg B$. Therefore, $w \neq w''$ and hence, by assumption, $w <_w w''$.

Solution 6.24. Suppose $M, w \models \Box(A \to B)$. If $M, w \models \neg \Diamond A$, then $M, w \models A \Box\!\!\!\to B$. So, suppose $M, w \models \Diamond A$, i.e., for some w' in W, wRw' and $M, w' \models A$. Since $M, w \models \Box(A \to B)$, it follows that wRw' and $M, w' \models A \land B$ (1). So, in order to show that $M, w \models A \Box\!\!\!\to B$ it suffices to prove that for all w'' in W, if $M, w'' \models A \land \neg B$, then $w' <_w w''$. So, suppose $M, w'' \models A \land \neg B$. Since $M, w \models \Box(A \to B)$ it follows that not wRw''. Then, by assumption, it follows from wRw' and not wRw'' that $w' <_w w''$.

Solution 6.25. The following search tree for the conjecture that $\vdash' \Box(P \to Q) \lor \Box(Q \to P)$ in $S4$ does not close and hence yields a Kripke counterexample in $S4$:

$$F \Box(P \to Q) \lor \Box(Q \to P)$$
$$F \Box(P \to Q), F \Box(Q \to P)$$

$F P \to Q$	$F Q \to P$		w_1	
TP, FQ	TQ, FP	$P\ w_2$		$w_3\ Q$

From this open search tree we can read off a Kripke counterexample in $S4$ to $P \mapsto Q \lor Q \mapsto P$. Let $M = \langle \{w_1, w_2, w_3\}, R, \models \rangle$ with R reflexive and transitive, $w_1 R w_2$, $w_1 R w_3$, $w_2 \models P$ and $w_3 \models Q$. Then $M, w_1 \not\models \Box(P \to Q) \lor \Box(Q \to P)$.

It happens that also $M, w_1 \not\models P \square\!\!\rightarrow Q$, since $M, w_1 \not\models \neg \Diamond P$ and for no w in $\{w_1, w_2, w_3\}$, $M, w \models P \wedge Q$. In a similar way one sees that $M, w_1 \not\models Q \square\!\!\rightarrow P$.

Solution 6.26. Deduction Theorem: $A^1, \ldots, A^n \vdash^* B$ iff $A^1, \ldots, A^{n-1} \vdash^* A^n \Rightarrow B$.
Proof: From right to left is trivial. From left to right: Suppose $A^1, \ldots, A^n \vdash^* B$. Replace in the given deduction of B from $A^1, \ldots, A^n$ each expression C_c with the natural number n occurring in the index c by the expression $(A^n \Rightarrow C)_{c-\{n\}}$, where the index $c - \{n\}$ results from c by leaving out n. The upper lines in the resulting schema may look as follows: $A^1_{\{1\}}, \ldots, A^{n-1}_{\{n-1\}}, A^n \Rightarrow A^n$, axiom.
The bottom line in the resulting schema just contains $A^n \Rightarrow B_{\{1,\ldots,n-1\}}$. Note that $A^n \Rightarrow A^n$ is an axiom. For Modus Ponens, $\dfrac{D_d \ (D \Rightarrow E)_e}{E_{d \cup e}}$ there are four possibilities:

i) n occurs in d, but not in e. Then we get $\dfrac{A^n \Rightarrow D_{d-\{n\}} \ (D \Rightarrow E)_e}{(A^n \Rightarrow E)_{d \cup e - \{n\}}}$ and using axiom 2 this is a derived rule.

ii) n occurs in e, but not in d. Then we get $\dfrac{D_d \ (A^n \Rightarrow (D \Rightarrow E))_{e-\{n\}}}{(A^n \Rightarrow E)_{d \cup e - \{n\}}}$ and by using axiom 4 this is a derived rule.

iii) n occurs both in d and in e. Then we get $\dfrac{(A^n \Rightarrow D)_{d-\{n\}} \ (A^n \Rightarrow (D \Rightarrow E))_{e-\{n\}}}{(A^n \Rightarrow E)_{d \cup e - \{n\}}}$ and by using axiom 3 this is a derived rule.

iv) In case n occurs neither in d nor in e, the application of Modus Ponens remains unchanged. So, the resulting schema can be extended - by using the axioms - to a deduction of $A^n \Rightarrow B$ from $A^1, \ldots, A^{n-1}$.

Solution 6.27. a) One easily checks that the axioms for weak implication are valid (1). For instance, let $M = \langle S, \emptyset, \cup, \models \rangle$ be a model; then $M \models_\emptyset A \Rightarrow A$, i.e., for all a in S, if $M \models_a A$, then $M \models_a A$. And the rule Modus Ponens preserves validity (2), more precisely: if $M \models_\emptyset B$ and $M \models_\emptyset B \Rightarrow C$, then $M \models_\emptyset C$.
Now suppose $A_1, \ldots, A_n \vdash^* B$. Then, by the deduction theorem, $\vdash^* A_1 \Rightarrow (\ldots \Rightarrow (A_n \Rightarrow B) \ldots)$, i.e., the latter formula can be obtained by a finite number of applications of Modus Ponens starting with the axioms for weak implication. So, by (1) and (2), for all models M, $M \models_\emptyset A_1 \Rightarrow (\ldots \Rightarrow (A_n \Rightarrow B) \ldots)$, i.e., $A_1, \ldots, A_n \models^* B$.
ii) Let $M = \langle \{\emptyset, \{1\}, \{2\}, \{1,2\}\}, \emptyset, \cup, \models \rangle$ be defined by $\{1\} \models Q$, $\{2\} \models P$ and $\{1,2\} \not\models P$. Then $M \not\models_{\{1\}} P \Rightarrow P$ and $M \not\models_\emptyset Q \Rightarrow (P \Rightarrow P)$.

Solution 6.28. 1. $(A \Rightarrow A) \Rightarrow (A \Rightarrow A)$, axiom 1 for weak implication.
2. $((A \Rightarrow A) \Rightarrow (A \Rightarrow A)) \Rightarrow (A \Rightarrow ((A \Rightarrow A) \Rightarrow A))$, axiom 4 for weak implication.
3. $A \Rightarrow ((A \Rightarrow A) \Rightarrow A)$, from 1 and 2 by MP.

Solution 6.29. $\square \forall x [A(x)] \vdash' \forall x [\square A(x)]$: $T \ \square \forall x[A(x)], F \ \forall x[\square A(x)]$
$T \ \square \forall x[A(x)], F \ \square A(a)$
$\overline{T \ \forall x[A(x)], F \ A(a)}$
$T \ A(a), F \ A(a)$ closure

Solution 6.30. Let $M = \langle W, R, \models \rangle$ be a Kripke model with W finite and non-empty, R transitive and irreflexive. Suppose $M, w \models \square(\square A \rightarrow A)$, i.e., for all w', if wRw'

and $M, w' \models \Box A$, then $M, w' \models A$. (1)

Next suppose that not $M, w \models \Box A$. Then there is $w_1 \in W$ such that wRw_1 and not $M, w_1 \models A$. From (1) it follows that not $M, w_1 \models \Box A$. Hence, there is $w_2 \in W$ such that w_1Rw_2 and not $M, w_2 \models A$. Because R is transitive, wRw_2. So, by (1), not $M, w_2 \models \Box A$. Consequently, there is $w_3 \in W$ such that w_2Rw_3 and not $M, w_3 \models A$. And so on.

So, we find a sequence $w = w_0, w_1, w_2, \ldots$ in W such that w_iRw_{i+1} and not $M, w_i \models A$. Because R is transitive and irreflexive it follows that $w_i \neq w_j$ for all i, j with $i \neq j$. So, W is infinite. Contradiction.

References

1. Ackermann, W., Begründung einer strengen Implikation. *Journal of Symbolic Logic*, 21, 113-128, 1956.
2. Anderson, A.R., and Belnap, N.D., *Entailment; the Logic of Relevance and Necessity*. Princeton University Press, 1975.
3. Benthem, J.F.A.K. van, *The Logic of Time*. Reidel, Dordrecht, 1983.
4. Blackburn, P., M. de Rijke and Y. Venema, *Modal Logic*. Cambridge Univ. Pr. 2001.
5. Boolos, G., J. Burgess and R. Jeffrey, *Computability and Logic*. Cambridge University Press, 5th edition, 2007.
6. Boolos, G., *The Unprovability of Consistency; an essay in modal logic*. Cambridge University Press, 1979.
7. Burgess, J.P., Logic and Time. *Journal of Symbolic Logic* 44, 566 - 582, 1979.
8. Church, A., The weak Theory of Implication. In: Menne-Wilhelmy-Angsil (eds.), *Kontrolliertes Denken, Untersuchungen zum Logikkalkül und der Logik der Einzelwissenschaften*. Kommissions-Verlag Karl Alber, Münich, pp. 22-37, 1951.
9. Donnellan, K., Reference and Definite Descriptions. *Philosophical Review* 75 (1966) 281-304.
10. Fagin, R., J.Y. Halpern, Y. Moses and M. Vardi, *Reasoning About Knowledge*. MIT Press, 2004.
11. Fitting, M., *Proof Methods for Modal and Intuitionistic Logics*. Reidel, 1983.
12. Føllesdal, D., Essentialism and Reference. In: Hahn, L.E., and P.A. Schilpp (eds.), *The Philosophy of W.V. Quine*. La Salle, Illinois, 1986.
13. Frege, G., *Begriffsschrift*. Halle, 1879.
14. Gent, I.P., A sequent system for Lewis's counterfactual logic VC. *Notre Dame Journal of Formal Logic*, 33 (1992) 369-382.
15. Goldblatt, R., *Logics of Time and Computation*. CSLI, Stanford, 1987.
16. Grice, P., Studies in the Way of Words. Harvard University Press, 1989.
17. Harper, W.L., R.C. Stalnaker, G. Pearce (eds.), *IFS,; Conditionals, Belief, Decision, Chance and Time*. Reidel, Dordrecht, 1976.
18. Hilpinen, R. (ed.), *Deontic Logic*. Introductory and Systematic Readings. Reidel, 1971.
19. Hubbeling, H.G., *Language, Logic and Criterion*. Van Gorcum, Assen, 1971.
20. Kneale, W. and M., *The Development of Logic*. Clarendon Press, Oxford, 1962.
21. Kripke, S.A., Semantical Analysis of Modal Logic. *Zeitschrift für Mathematische Logik und Grundlagen der Mathematik*, Band 9 (1963), 67-96.
22. Kripke, S.A., *Naming and Necessity*. Basil Blackwell, Oxford, 1980.
23. Kripke, S.A., *Identity and Necessity*. In: Schwartz, S.P., *Naming, Necessity and Natural Kinds*. Cornell University Press, 1977-1979, 66-101.
24. Lemmon, E.J., *An introduction to Modal Logic*. Basil Blackwell, Oxford, 1977.
25. Lewis, C.I., *A Survey of Symbolic Logic*, University of California Press, Berkeley, 1918.
26. Lewis, D., *Counterfactuals*. Harvard University Press, 1973.

27. Lewis, D., Counterfactuals and Comparative Possibility. *Journal of Philosophical Logic*, pp. 418-446, 1973.
28. Linsky, L. (ed.), *Reference and Modality*. Oxford University Press, London, 1971, 1979.
29. Meyer, J.-J. Ch. and W. van der Hoek, *Epistemic Logic for A.I. and Computer Science*. Cambridge University Press, 2004.
30. Montague, R., The Proper Treatment of Quantification in Ordinary English. Reprinted in R.H. Thomason, (ed.), *Formal Philosophy; Selected Papers of Richard Montague*. Yale University Press, London, pp. 247-270, 1974.
31. Perrick, M. and H.C.M. de Swart, Quantified Modal Logic, Reference and Essentialism. *Logique et Analyse* 143-144 (1993) 219-231.
32. Quine, W.V., Reference and Modality, *Journal of Symbolic Logic* (1953), 137-138.
33. Quine, W.V., *From a Logical Point of View*. Harvard University Press, 1980.
34. Searle, J.R. (ed.), *The Philosophy of Language*. Oxford University Press, 1971.
35. Solovay, R.M., Provability Interpretations of Modal Logic. *Journal of Symbolic Logic* 46, 661-662, 1981.
36. Swart, H.C.M. de, Systems of Natural Deduction for several systems of Modal Logic; constructive completeness proofs and effective decision procedures. *Logique et Analyse* 90-91, 263-284, 1980.
37. Swart, H.C.M. de, A System of Natural Deduction, an effective decision procedure and a constructive completeness proof for the counterfactual logics VC and VCS. *Journal of Symbolic Logic* 48, 1-20, 1983.
38. Urquhart, A., Semantics for Relevant Logics. *Journal of Symbolic Logic*, 37, 159-169, 1972.

Chapter 7
Philosophy of Language

Luc Bergmans, John Burgess, Amitabha Das Gupta and Harrie de Swart

Abstract This chapter aims to be an introduction to the philosophy of language and presents some major topics belonging to this field: the difference between use and mention, Frege's notions of Sinn (sense) and Bedeutung (reference), Mannoury's significs, speech acts, definite descriptions, Berry's and Grelling's paradox, the theory of direct reference, Kant's notions of analytic versus synthetic, logicism, logical positivism, presuppositions, Wittgenstein on meaning, syntax - semantics - pragmatics, conversational implicature, conditionals, Leibniz, de dicto - de re distinction, and grammars. It is fair to say that the Dutch mathematician Gerrit Mannoury (1867 - 1956) invented the notion of speech act long before Austin, Searle and others used this notion. In the subsection on Logicism we explain that - contrary to what many philosophers of science claim even nowadays - Kant was right in asserting that mathematical statements are not analytic, but synthetic.

7.1 Use and Mention

If we want to say something **about** an object, we use the name of that object. We are used to doing so when the object is a person, but one frequently gets confused when the object is a linguistic one. Names of linguistic objects can be formed by enclosing the linguistic object in single (or double) quotation marks. For instance, in the proposition

John is a teacher

we make a statement **about** a person using the name of that person; and, similarly, in the proposition

'man' is monosyllabic

we make a statement **about** the word (linguistic object) *man*, using the name of that word. Using the terminology of W.V. Quine, we say that in

Man is a rational animal

the word *man* is used, but not mentioned; and that in

>'Man' is monosyllabic

the word *man* is mentioned, but not used.

In practice the quotation marks are frequently suppressed, causing an equivocacy which is often convenient and harmless on the condition that one realizes what one is doing. So, instead of

>'man' is monosyllabic

one may come across

>man is monosyllabic.

Adopting Carnap's terminology, we say that the word *man* in the latter expression is *used autonymously*, i.e., as the name of that same word. So, in

>man is monosyllabic

the word *man* is both mentioned and used, though used in an anomalous manner, namely, autonymously. Some more examples: in

>The English translation of the French word *homme* has three letters

the word *man* is mentioned, but not used. In

>The second letter of man is a vowel, and in
>Man is a noun with a irregular plural,

the word *man* is mentioned and used autonymously.

The equivocacy, resulting from using the same word, *man*, both as a proper name of a linguistic expression and as a common name of certain mammals, may be removed by the use of added words in the sentence, or by the use of quotation marks, or of italics, as in

>The word man is monosyllabic,
>'Man' is monosyllabic,
>*Man* is monosyllabic.

The latter device has been used above several times. The reader is advised to do Exercise 7.1. The examples above are from Church [10].

Exercise 7.1. (B. Mates, *Elementary Logic*, 1972, pp. 40-41) Not using words autonymously, which of the following sentences are true?

1. 'The Iliad' is written in English.
2. 'The Iliad' is an epic poem.
3. 'The Morning Star' and 'The Evening Star' denote the same planet.
4. The Morning Star is the same as the Evening Star.
5. '7 + 5' = '12'
6. The expression ' 'The Campanile' ' begins with a quotation mark.
7. The expression ' 'der Haifisch' ' is suitable as the subject of an English sentence.
8. Saul is another name of Paul.
9. 'Mark Twain' was a pseudonym of Samuel Clemens.
10. 2 + 2 = 4 is synthetic.
11. Although 'x' is the 24th letter of a familiar alphabet, some authors have said x is the unknown.
12. We are using capital Roman letters 'A', 'B', 'C', ... to stand for any formulas.

7.2 Frege's Sinn und Bedeutung (Sense and Reference)

In his Begriffsschrift of 1879 Frege made a distinction between '-*A*' for 'the proposition that *A*' and '⊢ *A*' for 'it is a fact that *A*'. In '-*A*' and in '⊢ *A*' Frege calls '*A*' the conceptual content (begrifflichen Inhalt). Thus if '⊢ *A*' is an abbreviation for the statement 'unlike magnetic poles attract each other', '-*A*' is to convey only the thought of mutual attraction between unlike magnetic poles, without any judgment of the correctness of that thought (G. Frege [14], Section 2).

In section 8 of his Begriffsschrift Frege introduces ⊢ $a \equiv b$ as meaning: the sign *a* and the sign *b* have the same conceptual content so that *a* can always be replaced by *b* and conversely. However, if we consider

⊢ the morning star ≡ the evening star

it becomes clear that the definition just given must be wrong for the following two reasons: i) The expressions 'the morning star' and 'the evening star' have different conceptual contents, and ii) 'The morning star is identical with the evening star' has a meaning quite different from 'The morning star is identical with the morning star'. To verify the truth of the first sentence, astronomical observation is needed, but it is not necessary for the second one.

It is probably for these reasons that Frege, in his *Ueber Sinn und Bedeutung* of 1892, abandoned his talk of conceptual content and introduced a distinction between *sense (Sinn)* and *reference (Bedeutung)* instead. 'The morning star is identical with the evening star' then means: a) the expressions 'the morning star' and 'the evening star' refer to the same object, i.e., the planet Venus, called the reference (die Bedeutung), but b) they do so in different ways, because they have a different cognitive meaning or sense (Sinn).

The *reference (Bedeutung)* of an expression is what it 'stands for'. In the case of a proper name (Plato, France, the Titanic), it is the thing named; in the case of a singular definite description (Plato's father, the president of the United States), the object that fits the description.

In addition to words and the things (references) they stand for, Frege also insisted on taking into account the sense or cognitive meaning of words, since it is *through* its sense that an expression refers to an object. The sense provides the 'mode of presentation' of the object and referring to a reference is always achieved by way of sense.

Frege's *sense (Sinn)* includes the information content (cognitive meaning) of an expression, but not such features as (1) associations (emotional, literary; like the difference between 'horse' and 'steed'), (2) level of speech (formal, colloquial, slang, dialect, obsolescent, obscene; like the difference between 'regurgitate' and 'puke'), (3) indications of speaker's attitude (like the difference between 'but' and 'and' in 'he is a politician, but relatively honest', or the difference between 'they (still) have not arrived' with or without the 'still').

In poetry these other features are important, and a translation which merely preserved Frege's sense and lost these other features of 'meaning' would be a poor one.

In dry, objective scientific prose, only the sense is important. But for the study of literature, these extra features are of great importance. They distinguish 'I am determined/you are stubborn/he is pig-headed', which have, except for the change in personal pronoun, more or less the same information content or Fregean sense.

Frege emphasized the abstract nature of senses – they do not belong to any particular language (words in different languages can have the same cognitive meaning) and do not consist of individual psychological reactions in speakers of a language, but are something common to *all* speakers.

Obviously it is possible that two names or descriptions stand for the same object without being synonyms: we can have two expressions *a* and *b* with the same reference but with different senses. (The opposite cannot happen as we shall see below.) Indeed, this is Frege's explanation of how a statement '$a = b$' can be informative. In 'the Morning Star is the same thing as the Evening Star', for example, the two expressions 'Morning Star' and 'Evening Star' *refer to the same reference* (the planet Venus), but they *express a different sense*. And in '$2 + 2 = 4$' the names '$2 + 2$' and '4' refer to the same number, but they express a different sense.

Names for Frege include both proper names but also singular definite descriptions. Other writers use *singular term* or *designator* or *denoting-phrase* for Frege's *name*, which is less misleading, since we definitely want to include more than proper names.

The reference of a name is called an *object* (*ein Gegenstand*) by Frege. In other words, objects are anything which can be referred to by a name. This includes not just people and physical objects, but also abstractions (the Equator, numbers, justice) and events (the battle of Hastings). Frege has no special term for the sense of names. Carnap has called them *individual concepts*, not to be confused with Frege's *concepts* discussed below.

The expressions 'the greatest natural number' and 'the present king of France' do have a sense, but do not have a reference, because these expressions refer to nothing.

An expression is said to *express* its *sense* and *refer* to its *reference*. Other philosophers use the words *connotation* or *meaning* or *intension* for *sense* and *connote* or *mean* for *express*. *Denotation, designation, extension* and *signification* have all been used for *reference*, and *denote, designate* and *signify* for *refer*.

Frege goes on to argue that besides names and descriptions, predicates (or *general terms*) and sentences have both sense and reference.

Frege has no special term for the sense of a predicate ('is bald', 'lives in Princeton'). Others have said the predicate expresses a *property*, *attribute* or *quality*.

According to Frege, the reference (Bedeutung) of a predicate is a *concept* (*ein Begriff*). However, the nature of concepts is rather obscure. In addition to concepts, Frege recognizes classes. These are simply collections of objects. The class corresponding to a predicate, e.g., the class of all bald people corresponding to the predicate 'is bald', Frege calls the *extension* of the predicate. Most philosophers who follow Frege on the whole discard his concepts and simply speak of the class, and do not distinguish reference from extension.

7.2 Frege's Sinn und Bedeutung (Sense and Reference)

The two predicates 'has a heart' and 'has a liver' may be true of all the same things (*co-extensive*, as Frege says) without being synonymous: they can have the same reference and extension without having the same sense. (The opposite is impossible according to principle (i) below.) The *class* of all creatures with hearts may be exactly the same class as the *class* of all creatures with livers, but the *property* of having a heart is different from the *property* of having a liver: to say something has a heart does not mean the same as saying it has a liver.

In order to figure out what the *reference (Bedeutung)* of a sentence is, Frege seems to invoke two principles:

(i) expressions with the same sense have the same reference, and

(ii) (principle of *compositionality*:) the reference of a compound is entirely determined by the references of its parts. This implies that

(ii*) if we replace a name or description in a sentence by another name or description of the same object, the reference of the sentence is unchanged.

By the principles (i) and (ii*) all sentences below have the same reference:

Scott wrote Waverly;
Scott is the author of the 29 Waverly novels (i);
29 is the number of Waverly novels that Scott wrote (i);
29 is the number of counties in Utah (ii*);
Utah has 29 counties (i).

So, (i) and (ii) imply that seeming unrelated true statements 'Scott wrote Waverly' and 'Utah has 29 counties' have the same reference (and similarly for false sentences). Frege concludes that the *reference* of a sentence is just its truth value, either *true* or *false*. (The example is from Church, [10], pp. 24-25.)

According to the principle of compositionality, mentioned above, if a name or description has no reference, no sentence of which it is a part can have a reference (truth value). So, the sentence 'The king of France is bald' does not have a truth value (reference, Bedeutung), because its subject 'the king of France' has no reference. If there is no Pegasus, 'Pegasus is flying' can be neither true nor false.

Frege explained the *presuppositions* of a statement as those things which must be true if that statement is to have any truth value at all, and specifically stated that a statement involving a description like 'the present king of France' *presupposes* the existence of the thing satisfying this description. Later writers on presupposition (e.g., Strawson) take Frege as their starting point (see Section 7.11).

Frege calls the *sense* of a sentence a *proposition* or *thought* (*ein Gedanke*). Actually, as commentators on Frege have pointed out, sentences like 'it is raining', 'I have a headache' express different propositions (sometimes true, sometimes false) according to when, where, and by whom they are uttered, so that it is necessary to distinguish between the proposition expressed on any given occasion and the *meaning* of the sentence, which is always the same. (In mathematics words like 'I', 'here', 'now' seldom occur, and so this problem does not arise. Frege was mainly concerned with this area of discourse.)

Frege distinguishes between a *proposition* and an *assertion*. According to Frege, when I state, 'The door is open' or ask, 'Is the door open?' or request, 'Please open the door' or sigh, 'If only the door were open!' or command, 'Open the door!' or make a compound statement, 'If the door is open, then there'll be a draft', the same *proposition* is *expressed* in every case, but is only *asserted* in the first case. In the other cases no *assertion* is made; rather there is a question, request, etc.

Other philosophers have called a *proposition* a *phrastic* (Hare) or *locution* (Austin) and the element that must be added to make an assertion, or a question, or a request, etc., a *neustic* (Hare) or *illocutionary force* (Austin). The distinction between propositions and assertions is the origin of *speech act theory*, developed by Austin, Searle, and others (see Section 7.4 on speech acts).

Embedded in the aspects of Frege's theory of sense and reference, which have been dealt with so far, is the following contradiction. Consider the sentences:
(1) Somebody wonders whether Amsterdam is the capital of the Netherlands, and
(2) Somebody wonders whether Amsterdam is Amsterdam.
While (1) is probably true, (2) is false. So, (1) and (2) are likely to have different references (truth values). However, since the reference of 'Amsterdam' is the same as the reference of 'the capital of the Netherlands', the principle of compositionality seems to imply that (1) and (2) have the same truth value (reference).

Frege was aware of this problem and adapted his theory as follows. He postulated that in intensional contexts, created by phrases such as 'wonder whether', 'know that', and so on, expressions have an *indirect* (or *oblique*) *reference* and *sense* instead of their *direct* (or *ordinary*) *reference* and *sense*. The *indirect reference* of an expression is its ordinary sense and its indirect sense is something else. Consequently, the expressions 'Amsterdam' and 'the capital of the Netherlands' in the sentences (1) and (2) above have a different (indirect) reference, because both occur in the context 'wonders whether'. For that reason the principle of compositionality cannot be applied in order to derive that (1) and (2) would have the same reference (truth value).

The following schema gives a summary of Frege's theory of sense (Sinn) and reference (Bedeutung).

	proper names and singular definite descriptions	predicates	sentences
Examples	morning star; evening star; $2+2$; 4 the present king of France	has a heart has a liver	$2+2=4$; the morning star = the evening star. Scott wrote Waverly; Utah has 29 counties
Sinn sense		others: property, attribute, quality	ein Gedanke proposition, thought
Bedeutung reference	ein Gegenstand object	ein Begriff, concept extension of a predicate: class	truth value

'The morning star' and 'the evening star' express a different sense, but refer to the same reference (object). 'Has a heart' and 'has a liver' express a different sense (property), but refer to the same reference (class). 'Scott wrote Waverly' and 'Utah has 29 counties' express a different sense (proposition), but refer to the same reference (truth value).

'The present king of France' does have a sense, but does not have a reference. Hence, the sentence 'The present king of France is bald' does not have a reference (truth value).

As noted by J.R. Searle, although the distinction between sense and reference seems to be quite natural for names (proper names and singular definite descriptions), its extension to predicates and sentences is less compelling.

> To my mind it loses the most brilliant insight of the original distinction, an insight which reveals the connection between reference and truth: namely that an expression refers to an object only because it conveys something true of that object. But a predicate does not convey something true of a concept nor does a sentence convey something true of a truth value. (Searle, [47], p. 3)

Reading list on Frege: Carnap [9]; Church [10], Introduction; Dummett [13]; Frege [15]; Frege [16]; Heijenoort [23]; Searle [47]; Strawson [52].

7.3 Mannoury (1867-1956), Significs

> The language, which is used by all people as a means of understanding, is full of unclean elements that poison society, such as contaminated water poisons the population of a whole city. For that reason it is immediately needed to show that the water supply and the sources from which the city receives its drinking water, is contaminated by germs, and that it is most urgent to first purify these sources. [F. van Eeden in: Brouwer, L. E. J., F. Van Eeden, J. Van Ginneken en G. Mannoury, Signifische dialogen. 1939; translated from Dutch.]

Gerrit Mannoury's writings are likely to be enriching and thought-provoking for any student or scholar who takes a genuine interest in the phenomenon of language. This great Dutch thinker made many piercing remarks on the essential functions of language, on the nature of formalism, and on the connectedness of language-types that are generally considered incompatible. His views on meaning and the methods of describing it tended to be stated with refreshing and liberating relativism.

Mannoury was one of the founding members of the International Institute of Philosophy in Amsterdam (1917), which in many ways prepared the activities of the later Dutch Signific Circle (1922-1926) (for a history of Dutch significs from 1892 to 1926, see H.W. Schmitz [44]) and he remained the witful explainer and propagator of the signific ideas long after the circle had been dissolved (see, for example, Mannoury [34]).

Among the most prominent features of signific thought, and of Mannoury's thought in particular, was the idea of the *intentional* nature of language. This becomes apparent from the way in which Mannoury characterized *communicative acts*, of which *linguistic acts* form a subcategory:

> We shall call *communicative act* any act by which living beings (say human beings to simplify matters) try to influence directly the behavior or activity of other living beings. (G. Mannoury [33], p. 13.)

For Mannoury and his fellow significians, language was in the first place an *expression of the will* or, to quote L.E.J. Brouwer, another authority in the field:

> all utterances in words are more or less developed verbal imperatives, ... hence addressing always comes down to commanding or threatening, and understanding always comes down to obeying. (L.E.J. Brouwer [7], p. 333).

A language shared by a group serves to regulate and coordinate individual will or, to cite Brouwer again:

> to keep the movement of the Will of separate persons on one track. (L.E.J. Brouwer [6], p. 38).

The volitional function, which is primary from a signific point of view, can be illustrated particularly well through what Karl Bücher regarded as the historically primitive forms of poetry and music, namely the singing that accompanied manual labor. Wilhelm Wundt commented on this type of language use in his *Völkerpsychologie* (1900):

> Whenever several people join in the same work, the sounds which accompany the cadenced movements ... automatically bring about a pattern of co-operation which allows every participant to make the movements to the same rhythm. The resulting multiplication of rhythmic sounds increases the awakening of ardour. If in addition the shared labor is oriented towards one and the same object, such as in the case of the rowing of a boat or the joint hoisting or hauling of loads, the regular utterance of sounds again naturally becomes an expedient which rhythmically orders the singular powers synchronically or according to the sequence in which they mesh with one another. (Wundt [58] (included in the bibliography of G. Mannoury [34]), pp. 263-264.)

Here all the characteristics of what the significians considered to be the most original forms of linguistic usage are united: language which accompanies activity; language of people who focus their attention on one and the same object, or who pursue one and the same goal; language as mutual imposition of the will. In less primitive forms of language use than the ones mentioned above, other functions, such as the indicative or declarative ones, become more prominent, or are perceived as such. It was the merit of the significians to point to purposeful will and roots in human activity even there.

To Mannoury the meaning of any communicative act was composed of emotional or volitional elements on the one hand, and indicative or declarative ones on the other. The essential task of significs he held to be the disentangling and connecting of both types of elements (G. Mannoury [31], p. 113). He displayed great skill in uncovering the volitional aspect of utterances that seem purely indicative:

7.3 Mannoury (1867-1956), Significs

> Strictly speaking one cannot reasonably ask: 'is it true what you say?', but only: 'what do you want from me when you say this to me?', and 'can I agree with your goal?'. But of course such general remarks should always be taken 'cum grano salis', and it will be clear that nobody would be able to say what is really aimed at, or which kind of will is expressed in a sentence such as *there is a running horse*, but still I am sure that if someone all of a sudden made this important announcement to you, and you could not remotely guess what made him draw your attention to the movement of the Rosinante, you would be astonished and you would ask even without the slightest philosophical reflection: 'what do you mean, what are you getting at?', or, putting it more philosophically, 'what is the cause of your judgement?, which motive made you create this combination of thoughts?'. (G. Mannoury and D. Vuysje, archives, university library of Amsterdam; text of the lecture in file 14, p. 7; published version of the lecture: Mannoury [30]).

It is clear that finding dictionary-meanings of words or unalterable definitions of terms was not the main concern of the significians. Instead they were interested in the *use* of words in a particular context by specific people. The meaning of a communicative act was characterized as follows by Mannoury:

> the associations which link this act to the psychic complexes determined by the participants involved. (Mannoury [33], p. 13)

These participants are 'the speaker' and 'the listener' (in a very general sense, because communicative acts can also be wars, smiles and paintings).

Two main methods of empirical signific research into the meaning of linguistic acts were presented in Mannoury [34] (see also Schmitz [45]). The first, called *method of exhaustion*, consists of finding the range of situations to which a person reacts in the same verbal way; the second, termed *method of transformation*, aims at collecting the verbal reactions of various people to one particular situation (Mannoury [35], p. 44). These two methods are especially well adapted to the study of non-technical language.

However, the scope of signific analysis is by no means reduced to everyday language. As most of the significians were active in some other scientific field (mathematics, law, psychology, biology, ...) their signific writings displayed a marked interest in the communicative acts of science. Here too their main concern was detecting the emotional or volitional and delineating it from the indicative. Mannoury felt that every logical or physical formalism, and even purely mathematical communicative acts, encompassed an *empirical content* (on the level of indication) and an element of *belief* (on the level of emotion and volition). In the case of mathematics the empirical content of the theorems or demonstrations consists of the *knowledge of preceding formalisms* which speaker and listener share. The element of belief is to be identified with the *esthetic* or *sportive* aspect of mathematics, which is the key to its deepest truth (Mannoury [33], p. 46).

This should be understood as follows: Two mathematicians in the course of a discussion try to find the solution to a problem and join their efforts in a project of which words and signs on paper are only the external marks of progress. They develop, through corroboration, which equally gifted or equally trained people give each other when they strive for the same end, a feeling of *certainty* or *beauty*, which is nothing other than approached truth.

Mannoury distinguished between *active / speaking* mathematics and *passive / listening* mathematics and pointed to the tension between the two:

> It is the old song: speaking mathematics and listening mathematics are at loggerheads. ... Speaking mathematics searches, supposes, conjectures, guesses right or wrong, enjoys and suffers, gets dizzy and hits some nails, but listening mathematics remains calm and hides behind ready-made definitions and has logarithmic tables printed with typesetting plates. And does not want to know its mother any more! It has risen so high that it forgets where it came from ... (Mannoury [31], p. 31).

One of Mannoury's merits as a significian was that he showed with many examples that formalisms should not be considered in isolation, but need to be studied in relation to the intuitive insights from which they sprang, and to the purpose which they are supposed to serve.

According to the Dutch signific group it is possible to develop a classification of language-types (Mannoury ([33], pp. 19-20), a graded scheme, in which each type of language is situated on a particular level, and words or expressions of a higher type can be interpreted or replaced by words or expressions of a lower one, but never vice versa. In such a scheme the symbolic language of sciences displaying advanced formalization, for example mathematical logic, occupy the highest degree; primitive forms of language, in which the immediate expression of emotions prevails, belong to the lowest degree; and the language of daily social intercourse is situated somewhere in between.

The principle of *linguistic gradation* makes clear how even the most abstract systems of language (which by virtue of their rigidly regulated syntax or their constant 'word-word-associations' (Mannoury [36], p. 161), to use one of Mannoury's own terms, give the impression of complete independence and perfect self-sufficiency) remain anchored in the living language of emotion and intuition. Disrooted abstractions lead to the creation of false problems (for example, with regard to the void or the actual infinity; Mannoury [33], p. 53) and therefore have to be dismissed. Any language that loses contact with life should be shed off as a snake's skin of dead formula.

The significian has a role to play in this process of sloughing and renewal. A prerequisite for the success of his undertaking is a thorough understanding of the field to which the language in question applies. It is only through familiarity with the objects and approach roads that he is able to detect the flaws and imperfections of the existing means of expression. The significian will then break through the language, which is like a passive crust that is moulded to fit the terrain as discovered thus far. He will proceed to an active and synthetic refinement that matches the new needs and allows him and his fellow explorers to draw nearer to the objects that required redefining.

Progress through refinement of a language that is starting to flounder was exemplified by David van Dantzig, a second generation significian:

> Inasfar as progress of science consists of the discovery of new regularities of the *formal system*, the preceding formalization will be very useful, but it may be (even if one is willing to replace the old formalism by a new one) an impediment to the discovery of such new

properties of the objects under investigation, which require finer distinctions ('fine structure') of relations hitherto regarded (and formalized!) as 'identical'. It is to a large extent by such 'finer distinctions' and broader generalizations that progress of science proceeds, as numerous examples show. *After* they have been made, formalization may become useful again. Formalization therefore covers a small part of science only, in particular a part which to a certain extent is 'ready' or 'closed' at the moment, and therefore formalism is running *behind* actual science (van Dantzig [11], p. 515, quoted in Mannoury [36], p. 120).

The careful technical readjustment of formal or other language in order to serve modifying goals belongs to the synthetic activities of the significian.

Mannoury also tackled a synthetic project of a more general kind. In [33] (Polar Psychological Synthesis of Concepts) he developed a *unifying terminology* that aimed at bridging the sharp distinction between the *mathematical* way of thinking (characterized by formalism and specific objects of consideration) and the *ideological* way of thinking (characterized by metaphor and general points of view). In doing so Mannoury remained true to his relativist position, which allowed for polar opposition (in which each pole needs its opposite) rather than separated categories, and he avoided the false problems that arise from dualism.

In this same light one should view Mannoury's scheme which distinguishes and combines two types of negation: a *negation of choice* and an *exclusive negation* (Mannoury [32], pp. 333-334). The *negation of choice* is used in contexts where two alternatives clearly present themselves to the speaker's mind (e.g. It is raining or it is *not* raining; if this is *not* a big town, it must be a small town; etc). The *exclusive negation* is based on a negative volition, a refusal without a clear alternative. In natural language the exclusive negation is often marked by words such as 'not ... at all', which indicate a stronger emotional involvement on the part of the speaker.

Mannoury shows that it is possible to combine these two negations, and gives the following example: 'What is not a small town could be a big town, but also something quite different' (ibidem). The small town/big town dichotomy is governed by the negation of choice. However, the words 'something quite different' illustrate the effects of the exclusive negation, namely drawing the attention away from the given alternatives (small town/big town) without proposing another possibility.

Double negation also plays a crucial role in intuitionistic mathematics and logic, as developed by L.E.J. Brouwer and his students. The question of inspiration and transmission of ideas between G. Mannoury and L.E.J. Brouwer, who were fellow significians, fellow mathematicians and intimate friends, is far from resolved (see Schmitz [43]). However, in this treatment of negation, as in many other cases, one is not surprised to find analogies in their thinking.

Both Brouwer and Mannoury seem to start from a language of dichotomy and clearly perceived entities. These are Mannoury's given alternatives or Brouwer's constructions in the mathematician's mind. Into this language they insert specific expressions involving two negations. These expressions hint at what extends beyond dichotomy and the clearly perceived. In other words, the inserts call up Mannoury's undefined alternatives or Brouwer's as yet unfulfilled goals of construction projects in the mathematician's mind.

The task of significs has been defined as showing the link between indication and volition/emotion, between what we think we have, and what we reach for, whenever we communicate. Synthetic significs provides new language constructs that make explicit this connection. What Mannoury and the intuitionists do when combining and embedding the two negations, contributes to this explication. They make us see the link between the given and that what goes beyond, between specific objects of consideration and higher objectives, between mathematics and mysticism.

7.4 Speech Acts

According to J.L. Austin, any *speech act* comprises at least two, and typically three, sub-acts. These are what he calls the *locutionary*, the *illocutionary* and the *perlocutionary acts* involved in a total speech act.

The *locutionary act* includes the utterance of certain noises, the utterance of certain words in a certain construction, and the utterance of them with a certain meaning. *Locutionary acts* are acts of saying something and meaning it (and supplying a definite reference for any pronouns like 'this', 'he', etc.).

Most every time we say something and mean it – when we aren't just testing our voice, or acting in a play – we do in fact perform illocutionary acts.

Illocutionary acts are things we do *in* speaking like: requesting, welcoming, asking questions, demanding, inviting, giving orders, accusing, granting permission, asserting, promising, lying.

For more examples and a rough classification, see Austin [3] (*How to do things with words*, Lecture XII). Austin offers no precise definition of 'illocutionary act' (nor does anyone else for that matter), but one can pretty well agree how to extend the above list. The illocutionary act can be regarded as the force with which the sentence is employed. The distinction between locutionary and illocutionary acts recalls Frege's distinction between proposition (also called thought) and assertion (or question, command or whatever). As we have already had occasion to note in our discussion of Frege, the same proposition can be expressed in many different kinds of illocutionary acts:

> Please come (request or invitation);
> Will you come? (question);
> You will come (prediction);

all having the same propositional contents.

Some illocutionary acts (greeting, resigning, condoling) do not involve expressing propositions.

If there is any distinction between locution and illocution it is this: while the meaning of what we say severely restricts the range of illocutionary acts we can be performing (e.g., 'Get in here this instant, you S.O.B.' cannot be a polite request; nor of course can it be a question, an assertion, a promise, etc.), it may not suffice to determine completely the illocutionary force of what we say (e.g., 'Come in here'

7.4 Speech Acts

might, depending on the circumstances, be an invitation, a command, an official order, etc., 'I will come' might, depending on the context, be a promise, a statement of present intention, or a fatalistic prediction). Because the *conventional* meaning may not suffice to completely determine what illocutionary act is performed, it may happen that even when the hearer understands perfectly the meaning of speaker's words, there may be a gap between the speaker's *intentions* and how his utterance is *taken* by the hearer. (What is intended as a mere statement of present intention may be mistaken for a promise; what is intended as a polite request may be misinterpreted as a peremptory command.) The existence of a gap between locution and illocution means that the notion of illocutionary act belongs on the border between *semantics* (the theory of the meaning of words, what is conventional and common to all speakers independently of their particular circumstances) and *pragmatics* (the theory of the use of language by speakers taking into account not only the invariant meaning of the words but also aspects depending on the speaker's intentions and purposes in the particular speech situation).

Perlocutionary acts are things we can do *by* speaking like: persuading, perplexing, alarming, irritating, boring, convincing, deceiving, frightening.

In general it is possible to try and fail to perform a perlocutionary act (we can try to deceive someone but not succeed), whereas it hardly makes sense to speak of trying and failing in the case of an illocutionary act (like lying). Generally the illocutionary act is complete when we have spoken, so long as we have been understood, whereas the perlocutionary act requires our speech to have some kind of further *effect* on the hearer.

So, then 'I promise to come to dinner' will be the performance (1) of a locution – e.g., employing a certain grammatical construction, (2) of an illocution – that of making a promise, and (3) of a perlocution – e.g., cheering you up.

Rules. Illocutionary acts may be called a form of 'rule-governed behavior'. There are rules and procedures for how the act is to be performed, and rules saying what kind of further behavior on the part of the speaker and hearer is 'in order' once the act has been performed, and what kind of behavior is 'out of order'. (For example, a bigamist violates the procedural rules for getting married, one of which involves not being married already.) Breaking a promise, welcoming people and then treating them like unwanted intruders, etc., are violations of the rules about what is supposed to be done afterwards.

An important distinction between two kinds of rules has been made by J. Rawls and taken over by Searle and others: *regulative rules* prescribe how some form of behavior existing antecedently to and independently of the rules is to be carried out. Thus rules of table manners prescribe the manner in which people should eat, but they are going to eat anyhow whether or not anyone has thought up any rules of table manners. *Constitutive rules*, by contrast, create the very possibility of new forms of behavior which could not exist without the rules. The rules of a game like bridge or basketball constitute what it is to play bridge or basketball. Apart from the rules, these games have no existence. The rules governing illocutionary acts belong in the *constitutive* category.

In [48] (*What is a Speech Act*) Searle distinguishes two (not necessarily separate) parts in a sentence used to perform an illocutionary act: the proposition-indicating element and the function-indicating device. The latter indicates the illocutionary act the speaker is performing in the utterance of the sentence. J.R. Searle gives the following two examples.

(1) I promise that I will come.
(2) I promise to come.

The function-indicating device and the proposition-indicating element are separate in (1), but not so in (2). As function-indicating devices Searle mentions, among others, word order, stress, punctuation and performative verbs such as 'apologize', 'warn', 'state', etc. See J.R. Searle [48], pp. 43-44.

7.5 Definite Descriptions

Both Russell (1872-1970) and Wittgenstein (1889-1951), for different sets of reasons, rejected Frege's distinction between sense and reference. In 'Russell's Rejection of Frege's Theory of Sense and Reference' J.R. Searle critically examines Russell's reasons for doing so.

Frege's analysis of a sentence like 'The king of France is bald' would be that this sentence lacks a truth value (reference), because the subject expression has no reference, but that the lack of a truth value does not render the sentence meaningless, since this sentence does have a sense. Now, how does Russell, having already rejected Frege's theory of sense and reference, explain how sentences like this one can be meaningful, while there is nothing for the proposition, expressed by the sentence, to be about. In *On Denoting* (1905) Russell claims that the sentence in question appears to be in subject-predicate form, but is not really so. Its grammatical form is misleading as to its logical form. Russell's analysis of

$$\text{The king of France is bald}$$

is as follows:
$$\exists x [x \text{ is king of France} \land x \text{ is bald} \land \forall y [y \text{ is king of France} \to y = x]],$$
or equivalently, but shorter

$$\exists x [x \text{ is bald} \land \forall y [y \text{ is king of France} \rightleftarrows y = x]].$$

And since there is no king of France, this sentence is false.

Russell analyzed (say) 'The king of France is bald' as no simple subject-predicate statement but a far more complicated one, in which two different quantified variables occur. In Russell's theory, the deep structure of such statements is very different from what their surface grammar suggests.

So Russell does not give an explicit definition enabling one to replace a definite description by an equivalent wherever it appears, but a *contextual definition*, which

7.5 Definite Descriptions

enables one to replace sentences containing definite descriptions by equivalent sentences not containing definite descriptions.

Russell used the following 'iota' -notation

$$\iota x A(x): \text{the unique } x \text{ with property } A, \text{ and}$$

$$C(\iota x A(x)): \text{the unique } x \text{ with property } A \text{ has property } C$$

as shorthand for

$$\exists x[A(x) \wedge C(x) \wedge \forall y[A(y) \rightarrow y = x]].$$

Where the condition C is complex, the iota notation is ambiguous. Russell's simple example is well known:

$$\neg B(\iota x F(x)): \text{The king of France is not bald.}$$

Here the ambiguity of the iota notation corresponds to an ambiguity in the English, between these two:

1. $\neg(B(\iota x F(x)))$, i.e., $\neg \exists x[F(x) \wedge B(x) \wedge \forall y[F(y) \rightarrow y = x]]$. There is no object x such that x is king of France and x is bald and x is the only king of France. And this happens to be true.
2. $(\neg B)(\iota x F(x))$, i.e., $\exists x[F(x) \wedge (\neg B)(x) \wedge \forall y[F(y) \rightarrow y = x]]$. There is some object x such that x is king of France and x is not bald and x is the only king of France. And this happens to be false; so we have $\neg((\neg B)(\iota x F(x)))$.

Note that this latter expression is not equivalent to $B(\iota x F(x))$, i.e., $\exists x[F(x) \wedge B(x) \wedge \forall y[F(y) \rightarrow y = x]]$ (the king of France is bald): $\neg((\neg B)(\iota x F(x)))$ is true, while $B(\iota x F(x))$ is false. In Russell's jargon, the definite description $\iota x F(x)$ has *narrow scope* in version 1 and *wide scope* in version 2.

A less confusing notation for definite descriptions would result by treating them as a kind of quantifier:

$$(Ix)(F(x), B(x)) \text{ instead of } B(\iota x F(x)).$$

Then the sentence in version 1, $\neg(B(\iota x F(x)))$, would be rendered by $\neg(Ix)(F(x), B(x))$, and the sentence in version 2, $(\neg B)(\iota x F(x))$, by $(Ix)(F(x), \neg B(x))$. While it was somewhat strange to have both,

$$\neg(B(\iota x F(x))) \text{ and } \neg((\neg B)(\iota x F(x)))$$

in the new notation this would become

$$\neg(Ix)(F(x), B(x)) \text{ and } \neg(Ix)(F(x), \neg B(x))$$

which looks similar to

$$\neg\forall x[A(x)] \text{ and } \neg\forall x[\neg A(x)]$$

which does not look like a contradiction at all.

7.6 Berry's and Grelling's Paradox

In Subsection 2.10.1 we discussed the antinomy of the liar. This paradox results from considering a sentence which says of itself that it is not true. By making a sharp distinction between object-language and meta-language we could avoid this paradox. In this subsection two other antinomies are presented, those of the librarian G.G. Berry and of Kurt Grelling (1908), which can be avoided in a similar way by making the distinction between language and meta-language. While the paradox of the Liar is on the level of sentences, Berry's paradox is on the level of names/definite descriptions and Grelling's antinomy is on the level of predicates.

Berry's Paradox Consider the following definite description: *The least natural number not specifiable in less than twenty-two syllables.* (*)

First of all we should verify that such a natural number exists. That this actually is the case follows from the following observations: (i) There are only finitely many (different) syllables. (ii) Consequently, there are only finitely many phrases of less than 22 syllables. (iii) There are infinitely many natural numbers.

From (ii) and (iii) it follows that there is a least natural number that is not specifiable in less than twenty-two syllables. However, counting the number of syllables in (*) we find that we have specified that particular number in 21 syllables. Therefore, here is *Berry's paradox*:

The least natural number not specifiable in less than twenty-two syllables is specifiable in 21 syllables.

In order to avoid this paradox, one should realize that the expression 'specifiable' does not have a clear meaning. It must be supposed that we are talking with reference to the resources of some particular language, say L_0. 'Specifiable in terms of (expressions of) L_0', abbreviated by 'specifiable$_0$', does have a clear meaning. However, the expression 'specifiable$_0$', which is short for 'specifiable in terms of L_0', does not belong to L_0 itself, but to the meta-language L_1 of L_0. Keeping this in mind, we easily see that Berry's paradox is the result of a very loose usage of words and of identifying object-language and meta-language. Expressing ourselves more precisely, what we have actually found is that

The least natural number not specifiable$_0$ in less than twenty-two syllables (of L_0) is specifiable$_1$ in 21 syllables (of L_1).

In its specification we have used the expression *specifiable$_0$* which does not belong to the object-language L_0, but to the meta-language L_1 of L_0. So, making a clear distinction between object-language and meta-language and expressing ourselves precisely, the paradox simply disappears.

We are perhaps not accustomed to thinking of a natural language such as English as a sequence English$_0$, English$_1$, English$_2$, ..., where for each natural number n, English$_{n+1}$ is a meta-language of English$_n$. However, the paradox of the Liar, Berry's paradox and others force us to conceive of English in such a way and after a while the distinction between object-language and meta-language seems to be self-evident.

7.6 Berry's and Grelling's Paradox

That L_{n+1} is a meta-language of L_n means:
i) L_{n+1} contains L_n as a sublanguage ($L_n \subseteq L_{n+1}$), and
ii) L_{n+1} contains in addition means to talk **about** L_n.

Grelling's paradox Define the predicate 'autological' as 'being true of itself'. This predicate applies, for instance, to the adjectives 'short', 'English' and 'polysyllabic'. For example, the adjective 'short' is short and therefore, this adjective is autological.

Adjectives which are not autological are called heterological. The adjective 'long' is not long; the adjective 'German' is not German; and the adjective 'monosyllabic' is not monosyllabic. So, in Grelling's terminology, the adjectives 'long', 'German' and 'monosyllabic' are heterological.

Now consider the question whether the adjective 'heterological' is autological or not. If 'heterological' is autological, then it is true of itself, and hence it is heterological. Conversely, if 'heterological' is heterological, then it is true of itself and hence it is autological. So, this is *Grelling's paradox* (1908).

'heterological' is autological iff it is heterological (not autological).

This paradox is also the result of not making a sharp distinction between object-language and meta-language. Let 'true$_n$' belong to the language L_n. Then we can talk *about* expressions of language L_n, such as 'being not true$_n$ of itself (heterological$_n$)', in the meta-language L_{n+1} of L_n, but not in L_n itself. So, the question whether

$$\text{heterological}_n \text{ is autological}_n$$

does not make sense. What does make sense is the question whether

$$\text{heterological}_n \text{ is autological}_{n+1}. \qquad (*)$$

The answer to this question is no, since $(*)$ is equivalent to

$$\text{heterological}_n \text{ is a heterological}_n \text{ word}$$

which is meaningless.

We summarize this section in the schema below.

	proper names, definite descriptions	predicates	sentences
antinomy of	Berry (see this section)	Grelling (see this section)	the liar (see Subsection 2.10.1)
way out	distinction of object-language and meta-language: specifiable$_0$ specifiable$_1$ etc.	distinction of object-language and meta-language: true$_0$ of true$_1$ of etc.	distinction of object-language and meta-language: true$_0$ true$_1$ etc.
other way-outs			Namely-rider (see Subsection 2.10.1)

Another antinomy on the level of predicates is Russell's paradox (see Section 3.1). For further reading the reader is referred to Quine [39] and Kneale [27], Chapter XI.

Exercise 7.2. *Richard paradox, 1905* We may take the English alphabet as consisting of the blank space (to separate words), the 26 Latin letters, and the comma. By an 'expression' in the English language we may understand simply any finite sequence of these 28 symbols not beginning with a blank space. The expressions in the English language can then be enumerated by a simple device: first enumerate in alphabetical order all expressions of length 1, next all finitely many expressions of length 2, and so on.

Some English expressions, such as the expression 'the function which assigns to each natural number its square', define a number-theoretic function of one variable, i.e., a function $f : \mathbb{N} \to \mathbb{N}$. By striking out from the specified enumeration of all the expressions in the English language those which do not define a number-theoretic function, we obtain an enumeration, say $E_0, E_1, E_2, \ldots$, of those which do; say the functions defined are respectively $f_0, f_1, f_2, \ldots$. Now consider the function f defined by $f(n) = f_n(n) + 1$. This function f can be defined by an expression in the English language and hence should occur in the enumeration $f_0, f_1, f_2, \ldots$. (1)
On the other hand,
$f \neq f_1$ since $f(1) = f_1(1) + 1$,
$f \neq f_2$ since $f(2) = f_2(2) + 1$,
$f \neq f_3$ since $f(3) = f_3(3) + 1$,
and so on. Therefore, for all $i \in \mathbb{N}, f \neq f_i$. (2)
(1) and (2) are contradictory. Discover the flaw in the argument above.

7.7 The Theory of Direct Reference

According to the theory of *direct reference*, brought out by Keith Donnellan, Saul Kripke, Hilary Putnam and others, proper names ('Aristotle', 'Thales') and nouns standing for natural kinds ('gold', 'water', 'tiger') have no intension (Sinn) in the traditional sense, but only have reference; and this reference is established by a *causal chain* rather than by an associated description. For example, the reference to the person called 'Aristotle' is determined by a causal chain as follows. The person in question is given a name in a 'baptism' with the referent present. Next this name is handed on from speaker to speaker. It is in this way that we use the name 'Aristotle' referring to the person in question. We do not have to have any description of Aristotle; the information 'Aristotle was a philosopher' may be completely new to the one who is using the name 'Aristotle'.

There are at least two problems in the traditional theory of meaning:

1. In the traditional view, a proper name, like 'Jane', is identified with a description, such as 'the woman John is married to'. Now suppose that John is a bachelor. Then it would follow that Jane does not exist. This example makes clear that a

7.7 The Theory of Direct Reference

person can be referred to by his or her name even if the description of the person in question does not apply to that person.
2. According to the traditional theory, a tiger, for instance, is identified with an object which has certain properties, among which the property of having sharp teeth. Consequently, the statement 'tigers have sharp teeth' is analytic; this seems to be counter-intuitive.

In the traditional theory, the conjunction of properties which a tiger is supposed to have is called the intension of the word 'tiger' and is supposed to be the *essence* of tiger. In the traditional theory as well, intension determines extension. Similarly, in the traditional view, the proper name 'Aristotle' is identified with a description such as 'the most well-known man who studied under Plato'. As a consequence, the proposition 'Aristotle studied under Plato' would be an analytic truth. This is again against our intuition.

Typical of the theory of direct reference is the position, held by Kripke, Donnellan and others, that proper names and nouns standing for natural kinds refer independently of identifying descriptions.

In his paper [12] Donnellan distinguished between two kinds of use for definite descriptions – the *attributive*, and the *referential*. In order to make this distinction clear, Donnellan considered the use of the definite description 'Smith's Murderer' in the following two cases.

> Suppose first that we come upon poor Smith foully murdered. From the brutal manner of the killing and the fact that Smith was the most lovable person in the world, we might exclaim 'Smith's murderer is insane'. I will assume, to make it a simpler case, that in a quite ordinary sense we do not know who murdered Smith .... This, I shall say, is an *attributive use* of the definite description.

So, in the case of the attributive use, the speaker wants to say something about whoever or whatever fits the description even if he does not know who or what that is. On the other hand,

> Suppose that Jones has been charged with Smith's murder and has been placed on trial. Imagine that there is a discussion of Jones' odd behavior at his trial. We might sum up our impression of his behavior by saying 'Smith's murderer is insane'. If someone asks to whom we are referring by using this description, the answer here is 'Jones'. This, I shall say, is a *referential use* of the definite description. [K.S. Donnellan, [12], pp. 285-286.]

So, if the description 'Smith's murderer' is used referentially, the speaker is referring to Jones, even in the case that Jones turns out to be innocent. Note that in this case the description refers to Jones although it does not apply to Jones. To give another example, suppose someone asks me at a party who Mr. X is. I answer 'the man at the door with a glass of sherry in his hand'. Now suppose that the person referred to actually has a glass of white wine in his hand. Again the description may refer successfully without applying to the object referred to. These examples make clear that descriptions, when used referentially, do not always apply to the object they refer to. When using a description referentially, we have a definite object in mind whether or not it does fit the description.

It is typical of the theory of direct reference that proper names, like 'Jane', refer to some definite object, even when the description we supply, like 'the woman John is married to', does not apply to that object. This description may help us fix the reference, but it should not be taken to be the meaning of the name. And a similar view is held for nouns standing for natural kinds, like 'gold', 'water' and 'tiger'. The meaning of the word 'tiger' is its reference; identifying descriptions such as 'a tawny-coloured animal with sharp teeth, ...' only help us to fix the reference of this term.

Summarizing, according to the theory of direct reference, the meaning of a proper name or a natural kind term is its reference; the descriptions given in connection with these terms only help the hearer to pick out what the speaker has in mind.

In his paper [28] Kripke in addition holds the view that a proper name, like 'Aristotle', is a *rigid designator*, i.e., it designates the very same object in all possible worlds in which this object exists. Thus, in the sentence 'Aristotle might have been a carpenter', the proper name 'Aristotle' refers to the same individual referred to in the sentence 'Aristotle was the philosopher who was a pupil of Plato and taught Alexander'. The definite description 'the most well-known man who studied under Plato', though it designates Aristotle in the actual world, may designate other individuals in other possible worlds; for it is possible that Aristotle did not study under Plato. Contrary to the traditional theory of meaning, according to the theory of direct reference, the statement 'Aristotle studied under Plato' is not necessarily true (and hence not analytic).

Now, if a and b are rigid designators and $a = b$ is true (in this world), then $\Box(a = b)$ is true, i.e., $a = b$ is true in all possible worlds accessible from this one (see Exercise 7.3). So it follows from the thesis that proper names are rigid designators that all true identity statements of the form $a = b$, where a and b are proper names, are necessarily true. In particular, it follows that 'Hesperus is Phosphorus (the Morning Star is the Evening Star)' and 'Tully is Cicero', if true (in this world) are necessarily true. On the other hand, we do not know a priori that Hesperus (the Morning Star) is Phosphorus (the Evening Star); this was discovered by empirical observation. Therefore Kripke claims in his paper [29] that sentences like 'Hesperus is Phosphorus' and 'Tully is Cicero' if true (in this world) are *necessarily true and at the same time are a posteriori*.

Kripke extends his insights about proper names to nouns standing for natural kinds, such as 'gold', 'water' and 'tiger'. These nouns are rigid designators too, i.e., they refer to the same substance in all possible worlds in which this substance exists. Let us consider some interesting consequences of this point of view.

'Gold' being a rigid designator, the sentence 'gold is the element with atomic number 79', if true (in this world), will be true in all worlds (accessible from this one) and hence be necessarily true. Similarly, 'water' being a rigid designator, the sentence 'water has the chemical structure H_2O', if true (in this world), will be true in any world (accessible from this one) and hence be necessarily true. So both propositions, if true (in this world), are *necessarily true and* at the same time *a posteriori*.

7.8 Analytic - Synthetic

In Exercise 7.4 some examples are given of sentences which are *contingent*, i.e., not necessarily true, *and* at the same time *a priori*. Kripke defines a sentence *A* to be *analytic* if it is both necessary and a priori. Consequently, sentences like 'Hesperus is Phosphorus', 'Tully is Cicero', 'gold is the element with atomic number 79' and 'water is H_2O' are **not** analytic, since they are a posteriori, although necessarily true, if true (in this world).

Exercise 7.3. Suppose that '*a*' and '*b*' are rigid designators. Prove: if $a = b$ is true (in this world), then $\Box(a = b)$ is true. More precisely, for any Kripke model *M* and for any world *w* in *M*: if $M, w \models a = b$, then $M, w \models \Box(a = b)$.

Exercise 7.4. Regarding 'one meter' as a rigid designator, make clear that 'stick *S* is one meter long', where *S* is the standard meter in Paris, is a contingent and a priori truth. (See S.A. Kripke, [28] pp. 54-57.) Similarly, for 'water boils at hundred degrees Celsius', regarding '100 degree Celsius' as a rigid designator, and for 'I am here now'.

7.8 Analytic - Synthetic

In his *Critique of Pure Reason* (1781) Immanuel Kant [26] makes a distinction between analytic and synthetic judgments. Kant calls a judgment *analytic* if its predicate is contained (though covertly) in the subject, in other words, the predicate adds nothing to the conception of the subject. Kant gives 'All bodies are extended (Alle Körper sind ausgedehnt)' as an example of an analytic judgment; I need not go beyond the conception of *body* in order to find extension connected with it. If a judgment is not analytic, Kant calls it *synthetic*. So, a synthetic judgment adds to our conception of the subject a predicate which was not contained in it, and which no analysis could ever have discovered therein. Kant mentions 'All bodies are heavy (Alle Körper sind schwer)' as an example of a synthetic judgment.

Also in his *Critique of Pure Reason* Kant [26] makes a distinction between *a priori* knowledge and *a posteriori* knowledge. A priori knowledge is knowledge existing altogether independent of experience, while a posteriori knowledge is empirical knowledge, which has its sources in experience.

Sometimes one speaks of *logically necessary* truths instead of analytic truths and of *logically contingent* truths instead of synthetic truths, to be distinguished from physically necessary truths (truths which physically could not be otherwise, true in all physically possible worlds). The distinction between necessary and contingent truth is a *metaphysical* one. In her book [21], p. 170, S. Haack stresses that this distinction 'should be distinguished from the *epistemological* distinction between *a priori* and *a posteriori* truths'. Although these – the metaphysical and the epistemological – are certainly different distinctions, it is controversial whether they coincide in extension, that is, whether all and only necessary truths are *a priori* and all and only contingent truths are *a posteriori*.

In his *Critique of Pure Reason* Kant stresses that mathematical judgments are both a priori and synthetic. 'Proper mathematical propositions are always judgments *a priori*, and not empirical, because they carry along with them the conception of necessity, which cannot be given by experience.' Why are mathematical judgments synthetic? Kant considers the proposition $7+5 = 12$ as an example. 'The conception of twelve is by no means obtained by merely cogitating the union of seven and five; and we may analyse our conception of such a possible sum as long as we will, still we shall never discover in it the notion of twelve.' We must go beyond this conception of $7+5$ and have recourse to an intuition which corresponds to counting using our fingers: first take seven fingers, next five fingers extra, and then by starting to count right from the beginning we arrive at the number twelve.

```
    7:  1  1  1  1  1  1  1
    5:                       1  1  1  1  1
  7+5:  1  1  1  1  1  1  1  1  1  1  1  1
        1  2  3  4  5  6  7  8  9 10 11 12
```

'Arithmetical propositions are therefore always synthetic, of which we may become more clearly convinced by trying large numbers'. Geometrical propositions are also synthetic. As an example Kant gives 'A straight line between two points is the shortest', and explains 'For my conception of *straight* contains no notion of *quantity*, but is merely *qualitative*. The conception of the *shortest* is therefore wholly an addition, and by no analysis can it be extracted from our conception of a straight line'.

In more modern terminology, following roughly a 'Fregean' account of analyticity, one would define a proposition A to be *analytic* iff either

(i) A is an instance of a logically valid formula; e.g., 'No unmarried man is married' has the logical form $\neg\exists x[\neg P(x) \wedge P(x)]$, which is a valid formula, or

(ii) A is reducible to an instance of a logically valid formula by substitution of synonyms for synonyms; e.g., 'No bachelor is married'.

In his *Two dogmas of empiricism* W.V. Quine [40] is sceptical of the analytic/synthetic distinction. Quine argues as follows. In order to define the notion of analyticity we used the notion of synonymy in clause (ii) above. However, if one tries to explain this latter notion, one has to take recourse to other notions which directly or indirectly will have to be explained in terms of analyticity.

7.9 Logicism

Logicism dates from about 1900, its most important representatives being G. Frege [17] in his *Grundgesetze der Arithmetik* I, II (1893, 1903) and B. Russell [41] in his *Principia Mathematica* (1903), together with A.N. Whitehead. The program of the logicists was to reduce mathematics to logic. What do they mean by this? In his Grundgesetze der Arithmetik Frege defines the natural numbers in terms of sets as follows: 1 := the class of all sets having one element, 2 := the class of all sets having two elements, and so on. Next Frege shows that all kinds of properties of natural numbers can be logically deduced from a *naive comprehension principle*: if $A(x)$ is a property of an object x, then there exists a set $\{\,x \mid A(x)\,\}$ which contains precisely all objects x which have property A (see Section 3.1).

Logicism tried to introduce mathematical notions by means of explicit definitions; mathematical truths would then be logical consequences of these definitions. Mathematical propositions would then be reducible to logical propositions and hence mathematical truths would be analytical, contrary to what Kant said.

The greatest achievement of Logicism is that it succeeded in reducing great parts of mathematics to one single (formal) system, namely, set theory. The logicists believed that by doing this they reduced all of mathematics to logic without making use of any non-logical assumptions, hence showing that mathematical truths are analytic. However, they mistakenly held the naive comprehension principle for a logical axiom instead of a mathematical or set theoretical principle. So, what they actually did was reduce mathematics to logic **plus** set theory. And the axioms of set theory have a non-logical status! The axioms of set theory are – in Kant's terminology – synthetic, and surely not analytic. In his later years Frege came to realize that the axioms of set theory (see Chapter 3) are not a part of logic and gave up Logicism, which he had founded himself. The interested reader is referred to K. Gödel [19], *Russell's mathematical logic*.

Another way to see that a mathematical truth like $7+5=12$ is synthetic is to realize that $7+5=12$ is not a logically valid formula; it is true under the intended interpretation, but not true under all possible interpretations: if one interprets the + symbol as negation, the formula 5 + 7 = 12 yields a false proposition. $7+5=12$ can be logically deduced from the axioms of Peano for (formal) number theory (see Chapter 5), but it cannot be proved by the axioms and rules of formal logic alone.

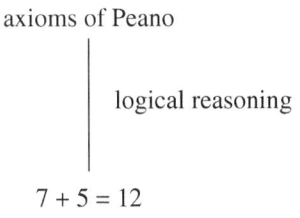

$$\begin{array}{c} \text{axioms of Peano} \\ | \\ \text{logical reasoning} \\ | \\ 7+5=12 \end{array}$$

Again, Peano's axioms are true under the intended interpretation, but are not (logically) valid and hence they do not belong to logic.

7.10 Logical Positivism

It is an old problem to draw the line between scientifically meaningful and meaningless statements. Consider the following quotation, taken from Hume's *Enquiry Concerning Human Understanding*:

> When we run over libraries, persuaded of these principles, what havoc must we make? If we take in our hand any volume; of divinity or school metaphysics, for instance; let us ask, *Does it contain any abstract reasoning concerning quantity of number?* No. *Does it contain*

any experimental reasoning concerning matter of fact and existence? No. Commit it then to the flames: for it can contain nothing but sophistry and illusion [David Hume, 1711-1776].

As we learn from A.J. Ayer [2], the quotation above is a good formulation of the positivist's position. In the 1930's the adjective *logical* was added, resulting in the term *Logical Positivism*, which underscored the successes of modern logic and the expectation that the new logical discoveries would be very fruitful for philosophy.

This logical positivism was typical of the *Vienna Circle*, a group of philosophers (among them Moritz Schlick, Rudolf Carnap and Otto Neurath), scientists and mathematicians (among them Karl Menger and Kurt Gödel). According to A.J. Ayer [2], Einstein, Russell and Wittgenstein had a clear kinship to the Vienna Circle and had a great influence upon it.

In order to draw a sharp distinction between scientifically meaningful statements and scientifically meaningless statements the *verification principle* was formulated: only those statements are scientifically meaningful which can be verified in principle; in other words, the meaning of a proposition is its method of verification. However, a proposition like 'all ravens are black', which has as logical form $\forall x[R(x) \to B(x)]$, cannot be verified due to the universal quantifier, $\forall$; at the same time we consider this proposition to be (scientifically) meaningful.

However, the proposition 'all ravens are black' can be conclusively falsified, since its negation 'not all ravens are black', being of the form $\neg\forall x[R(x) \to B(x)]$, is logically equivalent to 'some raven is not black', which has the logical form $\exists x[R(x) \land \neg B(x)]$, and hence can be verified. For this reason the *falsification principle* was formulated: only those statements are scientifically meaningful which can be falsified in principle. This principle seems to be more in conformity with scientific practice: hypotheses are set up and rejected as soon as experimental results force us to do so.

However, Otto Neurath himself soon realized that a slightly more complex proposition, like 'all men are mortal', which has the logical form $\forall x \exists y[P(x,y)]$ (for every person there is a moment of time such that ...), can neither be verified (due to the universal quantifier $\forall x$) nor falsified (due to the existential quantifier $\exists y$), since its negation 'not all men are mortal', being of the form $\neg\forall x \exists y[P(x,y)]$, is equivalent to 'some men are immortal', which has the logical form $\exists x \forall y[\neg P(x,y)]$, and hence – due to the universal quantifier $\forall y$ – cannot be verified. Falsification of $\forall x \exists y[P(x,y)]$ is equivalent to verification of $\neg \forall x \exists y[P(x,y)]$, i.e., verification of $\exists x \forall y[\neg P(x,y)]$, which is not possible in principle due to the universal quantifier $\forall y$. At the same time we want to consider a statement like 'all men are mortal' as (scientifically) meaningful. Therefore, we have to give up not only the verification principle, but also the falsification principle. This was already realized by Otto Neurath during his stay (1938-39) in the Netherlands (oral communication by Johan J. de Iongh).

Instead of the verification or falsification principle, a weaker criterion was formulated, called the *confirmation principle*: a statement is scientifically meaningful if and only if it is to some degree possible to confirm or disconfirm it. One way to confirm (increase the degree of credibility of) universal generalizations like 'all ravens are black' is to find things that are both ravens and black, and one way to disconfirm this proposition is to find things that are ravens but not black. The problem with this confirmation principle is that 'all ravens are black', $\forall x[R(x) \to B(x)]$, is logically equivalent to 'all non-black things are non-ravens', $\forall x[\neg B(x) \to \neg R(x)]$, and according to the confirmation principle, the latter proposition is confirmed by observations of non-black non-ravens; thus observations of brown shoes, white chalk, etc., would confirm the proposition 'all ravens are black'. Various attempts have been made to give the verification principle, in this weaker form, a precise expression, but the results have not been altogether satisfactory. For instance, a solution might be found by replacing the material implication $\to$ in $\forall x[R(x) \to B(x)]$ by the counterfactual implication $\square\!\!\to$ (see Section 6.9), for $\forall x[A(x) \,\square\!\!\to B(x)]$ is not logically equivalent to $\forall x[\neg B(x) \,\square\!\!\to \neg A(x)]$.

7.11 Presuppositions

Let us start this subsection with a quotation from Frege, *Ueber Sinn und Bedeutung*.

> If anything is asserted there is always an obvious presupposition that the simple or compound proper names used have reference. If therefore one asserts 'Kepler died in misery', there is a presupposition that the name 'Kepler' designates something; but it does not follow that the sense of the sentence 'Kepler died in misery' contains the thought that the name 'Kepler' designates something. If this were the case, the negation would have to run not 'Kepler did not die in misery' but 'Kepler did not die in misery, or the name 'Kepler' has no reference'. That the name 'Kepler' designates something is just as much a presupposition for the assertion 'Kepler died in misery' as for the contrary assertion. [G. Frege, 1892, in P. Geach and M. Black [18]].

Thus, according to Frege, the sentences
  (1) Kepler died in misery, and
  (2) Kepler did not die in misery
both presuppose that the name 'Kepler' has a reference. This presupposition is not part of the meaning of (1) or (2) respectively, since in that case the negation of (1) would not be (2), but
  (*) Kepler did not die in misery, or the name 'Kepler' has no reference.
If the presupposition is not satisfied, the speech act of asserting in (1) and (2) cannot be performed successfully.

As we have already seen in Section 7.5 on definite descriptions, Russell places the presupposition(s) of a sentence into an existentially quantified conjunction and by doing so he makes the presupposition part of the meaning of the sentence. For example, the sentence
  (1a) The king of France is bald
presupposes that
  (1b) There is a king of France,
but Russell translates (1a) by the expression $\exists x[x$ is king of France $\land\ x$ is bald $\land\ x$ is the only king of France], hence making the presupposition part of what is asserted.

As we have seen above, sentences containing proper names and sentences containing definite descriptions in subject position carry or induce presuppositions. R. van der Sandt gives in his book [42], on which this section is based, the following examples in which the (a)-sentences presuppose the corresponding (b)-sentences.

Quantifiers (all, a few, at least one, ...)
(2a) All John's children are asleep.
(2b) John has children.

Aspectual verbs (begin, stop)
(3a) Charles has stopped smoking.
(3b) Charles used to smoke.

Presuppositional adverbs (only, even, also, ...)
(4a) Only John voted for Harry.
(4b) John voted for Harry.

Contrastive stress
(5a) The *butcher* killed the goose.
(5b) Someone killed the goose.

Factive verbs (realize, regret, discover, ...)
(6a) Tom regrets that the goose has been killed.
(6b) The goose has been killed.

Cleft constructions
(7a) It was John who caught the thief.
(7b) Someone caught the thief.

It is widely accepted that presuppositions are characterized by the following three tests.

1. The *negation test*: presuppositions are preserved when the original sentence is embedded under negation. The negation of each (a)-sentence above still presupposes the corresponding (b)-sentence. For instance, the negation of (3a), 'Charles has not stopped smoking', still presupposes (3b), 'Charles used to smoke'.
2. The *modality test*: presuppositions are preserved when the original sentence is embedded under a possibility operator. For instance, 'It is possible that Charles has stopped smoking' still presupposes (3b), 'Charles used to smoke'.
3. The *antecedent test*: presuppositions are preserved when the original sentence is taken as the antecedent of a conditional statement. For instance, 'If Charles stopped smoking, his wife would be happy' still presupposes (3b) 'Charles used to smoke'.

(8) Charles managed to leave the country
entails
(9) Charles left the country.
But (8) does not presuppose (9), since (9) is not preserved in the application of the tests mentioned above.
(8i) Charles did not manage to leave the country,

7.11 Presuppositions

(8ii) Perhaps Charles managed to leave the country, and
(8iii) If Charles has managed to leave the country, then he will never come back,
do not suggest that (9) is true.

The verb 'manage' is called *implicative*, i.e., each sentence containing this verb entails the complement of that verb.

Next, consider sentence
(10) Charles was glad that he had left the country.
Applying the negation, modality and antecedent test to (10) teaches us that (9) is a presupposition of (10). These examples show (or at least suggest) that the tests mentioned above eliminate entailments, but preserve presuppositions. ([42], 2.3.)

When a presupposition is induced by a positive or negative *polarity element*, the application of the negation test is problematic.

Negative polarity elements, such as *at all, ever, anymore, matter that, mind that*. In order to form a grammatically correct expression, these elements are accompanied by a negation. When a presupposition is induced by a negative polarity element, the negation test cannot be applied, since the original sentence has no grammatical non-negated counterpart. The following examples are from van der Sandt [42].
(11a) Dick does not mind that his theory is wrong
presupposes
(11b) Dick's theory is wrong.
And
(12a) It does not matter that John was fired
presupposes
(12b) John was fired.
The presupposition-inducing elements *mind that* and *matter that* are negative polarity elements; so the non-negative versions of (11a) and (12a) are not grammatical. Thus the negation test cannot be applied in these cases. However, the modality test and the antecedent test can be applied successfully to (11a) and (12a). For this reason (11b) and (12b) are considered to be presuppositions of (11a) and (12a) respectively.

Positive polarity elements, such as *still, plenty of, perhaps, a lot, certainly, swarm with, be delighted*. When a presupposition is induced by a positive polarity element, the negation test may yield the wrong results since the negated sentence can evoke some kind of echo-effect. By this we mean that the negated sentence may suggest that the speaker rejects the original sentence, because he does not accept its presupposition. Consequently, when the original sentence is embedded under negation, the presupposition may get lost. Consider the following examples, again from [42].
(13a) John still believes in David's theory
presupposes
(13b) John believed in David's theory until recently.
And
(14a) Dick is delighted that his book is published
presupposes
(14b) Dick's book is published.

The presupposition inducing elements *still* and *be delighted* are positive polarity elements. Now, consider the negations of (13a) and (14a) respectively:
   John does not still believe in David's theory; (he never believed in it).
   Dick is not delighted that his book is published; (his book is not published).
On their most natural reading, the negations of (13a) and (14a) evoke the echo-effect that the speaker rejects the original sentence, not accepting its presupposition. So, application of the negation test would yield a negative result. On the other hand, if we apply the modality test and the antecedent test to (13a) and (14a) the truth of (13b) and (14b) is preserved. For this reason (13b) and (14b) are considered to be presuppositions of (13a) and (14a) respectively.

The negation, modality and antecedent test enable us to determine the elementary presuppositions of simple sentences. These elementary presuppositions are induced directly by certain lexical elements or syntactic constructions. The *projection problem* is the problem of finding out whether there is an algorithm – and if so, which one – which determines the presuppositions of a complex sentence on the basis of the presuppositions of its components.

In this connection it is important to realize that *presupposing is* **not** a binary relation between sentences and propositions, but *a ternary relation between sentences, propositions and contexts*. Whether a sentence presupposes a certain proposition or not may depend on the context! This can be illustrated by the following two examples, again taken from [42].
(15a) John will regret that there is a bouncer at the party
presupposes
(15b) There will be a bouncer at the party.
In the context that Charles is a competent bouncer and that problems are expected to arise, in sentence
(16) If Charles comes to the party, John will not regret that there is a bouncer at the party
presupposition (15b) is lost. In this context (16) does not presuppose (15b). However, in the context in which John gives a party and Charles, a notorious brawler, is one of the potential guests, the elementary presupposition (15b) is preserved. In such a context (16) does presuppose (15b).
(17a) Peter drinks too
presupposes
(17b) Someone other than Peter drinks.
Without any specification of a context, more precisely, in an empty context
(18) If Peter drinks too, the bottle is empty
preserves the presupposition (17b). However, in the context 'If John drinks, he drinks at least half a bottle', (18) does not presuppose (17b).

As is clear from these examples, in order to determine the presuppositions of a sentence one has to take into account not only its elementary presuppositions and its mode of composition, but also the relevant contextual information.

Exercise 7.5. The counterpart of the antecedent test for identifying elementary presuppositions would be the 'succedent test'. However, make clear that an elementary presupposition that is induced by the succedent need not be a presupposition of the entire sentence. Hint: consider the sentence 'If the king of France exists, then the king of France is bald'.

Exercise 7.6. Inspired by the negation test, 'A presupposes B' has been defined as: $A \models B$ and $\neg A \models B$. Make clear that this definition does not make sense. Hint: use the fact that $\models A \vee \neg A$.

7.12 Wittgenstein on meaning

> My *whole* task consists in explaining the nature of proposition. (Wittgenstein)

Wittgenstein (1889-1951) is probably unique in the history of philosophy having earned the name of being the author of two completely different, original and highly influential systems of philosophy. His early philosophy, represented in the *Tractatus* (1922) [55], and his later philosophy, represented mainly in the *Philosophical Investigations* [56] (1952), are regarded as two major classics in the philosophy of language. The present survey is chiefly concerned with Wittgenstein's later philosophy of language. However, reference to his earlier philosophy of language will be necessary since his later philosophy, though different from the earlier one, can never be viewed as separate from it. There is an underlying continuity between the two systems that unites them. This is his critique of language or the problem of meaning, a theme with which Wittgenstein was preoccupied in all his philosophical writings. Wittgenstein's remark quoted above 'My *whole* task consists in explaining the nature of proposition' (*Notebooks* 1914-1916 [54], p. 39) characterizes this central concern and also indicates the theme that dominates the different phases of his philosophical thought. To explain the nature of proposition is to explain the nature and function of language. That is why Wittgenstein was interested in questions like: what makes it possible to say something?; how can words in combination signify something?; how can the sense of an expression be communicated to others?. But the difference lies in his approach to these questions. The questions are the same, but the answers to these questions are different in the two phases of Wittgenstein's philosophy of language. The later Wittgenstein was not convinced of the answers he offered to these questions in the Tractatus period. As a result, there was a new set of answers offering a new understanding of the nature and function of language.

To put the difference in perspective, in the earlier period a proposition says something because it is a picture or model of reality. The reason for its being a picture of reality is that it has an isomorphic relation to reality. Further, it is held that the sense of a sentence is determined by its truth conditions. But when we come to the later period, this perspective is changed. A sentence is no longer thought to be the picture of reality nor its meaning to be determined by its truth conditions. A sentence, on the other hand, is compared with a tool to be used to perform various functions including that of describing reality. Its meaning is determined by the rule or the convention associated with it.

One of the most important thrusts of the earlier approach was to show that ultimately a proposition consists only of names and that every name stands for an object. Elementary propositions are thus placed at the very bottom of the system of the Tractatus and all other compound (molecular) propositions are truth-functionally related to elementary propositions. For an elementary proposition to be true, its pictorial form must be isomorphic to the state of affairs it represents. In other words, its structure must mirror the actual structure of the state of affairs. Failure to do so makes the proposition false (though the proposition will be meaningful because it represents a possible state of affairs). According to what has been called the *picture theory of meaning*, the sense of an elementary sentence is determined by its being true or false with respect to the reality of which it is a picture or model. The same holds true of molecular propositions since they are truth-functionally connected with elementary propositions. The truth or falsity of a molecular proposition is dependent on the truth or falsity of the elementary propositions. Similarly, the meaning of a molecular sentence is determined by the specific truth conditions which are related to the elementary propositions. The basic presupposition underlying all these theoretical moves is that the main function of language is to depict reality.

In Wittgenstein's later philosophy we are confronted with a different story regarding how linguistic expressions get their meaning. The aim of the analysis was not to arrive at elementary propositions conceived as the essence of language. The entire depth-grammar approach to language was abandoned. For the later Wittgenstein, language is ordinary language. The question of reducing it to something more basic, such as elementary propositions, does not arise. Language has to be described and understood the way it is found. Its function is not just to describe reality or to picture facts. It has varied functions to perform – it has multiple uses. As Wittgenstein said:

> It is wrong to say that in philosophy we consider an ideal language as opposed to our ordinary one. For this makes it appear as though we thought we could improve an ordinary language. But ordinary language is all right. [*The Blue and Brown Books* [57], p. 28]

The expression, 'ordinary language is all right' needs clarification. Wittgenstein does not mean that ordinary language is free of problems. It is all right in so far as it is used in the way it ought to be used. But very often it is found that language has not been used correctly. These are cases of misuse which give rise to a number of problems, including philosophical problems. Wittgenstein argued that many of the traditional disputes in philosophy arose precisely because of the misuse of language. Such disputes, therefore, do not have any real basis. What is required is a correct diagnosis which will reveal the pseudo nature of the philosophical problems arising due to the misuse of language. The need to reform ordinary language is ruled out since ordinary language is very rich in its content and also because it contains all the nuances regarding a particular concept. Later, Wittgenstein's entire approach to language took a radical turn, giving utmost importance to the notion of use. Language is to be understood not in terms of any predesigned fixed model but in the way it is used, i.e., the way it is used to perform various functions. This has far reaching consequences for semantics. The meaning of an expression is determined by its use and not by its truth conditions.

7.12 Wittgenstein on meaning

It may be of interest to mention the historical event that profoundly influenced Wittgenstein in this new direction of thinking. On March 10, 1928 in Vienna, Wittgenstein along with Herbert Feigl and Friedrich Waismann attended a lecture given by L.E.J. Brouwer on 'Mathematics, Science, and Language'. (See Hacker, *Insight and Illusion* [22], p. 120.) Feigl noted that this lecture made a tremendous impact on Wittgenstein and he was found to be visibly disturbed after the lecture was over. The theme of Brouwer's lecture was closer to the Kantian tradition of epistemology. Following Kant, Brouwer defends the constructive function of the human mind which provides the structure for organizing the data of experience. Both mathematics and language are examples of these constructive activities of the mind. Accordingly, as the argument goes, mathematical truths are not something to be discovered; they are to be invented. There are no independent, eternal truths that mathematics discloses. The *Tractatus* was not conceived in this constructivistic line of thinking. On the contrary, the essential thrust of the *Tractatus* was always realistic. The greatest evidence of this was found in the conception of logic and mathematics by the early Wittgenstein. Logic discloses the necessary structure which is inherent in all possible states of affairs and similarly mathematics, conceived as a set of tautologies, discloses the necessities inherent in the structure of reality. Finally, Wittgenstein comes to the determination of sense. The sense of all sentences must be determinate so that they can correspond with the objects of the world. Further, for the determination of sense, a well formulated language is required in which they can be completely articulated. The constructivistic theme of Brouwer's lecture, as Feigl reported, appealed to Wittgenstein so much that he thought of coming back to philosophy with a new approach and a new orientation to language. The result was the post-*Tractatus* writings of Wittgenstein upholding the constructivistic and conventionalistic view of language and meaning. Meaning of an expression does not correspond to any independent structure of reality. It is, on the other hand, determined by rules of use or conventions that people devise and adopt. With these introductory remarks explaining the transition of Wittgenstein from his early phase to his later phase, we now directly come to Wittgenstein's later view on meaning. In this attempt we shall be mainly concerned with the explication of the view that meaning is to be understood as use. This *'meaning as use'* view assumes a long chain of involved and interconnected arguments. In our presentation we shall follow an order which at the end will establish why meaning is to be understood as use.

Wittgenstein's revolt against essentialism was probably the first step towards his new approach to language and meaning. *Essentialism* is a view which says that there are common, uniform and essential properties. Acceptance of such essential properties becomes necessary, otherwise there cannot be any proper understanding of any thing. Accordingly, the search for common essences was a dominant theme in Western philosophy. Belief in essences was epitomized in Plato's doctrine of ideas. A similar tendency was found in Russell's attempt to discover the ultimate constituent of matter. The *Tractatus* is not an exception to this. Essentialism is ingrained in the very texture of the philosophical thinking of the *Tractatus*. One of the major goals of the *Tractatus* is to determine the limits of language which implies drawing the boundary that will separate sense from nonsense or what can be said from what cannot be said. In this task of separating the two, logic plays an important role. The task of logic is to make the distinction between sense and nonsense in such a way that the distinction becomes universal, necessary and a priori. In his search for universal essences Wittgenstein further talked about the general form of propositions and the essence of language. Wittgenstein tried to establish the essence of language by going back to the core from which all the propositions of language except those core propositions follow. These are elementary propositions and all other propositions are truth functions of these propositions. As stated earlier, elementary propositions consist of names, each of which designates a simple object in the world. Further, the configuration of names in a proposition depicts a state of affairs. This is, in brief, Wittgenstein's idea of the essence of all languages and also the essence of the relation between language and the world.

Later, Wittgenstein rejected this search for essences or 'craving for generality' as he called it. For him this whole search is illusory. But why do we look for such essences? Wittgenstein analysed it and offered some reasons for this craving for generality. In the following we shall present some of the reasons which will help us to understand why the search for essences is illusory.

There is a peculiar tendency in us that we always look for something common. We bring the particulars under a general term, for example, all houses are brought under the general term 'house', all tables are brought under the general term 'table' and so on. We believe that these general terms are meant to express the common properties which reside in the relevant particulars. But the falsity of this entire move is evident if we analyse what is meant by a common property as expressed by a general term. Wittgenstein (*The Blue and Brown Books* [57], pp. 17 to 20) explained this with reference to the analogy of games. In accordance with our belief in essentialism, we think that there must be something common to all games and the general term 'game' is meant to express this common property. But Wittgenstein says that the analysis of the term 'game' shows that it does not stand for any common property. The reason is that different games form a family in the sense that members 'exhibit family likeness', for example, 'some of them have the same nose, others the same eyebrows and others again the same way of walking'. What is important here is to note that these similarities or likenesses overlap, but by no means do they convey a general property. In other words, the overlapping likenesses cannot be mistaken for a general property.

Second, as Wittgenstein pointed out, there is a tendency to think that a man who has learnt a general term, say 'leaf', has come to possess a general picture of a leaf. This general picture is over and above the particular pictures of leaves. To have a general picture of leaf is to have a 'visual image' which 'contains what is common to all leaves'. Thus the meaning of a word is associated with an image which is correlated with the word. This is how essentialism arises and it gives rise to a questionable notion of meaning.

The next important source of our craving for generality is our preoccupation with the method of science. Philosophers are always influenced by the method of natural science which seeks to discover essences beneath the multiplicities. Natural phenomena are thus explained with reference to some universal laws. In this way the

7.12 Wittgenstein on meaning   361

method of natural science reduces multiplicities to some general patterns or regularities. Influenced by this method, philosophers have made similar attempts and offered generalisations in the same way as it is done in the sciences. Metaphysics is the prime example of this or as Wittgenstein put it 'this tendency is the real source of metaphysics' (*ibid*).

Finally, as Wittgenstein holds, this craving for generality also has its source in our 'contemptuous attitude towards the particular case' (*ibid*). We are averse to studying particular cases.

It has already been mentioned that for Wittgenstein the whole search for essences is illusory since there is nothing to be called a common property as has been shown in connection with the analysis of game. This calls for a drastic change in our philosophical method. To put it in concrete terms, Wittgenstein's suggestion is that our aim should be to study individual cases and while studying them we should take note of both their similarities and differences. A study of this sort will make us realise that there is no property common to all these cases. Instead of commonness, we will find that there are overlapping similarities on the basis of which we use general terms. As mentioned earlier, Wittgenstein compared this situation with a human family. The members of a family may resemble each other, but it is not the case that one particular characteristic is necessarily shared by all. What in traditional philosophy is taken as a common property expressing essence is in reality what Wittgenstein calls family resemblance (*Philosophical Investigations* [56], sec. 67). The same holds true of a game. The term 'game' covers the multitude of cases where all of them are found to have family resemblances. To put it in Wittgenstein's words:

> ... we see a complicated network of similarities overlapping and criss-crossing: sometimes overall similarities, sometimes similarities of detail (*ibid*, sec. 66).

This is how Wittgenstein moved from essentialism to the notion of family resemblance which provided the basis of his theory of language and meaning understood as use.

In the light of the above considerations, Wittgenstein made an analysis of language. Wittgenstein argued that we often make a common mistake by trying to discover certain fixed essences in language. According to St. Augustine's theory of language, the primary function of language is to use names to refer to entities or objects. Wittgenstein had a basic objection to any view which tries to define language in essentialistic terms. His objection is that language cannot be defined as having one single homogeneous nature. Language has a complex nature with an enormously complex variety of uses. Accordingly, to suggest that there is a certain use of language which is more fundamental than the other uses is wrong. There is no fundamental use of language. On the contrary, any attempt to do so will be a falsification of what language is.

That language has a variety of uses was explained by Wittgenstein with the help of a number of interesting and revealing analogies. To cite some of them: words in a language have been compared with tools in a tool box (*Philosophical Investigations* [56], Section 11). In a tool box there are different kinds of instruments, e.g., hammers, pliers, screwdrivers, etc. All of them have to perform different functions. In a similar way 'functions of words are as diverse as the functions of these objects'. Again, language has been compared with the inside of a cabin of a locomotive (*ibid*, Section 12). If you look into its cabin 'we see handles all looking more or less alike'. But each handle has its own function to perform. These analogies show why there cannot be any fundamental use of language. Language has a multiplicity of uses. A view which holds that all words are names for objects is wrong from the point of view of the actual functioning of language. To restrict language only to one such function will be like saying that the only role of money is to buy objects. But this is not true since money has many other uses (*ibid*, Section 120).

Wittgenstein further holds that many of the traditional disputes in philosophy arise because of this mistaken view that all words function like a proper name. As a result, philosophers have always been preoccupied with the task of showing what is designated by such abstract terms as 'mind', 'time', 'proposition', etc. That is why they ask questions like: 'What is time?', 'What is mind?', 'What is meaning?', 'What is length?', etc. These questions demand that corresponding to the terms expressed in these questions there must be objects existing outside. But at the same time we feel that there is nothing which we can point out as objects corresponding to these terms. Yet, there is pressure to say that 'we ought to point to something' (*The Blue and Brown Books* [57], p. 1). This produces in us a 'mental cramp' which becomes 'the great source of philosophical bewilderment' (*ibid*).

Wittgensteinian remedy to this bewilderment is that we must give up this fallacious view that words must name something. The very question: 'What is meaning?' or 'What is time?' is wrong. Instead, we should ask: 'What is an explanation of meaning?', 'How do we measure length?', 'How are numerical expressions used?'. Alternatively, to put it in a general form, meaning is to be seen as use.

The next crucially important concept that is intimately related to Wittgenstein's concept of meaning as use is the notion of a language game. A language game is a theoretical construct developed by Wittgenstein which offers a justification for the irreducible complexities that language exhibits. Language is not a monolithic system. On the contrary, it consists of different types of language games. These are as many language games as there are uses. St. Augustine's account of language, in spite of its limited application, is also considered to be a description of one kind of language game. It is a simple language game in which human beings communicate only by means of names where every name refers to some object. In Wittgenstein's famous example (*Philosophical Investigations* [56], sec. 2, 3) it is the language game played by the builder and the assistant in which the builder uses such words as 'slab' or 'block' to get the assistant to bring them out for him. This is a simple language game in which words are used as names for objects and thereby the purpose of communication is fulfilled, namely, giving the order and obeying it. There are also more complicated language games and Wittgenstein cites many instances of such games (*ibid*, sec. 23).

With the notion of language game, Wittgenstein introduced the notion of form of life. A language game is associated with a definite form of life. It is something which is embedded in the game itself. There is a controversy as to the exact meaning of form of life and its implications. Without going into these controversies, it is possible to offer a simple idea of what a form of life may mean. A form of life expresses the typical life situation within which a particular language game is played. It is the set of activities that define the form of life. Finally, coming back to use, to say that the meaning of an expression is to be understood in terms of its use is really to assert that it is the use of an expression in a particular language game that determines its meaning. In Wittgenstein's metaphor, a language game is the 'original home' of meaning and use.

In the above we have tried to show at a very rudimentary level the grounds that Wittgenstein offered to establish his new approach to meaning. However, his conception of meaning has various other dimensions and implications all of which together give rise to a philosophy of language which makes a new breakthrough in our understanding of language and meaning. It is also important to note in this connection that many of the implications of his theory of meaning are not restricted to the semantics of natural language. Wittgenstein extended his approach to some other frontiers of philosophical inquiry, such as the foundations of mathematics, philosophical psychology, philosophy of social sciences, etc. It is not possible in this brief survey to go into the various aspects of Wittgenstein's theory of meaning nor it is possible to show the application of this theory to other conceptual territory.

7.13 Syntax - Semantics - Pragnatics

Syntax is the study of sentences. It specifies the grammatical rules according to which well-formed expressions are built from basic expressions or from the letters of a given alphabet. The *syntax* of a language is concerned only with the *form* of the expressions, while the *semantics* is concerned with their *meaning*. So, the rules according to which the well-formed expressions of a language are formed and the rules belonging to a formal proof system, such as $\dfrac{A \quad A \to B}{B}$ and $\dfrac{A}{A \vee B}$, belong to the syntax of the language in question. These rules can be manipulated mechanically; a machine can be instructed to apply the rule Modus Ponens and to write down a B once it sees both A and $A \to B$, while the machine does not know the meanings of A, B and $\to$. The notions of (formal) proof and deduction, as well as the notions of (formal) provability and deducibility, clearly belong to the syntax: they are only concerned with the form of the formulas involved.

Semantics is the study of propositions. By a proposition we mean the cognitive meaning (Sinn, see Section 7.2) of a sentence. Semantics is the study of truth conditions for the sentences of certain languages in isolation from the context in which those sentences are uttered. Truth tables belong to the semantics, because they say how the truth value (meaning) of a composite proposition is related to the truth values (meanings) of the components from which it is built. The notions of validity and valid consequence also belong to the semantics: they are concerned with the meaning of the formulas in question.

Now consider the proposition expressed by 'Nixon is the president of the USA'. There is (was) a possible world in which this proposition is true, and there are possible worlds in which this proposition is false. Therefore, R.C. Stalnaker gave in his paper *Pragmatics* [50] the following mathematical definition of a proposition.

A *proposition* is a function from the set W of possible worlds to the set $\{0, 1\}$ of truth values. So a proposition assigns to each possible world a truth value. Stalnaker explains that under this definition, propositions have all the properties that have traditionally been ascribed to them:

1. A proposition is independent of the particular language and the linguistic formulation in which it is expressed. 'John writes a book', 'Johan schreibt ein Buch' and 'a book is written by John' all express the same proposition.
2. A proposition is independent of the speech act in which it figures. The same proposition 'I will come' may figure in the speech act of promising, in the speech act of asserting and in many others (see Section 7.4 on speech acts).

Pragmatics is the study of speech acts and the contexts in which they are performed. Typical examples of problems to be solved within pragmatics are:

- To find necessary and sufficient conditions for the successful performance of a speech act, such as promising and many others.
- To characterize the features of the context which help determine which proposition is expressed by a given sentence. For instance, the sentence 'I am here now' expresses different propositions in different contexts.
- To determine how the presuppositions of a given sentence depend on the context (see Section 7.11).

The syntactical and semantical rules for a language enable us to interpret a sentence like 'I am here now', although we do not have to know what the indexical expressions 'I', 'here' and 'now' stand for. This interpreted sentence will result in different propositions depending on the context: 'Harry is in Amsterdam on November 12, 1989', 'Mary is in New York on December 5, 1989', and so on. Therefore, the following mathematical definition (again from R.C. Stalnaker) of an interpreted sentence seems appropriate:

An *interpreted sentence* is a function from the set C of contexts to the set P of propositions. So an interpreted sentence assigns to each context a proposition. As we have seen above, the set P of all propositions is the set $\{0,1\}^W$ of all functions from the set W of possible worlds to the set $\{0, 1\}$ of truth values. Therefore, an interpreted sentence is a function from C to $\{0, 1\}^W$, where C is the set of contexts and W is the set of possible worlds. The overall picture is the following:

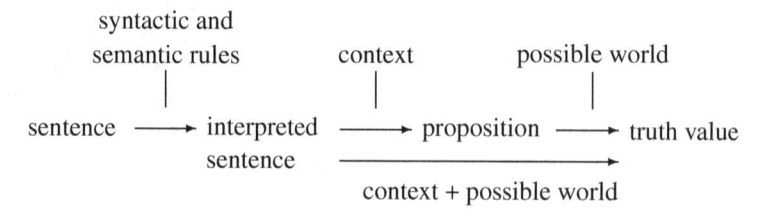

```
              syntactic and
              semantic rules        context        possible world
                    |                  |                  |
   sentence ──→ interpreted  ──→ proposition ──→ truth value
                 sentence       ─────────────────→
                                context + possible world
```

Stalnaker explains further that pragmatics-semantics could be treated as the study of the way in which truth values are dependent on context and a possible world in which case propositions are not explicitly taken into consideration on the road from sentences to truth values. However, propositions are of some independent interest: they are the objects of illocutionary acts, such as asserting, promising, questioning, etc., and of propositional attitudes, such as believing, knowing, hoping, wishing, and so on. This justifies the extra step on the road from sentences to truth values.

7.14 Conversational Implicature

P. Grice in the 1967 William James Lectures (published in 1989 in [20]) works out a theory in *pragmatics* which he calls the theory of *conversational implicature*. Generally speaking, in conversation we usually obey or try to obey rules something like the following:

> Quantity: Be informative
> Quality: Tell the truth
> Relation: Be relevant
> Mode: Avoid obscurity, prolixity, etc.

If the fact that *A* has been said, plus the assumption that the speaker is observing the above rules, plus other reasonable assumptions about the speaker's purposes and intentions in the context, logically entails that *B*, then we can say *A conversationally implicates B*.

It is possible for *A* to conversationally implicate many things which are in no way part of the *meaning* of *A*. For example, if X says 'I'm out of gas' and Y says 'there's a gas station around the corner', Y's remark conversationally implicates that the station in question is open. (Since the information that the station is there would be *irrelevant* to X's predicament otherwise.) If X says 'Your hat is either upstairs in the back bedroom or down in the hall closet', this remark conversationally implicates 'I don't know which', since if X did know which, this remark would not be the most *informative* one he could provide.

Grice shows how philosophers have sometimes mistaken conversational implicatures for elements of meaning. For instance, Strawson sometimes claims not-knowing-which must be part of the *meaning* of 'or' (and therefore the traditional treatment of disjunction in logic is misleading or false). Grice claims this is mistaking the conversational implicature cited above for an aspect of meaning.

Sometimes it is possible to *cancel* a conversational implicature by adding something to one's remark. For example, in the gas station case, 'I'm not sure whether it's open' and in the hat case, 'I know, but I'm not saying which' (one might say this if locating the hat was part of some sort of parlor game). The possibility of cancellation shows that the conversational implicatures definitely are not part of the *meaning* of the utterance.

7.15 Conditionals

In the examples below (from Adams, [1]) the conditional in (1) is in the *indicative* mood, while the conditional in (2) is a *subjunctive* one.

(1) If Oswald did not kill Kennedy, someone else did.

(2) If Oswald had not killed Kennedy, someone else would have.

(1) is true: someone killed Kennedy; but (2) is probably false. Therefore, different analyses are needed for indicative and for subjunctive conditionals.

A *counterfactual* conditional is an expression of the form 'if A were the case, then B would be the case', where A is supposed to be false. Not all subjunctive conditionals are counterfactual. Consider the argument: 'The murderer used an ice pick. But if the butler had done it, he wouldn't have used an ice pick. So the murderer must have been someone else'. If this subjunctive conditional were a counterfactual, then the speaker would be presupposing that the conclusion of his argument is true. (This example is from R.C. Stalnaker, [51].) Counterfactuals are discussed in Section 6.9. In this section we will restrict our attention from now on to indicative conditionals.

In Chapter 2 we have considered the so-called *paradoxes of material implication*: the following two inferences for material implication '$\to$' are valid, whereas the corresponding English versions seem invalid.

$$\frac{\neg A}{A \to B} \qquad \frac{\text{There is no oil in my coffee}}{\text{If there is oil in my coffee, then I like it}}$$

$$\frac{B}{A \to B} \qquad \frac{\text{I'll ski tomorrow}}{\text{If I break my leg today, then I'll ski tomorrow}}$$

(The latter example is from R. Jeffrey [25], p.74.)

So, the truth-functional reading of 'if ..., then ...', in which $A \to B$ is equivalent to $\neg A \lor B$, seems to conflict with judgments we ordinarily make. The paradoxical character of these inferences disappears if one realizes that

1. the material implication $A \to B$ has the same truth-table as $\neg A \lor B$,

2. speaking the truth is only one of the conversation rules one is expected to obey in daily discourse; one is also expected to be as relevant and informative as possible.

Now, if one has at one's disposal the information $\neg A$ (or B, respectively) and at the same time provides the information $A \to B$, i.e., $\neg A \lor B$, then one is speaking the truth, but a truth calculated to mislead, since the premiss $\neg A$ (or B, respectively) is so much simpler and more informative than the conclusion $A \to B$. If one knows the premiss $\neg A$ (or B, respectively), the conversation rules force us to assert this premiss instead of $A \to B$. Quoting R. Jeffrey [25], pp. 77-78:

> Thus defenders of the truth-functional reading of everyday conditionals point out that the disjunction '$\neg A \lor B$' shares with the conditional 'if A, then B' the feature that normally it is not to be asserted by someone who is in a position to deny 'A' or to assert 'B.'... Normally, then, conditionals will be asserted only by speakers who think the antecedent false or the consequent true, but do not know which. Such speakers will think they know of some connection between the components, by virtue of which they are sure (enough for the purposes at hand) that the first is false or the second is true. [R. Jeffrey, [25], pp. 77-78]

Summarizing in a slogan:

indicative conditional = material implication + conversation rules.

So H.P. Grice uses principles of conversation to explain facts about the use of conditionals that seem to conflict with the truth-functional analysis of the ordinary indicative conditional. In [50] R.C. Stalnaker follows another strategy, rejecting the material conditional analysis and in [49] Brian Skyrms also claims that the indicative conditional cannot be construed as the material implication '→' plus conversational implicature. The dispute between advocates of the truth-functional account of conditionals and the advocates of other, more complex but seemingly more adequate accounts is as old as logic itself. The truth-functional account is first known to have been proposed by Philo of Megara ca. 300 B.C. in opposition to the view of his teacher Diodorus Cronus. We know of this through the writings of Sextus Empiricus some 500 years later, the earlier documents having been lost. According to Sextus,

> Philo says that a sound conditional is one that does not begin with a truth and end with a falsehood. ... But Diodorus says it is one that neither could nor can begin with a truth and end with a falsehood. [W. & M. Kneale [27], p. 128]

There can be no doubt that what Sextus refers to is precisely the truth-functional connective that we have symbolized by the '→', for he says elsewhere,

> So according to him there are three ways in which a conditional may be true, and one in which it may be false. For a conditional is true when it begins with a truth and ends with a truth, like 'If it is day, it is light'; and true also when it begins with a falsehood and ends with a falsehood, like 'If the earth flies, the earth has wings'; and similarly a conditional which begins with a falsehood and ends with a truth is itself true, like 'If the earth flies, the earth exists'. A conditional is false only when it begins with a truth and ends with a falsehood, like 'If it is day, it is night'. [W. & M. Kneale [27], p. 130]

So Sextus reports Philo as attributing truth values to conditionals just as in our table for →, except for the order in which he lists the cases. Diodorus probably had in mind what later was called 'strict implication'; see Section 6.8. For relevant implication see Section 6.10.

7.16 Leibniz

We will here pay attention to only a few aspects of G. Leibniz (1646-1716). For more information the reader is referred to W. & M. Kneale, *The Development of Logic* [27] and to B. Mates, *Elementary Logic*, [37] , Chapter 12. What follows in this section is based on these works.

One of Leibniz' ideals was to develop a *lingua philosophica* or *characteristica universalis*, an artificial language that in its structure would mirror the structure of thought and that would not be affected with ambiguity and vagueness like ordinary language. His idea was that in such a language the linguistic expressions would be pictures, at it were, of the thoughts they represent, such that signs of complex thoughts are always built up in a unique way out of the signs for their composing parts. Leibniz believed that such a language would greatly facilitate thinking and communication and that it would permit the development of mechanical rules for deciding all questions of consistency or consequence. The language, when it is perfected, should be such that 'men of good will desiring to settle a controversy on any subject whatsoever will take their pens in their hands and say *Calculemus* (let us calculate)'. If we restrict ourselves to propositional logic, Leibniz' ideal has been realized: classical propositional logic is decidable (see Chapter 2). However, A. Church and A. Turing proved in 1936 that (classical) predicate logic is undecidable, i.e., there is no mechanical method to test logical consequence (in predicate logic), let alone philosophical truth.

Leibniz also developed a theory of identity, basing it on *Leibniz' Law*: *eadem sunt quorum unum potest substitui alteri salva veritate* – those things are the same if one may be substituted for the other with preservation of truth. Leibniz' Law is also called the substitutivity of identity and it is frequently formulated as follows.

$$a = b \to (\ldots a \ldots \rightleftarrows \ldots b \ldots),$$

where $\ldots a \ldots$ is a context containing occurrences of the name a, and $\ldots b \ldots$ is the same context in which one or more occurrences of a have been replaced by b; if $a = b$, then what holds for a holds for b and vice versa.

A consequence of Leibniz' law is that from 'it is necessary that $9 > 7$' and from 'the number of planets (in this world) $= 9$' it follows that 'it is necessary that the number of planets (in this world) > 7'. This result has been seen as problematic, in particular if one talks about 'the number of planets' instead of 'the number of planets in this world'. The definite description 'the number of planets' assigns different numbers to different possible worlds, but the phrase 'the number of planets in this world' is a rigid designator, referring to the number 9. So, the alleged problem is caused by a sloppy use of language and can be remedied by a careful and precise use of language. See Section 6.11.1.

Leibniz made a distinction between *truths of reason* and *truths of fact*. The truths of reason are those which could not possibly be false, i.e., – in modern terminology – which are *necessarily true*. Examples of such truths are: $2 + 2 = 4$, living creatures cannot survive fire, and so on. Truths of fact are called contingent truths nowadays; for example, unicorns do not exist, Amsterdam is the capital of the Netherlands, and so on. Leibniz spoke of the truths of reason as *true in all possible worlds*. He imagined that there are many *possible worlds* and that our actual world is one of them. '$2 + 2 = 4$' is true not only in this world, but also in any other world. 'Amsterdam is the capital of the Netherlands' is true in this world, but we can think of another world in which this proposition is false. In 1963, S. Kripke extended the notion of possible world with an *accessibility relation* between possible worlds, which enabled him to give adequate semantics for the different modal logics (see Chapter 6). The idea is that some worlds are accessible from the given world, and some are not. For instance, one could postulate (and one usually does) that worlds with different mathematical laws are not accessible from the present world.

7.17 De Dicto - De Re

If one wants to translate the sentence

It is possible that a Republican will win

into a logical formula, it becomes evident that this sentence is ambiguous. Using '$\Diamond$' for 'it is possible that', the predicate symbol R for 'being a Republican' and the symbol W for 'will win', there are two different translations of the sentence in question:

(1) $\exists x[R(x) \wedge \Diamond W(x)]$, and
(2) $\Diamond \exists x[R(x) \wedge W(x)]$.

(1) says, literally, that there is some particular individual (in the actual world) who is a Republican (in the actual world) and who may possibly win (in some imaginary world).

(2) says, literally, that it is possible that some Republican or other will win; more precisely, there is an imaginary world in which a person exists who is a Republican (in that world) en who wins (in that world)

(1) is called the *de re* or referential reading of the sentence above. Typical of the *de re* reading is that the possibility operator $\Diamond$ occurs within the scope of the (existential) quantifier. (2) is called the *de dicto* or non-referential reading of the sentence above. Typical of the *de dicto* reading is that the (existential) quantifier occurs within the scope of the possibility operator $\Diamond$.

The example above demonstrates that sentences containing modalities such as 'possibly', 'necessarily', 'John believes that ...', etc., in combination with existential or universal quantifiers may give rise to ambiguities. Speaking in terms of possible worlds (see Chapter 6) and interpreting '$\Diamond A$ (A is possible)' as 'there is some world accessible from the given world in which A holds', (1) says that in the given world there is a person who is a Republican and who will win in some world accessible from the given one, while (2) says that there is a world accessible from the given one in which there is a person who in that world is Republican and will win.

The proposition 'John finds a unicorn' can be properly translated as $\exists x[U(x) \wedge F(j,x)]$ where $U(a)$ stands for 'a is a unicorn', j stands for 'John' and $F(a,b)$ stands for 'a finds b'. But $\exists x[U(x) \wedge S(j,x)]$, where $S(a,b)$ stands for 'a seeks b' would be an improper translation of 'John seeks a unicorn', because the use of the existential quantifier commits us to an ontology in which unicorns do exist. In his '*The Proper Treatment of Quantification in Ordinary English*' R. Montague [38] develops a 'categorial' language in which 'John seeks a unicorn' can be properly translated.

7.18 Grammars

In the sixties Noam Chomsky developed the notion of grammar which turned out to be important not only for linguistics but also for computer science, for instance, in building parsers and compilers.

One of the main reasons for Chomsky to introduce the notion of grammar was to explain the linguistic competence of people; that they are able to produce new sentences they have never read or heard before. In order to do so, Chomsky assumed that everybody is equipped with certain grammatical rules which can be applied again and again to produce more and more linguistic expressions.

It is well-known that a sentence (S) can be built from a noun phrase (NP) and a verb phrase (VP). Chomsky's basic idea was to represent this fact as a *production rule* $S \rightarrow NP + VP$. This rule should be read as follows: whenever symbol S occurs, it is allowed to rewrite S as the string consisting of the symbols NP and VP. Similar rewrite or production rules, also called *phrase structure rules*, exist for NP and VP. For instance, $NP \rightarrow Art + N$, expressing that a noun phrase (NP) can be built from an article (Art) and a noun (N); and $VP \rightarrow Aux + V + NP$, expressing that a verb phrase (VP) may consist of an auxiliary verb (Aux), a main verb (V) and a noun phrase (NP).

In order to produce English sentences, we also need rewrite or production rules of the form $Aux \rightarrow$ can, $Aux \rightarrow$ may, $Aux \rightarrow$ will, $Aux \rightarrow$ must, usually represented by $Aux \rightarrow$ (can, may, will, must) for short. And we also need rules such as $V \rightarrow$ (read, hit, eat), $Art \rightarrow$ (a, the) and $N \rightarrow$ (boy, man, frog).

The expressions 'can', 'read', 'a', 'boy', etc., are called *terminals* because there are no production rules starting with these expressions. The symbols S, NP, VP, Art, N, Aux, V, on the other hand, are called *non-terminals* for obvious reasons.

Starting with the symbol S, we can, by repeated application of the production rules above, generate terminal strings to which no production rules can be applied. For instance, starting with S, the production rules mentioned above can generate the English sentence 'the boy must eat the frog'. The *derivation tree* or *phrase marker* for this sentence looks as follows.

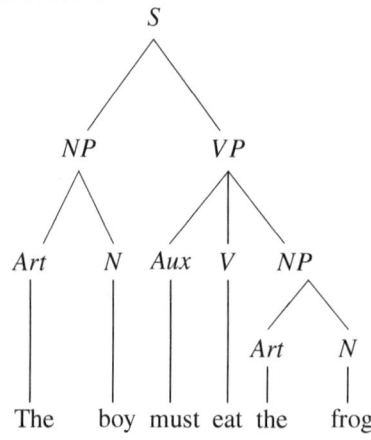

7.18 Grammars

In this way the production rules above can generate a small fragment of English. Together they form what Chomsky called a *grammar*. In his view, the production rules in a grammar represent the linguistic competence of a speaker and are part of everyone's innate mental equipment.

Definition 7.1 (Grammar). A (type 0) *grammar* G is a quadruple $\langle V_N, V_T, P, S \rangle$, where
1) V_N is a finite set; the elements of V_N are called *non-terminals*.
2) V_T is a finite set such that V_N and V_T have no elements in common; the elements of V_T are called *terminals*.
3) P is a finite set of expressions of the form $\alpha \to \beta$, where α and β are strings of finite length composed of symbols of V_N and/or V_T (i.e., $\alpha, \beta \in (V_N \cup V_T)^*$) and the length of α is at least 1 (i.e., $|\alpha| \geq 1$); the elements of P are called *productions*.
4) $S \in V_N$; S is called the sentence- or *start-symbol*.

Example 7.1. Let $G_1 = \langle V_N, V_T, P, S \rangle$ where
$V_N = \{S, NP, VP, Art, N, Aux, V\}$,
$V_T = \{$can, may, will, must, read, hit, eat, a, the, boy, man, frog$\}$, and
P consists of the following productions:

$S \to NP$ - VP $Aux \to$ (can, may, will, must)
$NP \to Art$ - N $V \to$ (read, hit, eat)
$VP \to Aux$ - V - NP $Art \to$ (a, the)
                         $N \to$ (boy, man, frog).

Example 7.2. Let $G_2 = \langle \{S\}, \{0,1\}, \{S \to 0S1, S \to 01\}, S \rangle$.
Here, S is the only non-terminal, 0 and 1 are terminals and there are two productions, $S \to 0S1$ and $S \to 01$.

By putting certain restrictions on the productions in P one obtains grammars of type 1, 2 and 3, respectively.

Definition 7.2 (Derivable from). Let $G = \langle V_N, V_T, P, S \rangle$ be a grammar and let α and β be finite strings composed of elements of V_N and/or V_T ($\alpha, \beta \in (V_N \cup V_T)^*$).
$\alpha \stackrel{*}{\underset{G}{\Rightarrow}} \beta$ (β is *derivable from* α in grammar G) := β can be obtained from α by application of one or more productions in P. By convention, $\alpha \stackrel{*}{\underset{G}{\Rightarrow}} \alpha$ for each string α.

Example 7.3. $S \stackrel{*}{\underset{G_1}{\Rightarrow}} Art$ - N - Aux - eat - NP and $VP \stackrel{*}{\underset{G_1}{\Rightarrow}} Aux$ - eat - the - frog.
$S \stackrel{*}{\underset{G_2}{\Rightarrow}} 00S11$.

Definition 7.3 (Language generated by a Grammar). Let $G = \langle V_N, V_T, P, S \rangle$ be a grammar. $L(G) := \{w \in V_T^* \mid S \stackrel{*}{\underset{G}{\Rightarrow}} w\}$, i.e., $L(G)$, called *the language generated by* G, is by definition the set of all strings (or words) w of terminals such that $S \stackrel{*}{\underset{G}{\Rightarrow}} w$.

So, a string w is in $L(G)$ iff 1) w consists solely of terminals and 2) w is derivable from S in G.

Example 7.4. The reader easily verifies that the sentences 'the boy must eat the frog' and 'a man may hit the boy' both are in $L(G_1)$, where G_1 is the grammar from Example 7.1. $L(G_2) = \{0^n 1^n \mid n \geq 1\}$, i.e., the language generated by G_2 (see Example 7.2) consists of all finite sequences composed of n 0's followed by the same number n of 1's ($n \geq 1$).

Consider the following three sentences.
1. Locusta is an alleged poisoner.
2. Locusta is a Roman poisoner.
3. Locusta is a skillful poisoner.

These three sentences have similar surface structures, but still intuitively we feel they are quite different in meaning. (An alleged poisoner is not another kind of poisoner along with Roman and Carthaginian or skillful and clumsy; a Roman poisoner is one who is both Roman and a poisoner, but a skillful strangler may be a clumsy poisoner.)

In order to explain this, Chomsky distinguishes the *deep structure* of a sentence, determined by the production rules of a grammar and the *surface structure* of a sentence which results by applying to the derivation tree of the sentence certain transformation rules. Corresponding with their quite different meanings, the sentences 1, 2 and 3 above have radically different deep structures. Transformation rules transform these different deep structures into similar surface structures. See Exercise 7.7.

Postulating that the deep structure determines the meaning of a sentence, Chomsky explains in this way that the sentences 1, 2 and 3 above have quite different meanings although they have a similar surface structure.

The sentences 'a man may hit the boy' and 'the boy may be hit by a man', on the other hand, have the same meaning, although they are syntactically different. This can also be explained in terms of deep and surface structures. These sentences have the same deep structure and hence the same meaning. And the surface structure of one of these sentences is obtained by applying certain transformation rules to its deep structure. See Exercise 7.8.

So, in Chomsky's view, the syntax of a language consists of two components:
1) a *base component*, containing the production (or phrase structure) rules. These rules generate the *deep structure* of each sentence.
2) a *transformational component*, containing the transformation rules which map derivation trees into (other) derivation trees. The transformation rules take as input a deep structure and generate as output a *surface structure*.

The deep structure determines the meaning of a sentence; the surface structure determines its sound.

In the case of the sentences 'a man will hit the boy', in the active mood, and 'the boy will be hit by a man', in the passive mood, two surface structures are derived from one deep structure. And in the case of 'Locusta is an alleged/Roman/skillful poisoner', similar surface structures are derived from several different deep structures.

One can show that the languages generated by a (type 0) grammar are precisely the languages recognized by a Turing machine; see, for instance, de Swart [53].

7.18 Grammars

Exercise 7.7. Let G be a grammar with the following phrase structure rules (productions):

$S \to S$ - [and] - S $\qquad$ $VP \to V$ - [that] - S
$S \to NP$ - VP $\qquad$ $VP \to (Adv)$ - V - N
$NP \to (Art)$ - (Adj) - N $\qquad$ $VP \to Copula$ - NP
$\qquad\qquad\qquad\qquad\qquad$ $VP \to Copula$ - Adj

$Art \to$ (a, an, the)
$Adj \to$ (roman, alleged, skilful)
$N \to$ (someone, Locusta, poisoner)
$V \to$ (allege, poison)
$Copula \to$ be
$Adv \to$ skilfully

Generate the phrase markers (derivation trees) in G for the deep structures of the following sentences:

1. Locusta is an alleged poisoner.
2. Locusta is a Roman poisoner.
3. Locusta is a skilful poisoner.

Various transformations give rise to the same surface structure.

Exercise 7.8. Let G be the same grammar as in Exercise 7.7 and consider the following transformation rule: N_1 - V - N_2 → N_2 - [be] - V - [ed] - [by] - N_1.
Check that this transformation rule, applied to the phrase marker (derivation tree) in G for the deep structure of 'someone is poisoned by Locusta', which is the same as the one of 'Locusta poisons someone', yields the phrase marker for the surface structure of 'someone is poisoned by Locusta'.

Exercise 7.9. Construct a grammar which generates precisely all formulas of propositional logic built from the atomic formulas Q, Q', Q'',... by means of the connectives $\wedge$, $\vee$ and $\neg$.

Exercise 7.10. Let $G = \langle \{S\}, \{0,1\}, \{S \to 0S1, S \to 01\}, S \rangle$. Show that $L(G) = \{0^n 1^n \mid n \geq 1\}$, where 0^n stands for 0 repeated n times and similarly for 1^n.

Exercise 7.11. Give a grammar generating the set of all finite strings w of 0's and 1's such that w does not contain two consecutive 1's.

Exercise 7.12. Give a grammar generating the set of all finite strings w of a's, b's and c's such that w consists of equal numbers of a's, b's and c's.

Exercise 7.13. Let $\{a,b\}^*$ be the set of all finite strings of a's and b's, including the empty string e of length 0. Let $L = \{w \in \{a,b\}^* \mid w$ contains an even number of b's$\}$. Check that $L - \{e\}$ is generated by the grammar $G = \langle \{S,B\}, \{a,b\}, P, S \rangle$ with $P = \{S \to aS, S \to a, S \to bB, B \to aB, B \to bS, B \to b\}$.

Exercise 7.14. With $\{a,b\}^*$ as in Exercise 7.13, let $L = \{w \in \{a,b\}^* \mid w$ does not contain three consecutive b's$\}$. Verify that $L - \{e\}$ is generated by the grammar $G = \langle \{S,B,C,D\}, \{a,b\}, P, S \rangle$ with P consisting of the following productions:

$$S \to aS, \quad B \to aS, \quad C \to aS, \quad D \to aD,$$
$$S \to a, \quad B \to a, \quad C \to a, \quad D \to bD.$$
$$S \to bB, \quad B \to bC, \quad C \to bD,$$
$$S \to b, \quad B \to b.$$

7.19 Solutions

Solution 7.1.
(a) true (b) false; The Iliad is an epic poem.
(c) true (d) true
(e) false; $7 + 5 = 12$. (f) true
(g) true (h) false; 'Saul' is another name of Paul.
(i) true (j) false; '$2 + 2 = 4$' is synthetic.
(k) true (l) true

Solution 7.2. The difficulty lies in our assumption that one can determine mechanically whether or not an alleged definition of a function is indeed such a definition. Since the function f, defined by $f(n) := f_n(n) + 1$, is not definable in the English language, our only recourse is to conclude: There is no algorithm that enables one to decide whether an alleged definition of a number-theoretic function is indeed such a definition. In other words, there is no algorithm that enables one to decide mechanically for any expression in the English language whether it defines a number-theoretic function or not.

Solution 7.3. Suppose that a and b are rigid designators. If '$a = b$' is true, so that 'a' and 'b' designate the same object in the actual world, then, since both names, being rigid designators, designate the same object in all possible worlds, '$a = b$' is true in all possible worlds, that is to say, it is necessarily true that $a = b$.

Solution 7.4. Since stick S is the standard meter in Paris, stick S is by definition one meter long. Therefore, the epistemological status of the statement 'stick S is one meter long' is that this statement is an *a priori* truth. Conceiving 'one meter' as a rigid designator, indicating the same length in all possible circumstances (worlds), the metaphysical status of 'stick S is one meter long' will be that of a *contingent* statement, since the length of stick S can vary with the temperature, humidity, etc.

Solution 7.5. 'The king of France is bald' induces the presupposition that there is a king of France. This presupposition is not induced by the sentence 'If the king of France exists, then the king of France is bald'.

Solution 7.6. Suppose we define 'A presupposes B' as: $A \models B$ and $\neg A \models B$. Then $A \vee \neg A \models B$. But $\models A \vee \neg A$. Therefore, $\models B$. So, 'A presupposes B' would mean that $\models B$. This is counter-intuitive.

7.19 Solutions

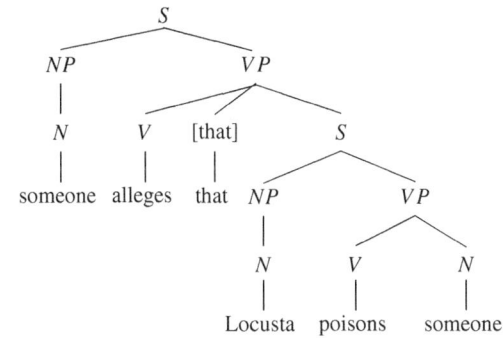

Solution 7.7. 1.

```
                    S
           ┌────────┴────────┐
          NP                 VP
           │      ┌──────┬───┴────┐
           N      V   [that]      S
           │      │     │    ┌────┴────┐
         someone alleges that NP       VP
                              │    ┌───┴───┐
                              N    V       N
                              │    │       │
                           Locusta poisons someone
```

2.
```
                         S
              ┌──────────┼──────────┐
              S        [and]        S
         ┌────┴───┐            ┌────┴────┐
        NP       VP           NP        VP
         │    ┌──┴──┐          │     ┌───┴──┐
         N    V     N          N   Copula   Adj
         │    │     │          │     │      │
      Locusta poisons someone and Locusta  is  Roman
```

3.
```
                   S
           ┌───────┴────────┐
          NP                VP
           │       ┌────────┼────────┐
           N      Adv       V        N
           │       │        │        │
        Locusta skilfully poisons  someone
```

Phrase-marker for the surface structure of the sentences in 1, 2 and 3:

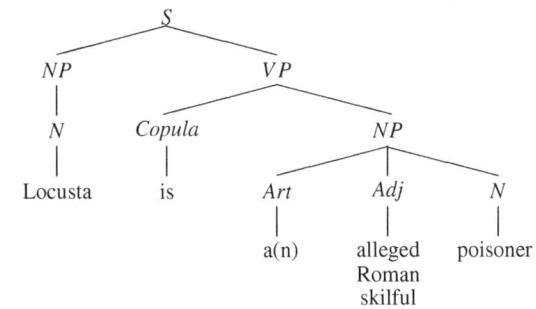

```
                       S
           ┌───────────┴──────────────┐
          NP                          VP
           │             ┌────────────┼────────┐
           N           Copula                  NP
           │             │         ┌────────┬──┴──┐
        Locusta          is       Art      Adj     N
                                   │        │      │
                                  a(n)   alleged poisoner
                                          Roman
                                          skilful
```

Solution 7.8.

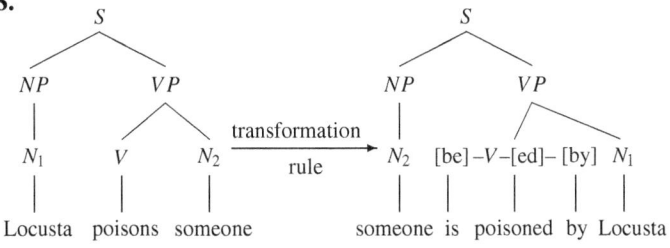

```
           S                                              S
      ┌────┴────┐                                   ┌─────┴─────┐
     NP        VP           transformation         NP          VP
      │     ┌───┴──┐             rule               │    ┌──┬───┼────┐
      N₁    V      N₂         ──────────→          N₂  [be]-V-[ed]-[by] N₁
      │     │      │                                │    │  │    │   │
   Locusta poisons someone                      someone is poisoned by Locusta
```

Solution 7.9. $G = \langle V_N, V_T, P, S \rangle$, where $V_T = \{Q, ', \wedge, \vee, \neg, (,)\}$, $V_N = \{S, A\}$, and $P = \{S \to A, S \to (S \wedge S), S \to (S \vee S), S \to (\neg S), A \to A', A \to Q\}$.

Solution 7.10. Each string of the form $0^n 1^n$, $n \geq 1$, is generated by G:
$$S \to 0S1 \to 00S11 \to 0^3 S 1^3 \to \ldots \to 0^{n-1} S 1^{n-1} \to 0^n 1^n.$$
Furthermore it is easy to see that these are the only strings in $L(G)$.

Solution 7.11. $G = \langle \{S, A\}, \{0, 1\}, P, S \rangle$, where P contains the following productions:
$$S \to 1 \; , \; A \to 0,$$
$$S \to 0 \; , \; A \to 0S,$$
$$S \to 0S \; , \; S \to 1A.$$

Solution 7.12. $G = \langle \{S, A, B, C\}, \{a, b, c\}, P, S \rangle$, where P consists of the following productions:
$$S \to ABC, \; A \to a, \; AB \to BA, \; BC \to CB,$$
$$S \to SS, \; B \to b, \; AC \to CA, \; CA \to AC,$$
$$C \to c, \; BA \to AB, \; CB \to BC.$$

Solution 7.13. We have to show that $L - \{e\} = L(G)$, i.e. that $L - \{e\}$ and $L(G)$ have the same elements.
1) So suppose $w \in L - \{e\}$, i.e., w contains an even number of b's and $w \neq e$. Then it is not hard to see that w can be generated by G, i.e., $S \overset{*}{\underset{G}{\Rightarrow}} w$, and hence $w \in L(G)$.
2) Conversely, suppose $w \in L(G)$. Then it follows from the definition of the productions in G that w contains an even number of b's and $w \neq e$, and hence that $w \in L - \{e\}$.

Solution 7.14. Similar to Solution 7.13. Note that if an expression u generated by the grammar ($S \overset{*}{\underset{G}{\Rightarrow}} u$) contains three or more consecutive b's, it must also contain the nonterminal D, and hence $u \notin V_T^*$, so $u \notin L(G)$.

References

1. Adams, E., Subjunctive and Indicative Conditionals. *Foundations of Language* 6, 89-94, 1970.
2. Ayer, A.J., (ed.), *Logical Positivism*. Editor's Introduction. Greenwood Press, Westport, C.T., 1959, 1978.
3. Austin, J.L., *How to Do Things with Words*. Oxford University Press, 1962.
4. Austin, J.L., Performative-Constative. In J.R. Searle, (ed.), *Philosophy of Language*.
5. Bergmans, L.J.M., Gerrit Mannoury. *Encyclopédie Philosophique Universelle*, vol. III Oeuvres. Les Presses Universitaire de France, Paris, 1992, columns 1778-1783.
6. Brouwer, L.E.J., *Leven, Kunst en Mystiek*. J. Waltman, Jr., Delft, 1905.
7. Brouwer, L.E.J., Boekbespreking van J.I. de Haan, Rechtskundige Significa en hare Toepassing op de begrippen: "Aansprakelijk, Verantwoordelijk, Toerekeningsvatbaar". Akademisch Proefschrift, Thesis, Amsterdam: W. Versluys, 1916. *Groot-Nederland* 14, 1916, pp. 333-336.
8. Bűcher, K., *Arbeit und Rhythmus*. (3rd edition) B.G.Teubner, Leipzig, 1902.
9. Carnap, R., *Meaning and Necessity*. University of Chicago Press, 1947, 1967, 2008.

References

10. Church, A., *Introduction to Mathematical Logic*. Princeton Univ. Press, 1956, 1996.
11. Dantzig, D. van, On the principles of intuitionistic and affirmative mathematics. *Indigationes mathematicae*, vol. IX, 4 & 5, 1947.
12. Donnellan, K., Reference and Definite Descriptions. *Philosophical Review* 75, 281-304, 1966.
13. Dummett, M., *Frege, Philosophy of Language*. Duckworth, 1981.
14. Frege, G., *Begriffsschrift und andere Aufsätze*. I. Angelelli (ed.), Olms, Hildesheim, 1964.
15. Frege, G., *Philosophical Writings*. Translated by P. Geach and M. Black. Basil Blackwell, Oxford, 1970.
16. Frege, G., *Funktion, Begriff, Bedeutung*. Fünf Logische Studien. G. Patzig (ed.), VandenHoeck & Ruprecht, Göttingen, 1975.
17. Frege, G., *Grundgesetze der Arithmetik*. 1893, 1903, 2009.
18. Geach, P. and M. Black (eds.) *Translations from the Philosophical Writings of Gottlob Frege*. Oxford University Press, 1980.
19. Gödel, K., *Russell's mathematical logic*. in P.A. Schilpp (ed.), *The philosophy of Bertrand Russell*, Tudor, N.Y., 1944; and in P. Benacerraf and H. Putnam, *Philosophy of Mathematics, Selected readings*, New York, 1964, 1983, 1998.
20. Grice, P., *Studies in the Way of Words*. Harvard University Press, 1989
21. Haack, S., *Philosophy of Logics*. Cambridge University Press, 1978, 1999.
22. Hacker, P.M.S., *Insight and Illusion: Wittgenstein on Philosophy and Metaphysics of Experience*. Oxford University Press, 1972.
23. Heijenoort, J. van, *From Frege to Gödel*. Begriffsschrift, a formula language, modeled upon that of arithmetic, for pure thought. Harvard University Press, Cambridge, 1967.
24. Heijerman, E. and Schmitz, H.W., (eds.), *Signifies, Mathematics and Semiotics. The Signific Movement in the Netherlands*. Proceedings of the International Conference held at Bonn, November 19-21, 1986. Modus Publikationen, Műnster, 1990.
25. Jeffrey, R., *Formal Logic: its scope and limits*. McGraw-Hill, New York, 1967, 1981, 2006.
26. Kant, I., *The Critique of Pure Reason*. William Benton, Publisher, Encyclopaedia Brittanica, in particular pp. 14-18.
27. Kneale, W. & M., *The Development of Logic*. Clarendon Press, Oxford, 1962, 2008.
28. Kripke, S.A., *Naming and Necessity*. Basil Blackwell, Oxford, 1980.
29. Kripke, S.A., *Identity and Necessity*. In [46], pp 66-101.
30. Mannoury, G., De filosofische grondslagen der verantwoordelijkheid. *Accountancy* 15 (12), december 1917, pp. 97-104.
31. Mannoury, G., *Mathesis en Mystiek: Een signifiese studie van kommunisties standpunt*. Maatschappij voor goede en goedkope lectuur, Amsterdam, 1925.
32. Mannoury, G., Die signifischen Grundlagen der Mathematik. *Erkenntnis*, Vol. IV, 1934.
33. Mannoury, G., *Les Fondements Psycho-Linguistiques des Mathématiques*. F.G. Kroonder, Bussum/Editions du Griffon, Neuchâtel, 1947.
34. Mannoury, G., *Handboek der Analytische Signifika; deel I: Geschiedenis der Begripskritiek*. F.G. Kroonder, Bussum, 1947.
35. Mannoury, G., *Handboek der Analytische Signifika; deel II: Hoofdbegrippen en Methoden der Signifika. Ontogenese en Fylogenese van het Verstandhoudingsapparaat*. F.G. Kroonder, Bussum, 1948.
36. Mannoury, G., *Polairpsychologische Begripssynthese*. F.G. Kroonder, Bussum, 1953.
37. Mates, B., *Elementary Logic*. Oxford University Press, 1965, 1972.
38. Montague, R., *The Proper Treatment of Quantification in Ordinary English*. Reprinted in R.H. Thomason (ed.), *Formal Philosophy; Selected Papers of Richard Montague*. Yale University Press, 1974.
39. Quine, W.V., *The Ways of Paradox and other Essays*. Harvard University Press, 1976.
40. Quine, W.V., *Two Dogmas of Empiricism*. In W.V. Quine, *From a Logical Point of View*. Harvard University Press, 1953, 1999.
41. Russell, B. and A.N. Whitehead, *Principia Mathematica*. 1903, 2011.
42. Sandt, R. van de, *Context and Presupposition*. Routledge, 1988.
43. Schmitz, H.W., Mannoury and Brouwer: Aspects of their relationship and cooperation. *Methodology and Science*, 20, 1987, pp. 40-62.

44. Schmitz, H.W., *De Hollandse Signifika: Een reconstructie van de geschiedenis van 1892 tot 1926*. Translated by J. van Nieuwstadt, Van Gorcum, Assen, 1990.
45. Schmitz, H.W., Empirical Methods of Signific Analysis of Meaning: Transformation and Exhaustion of Linguistic Acts. In Heijerman, E., and Schmitz, H.W., (eds.), 1990.
46. Schwartz, S.P., *Naming, Necessity and Natural Kinds*. Introduction, pp. 13-41. Cornell University Press, 1977-1979.
47. Searle, J.R. (ed.), *The Philosophy of Language*; Introduction. Oxford Univ. Pr., 1971.
48. Searle, J.R., What is a Speech Act. In: J.R. Searle, (ed.), *Philosophy of Language*.
49. Skyrms, B., *Causal Necessity*. Yale University Press, 1980.
50. Stalnaker, R.C., *Pragmatics*. In Davidson and Harman (eds.), *Semantics of Natural Language*. Reidel, Dordrecht, 1972.
51. Harper, W.L., R.C. Stalnaker, G. Pearce (eds.), *IFS*; Conditionals, Belief, Decision, Chance and Time. Reidel, Dordrecht, 1981.
52. Strawson, P.F., (ed.), *Philosophical Logic*; The Thought. Oxford Univ. Press, 1968.
53. Swart, H.C.M. de, *Logic and Computer Science*. Verlag Peter Lang, Frankfurt, 1994.
54. Wittgenstein, L., *Notebooks 1914-1916*. 2nd edition. Edited by G.H. von Wright and G.E.M. Anscombe. Translated by G.E.M. Anscombe. Basil Blackwell, Oxford 1961.
55. Wittgenstein, L., *Tractatus Logico-Philosophicus*. Translated by D.F. Pears and B.F. McGuiness. 2nd edition. Kegan Paul, London, 1961.
56. Wittgenstein, L., *Philosophical Investigations*. 3rd edition. Translated by G.E.M. Anscombe. Basil Blackwell, Oxford, 1967.
57. Wittgenstein, L., *The Blue and Brown Books*. 2nd edition. Basil Blackwell, Oxford, 1969.
58. Wundt, W., *Völkerpsychologie: Eine Untersuchung der Entwicklungsgesetze von Sprache, Mythus und Sitte; vol. I: Die Sprache*. W. Engelmann, Leipzig, 1900.

Chapter 8
Intuitionism and Intuitionistic Logic

H.C.M. (Harrie) de Swart

Abstract Brouwer's intuitionism is based on quite different philosophical ideas about the nature of mathematical objects than classical mathematics. This intuitionistic point of view results in a different use of language and in a corresponding different intuitionistic logic which is far more subtle than the classical use of language and corresponding classical logic. Nevertheless an intuitionistic deduction system and a notion of intuitionistic deducibility was developed by A. Heyting and it is amazing to see that a small change in the logical axioms, replacing the logical axiom $\neg\neg A \rightarrow A$ by $\neg A \rightarrow (A \rightarrow B)$, may have such far reaching consequences. Since finding (intuitionistic) formal deductions may be difficult, an intuitionistic tableaux based formal deduction system is presented in which the construction of intuitionistic deductions is rather straightforward. The semantics of intuitionistic logic and the notion of (intuitionistic) valid consequence are given in terms of (intuitionistic) Kripke models and it is shown that the three notions of intuitionistic valid consequence, intuitionistic deducibility and intuitionistic tableau-deducibility are equivalent. Intuitionistic sets are either finite constructions or otherwise, they are (subsets of) construction projects. Spreads are a particular kind of construction project, inducing specific principles which typically do not hold for other sets.

8.1 Intuitionism vs Platonism; basic ideas

A *classical mathematician* studies the properties of mathematical objects like an astronomer, who studies the properties of celestial bodies. Mathematical objects are like celestial bodies in the sense that they exist independently of us; they are created by God. And mathematicians are like astronomers who try to discover properties of these objects.

An *intuitionist* creates the mathematical objects himself. According to *Brouwer's intuitionism*, mathematical objects, like 5, 7, 12 and +, are mental constructions. A proposition about mathematical objects (like $5 + 7 = 12$) is true if one has a proof-construction that establishes it. Such a proof is again a mental construction.

> Mathematics is created by a free action, independent of experience. [L.E.J. Brouwer [3], p. 97]

In order to better understand the intuitionistic point of view, let us consider the classical or platonistic standpoint more closely. Using Dummett's terminology, one might say that from a classical or platonistic point of view mathematical objects exist in some external realm of mathematical reality; this realm of mathematical reality, existing objectively and independently of our knowledge, renders our mathematical statements (like $5 + 7 = 12$) true or false.

> From a classical or platonistic standpoint, the understanding of a mathematical statement consists in a grasp of what it is for that statement to be true, where truth may attach to it even when we have no means of recognizing the fact. [M. Dummett [4], p. 6-7]

The following quotation, translated from German, from P. Bernays [1] may further illuminate the classical or platonistic standpoint.

> One considers the objects of a theory as the elements of a totality and concludes from that: for every property, which can be expressed by means of the notions of the theory, it is an established fact whether there is an element in the totality which has this property or not. From this way of seeing things follow also the following alternatives: either all elements of a set have a given property or there is at least one which does not have this property.
>
> One finds in the axiomatics of geometry – in the form Hilbert has given to it – an example for this way of building a theory. When we compare Hilbert's axiomatics with those of Euclid, where we waive that in the case of the Greek mathematicians still some axioms are missing, we notice that Euclid talks about figures, which should be constructed, while for Hilbert the systems of points, of lines and of planes already exist right from the beginning. Euclid postulates: one can connect two points by a straight line; Hilbert, on the contrary, formulates the axiom: Given two arbitrary points, then there exists a straight line, on which both points are located. Existence refers here to the system of straight lines. This example already shows that the tendency, we are talking about, goes in the direction to consider all objects as detached from any connection with the thinking subject.
>
> Since this tendency has become valid most of all in the philosophy of Plato, let me be allowed, to call it Platonism. [P. Bernays [1], pp. 62-63]

The contrast between 'Platonism' and 'Intuitionism' already figures in essence in the comment of Proclos (450 A.D.) on the first book of Euclid's *Elements*, in which a distinction is made between Speusippos and his adherents (350 B.C.), who maintain that all construction problems are theorems, and Menaichmos and the people around him (350 B.C.), who maintain that all theorems are construction problems. (See Paul Ver Eecke [19], pp. 69-70.)

The different views underlying classical and intuitionistic mathematics also result in a different view of the infinite. In classical mathematics, the infinite (for instance, the set $\mathbb{N}$ of the natural numbers $0, 1, 2, 3, \ldots$) is treated as *actual* or *completed*. Quoting S.C. Kleene,

> An infinite set is regarded as existing as a completed totality, prior to or independently of any human process of generation or construction, and as though it could be spread out completely for our inspection. [S.C. Kleene [8], p. 48]

8.1 Intuitionism vs Platonism; basic ideas

Since for an intuitionist mathematical objects are mental constructions, in intuitionism the infinite is treated only as *potential* or *becoming* or *constructive*. Intuitionistically, the set $\mathbb{N}$ of natural numbers is identified with the *construction project* for its elements: start with 0, and add 1 to each natural number which has already been constructed before. And it was one of the main achievements of L.E.J. Brouwer (1881-1966) to solve the problem how we can talk constructively about the non-enumerable totality $\mathbb{R}$ of the real numbers (see Section 8.7).

For this purpose Brouwer introduced his notion of *spread*, which is again a construction project for producing the elements of the spread, the elements being (potentially) infinite sequences of natural numbers rather than natural numbers themselves. And since real numbers can be represented by infinite sequences of natural numbers, more precisely, by infinite sequences of intervals with rational endpoints, Brouwer's notion of spread enables us to talk constructively about the set $\mathbb{R}$ of the real numbers. (See Section 8.7 for a more elaborate discussion of sets in intuitionism and spreads in particular.)

Since, according to Brouwer, both the mathematical objects themselves and the proofs establishing properties of them are mental constructions, doing mathematics is in principle language-less. Nevertheless, language may be introduced for reasons of communication.

> People try by means of sounds and symbols to originate in others copies of mathematical constructions and reasonings which they have made themselves; by the same means they try to aid their own memory. In this way *mathematical language* comes into being, and as its special case *the language of logical reasoning*. [L.E.J. Brouwer, [3], p. 73]

8.1.1 Language

The following is a quotation from L.E.J. Brouwer [3].

> The immediate companion of the intellect is language. From life in the Intellect follows the impossibility to communicate directly, instinctively, by gesture or looks, or, even more spiritually, through all separation of distance. People then try and train themselves and their offspring in some form of communication by means of crude sounds, laboriously and helplessly, for never has anyone been able to communicate his soul by means of language. ...
>
> Only in those very narrowly delimited domains of the imagination such as the exclusively intellectual sciences – which are completely separated from the world of perception and therefore touch the least upon the essentially human – only there may mutual understanding be sustained for some time and succeed reasonably well. Little confusion is possible about the meaning of such words as 'equal' or 'triangle', but even then two different people will never think of them in exactly the same way. Even in the most restricted sciences, logic and mathematics (a sharp distinction between these two is hardly possible), no two different people will have the same conception of the fundamental notions of which these two sciences are constructed; and yet, they have a common will, and in both there is a small, unimportant part of the brain that forces the attention in a similar way. ...
>
> Language becomes ridiculous when one tries to express subtle nuances of will which are not a living reality to the speakers concerned, when for example so-called philosophers or metaphysicians discuss among themselves morality, God, consciousness, immortality or

the free will. These people do not even love each other, let alone share the same subtle movements of the soul. Sometimes they do not even know each other personally. They either talk at cross purposes or they each build their own little logical system that lacks any connection with reality. For logic is life in the human brain; it may accompany life outside it: it can never guide it by virtue of its own power. ...

Language by itself has no meaning; any philosophy, which in this way tried to find a firm foundation has come to grief. Lulled into sleep by the mistaken belief in its certainty one later hits upon deficiencies and contradictions. A language which does not derive its certainty from the human will, which claims to live on in the 'pure concept' is an absurdity. To be able to go on talking without being caught in contradiction or without making a silent assumption is an art to be valued only in an acrobat. [L.E.J. Brouwer [3], pp. 6-7]

8.1.2 *First Steps in Intuitionistic Reasoning*

Since, intuitionistically, the truth of a mathematical proposition is established by a proof – which is a particular kind of mental construction –, the meaning of the logical connectives has to be explained in terms of proof-constructions:

A proof of $A \wedge B$ is anything that is a proof of A and of B.

A proof of $A \vee B$ is, in fact, a proof either of A or of B, or yields an effective means, at least in principle, for obtaining a proof of one or other disjunct.

A proof of $A \to B$ is a construction of which we can recognize that, applied to any proof of A, it yields a proof of B. Such a proof is therefore an operation carrying proofs (of A) into proofs (of B).

Intuitionists consider $\neg A$ as an abbreviation for $A \to \bot$, postulating that nothing is a proof of $\bot$ (falsity).

Also the existential quantifier has a constructive meaning in intuitionism: $\exists x[P(x)]$ (there exists an x with the property P) means that I have an algorithm to construct an x in the given domain and next prove that this x has the property P. Consequently, $\exists x[P(x)]$ has a much stronger meaning than $\neg \forall x[\neg P(x)]$ (not every x has the property $\neg P$, i.e., the assumption $\forall x[\neg P(x)]$ yields a contradiction). The constructive reading of $A \vee B$ may be rendered by $\exists x[(x = 0 \wedge A) \vee (x = 1 \wedge B)]$.

The constructive meaning of the intuitionistic connectives causes that many familiar laws from classical logic are no longer valid. Below we shall make clear that the different philosophical points of view, the intuitionistic and the platonistic one, concerning the nature of mathematical objects, result in a quite different use of language.

1. It is reckless to affirm the validity of $A \vee \neg A$. Classically, the validity of $A \vee \neg A$ means that the state of affairs expressed by proposition A is either true or false, without necessarily having a method to decide which of these two. But intuitionistically the validity of $A \vee \neg A$ means that we have a method adequate in principle to solve *any* mathematical problem. Consider *Goldbach's conjecture* G, which states that each even number is the sum of two odd primes: $2 = 1 + 1$, $4 = 3 + 1$, $6 = 5 + 1$, $8 = 7 + 1$, $10 = 7 + 3$, $12 = 7 + 5$, $14 = 7 + 7$, $16 = 13 + 3$, $18 = 13 + 5$, ... One can check only finitely many individual instances,

8.1 Intuitionism vs Platonism; basic ideas

while Goldbach's conjecture is a statement about infinitely many (even) natural numbers. So far neither Goldbach's conjecture, G, nor its negation, $\neg G$, has been proved. We are therefore not in a position to affirm $G \vee \neg G$. Someone who does, claims that he or she can provide a proof either of G or of $\neg G$; such a person is called *reckless*. Of course, an intuitionist can prove that $(2+3=5) \vee \neg(2+3=5)$. But the validity of $A \vee \neg A$ means that he can give a proof of A or can give a proof of $\neg A$ for *any* mathematical proposition A. And this is a reckless statement. Note that the proposition $\neg(G \vee \neg G)$ implies a contradiction. From $G \to G \vee \neg G$ it follows that $\neg(G \vee \neg G) \to \neg G$. And from $\neg G \to G \vee \neg G$ it follows that $\neg(G \vee \neg G) \to \neg\neg G$. So, $\neg(G \vee \neg G)$ implies both $\neg G$ and $\neg\neg G$, i.e., a contradiction.

2. $\neg\neg(A \vee \neg A)$ is intuitionistically valid, and hence it is false to assert $\neg(A \vee \neg A)$.
   Proof: $\neg(A \vee B)$ is intuitionstically equivalent to $\neg A \wedge \neg B$. So, since $\neg(\neg A \wedge \neg\neg A)$ holds intuitionistically, it follows that $\neg\neg(A \vee \neg A)$ is intuitionistically valid.

3. It is reckless to affirm the validity of $\neg\neg E \to E$. For taking $E = A \vee \neg A$, we have seen in 2) that $\neg\neg(A \vee \neg A)$ is intuitionistically valid, while $A \vee \neg A$ is not, as we have seen in 1).

4. $E \to \neg\neg E$ is intuitionistically valid.
   Proof: We have a proof of $\neg\neg E$ when we can show that we shall never have a proof of $\neg E$, that is, when we show that we shall never have a proof that E will never be proved. Clearly, in general this does not amount to a proof of E itself, as we have seen in 3) by taking $E = A \vee \neg A$. On the other hand, a proof of E does count as a proof that E will never be disproved, for otherwise the possibility of deriving a contradiction would remain open; hence $E \to \neg\neg E$ is intuitionistically valid.

5. $\neg D \to \neg\neg\neg D$ is intuitionistically valid.
   Proof: From 4), taking $E = \neg D$.

6. $(A \to B) \to (\neg B \to \neg A)$ is intuitionistically valid. But the converse, $(\neg B \to \neg A) \to (A \to B)$ is not intuitionistically valid.
   Proof: Given $A \to B$, if we have a proof that B can never be proved, then clearly A can never be proved either, since we could transform any proof of A into a proof of B. We may thus always infer $\neg B \to \neg A$ from $A \to B$.
   From 6) follows immediately that $(\neg B \to \neg A) \to (\neg\neg A \to \neg\neg B)$ and hence also that $(\neg B \to \neg A) \to (A \to \neg\neg B)$ is intuitionistally valid, because $A \to \neg\neg A$ is intuitionistically valid. But because $\neg\neg B \to B$ is not intuitionistically valid, we may not conclude the validity of $(\neg B \to \neg A) \to (A \to B)$.

7. $\neg\neg\neg D \to \neg D$ is intuitionistically valid.
   Proof: By 4) $D \to \neg\neg D$ is intuitionistically valid. Hence by 6) $\neg\neg\neg D \to \neg D$ is intuitionistically valid. From 5) and 7) it follows that $\neg D$ and $\neg\neg\neg D$ are intuitionistically equivalent; in other words, three negation signs can be reduced to one, four negation signs can be reduced to two, five to one, and so on. However, recall that in general $\neg\neg E$ is not equivalent to E, as we have seen in 3).

8. It is reckless to assert the validity of $(A \to B) \vee (B \to A)$. Explanation: Let G be Goldbach's conjecture and F be another unsolved mathematical problem; then we are neither in the position to assert $F \to G$ nor in the position to assert $G \to F$. Notice that $(A \to B) \vee (B \to A)$ is classically valid; see Exercise 8.7.

In the preceding subsection we have seen how *a different philosophical point of view concerning the nature of mathematical objects, results automatically in a different use of language and in a different logic.*

The classical meaning of, for instance, $A \vee \neg A$, may be rendered intuitionistically by $\neg\neg(A \vee \neg A)$. In Exercise 8.12 we shall give a translation from classical formulas to intuitionistic formulas, preserving the actual meaning of the formulas.

While the classical connectives can all be defined in terms of $\neg$ and any one other (cf. Section 2.5), all three connectives listed above $\rightarrow$, $\wedge$ and $\vee$ are intuitionistically independent. Notice that $A \rightleftarrows B$ can be defined as $(A \rightarrow B) \wedge (B \rightarrow A)$.

In Section 8.2 a proof-theoretic formulation of intuitionistic propositional logic is given; we present a Hilbert-type proof system for intuitionistic propositional logic, consisting of (logical) axioms and one rule Modus Ponens (MP), and a tableaux system which given formulas $A_1, \ldots, A_n$ and B will give either an intuitionistic deduction of B from $A_1, \ldots, A_n$ or a counterexample showing that such a deduction cannot exist. In Section 8.4 we give a model-theoretic description of intuitionistic propositional logic in terms of Kripke models.

In classical propositional logic each formula built by means of connectives from only one atomic formula P is equivalent to either $P \wedge \neg P$, P, $\neg P$ or $P \rightarrow P$ (see Exercise 2.14). Nishimura [12] showed that in intuitionistic propositional logic, however, an infinite number of non-equivalent formulas can be built from only one atomic formula P (see Exercise 8.13). So by the intuitionistic refinement of the propositional language, formulas which are indistinguishable classically (i.e., equivalent in classical propositional logic) become different intuitionistically (i.e., non-equivalent in intuitionistic propositional logic). For instance, the formulas $\neg\neg A \rightarrow B$, $A \rightarrow B$ and $A \rightarrow \neg\neg B$ are classically equivalent, but not intuitionistically. Summarizing: *the language of the intuitionist is richer and more subtle than the language of the classical mathematician.*

Exercise 8.1. For the following pairs of formulas, which can be inferred from which intuitionistically?
(a) $A \rightarrow B$ and $\neg B \rightarrow \neg A$
(b) $\neg(A \wedge B)$ and $\neg A \vee \neg B$
(c) $A \rightarrow B$ and $\neg(A \wedge \neg B)$
(d) $A \rightarrow B$ and $\neg A \vee B$
(e) $A \rightarrow B \vee C$ and $(A \rightarrow B) \vee (A \rightarrow C)$
(f) $A \vee \neg A$ and $\neg\neg A \rightarrow A$

Note that the two formulas in each pair are classically equivalent (and hence classically indistinguishable), but not so intuitionistically. So, in this sense, the intuitionistic language is richer than the classical one.

Exercise 8.2. Verify that the following inferences are intuitionistically correct.

1. If $A \rightarrow C$, then $\neg\neg A \rightarrow \neg\neg C$.
2. If $A \rightarrow (B \rightarrow C)$ and B, then $\neg\neg A \rightarrow \neg\neg C$.
3. If $A \rightarrow (B \rightarrow C)$ and $\neg\neg A$, then $\neg\neg B \rightarrow \neg\neg C$.
4. If $\neg\neg(A \rightarrow B)$, then $\neg\neg A \rightarrow \neg\neg B$.
 (Hint: use 3 with $A \rightarrow B$, A, B instead of A, B, C, respectively.)

5. $\neg\neg A \to \neg\neg B$ iff $\neg\neg(A \to B)$.
6. If $\neg\neg A \land \neg\neg B$, then $\neg\neg(A \land B)$.
   (Hint: use 3 with A, B, $A \land B$ instead of A, B, C, respectively.)
7. $\neg\neg A \land \neg\neg B$ iff $\neg\neg(A \land B)$.
8. If $\neg\neg A \lor \neg\neg B$, then $\neg\neg(A \lor B)$.
9. Show that conversely *not* for all formulas A, B, if $\neg\neg(A \lor B)$, then $\neg\neg A \lor \neg\neg B$ intuitionistically.

Exercise 8.3. Note that if A is decidable, i.e., $A \lor \neg A$ is intuitionistically true, then also $\neg\neg A \to A$ is intuitionistically true.

8.2 Intuitionistic Propositional Logic: Syntax

The alphabet for intuitionistic propositional logic looks the same as the one for classical propositional logic, but the atomic formulas and the connectives now have a constructive interpretation, different from the classical interpretation.

Definition 8.1 (Alphabet). The alphabet for intuitionistic propositional logic consists of the following symbols:
1. $P_1, P_2, P_3, \ldots$, called atomic formulas or propositional variables, to be interpreted as (atomic) propositions.
2. $\to$, $\land$, $\lor$ and $\neg$, called connectives, to be interpreted constructively in terms of proofs.
3. (and), called brackets.

Definition 8.2 (Constructive interpretation of the connectives).
$A \to B$: I have a construction which transforms any proof of A into a proof of B.
$A \land B$: I can construct a proof of A and I can construct a proof of B.
$A \lor B$: I have an algorithm that yields a proof of A or a proof of B.
$\neg A$: $A \to \bot$, where $\bot$ is a special atomic formula, denoting falsity.
$\bot$:   falsity; a proof of this formula implies a proof of any formula.

Definition 8.3 (Formulas). 1. If P is any of the atomic formulas $P_1, P_2, P_3, \ldots$, then P is an (atomic) formula.
2. If A and B are formulas, then $(A \to B)$, $(A \land B)$, $(A \lor B)$ and $(\neg A)$ are (composite) formulas.

We apply the usual convention for leaving out brackets, see Section 2.1.

A proof theoretic formulation of *intuitionistic propositional logic* was given in 1928 by Arend Heyting (1898-1980) and is obtained by replacing axiom schema 8, $\neg\neg A \to A$ of classical logic (see Section 2.6) by axiom schema 8^i: $\neg A \to (A \to B)$. So, the axiom schemata for intuitionistic propositional logic are the following ones:

1. $A \to (B \to A)$
2. $(A \to B) \to ((A \to (B \to C)) \to (A \to C))$

3. $A \to (B \to A \wedge B)$
4a. $A \wedge B \to A$
4b. $A \wedge B \to B$
5a. $A \to A \vee B$
5b. $B \to A \vee B$
6. $(A \to C) \to ((B \to C) \to (A \vee B \to C))$
7. $(A \to B) \to ((A \to \neg B) \to \neg A)$
8^i. $\neg A \to (A \to B)$

In addition, like in classical propositional logic, Modus Ponens $\frac{A \quad A \to B}{B}$, is the only rule of inference for intuitionistic propositional logic.

Definition 8.4 (Intuitionistic Deducibility and Provability).
a) A *deduction* of B from $A_1, \ldots, A_n$ in intuitionistic propositional logic := a finite list of formulas with B as last one, such that each formula in the list is either:
1. one of the premises $A_1, \ldots, A_n$, or
2. an instance of one of the axiom schemata for intuitionistic propositional logic, or
3. obtained from two earlier formulas in the list by an application of Modus Ponens.
b) In case there are no premises $A_1, \ldots, A_n$, we speak of a (formal) *proof* of B.
c) B is *deducible* from $A_1, \ldots, A_n$ in intuitionistic propositional logic := there exists a deduction of B from $A_1, \ldots, A_n$ in intuitionistic propositional logic.
Notation: $A_1, \ldots, A_n \vdash_i B$. By $A_1, \ldots, A_n \nvdash_i B$ we mean: not $A_1, \ldots, A_n \vdash_i B$.
d) B is (formally) *provable* in intuitionistic propositional logic := there is a (formal) proof of B in intuitionistic propositional logic. **Notation**: $\vdash_i B$.

Since the intuitionistic axiom schema 8^i, $\neg A \to (A \to B)$ is provable in classical propositional logic, it follows that all propositional formulas provable intuitionistically are also provable classically; see Exercise 8.4. In order to formally prove $A \vee \neg A$ classically, we proved that $\neg\neg(A \vee \neg A)$ and next used $\neg\neg B \to B$ with $B = A \vee \neg A$. However, such a proof is no longer available in intuitionistic logic. In Section 8.4 we shall show that no intuitionistic formal proof of $A \vee \neg A$ can exist.

Since searching for deductions and proofs in terms of logical axioms and Modus Ponens may be difficult, we shall introduce a tableaux system for intuitionistic propositional logic in Section 8.3. In this system searching for an intuitionistic deduction of a putative conclusion from given premises is straightforward and a systematic search either yields such a deduction or provides us with a counterexample showing that such a deduction cannot exist.

Gentzen's natural deduction rules for *intuitionistic* propositional logic are obtained from Gentzen's natural deduction rules for classical propositional logic (see Section 2.7.2) by leaving out the double negation elimination rule $\frac{\neg\neg A}{A}$.

Exercise 8.4. (i) Show that all formulas provable in intuitionistic propositional logic are also provable in classical propositional logic.
(ii) Show that the *deduction theorem* also holds for intuitionistic propositional logic: if $A_1, \ldots, A_n, A \vdash_i B$, then $A_1, \ldots, A_n \vdash_i A \to B$.

8.3 Tableaux for Intuitionistic Propositional Logic

Exercise 8.5. Show that the deductions in Gentzen's system of natural deduction, found in Exercise 2.60 a i), b i) and c i) are intuitionistically correct, but not so are the deductions found in Exercise 2.60 a ii), b ii) and c ii).

8.3 Tableaux for Intuitionistic Propositional Logic

Definition 8.5 (Signed Formula). A *signed formula* is any expression of the form $T(A)$ or $F(A)$, where A is a formula.

Intuitionistically, $T(A)$ is read as: I have a proof of A; and $F(A)$ as: I do not have a proof of A (which is weaker than 'I have a proof of $\neg A$'!). If no confusion is possible. the brackets may be left out: so, we frequently write TA instead of $T(A)$ and FA instead of $F(A)$.

Definition 8.6 (Sequent). A *sequent* S is a finite set of signed formulas.

A *tableaux system* for *intuitionistic propositional logic* is obtained from the tableaux system in Section 2.8 for classical propositional logic by replacing the tableaux rules $F \to$ and $F \neg$ by

$$F \to_i \quad \frac{S, F\,B \to C}{S_T, TB, FC} \quad \text{and} \quad F \neg_i \quad \frac{S, F\,\neg B}{S_T, TB}.$$

respectively, where $S_T = \{TA \mid TA \in S\}$, i.e., S_T is the set of all T-signed formulas in S. We have drawn a line in the rules $F \to_i$ and $F\neg_i$ in order to stress that in the transition from S to S_T all F-signed formulas in S, if any, are lost.

For the sake of completeness we list here the *tableaux rules for intuitionistic propositional logic* (see also Fitting [5, 6]):

$$T\wedge \quad \frac{S, T\,B \wedge C}{S, TB, TC} \qquad F\wedge \quad \frac{S, F\,B \wedge C}{S, FB \mid S, FC}$$

$$T\vee \quad \frac{S, T\,B \vee C}{S, TB \mid S, TC} \qquad F\vee \quad \frac{S, F\,B \vee C}{S, FB, FC}$$

$$T\to \quad \frac{S, T\,B \to C}{S, FB \mid S, TC} \qquad F\to_i \quad \frac{S, F\,B \to C}{S_T, TB, FC}$$

$$T\neg \quad \frac{S, T\,\neg B}{S, FB} \qquad F\neg_i \quad \frac{S, F\,\neg B}{S_T, TB}$$

Notation: S, TA stands for $S \cup \{TA\}$, i.e., the set containing all signed formulas in S and in addition TA; and S, FA similarly stands for $S \cup \{FA\}$. Instead of $\{TB_1, \ldots, TB_m, FC_1, \ldots, FC_n\}$ we often simply write $TB_1, \ldots, TB_m, FC_1, \ldots, FC_n$. For example, by $\{TD, FE\}, TA$ we mean $\{TD, FE, TA\}$, but we will usually write TD, FE, TA.

Reading the tableaux rules from the top down, rule $F \to_i \dfrac{S, F B \to C}{S_T, TB, FC}$ is interpreted as follows: if I am in a proof-situation (this notion is analogous to the notion of possible world) in which I do not have a proof of $B \to C$, then there is a proof-situation accessible from the given one, in which I do have a proof of B without having a proof of C. The change from S to S_T (where S_T is the set of all T-expressions in S) is explained by noting that formulas of which I did not have a proof in the former situation may have been proved by me in the latter situation, while sentences once proved remain proved forever (an ideal mathematician does not forget); so F-signed formulas in S may not be copied down below the line.

In a similar way there is a shift of one proof-situation to another in the interpretation of rule $F \neg_i \dfrac{S, F \neg B}{S_T, TB}$, while in interpreting the other intuitionistic tableaux rules there is no such shift. For instance, rule

$$T \to \quad \dfrac{S, T B \to C}{S, FB \ | \ S, TC}$$

is read as follows: if I am in a (proof-)situation in which I have a proof of $B \to C$, then in that same (proof-)situation I do not have a proof of B or I do have a proof of C. So the intuitionistic tableaux rules in which there is a shift from S to S_T (the rules which have a bar in it) are precisely the rules the interpretation of which requires a shift from a former to a later proof-situation.

Notice that the rules for intuitionistic propositional logic still have the property that in any application of the rules all T-signed formulas in the upper half of the rule may be repeated in its lower half; because of the rules $F \to_i$ and $F \neg_i$ the same does not hold any more for the F-signed formulas.

Example 8.1. Below is an intuitionistic tableau-deduction of $\neg Q \to \neg P$ from $P \to Q$. The tableau starts with the premises T-signed and the putative conclusion F-signed. Informally, we check the possibility to have a proof of the premises without having a proof of the putative conclusion. Next we apply the tableaux rules and if all possibilities turn out to be closed, i.e., after all to be impossible, then we say that we have a tableau-deduction of the putative conclusion from the given premises.

$$\dfrac{T P \to Q, \ F \neg Q \to \neg P}{T P \to Q, \ T \neg Q, \ F \neg P}$$
$$T P \to Q, \ FQ, \ F \neg P$$
$$\dfrac{FP, \ FQ, \ F \neg P \ | \ TQ, \ FQ, \ F \neg P}{T P \to Q, \ T \neg Q, \ TP \ | \ \text{closure}}$$
$$T P \to Q, \ FQ, \ TP$$
$$FP, \ FQ, \ TP \ | \ TQ, \ FQ, \ TP$$
$$\text{closure} \ | \ \text{closure}$$

So, we start with the supposition that we have a proof of $P \to Q$ without having a proof of $\neg Q \to \neg P$. That might be possible in three different ways, but all of these three turn out to be impossible, in other words give closure. Therefore we shall say that $P \to Q \vdash'_i \neg Q \to \neg P$. The schema above is called a (closed) tableau $\mathcal{T}$ with initial branch $\mathcal{B}_0 = \{T\, P \to Q,\, F\, \neg Q \to \neg P\}$.

Let tableau $\mathcal{T}_0 = \{\mathcal{B}_0\}$ and $\mathcal{B}_1 = \mathcal{B}_0^* \cup \{T\neg Q, F\neg P\}$, where the $*$ in $\mathcal{B}^*$ indicates that only T-signed formulas in $\mathcal{B}$ may count towards closure. Then we call tableau $\mathcal{T}_1 = \{\mathcal{B}_1\}$ a *one-step expansion* of $\mathcal{T}_0$, corresponding with the application of rule $F \to_i$ to $F\ \neg Q \to \neg P$ in $\mathcal{B}_0$. Next let $\mathcal{B}_2 = \mathcal{B}_1 \cup \{FQ\}$, then we call $\mathcal{T}_2 = \{\mathcal{B}_2\}$ a *one-step expansion* of $\mathcal{T}_1$, corresponding with the application of rule $T\neg$ to $T\neg Q$ in $\mathcal{B}_1$. Next let branch $\mathcal{B}_{20} = \mathcal{B}_2 \cup \{FP\}$ and branch $\mathcal{B}_{21} = \mathcal{B}_2 \cup \{TQ\}$, then we call tableau $\mathcal{T}_3 = \{\mathcal{B}_{20}, \mathcal{B}_{21}\}$ a *one-step expansion* of $\mathcal{T}_2$, corresponding with the application of rule $T \to$ to $T\, P \to Q$ in $\mathcal{B}_2$. Let branch $\mathcal{B}_{200} = \mathcal{B}_{20}^* \cup \{TP\}$, then tableau $\mathcal{T}_4 = \{\mathcal{B}_{200}, \mathcal{B}_{21}\}$ is called a *one-step expansion* of $\mathcal{T}_3$, corresponding with the application of rule $F\neg_i$ to $F\neg P$ in $\mathcal{B}_{20}$. Let $\mathcal{B}_{2000} = \mathcal{B}_{200} \cup \{FQ\}$, then $\mathcal{T}_5 = \{\mathcal{B}_{2000}, \mathcal{B}_{21}\}$ is a one step expansion of $\mathcal{T}_4$.

Tableau $\mathcal{T}$ above consists of three branches and is closed since all of its branches are closed, i.e., contain for some formula A both TA and FA. Informally this means that the assumption that it is possible to have a proof of $P \to Q$ without having a proof of $\neg Q \to \neg P$ turns out to be untenable.

Definition 8.7 ((Tableau) Branch). (a) A *tableau branch* is a set of signed formulas. A branch is *closed* if it contains signed formulas TA and FA for some formula A. A branch that is not closed is called *open*.
(b) Let $\mathcal{B}$ be a branch and TA, resp. FA, a signed formula occurring in $\mathcal{B}$. TA, resp. FA, is *fulfilled* in $\mathcal{B}$ if (i) A is atomic, or (ii) $\mathcal{B}$ contains the bottom formulas in the application of the corresponding rule to A, and in case of the rules $T\vee$, $F\wedge$ and $T \to$, $\mathcal{B}$ contains one of the bottom formulas in the application of these rules.
(c) A branch $\mathcal{B}$ is *completed* if $\mathcal{B}$ is closed or every signed formula in $\mathcal{B}$ is fulfilled in $\mathcal{B}$.

Definition 8.8 (Tableau). (a) A set $\mathcal{T}$ of branches is a *tableau* with initial branch $\mathcal{B}_0$ if there is a sequence $\mathcal{T}_0, \mathcal{T}_1, \ldots, \mathcal{T}_n$ such that $\mathcal{T}_0 = \{\mathcal{B}_0\}$, each $\mathcal{T}_{i+1}$ is a one-step expansion of $\mathcal{T}_i$ ($0 \leq i < n$) and $\mathcal{T} = \mathcal{T}_n$.
(b) We say that a finite $\mathcal{B}$ has tableau $\mathcal{T}$ if $\mathcal{T}$ is a tableau with initial branch $\mathcal{B}$.
(c) A tableau $\mathcal{T}$ is *open* if some branch $\mathcal{B}$ in it is open, otherwise $\mathcal{T}$ is *closed*.
(d) A tableau is *completed* if each of its branches is completed, i.e., no application of a tableau rule can change the tableau.

Definition 8.9 (Tableau-deduction/proof). (a) A (logical) *tableau-deduction* of B from $A_1, \ldots, A_n$ (in intuitionistic propositional logic) is a tableau $\mathcal{T}$ with $\mathcal{B}_0 = \{TA_1, \ldots, TA_n, FB\}$ as initial branch, such that all branches of $\mathcal{T}$ are closed.
In case $n = 0$, i.e., there are no premises $A_1, \ldots, A_n$, this definition reduces to:
(b) A (logical) *tableau-proof* of B (in intuitionistic propositional logic) is a tableau $\mathcal{T}$ with $\mathcal{B}_0 = \{FB\}$ as initial sequent, such that all branches of $\mathcal{T}$ are closed.

Definition 8.10 (Tableau-deducible). (a) B is *tableau-deducible* from $A_1, \ldots, A_n$ (in intuitionistic propositional logic) := there exists a tableau-deduction of B from $A_1, \ldots, A_n$. Notation: $A_1, \ldots, A_n \vdash'_i B$. $A_1, \ldots, A_n \not\vdash'_i B$ means: not $A_1, \ldots, A_n \vdash'_i B$.

(b) B is *tableau-provable* (in intuitionistic propositional logic) := there exists a tableau-proof of B. **Notation**: $\vdash'_i B$.

(c) For Γ a (possibly infinite) set of formulas, B is *tableau-deducible from Γ* := there exists a finite list $A_1, \ldots, A_n$ of formulas in Γ such that $A_1, \ldots, A_n \vdash'_i B$.
Notation: $\Gamma \vdash'_i B$.

Example 8.2. We check whether we can show that $\neg Q \to \neg P \vdash'_i P \to Q$ or equivalently $\vdash'_i (\neg Q \to \neg P) \to (P \to Q)$. So, we start a tableau with $T \neg Q \to \neg P, F P \to Q$:

$$\frac{T \neg Q \to \neg P, F P \to Q}{\begin{array}{c} T \neg Q \to \neg P, TP, FQ \\ F \neg Q, TP, FQ \mid T \neg P, TP, FQ \\ F \neg Q, TP, FQ \mid FP, TP, FQ \\ T Q, TP \quad\quad \mid \text{closure} \end{array}}$$

The right branch does close, but the left branch does not; so we did not construct an intuitionistic tableau-deduction of $P \to Q$ from $\neg Q \to \neg P$. From this open branch we shall construct in Section 8.5 an intuitionistic Kripke counterexample in which $\neg Q \to \neg P$ is true, but in which $P \to Q$ is false. Since the intuitionistic proof system is sound, i.e., all formulas that are intuitionistically tableau-provable are also true in all intuitionistic Kripke models (cf. Theorem 8.2), it follows that there does not exist an intuitionistic tableau-deduction of $P \to Q$ from $\neg Q \to \neg P$.

In order to show that the intuitionistic notions of tableau-deducibility (Definition 8.10) and (Hilbert-type) deducibility (Definition 8.4) are equivalent, we first prove Theorem 8.1: if $A_1, \ldots, A_n \vdash'_i B$, then $A_1, \ldots, A_n \vdash_i B$. In Section 8.4 it is shown that the Hilbert type proof system for intuitionistic propositional logic is sound, i.e., if $A_1, \ldots, A_n \vdash_i B$, then also $A_1, \ldots, A_n \models_i B$ (B is an intuitionistically valid (or logical) consequence of $A_1, \ldots, A_n$). And in Section 8.5 we show completeness: if $A_1, \ldots, A_n \models_i B$, then $A_1, \ldots, A_n \vdash'_i B$.

Theorem 8.1. *(i) If B is tableau-deducible from $A_1, \ldots, A_n$ (in intuitionistic propositional logic), i.e., $A_1, \ldots, A_n \vdash'_i B$, then B is deducible from $A_1, \ldots, A_n$ (in intuitionistic propositional logic), i.e., $A_1, \ldots, A_n \vdash_i B$. In particular, for $n = 0$:*
(ii) If $\vdash'_i B$, then $\vdash_i B$.

Proof. Suppose $A_1, \ldots, A_n \vdash'_i B$, i.e., B is intuitionistically tableau-deducible from $A_1, \ldots, A_n$. It suffices to show:

for every sequent $S = \{TD_1, \ldots, TD_k, FE_1, \ldots, FE_m\}$ in an intuitionistic tableau-deduction of B from $A_1, \ldots, A_n$ it holds that $D_1, \ldots, D_k \vdash_i E_1 \vee \ldots \vee E_m$. (*)

Consequently, because $\{TA_1, \ldots, TA_n, FB\}$ is the first (upper) sequent in any given intuitionistic tableau-deduction of B from $A_1, \ldots, A_n$, we have that $A_1, \ldots, A_n \vdash_i B$. The proof of (*) is tedious, but has a simple plan: the statement is true for the closed sequents at the bottom of an intuitionistic tableau-deduction, and the statement remains true if we go up in the intuitionistic tableau-deduction via the T and F rules.
Basic step: Any closed sequent in an intuitionistic tableau-deduction of B from $A_1, \ldots, A_n$ is of the form $\{TD_1, \ldots, TD_k, TP, FP, FE_1, \ldots, FE_m\}$. So, we have to

8.3 Tableaux for Intuitionistic Propositional Logic

show that $D_1, \ldots, D_k, P \vdash_i P \vee E_1 \vee \ldots \vee E_m$. And this is straightforward: $D_1, \ldots, D_k, P \vdash_i P$ and $P \vdash_i P \vee E_1 \vee \ldots \vee E_m$.

Induction step: We have to show that for all rules the following is the case: if (*) holds for all lower sequent(s) in the rule (induction hypothesis), then (*) holds for the upper sequent in the rule. For convenience, we will suppose that $S = \{TD, FE\}$ in all rules.

Rule $T \rightarrow$:
$$\frac{TD, FE, FB \mid TD, FE, TC}{TD, FE, T\,B \rightarrow C}$$

Suppose $D \vdash_i E \vee B$ and $D, C \vdash_i E$ (induction hypothesis). To show: $D, B \rightarrow C \vdash_i E$. Because $E \vee B, B \rightarrow C \vdash_i E \vee C$ (see Exercise 2.50), by the first induction hypothesis,
$D, B \rightarrow C \vdash_i E \vee C$. (1)
From the second induction hypothesis, by the intuitionistic deduction theorem (see Exercise 8.4), $D \vdash_i C \rightarrow E$. (2)
Because $E \vee C, C \rightarrow E \vdash_i E \vee E$ (see Exercise 2.50), it follows from (1) and (2):
$D, B \rightarrow C \vdash_i E \vee E$. But (by $\vee$-elimination) $E \vee E \vdash_i E$. Hence $D, B \rightarrow C \vdash_i E$.

Rule $F \rightarrow_i$:
$$\frac{TD, TB, FC}{TD, FE, F\,B \rightarrow C}$$

Suppose $D, B \vdash_i C$ (induction hypothesis). Then by the (intuitionistic) deduction theorem (see Exercise 8.4), $D \vdash_i B \rightarrow C$ and hence $D \vdash_i E \vee (B \rightarrow C)$, what was to be shown.

The other tableaux rules are treated similarly, see Theorem 2.27. Notice that the proof for rule $F \rightarrow$ with $S = \{TD, FE\}$ instead of $S_T = \{TD\}$ in rule $F \rightarrow_i$, would not intuitionistically hold anymore: from $D, B \vdash_i E \vee C$ it does not follow that $D \vdash_i E \vee (B \rightarrow C)$, since in order to show the latter one needs the assumption $B \vee \neg B$. □

Exercise 8.6. a) Show that all axioms for intuitionistic propositional logic (see Section 8.2) are tableau-provable (in intuitionistic propositional logic). b) verify that:
1) $\vdash'_i A \rightarrow \neg\neg A$, but trying to show $\vdash'_i \neg\neg A \rightarrow A$ fails;
2) $\vdash'_i \neg\neg(A \vee \neg A)$, but trying to show $\vdash'_i A \vee \neg A$ fails;
3) $\neg A \vee B \vdash'_i A \rightarrow B$, but trying to show $A \rightarrow B \vdash'_i \neg A \vee B$ fails.
4) $\neg A \vee \neg B \vdash'_i \neg(A \wedge B)$, but trying to show $\neg(A \wedge B) \vdash'_i \neg A \vee \neg B$ fails.
5) $\vdash'_i \neg\neg\neg A \rightarrow \neg A$.
c) Check that it is not possible to construct an intuitionistic tableau-deduction of B from $A \rightarrow B$ and $\neg A \rightarrow B$; the intuitive reason for this is, of course, that $A \vee \neg A$ does not hold intuitionistically.

Exercise 8.7. Prove: $\vdash' (P \rightarrow Q) \vee (Q \rightarrow P)$ classically, but not intuitionistically.

Exercise 8.8. Prove the *Disjunction Property* for intuitionistic propositional logic: if $\vdash'_i B \vee C$, then $\vdash'_i B$ or $\vdash'_i C$. Notice that the corresponding statement for classical propositional logic does not hold.

Exercise 8.9. Show that the implications in the following diagrams all hold intuitionistically, but not in the converse direction. Note that the formulas in each diagram are classically equivalent.

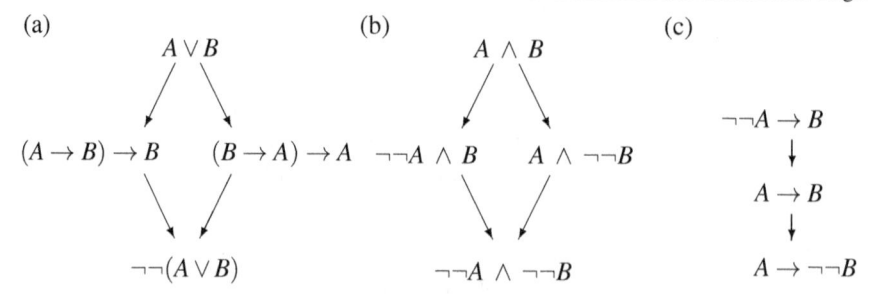

(a)

$A \vee B$
↙ ↘
$(A \to B) \to B$ $(B \to A) \to A$
↘ ↙
$\neg\neg(A \vee B)$

(b)

$A \wedge B$
↙ ↘
$\neg\neg A \wedge B$ $A \wedge \neg\neg B$
↘ ↙
$\neg\neg A \wedge \neg\neg B$

(c)

$\neg\neg A \to B$
↓
$A \to B$
↓
$A \to \neg\neg B$

Exercise 8.10. Show that the following pairs of formulas are intuitionistically equivalent. 1) $\neg\neg A \wedge \neg\neg B$ and $\neg\neg(A \wedge B)$; 2) $\neg\neg A \to \neg\neg B$, $\neg\neg(A \to B)$ and $A \to \neg\neg B$.

A is *stable* := $\vdash'_i \neg\neg A \to A$. So 1) and 2) above say that if both A and B are stable, then $A \wedge B$ and $A \to B$ are stable too.
3) Show that $\neg A$ is stable for each formula A.
4) For $A \veebar B := \neg(\neg A \wedge \neg B)$ show that $A \veebar B$ is stable, if both A and B are stable.
5) Show that $\neg\neg A \vee \neg\neg B$ and $\neg\neg(A \vee B)$ are *not* intuitionistically equivalent and hence we *cannot* conclude: if A and B are stable, then $A \vee B$ is stable too.

Exercise 8.11. A is *decidable* := $\vdash'_i A \vee \neg A$.
1) Prove that $\vdash'_i (A \vee \neg A) \to (\neg\neg A \to A)$. Hence, if A is decidable, then A is stable (see Exercise 8.10).
2) Prove that not $\vdash'_i (\neg\neg A \to A) \to (A \vee \neg A)$. Find a formula B which is stable such that we cannot say that it is decidable. (Hint: see Exercise 8.10, 3.)

Exercise 8.12. Let E^* come from E by replacing (or 'translating') each part of E of the form shown below in the first line by the respective expression shown in the second.
$$\begin{array}{ccccc} P & A \to B & A \wedge B & A \vee B & \neg A \\ \neg\neg P & A \to B & A \wedge B & \neg(\neg A \wedge \neg B) & \neg A \end{array}$$
1. Note that E^* is stable, i.e., $\vdash'_i \neg\neg E^* \to E^*$, for each formula E (see Exercise 8.10).
2. Using $\vdash_i \neg\neg E^* \to E^*$, prove that *classical propositional logic can be defined within the intuitionistic one*: if $A_1, \ldots, A_n \vdash B$ (classically), then $A_1^*, \ldots, A_n^* \vdash_i B^*$.

Exercise 8.13. Show that no two of the formulas $P \to P$, $P \wedge \neg P$, P, $\neg P$, $P \vee \neg P$, $(P \vee \neg P) \to P$ are intuitionistically equivalent. Confer this with the classical case where each formula built from only one atomic formula P is equivalent to either $P \to P$, $P \wedge \neg P$, P or $\neg P$ (see Exercise 2.14). In fact, the formulas mentioned above are just the initial formulas of an infinite list of formulas built from only one atomic formula P such that 1) no two of the formulas in the list are intuitionistically equivalent to each other, and 2) any formula built from only one atomic formula P is equivalent (intuitionistically) to one of the formulas in the list. See I. Nishimura [12].

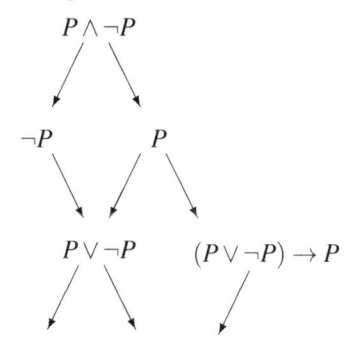

$$P \wedge \neg P$$
$$\swarrow \qquad \searrow$$
$$\neg P \qquad\qquad P$$
$$\swarrow \qquad\qquad \swarrow \searrow$$
$$P \vee \neg P \qquad (P \vee \neg P) \to P$$
$$\swarrow \searrow \qquad\qquad \swarrow$$

Exercise 8.14. * Let A be a formula and let Γ be a set of formulas of intuitionistic propositional logic. $\Gamma \mid A$ is defined by induction on A as follows.
$\Gamma \mid P \qquad := \Gamma \vdash_i P.$
$\Gamma \mid B \vee C \quad := \Gamma \Vdash B$ or $\Gamma \Vdash C$ where $\Gamma \Vdash A := \Gamma \mid A$ and $\Gamma \vdash_i A$.
$\Gamma \mid B \wedge C \quad := \Gamma \mid B$ and $\Gamma \mid C.$
$\Gamma \mid B \to C \;\; := $ if $\Gamma \Vdash B$, then $\Gamma \mid C.$
$\Gamma \mid \neg B \qquad := $ if $\Gamma \Vdash B$, then Γ is inconsistent.
Prove the following *theorem* (S.C. Kleene, 1962): if $\Gamma \mid C$ for all formulas C in Γ and $\Gamma \vdash_i A$, then $\Gamma \mid A$. Conclude the following *corollary*: If $H \mid H$ and $H \vdash_i B \vee C$, then $H \vdash_i B$ or $H \vdash_i C$.

Exercise 8.15. * Prove the following intuitionistic analogue of Theorem 2.18, introduced by S.A. Kripke (oral communication, August, 1977). Every consistent formula A is intuitionistically provably equivalent to a disjunction $A_1 \vee \ldots \vee A_n$, where each A_j, $1 \leq j \leq n$, is a consistent conjunction of atomic formulas, of negations $\neg B$, where not $A_j \vdash_i B$, and of implications $B \to C$, where not $A_j \vdash_i B$; hence for each such A_j, $A_j \mid A_j$ (see Exercise 8.14). Hint: if $A_j = (B \to C) \wedge A'_j$ and $A_j \vdash_i B$, then $\vdash_i A_j \rightleftarrows C \wedge A'_j$.

8.4 Intuitionistic Propositional Logic: Semantics

In this section we define a notion of *intuitionistically (Kripke-)valid consequence*, $A_1, \ldots, A_n \models_i B$, for intuitionistic propositional logic and we shall show that this (semantic) notion of valid consequence for intuitionistic propositional logic is equivalent to the (syntactic) notions of deducibility, $A_1, \ldots, A_n \vdash_i B$ and $A_1, \ldots, A_n \vdash'_i B$, for intuitionistic propositional logic.

In the intuitionistic tableaux rules (see Section 8.3) we interpret TA as: I have a proof of A, and FA as: I do not (yet) have a proof of A. Then, reading these rules from the top down, rule $F \to_i \dfrac{S,\; FB \to C}{S_T,\; TB,\; FC}$ may be interpreted as follows: if I am in a proof-situation (this notion is analogous to the notion of possible world in Chapter 6) in which I do not (yet) have a proof of $B \to C$, then there is a proof-situation accessible from the given one, in which I do have a proof of B without having a proof of C. The change from S to S_T, where S_T is the set of all T-signed formulas in S, is explained by noting that formulas of which I did not have a proof in the former situation may have been proved by me in the latter situation, while sentences once proved remain proved forever (an ideal mathematician does not forget); so F-signed formulas in S may not be copied down below the line.

In a similar way there is a shift of one proof-situation to another in the interpretation of rule $F\neg_i \frac{S, F\neg B}{S_T, TB}$, while in interpreting the other intuitionistic tableaux rules there is no such shift. For instance, rule $T \to_i \frac{S, T\ B \to C}{S, FB \mid S, TC}$ is read as follows: if I am in a (proof-)situation in which I have a proof of $B \to C$, then in that same (proof-)situation I may not have a proof of B or I may have a proof of C. So, the intuitionistic tableaux rules in which there is a shift from S to S_T (the rules which have a bar in it) are precisely the rules the interpretation of which requires a shift from a former to a later proof-situation.

These considerations form the basis for S.A. Kripke's semantics for intuitionistic logic; see Kripke [11]. In fact, this semantics has grown out of the possible world semantics for modal logics (see Chapter 6) developed by Kripke [10] two years earlier in 1963. And although E.W. Beth [2] in 1947 had already developed a semantics for intuitionistic logic which is very close to the later Kripke-semantics, the latter one has become more popular because it is easier to work with.

Definition 8.11 (Kripke model). $M = \langle S, R, \models_i \rangle$ is a *Kripke model* for intuitionistic propositional logic :=
1. S is a non-empty set; the elements of S are called possible proof-situations.
2. R is a binary relation on S, which is *reflexive*, i.e., for all s in S, sRs, and *transitive*, i.e., for all s, s', s'' in S, if sRs' and $s'Rs''$, then sRs''. sRs' is read as: the proof situation s' is accessible from and later in time than the proof-situation s; R is called the *accessibility relation*.
3. $\models_i$ is a relation between the elements of S and the atomic formulas (of intuitionistic propositional logic) such that for all s, s' in S, if $s \models_i P$ and sRs', then $s' \models_i P$. This condition is evident if one reads $s \models_i P$ as: P has been proved in the proof-situation s; once proved, it remains proved.

Definition 8.12 $(M, s \models_i A)$. Given a Kripke model $M = \langle S, R, \models_i \rangle$, we define $M, s \models_i A$, to be read as: A has been proved (or holds) in the situation s of the model M, for arbitrary s in S and for arbitrary formulas A as follows:
$M, s \models_i P \quad := s \models_i P$ (P atomic)
$M, s \models_i B \wedge C := M, s \models_i B$ and $M, s \models_i C$
$M, s \models_i B \vee C := M, s \models_i B$ or $M, s \models_i C$
$M, s \models_i B \to C :=$ for all t in S, if sRt, then not $M, t \models_i B$ or $M, t \models_i C$,
   or, equivalently, if sRt and $M, t \models_i B$, then $M, t \models_i C$,
$M, s \models_i \neg B \quad :=$ for all t in S, if sRt, then not $M, t \models_i B$.

Note that the definition of $M, s \models_i \neg B$ results from the one of $M, s \models_i B \to C$ by identifying $\neg B$ with $B \to \bot$ and postulating that for all s in S, not $M, s \models_i \bot$ ($\bot$ is the so-called false formula).

8.4 Intuitionistic Propositional Logic: Semantics

Lemma 8.1. *Let* $M = \langle S, R, \models_i \rangle$ *be a Kripke model, s and t elements of S and A a formula. If* $M, s \models_i A$ *and* sRt, *then* $M, t \models_i A$.

Proof. For atomic formulas P this follows immediately from condition 3 in Definition 8.11. Now suppose that the lemma has been proved for the formulas B and C (induction hypothesis), i.e., a) if $M, s \models_i B$ and sRt, then $M, t \models_i B$, and b) if $M, s \models_i C$ and sRt, then $M, t \models_i C$.
Then for $A = B \wedge C$ and $A = B \vee C$ the induction proposition follows immediately from the definition of $M, s \models_i A$ and from a) and b).
Now suppose $A = B \to C$, $M, s \models_i B \to C$ and sRt. We have to show that $M, t \models_i B \to C$, i.e., for all t' in S, if tRt' and $M, t' \models_i B$, then $M, t' \models_i C$. So suppose tRt' and $M, t' \models_i B$. Then since sRt and tRt', by the transitivity of R, sRt' and hence, since $M, s \models_i B \to C$ and $M, t' \models_i B$, it follows that $M, t' \models_i C$.
The case $A = \neg B$ is treated similarly. □

Definition 8.13 ($A_1, \ldots, A_n \models_i B$)**.**
1. Let $M = \langle S, R, \models_i \rangle$ be a Kripke model and A a formula. M is an *intuitionistic model* of A (or A *is true (intuitionistically) in* M) := for all s in S, $M, s \models_i A$.
Notation: $M \models_i A$. Otherwise M is called an *intuitionistic countermodel for A* (or an *intuitionistic counterexample to A*). **Notation**: $M \not\models_i A$.
2. A is *intuitionistically (Kripke-)valid* := for all Kripke models M, $M \models_i A$.
Notation: $\models_i A$.
3. B is an *intuitionistically (Kripke-)valid consequence* of $A_1, \ldots, A_n$:= for all Kripke models $M = \langle S, R, \models_i \rangle$ and for all s in S, if for all $j = 1, \ldots, n$, $M, s \models_i A_j$, then $M, s \models_i B$. **Notation**: $A_1, \ldots, A_n \models_i B$.
Note that $A_1, \ldots, A_n \models_i B$ iff $\models_i A_1 \wedge \ldots \wedge A_n \to B$.

Example 8.3. Let $M = \langle \{0, 1\}, R, \models_i \rangle$ be the intuitionistic (Kripke) model with $\{0, 1\}$ being the set consisting of the natural numbers 0 and 1 only, R being defined by $0R0$, $1R1$ and $0R1$ (and not $1R0$), and $\models_i$ being defined by $1 \models_i P$ (and not $0 \models_i P$). This model can be completely characterized by the following picture:

$$0$$
$$\downarrow$$
$$1\ P$$

Note that not $M, 0 \models_i P$ and not $M, 0 \models_i \neg P$ and hence that not $M, 0 \models_i P \vee \neg P$. So M is an intuitionistic counterexample to $P \vee \neg P$ and therefore $\not\models_i P \vee \neg P$. Also notice that for this Kripke model M, for any formula A, $M \models_i A$ iff $M, 0 \models_i A$.
Notice that $M, 0 \not\models_i \neg P$ and $M, 1 \not\models_i \neg P$ and therefore $M, 0 \models_i \neg\neg P$, while $M, 0 \not\models_i P$. So, M is an intuitionistic countermodel to $\neg\neg P \to P$. Hence $\neg\neg P \to P$ is not intuitionistically valid, i.e., $\not\models_i \neg\neg P \to P$.

Example 8.4. Let $M = \langle \{0, 1, 2\}, R, \models_i \rangle$ be the intuitionistic (Kripke) model with $\{0, 1, 2\}$ being the set consisting of the natural number 0, 1 and 2, R being defined by $0R0$, $1R1$, $2R2$; $0R1$, $0R2$ (and not $1R2$, not $2R1$, not $1R0$, not $2R0$), and $\models_i$ being defined by $1 \models_i P$, $2 \models_i Q$ (and not $1 \models_i Q$, not $2 \models_i P$, not $0 \models_i P$, not $0 \models_i Q$). This model can be completely characterized by the following picture:

```
        0
       ↙ ↘
    1 P   2 Q
```

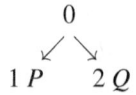

Note that $M,0 \models_i \neg(P \wedge Q)$, but not $M,0 \models_i \neg P$ and not $M,0 \models_i \neg Q$. Hence, $M,0 \not\models_i \neg(P \wedge Q) \rightarrow \neg P \vee \neg Q$. So M is a counterexample to $\neg(P \wedge Q) \rightarrow \neg P \vee \neg Q$ and therefore $\not\models_i \neg(P \wedge Q) \rightarrow \neg P \vee \neg Q$. Also notice that for this Kripke model M, for any formula A, $M \models_i A$ iff $M,0 \models_i A$.

Notice that $M,0 \not\models_i P \rightarrow Q$ and $M,0 \not\models_i Q \rightarrow P$. Therefore, M is an intuitionistic (Kripke) counterexample to $(P \rightarrow Q) \vee (Q \rightarrow P)$. Hence, $\not\models_i (P \rightarrow Q) \vee (Q \rightarrow P)$.

The reader can easily check that, for instance, $P \rightarrow \neg\neg P$ and $\neg P \vee \neg Q \rightarrow \neg(P \wedge Q)$ are Kripke valid, i.e., true in all intuitionistic Kripke models, in particular in the Kripke models of Example 8.3 and 8.4. Let us prove this for the formula $P \rightarrow \neg\neg P$. We have to show that $M,s \models_i P \rightarrow \neg\neg P$ for all Kripke models $M = \langle S, R, \models_i \rangle$ and for all s in S. So suppose sRt and $M,t \models_i P$. Then we must prove that $M,t \models_i \neg\neg P$, i.e., for all t' in S, if tRt', then $M,t' \not\models_i \neg P$. So suppose tRt'. Then, since $M,t \models_i P$, it follows from Lemma 8.1 that $M,t' \models_i P$ and consequently that $M,t' \not\models_i \neg P$. □

In Theorem 8.1 we have shown: if $A_1, \ldots, A_n \vdash'_i B$, then $A_1, \ldots, A_n \vdash_i B$, i.e., any formula which is intuitionistically tableau-deducible from given premises $A_1, \ldots, A_n$ is also intuitionistically deducible from these premises (in terms of the intuitionistic logical axioms and Modus Ponens).

Now we shall show the *soundness theorem for intuitionistic propositional logic*: if $A_1, \ldots, A_n \vdash_i B$, then $A_1, \ldots, A_n \models_i B$, i.e., each formula which is intuitionistically deducible from given premises $A_1, \ldots, A_n$ is also an intuitionistically (Kripke) valid consequence of these premises.

In Section 8.5 we shall close the circle and prove the *completeness of intuitionistic propositional logic*, that is, the intuitionistic logical axioms together with Modus Ponens are complete with respect to the intuitionistic (Kripke) semantics, i.e., if $A_1, \ldots, A_n \models_i B$, then $A_1, \ldots, A_n \vdash'_i B$.

Theorem 8.2 (Soundness). *If* $A_1, \ldots, A_n \vdash_i B$, *then* $A_1, \ldots, A_n \models_i B$.

Proof. It is easy to see that every intuitionistic logical axiom is intuitionistically (Kripke) valid. Let us check the intuitionistic logical axioms $A \rightarrow (B \rightarrow A)$ and $\neg A \rightarrow (A \rightarrow B)$. So, let $M = \langle S, R, \models_i \rangle$ be an intuitionistic (Kripke) model.
1. To show: for all s in S, $M,s \models_i A \rightarrow (B \rightarrow A)$. So, suppose sRt and $M,t \models_i A$ (1). Then we have to show that $M,t \models_i B \rightarrow A$. So suppose tRt' (2) and $M,t' \models_i B$. To show: $M,t' \models_i A$. This follows from (1), (2) and Lemma 8.1.
2. To show: for all s in S, $M,s \models_i \neg A \rightarrow (A \rightarrow B)$. So, suppose sRt and $M,t \models_i \neg A$, i.e., for all t' in S, if tRt', then $M,t' \not\models_i A$. Therefore, for all t' in S with tRt', if $M,t' \models_i A$, then $M,t' \models_i B$, i.e., $M,t \models_i A \rightarrow B$.
3. Next we have to show that Modus Ponens is sound with respect to the intuitionistic Kripke semantics, i.e., if $M,s \models_i A$ (1) and $M,s \models_i A \rightarrow B$ (2), then $M,s \models_i B$. So suppose (1) and (2). We have to show that $M,s \models_i B$. This follows from (1), sRs and (2). □

Exercise 8.16. Prove:
a) If M_1 is a Kripke counterexample to B and M_2 is a Kripke counterexample to C, then from M_1 and M_2 one can construct a Kripke counterexample to $B \vee C$.
b) Conclude from a): if $\models_i B \vee C$ ($B \vee C$ is intuitionistically Kripke valid), then $\models_i B$ or (in its classical meaning) $\models_i C$ (B is Kripke valid or C is Kripke valid).

8.5 Completeness of Intuitionistic Propositional Logic

We shall prove completeness of intuitionistic propositional logic, i.e., that any intuitionistically (Kripke-)valid consequence of given premises may be logically deduced by the intuitionistic tableaux rules from those premises: if $A_1, \ldots, A_n \models_i B$, then $A_1, \ldots, A_n \vdash'_i B$ (Theorem 8.5).

In order to prove completeness of intuitionistic propositional logic, we define a *procedure to construct a counterexample* to a given conjecture that $A_1, \ldots, A_n \vdash'_i B$ with the following property: if the procedure fails, i.e., does not yield a Kripke counterexample, we have in fact constructed an intuitionistic tableau-deduction of B from $A_1, \ldots, A_n$. The procedure makes use of the tableaux rules and produces 'trees' which we shall call *(intuitionistic) search trees*.

Definition 8.14 (Procedure to construct a counterexample). In order to construct a (Kripke-)counterexample to the conjecture that $A_1, \ldots, A_n \vdash'_i B$, we must construct an intuitionistic Kripke model M such that for some proof situation s in M, $M, s \models_i A_1 \wedge \ldots \wedge A_n$, but $M, s \not\models_i B$.

Step 1: Start with $\{TA_1, \ldots, TA_n, FB\}$ and apply all intuitionistic tableaux rules for the propositional connectives, except the rules $F \rightarrow_i$ and $F \neg_i$, as frequently as possible. However, in case one of the split-rules $T \rightarrow$, $T \vee$ and $F \wedge$ is applied, we make two search trees: one with the left split and one with the right split. Notice that for an intuitionistic tableau-deduction both search trees have to close.

For instance, consider the conjecture $\neg(P \wedge Q) \vdash_i \neg P \vee \neg Q$:

search tree (1)	search tree (2)
$T \neg(P \wedge Q)$, $F \neg P \vee \neg Q$	$T \neg(P \wedge Q)$, $F \neg P \vee \neg Q$
$F P \wedge Q$, $F \neg P \vee \neg Q$	$F P \wedge Q$, $F \neg P \vee \neg Q$
$F P \wedge Q$, $F \neg P$, $F \neg Q$	$F P \wedge Q$, $F \neg P$, $F \neg Q$
FP, $F \neg P$, $F \neg Q$	FQ, $F \neg P$, $F \neg Q$

In the transition from the third to the fourth line we apply the rule $F \wedge$ to $F P \wedge Q$, which causes a split. At that stage we make two search trees, one with the left split signed formula FP and one with the right split signed formula FQ. One continues to apply all possible rules, except the $F \neg_i$ and $F \rightarrow_i$ rules, as frequently as possible.

At this stage we have partially constructed one, two (or more) search trees, each consisting of one node labeled with signed formulas. A labeled node s in which all tableaux rules except the $F \neg_i$ and $F \rightarrow_i$ rules have been applied as frequently as possible will be called *logically complete*. Intuitively, this means that one has fully described which formulas have been proved and which formulas have not (yet) been proved in the present proof situation s. Next we continue to expand each search tree by one or more applications of the $F\neg_i$ and $F\to_i$ rules.

Step 2 Each labeled node s in a search tree τ which is logically complete may contain one or more signed formulas of the form $F\ \neg B$ or $F\ B\to C$. For each of the signed formulas of the form $F\ \neg B$ and $F\ B\to C$ in a labeled node s we construct a new node s', declare s' accessible from s in the given search tree τ, i.e., $sR_\tau s'$, and label this node s' with the formulas S_T, FB or S_T, TB, FC respectively, which result from applying the rule $F\neg_i$ to S, $F\ \neg B$ or the rule $F\to_i$ to S, $F\ B\to C$, respectively. It is important to copy all T-signed formulas occurring in s to the new node s' (formulas once proved remain proved). Notice that F-signed formulas that occur in labeled node s may not occur anymore in node s' and that for closure it suffices that one of the successor nodes contains TA and FA for some formula A.

Next we apply step 1 again, but now starting with S_T, FB or S_T, TB, FC, depending on whether rule $F\ \neg_i$ or $F\to_i$ has been applied, resulting in one or more logically complete nodes (proof situations) s'.

Step 1 and 2 are repeated as frequently as possible.

For search tree (1) above one may apply the $F\ \neg_i$ rule to $F\ \neg P$, losing all other F signed formulas, and one may apply the $F\ \neg_i$ rule to $F\ \neg Q$, again losing all other F signed formulas. Similarly for search tree (2) above.

search tree (1)
$T\ \neg(P\wedge Q),\ F\ \neg P\vee\neg Q$
$F\ P\wedge Q,\ F\ \neg P\vee\neg Q$
$F\ P\wedge Q,\ F\ \neg P,\ F\neg Q$
$FP,\ F\ \neg P,\ F\neg Q$
／＼
$T\ \neg(P\wedge Q),\ TP$ $T\ \neg(P\wedge Q),\ TQ$
$F\ P\wedge Q,\ TP$ $F\ P\wedge Q,\ TQ$

search tree (2)
$T\neg(P\wedge Q),\ F\ \neg P\vee\neg Q$
$F\ P\wedge Q,\ F\ \neg P\vee\neg Q$
$F\ P\wedge Q,\ F\ \neg P,\ F\ \neg Q$
$FQ,\ F\ \neg P,\ F\ \neg Q$
／＼
$T\ \neg(P\wedge Q),\ TP$ $T\ \neg(P\wedge Q),\ TQ$
$F\ P\wedge Q,\ TP$ $F\ P\wedge Q,\ TQ$

Application of rule $F\wedge$ to search tree (1) yields four different search trees; three of them are closed, i.e. contain a branch that is closed (a branch is *closed* if it contains TA and FA for some formula A), but one of them, search tree τ below, is not closed:

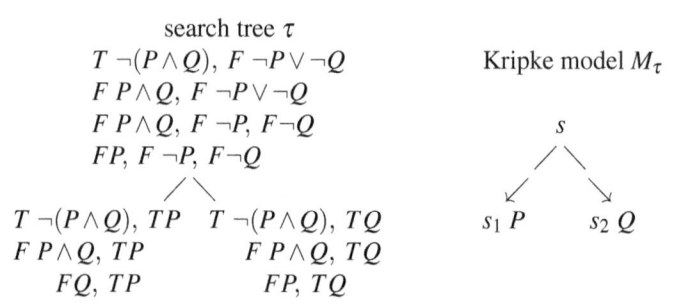

search tree τ
$T\ \neg(P\wedge Q),\ F\ \neg P\vee\neg Q$
$F\ P\wedge Q,\ F\ \neg P\vee\neg Q$
$F\ P\wedge Q,\ F\ \neg P,\ F\neg Q$
$FP,\ F\ \neg P,\ F\neg Q$
／＼
$T\ \neg(P\wedge Q),\ TP$ $T\ \neg(P\wedge Q),\ TQ$
$F\ P\wedge Q,\ TP$ $F\ P\wedge Q,\ TQ$
$FQ,\ TP$ $FP,\ TQ$

Kripke model M_τ

s
／＼
$s_1\ P$ $s_2\ Q$

The open (i.e., not closed) search tree τ yields a (Kripke-)counterexample M_τ to the conjecture that $\neg(P\wedge Q)\vdash'_i \neg P\vee\neg Q$, as depicted in the right column above.

The Kripke model $M_\tau = \langle\{s,s_1,s_2\}, R_\tau, \models_i\rangle$ is defined as follows: $sR_\tau s_1$, $sR_\tau s_2$, $s_1\models_i P$, corresponding with the fact that TP occurs in s_1, and $s_2\models_i Q$ corresponding

8.5 Completeness of Intuitionistic Propositional Logic

with the fact that TQ occurs in s_2. Clearly, $M_\tau, s \models_i \neg(P \land Q)$, corresponding with the fact that $T \neg(P \land Q)$ occurs in s, but $M_\tau, s \not\models_i \neg P$ and $M_\tau, s \not\models_i \neg Q$, corresponding with the fact that $F \neg P$, respectively $F \neg Q$, occurs in s.

Definition 8.15 (Search tree).
A search tree τ for the conjecture $A_1, \ldots, A_n \vdash'_i B$ is a set of nodes, labeled with signed formulas, with a relation R_τ between the nodes, such that:
0. The upper node contains $TA_1, \ldots, TA_n, FB$.
1. $sR_\tau s' := s' = s$ or s' is an immediate successor of s, i.e., s' results from applying the rule $F \neg_i$ or $F \to_i$ to a formula in s of the form $F \neg C$, respectively $F C \to D$.
2. For each node s in the search tree τ:
a) if TP occurs in s and $sR_\tau s'$, then TP occurs in s'.
b) if $T C \to D$ occurs in s, then for all s' in τ, if $sR_\tau s'$, then FC occurs in s' **or** TD occurs in s';
c) if $F C \to D$ occurs in s, then there is a node s' in τ with $sR_\tau s'$, TC occurs in s' **and** FD occurs in s';
d) if $T C \land D$ occurs in s, then TC occurs in s **and** TD occurs in s;
e) if $F C \land D$ occurs in s, then FC occurs in s **or** FD occurs in s;
f) if $T C \lor D$ occurs in s, then TC occurs in s **or** TD occurs in s.
g) if $F C \lor D$ occurs in s, then FC occurs in s **and** FD occurs in s;
h) if $T \neg C$ occurs in s, then for every node s' in τ, if $sR_\tau s'$, then FC occurs in s';
i) if $F \neg C$ occurs in s, then there is a node s' in τ with $sR_\tau s'$ and TC occurs in s'.

Definition 8.16 (Closed branch; closed search tree).
a) A branch in a search tree τ is *closed* if it contains at least one node labeled with TA and FA for some formula A. Otherwise, the branch is called *open*.
b) A search tree τ is *closed* if it contains at least one closed branch. Otherwise, the search tree is called *open*.

Theorem 8.3. *Let τ be an open search tree for the conjecture $A_1, \ldots, A_n \vdash'_i B$ with upper node s_0. Let S_τ the set of nodes in τ and let R_τ be defined as in Definition 8.15. Define $s \models_i P := TP$ occurs in s. Then $M_\tau = \langle S_\tau, R_\tau, \models_i \rangle$ is an (intuitionistic) Kripke countermodel to the conjecture that $A_1, \ldots, A_n \vdash'_i B$. More precisely, $M_\tau, s_0 \models A_1 \land \ldots \land A_n$, but $M_\tau, s_0 \not\models B$.*

Proof. Let τ be an open search tree with s_0 as upper node, containing $TA_1, \ldots, TA_n$, FB. Let $M_\tau = \langle S_\tau, R_\tau, \models_i \rangle$ be the corresponding Kripke model, as defined in the theorem. We shall prove by induction:
1) If TA occurs in s, then $M_\tau, s \models_i A$; and 2) If FA occurs in s, then $M_\tau, s \not\models_i A$.
Since $TA_1, \ldots, TA_n, FB$ occur in the top node s_0, it follows that $M_\tau, s_0 \models_i A_1 \land \ldots \land A_n$, but $M_\tau, s_0 \not\models_i B$. Therefore, $A_1, \ldots, A_n \not\models_i B$.
Induction basis Let $A = P$ be atomic. If TP occurs in s, then by definition $s \models_i P$, i.e., $M_\tau, s \models_i P$. If FP occurs in s, then - since τ is open - TP does not occur in s and hence by definition $s \not\models_i P$, i.e., $M_\tau, s \not\models_i P$.
Induction step Suppose 1) and 2) hold for C and D (induction hypothesis). We shall prove that 1) and 2) hold for $C \to D$, $C \land D$, $C \lor D$ and $\neg C$.
Let $A = C \to D$ and suppose $T C \to D$ occurs in s. Then according to Definition

8.15 b, for all s' in τ, if $sR_\tau s'$, then FC is in s' or TD is in s'. So, by the induction hypothesis, for all s' in τ, if $sR_\tau s'$, then $M_\tau, s' \not\models_i C$ or $M_\tau, s' \models_i D$. Consequently, $M_\tau, s \models_i C \to D$.

Let $A = C \to D$ and suppose $F\ C \to D$ occurs in s. Then according to Definition 8.15 c, there is a node s' in τ with $sR_\tau s'$ such that TC is in s' and FD is in s'. So, by the induction hypothesis, $M_\tau, s' \models_i C$ and $M_\tau, s' \not\models_i D$. Consequently, $M_\tau, s \not\models C \to D$. The cases that $A = C \wedge D$, $A = C \vee D$ and $A = \neg C$ are treated similarly. □

Example 8.5. We wonder whether $\vdash'_i P \vee \neg P$. So, in the left column below we start a search tree τ beginning with $F\ P \vee \neg P$:

$$
\begin{array}{cc}
F\ P \vee \neg P & s \\
FP, F\neg P & | \\
| & \downarrow \\
TP & s'\ P
\end{array}
$$

In the application of rule $F\neg_i$ to $F\neg P$ the F-signed formulas are not be copied to the next proof-situation. We do not find a tableau-proof of $P \vee \neg P$. Instead we have actually constructed a search tree τ for $P \vee \neg P$ which is open, i.e., no node in it contains both TB and FB for some formula B; and from this open search tree one can read off an intuitionistic Kripke counterexample to $P \vee \neg P$, as is shown in the right-column above. Confer Example 8.3.

Example 8.6. We wonder whether $\vdash'_i (P \to Q) \vee (Q \to P)$. So, in the left column below we start a search tree τ beginning with $F\ (P \to Q) \vee (Q \to P)$:

$$
\begin{array}{cc}
F\ (P \to Q) \vee (Q \to P) & s \\
FP \to Q, F\ Q \to P & \\
\diagup \diagdown & \diagup \diagdown \\
TP, FQ \quad TQ, FP & s_1\ P \quad s_2\ Q
\end{array}
$$

At the second line of the search tree in the left column above, we can apply rule $F \to_i$ to $F\ P \to Q$ losing the expression $F\ Q \to P$ or apply rule $F \to_i$ to $F\ Q \to P$ losing the expression $F\ P \to Q$. In neither case we find a tableau-proof of $(P \to Q) \vee (Q \to P)$. Instead we have actually constructed a search tree for $(P \to Q) \vee (Q \to P)$ which is open, i.e., no node in it contains both TB and FB for some formula B; and from this open search tree one can read off a Kripke counterexample to $(P \to Q) \vee (Q \to P)$. Confer Example 8.4.

Example 8.7. We wonder whether $P \vee \neg Q \vdash'_i (R \to P) \vee (Q \to R)$. Application of the procedure to construct a counterexample to this conjecture yields two different search trees which both turn out to be closed.

$$
\begin{array}{ll}
T\ P \vee \neg Q,\ F\ (R \to P) \vee (Q \to R) & T\ P \vee \neg Q,\ F\ (R \to P) \vee (Q \to R) \\
T\ P \vee \neg Q,\ F\ R \to P,\ F\ Q \to R & T\ P \vee \neg Q,\ F\ R \to P,\ F\ Q \to R \\
TP,\ T\ P \vee \neg Q,\ F\ R \to P,\ F\ Q \to R & T\neg Q,\ T\ P \vee \neg Q,\ F\ R \to P,\ F\ Q \to R \\
 & FQ,\ T\ P \vee \neg Q,\ F\ R \to P,\ F\ Q \to R
\end{array}
$$

$$
\begin{array}{ll}
\diagup \diagdown & \diagup \diagdown \\
TP,\ TR,\ FP \quad TP,\ TQ,\ FR & T\neg Q,\ TR,\ FP \quad T\neg Q,\ TQ,\ FR \\
\text{closed} & FQ,\ TR,\ FP \quad FQ,\ TQ,\ FR \\
 & \text{closed}
\end{array}
$$

8.5 Completeness of Intuitionistic Propositional Logic

Note that the two branches which give closure together yield an intuitionistic tableau-deduction of $(R \to P) \vee (Q \to R)$ from $P \vee \neg Q$:

$$T\ P \vee \neg Q,\ F\ (R \to P) \vee (Q \to R)$$
$$T\ P \vee \neg Q,\ F\ R \to P,\ F\ Q \to R$$

$T\ P \vee \neg Q,\ TP,\ \underline{F\ R \to P},\ F\ Q \to R$	$T\ P \vee \neg Q,\ T\ \neg Q,\ F\ R \to P,\ \underline{F\ Q \to R}$
$T\ P \vee \neg Q,\ TP,\ TR,\ FP$	$T\ P \vee \neg Q,\ T\ \neg Q,\ TQ,\ FR$
closure	$T\ P \vee \neg Q,\ FQ,\ TQ,\ FR$
	closure

The correctness of this remark is not accidental and follows immediately from the definition of a tableau-deduction and the structure of the procedure in Definition 8.14 to construct a counterexample.

Theorem 8.4. *If all search trees for the conjecture $A_1, \ldots, A_n \vdash'_i B$ are closed, i.e., contain closure in one of their branches, then $A_1, \ldots, A_n \vdash'_i B$.*

Proof. Suppose all search trees for the conjecture $A_1, \ldots, A_n \vdash'_i B$ are closed. Then it follows from the procedure to construct a counterexample to this conjecture that the closed branches together form an intuitionistic tableau-deduction of B from $A_1, \ldots, A_n$. □

Theorem 8.5 (Completeness). *If $A_1, \ldots, A_n \models_i B$, then $A_1, \ldots, A_n \vdash'_i B$.*

Proof. Suppose $A_1, \ldots, A_n \models_i B$. Apply the procedure to construct a counterexample to the conjecture $A_1, \ldots, A_n \vdash'_i B$. If one of the resulting search trees is open, say τ, then by Theorem 8.3, $M_\tau, s_0 \models_i A_1 \wedge \ldots \wedge A_n$, while $M_\tau, s_0 \not\models_i B$. This contradicts the assumption $A_1, \ldots, A_n \models_i B$. Hence, there can be no open search tree for the conjecture $A_1, \ldots, A_n \vdash'_i B$. That is, all search trees for this conjecture are closed. So, by Theorem 8.4, $A_1, \ldots, A_n \vdash'_i B$. □

The procedure to construct a counterexample to the conjecture $A_1, \ldots, A_n \vdash'_i B$ will stop after finitely many steps and then either yield an intuitionistic Kripke-counterexample or an intuitionistic tableau-deduction of B from $A_1, \ldots, A_n$. Therefore, intuitionistic propositional logic is *decidable*.

Theorem 8.6 (Decidability). *Intuitionistic propositional logic is decidable, i.e., there is a procedure to decide in finitely many steps whether $A_1, \ldots, A_n \vdash'_i B$.*

Proof. Given a conjecture $A_1, \ldots, A_n \vdash'_i B$, the procedure (in Definition 8.14) to construct a counterexample yields only finitely many different search trees, each of which will be completed in a finite number of steps. If they all close, by Theorem 8.4 they actually give us a tableau-deduction of B from $A_1, \ldots, A_n$, showing that $A_1, \ldots, A_n \vdash'_i B$ and hence $A_1, \ldots, A_n \models_i B$; if one of them is open, by Theorem 8.3 it actually yields a Kripke counterexample to the given conjecture, showing that $A_1, \ldots, A_n \not\models_i B$ and hence $A_1, \ldots, A_n \not\vdash'_i B$. □

Note that while the decision procedure for $A_1, \ldots, A_n \models B$ in classical propositional logic was immediately evident from the definition of $A_1, \ldots, A_n \models B$ (the truth table of B has 1 in all lines in which all of $A_1, \ldots, A_n$ are 1), this is not so for the notion of Kripke valid consequence ($A_1, \ldots, A_n \models_i B$) in intuitionistic propositional logic.

Exercise 8.17. Construct either an intuitionistic tableau-proof or an intuitionistic Kripke counterexample for the following formulas (confer Exercise 8.1):
(a) $(P \to Q) \to (\neg Q \to \neg P)$ and $(\neg Q \to \neg P) \to (P \to Q)$.
(b) $\neg(P \wedge Q) \to (\neg P \vee \neg Q)$ and $(\neg P \vee \neg Q) \to \neg(P \wedge Q)$.
(c) $(P \to Q) \to \neg(P \wedge \neg Q)$ and $\neg(P \wedge \neg Q) \to (P \to Q)$.
(d) $(P \to Q) \to (\neg P \vee Q)$ and $(\neg P \vee Q) \to (P \to Q)$.
(e) $(P \to Q \vee R) \to (P \to Q) \vee (P \to R)$ and $(P \to Q) \vee (P \to R) \to (P \to Q \vee R)$.
(f) $(P \vee \neg P) \to (\neg\neg P \to P)$ and $(\neg\neg P \to P) \to (P \vee \neg P)$.

Exercise 8.18. * (S.A. Kripke, oral communication, 1977)
Let $M = \langle S, R, \models_i \rangle$ be defined as follows: S is the set of all formulas A of intuitionistic propositional logic such that $A \mid A$ (see Exercise 8.14). For H, H' in S, let $HRH' := H' \vdash_i H$. For H in S let $H \models_i P := H \vdash_i P$ (P atomic).
Verify that $M = \langle S, R, \models_i \rangle$ is a Kripke model for intuitionistic propositional logic such that for all formulas A, $M, H \models_i A$ iff $H \vdash_i A$. Hint: use Exercise 8.14 and 8.15. As a corollary we have the *completeness theorem* for intuitionistic propositional logic: if $\models_i A$, then $\vdash_i A$.

8.6 Quantifiers in Intuitionism; Intuitionistic Predicate Logic

We have already seen in Section 8.1 that in classical mathematics the infinite – for instance, the set $\mathbb{N}$ of the natural numbers – is treated as actual or completed. On the other hand, since for an intuitionist mathematical objects are mental constructions, the set $\mathbb{N}$ of the natural numbers intuitionistically cannot be regarded as a completed totality, but only as potential or becoming or constructive. As explained in Section 8.1, the set $\mathbb{N}$ of the natural numbers is intuitionistically a construction project: start with 0 and for every natural number n already constructed earlier construct $n + 1$. As a consequence the quantifiers have intuitionistically a meaning different from the classical one.

The classical meaning of *there exists an n such that P(n)* ($\exists n[n \in \mathbb{N} \wedge P(n)]$), that somewhere in the completed infinite totality of the natural numbers there occurs an n such that $P(n)$, is not available to the intuitionist, since he does not conceive the set $\mathbb{N}$ of the natural numbers as a completed totality. The intuitionistic meaning of the proposition *there exists a natural number n such that P(n)* is that one can *construct* a natural number n which one can prove has the property P. So, an intuitionistic proof of the proposition in question must be *constructive*, i.e., it must indicate a concrete natural number with the property P, or at least indicate a method by which one can construct such a natural number.

The intuitionistic meaning of *all natural numbers n have the property P* ($\forall n[n \in \mathbb{N} \to P(n)]$), or briefly *for all n, P(n)*, is the following: I have a method (construction), which applied to any natural number n provides a proof of $P(n)$. Note that the classical concept of a completed infinity of the natural numbers does not occur in this intuitionistic interpretation of a universal quantification over the natural numbers.

8.6 Quantifiers in Intuitionism; Intuitionistic Predicate Logic

Propositions of the form *for all natural numbers n, P(n)* may be proved intuitionistically by using the principle of mathematical induction: if (1) $P(0)$ and (2) for all $n \in \mathbb{N}$, if $P(n)$, then $P(n+1)$, then for all n, $P(n)$. In order to arrive at an intuitionistic proof of the proposition *for all natural numbers n, P(n)*, the proofs of both the induction basis (1) and the induction step (2) should, of course, be intuitionistic too.

That intuitionistic methods are to be distinguished from non-intuitionistic ones is explained by S.C. Kleene [8] as follows:

> In classical mathematics there occur *non-constructive* or *indirect* existence proofs, which the intuitionists do not accept. For example, to prove *there exists an n such that P(n)*, the classical mathematician may deduce a contradiction from the assumption *for all n, not P(n)*. Under both the classical and the intuitionistic logic, by reductio ab absurdum this gives *not for all n, not P(n)*. The classical logic allows this result to be transformed into *there exists an n such that P(n)*, but not (in general) the intuitionistic. Such a classical existence proof leaves us no nearer than before the proof was given to having an example of a number n such that $P(n)$ (though sometimes we may afterwards be able to discover one by another method). [S.C. Kleene [8], p. 49]

Intuitionistic methods are to be distinguished from non-intuitionistic ones not only in the case of proofs, but also in the case of definitions. For instance, suppose one can show that the number 3 has a given property P if Goldbach's conjecture (G) is true and that if Goldbach's conjecture (G) is false, then the number 5 has the property P. From a classical point of view one may say that one has shown the existence of a natural number n with the property P. But because Goldbach's conjecture is an open, i.e., not solved, problem, from an intuitionistic point of view no construction of such a natural number n has been given. Neither 3 nor 5 is an example as long as Goldbach's conjecture has not been solved. From an intuitionistic point of view one has only proved the implication *if G or not G, then there exists an n such that P(n)*, where G is Goldbach's conjecture. From a classical point of view, the premiss $G \vee \neg G$ of this implication is available and consequently from a classical point of view one may infer its conclusion that there is a natural number n with the property P. However, in the present state of knowledge, an intuitionist does not accept the principle *G or not G* and hence *the number n which is equal to 3 if G, and equal to 5 if not G* is intuitionistically not a valid definition of a natural number n: one has no method to construct this natural number.

We have just seen that the quantifiers in intuitionism have a meaning quite different from their classical one. Let V be an (intuitionistic) set. $\forall x \in V[A(x)]$ (for every x in V, $A(x)$) means intuitionistically: I have a construction which assigns to each object a in V a proof of $A(a)$. And $\exists x \in V[A(x)]$ (for some x in V, $A(x)$) means intuitionistically: I can construct an object a in V and give a proof(-construction) of $A(a)$.

The reader should verify for himself that for the intuitionistic quantifiers we still have $\neg \exists x \in V[A(x)] \rightleftarrows \forall x \in V[\neg A(x)]$, and also $\exists x \in V[\neg A(x)] \rightarrow \neg \forall x \in V[A(x)]$, but not anymore the converse, $\neg \forall x \in V[A(x)] \rightarrow \exists x \in V[\neg A(x)]$: from the assumption that $\neg \forall x \in V[A(x)]$, i.e., that $\forall x \in V[A(x)]$ yields a contradiction, one can in general not construct a particular element x in V such that $\neg A(x)$. An intuitionistic (weak) counterexample to $\neg \forall x \in V[A(x)] \rightarrow \exists x \in V[\neg A(x)]$ is obtained as follows:

Let $C(k) :=$ in the decimal expansion of π a sequence of the form 0 1 2 3 4 5 6 7 8 9 occurs before the kth decimal. And let $\rho = 0. \, a_1 \, a_2 \, a_3 \ldots$, where $a_k = 3$ if $\neg C(k)$, and $a_k = 0$ if $C(k)$. Then if $\rho \neq \frac{1}{3}$, i.e., $\rho \neq 0.333\ldots$, then $\neg \forall k \in \mathbb{N}[\neg C(k)]$, and if $\rho \neq 0.33\ldots 0\,0\,0\ldots$, then $\neg \exists k \in \mathbb{N}[C(k)]$.

Let $\mathbb{Q}$ be the set of all rational numbers. Then
(1) $\neg \forall x \in \mathbb{Q}[x \neq \rho]$, i.e., it is not the case that ρ is irrational; for if ρ is irrational, then both $\rho \neq \frac{1}{3}$ and $\rho \neq 0.33\ldots 000\ldots$ and therefore both $\neg \forall k \in \mathbb{N}[\neg C(k)]$ and $\neg \exists k \in \mathbb{N}[C(k)]$ or equivalently $\forall k \in \mathbb{N}[\neg C(k)]$; contradiction.
(2) But it is reckless to assume that $\exists x \in \mathbb{Q}[\neg(x \neq \rho)]$: for indicating a rational number equal to ρ implies $\forall k \in \mathbb{N}[\neg C(k)] \vee \exists k \in \mathbb{N}[C(k)]$, or equivalently, $\neg \exists k \in \mathbb{N}[C(k)] \vee \exists k \in \mathbb{N}[C(k)]$, which clearly is a reckless statement since it states that I know whether in the decimal expansion of π a sequence of the form 0 1 2 3 4 5 6 7 8 9 occurs or not.

From (1) and (2) it follows that $\neg \forall x \in \mathbb{Q}[x \neq \rho] \to \exists x \in \mathbb{Q}[\neg(x \neq \rho)]$ is reckless. Note that $\neg \forall x \in V[A(x)] \to \exists x \in V[\neg A(x)]$ is a generalization of $\neg(P \wedge Q) \to \neg P \vee \neg Q$, of which we have already seen in Section 8.5 that it was intuitionistically reckless.

We are not able to give an intuitionistic proof of $\forall x \in V[\neg \neg A(x)] \to \neg \neg \forall x \in V[A(x)]$. In Section 8.7 we shall show that this formula does not hold intuitionistically in the case that $V = \{0,1\}^{\mathbb{N}}$: Let $A(\alpha)$ express that α is the infinite sequence consisting of only 0's, which we denote by $\alpha = \underline{0}$, i.e., $\forall n \in \mathbb{N}[\alpha(n) = 0]$. Then $\forall \alpha \in \{0,1\}^{\mathbb{N}}[\neg \neg (\alpha = \underline{0} \vee \alpha \neq \underline{0})]$, but $\neg \forall \alpha \in \{0,1\}^{\mathbb{N}}[\alpha = \underline{0} \vee \alpha \neq \underline{0}]$.
However, in the case that $V = \mathbb{N}$, whether $\forall x \in \mathbb{N}[\neg \neg A(x)] \to \neg \neg \forall x \in \mathbb{N}[A(x)]$ holds intuitionistically or not is still an open problem.

8.6.1 Deducibility for Intuitionistic Predicate Logic

The language of intuitionistic predicate logic is the same as the one for classical predicate logic (see Chapter 4), the difference being that the connectives and quantifiers now have another, constructive and intuitionistic meaning. For the sake of completeness we give the alphabet of predicate logic and mention the rules according to which the well formed expressions or formulas of predicate logic are formed.

Definition 8.17 (Alphabet for Predicate Logic).
The alphabet for (intuitionistic) predicate logic consists of the following symbols:
individual constants: $c_1, c_2, c_3, \ldots$
predicate symbols: $P_1, P_2, P_3, \ldots$, where P_i is supposed to be n_i-ary, i.e., taking n_i arguments.
free individual variables: $a_1, a_2, a_3, \ldots$; bound individual variables: $x_1, x_2, x_3, \ldots$
connectives: $\rightleftarrows, \to, \wedge, \vee, \neg$; quantifiers: $\exists, \forall$
brackets: (,), [,]

We shall use a, b to range over free individual variables, x, y, z to range over bound individual variables, and P, Q, R to range over predicate symbols.

8.6 Quantifiers in Intuitionism; Intuitionistic Predicate Logic

Definition 8.18 (Formulas of Predicate Logic).
If P is a n-ary predicate symbol and $t_1, \ldots, t_n$ are terms, i.e., individual constants or free individual variables, then $P(t_1, \ldots, t_n)$ is an (atomic) formula.
If B and C are formulas, then also $(B \rightleftarrows C)$, $(B \rightarrow C)$, $(B \wedge C)$, $(B \vee C)$ and $(\neg B)$ are formulas.
If $A(a)$ is a formula containing the free variable a, then $\forall x[A(x)]$ and $\exists x[A(x)]$ are formulas, where $A(x)$ results from $A(a)$ by replacing all occurrences of a by x.

A *Hilbert-type* proof-theoretic formulation of intuitionistic predicate logic is obtained by replacing axiom 8, $\neg\neg A \rightarrow A$ for classical predicate logic (see Section 4.4) by axiom 8^i, $\neg A \rightarrow (A \rightarrow B)$ (see Section 8.2). The other axioms and rules for the connectives and quantifiers remain unchanged and the reader should verify intuitively that they all hold true for the intuitionistic interpretation of $\rightleftarrows$, $\rightarrow$, $\wedge$, $\vee$, $\neg$, $\forall$ and $\exists$.

Definition 8.19 (Axioms and Rules for Intuitionistic Predicate Logic).
The (logical) axioms and rules for intuitionistic predicate logic consist of:
1. the axiom schemata for intuitionistic propositional logic, given in Section 8.2.
2. the axiom schemata for the quantifiers: $\forall x[A(x)] \rightarrow A(t)$ and $A(t) \rightarrow \exists x[A(x)]$, where t is a term, i.e., an individual constant or free individual variable.
3. the logical rules for $\rightarrow$, $\forall$ and $\exists$: $\dfrac{A \quad A \rightarrow B}{B}$ (MP), $\dfrac{C \rightarrow A(a)}{C \rightarrow \forall x[A(x)]}$, $\dfrac{A(a) \rightarrow C}{\exists x[A(x)] \rightarrow C}$, provided C does not contain a.

Definition 8.20 (Deduction; Deducible).
1. An intuitionistic (Hilbert-type) *deduction* of B from $A_1, \ldots, A_n$ (in predicate logic) is a finite list $B_1, \ldots, B_k$ of formulas, such that
(a) $B = B_k$ is the last formula in the list, and
(b) each formula in the list is either one of $A_1, \ldots, A_n$, or an axiom of intuitionistic predicate logic (i.e., an instance of one of the axiom schemata), or is obtained by application of one of the rules to formulas preceding it in the list, such that
(c) *all free variables of $A_1, \ldots, A_n$ are held constant*, i.e., the $\forall$-rule and the $\exists$-rule are not applied with respect to a free variable a occurring in $A_1, \ldots, A_n$, except preceding the first occurrence of $A_1, \ldots, A_n$ in the deduction.
2. B is intuitionistically *deducible from* $A_1, \ldots, A_n$:= there exists an intuitionistic (Hilbert-type) deduction of B from $A_1, \ldots, A_n$. **Notation**: $A_1, \ldots, A_n \vdash_i B$. The symbol $\vdash_i$ may be read 'yields intuitionistically'. $A_1, \ldots, A_n \nvdash_i B$ abbreviates: not $A_1, \ldots, A_n \vdash_i B$.
3. For Γ a (possibly infinite) set of formulas, B is intuitionistically *deducible from Γ* := there is a finite list $A_1, \ldots, A_n$ of formulas in Γ such that $A_1, \ldots, A_n \vdash_i B$. **Notation**: $\Gamma \vdash_i B$.

Example 8.8. $\exists x[\neg A(x)] \vdash_i \neg\forall x[A(x)]$. Proof: $\forall x[A(x)] \rightarrow A(a)$ is an (intuitionistic) axiom. Hence, $\vdash_i \neg A(a) \rightarrow \neg\forall x[A(x)]$, and hence, by application of the rule for $\exists$, $\vdash_i \exists x[\neg A(x)] \rightarrow \neg\forall x[A(x)]$. Consequently, $\exists x[\neg A(x)] \vdash_i \neg\forall x[A(x)]$.
However, conversely, the classical deduction of $\exists x[\neg A(x)]$ from $\neg\forall x[A(x)]$ is not intuitionistically valid: $A(a) \rightarrow \exists x[A(x)]$ is an axiom. Consequently, $\vdash \neg\exists x[A(x)] \rightarrow$

$\neg A(a)$ and hence $\vdash \neg \exists x[A(x)] \to \forall x[\neg A(x)]$ and $\vdash \neg \forall x[\neg A(x)] \to \neg\neg \exists x[A(x)]$. Now classically, but not intuitionistically, one may remove the double negation signs, obtaining $\vdash \neg \forall x[\neg A(x)] \to \exists x[A(x)]$ classically, but not intuitionistically.

Without proof we mention here the following generalization of Exercise 8.12, that *classical predicate logic can be defined within intuitionistic predicate logic*. For a proof of this theorem we refer the reader to S.C. Kleene [8], Section 82.

Theorem 8.7. *If $A_1, \ldots, A_n \vdash B$ (classically), then $A_1^*, \ldots, A_n^* \vdash_i B^*$ (intuitionistically), where A^* is defined as follows: For A atomic, $A^* := \neg\neg A$.*

$(B \wedge C)^* := B^* \wedge C^*$ $(\forall x[B(x)])^* := \forall x[B(x)^*]$,
$(B \vee C)^* := \neg(\neg B^* \wedge \neg C^*)$ $(\exists x[B(x)])^* := \neg \forall x[\neg B(x)^*]$,
$(A \to B)^* := A^* \to B^*$ $(\neg A)^* := \neg A^*$.

8.6.2 Tableaux for Intuitionistic Predicate Logic

A *tableaux* system for intuitionistic predicate logic is obtained by replacing the rules $F \to$, $F \neg$ and $F \forall$ for classical predicate logic (see Subsection 4.4.2) by the rules $F \to_i$, $F \neg_i$ and $F \forall_i$:

$$F \to_i \frac{S,\ F\ B \to C}{S_T,\ TB, FC} \qquad F \neg_i \frac{S,\ F\ \neg B}{S_T,\ TB} \qquad F \forall_i \frac{S,\ F\ \forall x[A(x)]}{S_T,\ FA(a)}$$

where a is new, i.e., does not occur in S, $F \forall x[A(x)]$, and where $S_T = \{TB \mid TB \in S\}$ is the set of all T-expressions in S. So, the tableaux rules for intuitionistic predicate logic are obtained by adding to the tableaux rules for the intuitionistic connectives, presented in Section 8.3, the intuitionistic tableaux rules for the quantifiers:

$T \exists$ $\dfrac{S,\ T\ \exists x[A(x)]}{S,\ TA(a)}$ $F \exists$ $\dfrac{S,\ F\ \exists x[A(x)]}{S,\ F\ \exists x[A(x)],\ FA(t)}$

a new: a does not occur in S, $T\exists x[A(x)]$ t being any term

$T \forall$ $\dfrac{S,\ T\ \forall x[A(x)]}{S,\ T\ \forall x[A(x)],\ TA(t)}$ $F \forall_i$ $\dfrac{S,\ F\ \forall x[A(x)]}{S_T,\ FA(a)}$

t being any term a new: a does not occur in S, $F\ \forall x[A(x)]$

The definitions of *an intuitionistic tableau-deduction of B from $A_1, \ldots, A_n$* and of *B is intuitionistically tableau-deducible from $A_1, \ldots, A_n$*, denoted by $A_1, \ldots, A_n \vdash'_i B$, are similar to the ones given in Section 8.3.

Example 8.9. $\exists x[\neg A(x)] \vdash'_i \neg \forall x[A(x)]$. Proof: the tableau in the left column below is an intuitionistic tableau-deduction of $\neg \forall x[A(x)]$ from $\exists x[\neg A(x)]$:

$T\ \exists x[\neg A(x)],\ F\ \neg \forall x[A(x)]$ $T\ \neg \forall x[A(x)],\ F\ \exists x[\neg A(x)]$
$T\ \exists x[\neg A(x)],\ T\ \forall x[A(x)]$ $F\ \forall x[A(x)],\ F\ \exists x[\neg A(x)]$
$T\ \neg A(a_1),\ T\ \forall x[A(x)]$ $F\ \forall x[A(x)],\ F\ \neg A(a_1)$
$F\ A(a_1),\ T\ \forall x[A(x)]$ $F\ \forall x[A(x)],\ T\ A(a_1)$
$F\ A(a_1),\ T\ A(a_1)$ $F\ A(a_2),\ T\ A(a_1)$
closure   no closure

8.6 Quantifiers in Intuitionism; Intuitionistic Predicate Logic 407

But, conversely, we do not find a tableau-deduction of $\exists x[\neg A(x)]$ from $\neg \forall x[A(x)]$: the tableau in the right column above fails to be an intuitionistic tableau deduction of $\exists x[\neg A(x)]$ from $\neg \forall x[A(x)]$.

Without proof we mention that Theorem 8.1 for intuitionistic propositional logic can be generalized to intuitionistic predicate logic.

Theorem 8.8. *(i) If B is tableau-deducible from $A_1, \ldots, A_n$ (in intuitionistic predicate logic), i.e., $A_1, \ldots, A_n \vdash'_i B$, then B is deducible from $A_1, \ldots, A_n$ (in intuitionistic predicate logic), i.e., $A_1, \ldots, A_n \vdash_i B$. In particular, for $n = 0$:*
(ii) If $\vdash'_i B$, then $\vdash_i B$.

8.6.3 Kripke Semantics for Intuitionistic Predicate Logic

In the definition of a Kripke model for intuitionistic predicate logic below, we shall, for the sake of simplicity in notation, assume that our language contains no individual constants. The definitions below generalize Definition 8.11 for intuitionistic propositional logic.

Definition 8.21 (Kripke Model for Intuitionistic Predicate Logic).
A *Kripke model M* for intuitionistic predicate logic is a quadruple $\langle S, R, U, \models_i \rangle$ such that
1. S is a non-empty set (of possible proof-situations);
2. R is a binary relation on S (regarded as the accessibility relation), which is reflexive and transitive (see Definition 8.11);
3. U assigns to each s in S a non-empty set $U(s)$, such that if sRs', then $U(s)$ is a subset of $U(s')$. $U(s)$ can be conceived as the Universe of the proof-situation s, i.e., the set of objects constructed in the situation s;
4. $\models_i$ is a relation between elements of S and expressions of the form $P(a_1, \ldots, a_k)[n_1, \ldots, n_k]$, such that
i) if $s \models P(a_1, \ldots, a_k)[n_1, \ldots, n_k]$, then $n_1, \ldots, n_k$ are elements of $U(s)$, and
ii) if $s \models P(a_1, \ldots, a_k)[n_1, \ldots, n_k]$ and sRs', then $s' \models P(a_1, \ldots, a_k)[n_1, \ldots, n_k]$.
$s \models P(a_1, \ldots, a_k)[n_1, \ldots, n_k]$ is to be read as: in the proof situation s it has been shown that $(n_1, \ldots, n_k)$ has the property P.

Definition 8.22. $(M, s \models_i A(a_1, \ldots, a_k)[n_1, \ldots, n_k])$
Given a Kripke model $M = \langle S, R, U, \models_i \rangle$, s in S, a formula $A(a_1, \ldots, a_k)$ and elements $n_1, \ldots, n_k$ in $U(s)$, $M, s \models_i A(a_1, \ldots, a_k)[n_1, \ldots, n_k]$ is defined as follows:
$M, s \models_i P(a_1, \ldots, a_k)[n_1, \ldots, n_k] := s \models_i P(a_1, \ldots, a_k)[n_1, \ldots, n_k]$;
$M, s \models_i B \wedge C[n_1, \ldots, n_k] := M, s \models_i B[n_1, \ldots, n_k]$ and $M, s \models_i C[n_1, \ldots, n_k]$;
The definition for $B \vee C$, $B \to C$ and $\neg B$ is analogous to the one (Definition 8.12) for intuitionistic propositional logic.
$M, s \models_i \forall x[B(x, a_1, \ldots, a_k)][n_1, \ldots, n_k] :=$ for all s' in S such that sRs' and for all n in $U(s')$, $M, s' \models_i B(a, a_1, \ldots, a_k)[n, n_1, \ldots, n_k]$ (where a is new);
$M, s \models_i \exists x[B(x, a_1, \ldots, a_k)][n_1, \ldots, n_k] := M, s \models B(a, a_1, \ldots, a_k)[n, n_1, \ldots, n_k]$ for at least one n in $U(s)$ (a being a new free variable).

Example 8.10. Let $M = \langle \{s,s'\}, R, U, \models_i \rangle$ be the Kripke model for intuitionistic predicate logic defined by sRs', $U(s) = \{1\}$, $U(s') = \{1,2\}$, $s \models_i P(a)[1]$, $s' \models_i P(a)[1]$, and $s' \not\models_i P(a)[2]$.

$$\{1\} \quad s \; P(a)[1]$$
$$\downarrow$$
$$\{1,2\} \; s' \; P(a)[1]$$

Then $M,s \models_i \neg\forall x[P(x)]$, because $M,s \not\models_i \forall x[P(x)]$ (since $M,s' \not\models_i P(a)[2]$) and $M,s' \not\models_i \forall x[P(x)]$. But $M,s \not\models_i \exists x[\neg P(x)]$, because $M,s \not\models_i \neg P(a)[1]$. Hence, $M,s \not\models_i \neg\forall x[P(x)] \to \exists x[\neg P(x)]$.

Example 8.11. Let $M = \langle S, R, U, \models_i \rangle$ be the Kripke model for intuitionistic predicate logic defined by:
$S = \{s_1, s_2\}$, s_1Rs_1, s_1Rs_2 and s_2Rs_2,
$U(s_1) = \{1\}$, $U(s_2) = \{1,2\}$,
$s_1 \models_i P(a)[1]$, $s_2 \models_i P(a)[1]$, $s_2 \models_i Q$, $s_2 \not\models_i P(a)[2]$,

$$\{1\} \quad \overset{s_1}{\underset{}{\big|}} P(a)[1]$$
$$\{1,2\} \; \big| \; P(a)[1], Q$$
$$s_2$$

Then $M,s_1 \models_i \forall x[P(x) \vee Q]$, but $M,s_1 \not\models_i \forall x[P(x)] \vee Q$ (because $M,s_1 \not\models_i \forall x[P(x)]$ and $M,s_1 \not\models Q$). Hence, $M,s_1 \not\models_i \forall x[P(x) \vee Q] \to \forall x[P(x)] \vee Q$.

Example 8.12. Let $M = \langle S, R, U, \models_i \rangle$ be the Kripke model for intuitionistic predicate logic defined by:
$S = \{s_1, s_2, \ldots\}$,
for all i, j, if $1 \leq i \leq j$, then s_iRs_j,
for all $i = 1, 2, \ldots, U(s_i) = \{1, \ldots, i\}$,
for all $i \geq 2, s_i \models P(a)[1], \ldots, s_i \models P(a)[i-1]$,
$s_i \not\models P(a)[i]$,

$$\{1\} \; s_1$$
$$\{1,2\} \; s_2 \; \big| \; P(1)$$
$$\{1,2,3\} \; s_3 \; \big| \; P(1), P(2)$$
$$\big|$$
$$\big|$$
$$\big|$$

Then for each s in S, $M,s \not\models_i \forall x[P(x) \vee \neg P(x)]$. Hence, $M,s_1 \models_i \neg\forall x[P(x) \vee \neg P(x)]$.

The reader can easily verify that the analogue of Lemma 8.1 holds again for intuitionistic predicate logic.

Lemma 8.2. *Let $M = \langle S, R, U, \models_i \rangle$ be a Kripke model, s and t elements of S and $A = A(a_1, \ldots, a_k)$ a formula.*
If $M,s \models_i A[n_1, \ldots, n_k]$ and sRt, then $M,t \models_i A[n_1, \ldots, n_k]$.

Definition 8.23 (Intuitionistically Kripke-valid Consequence).
Let $M = \langle S, R, U, \models_i \rangle$ be a Kripke model for intuitionistic predicate logic and $A = A(a_1, \ldots, a_k)$ a formula.
(a) M is a *model* for A, or A holds in M := for all s in S and for all $n_1, \ldots, n_k$ in $U(s)$, $M,s \models_i A(a_1, \ldots, a_k)[n_1, \ldots, n_k]$. **Notation**: $M \models_i A$.
(b) A is intuitionistically *Kripke-valid* := for all Kripke models M, $M \models_i A$.
Notation: $\models_i A$.
(c) B is an intuitionistically *Kripke-valid consequence* of $A_1, \ldots, A_n$:=
$\models_i A_1 \wedge \ldots A_n \to B$. **Notation**: $A_1, \ldots A_n \models_i B$.

Example 8.13. For the Kripke model M of Example 8.10, $M \not\models_i \neg\forall x[P(x)] \to \exists x[\neg P(x)]$. For the Kripke model M of Example 8.11, $M \not\models_i \forall x[P(x) \lor Q] \to \forall x[P(x)] \lor Q$. And for the Kripke model M of Example 8.12, $M \not\models \neg\neg\forall x[P(x) \lor \neg P(x)]$. Since $\models_i \forall x[\neg\neg(P(x) \lor \neg P(x))]$, it follows that $\forall x[\neg\neg A(x)] \not\models_i \neg\neg\forall x[A(x)]$.

8.6.4 Soundness and Completeness

The intuitionistic Hilbert-type proof system in Subsection 8.6.1 for intuitionistic predicate logic is sound with respect to the intuitionistic Kripke semantics given in Subsection 8.6.3.

Theorem 8.9 (Soundness). *If $A_1,\ldots,A_n \vdash_i B$, then $A_1,\ldots,A_n \models_i B$.*

Proof. The proof is a generalization of the proof of the soundness theorem (Theorem 8.2) for intuitionistic propositional logic.

The procedure to construct a counterexample to a given conjecture $A_1,\ldots,A_n \vdash'_i B$, given in Definition 8.14 for intuitionistic propositional logic, may be adapted to intuitionistic predicate logic, taking also the quantifiers into account. We shall illustrate this procedure in the Examples 8.14 and 8.15. Again, if this procedure yields an open search tree, then we have actually constructed an intuitionistic Kripke counterexample to the given conjecture. And if all search trees are closed, then the closed branches form together a tableau-deduction of B from $A_1,\ldots,A_n$. Hence, again we may conclude that the tableaux rules for intuitionistic predicate logic are complete with respect to the Kripke semantics for intuitionistic predicate logic:

Theorem 8.10 (Completeness). *If $A_1,\ldots,A_n \models_i B$, then $A_1,\ldots,A_n \vdash'_i B$.*

Finally, we may generalize the proof of Theorem 8.1 to intuitionistic predicate logic:

Theorem 8.11. *If $A_1,\ldots,A_n \vdash'_i B$, then $A_1,\ldots,A_n \vdash_i B$.*

Hence, the three notions of $A_1,\ldots,A_n \vdash_i B$, $A_1,\ldots,A_n \models_i B$ and $A_1,\ldots,A_n \vdash'_i B$ turn out to be equivalent.

Example 8.14. We wonder whether $\neg\forall x[P(x)] \vdash'_i \exists x[\neg P(x)]$. Our procedure to construct a counterexample to this conjecture yields the search tree in the left column below:

$T \neg\forall x[P(x)], F \exists x[\neg P(x)]$
$T \neg\forall x[P(x)], F \neg P(a_1)$
$T \neg\forall x[P(x)], T P(a_1)$ $\{1\}$ $s \ P(a)[1]$
$F \forall x[P(x)], T P(a_1)$
      | |
$T \neg\forall x[P(x)], F P(a_2), T P(a_1)$ $\{1,2\}$ $s' \ P(a)[1]$

Although we may proceed with developing this search tree, it is clear that we will never find closure. In fact, we have constructed the Kripke counterexample M described in Example 8.10, as depicted in the right column above. For instance, by definition $s \models_i P(a)[1]$, corresponding with the fact that $T \ P(a_1)$ occurs in s.

Example 8.15. We wonder whether $\forall x[P(x) \vee Q] \vdash_i' \forall x[P(x)] \vee Q$. Our procedure to construct a counterexample to this conjecture yields among others the search tree in the left column below:

$T \, \forall x[P(x) \vee Q], \, F \, \forall x[P(x)] \vee Q$
$T \, P(a_1) \vee Q, \, T \, \forall x[P(x) \vee Q], \, F \, \forall x[P(x)], \, FQ$ $\{1\} \, s_1 \,|\, P(a)[1]$
    $TP(a_1), \, T \, \forall x[P(x) \vee Q], \, F \, \forall x[P(x)], \, FQ$
                  |

$TP(a_1), \, T \, \forall x[P(x) \vee Q], \, FP(a_2)$
$TP(a_1), \, T \, P(a_2) \vee Q, \, T \, \forall x[P(x) \vee Q], \, FP(a_2)$ $\{1,2\} \, s_2 \,|\, P(a)[1], \, Q$
$TP(a_1), \, TQ, \, T \, \forall x[P(x) \vee Q], \, FP(a_2)$

From this open search tree one may easily read off the the Kripke counterexample described in Example 8.11, as depicted in the right column above. For instance, by definition $s_1 \models_i P(a)[1]$ corresponding to the fact that $T \, P(a_1)$ occurs in s_1 and $s_2 \models_i Q$ corresponding to the fact that TQ occurs in s_2.

The proofs of the *soundness* and *completeness* of intuitionistic predicate logic with respect to (intuitionistic) Kripke semantics are generalizations of the corresponding proofs for intuitionistic propositional logic (see the proofs of Theorem 8.2 and 8.5), and are analogous to the corresponding proofs for classical predicate logic (see Chapter 4).

There is however one complication. Also for intuitionistic predicate logic we have: if A is Kripke valid, then there is no open search tree starting with $F \, A$. Thus, if A is Kripke valid, we can conclude that each search tree starting with $F \, A$ is not open, i.e., each search tree starting with $F \, A$ is *not not* closed. Classically, we can conclude from this that each search tree starting with FA is closed (and hence that A is formally provable), but not so intuitionistically. So the completeness theorem for intuitionistic predicate logic with respect to (intuitionistic) Kripke semantics can only be proved if we use a classical metalanguage and not if we want to use intuitionistic metamathematics.

W. Veldman [16] has discovered a generalization of the notion of an intuitionistic Kripke model and hence a somewhat different notion of Kripke validity, such that completeness with respect to this generalized Kripke semantics can be proved intuitionistically. The essence is to allow that $\bot$ (falsum) is true in one or more proof situations and to demand that if $\bot$ is true in situation s ($s \models_i \bot$), then every formula A is true in s ($s \models_i A$). Next this idea was used by de Swart [14] to give another intuitionistic completeness proof with respect to a different semantics.

Note that the transition from 'not not closed' to 'closed' is intuitionistically correct in the case of intuitionistic propositional logic, because in that case we can for each search tree decide whether it is closed or not. And from $\forall t[C(t) \vee \neg C(t)]$ and $\forall t[\neg\neg C(t)]$, where $C(t)$ stands for 'the search tree t is closed', it follows intuitionistically that $\forall t[C(t)]$.

Like classical predicate logic, also intuitionistic predicate logic is *undecidable*. The search trees starting with $F \, A$ may become infinitely long, each time introducing new variables and we may not know whether we can stop at some stage.

Exercise 8.19. Verify (intuitively) that intuitionistically the following formulas hold true: a) $\neg \exists x \in V[\neg A(x)] \rightleftarrows \forall x \in V[\neg\neg A(x)]$; b) $\neg\neg \forall x \in V[A(x)] \to \forall x \in V[\neg\neg A(x)]$.
c) Verify (intuitively) that the following formula does not hold intuitionistically:
$\forall x \in V[\neg\neg A(x)] \to \forall x \in V[A(x)]$.
d1) Show that a) is intuitionistically tableau-provable and d2) that a) is also formally provable (in the intuitionistic Hilbert-type proof system).

Exercise 8.20. Prove that for A a formula in prenex normal form (see Subsection 4.3.5) $\vdash'_i A$ (in intuitionistic predicate logic) is decidable. Since intuitionistic predicate logic is undecidable, it follows that not every formula has a prenex normal form to which it is equivalent in intuitionistic predicate logic.

Exercise 8.21. Prove that if $\vdash'_i \exists x[A(x)]$ (intuitionistically) and in $A(a)$ there occur no other free variables than a, then $\vdash'_i \forall x[A(x)]$.

Exercise 8.22. Prove that $\forall x[P(x) \lor Q] \to \forall x[P(x)] \lor Q$ holds in all Kripke models (for intuitionistic predicate logic) $M = \langle S, R, U, \models_i \rangle$ with *constant domain*, i.e., $U(s) = U(s')$ for all s, s' in S. (Compare Example 8.11.)

Exercise 8.23. For each formula A of intuitionistic predicate logic we define a formula A' of modal predicate logic by induction as follows: $P' := \Box P$, $(B \to C)' := \Box(B' \to C')$, $(\neg B)' := \Box\neg(B')$, $(B \land C)' := B' \land C'$, $(B \lor C)' := B' \lor C'$, $(\forall x[B(x)])' := \Box\forall x[B(x)']$, $(\exists x[B(x)])' := \exists x[B(x)']$. Prove that $\models_i A$ iff $\models A'$ in S4, i.e., A is intuitionistically Kripke valid iff A' is valid in the modal logic S4 (see Chapter 6).

8.7 Sets in Intuitionism: Construction Projects and Spreads

Intuitionistically, a set – like any other mathematical object – should be a mental construction. Natural numbers can be conceived as objects which are finitely constructible. Intuitionistically, the set of all natural numbers is identified with the following *construction project*: a) 0 is a natural number, and b) if n is a natural number, then n' is a natural number too. (The term 'construction project' was coined by Johan J. de Iongh.)

The set $\mathbb{N}$ of the natural numbers is intuitionistically not regarded as a completed totality, but only as potential or becoming or constructive. The construction project can be stated in only two clauses, but it generates the potentially infinite set $\mathbb{N}$ of the natural numbers. At each stage only finitely many elements of $\mathbb{N}$ will actually have been constructed; but also at each stage the construction project tells us how to continue the construction of new natural numbers.

In classical mathematics one accepts the Powerset axiom: if V is a set, then $P(V)$ is a set too. It follows that $P(\mathbb{N})$, $PP(\mathbb{N})$, $PPP(\mathbb{N})$, ... are sets of ever increasing cardinality. However, these sets are not surveyable in the most literal meaning of the word; more precisely, no construction project is known of which we could reasonably say that it generates the elements of such a set in the course of time. For that reason, Brouwer rejected the powerset axiom and refused to accept the existence of these sets.

The notion of construction project is a primitive, i.e., undefined, one. But we certainly want to say that the clauses a) and b) above together define a construction project for $\mathbb{N}$. In what follows we introduce the notion of spread, which we want to consider as a particular kind of construction project. We do not exclude the possibility that one may discover other kinds of construction projects in the future, although we are not aware of them now.

The intuitionist constructs the integers from the natural numbers ($\mathbb{Z}$ is enumerable) and the rationals from the integers ($\mathbb{Q}$ is enumerable), just like this is done in classical set theory. A more difficult question is whether there is an intuitionistic set which can reasonably be called $\mathbb{R}$; in other words, if we can generate the elements of such an $\mathbb{R}$ by an appropriate construction project. Only if this is the case, does quantification over the reals ('for all $x \in \mathbb{R}\ldots$' and 'for some $x \in \mathbb{R}\ldots$') make sense. One could say that Brouwer invented the spread concept in order to answer this question.

Now the construction project for $\mathbb{R}$ is rather complicated. For that reason we first indicate a construction project for $\{0,1\}^{\mathbb{N}}$, i.e., the set of all (potentially) infinite sequences of zeros and ones. Schematically, the construction project for $\{0,1\}^{\mathbb{N}}$ looks as follows:

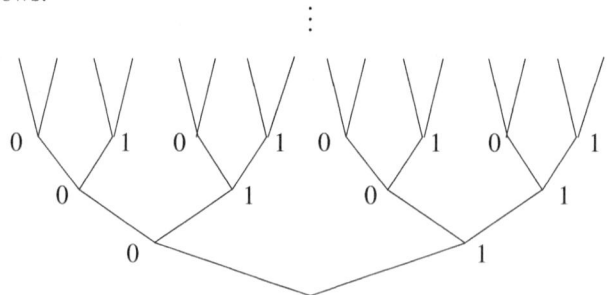

As an introduction one might think of a construction project as a mental project generating all possibilities to swim from Amsterdam to 'the end of the world', where at each stage one has the choice of going to the left or going to the right.

By choosing an element from $\{0,1\}$ at successive moments or stages, potentially infinite sequences of zero's and one's come into being. These sequences are generated in the course of time by a simple precept, called the choice-law: at each stage choose either a zero or a one. We identify the (intuitionistic) set $\{0,1\}^{\mathbb{N}}$ with this precept; and we call the potentially infinite sequences of zero's and one's the elements of this set, since they are generated in accordance with the corresponding choice-law. One does have an overall picture of how the elements of this set come into being.

The elements of a *spread* are generated by choosing natural numbers consecutively, with due observation of a *choice-law* (corresponding to the given spread),

8.7 Sets in Intuitionism: Construction Projects and Spreads

which prescribes which natural numbers may be chosen given the choices already made before.

In the case of $\{0, 1\}^{\mathbb{N}}$, the choice-law dictates that to each finite sequence of natural numbers already chosen before, we may choose only an element of $\{0, 1\}$. This spread is also called σ_{01} or the *binary spread*. A different choice-law is the one that dictates that given a finite sequence of natural numbers chosen before, one may choose an element of $\{0, 1\}$ if the given finite sequence does not contain 1 and that otherwise one has to choose an element of $\{1\}$. The spread belonging to this choice-law generates the monotone non-decreasing elements of σ_{01} and is called σ_{01mon}. Below is a picture of σ_{01mon}.

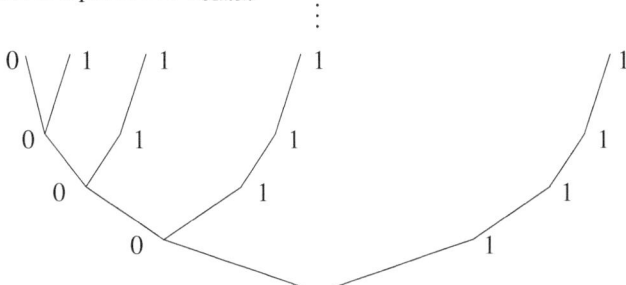

The elements of the spread, called *choice-sequences*, are the potentially infinite sequences of natural numbers which are admitted by the choice-law of the spread. So the elements of $\{0, 1\}^{\mathbb{N}}$ are the potentially infinite sequences of zero's and one's.

σ_ω is the *universal spread*, i.e., $\mathbb{N}^{\mathbb{N}}$. This choice-law dictates that to each finite sequence of natural numbers already chosen before, one may choose any natural number n in $\mathbb{N}$.

Some authors define a spread simply as a tree in which each node has at least one successor. This is not in the spirit of intuitionism, but rather in the spirit of classical mathematics.

We may consider every element of $\{0, 1\}^{\mathbb{N}}$ as the characteristic function of a subset of $\mathbb{N}$ (see Theorem 3.12). Note that from an intuitionistic point of view, the characteristic function K_U (Definition 3.27) of a subset U of $\mathbb{N}$ is only well defined if U is decidable, i.e., if for each $n \in \mathbb{N}$ one can decide whether $n \in U$ or not. So, the intuitionist also has at his disposal the set $P_{dec}(\mathbb{N})$, i.e., the set of all decidable subsets of $\mathbb{N}$.

The set containing just 0 if Goldbach's conjecture holds and containing just 1 if Goldbach's conjecture does not hold, is a subset of $\mathbb{N}$, but not a decidable one. In fact, the intuitionist does not know of any spread to which we could reasonably give the name $P(\mathbb{N})$. Up till now no one has succeeded to present a construction project which might be said to generate in the course of time *all* subsets of $\mathbb{N}$.

As he needed a construction project for $\mathbb{R}$, and not just for $\mathbb{N}^{\mathbb{N}}$, Brouwer generalized the notion of spread just introduced. A *dressed spread* consists of

1. a *choice-law*, prescribing – given a finite sequence of natural numbers already chosen before – which natural numbers may be chosen next, and

2. a *correlation-law*, which after each choice correlates effectively an object from a fixed countable set to the finite sequence of natural numbers chosen up till now.

The elements of a dressed spread, again called *choice-sequences*, are the potentially infinite sequences of objects which have been assigned by the correlation-law to the sequences of natural numbers chosen according to the choice-law.

In Brouwer's applications those correlated objects can be natural numbers, but also rational numbers and intervals with rational endpoints. Defining real numbers as infinite sequences of intervals with rational endpoints (for instance, $\sqrt{2} = [1,2], [1.4, 1.5], [1.41, 1.42], [1.414, 1.415], \ldots$), Brouwer indicated a specific choice-law and a specific correlation-law such that the corresponding spread has precisely the reals as elements.

A (dressed) spread is not thought of intuitionistically as the 'totality' of its elements, but rather as the pair consisting of the choice-law and the correlation-law, which together govern the generation process under which its elements grow. So a (dressed) spread is a construction project, which generates the elements of the spread in the course of time. These elements are potentially infinite sequences (for instance, of intervals with rational endpoints, in the case of $\mathbb{R}$), of which at each stage only a finite initial segment has been completed, but of which also is prescribed at each stage how the finite sequence already constructed can be continued. Among the elements of a spread there are choice sequences which are extensionally the same as individual choice sequences defined by a particular law or otherwise.

For sets which can be obtained by means of a construction project, in particular for spreads, some surprising axioms are defended. One of them is *Brouwer's Continuity Principle*, which we explain below.

Brouwer's Continuity Principle for natural numbers: Let σ be a spread. If $\forall \alpha \in \sigma \, \exists k \in \mathbb{N} \, [A(\alpha, k)]$, then

$$\forall \alpha \in \sigma \, \exists m \in \mathbb{N} \, \exists k \in \mathbb{N} \, \forall \beta \in \sigma \, [\bar{\beta}m = \bar{\alpha}m \to A(\beta, k)].$$

i.e., for each α in σ there is an initial segment of length m and a natural number k such that to all β in σ having the same initial segment of length m as α the same natural number k is correlated; $\bar{\beta}m = \bar{\alpha}m := \forall n < m[\beta(n) = \alpha(n)]$.

Justification: Although Brouwer used this principle without further justification, we now try to give a justification. Suppose $\forall \alpha \in \sigma \, \exists k \in \mathbb{N} \, [A(\alpha, k)]$. Because intuitionistically the elements of a spread are considered as continuously growing with new choices and not as being completed, and because natural numbers themselves are finite constructions, the correlation that associates with each α in σ a natural number k can intuitionistically only consist in such a way that the correlated natural numbers will be determined effectively at a certain finite stage in the growth of the choice sequences. That is, intuitionistically the correlated natural numbers will have to be determined by finite initial segments of the choice sequences. The justification of Brouwer's principle ultimately rests on the insight that each element α in σ can be thought of as being given step by step, also in the case that some particular α is determined by a finite law. And the truth of $A(\alpha, k)$ does not depend on the manner in which α has been generated. □

8.7 Sets in Intuitionism: Construction Projects and Spreads

We now derive a consequence from Brouwer's principle.

Theorem 8.12. *Let $\sigma = \sigma_{01}$ and $\underline{0}$ be the element in σ_{01} such that $\forall n \in \mathbb{N} \; [\underline{0}(n) = 0]$. Then $\neg \forall \alpha \in \sigma \; [\alpha = \underline{0} \vee \neg(\alpha = \underline{0})]$.*

Proof. Suppose $\forall \alpha \in \sigma \; [\alpha = \underline{0} \vee \neg(\alpha = \underline{0})]$, i.e.,

$$\forall \alpha \in \sigma \; \exists k \in \mathbb{N} \; [(k = 0 \wedge \alpha = \underline{0}) \vee (k = 1 \wedge \neg(\alpha = \underline{0}))].$$

By Brouwer's principle, $\forall \alpha \in \sigma \; \exists m \in \mathbb{N} \; \exists k \in \mathbb{N} \; \forall \beta \in \sigma \; [\bar{\beta}m = \bar{\alpha}m \rightarrow$

$$(k = 0 \wedge \beta = \underline{0}) \vee (k = 1 \wedge \neg(\beta = \underline{0}))].$$

Now consider $\alpha = \underline{0}$. Then there is $m \in \mathbb{N}$ such that $\forall \beta \in \sigma \; [\bar{\beta}m = \bar{\alpha}m \rightarrow \beta = \underline{0}]$. However, let β be such that $\bar{\beta}m = \bar{\alpha}m = \bar{\underline{0}}m$ and $\beta(m) = 1$. Then $\beta \neq \underline{0}$. Contradiction. Therefore $\neg \forall \alpha \in \sigma \; [\alpha = \underline{0} \vee \neg(\alpha = \underline{0})]$. □

Notice that although for each $\alpha \in \sigma_{01}$ the statement $\alpha = \underline{0} \vee \alpha \neq \underline{0}$ itself does not yield a contradiction – in fact, $\neg\neg(\alpha = \underline{0} \vee \alpha \neq \underline{0})$ –, the simultaneous quantification over all $\alpha \in \sigma_{01}$ of the expression $\alpha = \underline{0} \vee \alpha \neq \underline{0}$ does lead to a contradiction. Summarizing: $\neg\neg(\alpha = \underline{0} \vee \alpha \neq \underline{0})$, but $\neg \forall \alpha \in \sigma_{01} \; [\alpha = \underline{0} \vee \alpha \neq \underline{0}]$.

From Brouwer's principle it also follows that there do not exist bijective mappings from either σ_{01} or σ_{01mon} to $\mathbb{N}$ (Brouwer 1918; see Exercise 8.24). Note that from a classical point of view σ_{01mon} is an enumerable set.

Given a construction project which results in an intuitionistic set V and given a well defined extensional property $A(x)$ concerning the elements x of V, an intuitionist also accepts $W := \{x \in V \mid A(x)\}$ as a set, for quantification over W may be explained in the usual way as a restricted quantification over V: $\forall x \in W \; [E(x)] := \forall x \in V \; [A(x) \rightarrow E(x)]$ and $\exists x \in W \; [E(x)] := \exists x \in V \; [A(x) \wedge E(x)]$.

For a more extensive treatment of spreads and the axioms holding for them see Gielen, Veldman and de Swart [7].

Choice Sequences *Choice-sequences* are the potentially infinite sequences of natural numbers which are generated by the choice-law of the spread. And in the case of a dressed spread *choice sequences* are the potentially infinite sequences of objects which have been assigned by the correlation-law to the sequences of natural numbers chosen according to the choice-law.

Some particular choice sequences may be called *lawlike*, for instance, the sequence $\underline{0}$ which is generated by the choice-law that dictates that given a finite sequence of choices already made before one has to choose a 0 and nothing else. Other particular choice sequences may be called *lawless*, for instance, the sequence which is generated by the choice-law that dictates that given any finite sequence of choices already made before one has to choose any natural number and of which we have determined in advance that the choice-law will never impose any restriction on further choices.

However, the expression 'α is lawlike' is not a well defined propositional function for $\alpha \in \sigma_\omega$ for the following reasons:
1. The notion of finite law has not been defined precisely.
2. Let $A(\alpha) := \alpha$ is lawlike. Then $A(\underline{0})$ is true. But $A(0,0,0,\ldots)$ is not true for the

sequence 0, 0, 0, ... which 'accidentally' only contains zero's. But a well defined propositional function should be a function in the sense that if α and β are extensionally equal, then $A(\alpha)$ and $A(\beta)$ should be the same well defined assertion.

'$\underline{0}$ is lawlike' is a well defined proposition (what-ever 'lawlike' may mean precisely). But as long as α has not been specified by some specific finite law, the expression 'α is lawlike' has no clear meaning, because the notion of 'finite law' has not been defined. Consequently, we are not able to speak about the set of all lawlike sequences and hence we cannot quantify over them. Similar observations hold for the expressions 'α is lawless' and 'the set of all lawless sequences'. We have no construction project that generates in the course of time all lawlike (respectively, all lawless) sequences! See de Swart [15].

$A(\beta)$, e.g., $\forall n[\beta(n) = 0]$, is a propositional function, rather than a well defined proposition. If we have a construction determining all values of β, then this construction together with our understanding of $A(\beta)$ would result in a well defined proposition. But as long as such a finite law for β is not given, we have neither a proof of $A(\beta)$, nor an insight in the impossibility of experiencing the truth of $A(\beta)$.

Summarizing: 1. Intuitionistic objects, in particular sets, are finite constructions or *construction projects*. 2. Construction projects for non-denumerable sets technically boil down to spreads. A *dressed spread* consists of i) a choice law and ii) a correlation law. 3. *Choice sequences* are just the elements of a spread. 4. Brouwer defines the set $\mathbb{R}$ as a dressed spread whose choice sequences are infinite sequences of (decreasing) intervals with rational endpoints. 5. Brouwer's principle is proved by reflection on what it means to have a proof of $\forall \alpha \in \sigma \exists k \in \mathbb{N} \, [\, A(\alpha, k) \,]$, rather than the result of the peculiar epistemological status of choice sequences. 6. Quantification only makes sense if it is a quantification over an intuitionistic set, i.e., a set for which one has a construction(-project).

Starting from these philosophically sound principles it is quite possible to develop intuitionistic mathematics, enough for the purposes of science (physics, economics, etc.). See, for instance, Veldman [17, 18] and de Swart [13].

Exercise 8.24. Using Brouwer's principle, prove that there is no bijection from σ_{01} to $\mathbb{N}$, neither from σ_{01mon} to $\mathbb{N}$. However, from a classical point of view σ_{01mon} is enumerable; explain the difference.

8.8 The Brouwer Kripke axiom

Brouwer-Kripke axiom (Brouwer, 1948) Let P be a *determinate* proposition. Then there is an α in σ_{01} such that $P \leftrightarrows \exists n[\alpha(n) = 1]$.

Justification Given a determinate proposition P it can be pondered again and again in my mathematical life. We construct α as follows: $\alpha(n) = 1$ if at stage n I did succeed in proving P; otherwise, $\alpha(n) = 0$. □

Johan de Iongh stressed that P should be *determinate*, i.e., P should not depend on infinite objects which are still under construction. As long as information about P has not yet been completed, I cannot really start to think about its truth. In particular, P should not be of the form $\forall n[\beta(n) = 0]$, in which case one would obtain a contradiction from the Brouwer-Kripke axiom and what Kleene [9] called Brouwer's principle for functions. Wim Veldman [17] calls this principle AC_{11}, where AC stands for Axiom of Choice.

Theorem: The Brouwer Kripke axiom and AC_{11} (Brouwer's principle for functions) are contradictory when applied to the expression $\forall n[\beta(n) = 0]$ with β in σ_{01}.

For a precise formulation of AC_{11} and for a proof of this theorem we refer the reader to [17] or [7]. Several authors have blamed this contradiction on AC_{11} (Brouwer's principle for functions); however, the restriction proposed by Johan de Iongh that P should be determinate seems rather natural and self-evident, while AC_{11} has a good justification.

Application of the Brouwer-Kripke axiom: Let $\alpha(n) = 1$ if at stage n I have a proof of $G \vee \neg G$, where G is Goldbach's conjecture. Then $G \vee \neg G \leftrightarrow \exists n[\alpha(n) = 1]$. Because of $\neg\neg(G \vee \neg G)$ we know $\neg\neg\exists n[\alpha(n) = 1]$. But $G \vee \neg G$, i.e., $\exists n[\alpha(n) = 1]$ cannot be asserted.

8.9 Solutions

Solution 8.1. (a) If $A \to B$, then $\neg B \to \neg A$. Conversely, if $\neg B \to \neg A$, then $A \to \neg\neg B$. But we are not entitled to infer $A \to B$ from $\neg B \to \neg A$. For $\neg B \to \neg(\neg\neg B)$ is intuitionistically valid and $\neg\neg B \to B$ is not.
(b) $(\neg A \vee \neg B) \to \neg(A \wedge B)$ is intuitionistically valid: $(\neg A \vee \neg B)$ and $(A \wedge B)$ together yield a contradiction. The converse formula $\neg(A \wedge B) \to (\neg A \vee \neg B)$ is not valid intuitionistically: interpreting A as Goldbach's conjecture and B as $\neg A$ we have $\neg(A \wedge \neg A)$, but not $\neg A \vee \neg\neg A$.
(c) $(A \to B) \to \neg(A \wedge \neg B)$ is intuitionistically valid: $(A \to B)$ and $(A \wedge \neg B)$ together yield a contradiction. But $\neg(A \wedge \neg B) \to (A \to B)$ is not valid intuitionistically: interpreting A as $\neg\neg B$ we have $\neg(\neg\neg B \wedge \neg B)$, but not $\neg\neg B \to B$, as we have seen before.
(d) $(\neg A \vee B) \to (A \to B)$ is intuitionistically valid: $\neg A \to (A \to B)$ and $B \to (A \to B)$, hence $(\neg A \vee B) \to (A \to B)$. But $(A \to B) \to (\neg A \vee B)$ is not valid intuitionistically: interpreting B as $\neg\neg A$ we have $A \to \neg\neg A$, but not $\neg A \vee \neg\neg A$.
(e) $((A \to B) \vee (A \to C)) \to (A \to B \vee C)$ is intuitionistically valid: $(A \to B) \to (A \to B \vee C)$ and $(A \to C) \to (A \to B \vee C)$, hence $((A \to B) \vee (A \to C)) \to (A \to B \vee C)$. But the converse formula is not valid intuitionistically: interpreting A as $B \vee C$ we have $B \vee C \to B \vee C$, but not $(B \vee C \to B) \vee (B \vee C \to C)$.
(f) $(A \vee \neg A) \to (\neg\neg A \to A)$ is intuitionistically valid: for if $A \vee \neg A$, then $\neg\neg A$ rules out the second possibility $\neg A$; so, only the first possibility A is left. $(\neg\neg A \to A) \to (A \vee \neg A)$ is not intuitionistically valid: interpreting A as $\neg\neg P$ we have $\neg\neg\neg\neg P \to \neg\neg P$, but not $\neg\neg P \vee \neg\neg\neg P$ which is equivalent to $\neg\neg P \vee \neg P$.

Solution 8.2. 1. If $A \to C$, then $\neg C \to \neg A$ and hence $\neg(\neg A) \to \neg(\neg C)$.
2. If $A \to (B \to C)$ and B, then $A \to C$ and so by 1) $\neg\neg A \to \neg\neg C$.
3. If $A \to (B \to C)$ and $\neg\neg A$, then by 2) $B \to \neg\neg C$ and hence by 1) $\neg\neg B \to \neg\neg\neg\neg C$, i.e., $\neg\neg B \to \neg\neg C$.
4. From 3) with $A \to B$, A, B instead of A, B, C respectively, if $(A \to B) \to (A \to B)$ and $\neg\neg(A \to B)$, then $\neg\neg A \to \neg\neg B$.
5. Suppose $\neg\neg A \to \neg\neg B$ and $\neg(A \to B)$. Then $\neg\neg A$ and $\neg B$, for $A \to B$ follows both from $\neg A$ and from B. So, $\neg\neg A$ and $\neg\neg A \to \neg\neg B$. Therefore, $\neg\neg B$. Contradiction with $\neg B$. So, if $\neg\neg A \to \neg\neg B$, then $\neg\neg(A \to B)$.
6. From 3) with A, B, $A \wedge B$ instead of A, B, C respectively, if $A \to (B \to A \wedge B)$ and $\neg\neg A$, then $\neg\neg B \to \neg\neg(A \wedge B)$. Hence, if $\neg\neg A \wedge \neg\neg B$, then $\neg\neg(A \wedge B)$.
7. $A \wedge B \to A$. Hence by 1), $\neg\neg(A \wedge B) \to \neg\neg A$. Similarly, $\neg\neg(A \wedge B) \to \neg\neg B$.
8. $A \to A \vee B$. Hence by 1), $\neg\neg A \to \neg\neg(A \vee B)$. Similarly, $\neg\neg B \to \neg\neg(A \vee B)$. Therefore, if $\neg\neg A \vee \neg\neg B$, then $\neg\neg(A \vee B)$.
9. Take $B = \neg A$. Then $\neg\neg(A \vee \neg A)$, but $\neg\neg A \vee \neg A$ is not intuitionistically valid.

Solution 8.3. If A is decidable, i.e., $A \vee \neg A$, then $\neg\neg A$ eliminates the second option $\neg A$ and consequently only the first option A is left.

Solution 8.4. Axiom 8^i, $\neg A \to (A \to B)$, of intuitionistic propositional logic is (formally) provable in classical propositional logic. This follows from $\neg A, A \vdash B$ (weak negation elimination) by two applications of the deduction theorem. Hence, we have (i): all formulas provable in the intuitionistic system are provable in the classical system. Since the proof of the deduction theorem only uses the axiom schemas 1 and 2 and applications of Modus Ponens and since all these tools are available in intuitionistic propositional logic, we have (ii): the deduction theorem also holds for intuitionistic propositional logic.

Solution 8.5. Note that the deductions found in Exercise 2.60 a ii), b ii) and c ii) do use the rule of double negation elimination $(d\neg E)$, while those in a i), b i) and c i) do not.

Solution 8.6. a) We restrict ourselves to a tableau-proof of $A \to (B \to A)$ and of $\neg A \to (A \to B)$:

$F\ A \to (B \to A)$	$F\ \neg A \to (A \to B)$
$TA, F\ B \to A$	$T\ \neg A, F\ A \to B$
TA, TB, FA	$T\ \neg A, FA, F\ A \to B$
closure	$T\ \neg A, TA, FB$
	$T\ \neg A, FA, TA, FB$
	closure

b) 1) In the left column below is an intuitionistic tableau-proof of $A \to \neg\neg A$, while in the right column there is a failed attempt to give an intuitionistic tableau-proof of $\neg\neg A \to A$:

$F\ A \to \neg\neg A$	$F\ \neg\neg A \to A$
$TA, F\ \neg\neg A$	$T\ \neg\neg A, FA$
$TA, T\ \neg A$	$T\ \neg\neg A, F\ \neg A, FA$
TA, FA	$T\ \neg\neg A, TA$
closure	no closure

8.9 Solutions

b) 2) Below is in the left column an intuitionistic tableau-proof of $\neg\neg(A \vee \neg A)$, while in the right column there is a failed attempt to give an intuitionistic tableau-proof of $A \vee \neg A$:

$$\begin{array}{ll}
F \neg\neg(A \vee \neg A) & F\ A \vee \neg A \\
\hline
T \neg(A \vee \neg A) & FA,\ F\neg A \\
\hline
T \neg(A \vee \neg A),\ F\ A \vee \neg A & TA \\
T \neg(A \vee \neg A),\ FA,\ F\neg A & \text{no closure} \\
\hline
T \neg(A \vee \neg A),\ TA & \\
T \neg(A \vee \neg A),\ F\ A \vee \neg A,\ TA & \\
T \neg(A \vee \neg A),\ FA,\ F\neg A,\ TA & \\
\text{closure} &
\end{array}$$

c) It is not possible to construct an intuitionistic tableau-deduction of B from $A \to B$ and $\neg A \to B$:

$T\ A \to B,\ T \neg A \to B,\ FB$

$FA,\ T\ A \to B,\ T \neg A \to B,\ FB$ $\quad\mid\quad$ $TB,\ T \neg A \to B,\ FB$

$FA,\ T\ A \to B,\ F\neg A,\ FB$ $\mid$ $FA,\ T\ A \to B,\ TB,\ FB$ $\mid$ $TB,\ T \neg A \to B,\ FB$

$T\ A \to B,\ TA$

$FA,\ TA$ $\mid$ $TB,\ TA$

$\quad\quad$ no closure

Solution 8.7. Below in the left column there is a classical tableau-proof of $(P \to Q) \vee (Q \to P)$, and in the right column there are two failed attempts to construct an intuitionistic tableau proof of $(P \to Q) \vee (Q \to P)$.

$$\begin{array}{ll}
F\ (P \to Q) \vee (Q \to P) & F\ (P \to Q) \vee (Q \to P) \\
F\ P \to Q,\ F\ Q \to P & F\ P \to Q,\ F\ Q \to P \\
TP,\ FQ,\ F\ Q \to P & \swarrow\quad\searrow \\
TP,\ FQ,\ TQ,\ FP & TP,\ FQ \quad\quad TQ,\ FP \\
& \text{no closure}\quad\text{no closure}
\end{array}$$

Solution 8.8. In all F-rules, except in the rules $F \to$ and $F\neg$, going from the top downwards, only F-formulas are introduced, while in the intuitionistic rules $F \to_i$ and $F\neg_i$, S is replaced by S_T. So: if an intuitionistic tableau-proof starts with

$\quad\quad F\ B \vee C$
$\quad\quad FB,\ FC$

then it is impossible that in the lowest sequents – which are of the form S, TA, FA – in the tableau-proof of $B \vee C$, TA results from FB and FA results from FC, by application of the rules. Hence, if $\vdash'_i B \vee C$, then $\vdash'_i B$ or $\vdash'_i C$.

The classical variant does not hold; for instance, $\vdash' P \vee \neg P$, but $\nvdash' P$ and $\nvdash' \neg P$.

Solution 8.9. $A \vee B \vdash'_i (A \to B) \to B$, but not $(A \to B) \to B \vdash'_i A \vee B$:

$T\ A \vee B,\ F\ (A \to B) \to B$ $\quad\quad\quad$ $T\ (A \to B) \to B,\ F\ A \vee B$

$T\ A \vee B,\ T\ A \to B,\ FB$ $\quad\quad\quad\quad\ $ $T\ (A \to B) \to B,\ FA,\ FB$

$TA,\ T\ A \to B,\ FB$ $\mid$ $TB,\ T\ A \to B,\ FB$ $\quad$ $F\ A \to B,\ FA,\ FB$ $\mid$ $TB,\ FA,\ FB$

$TA,\ FA,\ FB$ $\mid$ $TA,\ TB,\ FB$ $\mid$ closure $\quad\quad$ $TA,\ FB$

closure $\quad\quad$ closure $\quad\quad\quad\quad\quad\quad\quad\quad\quad$ no closure

$(A \to B) \to B \vdash'_i \neg\neg(A \lor B)$, but conversely not $\neg\neg(A \lor B) \vdash'_i (A \to B) \to B$:

$T (A \to B) \to B, F \neg\neg(A \lor B)$	$T \neg\neg(A \lor B), F (A \to B) \to B$
$T (A \to B) \to B, T \neg(A \lor B)$	$T \neg\neg(A \lor B), T A \to B, FB$
$T (A \to B) \to B, F A \lor B$	$T \neg\neg(A \lor B), F A, FB \mid T \neg\neg, T B, FB$
$T (A \to B) \to B, FA, FB$	$F \neg(A \lor B), FA, FB \quad \mid$ closure
$F A \to B, FA, FB \mid TB, FA, FB$	$T A \lor B, T A \to B$
$TA, FB, T\neg(A \lor B) \mid$ closure	$TA, T A \to B \mid TB, T A \to B$
$TA, FB, F A \lor B$	$TA, FA \mid TA, TB \mid TB, FA \mid TB, TB$
TA, FB, FA, FB	closure $\mid$ no closure $\mid$ no closure $\mid$ no closure
closure	

Solution 8.10. 1) $\neg\neg A \land \neg\neg B \vdash'_i \neg\neg(A \land B)$ and $\neg\neg(A \land B) \vdash'_i \neg\neg A \land \neg\neg B$.
2) $\neg\neg A \to \neg\neg B \vdash'_i \neg\neg(A \to B)$ and conversely; $\neg\neg(A \to B) \vdash'_i A \to \neg\neg B$ and conversely.
3) for each formula A, $\vdash'_i \neg\neg\neg A \to \neg A$.
4) It follows from 3) that (if A and B are stable, then) $A \lor B$ is stable.
5) $\neg\neg A \lor \neg\neg B \vdash'_i \neg\neg(A \lor B)$, but conversely not $\neg\neg(A \lor B) \vdash'_i \neg\neg A \lor \neg\neg B$.

Solution 8.11. 1) Here is an intuitionistic tableau-proof of $(A \lor \neg A) \to (\neg\neg A \to A)$:

$F (A \lor \neg A) \to (\neg\neg A \to A)$
$T A \lor \neg A, F \neg\neg A \to A$
$T A \lor \neg A, T \neg\neg A, FA$
$TA, T \neg\neg A, FA \mid T \neg A, T \neg\neg A, FA$
closure $\qquad \mid T \neg A, F \neg A, FA$
$\qquad\qquad\qquad$ closure

2) $\neg P$ (P atomic) is stable, i.e., $\vdash'_i \neg\neg\neg P \to \neg P$, but $\neg P$ is not decidable, i.e., not $\vdash'_i \neg P \lor \neg\neg P$.

Solution 8.12. 1. If $E = P$ (atomic), then $E^* = \neg\neg P$ and $\vdash'_i \neg\neg\neg\neg P \to \neg\neg P$, i.e., $E^* = \neg\neg P$ is stable. Suppose that A^* and B^* are stable (induction hypothesis).
If $E = A \land B$, then $E^* = A^* \land B^*$; and by Exercise 8.10 (1) E^* is stable.
If $E = A \to B$, then $E^* = A^* \to B^*$; and by Exercise 8.10 (2) E^* is stable.
If $E = \neg A$, then $E^* = \neg A^*$; and E^* is stable by Exercise 8.10 (3).
If $E = A \lor B$, then $E^* = \neg(\neg A^* \land \neg B^*)$; and by Exercise 8.10 (4) E^* is stable.
2. Suppose $A_1, \ldots, A_n \vdash B$ (classically), i.e., there is a schema of the form

$$\frac{\text{axiom 1}, \ldots, \text{axiom 8}, A_1, \ldots, A_n}{\dfrac{C \quad C \to D}{D}}$$
$$B$$

Replace each formula E in this schema by E^*.
(axiom 1)$^* = (A \to (B \to A))^* = A^* \to (B^* \to A^*)$ is again an instance of axiom schema 1. (axiom 5a)$^* = (A \to A \lor B)^* = A^* \to (A \lor B)^* = A^* \to \neg(\neg A^* \land \neg B^*)$ is intuitionistically provable.

(axiom 6)$^* = ((A \to C) \to ((B \to C) \to (A \vee B \to C)))^* = (A^* \to C^*) \to ((B^* \to C^*) \to (\neg(\neg A^* \wedge \neg B^*) \to C^*))$. Now, $\vdash_i (A^* \to C^*) \to ((B^* \to C^*) \to (\neg(\neg A^* \wedge \neg B^*) \to \neg\neg C^*))$ and, since C^* is stable, $\vdash_i \neg\neg C^* \to C^*$. Therefore, $\vdash_i$ (axiom 6)*.
(axiom 8)$^* = (\neg\neg A \to A)^* = \neg\neg A^* \to A^*$ is intuitionistically provable (since A^* is stable).

Since $(C \to D)^* = C^* \to D^*$, $\dfrac{C^* \quad (C \to D)^*}{D^*}$ is again an application of Modus Ponens. Therefore, the schema above can be transformed into an intuitionistic deduction of B^* from $A_1^*, \ldots, A_n^*$.

Solution 8.13. While classically the formulas $P \to P$ and $P \vee \neg P$ are equivalent and both classically valid (always true), because of the stronger meaning of the $\vee$-connective in intuitionism, the formula $P \vee \neg P$ is intuitionistically no longer valid; however, the formula $P \to P$ is also intuitionistically valid.

Solution 8.14. * Suppose $\Gamma \mid C$ for all formulas C in Γ and let $\Gamma \vdash_i A$. Then either
1. $A \in \Gamma$, or 2. A is an axiom, or 3. there are formulas B and $B \to A$ such that $\Gamma \vdash_i B$
(1), $\Gamma \vdash_i B \to A$ and A is deduced from B and $B \to A$ by Modus Ponens.
In case 1, $\Gamma \mid A$ by hypothesis. In case 2, one easily checks that $\Gamma \mid A$ for each intuitionistic axiom A. In case 3, by induction hypothesis, $\Gamma \mid B$ (2) and $\Gamma \mid B \to A$ (3). From (1) and (2), $\Gamma \Vdash B$ and hence, by (3), $\Gamma \mid A$.
Suppose $H \mid H$ and $H \vdash_i B \vee C$. Then, by the Theorem just proved, $H \mid B \vee C$, i.e., $H \Vdash B$ or $H \Vdash C$. Hence, $H \vdash_i B$ or $H \vdash_i C$.

Solution 8.15. * For $A = P$ (atomic), the theorem is trivial. Induction Hypothesis: $\vdash_i A_1 \rightleftarrows B_1 \vee \ldots \vee B_m$ and $\vdash_i A_2 \rightleftarrows C_1 \vee \ldots \vee C_n$, where each B_i and C_j satisfies the conditions specified.
Case 1: $A = A_1 \vee A_2$; then $\vdash_i A \rightleftarrows B_1 \vee \ldots \vee B_m \vee C_1 \vee \ldots \vee C_n$.
Case 2: $A = A_1 \wedge A_2$; then $\vdash_i A \rightleftarrows (B_1 \wedge C_1) \vee \ldots \vee (B_m \wedge C_1) \vee \ldots \vee (B_m \wedge C_n)$, leaving out those $B_i \wedge C_j$ which are inconsistent.
Case 3: $A = A_1 \to A_2$. If $A_1 \to A_2 \vdash_i A_1$, then $\vdash_i A \rightleftarrows A_2$; hence $\vdash_i A \rightleftarrows C_1 \vee \ldots \vee C_n$. If $A_1 \to A_2 \nvdash_i A_1$, then $\vdash_i A \rightleftarrows B_1 \vee \ldots \vee B_m \to C_1 \vee \ldots \vee C_n$ and hence $\vdash_i A \rightleftarrows (B_1 \to C_1 \vee \ldots \vee C_n) \wedge \ldots \wedge (B_m \to C_1 \vee \ldots \vee C_n)$, where for all k, $1 \leq k \leq m$, $A \nvdash_i B_k$ because we have supposed that $A \nvdash_i A_1$ and therefore $A \nvdash_i B_1 \vee \ldots \vee B_m$. $(B_1 \to C_1 \vee \ldots \vee C_n) \wedge \ldots \wedge (B_m \to C_1 \vee \ldots \vee C_n)$ is consistent since, by hypothesis, $A \nvdash_i A_1$.
Case 4: $A = \neg A_1$; then $\vdash_i A \rightleftarrows \neg B_1 \wedge \ldots \wedge \neg B_m$.
Proof of $A_j \mid A_j$: Let $A_j = P \wedge \neg B \wedge (C \to D)$, where $A_j \nvdash_i C$ and P atomic. To show: $A_j \mid P$ and $A_j \mid \neg B$ and $A_j \mid C \to D$, i.e., $A_j \vdash_i P$ and not $A_j \Vdash B$ and (if $A_j \Vdash C$, then $A_j \mid D$). $A_j \vdash_i P$ is trivial. Because $A_j \vdash_i \neg B$ and A_j is consistent, it follows that not $A_j \Vdash B$. And because of $A_j \nvdash_i C$ it follows that (if $A_j \Vdash C$, then $A_j \mid D$).

Solution 8.16.

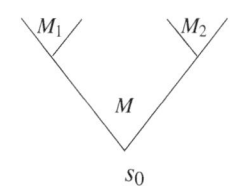

a) Let $M_1 = \langle S_1, R_1, \models_1\rangle$ be an intuitionistic Kripke model such that not $M_1, s_1 \models_1 B$ for some s_1 in S_1. And let $M_2 = \langle S_2, R_2, \models_2\rangle$ be a Kripke model such that not $M_2, s_2 \models_2 C$ for some s_2 in S_2. Let $M = \langle S, R, \models_i\rangle$ be the following Kripke model:
1) S contains all nodes of S_1 and of S_2 and in addition one extra node s_0.
2) $s_0 R s$ for all s in S_1; $s_0 R s$ for all s in S_2. R restricted to S_1 equals R_1 and R restricted to S_2 equals R_2, i.e., for all s, t in S_1, $sRt := sR_1 t$ and for all s, t in S_2, $sRt := sR_2 t$.
3) For s in S_1, $s \models_i P := s \models_1 P$ and for s in S_2, $s \models_i P := s \models_2 P$. For all atomic formulas P, by definition, $s_0 \not\models_i P$. One easily checks that for all formulas A, $M, s \models_i A$ iff $M_1, s \models_1 A$, if s in S_1, and $M, s \models_i A$ iff $M_2, s \models_2 A$, if s in S_2.
Now suppose $M, s_0 \models_i B \vee C$. Then $M, s_0 \models_i B$ or $M, s_0 \models_i C$.
Case 1: $M, s_0 \models_i B$; then $M, s_1 \models_i B$ and hence, $M_1, s_1 \models_1 B$; contradiction.
Case 2: $M, s_0 \models_i C$; then $M, s_2 \models_i C$ and hence, $M_2, s_2 \models_2 C$; contradiction.
Therefore, not $M, s_0 \models_i B \vee C$.
b) Suppose $\models_i B \vee C$, not $\models_i B$ and not $\models_i C$. Then there is a Kripke counterexample M_1 to B and a Kripke counterexample M_2 to C. By a) it follows that there is a Kripke counterexample M to $B \vee C$, contradicting $\models B \vee C$.

Solution 8.17. (a) The following schema is an intuitionistic tableau-proof of $(P \to Q) \to (\neg Q \to \neg P)$:

$$\begin{array}{c} F\ (P \to Q) \to (\neg Q \to \neg P) \\ \hline T\ P \to Q,\ F\ \neg Q \to \neg P \\ \hline T\ P \to Q,\ T\ \neg Q,\ F\ \neg P \\ \hline T\ P \to Q,\ T\ \neg Q,\ TP \\ \hline T\ P \to Q,\ FQ,\ TP \\ \hline FP,\ FQ,\ TP\ |\ TQ,\ FQ,\ TP \end{array}$$

Applying our procedure to construct a Kripke counterexample for $(\neg Q \to \neg P) \to (P \to Q)$ we find two different search trees, one of which is open:

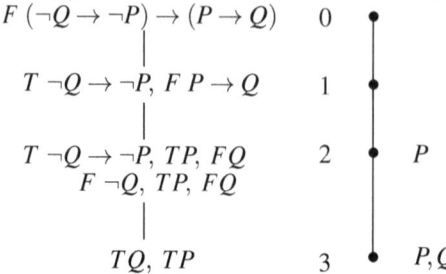

$F\ (\neg Q \to \neg P) \to (P \to Q)$ 0

$T\ \neg Q \to \neg P,\ F\ P \to Q$ 1

$T\ \neg Q \to \neg P,\ TP,\ FQ$
$F\ \neg Q,\ TP,\ FQ$ 2 P

$TQ,\ TP$ 3 P, Q

From this open search tree one can read off a Kripke counterexample to $(\neg Q \to \neg P) \to (P \to Q)$: $M = \langle \{0, 1, 2, 3\}, R, \models_i\rangle$, where R is reflexive and transitive such that $0R1$, $1R2$ and $2R3$; and $2 \models_i P$, $3 \models_i P$ and $3 \models_i Q$. Then $M, 0 \not\models_i (\neg Q \to \neg P) \to (P \to Q)$.
(b) ... (f) are treated similarly.

Solution 8.18. * $M = \langle S, R, \models_i\rangle$ is a Kripke model, for:
1. R is reflexive, i.e., $H \vdash_i H$,
2. R is transitive, i.e., if $H' \vdash_i H$ and $H'' \vdash_i H'$, then $H'' \vdash_i H$,

8.9 Solutions

3. if $H \models_i P$ and HRH', then $H' \models_i P$, i.e., if $H \vdash_i P$ and $H' \vdash_i H$, then $H' \vdash_i P$.
Proof of: $M, H \models_i A$ iff $H \vdash_i A$, for $H \in S$, i.e., $H \mid H$.
1. For $A = P$ (atomic), by definition.
2. Induction hypothesis: $M, H \models_i B$ iff $H \vdash_i B$, and $M, H \models_i C$ iff $H \vdash_i C$.
a) $A = B \land C$: $M, H \models_i B \land C$ iff $M, H \models_i B$ and $M, H \models_i C$,
             (ind. hyp.)   iff $H \vdash_i B$ and $H \vdash_i C$,
                             iff $H \vdash_i B \land C$.
b) $A = B \lor C$: $M, H \models_i B \lor C$ iff $M, H \models_i B$ or $M, H \models_i C$,
             (ind. hyp.)   iff $H \vdash_i B$ or $H \vdash_i C$,
             (Exercise 8.14) iff $H \vdash_i B \lor C$.
c) $A = B \to C$: $M, H \models_i B \to C$ iff for all H' in S such that HRH',
                                if $M, H' \models_i B$, then $M, H' \models_i C$.
             (ind. hyp.)   iff for all H' in S such that $H' \vdash_i H$,
                                if $H' \vdash_i B$, then $H' \vdash_i C$.   (†)

To show: (†) iff $H \vdash_i B \to C$. i) Suppose $H \vdash_i B \to C$. Then (†) easily follows.
ii) Suppose (†). By Exercise 8.15 $H \land B$ is intuitionistically provably equivalent to a disjunction $H_1 \lor \ldots \lor H_n$ such that for all H_j, $1 \leq j \leq n$, $H_j \mid H_j$, i.e., H_j is an element of S. Now $H_j \vdash_i H \land B$. So, $H_j \vdash_i H$ and $H_j \vdash_i B$. Hence, by (†), for all j, $1 \leq j \leq n$, $H_j \vdash_i C$. Therefore, by $\lor$-elimination, $H_1 \lor \ldots \lor H_n \vdash_i C$; so, $H \land B \vdash_i C$. Consequently, $H \vdash_i B \to C$.
Now suppose $\models_i A$. One easily checks that $P \to P \mid P \to P$, i.e., $P \to P$ is an element of S. So, $M, P \to P \models_i A$. Therefore, $P \to P \vdash_i A$, i.e., $\vdash_i A$.

Solution 8.19. a) $\neg \exists x \in V[\neg A(x)] \to \forall x \in V[\neg\neg A(x)]$: Suppose $\neg \exists x \in V[\neg A(x)]$, a in V and $\neg A(a)$. Then $\exists x \in V[\neg A(x)]$. Contradiction. Therefore, $\forall x \in V[\neg\neg A(x)]$.
$\forall x \in V[\neg\neg A(x)] \to \neg \exists x \in V[\neg A(x)]$: Suppose $\forall x \in V[\neg\neg A(x)]$ and $\exists x \in V[\neg A(x)]$. Then for some a in V, both $\neg A(a)$ and $\neg\neg A(a)$. Contradiction. Therefore, $\neg \exists x \in V[\neg A(x)]$.
b) Suppose $\neg\neg \forall x \in V[A(x)]$, a in V and $\neg A(a)$. Then $\neg \forall x \in V[A(x)]$. Contradiction. Therefore $\forall x \in V[\neg\neg A(x)]$.
c) Let $V = \{a\}, A(a) := P \lor \neg P$. Then $\forall x \in V[\neg\neg A(x)]$ iff $\neg\neg(P \lor \neg P)$ and $\forall x \in V[A(x)]$ iff $P \lor \neg P$. Now $\neg\neg(P \lor \neg P)$ is intuitionistically valid, while $P \lor \neg P$ is intuitionistically invalid (see Section 8.1.2).

d1)

$F \ \neg\exists x[\neg A(x)] \to \forall x[\neg\neg A(x)]$
$\overline{T \ \neg\exists x[\neg A(x)], \ F \ \forall x[\neg\neg A(x)]}$
$\overline{T \ \neg\exists x[\neg A(x)], \ F \ \neg\neg A(a_1)}$
$\overline{T \ \neg\exists x[\neg A(x)], \ T \ \neg A(a_1)}$
$\overline{F \ \exists x[\neg A(x)], \ T \ \neg A(a_1)}$
$\overline{F \ \neg A(a_1), \ T \ \neg A(a_1)}$
$\overline{T A(a_1), \ T \ \neg A(a_1)}$
$T A(a_1), \ F A(a_1)$
       closure
Hence, $\vdash'_i \neg\exists x[\neg A(x)] \to \forall x[\neg\neg A(x)]$

$F \ \forall x[\neg\neg A(x)] \to \neg\exists x[\neg A(x)]$
$\overline{T \ \forall x[\neg\neg A(x)], \ F \ \neg\exists x[\neg A(x)]}$
$\overline{T \ \forall x[\neg\neg A(x)], \ T \ \exists x[\neg A(x)]}$
$\overline{T \ \forall x[\neg\neg A(x)], \ T \ \neg A(a_1)}$
$\overline{T \ \neg\neg A(a_1), \ T \ \neg A(a_1)}$
$\overline{F \ \neg A(a_1), \ T \ \neg A(a_1)}$
$\overline{T A(a_1), \ T \ \neg A(a_1)}$
$T A(a_1), \ F A(a_1)$
       closure
Hence, $\vdash' \forall x[\neg\neg A(x)] \to \neg\exists x[\neg A(x)]$

d2) By $\exists$-introduction, $\neg\exists x[\neg A(x)], \neg A(a) \vdash \exists x[\neg A(x)]$. Also $\neg\exists x[\neg A(x)], \neg A(a) \vdash \neg\exists x[\neg A(x)]$. Therefore, by $\neg$-introduction, $\neg\exists x[\neg A(x)] \vdash \neg\neg A(a)$. Hence, by $\forall$-introduction, $\neg\exists x[\neg A(x)] \vdash \forall x[\neg\neg A(x)]$.
$\forall x[\neg\neg A(x)], \neg A(a) \vdash \neg A(a) \wedge \neg\neg A(a)$, and hence, by weak negation elimination, $\forall x[\neg\neg A(x)], \neg A(a) \vdash B \wedge \neg B$ for any B. Hence, by $\exists$-elimination, $\forall x[\neg\neg A(x)], \exists x[\neg A(x)] \vdash B \wedge \neg B$. So, by $\neg$-introduction, $\forall x[\neg\neg A(x)] \vdash \neg\exists x[\neg A(x)]$.

Solution 8.20. $Z(\exists x[B(x, a_1, \ldots, a_m)]) := \{B(a_k, a_1, \ldots, a_m), B(a_1, a_1, \ldots, a_m), \ldots, B(a_m, a_1, \ldots, a_m)\}$, and let $Z(\forall x[B(x, a_1, \ldots, a_m)]) := \{B(a_k, a_1, \ldots, a_m)\}$, where a_k is the first free variable not occurring in $\exists x[B(x, a_1, \ldots, a_m)], \forall x[B(x, a_1, \ldots, a_m)]$ resp. If V is a set of formulas of the form $\exists x[B(x, a_1, \ldots, a_m)]$ or $\forall x[B(x, a_1, \ldots, a_m)]$, then $Z(V) :=$ the union of all $Z(C)$ for C in V. Clearly, if V is finite, then $Z(V)$ is finite and the number of elements of $Z(V)$ can easily be estimated from the definitions above.
Now suppose A is a prenex formula $Q^1 x_1 \ldots Q^n x_n[B]$, where $Q^i = \forall$ or $Q^i = \exists$. Let $V_1 := Z(A)$ and by induction $V_k := Z(V_{k-1})$, $k = 2, \ldots, n$. Note that $\vdash' A$ iff V_1 contains at least one tableau-provable formula. By an easy induction with respect to k, $k = 2, \ldots, n$, we find that $\vdash' A$ iff V_k contains at least one tableau-provable formula. Consequently, A is tableau-provable iff V_n contains at least one tableau-provable formula. However, all formulas in V_n are quantifier-free and hence are decidable. And since V_n is finite, we can decide by a finite method whether there is a tableau-provable formula in V_n.

Solution 8.21. Suppose $\vdash'_i \exists x[A(x)]$ and a_k is the only free variable in $A(a_k)$. A tableau-proof of $\exists x[A(x)]$ starts with
$$\frac{F \exists x[A(x)]}{F A(a_k)}$$
and then proceeds to closure. By replacing in a given tableau-proof of $\exists x[A(x)]$ the upper sequent, $F\exists x[A(x)]$, by $F\forall x[A(x)]$, one obtains a tableau-proof of $\forall x[A(x)]$.

Solution 8.22. Let $M = \langle S, R, U, \models_i \rangle$ be a Kripke model (for intuitionistic predicate logic) with constant domain U. $M, s \models_i \forall x[P(x) \vee Q] :=$ for all s' in S such that sRs' and for all $n \in U(s')$, $M, s' \models_i P(a)[n]$ or $M, s' \models_i Q$. (1)
$M, s \models_i \forall x[P(x)] \vee Q := M, s \models_i Q$ or for all s' in S such that sRs' and for all $n \in U(s')$, $M, s' \models_i P(a)[n]$. (2)
(1) $\to$ (2): If $M, s \models_i Q$, we are done. If $M, s \not\models_i Q$, then by (1) for all $n \in U(s)$, $M, s \models_i P(a)[n]$. Now suppose sRs' and $n \in U(s')$; then, since U is constant, $n \in U(s)$ and, since $M, s \models_i P(a)[n]$ and sRs', $M, s' \models_i P(a)[n]$, which was to be shown.

Solution 8.23. Suppose $\models A'$ in S4. We want to show that $\models_i A$ (intuitionistically). So, let $M = \langle S, R, U, \models_i \rangle$ be a Kripke model (for intuitionistic predicate logic). Define the Kripke model $M' = \langle S, R, U, \models \rangle$ for S4 as follows: $s \models P := s \models_i P$. Then, since R is reflexive and transitive, M' is a Kripke model for the modal logic S4 (satisfying the extra condition: if $M', s \models A$ and sRt, then $M', t \models A$).
Claim: $M, s \models_i A$ intuitionistically iff $M', s \models A'$ in S4. Now, since $\models A'$ in S4, $M' \models A'$ in S4 and hence, $M \models_i A$ (intuitionistically).
Proof of claim: for $A = P$ (atomic), $M, s \models_i P$ (intuitionistically) iff

8.9 Solutions

for all t in S, if sRt, then $M,t \models_i P$, iff
for all t in S, if sRt, then $M',t \models P$, iff
$M',s \models \Box P$ in S4, iff
$M',s \models P'$ in S4.
Induction hypothesis: the claim is correct for B and C. We shall show that the claim also holds for $B \to C$ and for $\forall x[B(x)]$, leaving the other cases to the reader.
$M,s \models_i B \to C$ (intuitionistically) iff for all t in S, if sRt and $M,t \models_i B$, then $M,t \models_i C$. By induction hypothesis, this is equivalent to: for all t in S, if sRt and $M',t \models B'$ in S4, then $M',t \models C'$ in S4. And this latter expression is equivalent to $M',s \models \Box(B' \to C')$, i.e., $M',s \models (B \to C)'$ in S4.
$M,s \models_i \forall x[B(x)]$ iff for all t with sRt and for all n in $U(t)$, $M,t \models_i B(a)[n]$,
(ind. hyp.) iff for all t with sRt and for all n in $U(t)$, $M',t \models B(a)'[n]$ in S4,
iff for all t with sRt, $M',t \models \forall x[B(x)']$ in S4,
iff $M',s \models \Box \forall x[B(x)']$ in S4
iff $M',s \models (\forall x[B(x)])'$ in S4.

Conversely, suppose $\models_i A$ intuitionistically. To show: $\models A'$ in S4. So, let $M = \langle W, R, U, \models \rangle$ be a Kripke model for S4, i.e., R is reflexive and transitive. Define $M_i = \langle W, R, U, \models_i \rangle$ as follows: $w \models_i P :=$ for all w' in W, if wRw', then $w' \models P$. Then M_i is a Kripke model for intuitionistic logic, since from the transitivity of R it follows that: if $w \models_i P$ and wRw', then $w' \models_i P$.
Claim: $M_i, w \models_i A$ (intuitionistically) iff $M,w \models A'$ in S4. So, since $\models_i A$ (intuitionistically), $M_i \models_i A$ intuitionistically and hence, $M \models A'$ in S4.
Proof of claim : For $A = P$ (atomic), $M_i, w \models_i P$ (intuitionistically) iff $w \models_i P$,
iff for all w' in W, if wRw', then $w' \models P$,
iff $M,w \models \Box P$,
iff $M,w \models P'$ in S4.
Induction hypothesis: the claim holds for B and C. We shall show that the claim holds for $B \to C$ and for $\forall x[B(x)]$, leaving the other cases to the reader.
$M_i, w \models_i B \to C$ (intuitionistically) := for all w' in W, if wRw' and $M_i, w' \models_i B$, then $M_i, w' \models_i C$. By the induction hypothesis this is equivalent to: for all w' in W, if wRw' and $M,w' \models B'$, then $M,w' \models C'$. And this latter expression is equivalent to $M,w \models \Box(B' \to C')$, in other words, $M,w \models (B \to C)'$ in S4.
$M_i, w \models_i \forall x[B(x)] :=$ for all w' with wRw' and for all n in $U(w')$, $M_i, w' \models_i B(a)[n]$,
iff (ind. hyp.) for all w' with wRw' and for all n in $U(w')$, $M,w' \models B(a)'[n]$,
iff for all w' with wRw', $M,w' \models \forall x[B(x)']$. And the latter expression is equivalent to $M,w \models \Box \forall x[B(x)']$ and hence to $M,w \models (\forall x[B(x)])'$ in S4.

Solution 8.24. Suppose $f : \sigma_{01} \to \mathbb{N}$ were a bijection. Then $\forall \alpha \in \sigma_{01} \; \exists k \in \mathbb{N} \; [k = f(\alpha)]$. By Brouwer's principle, $\forall \alpha \in \sigma_{01} \; \exists m \in \mathbb{N} \; \exists k \in \mathbb{N} \; \forall \beta \in \sigma_{01} \; [\bar{\beta}m = \bar{\alpha}m \to k = f(\beta)]$. Let $\alpha = \underline{0}$. Then there are $m \in \mathbb{N}$ and $k \in \mathbb{N}$ such that

$$\forall \beta \in \sigma_{01} \; [\bar{\beta}m = \bar{\alpha}m = \underline{\bar{0}} \to k = f(\beta)].$$

Now, take β such that $\bar{\beta}m = \bar{\alpha}m = \underline{\bar{0}}m$ and $\beta(m) \neq \alpha(m)$. Then $\beta \neq \alpha$, but $k = f(\beta) = f(\alpha)$. So, f is not injective. Contradiction. The proof for σ_{01mon} is similar. The classical function $f : \sigma_{01mon} \to \mathbb{N}$, defined by $f(\underline{0}) = 0, f(\underline{1}) = 1, f(0\underline{1}) = 2, f(00\underline{1}) = 3$, etc. is intuitionistically not well defined: we cannot determine the value of $f(\alpha)$ for the sequence α defined by $\alpha(n) = 0$ if at stage n I do not have a proof of Goldbach's conjecture G and $\alpha(n) = 1$ if at stage n I do have a proof of G.

References

1. Bernays, P. *Abhandlungen zur Philosophie der Mathematik*. Wissenschaftliche Buchgesellschaft, Darmstadt, 1976.
2. Beth, E.W., Semantical Considerations on Intuitionistic Mathematics. *Indagationes Mathematicae* 9, pp. 572-577, 1947.
3. Brouwer, L.E.J., *Collected Works*, Volume I. A. Heyting (ed.), North-Holland, 1975.
4. Dummett, M., *Elements of Intuitionism*. Clarendon Press, Oxford, 1977, 2000.
5. Fitting, M., *Intuitionistic Logic, Model Theory and Forcing*. North-Holland, 1969.
6. Fitting, M., *Proof Methods for Modal and Intuitionistic Logics*. Reidel, Dordrecht, 1983.
7. Gielen, W., H. de Swart and W. Veldman, The Continuum Hypothesis in Intuitionism. *Journal of Symbolic Logic*, 46, pp. 121-136, 1981.
8. Kleene, S.C., *Introduction to Metamathematics*. Elsevier 1952, 2009.
9. Kleene, S.C. and R.E. Vesley, *The Foundations of Intuitionistic Mathematics*. North-Holland, 1965.
10. Kripke, S.A., Semantical Analysis of Modal Logic, *Zeitschrift für Mathematische Logik und Grundlagen der Mathematik*, Band 9, pp. 67-96, 1963.
11. Kripke, S.A., Semantical Analysis of Intuitionistic Logic. In: Crossley, J.N., and Dummett, M.A.E. (eds.), *Formal Systems and Recursive Functions*, North-Holland, pp. 92-130, 1965.
12. Nishimura, I., On formulas of one variable in intuitionistic propositional calculus. *Journal of Symbolic Logic* 25, 327-332, 1960.
13. Swart, H.C.M. de, Elements of Intuitionistic Analysis. *Zeitschrift für Mathematische Logik und Grundlagen der Mathematik*, Band 22, pp. 289-298 and 501-508, 1976.
14. Swart, H.C.M. de, Another intuitionistic completeness proof. *Journal of Symbolic Logic*, 41, pp. 644-662, 1976.
15. Swart, H.C.M. de, Spreads or Choice Sequences. *History and Philosophy of Logic*, 13, pp. 203-213, 1992.
16. Veldman, W., An intuitionistic completeness theorem for intuitionistic predicate logic. *Journal of Symbolic Logic* 41, pp. 159-166, 1976.
17. Veldman, W., *Investigations in Intuitionistic Hierarchy Theory*. PhD thesis, Nijmegen, 1981.
18. Veldman, W., The Borel Hierarchy Theorem from Brouwer's Intuitionistic Perspective. *Journal of Symbolic Logic*, 73, pp. 1-64, 2008.
19. Proclus de Lycie. Les Commentaires sur le premier livre des Eléments d'Euclide : Traduits pour la première fois du grec en français avec une introduction et des notes par Paul Ver Eecke, 1948.

Chapter 9
Applications: Prolog; Relational Databases and SQL; Social Choice Theory

H.C.M. (Harrie) de Swart

9.1 Programming in Logic

Abstract The language of logic can be used as a declarative programming language, i.e., the programmer has to describe *what* the problem is, not *how* it should be solved. We introduce logic programming by means of an example and explain how the system answers questions given a certain program. The possibility of recursive definitions is one of the cornerstones of logic programming. Prolog is a particular form of logic programming; it has been implemented in a certain way. As a consequence, although declarative in principle, Prolog also has certain procedural aspects. The syntax of logic programming in general and of Prolog in particular is very simple. Although the reasoning mechanism should use unification, many systems work with a simpler form, called matching, for reasons of efficiency. Lists are important terms in logic programming. Cut is a procedural device needed to keep programs efficient. Negation is implemented by means of cut and hence differs from logical negation. Logic programming has many applications in (deductive) databases and in Artificial Intelligence. We discuss the most important pitfalls.

Example 9.1. The best way to introduce the subject of logic programming seems to be to give an example of a concrete logic program. The following example is from I. Bratko [7].

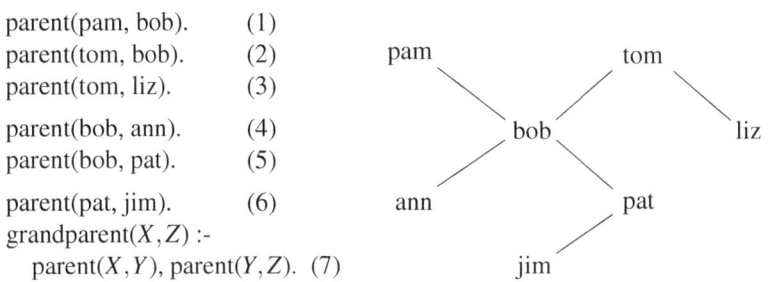

parent(pam, bob).   (1)
parent(tom, bob).   (2)
parent(tom, liz).   (3)

parent(bob, ann).   (4)
parent(bob, pat).   (5)

parent(pat, jim).   (6)
grandparent(X, Z) :-
   parent(X, Y), parent(Y, Z).   (7)

This logic program consists of seven *clauses*. The first six of them are called *facts*; They express that pam is a parent of bob, tom is a parent of bob, etc. The last clause is called a *rule*; it expresses that X is a grandparent of Z if X is a parent of Y and Y is a parent of Z. The symbol :- is to be read as 'if' and the comma between 'parent(X,Y)' and 'parent(Y,Z)' is to be read as 'and'. 'X', 'Y' and 'Z' are called *variables*; it is allowed to replace them by the names of arbitrary individual objects. In the rule 'L :- L_1, L_2', 'L' is called the *head* of the rule and 'L_1, L_2' is called the *body*.

Given a logic program P, the user may ask questions. Given the logic program just presented one might ask who are the parents of bob; in other words, for which X is it true that X is a parent of bob? This question is formulated as follows:

:- parent(X, bob).

or

?- parent(X, bob).

The question whether bob and liz have a common parent is formulated as follows:

:- parent(X, bob), parent(X, liz).

or

?- parent(X, bob), parent(X, liz).

In *Prolog*, which stands for PROgramming in LOGic and which is just a particular form of logic programming, the answers to these questions are found as follows. Given the logic program mentioned above and given the question

?- parent(X, bob).

the Prolog system tries to *match* or *unify* the clause 'parent(X, bob)' with the first fact 'parent(pam, bob)' in the given program. This matching or unifying succeeds by replacing the variable 'X' by 'pam'. So, 'X = pam' is the first answer to this question. Next, the Prolog system tries to match or unify the clause 'parent(X, bob)' with the second fact 'parent(tom, bob)' in the given program, yielding the second answer 'X = tom'. Since the Prolog system cannot succeed in unifying or matching the clause 'parent(X, bob)' with the other facts in the program, there are no more answers to this question.

The following picture describes graphically how the Prolog system finds the answers to the question '?- parent(X, bob).' given the program in Example 9.1.

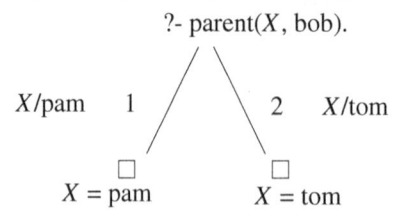

?- parent(X, bob).

X/pam   1   2 X/tom

□   □
X = pam X = tom

This picture is called the *search tree* for the question '?- parent(X, bob).' given the program above. The numbers 1 and 2 refer to the first and second facts in the program. 'X/pam' indicates that the variable 'X' is substituted by 'pam' in order to match the clause 'parent(X, bob)' with fact (1) in the program. The symbol '□' indicates that no other questions remain to be answered.

9.1 Programming in Logic

The search tree for the question 'do bob and liz have a common parent?' given the logic program in Example 9.1 looks as follows.

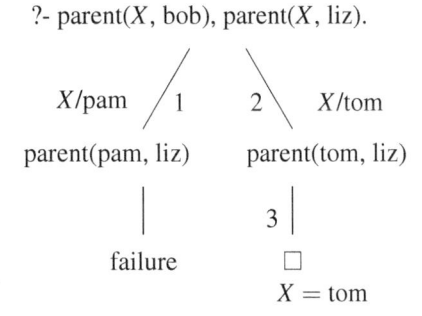

$$\text{?- parent}(X, \text{bob}), \text{parent}(X, \text{liz}).$$

```
                    X/pam  / 1      2 \  X/tom
                  parent(pam, liz)   parent(tom, liz)
                         |              3 |
                       failure           □
                                       X = tom
```

In this example there are two simultaneous questions or *goals*: parent(X, bob), parent(X, liz). The Prolog system selects the *left-most* goal first and tries to match it with the facts in the program. The first possibility to do so is by matching it with fact (1) in the program via the *substitution* X/pam. Then only the goal 'parent(X, liz)' remains with 'X' replaced by 'pam. However, this goal cannot be realized, in the sense that there is no such fact in the program. Hence, this trial to realize the two goals fails. Then the Prolog system *backtracks* and tries to realize the first goal in another way. This can be done by matching it with fact (2) in the program via the substitution X/tom. The second goal 'parent(X, liz)' remains with 'X' replaced by 'tom'. Since 'parent(tom, liz)' occurs as fact (3) in the program, this goal can be realized by the program and no other goals remain.

Given the logic program in Example 9.1, the search tree for the question 'who are pat's grandparents?', i.e., 'for which X is it true that X is a grandparent of pat', looks as follows.

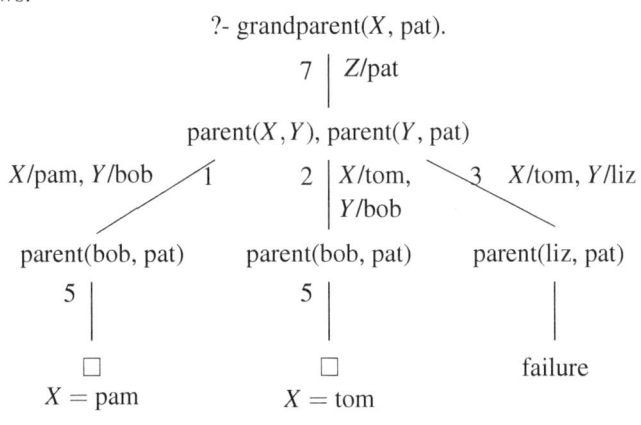

$$\text{?- grandparent}(X, \text{pat}).$$

```
                              7 | Z/pat

                        parent(X, Y), parent(Y, pat)
     X/pam, Y/bob  / 1      2 | X/tom,    \ 3  X/tom, Y/liz
                               | Y/bob
     parent(bob, pat)        parent(bob, pat)        parent(liz, pat)
          5 |                     5 |                     |
            □                       □                   failure
         X = pam                  X = tom
```

Given the question or goal '?- grandparent(X, pat).', the Prolog system looks for facts of this form in the given program, but does not find any. It also tries to match or unify the goal with the head of a rule in the program. This succeeds: the goal 'grandparent(X, pat)' can be unified with the head of clause (7) in the program via the substitution Z/pat. Then the original goal is replaced by two new goals:

parent(X,Y), parent(Y, pat). The *left-most* goal is selected first. Looking at the program, we see that there are six different ways to realize this goal. The first possibility is to use the first clause in the program via the substitution X/pam, Y/bob. The second goal 'parent(Y, pat)' remains with 'Y' substituted by 'bob'. Since 'parent(bob, pat)' is the 5th clause in the program, this goal is realized and no other goals remain. The Prolog system answers 'X = pam'. Next the system *backtracks* and tries to realize the left-most goal 'parent(X,Y)' in another way. This can be done by using the second clause in the program via the substitution X/tom, Y/bob. Then the second goal 'parent(Y, pat)' remains with 'Y' replaced by 'bob'. Since this is the 5th fact in the program, the second branch in the search tree is completed successfully and the system answers 'X = tom'. Again, backtracking takes place and Prolog realizes the left-most goal 'parent(X,Y)' by using the third fact in the program via the substitution X/tom, Y/liz. Then the goal 'parent(liz, pat)' remains. Looking at the program, the Prolog system discovers that this goal cannot be realized. It backtracks and tries to realize the left-most goal 'parent(X,Y)' in a fourth way, and so on.

Summarizing: In logic programming a program consists of *facts* and *rules*. A logic program is a kind of database. A question is a finite sequence of one or more goals. Given a logic program, questions are answered by trying exhaustively to realize the goals by *matching* (or *unifying*) them with the facts and/or the heads of the rules in the program, possibly via *substitution* of the variables. A logic programming system accepts facts and rules as a set of (non-logical) axioms and a question as a putative theorem or conclusion. The logic programming system tries to deduce this putative conclusion logically from the axioms.

Typical features of Prolog are that it has a *left-most selection rule*, that the search is *depth-first* (not breadth-first) and that after successful or unsuccessful termination of a branch in the search tree, *backtracking* takes place in order to find alternative solutions.

The reader is advised to do Exercise 9.1.

9.1.1 Recursion

The use of recursive definitions (of predicates or relations) is typical in logic programming. As an example we might add the following two clauses to the logic program in Example 9.1:

```
pred(X,Z) :- parent(X,Z).
pred(X,Z) :- parent(X,Y), pred(Y,Z).
```

```
X
|    parent
Y
⋮    pred
Z
```

These two clauses define the predecessor relation, abbreviated by 'pred'. The first clause expresses that X is a predecessor of Z if X is a parent of Z. And the second one expresses that X is a predecessor of Z if (for some Y) X is a parent of Y and

9.1 Programming in Logic

Y is a predecessor of Z. So, in the second clause the relation 'pred' recurs in the definition of 'pred(X, Z)'. For this reason one speaks of a *recursive definition*.

In order to understand the role of recursive definitions in logic programming, let us ask the question 'is tom a predecessor of pat?', given the program which results from adding the two clauses for 'pred' to the program in Example 9.1. This program then looks as follows:

Example 9.2.

parent(pam, bob).	(1)
parent(tom, bob).	(2)
parent(tom, liz).	(3)
parent(bob, ann).	(4)
parent(bob, pat).	(5)
parent(pat, jim).	(6)
grandparent(X, Z) :- parent(X, Y), parent(Y, Z).	(7)
pred(X, Z) :- parent(X, Z).	(8)
pred(X, Z) :- parent(X, Y), pred(Y, Z).	(9)

Given this program, the search tree for the question (or goal) 'is tom a predecessor of pat?' looks as follows:

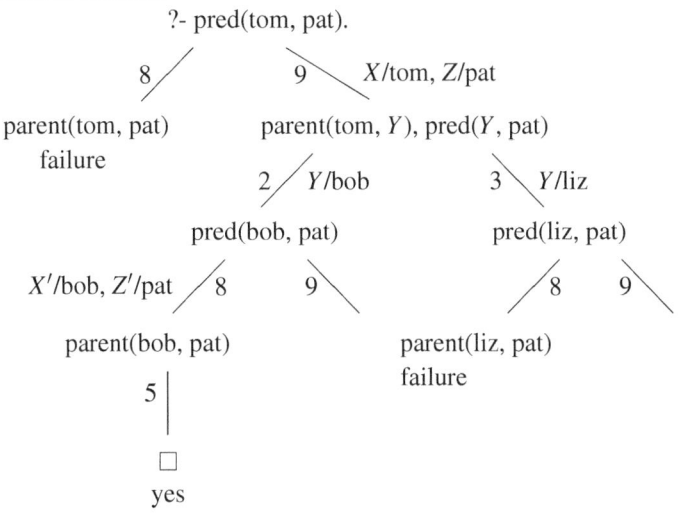

```
                    ?- pred(tom, pat).
              8 /                    9 \   X/tom, Z/pat
      parent(tom, pat)       parent(tom, Y), pred(Y, pat)
         failure           2 / Y/bob         3 \ Y/liz
                         pred(bob, pat)      pred(liz, pat)
            X'/bob, Z'/pat / 8      9 \       / 8        9 \
                 parent(bob, pat)        parent(liz, pat)
                       5 |                   failure
                       □
                      yes
```

In order to answer the question 'pred(tom, pat)' the Prolog system tries to unify this goal with a fact or the head of a rule in the program. The first possibility is to unify 'pred(tom, pat)' with the head of clause (8) via the substitution X/tom, Z/pat. The goal 'parent(tom, pat)' is the result. Since there is no such fact in the given program, the left-most branch in the search tree terminates unsuccessfully and backtracking takes place. The original goal can also be unified with the head of clause (9) in the program via the substitution X/tom, Z/pat. Then the original goal is replaced by two new goals: parent(tom, Y), pred(Y, pat). Prolog selects the left-most goal first. It can be unified with clause (2) in the program via the substitution Y/bob. Then the goal 'pred(Y, pat)' remains with 'Y' substituted by 'bob'. In order to realize this latter goal, Prolog first matches it with the head of clause (8). Since 'X' and 'Z'

have already been substituted by 'tom' and 'pat' respectively, the Prolog system has replaced the variables X and Z in clause (8) by X' and Z' respectively. Clause (8) now looks as follows:

$$\text{pred}(X', Z') \text{ :- parent}(X', Z'). \tag{8'}$$

The goal 'pred(bob, pat)' can now be unified with the head of clause (8'). This yields the new goal 'parent(bob, pat)'. Because of the 5th clause in the program, the second branch (from the left) in the search tree terminates successfully, and the answer to the original question is 'yes'. Backtracking takes place in order to see whether the goal 'pred(bob, pat)' can be realized in other ways. In our search tree we have not worked this out. But it turns out that the goal 'pred(bob, pat)' cannot be realized in another way. Further backtracking takes place in order to see whether the leftmost goal in 'parent(tom, Y), pred(Y, pat)' can be realized in another way. This may indeed be the case. Applying clause (3) in the program and substituting 'liz' for 'Y', the goal 'pred(Y, pat)' remains with 'Y' replaced by 'liz', and so on.

9.1.2 Declarative versus Procedural Programming

Logic programming is in principle *declarative*: the programmer only has to describe the problem, in other words, he must formulate *what* the problem is; but he does not have to specify *how* the problem has to be solved. The logic programmer is more concerned with *knowledge* than with *algorithms*.

In order to answer certain questions, the logic programmer has to formulate all relevant information, consisting of facts and rules, in a logic program. Given a certain program, the logic programming system will try systematically to deduce the answer to any question from the facts and rules in the program. This is done by an exhaustive search, which can be represented in a *search tree*.

Given the program in Example 9.2, the answer to the question 'is tom a predecessor of pat?' is 'yes'. This means that pred(tom, pat) logically follows from (is a valid consequence of) the facts and the rules in the program. The answer to the question 'is pam a predecessor of liz?' will be 'no', meaning that pred(pam, liz) is not a logical consequence of the (facts and rules in the) program. This does **not** mean that ¬ pred(pam, liz) is a logical consequence of the program in question! And the answer to the question

$$\text{?- parent}(X, \text{bob}), \text{parent}(X, \text{liz}).$$

was 'X = tom'. This means that both parent(tom, bob) and parent(tom, liz) logically follow from the given program.

One can prove what is called *soundness*: given any logic program P and question or goal G every computed answer logically follows from P. The converse problem is *completeness*: given any logic program P, is the logic programming system able to compute any goal which logically follows from P? This problem is more difficult and cannot be answered with a simple 'yes' or 'no'.

9.1 Programming in Logic

Programming languages like Pascal, Algol and C are *procedural* languages: the programmer has to specify *how* the problem has to be solved. Although logic programming is in principle declarative and not procedural, the logic programmer has to take into account certain procedural aspects. We have already noticed above that Prolog, being a particular – but most popular – logic programming system, has a *left-most selection rule* and a *depth-first search* strategy. This strategy first develops the left-most branch in the search tree. As long as this branch has not been terminated, no other branches are developed. It also searches for facts and rules in the program in the order they have been programmed. The programmer, who writes a logic program in Prolog, has to take the procedural aspects of the Prolog system into account. The order of the facts and rules in his program may be important and even the order of the goals in the body of a rule may be important. In order to make this clear, look at the following four definitions of the predecessor relation.

pred(X,Z) :- parent(X,Z).
pred(X,Z) :- parent(X,Y), pred(Y,Z).

        pred2(X,Z) :- parent(X,Y), pred2(Y,Z).
        pred2(X,Z) :- parent(X,Z).

pred3(X,Z) :- parent(X,Z).
pred3(X,Z) :- pred3(X,Y), parent(Y,Z).

        pred4(X,Z) :- pred4(X,Y), parent(Y,Z).   (I)
        pred4(X,Z) :- parent(X,Z) .   (II)

In the definition of 'pred2' and 'pred4' the order of the clauses is reversed with respect to the definition of 'pred'. And in the definition of 'pred3' and 'pred4' the goals in the body of the recursion clause are reversed with respect to the definition of 'pred'. This may have disastrous consequences. Consider the search tree for ?- pred4(tom, pat).

    ?- pred4(tom, pat).

    I | X/tom, Z/pat

    pred4(tom, Y), parent(Y, pat)

    I | X'/tom, Z'/Y

    pred4(tom, Y'), parent(Y',Y), parent(Y, pat)

    I |

    .
    .
    .

    ad infinitum

The left-most branch in this search tree will be infinitely long since clause I will be applied again and again. And the other branches in the search tree will not be developed. So, the Prolog system will give no answer to the question '?- pred4(tom, pat).' and although the Prolog system does answer the questions '?- pred2(tom, pat).' and '?- pred3(tom, pat).' positively, Exercise 9.2 makes clear that from a procedural point of view, pred2 and pred3 do not give an adequate description of the predecessor relation.

To summarize, although the definitions of 'pred', 'pred2', 'pred3' and 'pred4' are equivalent from a logical or declarative point of view, they are quite different from a procedural point of view. Consequently, the logic programmer has to take into account the particular procedural aspects of the logic programming system he or she is working with. For recursive definitions there is a simple rule:

i) The more simple clause is put first (supposing that the logic programming system searches through the program from top to bottom). In the case of the predecessor relation this is the clause 'pred(X,Z) :- parent(X,Z).'

ii) The goals in the body of the recursion clause should be ordered from simple to more complex (supposing a left-most selection rule in the logic programming system). So, the recursion clause for the predecessor relation should be

$$\text{pred}(X,Z) :\text{- parent}(X,Y), \text{pred}(Y,Z).$$

since 'parent' is a more simple relation than 'pred'.

Another procedural aspect of Prolog is the cut, denoted by '!'. The cut prunes part of the search tree. This enhances efficiency, but may be dangerous if the pruned part contains successful branches. We will discuss this feature of Prolog further on in Subsection 9.1.6.

9.1.3 Syntax

The syntax of a logic programming language is that of first-order predicate logic (see Chapter 4) with possibly some modifications in notation.

Definition 9.1 (Alphabet of Prolog). The *alphabet* consists of:
a) *Individual variables*, such as $X1, X2, \ldots, X, Y$, Person, ...
b) *Individual constants*, such as pam, bob, liz, ... ; 0, 1, 2, ...
c) *Function symbols*, such as father_of, mother_of, ... ; $+, *, \ldots$
d) *Predicate symbols*, such as parent, male, female; $=, <, \ldots$

The programmer is free to choose his own symbols in the alphabet. However, in Prolog any expression starting with a capital is a variable. For instance, the expressions 'Pam' and 'Bob' are variables in Prolog, while 'pam' and 'bob' are individual constants. Each function and predicate symbol is k-ary for some k; the arity is chosen by the programmer.

Definition 9.2 (Terms). *Terms* are defined as in Chapter 4:
a) Each individual variable and each individual constant is a term.
b) If f is a k-ary function symbol and $t_1, \ldots, t_k$ are terms, then $f(t_1, \ldots, t_k)$ is a term.

9.1 Programming in Logic

Examples of terms are: 1) Person, tom, bob; 1, 2.
2) father_of(Person), mother_of(bob); +(1, 2), usually written as $1+2$.
3) mother_of(father_of(Person)); *(1, +(1, 2)), usually written as $1*(1+2)$.

Definition 9.3 (Atomic Formulas). *Atomic formulas* are defined as in Chapter 4: if p is an n-ary predicate symbol and $t_1, \ldots, t_n$ are terms, then $p(t_1, \ldots, t_n)$ is an atomic formula.

Examples of atomic formulas: parent(tom, bob), parent(X, liz), female(X), male(Y), male(bob); = (X, 1), usually written as $X = 1$, and <(2, 3), usually written as $2 < 3$.

Definition 9.4 (Definite Program Clause, Goal, Program, Clause).
a) A *definite program clause* is any expression of the form B :- $A_1, \ldots, A_m$ ($m \geq 0$), where B and $A_1, \ldots, A_m$ are atomic formulas (Cf. Definition 2.10). If $m = 0$, one writes simply 'B' instead of 'B :- '. 'B :- $A_1, \ldots, A_m$' is to be read as: B if A_1 and $\ldots$ and A_m; in the notation of Chapter 4: $B \leftarrow A_1 \wedge \ldots \wedge A_m$. B is called the *head* and $A_1, \ldots, A_m$ is called the *body* of the clause B :- $A_1, \ldots, A_m$. If $m = 0$, one calls the definite program clause a *fact*, otherwise a *rule*.
b) A *definite goal* is any expression of the form :- $A_1, \ldots, A_m$ ($m \geq 0$), where $A_1, \ldots, A_m$ are atomic formulas. Each A_i is called a *subgoal* of the goal. If $m = 0$, one speaks of the *empty goal* or *empty clause*, denoted by '□'.
c) A *definite program* is a finite set of definite program clauses. See, for instance, Example 9.1.
d) A *definite clause* or *Horn clause* is either a definite program clause or a definite goal.

Definition 9.5 (Literal, Normal Program Clause, Goal, Program).
a) A *literal* is an atomic formula A or the negation 'not A' of an atomic formula A. In the notation of Chapter 4 'not A' was written as '$\neg A$'.
b) A *normal program clause* is any expression of the form B :- $L_1, \ldots, L_m$, where B is an atomic formula and $L_1, \ldots, L_m$ are literals.
Example: sister(X, Y) :- parent(Z, X), parent(Z, Y), female(X), not $X = Y$.
c) A *normal goal* is any expression of the form :- $L_1, \ldots, L_m$, where $L_1, \ldots, L_m$ are literals.
d) A *normal program* is a finite set of normal program clauses.

From a theoretical point of view, in particular with respect to completeness, definite logic programs are to be preferred to normal logic programs. However, for practical purposes the latter ones are often needed. We return to the problems connected with negation in Subsection 9.1.7 and 9.1.9.

Definite and normal program clauses are formulas of a special kind. In order to see this, let us repeat the definition of *formulas* as given in Chapter 4.

Definition 9.6 (Formulas).
a) Any atomic formula is a formula.
b) If A and B are formulas, then so are $(A \rightleftarrows B)$, $(A \rightarrow B)$, $(A \wedge B)$, $(A \vee B)$ and $(\neg A)$, to be read as 'A if and only if B', 'if A, then B', 'A and B', 'A or B' and 'not A', respectively.

c) If A is a formula and x is an individual variable, then $\forall x[A]$ and $\exists x[A]$ are formulas, to be read as 'for all x, A' and 'there is at least one x such that A', respectively. (See Definition 4.5 for more details.)

In Theorem 2.18 we have shown that any formula not containing the quantifiers $\forall$ and $\exists$ is equivalent to a finite conjunction of clauses, where a *clause* is a formula of the form $A_1 \wedge \ldots \wedge A_m \to B_1 \vee \ldots \vee B_n$ or, in different notation, $B_1 \vee \ldots \vee B_n$:- $A_1,\ldots,A_m$, the A's and B's being *atomic* formulas. (See the topic on Knowledge Representation and Prolog in Subsection 2.5.2.) Note that $B_1 \vee \ldots \vee B_n$:- $A_1,\ldots,A_m$ is equivalent to $B_1 \vee \ldots \vee B_n \vee \neg A_1 \vee \ldots \vee \neg A_m$.

However, in Prolog only *definite clauses* are allowed, i.e., clauses with $n \leq 1$, for reasons of efficiency. For instance, suppose we had a program containing the clause $p(1) \vee q(2)$. Now the question ?- $p(X) \vee q(X)$ should be answered as follows: $X = 1$ if $p(1)$ and $X = 2$ if $q(2)$. It is hard to implement a system that is able to give such conditional answers.

In Subsection 4.3.6 on Skolemization and Clausal Form we have defined the *clausal form* $C(A)$ of any formula A, such that 1) $C(A)$ is a conjunction of clauses, and 2) A is satisfiable iff $C(A)$ is satisfiable. And in Subsection 4.7.3 on Logic and Artificial Intelligence we have shown that any (definite) logic program is a formula in clausal form.

9.1.4 Matching versus Unification

In the preceding examples we have seen that a Prolog system, given a certain program and answering a certain question, makes use of what is called matching or unification. In this subsection we want to describe matching and unification more precisely and to point out the difference between them.

Definition 9.7 (Matching). *Matching* is a process that takes as input two terms or atomic formulas and checks whether they match. The rules governing this process are the following:
1. Two individual constants match only if they are syntactically the same.
2. If X is a variable and t a term, then they match and X is instantiated to, or substituted by, t.
3. Two terms match only if they have the same principal function symbol and all their arguments match.
4. Two atomic formulas match only if they have the same principal predicate symbol and all their arguments match.

Example 9.3. a) The pair $\{parent(X, bob), parent(pam, bob)\}$ can be matched.
b) The pair $\{parent(X, bob), parent(pat, jim)\}$ cannot be matched, because the second arguments cannot be matched.
c) The pair $\{p(f(X),Z), p(Y,c)\}$ can be matched via the substitution $Y/f(X)$, Z/c.
d) The pair $\{p(f(X),c), p(Y,f(Z))\}$ cannot be matched, since the second arguments cannot be matched.

9.1 Programming in Logic

e) The pair $\{p(X,X), p(Y,f(Y))\}$ can be matched. The first arguments can be matched via the substitution X/Y. The resulting pair is $\{p(Y,Y), p(Y,f(Y))\}$. The second arguments can be matched via the substitution $Y/f(Y)$.

The following logic program shows that in some cases matching yields undesirable results. Let P be the following logic program:

$$\text{parent}(Y, \text{child\_of}(Y)). \tag{1}$$

Note that 'parent(X,X)' is similar to '$p(X,X)$' and that 'parent$(Y, \text{child\_of}(Y))$' is similar to '$p(Y,f(Y))$' in Example 9.3 e). Given this program, the search tree for the question '?- parent(X,X).' looks as follows.

$$?\text{- parent}(X,X).$$

(1) | X/Y, $Y/\text{child\_of}(Y)$

□
yes

However, this result is undesired, since 'parent(X,X)' does *not* logically follow from program P. In the intended interpretation the only formula (1) in P expresses a true proposition, while 'parent(X,X)' expresses a false proposition for any value of X.

For reasons of efficiency, most logic programming systems make use of matching and take for granted that in some cases this may yield the wrong results. What they should do, however, is to replace matching by unification and take for granted that in some cases this may be inefficient.

Definition 9.8 (Unification). *Unification* is characterized by the following slogan:

$$\text{Unification} = \text{matching} + \text{occur check}.$$

The *occur check* involves checking whether in the substitution of a term t for a variable X (clause 2 in the definition of matching), the variable X does not occur in t. If X does occur in t, then unification fails, while matching may succeed.

In Example 9.3 e) we have seen that the pair $\{p(X,X), p(Y,f(Y))\}$ can be matched. However, this pair cannot be unified. After the substitution X/Y the resulting pair is $\{p(Y,Y), p(Y,f(Y))\}$. And although the second arguments in this pair can be matched, they cannot be unified since the variable Y does occur in $f(Y)$.

The *unification algorithm* is like the matching algorithm given above, except that the occur check is added to clause 2. Below, we demonstrate how the unification algorithm works in a few examples.

Example 9.4. (Lloyd [16]):
Is it possible to unify $p(f(c), g(X))$ and $p(Y,Y)$?
1) The predicate symbols are identical.
2) The left-most arguments that differ are $f(c)$ and Y. *Occur check*: Y does not occur in $f(c)$. So, replace Y by $f(c)$. Result: $p(f(c), g(X))$ and $p(f(c), f(c))$.
3) The left-most arguments that differ are $g(X)$ and $f(c)$. These terms have different principal function symbols and hence cannot be unified.
Conclusion: $p(f(c), g(X))$ and $p(Y,Y)$ cannot be unified. Nor can they be matched.

Example 9.5. (Lloyd [16]):
Can $p(c,X,h(g(Z)))$ and $p(Z,h(Y),h(Y))$ be unified?
1) The predicate symbols are identical.
2) The left-most arguments that differ are c and Z. *Occur check*: Z does not occur in c. So, replace Z by c. Result: $p(c,X,h(g(c)))$ and $p(c,h(Y),h(Y))$.
3) The left-most arguments that differ now are X and $h(Y)$. *Occur check*: X does not occur in $h(Y)$. So, replace X by $h(Y)$. Result: $p(c,h(Y),h(g(c)))$ and $p(c,h(Y),h(Y))$.
4) The left-most arguments that differ now are $h(g(c))$ and $h(Y)$. The principal function symbols are identical. The arguments $g(c)$ and Y are different. *Occur check*: Y does not occur in $g(c)$. So, replace Y by $g(c)$. Result: $h(g(c))$ and $h(g(c))$.
Conclusion: $p(c,X,h(g(Z)))$ and $p(Z,h(Y),h(Y))$ are unifiable via the substitutions Z/c, $X/h(Y)$ and $Y/g(c)$.

For more details about substitution and unification the reader is referred to Lloyd [16]. See also Exercise 9.3.

Note that both the matching and the unification algorithms result in a *most general unifier* (substitution) in the sense that no more is substituted than strictly necessary. For instance, unifying date(D, M, 1983) and date($D1$, may, Y) results in the substitution $D/D1$, M/may and Y/1983. The substitutions D/3, $D1$/3, M/may and Y/1983 also unify the two terms, but are less general.

One can show that given a logic program P any answer to a question is correct in the sense that the computed answer is a logical consequence of the given program, provided the system uses unification instead of matching. For instance, given the program of Example 9.1 the Prolog system computes two answers to the question ?- grandparent(X, pat): X = pam; X = tom. The theorem just mentioned, called the *soundness* theorem, then says that 'grandparent(pam, pat)' and 'grandparent(tom, pat)' are logical consequences of the given program. For more details we refer the reader to Lloyd [16] where also the converse problem is discussed whether any logical consequence of a given program can be computed by the Prolog system (*completeness*).

9.1.5 Lists, Arithmetic

Lists are very important terms in the practice of logic programming. For instance, if one wants to represent information about families in a logic program, one has the problem that different families have different numbers of children. By putting the children of any family in a list, one can represent any family in a uniform way:

family(Father, Mother, List_of_Children).

Here, 'family' is a predicate symbol taking three arguments, no matter how many children there are.

Lists are terms of a special kind, defined as follows.

9.1 Programming in Logic

Definition 9.9 (Lists).
1) [ ] is a list, called the *empty list*.
2) If t is any term and L is a list, then $[t \mid L]$ is a list.
In $[t \mid L]$, t is called the *head* and L is called the *tail* of the list $[t \mid L]$.

Example 9.6. Examples of lists:
1) [ ]
2) $[c \mid [\,]]$, usually rendered as $[c]$.
3) $[b \mid [c]]$, usually rendered as $[b,c]$.
4) $[a \mid [b,c]]$, usually rendered as $[a,b,c]$.

Working with lists, one needs a program that determines the elements or members of a given list. Reading 'member(X,L)' as 'X is a member of list L', the membership relation is defined recursively as follows.

Definition 9.10 (Member). member$(X,[X \mid L])$. (1)
member$(X,[Y \mid L])$:- member(X,L) . (2)

In words: X is a member of a given list if (1) X is the head of the list, or (2) X is a member of the tail of the list. Given this program, the search tree for the question '?- member$(X,[b,c])$.' (what are the members of the list $[b,c]$?) looks as follows.

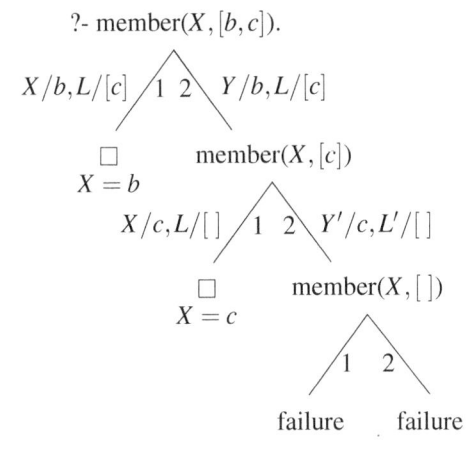

$$?\text{- member}(X,[b,c]).$$

```
          ?- member(X,[b,c]).
         /                  \
  X/b,L/[c] 1              2  Y/b,L/[c]
       /                         \
      □                      member(X,[c])
    X = b                    /           \
                    X/c,L/[] 1          2  Y'/c,L'/[]
                         /                    \
                        □                 member(X,[])
                      X = c                 /      \
                                           1        2
                                        failure  failure
```

In Exercise 9.4 the reader is invited to define a concatenation relation for lists and in Exercise 9.5 to define a relation for deleting members from a given list.

Prolog contains some built-in arithmetic operations which can be used in the infix notation, i.e., $2+3$ instead of $+(2,3)$, etc. Among them are

$+$ for addition, $*$ for multiplication,
$-$ for subtraction, $/$ for division.

When doing arithmetic in Prolog it is important to realize that '=' is a built-in matching operator, while 'is' is a built-in operator that forces the evaluation of the term in question. In order to make the difference clear, consider the following examples.

?- $X = 2+3$.   ?- X is $2+3$.   ?- $2+3 = 3+2$.
$X = 2+3$ $X = 5$ no

Other built-in operators which force the evaluation of the terms in question are:
> is greater than;
>= is greater than or equal to;
=:= the values of the left and right terms are equal;
=\= the values of the left and right terms are not equal.

Example 9.7.
?- 2+3 =:= 3+2.   ?- 2*3 > 5.
yes   yes

It is important to realize that all arguments must be instantiated to numbers at the time that the evaluation is carried out. Examples:

?- X is 2*3, X > 5.   ?- X > 5, X is 2*3.
X = 6   control error

9.1.6 Cut

max(X, Y, Z), to be read as 'Z is the maximum of X and Y', can be defined as follows.

$$\text{max}(X,Y,X) :\!- X >= Y. \quad (1)$$
$$\text{max}(X,Y,Y) :\!- Y > X. \quad (2)$$

Now the programmer, but not the logic programming system, knows that if the goal $X >= Y$ succeeds, then the goal $Y > X$ is bound to fail. So, given this program and the question ?- max(3, 2, Z), it is a waste of time and energy to try to apply the second clause via backtracking, once the left-most branch in the search tree has terminated successfully.

?- max(3, 2, Z).

X/3, Y/2, Z/X/1   2 X/3, Y/2, Z/Y

3 >= 2   2 > 3

□   failure
Z = 3

It is attractive to have a control facility that prunes that part of a search tree that only contains unsuccessful branches. Prolog has such a control facility, called *cut* and denoted by '!'. The cut ! can be conceived of as a true atomic formula or as a goal that always succeeds. However, while the declarative or logical meaning of '!' is 'true', the procedural meaning of '!' is the pruning of the search tree. Given the program

$$\text{max1}(X,Y,X) :\!- X >= Y, !. \quad (1)$$
$$\text{max1}(X,Y,Y). \quad (2)$$

9.1 Programming in Logic

the search trees for the questions ?- max1(3, 2, Z) and ?- max1(2, 3, Z) look as follows:

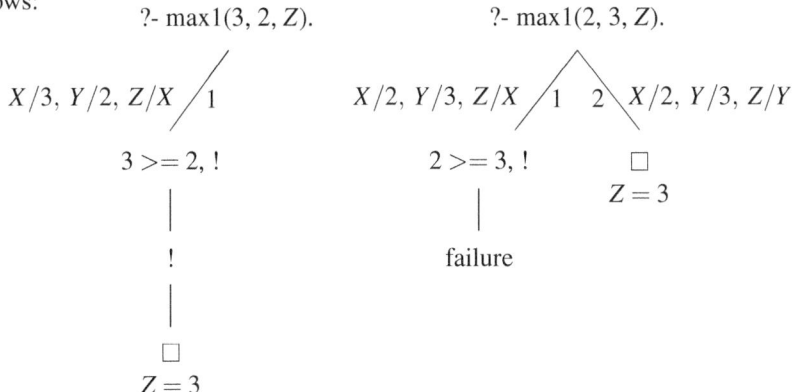

```
            ?- max1(3, 2, Z).                    ?- max1(2, 3, Z).

    X/3, Y/2, Z/X  / 1            X/2, Y/3, Z/X  / 1    2 \ X/2, Y/3, Z/Y
           3 >= 2, !                   2 >= 3, !              □
              |                           |                 Z = 3
              !                        failure
              |
              □
            Z = 3
```

The right-most branch in the search tree for ?- max1(3, 2, Z) is pruned because first the goal $3 >= 2$ succeeds and next the cut is passed. The goal $2 >= 3$ in the left-most branch in the search tree for ?- max1(2, 3, Z) fails, hence the cut is not passed and backtracking takes place as usual.

From a procedural point of view, the programs for max(X,Y,Z) and for max1(X,Y,Z) yield the same results. However, from a declarative or logical point of view the program for max1(X,Y,Z) is not an adequate description of the maximum relation. Since the declarative meaning of '!' is 'true', from a declarative point of view the program for max1 is equivalent to the following one.

$$\text{max2}(X,Y,X) :\!-\ X >= Y \qquad (1)$$
$$\text{max2}(X,Y,Y). \qquad (2)$$

But the question ?- max2(3, 2, Z) yields two answers: $Z = 3$ and $Z = 2$.

```
            ?- max2(3, 2, Z).

    X/3, Y/2, Z/X / 1    2 \ X/3, Y/2, Z/Y
           3 >= 2              □
              |              Z = 2
              □
            Z = 3
```

So, if one wants a program for the maximum relation that is both correct from a declarative point of view and efficient from a procedural point of view, the following program is to be preferred.

$$\text{max3}(X,Y,X) :\!-\ X >= Y, !. \qquad (1)$$
$$\text{max3}(X,Y,Y) :\!-\ Y > X. \qquad (2)$$

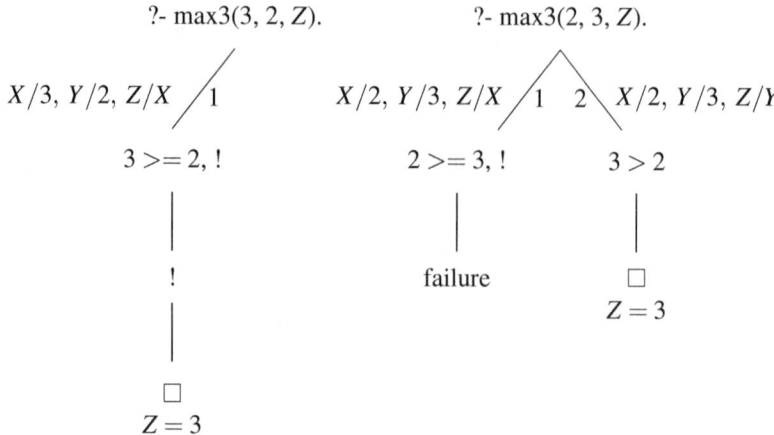

```
        ?- max3(3, 2, Z).                    ?- max3(2, 3, Z).
       /                                    /        \
  X/3, Y/2, Z/X  / 1           X/2, Y/3, Z/X / 1   2 \ X/2, Y/3, Z/Y
       |                              |                  |
     3 >= 2, !                      2 >= 3, !          3 > 2
       |                              |                  |
       !                           failure              □
       |                                              Z = 3
       □
     Z = 3
```

So, a cut prunes the search tree. This is safe if the pruned part contains no successful branches. In that case the cut merely enhances efficiency; it saves time. However, if the pruned part contains successful branches, the use of cut may have disastrous consequences. For that reason the programmer should be very careful in using this control facility. Unfortunately, more complicated programs often require the use of cuts in order to keep the program efficient.

What part of the search tree is pruned by using cut? In order to answer this question more precisely, consider the following program P.

$$p(X) :\!\!- q(X), r(X).$$
$$\vdots$$
$$q(X) :\!\!- s(X), t(X), !, u(X).$$
$$q(X) :\!\!- v(X).$$
$$\vdots$$
$$r(1).$$
$$s(1).$$
$$t(1).$$
$$u(X) :\!\!- X = 5.$$
$$u(X) :\!\!- X > 2.$$

The following picture shows the effect of cut on the search tree for the question '?- $p(X)$.', given program P above. Given the program P above, the goal '?- $p(X)$.' will be answered with 'no', even if we add the fact $v(1)$ to P. In that case there is a successful branch in the search tree, namely, the branch with $v(X)$, $r(X)$. However, this branch will be pruned because of the cut. If we add a second rule to P, '$p(X) :\!\!- X = 2$.', the goal '?- $p(X)$.' will have '$X = 2$' as its only solution.

In order to formulate precisely what the effect of cut on the search tree is, we have to define the notion of *parent* goal. The *parent* goal is the goal that causes the clause containing the cut to be activated. In our example this is $q(X)$. The cut commits the system to all choices made between the time the parent goal was involved and the

9.1 Programming in Logic   443

time the cut was encountered. The remaining alternatives between the parent goal
and the cut are discarded.   ?- $p(X)$.

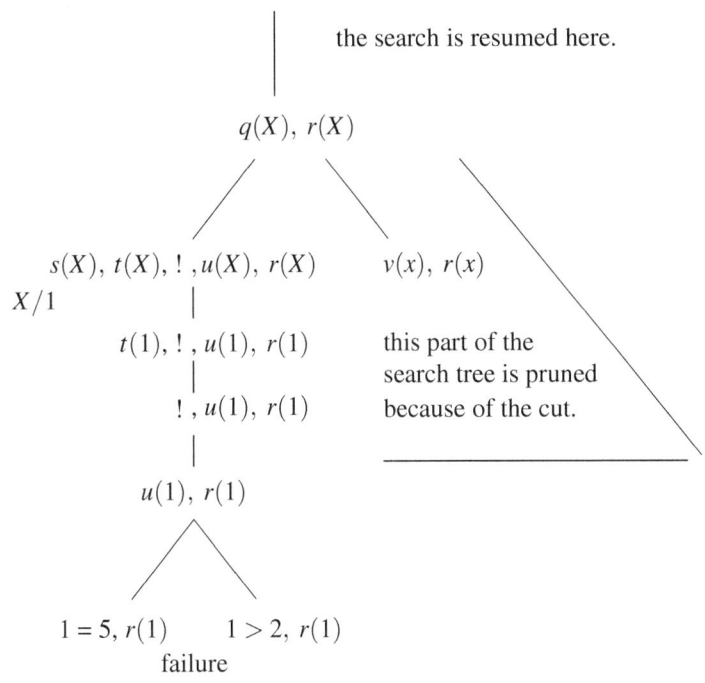

$$
\begin{array}{c}
\text{the search is resumed here.} \\
q(X),\ r(X)
\end{array}
$$

```
                        q(X), r(X)
                       /     |        \
        s(X),t(X),!,u(X),r(X)   v(x),r(x)
    X/1              |
                t(1),!,u(1),r(1)       this part of the
                     |                 search tree is pruned
                !,u(1),r(1)            because of the cut.
                     |
                  u(1), r(1)
                   /      \
           1=5, r(1)    1>2, r(1)
                      failure
```

Exercises 9.7 and 9.8 give some other programs containing cuts.

9.1.7 Negation as Failure

Prolog has a built-in operator 'not', which has been defined, using cut, as follows.

$$\text{not}\,(A) :\!\!- A,\ !,\ \text{fail}.$$
$$\text{not}\,(A) :\!\!- \text{true}.$$

In order to understand this definition, the reader should know that 'fail' and 'true'
are built-in expressions which always fail or succeed respectively, when they are
invoked. From this definition it follows immediately that

(i) the goal 'not (A)' fails if the search tree for '?- A.' is finite and has a successful
branch, and

(ii) the goal 'not (A)' succeeds if the search tree for '?- A.' is finite and has no
successful branches.

Note that if the search tree for '?- A.' contains no successful branches and has
at least one infinite branch, then the Prolog system cannot answer the question '?-
not (A).'. In order to see how Prolog handles negation, let us consider the following
program P.

student(tom).
student(jane).
teacher(mary).

It is important to note that neither 'student(mary)' nor 'not student(mary)' are logical consequences of *P*.

Given this program, the question '?- not student(mary).' is answered by the Prolog system as follows:

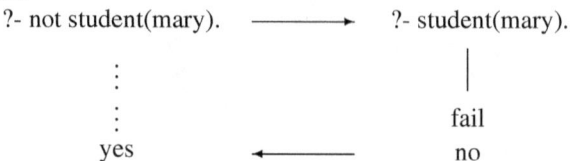

```
?- not student(mary).    ⟶    ?- student(mary).
       ⋮                              |
       ⋮                             fail
      yes            ⟵                no
```

The Prolog system uses what is called the *Negation as Failure (NF)* rule: if the search tree for *A*, given a certain program, is finite and has no successful branches, then conclude 'not *A*'.

The Negation as Failure rule is *non-monotonic*, i.e., adding new facts and/or rules to the given program may eliminate some former conclusions. For instance, if we add the fact 'student(mary)' to the program *P* above, the conclusion 'not student(mary)' can no longer be drawn. More information may lead to different (and other) conclusions.

Program *P* above is equivalent to the following one:
student(X) :- X = tom.
student(X) :- X = jane.
teacher(X) :- X = mary.

In most cases what the programmer has in mind is not the program *P* itself, but what is called the *completion* of *P*:
student(X) iff X = tom or X = jane.
teacher(X) iff X = mary.

The completion of *P* is obtained by replacing the if's in program *P* by iff's. And although 'not student(mary)' is not a logical consequence of *P*, it is a logical consequence of the completion of *P*. Both the Negation as Failure rule and the process of completion capture the idea that information not given by the program is taken to be false.

Exercises 9.9 and 9.10 make clear that for programs which contain negation, the use of cut may affect the soundness of the system.

9.1.8 Applications: Deductive Databases and Artificial Intelligence

In Example 9.1 we have given a very simple application of logic programming to databases. In this example a database is given containing facts or data concerning who is a parent of whom. This database has been extended with rules stating under what conditions the grandparent relation applies. We have seen that one can add other rules such as rules for the predecessor relation, the mother relation, etc. (see also Exercise 9.1). These rules enable the user to derive conclusions from the

9.1 Programming in Logic

database which are not explicitly present in the database (as static facts), but which can be logically deduced from the facts in the database by means of application of the rules. For this reason one speaks of *deductive databases*. A logic program can be viewed as a (deductive) database, consisting of facts and rules. *Relational databases* correspond to logic programs consisting only of facts.

Prolog contains a number of facilities for updating databases. For instance, the goal 'assert(C)' will always succeed and will result in adding the program clause C to the database. The goal 'asserta(C)' adds C at the beginning of the database and the goal 'assertz(C)' adds C at the end of the database. The goal 'retract(C)' deletes a program clause that matches C.

Example 9.8. The following non-trivial example of a deductive database is from Bratko [7], Section 4.1. The database or logic program contains facts of the following form:

family(
person(tom, fox, date(7, may, 1950), works(bbc, 15200)),
person(ann, fox, date(9, may, 1951), unemployed),
[person(pat, fox, date(5, may, 1973), unemployed),
person(jim, fox, date(5, may, 1973), unemployed)]
).

These atomic formulas are built from a ternary predicate symbol 'family', a 4-ary function symbol 'person', a ternary function symbol 'date', a binary function symbol 'works' and a number of individual constants. The overall structure of these facts is: family (Father, Mother, List_of_Children). Now, given a database of the type above, the question 'give name and surname of all married woman who have at least two children' can be formulated in Prolog as follows:

?- family(_, person(Name, Surname, _, _), [_, _| _]).

In order to understand this formulation the reader should know that '_' is a so-called *anonymous variable*, i.e., a variable whose value is not given when Prolog answers the question. Among the answers to this question would be:

Name = ann,
Surname = fox.

In Exercise 9.11 the reader is invited to add a number of rules to the database such that many other questions can be asked in a straightforward manner.

Since any logic programming system is equipped with a reasoning mechanism, one might say that any such system is able to simulate reasoning and hence disposes of *Artificial Intelligence (AI)*. This makes logic programming a very appropriate tool for solving many problems, which are generally considered to belong to the field of Artificial Intelligence. Many puzzles can be solved by appropriate logic programs. A nice example is *cryptarithmetic puzzles*, such as

$$\begin{array}{r} S\,E\,N\,D \\ \underline{M\,O\,R\,E} \\ M\,O\,N\,E\,Y \end{array} +$$

where the problem is to assign decimal digits to the letters of the alphabet such that the above sum is correct. Bratko's book [7] contains in Section 7.1 a Prolog program for solving cryptarithmetic puzzles.

Example 9.9. We give a simple logic program for colouring a given map, such that the colour in each region is different from the colours in all its adjacent regions.

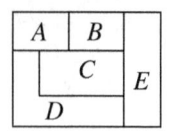

A	B	
	C	E
	D	

color(X) :- X = red.
color(X) :- X = blue.
color(X) :- X = green.
color(X) :- X = black.
next(X,Y) :- color(X), color(Y), not (X = Y).
colormap([A,B,C,D,E]) :- next(A,B), next(A,C), next(A,D), next(B,C),
                        next(B,E), next(C,D), next(C,E), next(D,E).

Given this program, the appropriate question to ask is

$$?\text{- colormap}(Z).$$

Example 9.10. Another example of the use of logic programming in the domain of Artificial Intelligence is for *parsing* sentences. The following program is for parsing sentences in a very simple and small fragment of English.

np([john]).	'john' is a noun phrase	tv([loves]).	'loves' is a transitive verb
np([mary]).		tv([hates]).	
np([bill]).		det([a]).	'a' is a determiner.
cn([dog]).	'dog' is a common noun	det([the]).	
cn([woman]).		vp([walks]).	'walks' is a verb phrase.
cn([man]).		vp([talks]).	

np(L) :- conc(L1, L2, L), det(L1), cn(L2).
vp(L) :- conc(L1, L2, L), tv(L1), np(L2).
s(L) :- conc(L1, L2, L), np(L1), vp(L2).
conc([ ], L, L).
conc([X | L1], L2, [X | L3]) :- conc(L1, L2, L3).

In this program 'conc(L1, L2, L)' should be read as 'L is the concatenation of $L1$ and $L2$', and 's(L)' should be read as 'L is a sentence'. Given this program, questions one might ask are:

?- s([john, hates, the, dog]).   ?- s([john, hates, the, walks]).
yes   no

The question '?- s(S).' will generate all syntactically correct sentences in the given fragment of English.

9.1 Programming in Logic

When a logic program behaves like an expert in some specific domain such as medical diagnosis or system break-down diagnosis, the logic program is called an *expert system* or a *knowledge-based-system*. By 'behaving like an expert' we mean that 1) the logic program contains some expertise information concerning a specific domain, 2) that the program must be able to ask certain questions to the user and 3) that the program must be able to indicate in a user friendly manner how it has derived the answer(s) to a given question. Logic programs which satisfy these conditions become rather complex. Relatively simple examples of such expert systems can be found, among others, in Bratko [7], Chapter 14.

9.1.9 Pitfalls

There are at least four pitfalls the logic programmer should be aware of.

1. We have already mentioned that most actual logic programming systems use matching instead of unification for reasons of efficiency. However, as indicated in Subsection 9.1.4 on Matching versus Unification, it may happen as a consequence that some goal is answered affirmatively, while the goal does not logically follow from the given program. In other words, the lack of the occur check destroys the soundness of the system.

2. The occurrence of a cut in a *definite* program does not affect the soundness of the system, although it may affect the completeness of the system by pruning successful branches. However, Exercise 9.10 makes clear that the use of cut in a *normal* program may even destroy the soundness of the system.

3. Consider the following program P (from Lloyd [16], Section 10).

$$\begin{aligned}
&p(a, b). &&(1)\\
&p(c, b). &&(2)\\
&p(X, Z) \text{ :- } p(X, Y), p(Y, Z). &&(3)\\
&p(X, Y) \text{ :- } p(Y, X). &&(4)
\end{aligned}$$

Now it is easy to see that $p(a, c)$ is a logical consequence of P. From (2) and (4) it follows that $p(b, c)$ (5). And from (1), (5) and (3) it follows that $p(a, c)$.

However, given this program, the question '?- p(a, c).' will not be answered by any of the existing Prolog systems. In order to see why, let us consider the search tree for this question.

Any logic programming system that uses a depth-first search, combined with a fixed order for trying clauses given by their ordering in the program, will never find the success branch, because the left-most branch in the search tree is infinite. We have seen that all the clauses (1), (2), (3) and (4) were used in concluding $p(a, c)$ from P. However, in the left-most branch of the search tree for '?- p(a, c).', clause (4) will never be applied. Interchanging clauses (3) and (4) in the program P would result in a left-most branch in which clause (3) is never applied, while all the clauses in P are necessary to deduce $p(a, c)$.

The solution to this problem would be a logic programming system with a breadth-first search rule. However, it is unlikely that such a system can be implemented efficiently.

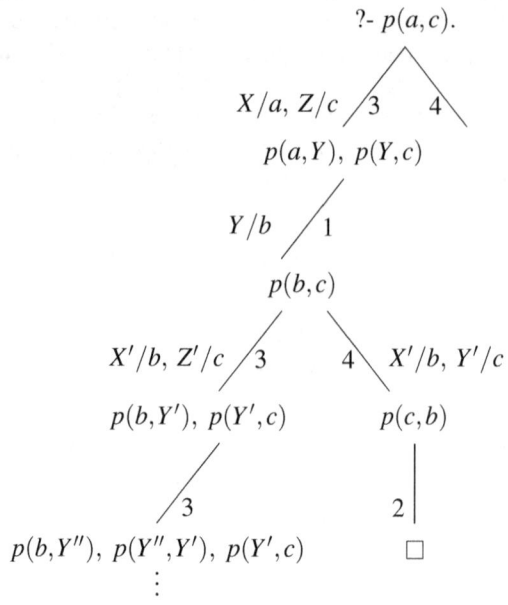

$$?\text{-}\ p(a,c).$$

```
                    ?- p(a,c).
                  X/a, Z/c ╱3    4╲
                    p(a,Y), p(Y,c)
                       Y/b ╱ 1
                       p(b,c)
              X'/b, Z'/c ╱3    4╲ X'/b, Y'/c
              p(b,Y'), p(Y',c)    p(c,b)
                    ╱3                │2
       p(b,Y''), p(Y'',Y'), p(Y',c)   □
                    ⋮
```

4. Many logic programming systems do not satisfy the *safeness condition*: negative literals are only allowed to be selected if they do not contain any variables. The safeness condition can be implemented by delaying the treatment of negative subgoals until any variable in the subgoal has been substituted by a term not containing variables.

Violation of the safeness condition affects the soundness of the system. Consider, for instance, the following program P:

$$\begin{aligned}&\text{bachelor}(X) \text{:- not married}(X), \text{man}(X). &(1)\\ &\text{man(bob).} &(2)\\ &\text{married(alice).} &(3)\end{aligned}$$

What the programmer actually has in mind is not P itself, but the *completion* of P, consisting of the following formulas:

$$\begin{aligned}&\text{bachelor}(X) \rightleftarrows \text{not married}(X), \text{man}(X).\\ &\text{man}(X) \rightleftarrows X = \text{bob.}\\ &\text{married}(X) \rightleftarrows X = \text{alice.}\end{aligned}$$

From the completion of P it logically follows that for some X, bachelor(X), namely X = bob. A logic programming system that delays the treatment of a negative subgoal, until all variables in the subgoal have been replaced by terms not containing variables, will answer the question '?- bachelor(X).' with 'X = bob'.

9.1 Programming in Logic

```
            ?- bachelor(X).              The goal printed in italics is
                  |                      the goal selected by a system
                  | (1)                  satisfying the safeness condition.
          not married(X), man(X)
                  |
          X/bob   | (2)
          not married(bob)    ─────────────▶  ?- married(bob).
                  |                                  |
                  □  ◀─────────────────────────── failure
               success
               X = bob
```

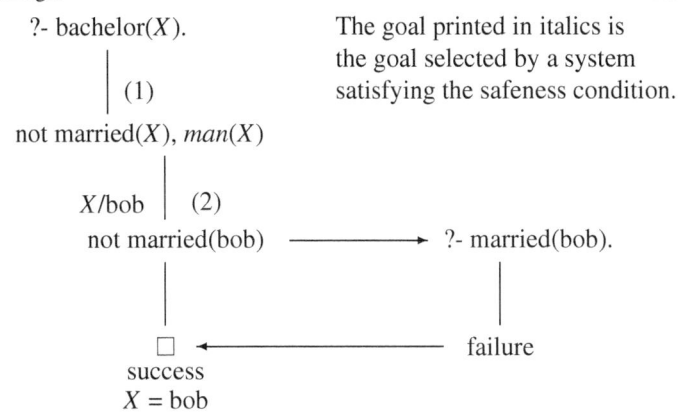

However, a logic programming system that does not satisfy the safeness condition will answer the question '?- bachelor(X).' with 'no'.

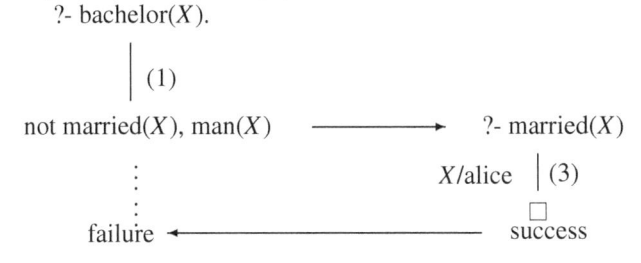

```
            ?- bachelor(X).
                  |
                  | (1)
          not married(X), man(X)    ─────────────▶  ?- married(X)
                  ⋮                                 X/alice │ (3)
                  ⋮                                         □
               failure  ◀───────────────────────────     success
```

Exercise 9.1. Extend the program concerning the parent relation in Example 9.1 with rules which define the offspring, the father, the mother, the sister and the brother relation. It will be necessary to introduce unary predicate symbols 'male' and 'female' and to add some facts about the sex of the persons whose names occur in the program. Given the extended program, construct the search trees for the following questions.
  ?- mother(tom, liz).   ?- mother(X, bob).
  ?- sister(ann, pat).   ?- father(bob, Y).

Exercise 9.2. Let the predecessor relation be added to the program in Example 9.1 (concerning the parent relation) in the following ways.

a) pred2(X,Z) :- parent(X,Y), pred2(Y,Z).
   pred2(X,Z) :- parent(X,Z).

b) pred3(X,Z) :- parent(X,Z).
   pred3(X,Z) :- pred3(X,Y), parent(Y,Z).

Construct the search trees for the following questions: ?- pred2(tom, pat). ?- pred3(tom, pat). and ?- pred3(liz, jim). Conclude that from a procedural point of view, pred2 and pred3 do not describe the predecessor relation in an adequate way. (The examples are from Bratko [7], Section 2.6.2.)

Exercise 9.3. Determine whether the following pairs can be matched or unified:
a) p($f(X),Z$) and p(Y,c); b) X and $f(X)$ and c) p($f(X),c$) and p($Y,f(Z)$).

Exercise 9.4. Give a recursive definition of the concatenation relation, reading 'conc($L1, L2, L$)' as 'L is the concatenation of the lists $L1$ and $L2$'.

Exercise 9.5. Give a recursive definition of the deletion relation, with 'del($X, L, L1$)' read as '$L1$ results from the list L by deleting one occurrence of X'.

Exercise 9.6. Give a recursive definition for establishing the length of a list, reading 'length(L, N)' as 'N is the number of elements in the list L'.

Exercise 9.7. (Bratko [7]) Let P be the following program: p(1). p(2) :- ! . p(3). Construct the search trees for the following goals: a) ?- p(X). b) ?- p(X), p(Y). c) ?- p(X), !, p(Y).

Exercise 9.8. Let P be the following program:
$$p(X, 0) :- X < 1.$$
$$p(X, 1) :- X >= 1, X < 2.$$
$$p(X, 2) :- X >= 2.$$
Using cuts, change P into a program which is declaratively equivalent, but procedurally more efficient.

Exercise 9.9. (Lloyd [16]) Consider the following program P for the subset relation, representing sets by lists, where p(X, Y) expresses that X is not a subset of Y.

subset(X, Y) :- not p(X, Y).	(1)
p(X, Y) :- member(Z, X), not member(Z, Y).	(2)
member(X, [$X \mid L$]).	(3)
member(X, [$Y \mid L$]) :- member(X, L).	(4)

Make clear how the Prolog system answers the question: ?- subset([1, 2], [1, 2, 3]).

Exercise 9.10. If we replace clause (3) in the program of Exercise 9.9 by the clause

member(X, [$X \mid L$]) :- ! .   (3$'$)

then the membership program will generate just one solution and not all possible solutions. Verify that if we do so, the question '?- subset([1, 2, 3], [1]).' will be answered affirmatively, while 'not subset([1, 2, 3], [1])' logically follows from P. So, the use of cut in combination with negation may affect the soundness of the system!

Exercise 9.11. Extend the database in Example 9.8 with appropriate rules, such that the following questions can be formulated in Prolog in an adequate way. (Confer Bratko [7], Section 4.1.)
1. Give the names and surnames of all people in the database.
2. Give all children born in 1973.
3. Give the names and surnames of all employed wives.
4. Give the names and surnames of all unemployed people born before 1960.
5. Give all people born before 1960 whose salary is more than 10000.
6. Give the surnames of all families with at least two children.
7. Give the surnames of all families without children.

Exercise 9.12. Instead of $f(t1, t2)$, Prolog also allows the *infix notation* $t1 \; f \; t2$. For that purpose it is necessary to define f as an operator with a given precedence. The *precedence* of an arbitrary term is then defined as follows.
1. The precedence of individual variables and individual constants is 0.
2. The precedence of $f(t_1,\ldots,t_n)$ is the precedence of f.
3. The precedence of (t), t a term, is 0.

In order to ensure that $a+b*c$ is interpreted as $a+(b*c)$ and not as $(a+b)*c$, the operators $+$ and $*$ may be defined as follows.

$$\text{op}(500, yfx, +). \quad (1)$$
$$\text{op}(400, yfx, *). \quad (2)$$

In (1) the operator $+$ is defined as an infix operator (i.e., occurring between its arguments) with precedence 500. 'y' represents an argument whose precedence must be lower than or equal to that of the operator, and 'x' represents an argument whose precedence must be strictly lower than that of the operator.
1. Check that under the definitions (1) and (2) $a+b*c$ is understood as $a+(b*c)$ and not as $(a+b)*c$.
2. Defining '$-$' by 'op(500, yfx, $-$).', check that $a-b-c$ is read as $(a-b)-c$ and not as $a-(b-c)$.
3. Defining 'has' by 'op(600, xfx, has).', check that instead of 'has(peter, information).' the programmer can write 'peter has information.'.

9.2 Relational Databases and SQL

Abstract In this section we shall concentrate on the conceptual schema, i.e., the description of a database on a logical level. Only the relational model of databases will be discussed, because this model is most interesting from a logical and set-theoretical point of view. The description of the logical structure of relational databases in set theoretic terms shows that a Query Language such as SQL is a very natural one. Tuple-, table- and database- *constraints* are discussed. The notion of *key* is introduced and we also discuss the *Boyce-Codd Normal Form*, the *projection* of a table and the (natural) *join* of two tables. The material presented in this section is based on F. Remmen's book *Databases* (in Dutch) and on de Brock [8].

By a database we mean a class of permanent data, which is available to all users of an information system. These data relate to the objects which are relevant to the information system and to the attributes which are relevant to these objects. For instance, the permanent data of a hospital organisation include, among other things, the name, address and residence of each patient.

These permanent data should be available to all users of an information system. This availability for many users has important consequences as different groups of users will be interested in the data in different manners. For instance, an administrator in a hospital organisation will be in need of financial data about persons and rates, while a specialist needs to have at his disposal all medical data of persons and of all treatments to be applied.

Which objects with what properties are relevant to the information system can only be determined by the users. The design and implementation of a database will be a compromise between the different and partly clashing desires of the different users. It is the task of the Data Base Administrator (DBA) to bring about such a compromise.

In current database terminology the difference between species and individual is usually indicated by the difference between *type* and *occurrence*. So one can speak of the (object) type patient, and of the (object) occurrence of a patient in a hospital-organisation. In this example we have one type (species) with – in general – many occurrences (individuals).

Each user communicates with the database via the Data Base Management System (DBMS). In fact, a DBMS can be considered as a special expansion of the operating system.

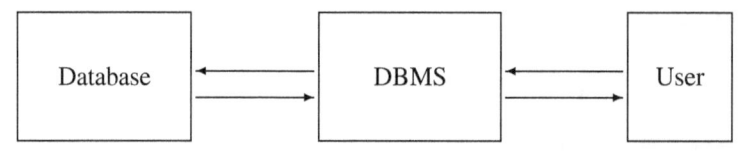

```
Database <---> DBMS <---> User
```

The users of an information-system are interested in information about individual objects and information about an individual object can only be provided in the form of values of one or more attributes. In general, many possible values will be available for each attribute. For instance, the attribute 'pnr' (short for 'patientnumber') of the object 'patient' may have a value between 1 and 100000, and the attribute 'pnm' (patient-name) may have a value consisting of a combination of at most 25 characters. In general, we demand that the values of an attribute form a set.

The set of attributes of an object together with the sets of values belonging to them is called the *object-characterisation* of that object. In the following examples it is made clear how we shall render an object-characterisation.

obchar patient =
    attrib   pnr : $\{1, \ldots, 100000\}$, patientnummer
             pnm : chs25   , name
             padr : chs20   , address
             pres : chs20   , residence
             db   : $\{18800101, \ldots, 19991231\}$, date of birth
             sex : $\{m, f\}$, sex
  endobchar

By chs25 (character string 25) we mean the set of all strings of at least one and at most 25 signs (letters, figures). In the following example – an object-characterisation of the object 'admission' (into a hospital) – we use the abbreviation 'dat' for the set $\{19500101, \ldots, 19991231\}$, i.e., the set of all natural numbers between 19500101 and 19991231.

9.2 Relational Databases and SQL

obchar admission =
    attrib
pnr	: $\{1,\ldots,100000\}$	,	
pnm	: chs25	,	
padr	: chs20	,	
pres	: chs20	,	
indat	: dat	,	date of admission
outdat	: dat	,	date of discharge
reas	: chs25	,	reason of admission
snr	: $\{1,\ldots,100000\}$	,	number of specialist
snm	: chs25	,	name of specialist
rnr	: $\{1,\ldots,1000\}$	,	number of nursing-room
wnr	: $\{1,\ldots,15\}$	,	number of ward

endobchar

Lastly, we give as an example an object-characterisation of the object 'specialist'.

obchar specialist =
    attrib
snr	: $\{1,\ldots,100000\}$	,	registration-number
snm	: chs25	,	name
sadr	: chs20	,	address
sres	: chs20	,	residence
wnr	: $\{1,\ldots,15\}$	,	number of ward
nbd	: $\{1,\ldots,100\}$	,	number of beds

endobchar

Definition 9.11 (Object-characterization; Tuple). Let O be an object with attributes $A_1,\ldots,A_m$ and let $W_1,\ldots,W_m$ be the sets of values belonging to $A_1,\ldots,A_m$ respectively. Then we define

$$F_O := \{(A_1,W_1),\ldots,(A_m,W_m)\},$$

and call it the *object-characterisation* of O.

Next we define $\pi(F_O) :=$

$$\{t \mid t = \{(A_1,w_1),\ldots,(A_m,w_m)\} \text{ for some } w_1 \in W_1,\ldots,w_m \in W_m\}.$$

The elements of $\pi(F_O)$ are called *tuples* for O. A tuple for O is a function (and hence a relation) with domain $\{A_1,\ldots,A_m\}$. If t is a tuple for O and $(A_i,w_i) \in t$, then we write $t(A_i)$ for w_i.

So a *tuple* t for O is a set $\{(A_1,w_1),\ldots,(A_m,w_m)\}$ with $w_1 \in W_1, \ldots, w_m \in W_m$. We write $w_1 = t(A_1),\ldots,w_m = t(A_m)$. Each tuple for O represents one object-occurrence. By mentioning the attributes at the head of columns, we can list the tuples for a given object O in a table, each row in the table corresponding to a tuple for O. For instance, below we give a partial table for the object 'patient'.

pnr	pnm	padr	pres	db	sex
537	Blunt	36 Evans Drive	Cranbury	19080527	m
498	Kiviat	67 Main Street	Newark	19090730	f

In general it holds that not all possible combinations of values of the attributes $A_1, \ldots, A_m$ will be allowed. In the literature on databases these restrictions are called *constraints*. We distinguish constraints on tuples, constraints on tables, and constraints on databases.

Below we give some examples of tuple-constraints, i.e., constraints on tuples.
$C_1(t)$: if $t(\text{pnr}) < 200$, then $t(\text{db}) < 19000101$; t being a tuple for object 'patient'.
$C_2(t)$: $t(\text{indat}) < t(\text{outdat})$; t being a tuple for object 'admission' (into hospital).
$C_3(t)$: if $t(\text{wnr}) = 9$, then $t(\text{sres}) = $ Princeton; and if $t(\text{wnr}) = 7$, then $t(\text{nbd}) \leq 2$; t being a tuple for object 'specialist'.

So a *tuple-constraint* is a condition on tuples for a given object O, such that it can be determined whether the condition holds for a given tuple t or not, completely independent of the other tuples.

Definition 9.12 (Tuple-type). Given an object O and a tuple-constraint C

$$T\text{-}O := \{t \in \pi(F_O) \mid C(t)\}.$$

T-O is called the *tuple-type* for O (determined by the constraint C) and is the set of all tuples t for O satisfying the condition C.

If the number of tuples t for O satisfying a condition C is finite (and in practice not too large), then the tuple-type T-O for O can be rendered by an exhaustive list of all object-occurrences satisfying condition C.

A *table for O* is by definition a subset of T-O. There may also be constraints on such tables. We give some examples below.
$TC_1(D) : \forall t_1, t_2 \in D \ [\ t_1(\text{pnr}) = t_2(\text{pnr}) \to t_1 = t_2\ ]$; D being a table for the object 'patient'. $\forall t_1, t_2 \in D$ stands for 'for all t_1 and t_2 in D'. This table-constraint TC_1 is also formulated as follows: {pnr} is *uniquely identifying*, or: {pnr} uni, for short.
$TC_2(D) : \forall t_1, t_2 \in D \ [\ t_1(\text{pnr}) = t_2(\text{pnr}) \land t_1(\text{indat}) = t_2(\text{indat}) \to t_1 = t_2]$; D being a table for the object 'admission'. This table-constraint TC_2 is also formulated as follows: {pnr, indat} is *uniquely identifying*, or: {pnr, indat} uni, for short.
$TC_3(D) :$ {snr} uni, and {snm, sadr, sres} uni, and the number of specialists at ward 9 is at least 2; D being a table for the object 'specialist'.

So, a *table-constraint* indicates which subsets of a tuple-type are allowed. The set of all tables allowed, given an object O, is called a table-type for O.

Definition 9.13 (Table-type). Let O be an object, T-O a tuple-type for O and TC a table-constraint for O. Then

$$TT\text{-}O := \{D \in P(T\text{-}O) \mid TC(D)\}$$

is called a *table-type* for O. If $D \in TT$-O, we say D is a *table of type O*.

Definition 9.14 (Functional Dependence; Uniquely Identifying; Key).
Let O be an object with attributes $A_1, \ldots, A_m$, and let D be a table of type O. Let $V, W \subseteq \{A_1, \ldots, A_m\}$.

9.2 Relational Databases and SQL

1. $V \to W$ in $D := \forall t_1, t_2 \in D \ [\, t_1 \lceil V = t_2 \lceil V \to t_1 \lceil W = t_2 \lceil W\,]$, where $t \lceil V$ is the restriction of t to V. In words: V *functionally determines* W *in* D, or W is *functionally dependent on* V *in* D.
2. V is *uniquely identifying within* $D := V \to \{A_1, \ldots, A_m\}$ in D, i.e.,
$\forall t_1, t_2 \in D \ [\, t_1 \lceil V = t_2 \lceil V \to t_1 = t_2\,]$.
3. $V \to W$ for $O :=$ for every table D of type O, $V \to W$ in D.
V is a *key for* $O := V$ is uniquely identifying within every table of type O.

Example 9.11. {pnr} is a key for 'patient'; {pnr, indat} is a key for 'admission'; {pnm, padr, pres} is a key for 'patient'; {snr} is a key for 'specialist'.

Within the framework of an information-system one usually will be interested in more than only one table-type. For instance, in a hospital-organisation one may be interested in patients, admissions (into the hospital) and specialists and hence also in the table-types TT-patient, TT-admission and TT-specialist belonging to them. At a certain moment the situation of the hospital, at least with respect to patients, admissions and specialists, can be summed up by three tables, one of type TT-patient, one of type TT-admission and one of type TT-specialist. Such a triple of tables is called a *relational database*.

Definition 9.15 (Database-characterisation). A set of objects together with table-types belonging to them is called a *database-characterisation*. More precisely, let $O_1, \ldots, O_n$ be objects, together with table-types TT-$O_1, \ldots, TT$-O_n belonging to them. Then

$$F_{DB} := \{(O_1, TT\text{-}O_1), \ldots, (O_n, TT\text{-}O_n)\}$$

is a database-characterisation. In the following example it is made clear how we shall render a database-characterisation.

Example 9.12. The database-characterisation for the combination of the objects patient, admission and specialist looks as follows:
  dbchar hospital =
    obj   pat : TT-patient ,
         adm : TT-admission,
         spec : TT-specialist
  enddbchar

Note the analogy between an object-characterisation and a database-characterisation: the attributes are replaced by objects and the sets of values by table-types.

Definition 9.16 (Relational Databases).
$\pi(F_{DB}) := \{\{(O_1, D_1), \ldots, (O_n, D_n)\} \mid D_1 \in TT\text{-}O_1, \ldots, D_n \in TT\text{-}O_n\}$. The elements of $\pi(F_{DB})$ are called *relational databases*.

Given a database-characterisation with objects $O_1, O_2, \ldots, O_n$ and table-types TT-O_1, TT-$O_2, \ldots, TT$-O_n belonging to them, in general not all databases in $\pi(F_{DB})$ will be allowed. This brings us to the last class of constraints, the so-called *database-constraints*. The set of all databases satisfying a certain database-constraint is called a *database-type*.

An important subclass of database-constraints is formed by the so-called *subset-requirements*. We make this notion clear by means of the following two database-constraints.

$DC_1(D_1,D_2,D_3) := \forall t_2 \in D_2 \; \exists t_1 \in D_1 \; [\; t_2(\text{pnr}) = t_1(\text{pnr}) \;]$, in words: for each admission-tuple t_2 there is a patient-tuple t_1 such that the value of pnr in t_2 is equal to the value of pnr in t_1.

$DC_2(D_1,D_2,D_3) := \forall t_2 \in D_2 \; \exists t_3 \in D_3 \; [\; t_2(\text{snr}) = t_3(\text{snr}) \;]$, in words: for each admission-tuple t_2 there is a specialist-tuple t_3 such that the value of snr in t_2 is equal to the value of snr in t_3.

The constraint DC_1 means that for any database allowed the set of pnr-values in the admission-table is a subset of the set of pnr-values in the patient-table. A similar remark is to be made for DC_2. For these subset-requirements the following notation is used: for DC_1, ssr(adm.pnr, pat.pnr); for DC_2, ssr(adm.snr, spec.snr).

Below we give an example of a database-type. The symbols *tatp, tutp, obchar, attrib* stand for table-type, tuple-type, object-characterisation and attribute respectively. The symbols *tuc, tac, dbc* stand for tuple-constraint, table-constraint and database-constraint respectively.

Type nr   : {1,..., 100000}
    hoev : {1,..., 100}
    dat   : {19000101,..., 19991231}
    tatp TT-patient =
        tutp T-patient =
            obchar patient =
                attrib pnr   : nr   ,
                        pnm : chs25   ,
                        padr : chs20   ,
                        pres : chs20   ,
                        db   : {18800101, ..., 19991231} ,
                        sex : {m,f}
            endobchar;
            tuc   pnr < 200 → db < 19000101
        endtutp;
        tac   {pnr} uni,
              {pnm, padr, pres} uni
    endtatp,
    tatp TT-admission =
        tutp T-admission =
            obchar admission =
                attrib pnr   : nr   ,
                        pnm : chs25 ,
                        padr : chs20 ,
                        pres : chs20 ,
                        indat : dat   ,
                        outdat : dat   ,

9.2 Relational Databases and SQL

       attrib reas : chs25 ,
          snr : nr   ,
           snm : chs25 ,
           rnr : $\{1, \ldots, 1000\}$,
           wnr : $\{1, \ldots, 15\}$
      endobchar;
      tuc indat $<$ outdat,
         reas = 'informaritis' $\rightarrow$ rnr = 5
    endtutp ;
    tac {pnr, indat} key
 endtatp,
 tatp TT-specialist =
    tutp T-specialist =
      obchar specialist =
        attrib snr : nr   ,
           snm : chs25 ,
           sadr : chs20 ,
           sres : chs20 ,
           wnr : $\{1, \ldots, 15\}$,
           nbd : hoev
      endobchar;
      tuc wnr = 9 $\rightarrow$ sres = 'Princeton',
         wnr = 7 $\rightarrow$ nbd ≤ 2
    endtutp;
    tac keys {{snr}, {snm, sadr, sres}},
       'at least two specialists at ward 9'
 endtatp,
  dbtype DT-hospital =
    dbchar hospital =
      obj pat : *TT*-patient ,
        adm : *TT*-admission ,
        spec : *TT*-specialist
    enddbchar;
    dbc ssr (adm.pnr, pat.pnr),
       ssr (adm.snr, spec.snr)
  enddbtype,
endtype

Looking at the database-type given above, we see that some attributes are *redundant*: pnm, padr and pres in the table for 'admission' are uniquely determined by pnr and already occur in the table for 'patient'; snm in the table for 'admission' is uniquely determined by snr and also occurs in the table for 'specialist'; and wnr in the table for 'admission' is uniquely determined by rnr, although a table in which both rnr and wnr already occur is not (yet) available. For these reasons we say that the table-type *TT*-admission given above is not *normal*. In concrete cases this means that in case of a change of address of a patient not only the table for 'patient' has to be updated, but also the table for 'admission'; otherwise, an inconsistent database would result.

Definition 9.17 (Boyce-Codd Normal Form). Let O be an object with attributes $A_1, \ldots, A_m$ and let TT-O be a table-type for O. TT-O is in *Boyce-Codd Normal Form* (BCNF) := for all $V \subseteq \{A_1, \ldots, A_m\}$ and for all $A \in \{A_1, \ldots, A_m\}$: if $V \to \{A\}$ for O and $A \notin V$, then V is a key for O. Informally: TT-O is in BCNF if every set V of attributes which determines an attribute outside of V is a key for O.

Since $\{pnr\} \to \{pnm\}$ for 'admission', $pnm \notin \{pnr\}$, but $\{pnr\}$ is not a key for 'admission', it follows that TT-admission is not in Boyce-Codd Normal Form. (Remember that $\{pnr, indat\}$ is a key for 'admission'.)

In the literature one also finds various other normal forms including the first, second, third and fourth normal forms. If TT-O is in BCNF, then TT-O is also in 3NF (Third Normal Form).

Definition 9.18 (Normal Database-type). A *normal database-type* is a database-type in which each table-type is normal.

Example 9.13. We can convert the database-type given above into a normal database-type by applying the following two operations to the given database-type:
1. In the table-type TT-admission leave out the attributes pnm, padr, pres, snm and wnr.
2. Add a table-type TT-room as follows:

    tatp TT-room =
        tutp T-room =
            obchar room =
                attrib rnr : $\{1, \ldots, 1000\}$,
                    wnr : $\{1, \ldots, 15\}$
            endobchar;
        endtutp;
        tac $\{rnr\}$ uni
    endtatp.

The result is a normal database-type, in which redundancies are avoided, while all information has been saved. (However, in practice, redundant *storage* of data may be necessary, for instance, because of the required time of response.)

Of course, a database-type for any actual hospital organisation will be much more complex than the simple example considered here.

Definition 9.19 (Projection). Let O be an object with attributes $A_1, \ldots, A_m$, $V \subseteq \{A_1, \ldots, A_m\}$ and let D be a table of type O. $D \parallel V$, the *projection* of D on V, is by definition $\{t \lceil V \; ; \; t \in D\}$.

Example 9.14. For instance, for the following table D_1:

nr	name	sal	sex	dept
8	Johnson	2200	male	1
7	Johnson	3100	female	2
9	Kiviat	2900	male	1

$D_1 \parallel \{sex, dept\}$ is the table

sex	dept
male	1
female	2

.

9.2 Relational Databases and SQL

Definition 9.20 (Compatible tuples). Let O_1 be an object with attributes $A_1,\ldots,A_m$ and O_2 an object with attributes $B_1,\ldots,B_n$. Let t_1 be a tuple for O_1 and t_2 a tuple for O_2. t_1 and t_2 are *compatible* := $t_1\lceil\{A_1,\ldots,A_m\}\cap\{B_1,\ldots,B_n\} = t_2\lceil\{A_1,\ldots,A_m\}\cap\{B_1,\ldots,B_n\}$.

Definition 9.21 (Join). Let D_1 be a table of type O_1 and D_2 a table of type O_2. $D_1 \bowtie D_2 := \{t_1 \cup t_2 \mid t_1 \in D_1 \text{ and } t_2 \in D_2 \text{ and } t_1 \text{ and } t_2 \text{ are compatible}\}$.
$D_1 \bowtie D_2$ is called the (natural) *join* of D_1 and D_2.

Example 9.15. For instance, let D_2 be the table:

anr	name	man
2	planning	7
1	production	9

and let D_3 result from table D_1 in Example 9.14 by replacing 'dept' by 'anr'. Then $D_3 \bowtie D_2$ is the table:

nr	name	sal	sex	anr	name	man
8	Johnson	2200	male	1	production	9
7	Johnson	3100	female	2	planning	7
9	Kiviat	2900	male	1	production	9

9.2.1 SQL

The purpose of a *query-language* is to enable the user to make use of the data stored in the database in an user-friendly manner. In order to give a more concrete idea of a query-language, we shall treat some elements of the query-language SQL (Structured Query Language; 1980) on the basis of some examples. Having understood the logical structure of a relational database, query-languages such as SQL become very perspicuous.

The terminology of SQL is familiar to the terminology of set theory, as will become clear from the examples below. In these examples P, ADM, SP and R stand for the set (or table) of all patients, the set of all admissions, the set of all specialists and the set of all rooms, respectively. The examples all refer to the objects described in the *normalized database* given in Example 9.13.

Example 9.16. Describe the set of numbers, names and addresses of all patients who live in Princeton and were born before 1960.
Answer: a) $\{t\lceil\{\text{pnr, pnm, padr}\} \mid t \in P \mid t(\text{pres}) = \text{'Princeton'} \land t(\text{db}) < 19600101\}$.
b) Now, in SQL the query 'give number, name and address of all patients who live in Princeton and were born before 1960' is formulated as follows:

```
SELECT   t.pnr, t.pnm, t.padr
FROM     P t
WHERE    t.pres = 'Princeton'
AND      t.db < 19600101
```
Here t.pnr corresponds to t(pnr).

Example 9.17. Describe the set of numbers, names and addresses of all patients who were admitted into hospital in the period between May 26 and July 11, 1981.
Answer: a) $\{t \lceil \{\text{pnr, pnm, padr}\} \mid t \in P \mid \exists s \in \text{ADM}$
$[s(\text{pnr}) = t(\text{pnr}) \land s(\text{indat}) \geq 19810526 \land s(\text{indat}) \leq 19810711]\}$
or, equivalently,
$\{t \lceil \{\text{pnr, pnm, padr}\} \mid t \in P \mid t(\text{pnr}) \in \{s(\text{pnr}) \mid s \in \text{ADM} \mid s(\text{indat}) \geq 19810526 \land s(\text{indat}) \leq 19810711\}\}$.
b) Now, in SQL the query 'give number, name and address of all patients who were admitted into hospital in the period between May 26 and July 11, 1981' is formulated as follows:

SELECT t.pnr, t.pnm, t.padr
FROM   P t
WHERE t.pnr IN
       (SELECT s.pnr
       FROM   ADM s
       WHERE s.indat $\geq$ 19810526
       AND   s.indat $\leq$ 19810711)

Example 9.18. Describe the set of names, addresses and residences of all specialists who were responsible for an admission in August 1977 of a patient from Princeton for reason 034.
Answer: a) $\{t \lceil \{\text{snm, sadr, sres}\} \mid t \in SP \mid \exists s \in \text{ADM} \, [\, s(\text{snr}) = t(\text{snr}) \land s(\text{reas}) = 034 \land 19770801 \leq s(\text{indat}) \leq 19770831 \land \exists u \in P \, [\, u(\text{pnr}) = s(\text{pnr}) \land u(\text{pres}) = \text{'Princeton'}]]\}$
       or, equivalently,
$\{t \lceil \{\text{snm, sadr, sres}\} \mid t \in SP \mid t(\text{snr}) \in \{s(\text{snr}) \mid s \in \text{ADM} \mid s(\text{reas}) = 034 \land 19770801 \leq s(\text{indat}) \leq 19770831 \land s(\text{pnr}) \in \{u(\text{pnr}) \mid u \in P \mid u(\text{pres}) = \text{'Princeton'}\}\}\}$.
b) Now, in SQL the query 'give name, address and residence of all specialists who were responsible for an admission in August 1977 of a patient from Princeton for reason 034' is formulated as follows.

SELECT t.snm, t.sadr, t.sres
FROM   SP t
WHERE t.snr IN
       (SELECT s.snr
       FROM   ADM s
       WHERE s.reas = 034
       AND   s.indat $\leq$ 19770831
       AND   s.indat $\geq$ 19770801
       AND   s.pnr IN
             (SELECT u.pnr
             FROM   P u
             WHERE u.pres = 'Princeton'))

Example 9.19. Describe the set of numbers and names of all patients, reason of admission and number of nursing-room, who were admitted into hospital between

9.2 Relational Databases and SQL

September 1 and 5, 1977, in ward number 9.
Answer: a) $\{t\lceil\{\text{pnr, pnm, reas, rnr}\} \mid t \in P \bowtie \text{ADM} \mid 19770901 \leq t(\text{indat}) \leq 19770905 \wedge \exists s \in R \, [s(\text{rnr}) = t(\text{rnr}) \wedge s(\text{wnr}) = 9]\}$
or, equivalently,
$\{t\lceil\{\text{pnr, pnm, reas, rnr}\} \mid t \in P \bowtie \text{ADM} \mid 19770901 \leq t(\text{indat}) \leq 19770905 \wedge t(\text{rnr}) \in \{s(\text{rnr}) \mid s \in R \mid s(\text{wnr}) = 9\}\}$, where $P \bowtie \text{ADM}$ is the join of P and ADM.
b) Now, in SQL the query 'give number and name of all patients, reason of admission and number of nursing-room, who were admitted into hospital in the period between September 1 and 5, 1977, in ward number 9' is formulated as follows.

```
SELECT  t1.pnr, t1.pnm, t2.reas, t2.rnr
FROM    P t1, ADM t2
WHERE   t1.pnr = t2.pnr
AND     t2.indat ≥ 19770901
AND     t2.indat ≤ 19770905
AND     t2.rnr IN
           (SELECT s.rnr
            FROM   R s
            WHERE  s.wnr = 9)
```

Example 9.20. Describe the set of all room-numbers, in which no patients from Cranbury were hospitalized in the period between August 11 and 17, 1977.
Answer: a) $\{s(\text{rnr}) \mid s \in R \mid \neg\exists t \in \text{ADM} \, [\, t(\text{rnr}) = s(\text{rnr}) \wedge 19770811 \leq t(\text{indat}) \leq 19770817 \wedge A(t)]\}$ where $A(t) :=$
  i) $\exists u \in P \, [u(\text{pnr}) = t(\text{pnr}) \wedge u(\text{pres}) = \text{'Cranbury'}]$ or, equivalently,
  ii) $t(\text{pnr}) \in \{u(\text{pnr}) \mid u \in P \mid u(\text{pres}) = \text{'Cranbury'}\}$.
Note that $\neg\exists t \in \text{ADM} \, [t(\text{rnr}) = s(\text{rnr}) \wedge \ldots \wedge A(t)]$ is equivalent to

$$s(\text{rnr}) \notin \{t(\text{rnr}) \mid t \in \text{ADM} \mid \ldots \wedge A(t)\}.$$

b) Now, in SQL the query 'give the numbers of all rooms, in which no patients from Cranbury were hospitalized in the period between August 11 and 17, 1977' can be formulated as follows.

```
SELECT  s.rnr
FROM    R s
WHERE   s.rnr NOT IN
           (SELECT t.rnr
            FROM   ADM t
            WHERE  t.indat ≤ 19770817
            AND    t.indat ≥ 19770811
            AND    t.pnr IN
                      (SELECT u.pnr
                       FROM   P u
                       WHERE  u.pres = 'Cranbury'))
```

For further reading, the reader is referred to E. O. de Brock [8].

Exercise 9.13. The following queries all refer to the normalized database given in Example 9.13. Formulate these queries into SQL.
a) Give name, address and residence of all specialists from ward number 9, having more than two beds.
b) Give number and name of all specialists who were responsible for admission on March 3, 1980, because of informaritis.
c) Give number of all rooms in which no patients from Princeton were hospitalized in the period between May 9 and 18, 1980.
d) Give number, name, address and residence of all patients who were hospitalized by a specialist of ward number 9.

Exercise 9.14. Let D_1 be the table

nr	name	sal	sex	dept
8	Johnson	2200	male	1
7	Johnson	3100	female	2
9	Kiviat	2900	male	1

and let D_2 be the table

anr	name	man
2	planning	7
1	production	9

Determine $D_1 \bowtie D_2$. Let D_3 result from D_1 by replacing 'dept' by 'anr' and 'name' by 'wnm'. Determine $D_3 \bowtie D_2$.
Let D_4 result from D_2 by replacing 'man' by 'nr' and 'name' by 'anm'. Determine $D_1 \bowtie D_4$ and $D_3 \bowtie D_4$.

Exercise 9.15. Make clear why the following set does not describe the set of all room-numbers in which no patients from Cranbury were hospitalized in the period between August 11 and 17, 1977 (compare Example 9.20).
$\{s(\text{rnr}) \mid s \in R \mid \exists t \in \text{ADM} [\ t(\text{rnr}) = s(\text{rnr}) \land 19770811 \leq t(\text{indat}) \leq 19770817 \land \neg \exists u \in P [\ u(\text{pnr}) = t(\text{pnr}) \land u(\text{pres}) = \text{'Cranbury'}]]\}$.
Hint: Consider the following tables.

R	rnr	wnr
s1	11	5
s2	12	5
s3	13	6

ADM	pnr	indat	rnr
t1	400	19770812	11
t2	500	19770813	11
t3	600	19770814	12

P	pnr	pres
u1	400	Princeton
u2	500	Cranbury
u3	600	Cranbury

9.3 Social Choice Theory; Majority Judgment

Abstract We show that most well-known and most frequently used voting rules have a number of unacceptable properties. The hope for a voting rule with only nice properties seemed to be vanished when Kenneth Arrow [1] proved his impossibility theorem in 1951. However, in 2010 Michel Balinski and Rida Laraki made clear that – by asking voters for their evaluations of the candidates instead of their preferences over the candidates – a nice voting rule does exist: Majority Judgment (MJ). They show how poorly the existing voting rules perform in the French and American presidential elections and how Majority Judgment would lead to other and more plausible results.

9.3.1 Introduction

When choosing a mayor, president, chairman, etc., usually the first thought is: most votes count. Many people think that democracy is more or less identical to application of 'most votes count', in other words, the Plurality Rule (PR). However, this procedure to choose a winner or a common (or social) preference over the candidates or alternatives has many defects. This rule takes only the top preference of the voters into account, ignores the second, third, etc. preferences of the voters and hence causes serious *loss of information*. In technical terms, this procedure is not Independent of Irrelevant Alternatives (not IIA), as we shall see in Section 9.3.2.

Is then pairwise comparison, in other words Majority Rule (MR), a good alternative? This procedure does take the individual preference orderings of the voters over the alternatives into account and is Independent of Irrelevant Alternatives. However, it is not transitive and hence does not in all cases yield a feasable outcome, as we shall see in Section 9.3.3. By the way, 'most votes count' and 'pairwise comparison' coincide in the case of only two alternatives, i.e., with only two candidates Plurality Rule and Majority Rule give the same outcome.

In 1951 K. Arrow [1] proved that any voting rule which takes as input the individual preference orderings (over the candidates or alternatives) of the voters and which is transitive and Independent of Irrelevant Alternatives (together with some other natural properties like anonymity and neutrality) is dictatorial, i.e., there will be a voter whose preference is always the outcome of the voting rule, no matter what the preferences of the other voters are. In Section 9.3.6 we shall give a simple proof of (a version of) Arrow's theorem, due to Balinski and Laraki [3].

Recently, Balinski and Laraki [3] showed that even with only two candidates 'most votes count' in many cases may give an unnatural or counterintuitive outcome, i.e., it may select a candidate as winner who in fact has lower evaluations than his competitor. In their words: Majority Rule does not respect domination. Consequently, Majority Rule and Plurality Rule are disqualified as good voting rules for determining a winner or a common preference ordering over the alternatives. We shall elaborate this in Section 9.3.7.

Considering all this, the conclusion seems to be inevitable: there is no 'good' voting rule to determine a winner (or a common preference over the candidates) in an election, where we mean by 'good' that the voting rule is transitive, Independent of Irrelevant Alternatives and in addition respects domination.

However, already in 2010 Balinski and Laraki [2] presented their Majority Judgment (MJ). This voting rule takes as input not the individual preference orderings (over the alternatives) of the voters, but the evaluations by the voters of the different candidates in sufficiently varied terms, like for instance: excellent (*ex*), very good (*vg*), good (*go*), acceptable (*ac*), poor (*po*) and reject (*re*). It turns out that this voting rule, Majority Judgment, is IIA, transitive and does respect domination and nevertheless is not dictatorial. In addition, this Majority Judgment contains certain safeguards to prevent successful manipulation by the voters. We describe this voting rule in Section 9.3.8.

How is it possible that Majority Judgment escapes the curse of Arrow's theorem? Because MJ takes the evaluations of the candidates by the voters as input and not the individual preferences over the candidates. Here it is important to notice that from the evaluations of the candidates by a voter one may deduce the individual preference ordering of this voter, but that conversely, from the individual preference ordering over the candidates one cannot deduce the evaluations of the candidates by the voter in question. So, an evaluation of all candidates by a voter is much *more informative* than his preference ordering over the candidates. In addition, if two voters say that they prefer candidate A to candidate B, they may mean quite different things: one that he judges A as excellent and B as acceptable, the other that he judges A as poor and B as even more poor. In other words, individual preference orderings over the candidates lead to a babylonian confusion of tongues and one should not be surprised that this yields problems, as becomes evident from Arrow's theorem.

Balinski and Laraki show on the basis of the presidential elections in the USA [4] and in France [5] how poorly our familiar ways of choosing a president may work out and illustrate with these examples from real life how their Majority Judgment would lead to other and more plausible outcomes. We discuss this in Section 9.3.12 (USA) and 9.3.13 (France). In Section 9.3.14 we pay attention to the situation in the Netherlands.

9.3.2 *Plurality Rule (PR): most votes count*

In the year 2000 there were presidential elections in the USA with Bush, Gore and Nader as the most important candidates. In Florida the result of the ballot was approximately as follows:

41% Bush
39% Gore
20% Nader

Because 'most votes count' or Plurality Rule (PR) is applied, Bush was the winner in Florida (with in fact only a few hundred votes more than Gore). But most votes count? Or rather not? The individual preferences of the voters were approximately as given in the following *profile p*.

41% Bush Gore Nader
39% Gore Nader Bush
20% Nader Gore Bush

Notice that (39 + 20) = 59% of the voters, hence a majority, has Bush as last preference. But Plurality Rule chooses Bush as the winner. How can this be? Because the Plurality Rule causes *loss of information*: only the first preferences of the voters

9.3 Social Choice Theory; Majority Judgment

are taken into account, the second, third, etc. preferences of the voters are left out of consideration.

Taking this extra information into account, pairwise comparison, in other words Majority Rule (MR), yields the following result: both Gore and Nader beat Bush with 39 + 20 = 59% against 41. And Gore beats Nader with 41 + 39 = 80% against 20. So, the outcome under pairwise comparison (MR) would be: *Gore Nader Bush*, in this order, while the outcome under Pluraiity Rule was: *Bush Gore Nader*.

>   PR   Bush   Gore   Nader
>   MR   Gore   Nader   Bush

In a pairwise comparison Bush loses of every other candidate, and is therefore called a *Condorcet loser*, but he becomes the winner under 'most votes count'. Gore beats every other candidate in a pairwise comparison and is therefore called the *Condorcet winner*.

Candidate Nader was irrelevant in the sense that he did not have a chance to become president. For that reason he could have withdrawn his candidacy. One might think, no problem, because Nader was not chosen anyway. However, without Nader the profile above looks like this:

>   41%   Bush   Gore
>   39 + 20 = 59%   Gore   Bush

Now, when applying 'most votes count', Gore would win instead of Bush. So, under 'most votes count' the choice between Bush and Gore is determined by the participation or non-participation of a third (irrelevant) candidate. In other words, 'most votes count' (PR) is *not Independent of Irrelevant Alternatives* (not IIA). Notice that Majority Rule (MR), or pairwise comparison, is (by definition) IIA.

Related to this, the 20% voters with preference ordering *Nader Gore Bush* prefer Gore to Bush. By giving an improper order of preference *Gore Nader Bush* they can ensure that under 'most votes count' Gore becomes the winner with 39 + 20 = 59% of the votes, which is a better outcome for them. In other words, 'most votes count' (PR) is *not strategy-proof*, i.e., cheating may pay off.

Another objection against Plurality Rule (PR) has been pointed out by Donald Saari [13, 14, 15]. Profile *p* above contains what Saari calls a *reversal portion*:

>   20   Bush   Gore   Nader
>   20   Nader   Gore   Bush

These 20 + 20 voters have diametrically opposed preferences and hence cancel each other out. One would intuitively expect that adding a reversal portion to or subtracting it from a given profile does not change the outcome. However, subtracting the reversal portion in question from the original profile *p* yields:

>   21   Bush   Gore   Nader
>   39   Gore   Nader   Bush

Now, under 'most votes count' Gore instead of Bush would become the winner, while one would expect intuitively that the outcome does not change.

9.3.3 Majority Rule (MR): pairwise comparison

As we have remarked earlier, Majority Rule (MR), or pairwise comparison, is Independent of Irrelevant Alternatives (IIA). This follows immediately from the definition of Majority Rule: in a competition between two candidates A and B only the relative positions of A and B in the given profile are compared and a third alternative C has no influence on that. Related to this is that Majority Rule is also strategy-proof; see Exercise 9.16. This might suggest that Majority Rule is a perfect voting rule to aggregate the individual preference orderings of the voters to a common or social ordering of the candidates. However, this is not the case, because in some cases Majority Rule does not yield a feasible outcome, as illustrated by the following so called *Condorcet profile q*:
$$\begin{array}{cccc} 1/3 & a & b & c \\ 1/3 & b & c & a \\ 1/3 & c & a & b \end{array}$$
A majority (group 1 and 3) prefers a to b, another majority (group 1 and 2) prefers b to c and again another majority (group 2 and 3) prefers c to a. So, a beats b and b beats c, but not a beats c. On the contrary, c beats a. In other words: Majority Rule is *not transitive*. The outcome under Majority Rule may be cyclic: $a\ b\ c\ a$. This is called *Condorcet's paradox*.

Notice that with only two alternatives violation of transitivity cannot occur because transitivity refers to three alternatives. Transitivity of a relation R on a set V means by definition: if aRb and bRc, then aRc for all elements a,b,c in V.

In the case of three alternatives and a great number of voters, supposing that every individual preference ordering is equally likely, the probability of the occurrence of the *Condorcet paradox*, i.e., the probability of a cyclic outcome, is 1 out of 16, a number which is not negligible small; see Gehrlein [11].

As pointed out by Saari [13, 14, 15], the outcome under Majority Rule may change when we add a *Condorcet portion* to, or subtract it from, a given profile. For instance, consider the following profile r:
$$\begin{array}{cccc} 1: & a & c & b \\ 2: & b & a & c \end{array}$$
If we apply Majority Rule to this profile r the outcome is: $b\ a\ c$. But if add to profile r the Condorcet portion s:
$$\begin{array}{cccc} 2: & a & b & c \\ 2: & b & c & a \\ 2: & c & a & b \end{array}$$
and next apply Majority Rule to the profile $r+s$ the outcome will become $a\ b\ c$. This is counterintuitive: a Condorcet portion represents voters whose collective advice with regard to social choice is confused and hence should be ignored. Note that in a Condorcet portion each candidate is an equal number of times first, second and third choice. So, intuitively, nobody is preferred. A Condorcet portion should give a tie. But it does not necessarily so under Majority Rule, as we have just seen.

9.3.4 Borda Rule (BR)

The French mathematician and political scientist Jean-Charles de Borda (± 1750) proposed to count the number of candidates beaten by a given candidate. That is, if a voter gives an order of preference *Bush Gore Nader*, Bush gets 2 (Borda) points, because he beats both Gore and Nader, Gore gets 1 (Borda) point because he beats only one candidate and Nader gets 0 (Borda) points. Given profile p above

$$\begin{array}{llll} 41\% & \text{Bush} & \text{Gore} & \text{Nader} \\ 39\% & \text{Gore} & \text{Nader} & \text{Bush} \\ 20\% & \text{Nader} & \text{Gore} & \text{Bush} \end{array}$$

the Borda score of Bush is: $(41 \times 2) + (39 \times 0) + (20 \times 0) = 82$,
the Borda score of Gore is: $(41 \times 1) + (39 \times 2) + (20 \times 1) = 139$, and
the Borda score of Nader is: $(41 \times 0) + (39 \times 1) + (20 \times 2) = 79$.
So, the outcome under the Borda Rule (BR) would be: *Gore Bush Nader*, in this order.

Although the Borda Rule takes the individual preference orderings of the voters into account, the Borda Rule still causes *loss of information*: it does not take into account the intensity with which one candidate is preferred to the next one. If a voter indicates that he prefers candidate A to B he may mean quite different things: he may evaluate A as excellent and B as very good, he may evaluate A as excellent and B as poor, or he may evaluate A as poor and B as reject.

Like Plurality Rule, also the Borda Rule is not Independent of Irrelevant Alternatives (not IIA), as illustrated by the following profile:
$$\begin{array}{ll} 3: & c \quad a \quad b \\ 2: & a \quad b \quad c \\ 1: & a \quad c \quad b \\ 1: & b \quad c \quad a \end{array}$$

Given this profile, c is the Condorcet winner, i.e., c beats all other candidates in a pairwise comparison, but a is the Borda winner with $(3 \times 1) + (2 \times 2) + (1 \times 2) + (1 \times 0) = 9$ Borda points against 8 Borda points for c. In a competition between a and c under application of the Borda Rule the third alternative b turns out to be decisive: without the participation of b the Borda winner would become c with 4 Borda points against only 3 for a.

A serious disadvantage of the Borda Rule is that voters can rather easily act strategically: by giving an improper order of preference they may be able to achieve an outcome which is better for them. The three voters with preference $c\ a\ b$ who want c to win, can easily pretend that a is their last preference and pretend that their order of preference is $c\ b\ a$. In this way they achieve that a gets 3 Borda points less, hence $9 - 3 = 6$, the number of Borda points for c remains 8 and the Borda score of b becomes $4 + 3 = 7$. So, by giving an improper order of preference these three voters can achieve an outcome c which they prefer to the outcome a when they give their proper order of preference. In other words, the Borda Rule is *not strategy-proof*.

Another objection against the Borda Rule has been pointed out by Balinski and Laraki [2]: if one removes the Borda winner Gore from the given profile p, the order of the remaining candidates may change under the Borda Rule: leaving out the winner Gore from profile p we get:

41 Bush Nader
39 Nader Bush
20 Nader Bush

Applying the Borda Rule to this profile yields *Nader Bush* as social outcome, while with the winner Gore present the social order between these two candidates was just the opposite: *Bush Nader*.

As pointed out by Saari [13, 14, 15], the outcome under the Borda Rule remains unaffected by adding a reversal portion to, or subtracting it from, a given profile. Why is this so? A reversal portion has the following structure:

$$\begin{array}{ccc} a & b & c \\ c & b & a \end{array}$$

Applying the Borda Rule to this reversal portion, a gets $2 + 0 = 2$ Borda points, b gets $1 + 1 = 2$ Borda points and c also gets $0 + 2 = 2$ Borda Points. So, when we add or subtract a reversal portion, the alternatives get the same number of Borda points more or less. A similar result holds for Majority Rule, but not for Plurality Rule, as we have seen in Section 9.3.2.

The outcome under the Borda Rule also remains unaffected by adding a Condorcet portion to, or subtracting it from, a given profile. The reason is simple: all alternatives in a Condorcet portion get the same number of Borda points. A similar result holds for Plurality Rule, but not for Majority Rule as we have seen in Section 9.3.3.

9.3.5 Outcome depends on the Voting Rule

In the preceding subsections we have seen that the outcome of an election does not depend so much on the preferences of the electorate, but rather on the voting rule which aggregates the individual preferences of the voters to a common or social order of preference. Given profile p above, the outcome

under Plurality Rule is: Bush Gore Nader
under Majority Rule is: Gore Nader Bush
and under the Borda Rule is: Gore Bush Nader

Notice that with only two alternatives, Plurality Rule, Majority Rule and the Borda Rule are equivalent, i.e., for all profiles they yield the same outcome; see Exercise 9.17.

9.3.6 Arrow's Impossibility Theorem

In the preceeding subsections we have seen that Plurality Rule (PR) or 'most votes count' and the Borda Rule are not Independent of Irrelevant Alternatives (not IIA), but they are transitive. On the other hand, Majority Rule (MR) or pairwise compari-

9.3 Social Choice Theory; Majority Judgment

son is IIA, but not transitive. The question remains whether one can devise a voting rule which is both IIA and transitive. In 1951 K. Arrow made an abrupt end to this hope by publishing his so called *impossibility theorem* [1]: for three or more alternatives every voting rule which takes as input the individual preference orderings of the voters and which satifies IIA and transitivity (together with some other elementary properties like anonymity and neutrality) is dictatorial, i.e., there will be a voter whose preference is always the social or common preference, no matter what the preferences of the other voters are. Such a voter is called a *dictator*.

First some definitions.

Definition 9.22 (Profile). A profile p associates with every voter a (linear or weak) ordering of the candidates or alternatives.

Definition 9.23 (Voting Rule). A voting rule or voting method M assigns to every profile a common (or social) (weak) ordering $\succeq_M$ of the candidates. The ordering $\succeq_M$ may be weak, i.e., indifferences ($A \approx_M B$, i.e., $A \succeq_M B$ and $B \succeq_M A$) may occur.

There are many proofs of Arrow's theorem. Below we present a simple proof of (a version of) Arrow's theorem, recently published by Balinski and Laraki [3]. They start with listing *May's axioms* [12] *for a voting method M in the case of two candidates*:

Definition 9.24 (May's axioms for a voting method M in the case of two alternatives). 1. *Based on comparisons* The input of the voting method M consists of the individual preference orderings of the voters over the candidates or alternatives.
2. *Unrestricted domain* Every vote configuration (profile) is allowed, in other words, the voting method M should assign a social ordering to every profile p.
3. *Anonymity* Interhanging the names of the voters does not change the outcome.
4. *Neutrality* Interchanging the names of the alternatives does not change the outcome.
5. *Monotonicity* If A wins or is socially indifferent to B ($A \succeq_M B$) and one or more voters change their preference in favor of A, then the voting method M will put A above B ($A \succ_M B$).
6. *Completeness* Given a pair of candidates A and B, the voting method M will put A above B ($A \succ_M B$) or B above A ($B \succ_M A$) or declare them indifferent ($A \approx_M B$).

Theorem 9.1 (May [12]). *In the case of only two alternatives the only voting method which satisfies May's axioms is Majority Rule.* (Remember that in the case of two alternatives Majority Rule, Plurality Rule and the Borda Rule are equivalent!)

Proof. (Balinski and Laraki [3]) Suppose two alternatives A and B and the voting method M satisfies May's axioms. Anonymity implies that only the numbers count: the number n_A of voters who prefer A to B, the number n_B of voters who prefer B to A and the number n_{AB} of voters who are indifferent between A and B. Completeness guarantees that there must be an outcome.

Suppose $n_A = n_B$ and $A \succ_M B$. Because of neutrality changing the names of A and B results in $B \succ_M A$. But the new profile is identical to the original profile. Contradiction. Hence, by completeness, $A \approx_M B$ when $n_A = n_B$.

Suppose $n_A > n_B$. Change the preferences of $n_A - n_B$ voters who prefer A to B in indifferences. By May's axiom of unrestricted domain this profile is allowed and given this profile $A \approx_M B$, as we have just seen. Changing this profile back to the original profile yields $A \succ_M B$ according to May's monotonicity axiom. □

For the case of an arbitrary number of candidates Balinski and Laraki [3] add to May's axioms the following two axioms:
7. *Transitivity* If $A \succeq_M B$ and $B \succeq_M C$, then $A \succeq_M C$.
8. *Independence of Irrelevant Alternatives* (IIA) If $A \succeq_M B$ and other candidates are dropped or adjoined, then again $A \succeq_M B$.
Next Balinski and Laraki prove the following version of Arrow's impossibility theorem.

Theorem 9.2 (Arrow's impossibility theorem [1]**).** *For $n \geq 3$ candidates there is no voting method M which satisfies all eight axioms.*

Proof. (Balinski and Laraki [3]) Consider any two candidates A and B. According to IIA it is sufficient to consider only these two. By Theorem 9.1 axioms 1 till 6 imply that the voting method M is Majority Rule. Because of the axiom of unrestricted domain, Condorcet's paradoxical profile is admitted and hence transitivity is violated. Hence, there can be no voting method which satisfies all eight axioms. □

The question whether it is possible to escape from Arrow's impossibility theorem has kept many scientists busy for more than 60 years: mathematicians, economists, political scientists and philosophers. Notice that when two people say that they prefer A to B they may mean quite different things: one may mean that A is excellent and B is (very) good, while the other may mean that A is acceptable and B should be rejected. With many voters a babylonian confusion of tongues is the result and it should not come as a surprise that problems like the impossibility theorem show up.

Already in the first half of last century people like Gerrit Mannoury, L.E.J. Brouwer, David van Dantzig, Frederik van Eeden and some other like minded, unified in the *Signific Circle* in the Netherlands, have pointed to the importance of a careful use of language. We quote Mannoury:

> Who wants to control his feelings must first analyze them and the traditional language forms are utterly insufficient for this purpose. [Mannoury 1917]

> To the further development of philosophical thoughts an impediment stands in the way. ... I know of no image that gives a clearer idea of this impediment than that of the tower of Babel, symbol of the confusion of tongues. [Mannoury 2017]

This is precisely what happens if different people say that they prefer A to B. They all mean something else!

9.3.7 Domination

In their book [2] Balinski and Laraki present a solution: instead of asking voters their preference ordering over the candidates, one should ask them to give an *evaluation*

9.3 Social Choice Theory; Majority Judgment

of all candidates in terms which are well understood by everyone involved. For instance in terms of: excellent (*ex*), very good (*vg*), good (*go*), acceptable (*ac*), poor (*po*) and reject (*re*). The range of evaluations should be sufficiently large such that every voter can express his distinction of the candidates.

Notice that evaluations are much *more informative* than preference orderings: from the evaluations of the candidates by a voter one can easily deduce his preference ordering over the candidates, but not vice versa! From a preference ordering over the candidates one cannot deduce the evaluations by the voter in question.

By a more precise use of language, evaluations instead of orderings, Balinski and Laraki [3] do an astonishing, if not shocking, discovery: **Majority Rule does not respect domination!** Let us illustrate what we mean by an example. Consider two candidates A and B who are evaluated by five voters as rendered in the following *opinion profile*:

voter	1	2	3	4	5
candidate A	go	ac	po	ex	vg
candidate B	vg	go	ac	po	re

The first three voters slightly prefer B to A, while the last two voters strongly prefer A to B. According to Majority Rule A is beaten by B with 2 against 3: $B \succ_{MR} A$.

However, if we look at the evaluations of A and B, ordered from high till low, then the following *merit-profile* results:

$$A \quad ex \quad vg \quad go \quad ac \quad po$$
$$B \quad vg \quad go \quad ac \quad po \quad re$$

It is A who has the better evaluations, in other words, the evaluations of A *dominate* those of B. Hence, A instead of B should be the winner! Majority Rule does not respect domination. On the other hand, any reasonable voting rule should respect domination. Question is whether there exists such a voting rule. And yes, there is: Majority Judgment (MJ) of Balinski and Laraki [2, 3]. Let us illustrate how Majority Judgment works by applying it to the situation just given.

There is a majority of 3 voters who think that A deserves at least a *go*, and there is another majority of 3 voters who think that A deserves at most a *go*. For that reason the *majority grade* of A is by definition *go*. For B there is a majority of 3 voters who think that B deserves at least an *ac*, and another majority of 3 voters who think that B deserves at most an *ac*. Hence, the *majority grade* of B is by definition *ac*. The majority grade of A is higher than the one of B and hence, according to Majority Judgment, A is the winner: $A \succ_{MJ} B$.

Majority Judgment (MJ) looks horizontally for majorities in the merit-profile, while Majority Rule (MR) looks vertically for majorities in the opinion-profile. Majority Judgment (MJ) respects domination, however, Majority Rule (MR) does not.

9.3.8 Majority Judgment (MJ)

Balinski and Laraki develop in their book [2] and in their article [3] a theory, called Majority Judgment (MJ), to aggregate the evaluations (instead of the preference orderings) of the candidates by the voters to a common or social (weak) preference ordering $\succeq_{MJ}$ over the candidates. As suggested by the name Majority Judgment, majorities play an essential role in this aggregation method. Majority Judgment (MJ) is Independent of Irrelevant Alternatives (IIA), transitive and does respect domination.

To explain how Majority Judgment works, let us consider an example with three candidates *A*, *B* and *C* and six voters or judges. The evaluations of the candidates by the voters are given in the following *opinion-profile*:

voter	1	2	3	4	5	6
A:	ex	ex	vg	ex	ex	ex
B:	ex	vg	vg	vg	go	vg
C:	ac	ex	go	vg	vg	ex

Anonymity requires that only the judgments or grades count. The number of times that each grade occurs, from high till low, is rendered in the *merit-profile* of the candidates:

A: ex ex **ex** **ex** ex vg
B: ex [vg **vg** **vg** vg] go
C: ex [ex **vg** **vg** go] ac

There is a 4/6 majority of voters who think that *C* deserves at least a *vg* and there is another 4/6 majority of voters who think that *C* deserves at most a *vg*. So, for *C* there is a 4/6 majority for [*vg*, *vg*]. The *majority grade* of *C* is therefore by definition *vg*. It is the most accurate possible majority decision about the evaluations of *C*. In a similar way the 4/6 majorities for *A* and *B* have been indicated in boldface.

The merit-profile $\alpha = (\alpha_1, \alpha_2, \ldots, \alpha_n)$ of candidate *A dominates* the merit-profile $\beta = (\beta_1, \beta_2, \ldots, \beta_n)$ of candidate *B* iff for every *i*, $\alpha_i \geq \beta_i$ and for at least one *k*, $\alpha_k > \beta_k$. Every reasonable voting method should respect domination. In our example the merit-profile of *A* dominates the one of *B* and the one of *C*. Therefore, Majority Judgment (MJ) will make *A* the winner: $A \succ_{MJ} B$ and $A \succ_{MJ} C$.

How should Majority Judgment (MJ) rank *B* and *C*? The 4/6 majorities for *B* and *C* are identical: [*vg*, *vg*]. But for *B* the 5/6 majority (indicated by the square brackets) is for [*vg*, *vg*], while for *C* the 5/6 majority is for [*ex*, *go*]. Because none of these pairs dominates the other and because there is more consensus in the evaluations of *B* than in those of *C*, Majority Judgment (MJ) will rank *B* above *C*. So, the social or common preference ordering under Majority Judgment will be: $A \succ_{MJ} B \succ_{MJ} C$. Notice that Majority Rule (MR), applied to the opinion-profile in our example, will rank *C* above *B*: *C* beats *B* with 3 against 2, so $C \succ_{MR} B$.

More generally, suppose the evaluations of *B* are $\beta = (\beta_1, \beta_2, \ldots, \beta_n)$ and those of *C* are $\gamma = (\gamma_1, \gamma_2, \ldots, \gamma_n)$, both from high till low, and suppose the most accurate majority where the candidates *B* and *C* differ is the majority for $[\beta_k, \beta_{n-k+1}] \neq [\gamma_k, \gamma_{n-k+1}]$. We call $[\beta_k, \beta_{n-k+1}]$ *B*'s *middle-most block* with respect to *C* and $[\gamma_k, \gamma_{n-k+1}]$ *C*'s middle-most block with respect to *B*. Majority Judgment'(MJ) ranks *B* above *C* iff (a) the middle-most block of *B* with respect to *C* dominates the middle-most block of *C* with respect to *B*, or (b) the middle-most block of *B* with respect to *C* shows more consensus than the one of *C* with respect to *B*.

So, $B \succ_{MJ} C$ iff (a) $\beta_k \succeq \gamma_k$ and $\beta_{n-k+1} \succeq \gamma_{n-k+1}$, with at least one $\succeq$ strict, or (b) $\gamma_k \succ \beta_k \succeq \beta_{n-k+1} \succ \gamma_{n-k+1}$. In all other cases the collections of evaluations are identical and $B \approx_{MJ} C$.

9.3.9 Properties of Majority Judgment

From the definition of Majority Judgment (MJ) follows immediately:

Theorem 9.3. *(Balinski and Laraki) Majority Judgment takes as input the evaluations of the candidates by the voters and satisfies all axioms 2 till 8 in subsection 9.3.6.*

In addition, Majority Judgment (MJ) has among others the following properties.

1. Majority Judgment (MJ) gives a social preference ordering $\succeq_{MJ}$ of the candidates or alternatives and society is indifferent between two candidates A and B, $A \approx_{MJ} B$, precisely when they have the same evaluations. Majority Judgment measures the support of the electorate for the candidates and orders them in proportion to their support. With Majority Rule the voters cannot express their opinions about the candidates, every voter is restricted to supporting one candidate at the expense of all others.
2. From the definitions it is evident that Majority Judgment (MJ) is *Independent of Irrelevant Alternatives* (IIA): whether $A \succeq_{MJ} B$ or $B \succeq_{MJ} A$ does not depend on a third alternative C. As we saw, Plurality Rule and the Borda Rule are not IIA.
3. With more than two candidates, $\succeq_{MJ}$ is *transitive*: if $A \succeq_{MJ} B$ and $B \succeq_{MJ} C$, then $A \succeq_{MJ} C$. As we have seen, Majority Rule (MR) is not transitive.
4. Majority Judgment (MJ) respects domination: if the evaluations of A dominate those of B, then $A \succ_{MJ} B$. Majority Rule (and hence also Plurality Rule and the Borda Rule) do not respect domination.
5. Majority Judgment is *strategy-proof in grading*: a group of voters whose input is higher (respectively, lower) than the majority grade cannot raise (respectively, lower) the majority grade. For instance, suppose candidate A receives the following grades: *good acceptable poor*. The majority grade of A is *acceptable*. The voter who gave A a *good* thinks the majority grade *acceptable* is too low, but he cannot raise the majority grade of A; giving an *excellent* instead of a *good* does not raise the majority grade of A.
   This property certainly does not hold for mechanisms based on adding numbers or taking averages of numbers, neither for the Borda Rule and its variants.
6. Majority Judgment (MJ) is *partially strategy-proof in ranking*: if a voter who prefers B to A, can raise the majority grade of B, then he cannot lower the majority grade of A; and if he can lower the majority grade of A, then he cannot raise the majority grade of B. For instance, suppose voter i gives B a higher evaluation than A and A has the same majority grade as B.

$$\begin{array}{c} B \\ \text{majority grade} \\ A \end{array} \quad \begin{array}{c} \leftarrow \\ | \\ | \end{array} \quad \begin{array}{c} \cdot \\ i \\ | \end{array} \quad \begin{array}{c} \\ i \\ \cdot \end{array}$$

The only way in which voter i can raise the majority grade of B is by giving B a grade higher than its majority grade instead of a grade lower than B's majority grade. But because i gave a lower grade to A than to B, he cannot lower A's majority grade.

This property certainly does not hold for mechanisms based on adding numbers or taking averages of numbers, neither for the Borda Rule and its variants.

7. The majority grade of a candidate is an important signal both to the candidate and to the electorate.
8. Majority Judgment stimulates candidates to get the highest possible grades of as many voters as possible; every grade contributes to the final judgment.
9. Candidates cannot focus on 51% of the electorate and, once the winner, claim to represent the whole electorate.

9.3.10 Point Summing and Approval Voting

One should notice that voting methods, where voters give points to candidates and where candidates are ordered according to the number of points they have collected, like Majority Judgment also satisfy the axioms 2 till 8 in Section 9.3.6. However, such methods are not strategy-proof neither in grading nor in ranking. In addition, a voting method based on giving points to the candidates is not consistent with Majority Judgment, neither with Majority Rule. Consider the following example:

	1	2	3	4	5	6	7
A:	ex	ex	ex	ac	ac	ac	ac
B:	po	po	po	go	go	go	go

Looking at this opinion profile vertically, we see that B beats A with 4 against 3, so B is the Majority Rule winner: $B \succ_{MR} A$. Looking at this profile horizontally, we see that the majority grade of B is go and the majority grade of A is only ac; so, in this example B is also the Majority Judgment winner: $B \succ_{MJ} A$. However, with 5 points for ex, 4 for vg, 3 for go, 2 for ac, 1 for po and 0 for re, A wins with 23 points against 15 for B. So, adding points is not consistent with Majority Judgment, neither with Majority Rule.

The idea of Approval Voting [6] is that every voter gives 1 point to each candidate he or she approves of and 0 points to every candidate he or she disapproves of. With 1 point for go or higher, B wins with 4 points against 3 for A. But with 1 point for ac or higher, A wins with 7 points against 4 for B. So, Approval Voting yields arbitrary outcomes and is not consistent with Majority Judgment, neither with Majority Rule.

9.3.11 Majority Judgment with many Voters

Consider the following merit profile for two candidates A and B:

	ex	vg	go	↕	go	ac	po	re
A:	28.63	16.42	04.95	↕	06.72	14.79	14.25	14.24
B:	12.35	21.71	15.94	↕	09.30	20.08	11.94	08.69

Left and right of the middle one finds 50% of the number of evaluations. For $\varepsilon \leq 4.95$, A and B have a $(50+\varepsilon)\%$ majority for [go, go]. But for $\varepsilon < 6.72 - 4.95 = 1.77$, A has a $(54.95+\varepsilon)\%$ majority for [vg, go], while B has a $(54.95+\varepsilon)\%$ majority for [go, go]. Because A's middlemost block dominates the one of B, $A \succ_{MJ} B$. This is the case because $4.95 < \min\{6.72, 15.94, 9.30\}$. Finding the smallest of these four numbers is the same as finding the highest percentage of each candidate's grades strictly above and strictly below their majority grades.

Let p_A be the percentage of A's grades strictly above the majority grade α_A of A and q_A the percentage of A's grades strictly below α_A. A's *majority gauge* is by definition (p_A, α_A, q_A). So, in our example the majority gauge of A is (45.05, go, 43.28) and the majority gauge of B is (34.06, go, 40.71).

The *majority-gauge rule* $\succ_{MG}$ ranks A above B, $A \succ_{MG} B$, iff $\alpha_A \succ \alpha_B$ or ($\alpha_A = \alpha_B$ and $p_A > \max\{q_A, p_B, q_B\}$) or ($\alpha_A = \alpha_B$ and $q_B > \max\{p_A, q_A, p_B\}$).

In our example: $p_A = 45.05 > \max\{43.28, 34.06, 40.71\}$, therefore $A \succ_{MG} B$. If $\succeq_{MG}$ is decisive (written as $\succ_{MG}$), then its ordering $\succ_{MG}$ is identical to the one of $\succ_{MJ}$. So, in our example it also follows that $A \succ_{MJ} B$, as we already saw above.

9.3.12 Presidential Elections in the USA

In [4] Balinski and Laraki give an analysis of the recent (2016) presidential elections in the USA. Their conclusion is unambiguous: the voting method in the USA does not work, more precisely, it does not select the candidate who gets globally the highest evaluation of the electorate. To illustrate this, they use the results mentioned below of a poll by the Pew Research Center in March 2016 among 1787 voters from all political stripes.

candidate	great	good	average	poor	terrible
A	05	28	**39**	13	15 %
B	10	26	**26**	15	23 %
C	07	22	**31**	17	23 %
D	11	22	20	16	**31** %
E	10	16	12	**15**	47 %

For candidate A there is a majority of $05 + 28 + 39 = 64\%$ who thinks that he deserves at least an *average* and there is another majority of $15 + 13 + 39 = 67\%$ who thinks that he deserves at most an *average*. Therefore, *average* is by definition the majority grade of candidate A. In the table the majority grades of the different candidates have been indicated by bold face letters. Notice that the opinions of the voters are clearly much more detailed than can be expressed by Majority Rule. Also the percentages of voters who think that candidates D and E would be bad presidents is relatively high.

Next Balinski and Laraki determine how, given these judgments, Majority Judgment would rank the candidates. The majority grade of candidate A, B, C and D is *average*, the one of candidate E is *poor*. The majority gauge of candidate A is (33, average, 28), because $p_A = 5 + 28 = 33$ and $q_A = 13 + 15 = 28$. In the table below the majority gauges of all candidates are listed, from which one may derive a ranking of the candidates according to the majority-gauge rule, which is also the Majority Judgment ranking.

	majority grade	majority gauge
1. A	average	(33, average, 28)
2. B	average	(36, average, 38)
3. C	average	(29, average, 40)
4. D	average	(33, average, 47)
5. E	poor	(38, poor, 47)

Because $q_B = 38 > \max\{33, 28, 36\}$ it follows that $A \succ_{MG} B$; because $q_C = 40 > \max\{36, 38, 29\}$ it follows that $B \succ_{MG} C$ and because $q_D = 47 > \max\{29, 40, 33\}$ it follows that $C \succ_{MG} D$. Finally, because the majority grade *average* of D is higher than the majority grade *poor* of E, it follows that $D \succ_{MG} E$.

Amazingly, at election day the two main candidates were D and E, of which E won the election, because he won in most states, although he did not get most votes.

The Majority Judgment ranking is the logical result of majorities which decide about the judgments of the candidates instead of Majority Rule which ranks candidates according to the number of votes they get. *Majority Judgment measures the support of the electorate for the different candidates and ranks them according to their support.* With Majority Rule the voters cannot express their opinions about the candidates; every voter is restricted to supporting one candidate at the same time excluding all others.

Why can Majority Rule work out so poorly? To make this clear, Balinski and Laraki [4] consider the merit-profile of candidates D and E:

	great	good	average	poor	terrible
D	11	22	**20**	16	31 %
E	10	16	12	**15**	47 %

Notice that the evaluations of D dominate those of E. Hence, D should win, as also becomes clear from the following table:

at least	great	good	average	poor	terrible
D	11	33	53	69	100 %
E	10	26	38	53	100 %

Any decent voting method should rank D above E. But Majority Rule can easily fail to make D the winner: suppose that underlying the merit-profile for D and E is the following opinion-profile for these candidates:

9.3 Social Choice Theory; Majority Judgment

	10	16	12	15	14	11	12	04	04	02
D	go	av	po	te	te	gr	go	av	po	te
E	gr	go	av	po	te	te	te	te	te	te

The individual vote percentages in this opinion-profile are in accordance with the degrees that each candidate received in the merit-profile. For instance, the 22% voters who gave a *good* to D are now divided in two groups: a group of 10% voters who gave a *good* to D and a *great* to E and a group of 12% voters who evaluated D as *good* and E as *terrible*. Applying Majority Rule to this opinion-profile, E will beat D with $10 + 16 + 12 + 15 = 53\%$ against $11 + 12 + 4 + 4 = 31\%$, while D's evaluations dominate those of E. Notice that in this opinion-profile the 53% voters who prefer E to D only slightly do so, while most voters who prefer D to E do so strongly.

9.3.13 Presidential Elections in France

In [5] Balinki and Laraki take a look at the French presidential elections. Their conclusion is again extremely negative: the French election system can easily select a winner who is rejected by a vast majority of the voters. The French presidential election is in two rounds: 1. If in the first round a candidate has more than half of the votes, then he or she is elected. 2. Otherwise, there is a second round between the two candidates with most votes in the first round.

Let us start with having a look at the presidential elections of 2007 with twelve candidates of which Sarkozy, Royal and Bayrou were the most important ones. The results of the first round were as follows:

$$\begin{array}{lll} 31.2\% & \text{Sarkozy} & \text{Bayrou Royal} \\ 25.9\% & \text{Royal} & \text{Bayrou Sarkozy} \\ 18.6\% & \text{Bayrou} & \\ xy.z\% & ??? & \text{Bayrou Sarkozy/Royal} \end{array}$$

In the first round Sarkozy and Royal had most votes, but less than 50%. Therefore, there was a second round between them, in which Sarkozy won. But the polls showed clearly that a majority of $25.9 + 18.6 + xy.z$ % of the voters preferred Bayrou to Sarkozy and that another majority of $31.2 + 18.6 + xy.z$ % preferred Bayrou to Royal. As we shall see further on in this subsection applying Majority Judgment would most likely have chosen Bayrou as the winner.

At the French presidential elections of April 21, 2017, there were initially three major candidates, say A, B and C. Suppose the preference orderings of the voters were as follows:

$$\begin{array}{llll} 34\% & A & B & C \\ 32\% & B & A & C \\ 34\% & C & B & A \end{array}$$

In this case nobody has more than 50% of the votes and B, who has least votes, is eliminated. The second round is then between A and C, in which A gets $34 + 32 =$

66% of the votes and wins. Next suppose that in the first round A gets more support at the expense of candidate C:

$$\begin{array}{ccc} 37\% & A\ B\ C \\ 32\% & B\ A\ C \\ 31\% & C\ B\ A \end{array}$$

Then after the first round C is eliminated and B wins in the second round with $32 + 31 = 63\%$ of the votes. More support for the winning candidate A in the first round causes that he becomes a loser instead of a winner. In other words: *the French election mechanism is not monotonic*: more support may mean losing instead of winning.

On April 22, 2007, Balinski and Laraki did an experiment among 1752 voters in three districts of Orsay. These voters were asked to fill in, apart from the official voting ballot, also the following voting ballot.

Pour présider la France, ayant pris tous les éléments en compte, je juge en conscience que ce candidat serait:

	très bien	bien	asses bien	passable	insuffisant	à rejeter
Besancenot						
Buffet						
Schivardi						
Bayrou						
Bové						
Voynet						
Villiers						
Royal						
Nihous						
Le Pen						
Laguiller						
Sarkozy						

Attribuer à chaque candidat une évaluation parmi les mentions.

The results for the three most important candidates were:

	exc	very good	good	acc	poor	reject
Bayrou	13.6	30.7	**25.1**	14.8	8.4	7.4
Royal	16.7	22.7	**19.1**	16.8	12.2	12.6
Sarkozy	19.1	19.8	**14.3**	11.5	7.1	28.2

All three candidates have majority grade *good*. Let p, resp. q be the percentage strictly above, resp. strictly below the majority grade.

majority rank	p	majority grade	q	national rank
1 Bayrou	44.3	*good*	30.6	3
2 Royal	39.4	*good*	41.5	2
3 Sarkozy	38.9	*good*	46.9	1

The majority gauge rule yields the ranking: 1 Bayrou, 2 Royal en 3 Sarkozy. One may easily motivate this outcome by looking at the cumulative table below. With the exception of the *exc* column it holds for every column that Bayrou scores better than Royal and Royal better than Sarkozy.

at least	exc	very good	good	acc	poor	reject
Bayrou	13.6	44.3	69.4	84.2	92.6	100
Royal	16.7	39.4	58.5	75.3	87.5	100
Sarkozy	19.1	38.9	53.2	64.7	71.8	100

9.3 Social Choice Theory; Majority Judgment

9.3.14 Elections for Parliament in the Netherlands

Majority Judgment may be used to determine a common or social preference ordering over the candidates. Those candidates may be political parties. But Majority Judgment does not yield a seat distribution among the parties. However, Majority Judgment might be used in the Netherlands to choose a mayor, a prime minister, a chairman, etc. That there is a need in the Netherlands for a better election mechanism may become evident from the following examples.

In the table below one finds the vote and seat distribution after the elections for parliament on September 6, 1989:

Party	% of votes	number of seats
CDA	35.3	54
PvdA	31.9	49
VVD	14.6	22
D66	07.9	12
GL	04.1	06
SR	05.0	07

Suppose the following plausible profile is underlying the seat distribution above:

35.3	CDA	D66	VVD	SR	PvdA	GL
31.9	PvdA	GL	D66	CDA	VVD	SR
14.6	VVD	PvdA	D66	SR	CDA	GL
07.9	D66	PvdA	CDA	VVD	GL	SR
04.1	GL	PvdA	D66	CDA	VVD	SR
05.0	SR	VVD	CDA	D66	PvdA	GL

Notice: VVD beats PvdA with 35.3 + 14.6 + 05.0 = 54.9 against 31.9 + 07.9 + 04.1 = 43.9, but PvdA gets 49 seats and VVD only 22. Similarly: D66 beats CDA with 31.9 + 14.6 + 07.9 + 04.1 = 58.5 against 35.3 + 05.0 = 40.3, but CDA gets 54 seats and D66 only 12. Van Deemen [9] calls this phenomenon: the *more preferred, but less seats* paradox.

The situation may be even worse: a party may beat every other party in a pairwise comparison (Majority Rule) and still get less or no seats at all, as becomes clear from the example below. On September 6, 1989, the Greens (G) were participating, but did not get any seat. Suppose G was for all voters the second choice:

35.3	CDA	G	D66	VVD	SR	PvdA	GL
31.9	PvdA	G	GL	D66	CDA	VVD	SR
14.6	VVD	G	PvdA	D66	SR	CDA	GL
07.9	D66	G	PvdA	CDA	VVD	GL	SR
04.1	GL	G	PvdA	D66	CDA	VVD	SR
05.0	SR	G	VVD	CDA	D66	PvdA	GL

Under pairwise comparison (Majority Rule) G beats every other party and hence is the Condorcet winner. But G gets no seat at all in the Dutch system. At another occasion, a similar fate struck party DS70, which was second or third choice for many voters. Van Deemen [9] calls this phenomenon: the *Condorcet winner, but no or less seats* paradox.

From empirical research [10] it turns out that the 'more preferred, but less seats' paradox occurs abundantly. And from empirical research it also becomes clear that

D66 in 1994 was the Condorcet winner, but got less seats than PvDA, CDA and VVD. In 1982 PvdA was the Condorcet winner, but got less seats than CDA.

Exercise 9.16. Prove that pairwise comparison is *strategy-proof* in the following sense: Let S be a set of voters and p, q profiles such that $p(i) = q(i)$ for all voters i not in S (the individuals in S give in q a dishonest preference). Let x be the Condorcet winner given p and y the Condorcet winner given q. Suppose $x \neq y$. Then there is an individual $i \in S$ who in his honest individual preference ordering $p(i)$ strictly prefers x to y. So, for that individual the strategic change towards $q(i)$ is a disadvantage.

Exercise 9.17. Prove that for two alternatives 'most votes count' (Plurality Rule), pairwise comparison (Majority Rule) and the Borda Rule give the same results. Conclude that Arrow's theorem does not hold for the case of two alternatives.

Exercise 9.18. *Agenda's: Berlin versus Bonn*
At June 20, 1991, the German parliament had to make a choice among the following three alternatives:
(a) the parliament moves to Berlin, but the ministries stay in Bonn;
(b) both the parliament and the ministries move to Berlin;
(c) both the parliament and the ministries stay in Bonn.
The council of elderly had made an agenda, which was essentially as follows: in the first round the representatives have to make a choice between (a) and not (a). In the second round: if (a) is accepted, then the final choice is (a); if not, then the representatives have to choose between (b) and (c).
From a reconstruction it has become pretty evident that the preferences of the 660 representatives were given in the following profile p:

077: a b c
070: a c b
178: b a c
083: b c a
190: c a b
062: c b a

i) Check that the outcome will be (b), in accordance with the real state of affairs. Verify that given profile p there is no Condorcet winner.
ii) Why is the agenda set by the council of elderly not fair?
iii) Check that if the 83 representatives change their preference ordering b c a into b a c and the preference orderings of the other representatives remain the same, then (a) will be the Condorcet winner. Nevertheless, in this case (b) will again be the outcome under the agenda devised by the council of elderly.
iv) A more fair agenda than the one above would be agenda I: in the first round choose between (a) and (b), and in the second round choose between the winner of the first round and (c). Why is this agenda more fair? Check that if a Condorcet winner exists, it will always be the outcome under this agenda I. Check that the outcome under agenda I given profile p will be (c).
v) Devise an agenda II, respectively III, such that the outcome under agenda II, respectively III, given profile p, will be (a), respectively (b).

Exercise 9.19. *District Paradox: more votes, but less seats.*
Suppose there are three districts and two parties, twenty voters in each district and

9.3 Social Choice Theory; Majority Judgment

in each district the Plurality Rule is used to determine the winner. Suppose the ballot yields the following results:

	Candidate of party A	Candidate of party B	Elected candidate
district 1	11 votes	09 votes	A
district 2	11 votes	09 votes	A
district 3	05 votes	15 votes	B

Party A gets a majority in the House of Commons and will form the cabinet. But party B receives more votes (33) than party A (27). So, if the government would be chosen directly, it would be composed by party B. The majority attributed to party A is called a *manufactured majority*: a majority of the seats obtained by a minority of the voters.

Exercise 9.20. *Discursive Paradox in judgment aggregation*
We explain the *discursive paradox* using the following example, due to Saari: A three member faculty committee must determine whether or not a student should be advanced to Ph.D. candidacy. A majority vote is required to advance. Each faculty member's decision is based on the student's performance on both a written and an oral exam. If a faculty member feels that the student failed one or both of these exams, she is instructed to fail the student. The results follow, where a 'yes' or 'no' indicates the judge's opinion on an exam and whether to advance.

Judge	written	oral	decision
1	yes	yes	yes
2	no	yes	no
3	yes	no	no
Outcome	yes	yes	no

Exercise 9.21. Consider the following Condorcet table of D. Saari:

Ranking	$\{A,B\}$	$\{B,C\}$	$\{A,C\}$
$A > B > C$	$A > B$	$B > C$	$A > C$
$B > C > A$	$B > A$	$B > C$	$C > A$
$C > A > B$	$A > B$	$C > B$	$C > A$
Outcome	$A > B$	$B > C$	$C > A$

Verify that by replacing $A > B$, $B > C$ and $A > C$ by 'yes', $B > A$, $C > B$ and $C > A$ by 'no', the discursive paradox in Exercise 9.20 is a special case of the Condorcet paradox. Notice that the table below, in which the individual preferences are not transitive but cyclic, gives under Majority Rule the same result as the table above:

Ranking	$\{A,B\}$	$\{B,C\}$	$\{A,C\}$
$A > B > C > A$	$A > B$	$B > C$	$C > A$
$B > C > A > B$	$A > B$	$B > C$	$C > A$
$C > B > A > C$	$B > A$	$C > B$	$A > C$
Outcome	$A > B$	$B > C$	$C > A$

So, pairwise comparison ignores the rationality of the voters, i.e., that voters are transitive. Similarly, the IIA condition ignores the rationality of the voters.

Exercise 9.22. *Sen's Paradox: even a minimal form of Liberalism is impossible.*
A voting rule satisfies the *Pareto condition* := if all voters prefer x to y, then also society should prefer x to y.

Assuming that voter 1, respectively 2, is decisive over the pair $\{A,B\}$, respectively $\{C,D\}$ and that the voting rule satisfies the Pareto condition, determine in the table below the outcome for each pair and notice that a cyclic outcome results. This is *Sen's paradox*: even a minimal form of Liberalism is impossible.

Voter	Preference	$\{A,B\}$	$\{B,C\}$	$\{C,D\}$	$\{A,D\}$
1	$D > A > B > C$	$A > B$	$B > C$	–	$D > A$
2	$B > C > D > A$	–	$B > C$	$C > D$	$D > A$
	outcome				

In this table a dash indicates a ranking that is irrelevant for the decision rule because another agent is decisive over that pair.

Notice that for instance for voter 2 it is immaterial whether his $\{A,B\}$ preference is $A > B$ or $B > A$ (because voter 1 is decisive over this pair). But the first choice makes his preferences cyclic, while the second choice makes them transitive - a huge difference. So, the assumptions imposed on the voting rule dismiss the individual rationality assumption (that a voter's preferences are transitive).

9.4 Solutions

Solution 9.1. Extend the program in Example 9.1 as follows.
(8) male(bob).   (11) female(pam).   (14) female(ann).
(9) male(tom).   (12) female(liz).   (15) offspring(X,Y) :- parent(Y,X).
(10) male(jim).   (13) female(pat).   (16) father(X,Y) :- parent(X,Y), male(X).
(17) mother(X,Y) :- parent(X,Y), female(X).
(18) sister(X,Y) :- parent(Z,X), parent(Z,Y), female(X).
(19) brother(X,Y) :- parent(Z,X), parent(Z,Y), male(X).

```
     ?- mother(tom, liz).                    ?- sister(ann, pat).
           (17) |                                  (18) |
     parent(tom, liz), female(tom)      parent(Z, ann), parent(Z, pat), female(ann)
            (3) |                                   (4) |
          female(tom)                      parent(bob, pat), female(ann)
               |                                    (5) |
            failure                              female(ann)
              no                                   (14) |
                                                     □
                                                    yes
```

9.4 Solutions

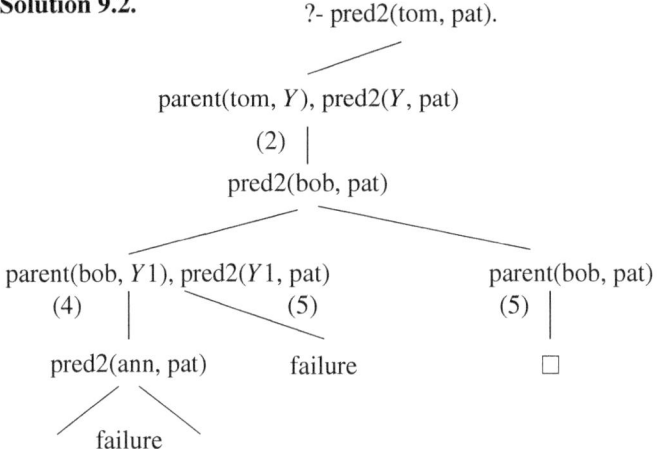

Solution 9.2.

```
                        ?- pred2(tom, pat).
                              /
              parent(tom, Y), pred2(Y, pat)
                         (2) |
                      pred2(bob, pat)
                      /            \
   parent(bob, Y1), pred2(Y1, pat)   parent(bob, pat)
        (4)  |         (5)               (5) |
     pred2(ann, pat)    failure              □
        /      \
    failure
```

The system has to backtrack many times before it finds a successful branch in the search tree.

Solution 9.3. a) Replace Y by $f(X)$; Y does not occur in $f(X)$; result: $p(f(X),Z)$ and $p(f(X),c)$. Next replace Z by c. Consequently, $p(f(X),Z)$ and $p(Y,c)$ can be matched and unified. Result: $p(f(X),c)$.
b) X and $f(X)$ can be matched, but not unified: replace X by $f(X)$; but X does occur in $f(X)$.
c) Replace Y by $f(X)$. Result: $p(f(X),c)$ and $p(f(X),f(Z))$. c and $f(Z)$ cannot be matched.

Solution 9.4. conc([ ], L, L).   conc([X | L1], L2, [X | L]) :- conc(L1, L2, L).

Solution 9.5. del(X, [X | L], L).   del(X, [Y | L], [Y | L1]) :- del(X, L, L1).

Solution 9.6. length([ ], 0).   length([X | L], N) :- length(L, M), N is M + 1.

Solution 9.7.

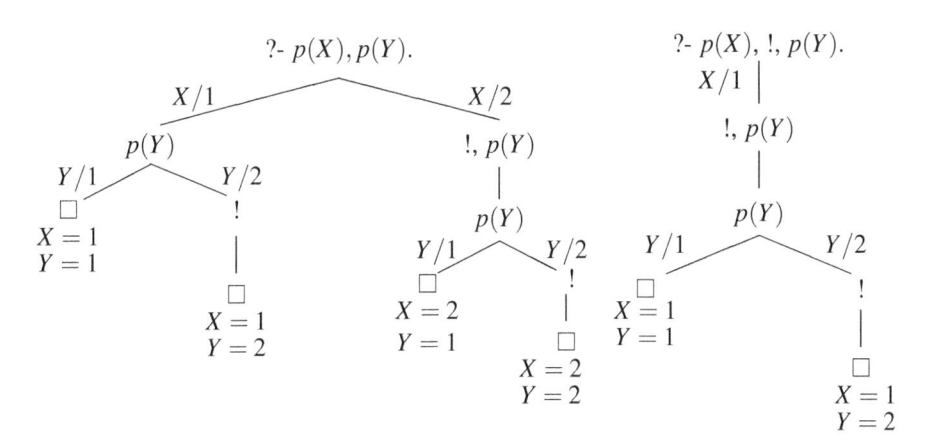

```
              ?- p(X), p(Y).                          ?- p(X), !, p(Y).
         X/1  /        \ X/2                           X/1 |
          p(Y)          !, p(Y)                         !, p(Y)
       Y/1 / \ Y/2         |                               |
          □   !           p(Y)                            p(Y)
        X=1  |          Y/1 / \ Y/2              Y/1 /         \ Y/2
        Y=1  □           □    !                    □             !
            X=1         X=2   |                   X=1            |
            Y=2         Y=1   □                   Y=1            □
                            X=2                                 X=1
                            Y=2                                 Y=2
```

Solution 9.8.
$$p(X,0) :\text{-} X < 1, ! .$$
$$p(X,1) :\text{-} X >= 1, X < 2, ! .$$
$$p(X,2) :\text{-} X >= 2 .$$

Solution 9.9.

?- subset([1, 2], [1, 2, 3]).

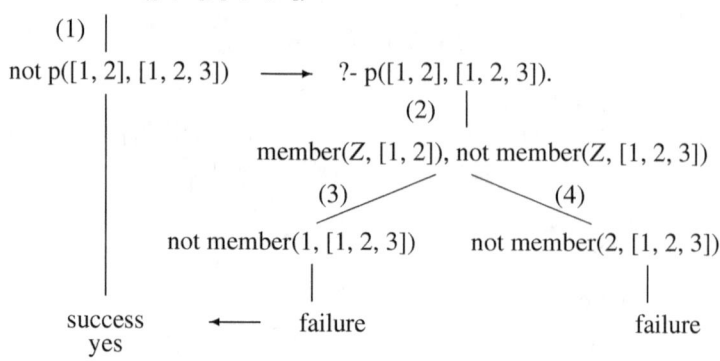

(1) |

not p([1, 2], [1, 2, 3]) ⟶ ?- p([1, 2], [1, 2, 3]).

(2) |

member(Z, [1, 2]), not member(Z, [1, 2, 3])

(3) ╱   ╲ (4)

not member(1, [1, 2, 3])   not member(2, [1, 2, 3])

| | |

success   ⟵   failure   failure
yes

Solution 9.10.

?- subset([1, 2, 3], [1]).

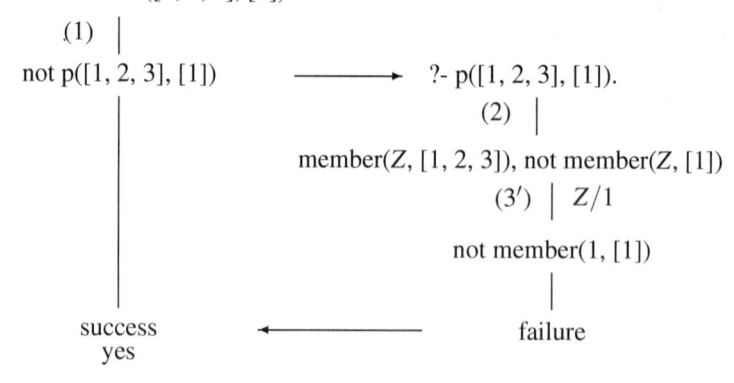

(1) |

not p([1, 2, 3], [1]) ⟶ ?- p([1, 2, 3], [1]).

(2) |

member(Z, [1, 2, 3]), not member(Z, [1])

(3') | Z/1

not member(1, [1])

| |

success   ⟵   failure
yes

Solution 9.11.

husband(X) :- family(X, _, _).
wife(X) :- family(_, X, _).
child(X) :- family(_, _, L), member(X, L).
exists(X) :- husband(X); wife(X); child(X).

1. ?- exists(person(N, S, _, _)).
2. ?- child(person(N, S, date(_, _, 1973), _)).
3. ?- wife(person(N, S, _, works(_, _))).
4. ?- exists(person(N, S, date(_, _, Y), unemployed)), Y < 1960.
5. ?- exists(person(N, S, date(_, _, Y), works(_, Sal))), Y < 1960, Sal > 10000.
6. ?- family(person(_, S, _, _), _, [_, _ | _]).
7. ?- family(person(_, S, _, _), _, [ ]).

Solution 9.12. 1. The reading $(a+b)*c$ has the following structure:

9.4 Solutions

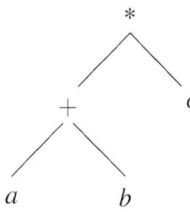

Now the precedence of $a+b$ is 500, which is greater than the precedence of $*$. Therefore this reading is rejected.

2. The reading $a-(b-c)$ has the following structure:

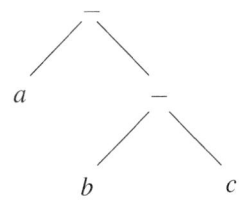

The precedence of $b-c$ is 500, which is not strictly smaller than that of the operator $-$. Since the operator $-$ has been defined to be of type yfx, this reading is impossible

3. Since 'has' has been defined as an infix operator and the arguments 'peter' and 'information' have precedence 0, which is strictly smaller than the precedence 600 of 'has', 'peter has information' will be read as 'has(peter, information)'.

Solution 9.13.

a)
```
SELECT  t.snm, t.sadr, t.res
FROM    SP t
WHERE   t.wnr = 9
AND     t.nbd > 2
```

b)
```
SELECT t.snr, t.snm
FROM SP t
WHERE t.snr IN
    (SELECT u.snr
     FROM ADM u
     WHERE u.indat = 19800303
     AND u.reas = 'informaritis')
```

c)
```
SELECT  s.rnr
FROM    R s
WHERE   s.rnr NOT IN
    (SELECT t.rnr
     FROM ADM t
     WHERE t.indat ≤ 19800518
     AND t.indat ≥ 19800509
     AND t.pnr IN
        (SELECT u.pnr
         FROM P u
         WHERE u.pres = 'Princeton'))
```

d)
```
SELECT t.pnr, t.pnm, t.padr, t.pres
FROM P t
WHERE t.pnr IN
    (SELECT u.pnr
     FROM ADM u
     WHERE u.snr IN
        (SELECT s.snr
         FROM SP s
         WHERE s.wnr = 9))
```

Solution 9.14. $D_1 \bowtie D_2$ is the empty set since no tuples agree on the common attribute 'name'.

$D_3 \bowtie D_2$

nr	wnm	sal	sex	anr	name	man
8	Johnson	2200	male	1	production	9
7	Johnson	3100	female	2	planning	7
9	Kiviat	2900	male	1	production	9

486   9 Applications: Prolog; Relational Databases and SQL; Social Choice Theory

$D_1 \bowtie D_4$

nr	name	sal	sex	dept	anr	anm
7	Johnson	3100	female	2	2	planning
9	Kiviat	2900	male	1	1	production

$D_3 \bowtie D_4$

nr	wnm	sal	sex	anr	anm
7	Johnson	3100	female	2	planning
9	Kiviat	2900	male	1	production

Solution 9.15. The patient with number 500 is from Cranbury and has been hospitalized in the period in question in room number 11. So, 11 should not occur in the set of all room-numbers in which no patients from Cranbury were hospitalized between August 11 and 17, 1977. However, 11 is an element of the indicated set. For consider $s1$; $s1(\text{rnr}) = 11$. Then there is $t \in \text{ADM}$, namely $t1$, such that $t(\text{rnr}) = s1(\text{rnr})$ and $19770811 \le t(\text{indat}) \le 19770817$ and $\neg \exists u \in P\ [\ u(\text{pnr}) = t(\text{pnr}) = 400 \land u(\text{pres}) = \text{'Cranbury'}\ ]$.

Solution 9.16. Let S be a set of voters (who manipulate) and p, q profiles such that for all i not in S, $p(i) = q(i)$. Let x be the Condorcet winner at p and let y be the Condorcet winner at q. Suppose that $x \ne y$.
Because x is the Condorcet winner at p, we know for profile p that x beats y in a pairwise comparison. And because y is the Condorcet winner at q, we know for profile q that y beats x in a pairwise comparison. So, there must be at least one individual i such that
1. i prefers x to y in p, and
2. i prefers y to x in q.
Because only voters in coalition S give another (dishonest) preference order, individual i must be in coalition S. Because in the real (honest) profile p, i prefers x to y, i is punished for the strategic behaviour of the coalition S he or she belongs to.

Solution 9.17. a) Call the alternatives x en y and suppose:

m: $x\ y$
n: $y\ x$

Then the Borda score of x equals m en the one of y equals n. Therefore: the outcome under the Borda Rule is $x\ y$ if and only if (iff) $m > n$. But also: the outcome under Plurality Rule is $x\ y$ iff $m > n$. And similarly: the outcome under Majority Rule is $x\ y$ iff $m > n$. Hence, with two alternatives, the Borda Rule, Plurality Rule and Majority Rule yield the same outcome.
b) Because Majority Rule is Independent of Irrelevant Alternatives and in the case of two alternatives trivially is transitive (transitivity says something about 3 alternatives), this makes clear that in the case of two alternatives the theorem of Arrow does not apply: Majority Rule is not dictatorial.

Solution 9.18. i) The first round is between (a) and not (a); $77 + 70 = 147$ voters vote for (a), all others vote for not (a) So, the second round is between (b) and (c). $77 + 178 + 83 = 338$ representatives vote for (b) and $70 + 190 + 62 = 322$ vote for (c). Therefore (b) wins.
  Given profile p, (a) beats (b) with $77 + 70 + 190 = 337$ votes against 323; (b) beats (c) with $77 + 178 + 83 = 338$ votes against 322; and (c) beats (a) with $83 +$

190 + 62 = 335 votes against 325. So, given p there is no Condorcet winner.

ii) The agenda set by the council of elderly is not fair, because (a) has to compete with both (b) and (c) simultaneously.

iii) If the 83 representatives change their preference ordering b c a into b a c and the preference orderings of the other representatives remain the same, then one easily checks that (a) beats (b) with 77 + 70 + 190 = 337 against 323 votes and that (a) beats (c) with 77 + 70 + 178 + 83 = 408 against 252 votes. So, in this new configuration (a) is the Condorcet winner. But according to the agenda set by the council of elderly (b) would again become the winner.

iv) Agenda I is more fair than the agenda set by the council of elderly because according to this agenda in every round only two alternatives are compared and every alternative is compared with at least one other alternative. If given a profile there is a Condorcet winner, this Condorcet winner will also win using agenda I, because the Condorcet winner will in the first or the second round be compared with another alternative and from that moment on be the winner in every next round. Given profile p and using agenda I, in the first round (a) beats (b) and in the second round (c) beats (a). So, the outcome under agenda I given profile p will be (c).

v) Agenda II: first round between (b) and (c); second round: between (a) and the winner of the first round. Given profile p, the outcome under agenda II will be (a). Agenda III: first round between (a) and (c); second round: between (b) and the winner of the first round. Given profile p, the outcome under agenda III will be (b).

Solution 9.19. According to Plurality Rule party A wins in district 1 and 2, while party B only wins in district 3. So, party A gets 2/3 of the seats in parliament. But the total number of votes for party A is 11 + 11 + 5 = 27, while party B has 9 + 9 + 15 = 33 votes.

Solution 9.20. A 2/3 majority of the judges gives a 'yes' for the written exam, another 2/3 majority of the judges gives a 'yes' for the oral exam, but another 2/3 majority of the judges gives a 'no' for the final decision. So, judgment aggregation with Majority Rule is problematic.

Solution 9.21. Majority Rule looks only at pairs of candidates. Transitivity concerns three or more candidates. By looking only at pairs of candidates, as required by Independence of Irrelevant Alternatives, transitivity, and hence the rationality of the voters, cannot be taken into account.

Solution 9.22. The outcome is: $A > B$ (1), $B > C$ (2), $C > D$ (3) and $D > A$ (4). (1) because 1 is decisive over the pair $\{A,B\}$, (2) because of the Pareto condition, (3) because 2 is decisive over the pair $\{C,D\}$ and (4) because of the Pareto condition.

References

1. Arrow, K., *Social Choice and Individual Values*. Yale University Press, 1951.

2. Balinski, M., and R. Laraki, *Majority Judgment; Measuring, Ranking and Electing*. MIT Press, Cambridge; MA, 2010.
3. Balinski, M., and R. Laraki, *Majority Judgment vs Majority Rule*. Cahier 2016-4, Ecole Polytechnique, Paris, 2016.
4. Balinski, M., and R. Laraki, *Trump and Clinton victorious: proof that US voting system does not work*. The Conversation 58752.
5. Balinski, M., and R. Laraki, *Pour éviter un nouveau 21 Avril instaurons le jugement majoritaire*. The Conversation 58178.
6. Brams, S. and Peter C. Fishburn, *Approval Voting*. Springer, 2007.
7. Bratko, I., *PROLOG, Programming for Artificial Intelligence*. Addison Wesley, 1986, 2011.
8. Brock, E.O. de, *The Foundations of Semantic Databases*. Prentice Hall, 1993.
9. Deemen, A. van, Paradoxes of Voting in List Systems of Proportional Representation. *Electoral Studies*, 12:3, 234-241, 1993.
10. Deemen, A. van, Empirical evidence of paradoxes of voting in Dutch elections. *Public Choice* 97: 475-490, 1998.
11. Gehrlein, W.V., Condorcet's paradox and the likelihood of its occurrence: different perspectives on balanced preferences. *Theory and Decision* 52, pp. 171-199, 2002.
12. May, K.O., A set of independent, necessary and sufficient conditions for simple majority decision. *Econometrica* 20, 680-684, 1952.
13. Saari, D., *Chaotic Elections! A mathematician looks at voting*. American Mathematical Society, 2001.
14. Saari, D., *Decisions and Elections; explaining the unexpected*. Cambridge University Press, 2001.
15. Saari, D., *Disposing Dictators, Demystifying Voting Paradoxes*. Cambridge University Press, 2008.
16. Lloyd, J.W., *Foundations of Logic Programming*. Springer Verlag, Berlin, 1987.
17. Sterling, L. and Shapiro, E., *The Art of Prolog: Advanced Programming Techniques*. MIT Press, 1986, 1994.

Chapter 10
Fallacies and Unfair Discussion Methods

H.C.M. (Harrie) de Swart

Abstract Many discussions and meetings are led perfectly from a formal and procedural perspective, but the quality of the in-depth discussion is nevertheless poor. The cause of poor thinking should be sought in the weakness of human nature, rather than in the limitations of our intelligence. Among the weaknesses of human nature are ambitions, emotions, prejudices and laziness of thinking. The goal of a discussion is not to be right or to overplay or mislead the other, but to discover the truth or to come to an agreement by common and orderly thinking. In Section 10.2 we discuss a dozen fallacies and in Section 10.3 a dozen unfair discussion methods. This chapter follows - broadly speaking - the nice arrangement of fallacies and unfair discussion methods of a Dutch booklet from the 1950s, Zindelijk denken [Thinking clearly], by A.F.G. van Hoesel [2]. Many examples in this chapter also come from this booklet.

10.1 Introduction

Ideally, an argument consists of carefully specified premises or assumptions and a conclusion which logically follows from the premises. Logical validity of an argument means that *if* the premises are true, *then* the conclusion must also be true. In Chapter 1 we have already seen that logical validity of an argument does not mean that the premises are true, nor that the conclusion is true. We may have a logically valid argument with a false conclusion when at least one of the premises is false. And a logically invalid argument may have a conclusion that is true, when its truth is not based on the given premises but on other grounds. One should also realize that from a set of inconsistent premises one may conclude anything one wants: ex falso sequitur quod libet; a principle popular among many politicians.

In Subsection 2.3.2 we already mentioned that in real life premises and even the conclusion may be tacit, in which case one speaks of *enthymemes*. Premises may be left implicit for practical reasons or because the speaker is not aware of them himself, but might also be omitted in order to mislead the audience.

One may distinguish formal and informal fallacies. A *formal fallacy* is an invalid argument whose incorrectness can be established via a formal representation in an appropriate logical system. A simple example is: A implies B $(A \rightarrow B)$ and B; hence A. For instance: if the weather is nice, then John will come. John comes; hence the weather is nice. That this argument is incorrect may become clear from the following example which has exactly the same structure: if Bill Gates owns all the gold in Fort Knox, then he is rich. Bill Gates is rich; hence Bill Gates owns all the gold in Fort Knox. We discussed a number of such formal fallacies in Chapter 1.

In this Chapter we want to focus on *informal fallacies* in which the putative conclusion is not supported by the content of the premises, but is based on the ambitions, emotions, prejudices and/or laziness of thinking of the people involved. In real life, these weaknesses of human nature play a major role in argumentation, debating and discussions. Quoting Jean de Boisson: 'It is difficult to take someone who has a different opinion for a wise person'. A speaker may be too proud to admit that he is wrong, he may be irritated by his opponent and consequently say more than he can justify, he may have prejudices which he does not want to give up and/or he may be too lazy to study an issue carefully and for that reason oversimplify it.

So, in real life discussions and debating it is important that one is aware of all kinds of tricks which are used, consciously or unconsciously, by one's opponent to suggest that you are wrong, while in fact your opponent is wrong. In this Chapter we give a classification of fallacies and unfair discussion methods, which is based on the Dutch booklet by A.F.G. van Hoesel [2]. This classification is not meant to be exhaustive, and the different categories are not necessarily mutually exclusive.

Quoting Arthur Schopenhauer in his booklet 'The Art of Always Being Right' [4]:

> A man may be objectively in the right, and nevertheless in the eyes of bystanders, and sometimes in his own, he may come off worst. For example, I may advance a proof of some assertion, and my adversary may refute the proof, and thus appear to have refuted the assertion. There may, nevertheless, be other proofs. In this case ... he comes off best, although, as a matter of fact, he is in the wrong. [p. 23]
> If the reader asks how this is, I reply that it is simply the natural baseness of human nature. If human nature were ... thoroughly honourable, we should in every debate have no other aim than the discovery of truth. We should not in the least care whether the truth proves to be in favour of the opinion which we had begun by expressing, or of the opinion of our adversary. That we should regard as a matter of no importance ... . But, as things are, it is the main concern. Our innate vanity will not allow that our first position was wrong and our adversary's right. [p. 24]
> The way out of this difficulty would be simply to take the trouble always to form a correct judgement. For this a man would have to think before he spoke. But, with most men, innate vanity is accompanied by loquacity and innate dishonesty. They speak before they think; and even though they may afterwards perceive that they are wrong they want it to seem the contrary. The interest in truth, which may be presumed to have been their only motive when they stated the proposition alleged to be true, now gives way to the interests of vanity. So, for the sake of vanity, what is true must seem false, and what is false must seem true. [p.25]

The topic and purpose of this Chapter is best formulated by Schopenhauer [4], p. 29: 'Even when a man has truth on his side, he needs dialectic in order to defend and maintain it; he must know what the dishonest tricks are, in order to meet them, so as to beat the enemy with his own weapons.'

10.2 Fallacies

A *fallacy* or *sophism* is a reason or reasoning which sounds plausible, but actually is not adequate. The oldest known treatises are:
1. the dialogue *Euthydemos* of Plato, written about 384 BC, in which he satirizes what he presents as the logical fallacies of the Sophists, Euthydemos among them;
2. *Sophistikoi elenchoi* (sophistical refutations) of his pupil Aristotle, in which the emphasis is on semantic and rhetorical matters having to do with argumentation.

10.2.1 Clichés and Killers

A **cliché** is a frequently used expression that has lost its freshness and descriptive power. It refers to a saying or expression that, upon its inception, was striking and thought-provoking, but has been so overused that it has become boring and unoriginal. The French poet Gérard de Nerval said: 'The first man who compared a woman to a rose was a poet, the second, an imbecile'. Synonyms for the word cliché are: platitude, commonplace, saying.

Example 10.1 (Clichés). a) Opposites attract;
b) Woke up on the wrong side of the bed.

Clichés frequently express experiences of many generations in a compact way and hence contain a core of truth. Such expressions are easy to handle in a debate and meet the laziness of thinking of both speaker and listener, because they are nice to hear. Statements like 'time is money' and 'if the need is the highest, the rescue is near' - although not true - are generally considered to be true and do not attract scrutiny from the listener.

Many clichés have meanings that are obvious; others have meanings that are only clear if you know the context. For instance, the obvious meaning of 'any port in a storm' is that in a bad situation anything will do. However, this cliché can also be used when talking about someone who has many lovers.

Example 10.2 (Clichés). Some more examples of clichés are:

I thank you from the bottom of my heart	It's only a drop in the bucket
Do not play with fire	Beauty is skin deep
All that glitters isn't gold	He has his tail between his legs
Had nerves of steel	The time of my life
The calm before the storm	Laughter is the best medicine
Time heals all wounds	Frightened to death
Read between the lines	Only time will tell
All is fair in love and war	Haste makes waste

A **killer** or silencer is a meaningless argument to divert a conversation from the subject, hence cutting off a further exchange of views. In some contexts these arguments may be appropriate and true, in others they are only meant to finish the discussion without further arguments.

Example 10.3 (Killers). a) The truth is in the middle;
b) The exception proves the rule.

For instance, if in a discussion someone says that all football players have a high salary and his opponent argues that he knows some amateur players who get nothing, the answer that this exception proves the rule is simply misleading. The exception just shows that the original statement was too general and that it would have been more appropriate to state that many or most football players earn a high salary. In which case the opponent would certainly have agreed.

When two persons have opposite views concerning a certain item, frequently a third person tries to make a wise impression by stating: 'gentlemen, would not the truth be in the middle'. However, when one person says '2 + 2 = 4' and the other says '2 + 2 = 6', then the truth is certainly not in the middle. This killer argument of the middle way is not to be confused with a compromise where one tries to unite what is acceptable to both parties, in order to be able to proceed.

If in a discussion about improvements in the cafeteria of a company one of the engineers states 'let us be realistic; the first mission of the company is production', this argument looks like a down-to-earth argument, but it ignores the fact that a better canteen may result in a better production. And suggestions of employees to improve the production process are frequently dismissed by statements as 'Tell me something I don't know' or 'since when are you the expert'.

If in a political discussion someone claims that there are good arguments for immigration restrictions, a liberal who dismisses the speaker on the basis of her being a conservative, ends the discussion without asking for clarification. Similarly, if a person says he has strong arguments in favor of nuclear energy, someone might immediately use a killer argument like 'that is just your opinion' to finish the discussion and most likely no one will ask for the announced arguments.

One may also kill a discussion by using body language, a facial expression or by raising one's eyebrows.

Example 10.4 (Killers). Some more examples of killers:

It is only a matter of taste	Do not worry; it is as it is
Impossible!	That is nothing for our clients
It is too difficult to handle	Too expensive!
That is illogical	More research is needed
The management will not like the idea	There is no budget for it
Not my responsibility	That is too great a change
Let's keep it under consideration	We do not have time for that
The market is not yet ripe	We are too small for that
I have never heard of this	We will put someone on it later
Practice is always different	There he goes again
I already know what you are going to say	You are a right wing zealot

10.2.2 Improper or hasty Generalizations

An improper generalization is a general statement based on frequently emotional experiences with only a small number of particular instances.

Example 10.5 (Improper generalizations). a) Civil servants are lazy; b) Juvenile delinquents are psychopaths; c) Women are vain; d) Blondes are stupid.

When someone has met two or three civil servants whom he viewed as being lazy, he will be inclined to generalize his limited experience to: civil servants are lazy. This latter expression will be understood by most people as: *all* civil servants are lazy. However, if the person in question would generalize his experience with two lazy officers to 'all civil servants are lazy', it would become easy to reject his statement. So, the person in question will say 'civil servants are lazy', while the only thing he is entitled to say would be something like 'some civil servants are lazy'. However, this statement is so weak that it looks completely uninteresting. That is why one will usually say 'civil servants are lazy'.

Similar stories may be told about expressions like 'women are vain', 'children are difficult to handle', 'specialists are expensive', 'men are egoistic', 'people from Morocco cannot be trusted', etc. In general, there is no proof at all to suppose that among civil servants there is a higher percentage of lazy ones than among masons, carpenters or gardeners. Frequently, improper generalizations, like 'politicians are unreliable' and 'blondes are stupid', are the consequence of emotional experiences with some particular instances, which for convenience are generalized, even when counterexamples are known.

Consider the following four statements (van Hoesel [2]):
1. All juvenile delinquents are psychopaths.
2. Juvenile delinquents are psychopaths.
3. The juvenile delinquents I have had in my practice are psychopaths.
4. The juvenile delinquents I have had in my practice are psychopaths; but I have to add that I only had two.

Notice that the third sentence looks as a scientific generalization and suggests a sufficient number of observations. The craftiness of the third sentence lies in the fact that, on the one hand, a fair restriction is made by saying 'that I have had in my practice' (a restriction that undoubtedly inspires confidence), while on the other hand it fails to indicate on how many practical cases the judgment is based.

Notice that in some cases it is completely justified to draw a general conclusion from a single observation. For instance, if a scientist in one experiment determines the melting point of some substance. Experience has learned us that the melting point of a substance is invariable (all other things, such as air pressure, being equal). So, in this case one single observation justifies the generalization. On the other hand, suppose that for a long time one has thought that swans are white, because one has never seen a swan with a different colour. But this could be simply because the person has never been to a different part of the continent where there are black swans. In this case the thousands and thousands of observations did not justify the absolute generalization 'all swans are white'.

Example 10.6 (Improper generalizations). Some more examples:
My grandfather smoked all day and he made it to 95, so smoking is not bad!
My friends all study law and I never saw them reading a book. So, it seems to me that law students do not read books.
Most employers are too picky; I have applied for three different jobs and have not been hired.
The last five years were very warm, so the climate has changed.
Last spring we stayed in a hotel in Germany and everything was extremely clean; so, you see, Germans are very neat and hygienic.
Today 50% of the women who took the driving test failed. Women must be incompetent drivers. (But the speaker does not mention that only two women took the test today.)

One makes a *slippery slope argument* when one takes several related ideas and inappropriately makes a generalization about them all.

Example 10.7 (Slippery slope arguments).
If we stop insisting that students wear button-up shirts to class, next thing you know, they will be coming to class in pajamas.
If the border of Europe is not at the border of Turkey, then one may equally well form a union with China.
If we allow him to smoke a cigarette now, he will become addicted to cocaine.
If the health insurance company were to start paying for viagra, by tomorrow people will expect them to start reimbursing BMWs.

Another type of improper generalization is the *questionable analogy* which takes an analogy and inappropriately generalizes the relationship between the two items. See also Subsection 10.3.4.3.

Example 10.8 (Questionable analogy).
Forcing people to pay taxes is like cornering them in a dark alley and demanding their money.
You can not fold that book as the back of the book cannot stand it. I do not fold you in half either.
Education is like cake. A small amount tastes sweet, but eat too much and it will spoil your teeth. Likewise, too much education is not good.

10.2.3 Thinking simplistically

When one is confronted with large complex problems or theories which require a lot of knowledge, effort and thinking in order to understand them, our laziness of thinking frequently leads us to leave out the nuances. One may simplify Einstein's theory of relativity to 'everything is relative', Freud's theory about subconsciousness to 'everything is sexuality' and one may dismiss a person who is concerned about overpopulation by calling him a misanthrope. Frequently one does not (want

10.2 Fallacies

to) take the time nor the effort to study the problem in depth, while on the other hand one wants to participate in the discussion, resulting in an oversimplification of the problem or theory in question. Questions like 'can you explain to me in five minutes what philosophy is all about' are typical examples of our laziness of thinking. When the discussion takes place among people with limited competence, the one who simplifies will frequently have the sympathy of the others, because the only specialist in the group is hard to understand and seems to make things more complicated than necessary. With slogans as 'simplicity is the hallmark of truth' the one who simplifies may defend his position by suggesting that his opponent, the specialist, makes things too complicated. If a child asks his mother what Jehovah's witnesses stand for, the mother may give the following oversimplified answer: they are people who do not accept blood transfusions when they need it. Such an answer ignores completely the essence that Jehovah's witnesses take the Bible as their source of inspiration.

Example 10.9 (Thinking simplistically). Arthur Schopenhauer [4] gives a nice example in his Chapter 28: Persuade the audience, not the opponent.

> This is chiefly practicable in a dispute between scholars in the presence of the unlearned. If you have no refutation *whatsoever*, you can make one *aimed at the audience*; that is to say, you can start some invalid objection, which only an expert sees to be invalid. Though your opponent is an expert, those who form your audience are not, and accordingly, in their eyes, he is defeated, particularly if the objections which you make places him in any ridiculous light. People are ready to laugh, and you have the laughers on your side. To show that your objection is an idle one, would require a long explanation on the part of your opponent, and a reference to the principles of the branch of knowledge in question, or to the elements of the matter which you are discussing; and people are not disposed to listen to it.
>
> For example, your opponent states that in the original formation of a mountain-range the granite and other elements in its composition were, by reason of their high temperature, in a fluid or molten state; that the temperature must have amounted to some 480 degrees Fahrenheit; and that when the mass took shape it was covered by the sea. You reply that at that temperature – indeed, long before it had been reached, namely, at 212 degrees Fahrenheit – the sea would have been boiled away; and spread through the air in the form of steam. At this the audience laughs. To refute the objection, your opponent would have to show that the boiling-point depends not only on the degree of warmth, but also on the atmospheric pressure, and that as soon as about half the seawater had gone off in the shape of steam, this pressure would be so greatly increased that the rest of it would fail to boil even at a temperature of 480 degrees. He is debarred from giving this explanation, as it would require a treatise to demonstrate the matter to those who had no acquaintance with physics.

In daily life one may not be able to avoid simplistic thinking completely, because one cannot be an expert in all fields. A good example is when a doctor has to explain to a patient what is wrong with him or her. He cannot expect that the patient has the knowledge he has himself, so he must resort to simplifications that are hopefully understood by the patient. When one has to choose between two or three cars or insurances, one is not able to take all aspects and details into account. In such cases one has to act at a certain moment and make the choice which seems overall best at that moment.

If one wants to become a member of a political party and one wavers between two of them because both have more attractive and less attractive elements, then opting for one of them will make one understand and respect people who opted for the other party. And based on new facts and experiences one may change one's mind later on.

10.2.4 Appeal to ignorance

A particular form of simplistic thinking is the appeal to ignorance. The speaker shifts the burden of proof to his opponent instead of offering an argument for his own claim. For example, if the speaker claims that someone is guilty by saying to him: prove to me that you are innocent.

Example 10.10 (Appeal to ignorance). No one has ever been able to prove that ghosts do exist, so they must not be real.

However, the same argument strategy may be used to support the opposite claim: No one has ever been able to prove that ghosts do not exist, so they must be real. Ignorance is not proof of anything except that one does not know something.

A more relevant example is from a discussion in a city council:

Example 10.11 (Appeal to ignorance). No one has been able to prove that radiation from transmission masts is safe; therefore, we should not allow them in our city.

However, similar reasoning may be used to allow them: No one has been able to prove that radiation from transmission masts is dangerous; therefore, they are safe.

Example 10.12 (Appeal to ignorance). Newton's theory of classical mechanics is not one hundred percent accurate. Therefore, Einstein's theory of relativity must be true.

Perhaps the theory of quantum mechanics is more accurate and Einstein's theory is flawed. Perhaps all theories in question are wrong. If one disproves someone's claim that $2 + 2 = 5$, it does not mean that my claim that $2 + 2 = 7$ is true.

The term *argumentum ad ignorantiam* was introduced by John Locke in his Essay Concerning Human Understanding (1690). This fallacy essentially boils down to the following two variants:
- Inferring that something is true from the fact that it has not been proven to be false;
- Inferring that something is false from the fact that it has not been proven to be true.
In the context of science, the mistake in the first variant is that a model can be false even though there are to date no known experimental falsifications – that is, even though the model is thus far in agreement with experimental data. The mistake in the second variant is that a model can be true even though it has not yet been tested.

As to the first variant, here are some historical examples that date from the time that Newtonian mechanics (now proven to be false on a micro and on a macro level) was still in agreement with all experiments:
- 'We are probably nearing the limit of all we can know about astronomy.' (Simon

10.2 Fallacies

Newcomb, astronomer, 1888)
- 'The more important fundamental laws and facts of physical science have all been discovered .... Our future discoveries must be looked for in the sixth place of decimals.' (Physicist Albert. A. Michelson, 1894)
- 'There is nothing new to be discovered in physics now. All that remains is more and more precise measurement.' (Lord Kelvin, 1900)

Also, currently, the adjective standard in the 'standard model of particle and interactions' (the name for a body of theories in particle physics) reflects the confidence of the physics community that this is basically the correct picture. But, truth be told: this has not been refuted yet.

As to the second variant, we have this interesting quote: 'Third-rate scientists cry that everything has to be proven and mistake not being proven to be true as proven to be false or at least not worthy of further consideration. (Hans Ten Dam, Journal of Regression Therapy, VIII(1), 1994)

And so, this fallacy lies at the very basis of the fact that anyone who comes up with a new theory will have a hard time getting it published in a recognized journal. It is virtually a certainty that he will stumble on a referee report recommending rejection along these lines:
- the author comes up with a new theory;
- this new theory is not proven to be correct in every aspect;
- therefore, the theory should be rejected, i.e., is not worthy of further consideration.

Practically every professional scientist who works on new theories will have had a rejection along these lines at least once in his career. The mistake is thus to think that a theory that has not been proven to be true in every aspect is not worthy of further consideration. Of course, there may be good reasons to reject a new theory, but the point is that it is a mistake to reject it as unworthy of further consideration because it has not been proven to be true. The key is to remain impartial. That is actually another one of the so-called principles of good scientific practice that are widely agreed upon: the principle of impartiality. This implies, among other things, that a different intellectual stance must be respected.

10.2.5 Speculative Thinking

Opinions should be based on facts, not on speculations. Speculating may be interesting at the stock market, sometimes yielding profit and sometimes yielding loss. Speculations may be useful because they suggest what might be the case or what might happen. But only facts can tell us what actually is the case or what actually happens. Nevertheless, speculative arguments are frequently used in discussions among people. Here are some examples: every right-minded person knows that it must be like that; it cannot be otherwise; it has always been the case; it cannot be that that's right. Frequently one argues that things are the way they are because it always was the case or because it should be this way. But to quote Johan de Iongh:

'One of the most important tasks of a philosopher is to make clear that things do not have to be the way they are, that they might be different and in some cases even should be different.'

Example 10.13 (Speculative thinking). Here are three examples, all from [2].

A good and simple example is the following discussion. Based on the results of some tests, a doctor prescribes a patient a diet without salt. When his wife is informed about this, she reacts as follows: no salt at all? That can never be good! Asking this woman on which facts or arguments her statement is based, she will probably look at you in amazement and say: it cannot be that that is right.

In a discussion between a biologist who is enthusiastic about Darwin's theory of evolution and a skeptic, the latter might bring in the following arguments against Darwin's theory, all of them speculative and not based on facts: 1. It may never have been God's intention to let the most beautiful part of His creation originate from a being equipped with only instincts; 2. It must be excluded that mankind descends from such a stinking monkey; 3. For me it is certain that the higher can never have evolved from the lower.

Another example is the discussion between two non-American managers with opposite views about some new method introduced in the United States. The one opposed to the method might use the following arguments, again all of them speculative and not based on any facts: 1. It can never be good to always emulate America; 2. We have everything we need for our company; you may be able to put something else in its place, but certainly not something better; 3. A system that has proven its practicality for so long has to be much better than such a newfangled American theory. Maybe, it will turn out that the new policies should be rejected, but these arguments are purely emotional and not based on facts.

Strikingly, people using speculative arguments frequently do so with great self-consciousness and without showing any doubts about their own points of view. They tend to react very emotionally to objections with expressions like: crazy to run loose to assume that ...; for everyone with a little sense, it is obvious that ...; every right-minded person knows that this has to be the case. See Section 10.3.4.

One might think that speculative argumentation does not occur in a purely scientific environment. Unfortunately, this is too good to be true. An example is the election of a president, mayor or chairman. We have been holding elections already many years in the familiar way, but from social choice theory it is evident that practically all existing election methods are seriously defective. Nevertheless, a scientifically well defended proposal for another completely new election method, namely Balinski and Laraki's Majority Judgment, is generally met with great skepticism, also among specialists in social choice theory. Similarly, Einstein's Relativity Theory was originally met with great scepticism. See Section 10.2.6 for more examples in the history of science.

And although organizations funding scientific research claim that they select the best projects, their arguments to fund or not fund particular projects are in fact frequently of a speculative nature. One also sees the phenomenon that scientists have prejudices or presuppositions they are not aware of and consequently proceed down a dead alley. Giving up the original prejudices or presuppositions might harm their reputation or might mean the end of their funding.

10.2.6 Incredulity

This fallacy essentially boils down to this: what I don't believe cannot be true. A weaker form is this: what I don't believe is not worthy of further consideration.

In the history of science there have been numerous occasions where scientists have been collectively mistaken in their rejection of a new idea: often the mistake then stems from this fallacy. It is thus a mistake to think that something cannot be true (or valuable) if you don't believe it: the opposite is true – that is, something can be true even if you don't believe it. Below are some historical examples that are based on this fallacy:

'... so many centuries after the Creation it is unlikely that anyone could find hitherto unknown lands of any value.' (committee advising Ferdinand and Isabella regarding Columbus' proposal, 1486)

'Drill for oil? You mean drill into the ground to try and find oil? You're crazy.' (drillers who Edwin L. Drake tried to enlist to his project to drill for oil in 1859)

'Louis Pasteur's theory of germs is a ridiculous fiction.' (Pierre Pachet, Professor of Physiology at Toulouse, 1872)

'Fooling around with alternating current is just a waste of time. Nobody will use it, ever.' (Thomas Edison, 1889)

'Heavier-than-air flying machines are impossible.' (Lord Kelvin, president Royal Society, 1895)

'Airplanes are interesting toys but of no military value.' (Marechal Ferdinand Foch, Professor of Strategy, Ecole Superieure de Guerre, 1911)

'All a trick.' 'A Mere Mountebank.' 'Absolute swindler.' (members of Britain's Royal Society, 1926, after a demonstration of television)

'Space travel is bunk.' (Sir Harold Spencer Jones, Astronomer Royal of Britain, 1957, two weeks before the launch of Sputnik)

Besides that, this fallacy reflecting a standard response of the human mind has been used in politics by a variety of governments, who very well know that they will easily get away with colossal lies because the people simply cannot believe that their own government would have the impunity to resort to such large-scale falsehoods. Concluding, the truth of the matter is that only very few people are able to consider the situation that their own belief about something is wrong. The famous Russian novelist Leo Tolstoy expressed this as follows:

> I know that the majority, not only of those that are considered intelligent people, but even of the really very intelligent people that are able to understand the most difficult scientific, mathematical, philosophical, problems, only very rarely can comprehend even the most simple and evident truth, if it is such that as a result thereof they would have to admit that their own, sometimes difficultly acquired opinion about things, which they are proud of, which they have taught others, and which they have based their entire lives on, might be false. [Leo Tolstoy, What is Art?, Ch. XIV (1897) (translation by M. Cabbolet)]

The *fallacy of incredulity* applies when a scientist spontaneously and fiercely rejects ideas which are inconsistent with what he has believed himself all his life. A kind of reverse fallacy of incredulity is when a scientist uses any piece of evidence as proof for his favored claim. A recent example is the claim that the Higgs boson exists. In the literature it is even stated that scientists have observed the Higgs Boson. But what one has actually observed are the decay products of the Higgs boson during a very small fraction of a second.

A particular form of the fallacy of incredulity frequently occurs when someone questions a widely accepted model. It has virtually become the standard reaction of 'experts' to any dissenting paper that questions a widely accepted model, to (often publicly) denounce its author as incompetent. According to Brian Martin, who has devoted his career to the study of the suppression of dissent in modern times, the reasoning is as follows:
- Observation: an author criticizes a widely used model.
- (Tacit) assumption: the author in question is not aware of the reasons why the model has become widely used.
- Conclusion: the author is incompetent.

This is a clear-cut case of jumping to conclusions. The mistake is thus to think that when someone criticizes an accepted model, he or she is therefore unaware of the reasons why that model has become accepted. However, the opposite is frequently the case: an author may criticize a widely used model, even though he or she is competent in the relevant field. Of course, an author who criticizes an accepted model may indeed be incompetent, but the point is that this incompetence cannot be deduced immediately from the sheer fact alone that he or she criticizes the model. Unfortunately, this is what frequently happens in scientific discourse!

Example 10.14 (Incredulity). 'Professor Goddard does not know the relation between action and reaction and the need to have something better than a vacuum against which to react. He seems to lack the basic knowledge ladled out daily in high schools.' (1921, New York Times editorial about Robert Goddard's revolutionary rocket work)

The observation is that Goddard comes up with an idea for a rocket. At the time this was considered impossible within the framework of Newtonian mechanics: the tacit assumption is thus that anyone who nevertheless suggests that rockets are possible does not know Newtonian mechanics.

10.2.7 The use of Terms with a vague Meaning

An essential ingredient for a good discussion is that all discussants involved know what they are talking about. Nevertheless it rather frequently happens that people talk past each other. The cause is then that the topic of the discussion is extremely vague and therefore has a different meaning for everyone involved. Examples of words with a vague meaning are: democracy, slavery, intelligence, socialism, capitalism, power, green, sustainable. In a discussion with an alderman I heard him say: 'that is democracy: most votes count'. But from social choice theory we know that there are many ways to aggregate the preferences of the people into a social or common preference and that 'most votes count' is one of the worst ways to do so.

'I love you' is another example of an expression with a vague meaning. It may mean: I will take care of you, I find you attractive, I want to make love to you, I will be faithful to you, I want to marry you, and all kinds of other things in between.

Example 10.15 (Vague terms).
A man after visiting a modern production facility might argue that the employees in the factory have become slaves, while his opponent might counter argue that the employees are allowed to complain about their circumstances, that they can quit their job, that they have a nice canteen, vacation days etc. The first person, however, may talk about slavery in the sense that the machine rules over the human being, controls his pace and his actions and deprives him of his initiative, while for his opponent the word slavery means quite something else.

Someone argues that John will almost surely vote for the socialistic party, because John is a very social person. However, socialism is a political doctrine, which has nothing to do with the property of John's being a social person.

How is it possible that one so frequently does not realize the vagueness of the terms used and does not take the trouble to make the terms in question more precise? The answer is simple: laziness in general and laziness of thinking in particular. We hear many people talk about democracy, socialism, etc. and they all make the impression that they know what they are talking about, which most likely actually is not the case. Consequently, different people give different meanings to the same words, in this way laying the foundations for many confusing discussions.

Already in the first half of the 20th century the Dutch significists, among them Gerrit Mannoury and Frederik van Eeden, warned for an imprecise use of language resulting in a Babylonian confusion of tongues. See Section 7.3.

> There is an obstacle in the way of the further development and impact of philosophical thought. ... I know of no image that may give a clearer idea of the obstacle I have in mind than the one of the Tower of Babylon, a symbol of the confusion of languages. [Mannoury, 1917; translated from Dutch.]

> The language, which is used by all people as a means of understanding, is full of unclean elements that poison society, such as contaminated water poisons the population of a whole city. For that reason it is important to immediately show that the water supply and the sources from which the city receives its drinking water is contaminated by germs, and it is most urgent to first purify these sources. [F. van Eeden in: Brouwer, L. E. J., F. Van Eeden, J. Van Ginneken en G. Mannoury, Signifische dialogen. 1939; translated from Dutch.]

If in a discussion about psychopaths one realizes that one does not know the content of this term, one might start by looking up the meaning of this word in a dictionary or encyclopedia. But a description of the word psychopath found in the encyclopedia will not suffice and still remains vague. In order to grasp the relevant concept, we need to know a number of examples of psychopaths. It is important that we cannot only verbalize what a psychopath is, but that we also know the living reality that lies behind it. The latter can be achieved by giving clear examples, such as querulants, kleptomaniacs, criminals, intrigants, fanatics and bigots, giving concrete examples of each of them. In this way we prevent our mind from being filled up with vague or empty notions which say nothing about the world around us.

New words and expressions enter the discussion arena now and then. A modern example is the notion of sustainability. Everyone seems to understand what this word means, i.e., pretends to understand this notion. But in all honesty, this notion is still unclear to the present author.

10.2.8 The Danger of Words with more than one Meaning

Some words do have more than one meaning. A good example is the word nature. It may mean: character; for instance when one speaks about the stubborn nature of John. It may mean: creation; for instance when one speaks of human intervention in nature. It may mean: the status in which primitive people live; for instance when one talks about primitive peoples. By itself it is not a real problem that one and the same word may have different meanings depending on the context. But it becomes problematic when in the same conversation the word is used with quite different meanings, causing a Babylonian confusion of tongues. This may be illustrated by the following conversation between a teacher and the father of one her pupils.

Example 10.16 (Words with more than one meaning). (van Hoesel [2])
Teacher You should talk with your son; a boy with such a stubborn nature must be dealt with firmly.
Father I am not so sure. I doubt whether we are allowed to intervene in nature. Nature is the creation of God and hence is not only beautiful but also perfect.
Teacher Of course, but you do not want to claim that the stubbornness of John is completely natural and should be accepted.
Father What should I say? Nature is nature. Look at the primitive peoples. We find cannibalism there. But because nature is the creation of God, it is perfect. For the same reason there is little to argue against the stubbornness of John.

Synonyms are two words for the same conception; for instance, 'honorable' and 'honest'. *Homonyms* are two conceptions which are covered by the same word. For instance, 'deep' and 'high' used at one moment for bodies, at another moment for tones. Schopenhauer [4] gives the following examples.

Example 10.17 (Words with more than one meaning).
1. Every light can be extinguished. The intellect is a light. Therefore, it can be extinguished.
2. A: You are not yet initiated into the mysteries of the Kantian philosophy.
B: Oh, if it is mysteries you are talking of, I'll have nothing to do with them.

Another example of an expression with more than one meaning is: do not shoot, please. It may be used by someone who does not want photographers to take a picture of him. But in the newspaper of the next day it may be reported that there was an attack on the person in question.

Example 10.18 (Words with more than one meaning).
According to Plato the end of a thing is its perfection. But death is the end of life. Hence, death is the perfection of life.

In Plato's usage the word end means: goal. But in 'death is the end of life' the word end means quite something else: termination.

Example 10.19 (Words with more than one meaning).
Giving money to charity is the right thing to do. So, charities have a right to our money.

The first time the word right is used in the sense of correct or good, but the second time it is used in the sense of a claim. Two completely different things.

The words 'true' and 'truth' should be avoided as much as possible. That a statement is true may mean that I have a (mathematical) proof of it; for instance, when I say that '5 + 7 = 12' is true. That a statement is true may also mean that it is in accordance with (empirical) facts; for instance, when I say that it is true that the earth revolves around the sun. But in a social context the word true may also mean that the speaker agrees with what is said; for instance, when I say that orchids are beautiful and you react with 'that is true'. Mathematicians avoid the word true altogether and simply say that 5 + 7 = 12.

The word automation may also have different meanings: self-regulating, mechanization, computerization. A psychologist will most likely use this word in another meaning than a technical engineer. Similarly, the word capital may have quite different meanings: 1. the most important city or town of a country or region; 2. wealth in the form of money or other assets owned by a person or organization; 3. a letter of the size and form used to begin sentences and names; 4. the distinct, typically broader section at the head of a pillar or column.

Example 10.20 (Words with more than one meaning).
The constitution says that all men are equal. But this is clearly not true, because there are rich and poor people, wise and stupid people.

The constitution stipulates that all citizens are equal for the law, i.e., that everyone will be treated in the same way and that no one will be privileged. This has nothing to do with economic equality or equality of intelligence.

Similarly, the word 'complete' has entirely different meanings in theories about mathematics and physics, which makes the following argument misleading.

Example 10.21 (Words with more than one meaning).
Gödel has proved that (formal) mathematics (including elementary number theory) is not complete. Einstein's relativity theory is expressed in mathematics. Therefore, Einstein's relativity theory cannot be complete.

10.2.9 Aprioristic Reasoning

Someone claims that all tables have four legs. You realize you have seen a table with only three legs and you present this counterexample to your opponent. To which he responds: Sorry, such a thing I do not call a table. What is happening here is that the property of having four legs is made part of the definition of the notion of table. Consequently, the proposition 'all tables have four legs' is what Kant would call an analytic statement: the predicate 'having four legs' is contained in the subject concept (tables) of the sentence. In fact, in this way the content of the sentence 'all tables have four legs' is completely empty and the speaker is always right, an undoubtedly desirable situation. A. Schopenhauer discusses in [4] that there are many other tricks for always being right. The situation is similar to the one in which a magician pulls a rabbit out of his hat: everyone knows he has put the rabbit in the hat before.

Example 10.22 (Aprioristic reasoning). Some more examples:
The director of a company argues that all his managers are high level and his co-director notices surprised that at least two of them are of questionable level. Then the director might react with something like: I do not call these guys managers; they should never have been appointed as such. Again, the director makes the property of being high level part of his definition of manager.

A priest argues that Christians are living a more decent life than non-Christians. His opponent mentions some persons which go to church every Sunday, but are drunk the same evening, beat their wife and neglect their children. To which the priest reacts with: sorry, I do not call such people (real) Christians.

Little John claims that all cars have four wheels. His little sister objects that she has seen a car with only three wheels. But John replies with: That's not a car.

All Scottish men love whisky. John is a Scott, but he does not like whisky. So, John is no real Scott!

10.2.10 Circular Reasoning

A circular argument is like a revolving door that one cannot get out of. Its general structure is: A because of B and B because of A. Consider for instance the following conversation (van Hoesel [2]):

Example 10.23 (Circular reasoning).
John I believe that nowadays all young people are lazy.
Codd What might be the reason for this?
John I think they never learned to work.
Codd How could this happen?
John It seems to me because they are simply lazy.

The circular argument becomes less perspicuous when it is of type: A because of B, B because of C, C because of D, D because of E and E because of A.

10.2 Fallacies

In rhetoric too people are often guilty of circular reasoning. For instance, someone argues in the heat of his argument: Why does it have to be? Because it's possible! And why is it possible? Because it has to be!

The conversation below also illustrates circular reasoning (van Hoesel [2]):
Teacher Children, do you know that a human being has s soul?
Children Yes, we know.
Teacher But can you also prove this?
Children No, we cannot.
Teacher I will explain. You have all seen an obituary card. If you looked carefully, then you have seen that it mentioned 'pray for the soul of the dead person'. Well, you understand they would not have written this if the human being would not have a soul. Do you understand?
Children Yes!

In circular reasoning, also called *begging the question*, the same proposition is formulated in different words, obscuring the fact that the same proposition is used both as a premiss and a conclusion. In the following examples, the author is repeating the same assertion in different words and then attempting to 'prove' the first assertion with the second one.

Example 10.24 (Circular reasoning).
God exists because it is mentioned in the Bible. What is mentioned in the Bible is true, because it is God's word.
Of course, freedom of speech is important. Everyone must be able to say what he wants.
I am no kleptomaniac, for I do not steal.
I am the director since I have the final word here.

10.2.11 Applying double Standards

It is amazing to see how people use arguments in one context, but refuse to use the same argument in another context. Usually, such an argument is used when it is beneficial to oneself, but not when it is beneficial to others.

Politicians in Western Europe are very strict in condemning what they call expansion of Russia, pointing for instance to the Crimea, but the same politicians consider NATO's enlargement of its territory into many former Russian states to be no issue.

Another example from real life: a jealous husband and his wife, where the husband is always trying to seduce other women, while he does not even allow his wife to dance with another man. Even worse: he refuses to dance, but does not allow his wife to dance with somebody else.

Example 10.25 (Applying double standards). (van Hoesel [2])
A father to his son: you pay too much attention whether your girlfriend is beautiful; the appearance is not important, only the inner self is. The son answers: I find her so charming! The father replies: That is because of the make-up she is using. The son:

But you said that only the inner self is important and not the appearance. So, the father argues that his son's girlfriend is charming because of her appearance, using make-up, while he just said that the appearance is not important.

Two friends decided to go to a football match, but forgot to decide who would buy the tickets. So it happened that each of them bought two tickets. When they discovered their mistake, they blamed each other for not having informed the other about buying the tickets. None of them saw that the argument could be reversed against themselves.

In one and the same conversation the director of a company, in discussion with his wife, argues that the fact that they spend a lot of money is useful because it stimulates the economy and provides employment opportunities. But when his wife argues that the employees should have a higher salary, the same man argues that this would only mean that they will waste their money.

I have experienced several times in a city council that one has wasted lots of money for projects which were doomed to failure, as has become clear afterwards, while refusing to spend money for useful projects on the basis that there was no money for it.

Applying double standards is even evident in daily language, as shown by the following examples:
When a man dates many women, he is an interesting Don Juan, a womanizer. But when a woman dates many men, she is immoral and a slut.
A man who is not married is a bachelor. But when a woman is not married, she is an old spinster.
A man in his forties is in the prime of his life. But a woman of that age is already an older lady.
A man who spends much money is called generous. When a woman does the same she is called wasteful.
If a man argues strongly in an exalted tone, he is called masculine. But a woman doing the same is called quarrelsome.

When one hears the production of atomic bombs defended by the argument that it gives employment to many people, this argument does not contain an inconsistency. That this argument is not convincing may be made clear by applying the same argument to another situation: destroying whole cities is useful because it gives employment to many people. In this way it hopefully becomes clear that the person in question is applying double standards.

10.2.12 Rationalizing

People want something, frequently based on unconscious premature judgments or habits, and next try to give more or less good reasons to support this position; however, these reasons are not convincing or are not the real motives. The notion of rationalizing is best explained by the following anecdote: there once was a fox that

10.2 Fallacies 507

lost its tail and then told itself and the world that tailless foxes are much more fashionable.

Example 10.26 (Rationalizing). A simple example is the following: a husband is pretty lazy and likes to read the newspaper and watch TV when he comes home. His wife is tired and asks him to do some shopping. The man reacts by saying: my darling, you look a bit pale today, I think it would be good for you to make a small walk to the shopping center. His wife replies: yes, you might be right.

Needless to say that this fallacy frequently occurs in political decision making. Politicians want something, frequently based on private hobbies and premature judgments, and do their very best to find all kinds of more or less reasonable arguments to motivate their proposal, usually carefully remaining silent about their real motives. One frequently sees that they, confronted with new facts and counter-arguments, do not want to give up their premature judgments and do everything to spasmodically maintain their original position. By doing this their premature judgment becomes a prejudice.

Prejudices are the result of emotional and practical needs such as certainty, safety, security, appreciation, physical well-being and to preserve what is familiar. These needs and desires bring us as it were automatically to accepting certain viewpoints and opinions, which are certainly not the result of critical analysis. In this context one may be reminded of the saying: the wish is father to the thought.

Thinking is not a matter of our intelligence alone, but the whole human being is involved with all his emotions and premature judgments. As a member of a certain class, religion or group everyone has unconsciously built up certain premature judgments which seem to be self evident and have never been submitted to critical analysis.

Strong prejudices are even able to reduce or eliminate critical thinking of (very) intelligent persons, as becomes clear from the following little experiment. A group of students is asked to judge the correctness of the following two arguments:
1. Because many people from Israel are hospitable and many hospitable people have a good character, many people from Israel have a good character.
2. Because many Jews are warlike and many warlike people are slavish, many Jews are slavish.
Both arguments have the same structure and are evidently not correct.

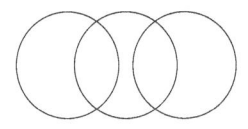

The left circle represents the people from Israel, respectively the Jews; the middle circle represents the hospitable, respectively the warlike people; and the right circle represents the people with a good character, respectively the slavish people. Clearly, the two outer circles may have nothing in common.

However, many people who are sympathetic towards Israel will judge the first argument as correct and the second one as incorrect, while many people who have a prejudice against Israel will judge the first argument as incorrect and the second one as correct.

So, the human being with a prejudice is not aware that his conviction is the result of his own desires and needs. Since the real reasons for his opinion remain hidden for himself, he will unconsciously create certain reasons or arguments. This process is called *rationalizing*: the rational or reasonable foundation of an opinion or conviction, which is essentially based on irrational grounds.

Example 10.27 (Rationalizing). (van Hoesel [2])
Sometimes the prejudiced person will try to maintain his prejudice with the most contradictory arguments, as for instance in the following example which illustrates the saying: it is an easy thing to find a staff to beat a dog.
X What I do not like about Jews is that they only look at their own group.
Y I doubt whether you are right. It turns out that they give relatively more money to charities than non-Jews.
X This only proves that they try to buy the favor of mankind by giving money. Jews only think of money which is the reason that so many Jews are bankers.
Y Recent research has shown that the number of Jews in the banking world is negligible.
X That is the point. These people are not concerned with respectable matters.

Example 10.28 (Rationalizing). (van Hoesel [2])
When a large company wanted to introduce clocking (on/off), one of the employees came with a number of fundamental objections: 1. impairment of personal freedom; 2. people should be trusted; 3. to gain trust you first have to give confidence; 4. employees will also leave exactly in time. All these arguments against clocking look reasonable, but, no surprise, the employee in question was always too late, because he had problems leaving his bed in time.

10.2.13 After this, therefore because of this

A simple example of this fallacy is provided by people who argue that their headache has disappeared due to taking a paracetamol tablet. After taking the paracetamol, the headache disappeared and one concludes that it disappeared because of taking this medicine. The idea that the headache might have disappeared without taking paracetamol does not occur to these people.

This fallacy, in Latin called '*post hoc, ergo propter hoc*' (after this, therefore because of this) consists of assuming that a certain fact is a consequence of another fact, only on the basis that the one fact is chronologically later than the other fact. It occurs very frequently, also in modern times.

Example 10.29 (Post hoc, ergo propter hoc). Some more examples:
'Last ten years climate has changed; that must be a consequence of CO_2 emissions.'
That CO_2 emissions were earlier than climate change is hard to refuse, but that they are the cause of climate change is another question.

10.2 Fallacies

The sun always comes up after the cock has crowed, so the sun rises because the cock has crowed.

The inhabitants of some islands in the Pacific were convinced that lice keep people healthy. They had observed that many healthy people had lice, while sick people frequently do not have them. What they did not realize is that the lice run away from sick people because due to fever their temperature is too high for them.

Also commercials frequently suggest a causal relation only on the basis that the one follows chronologically on the other:

Example 10.30 (Post hoc, ergo propter hoc).
She was a wallflower, now she is engaged. She uses Lucia soap.

He was tired of being alone; now he is happily married. He signed up for our dating site.

You want to be happy too? Our car is the perfect one for you.

Many people, among them many doctors, believe that injections against influenza prevent them from having this disease, although many controlled experiments have shown that they were useless.

Example 10.31 (Post hoc, ergo propter hoc).
Smith became president. Next the economy flourished. So, the presidency of Smith was good for the economy.

Possibly the presidency of Smith was beneficial for the economy, but not necessarily so. The effect of politicians on the economy should not be overestimated. The economy may flourish for many other reasons, under any president.

Every cause always precedes its consequence, but not everything that precedes a result is a cause!

A similar mistake is when one concludes from the parallel occurrence of phenomena that one is causing the other. In Latin this fallacy is called *'cum hoc, ergo propter hoc'* (together with this, therefore because of this). A good example is the following one (van Hoesel [2]). Reliable statistics show that students who smoke in general have lower grades than students who do not smoke. Opponents against smoking will gratefully conclude from this that smoking is harmful for learning. However, one may also reverse this conclusion: lower grades are causing students to smoke. A third even more likely conclusion might be that students who like to be popular and to make a social impression will for that reason smoke and will avoid everything that might lead them to being mistaken for an eager beaver.

Example 10.32 (Cum hoc, ergo propter hoc). Some more examples:
When in a certain village some form of cancer statistically occurs more frequently than elsewhere, people may suggest that a particular factory in the neighborhood of the village is responsible for it. However, it might well be that the real cause is that the people in the village do not eat healthy for whatever reason.

The last 200 years the number of pirates has decreased and global warming has increased. So climate change is due to the fact that there are fewer pirates.

I was just thinking about you when the phone rang. That cannot be a coincidence.

10.3 Unfair Discussion Methods

Once more: the purpose of a discussion is not to be proved right, or to outdo, to force or to mislead the other, but to discover the truth or to reach an agreement through joint and ordered thinking. In this section we will point out and distinguish a number of unfair discussion methods in the hope of making the reader aware of them and to help the reader not to become the victim of so many unfair tricks that are used, consciously or unconsciously, in local councils, parliaments and other meetings.

10.3.1 Pushing someone into an extreme corner

There is a well known Dutch saying: whoever claims a lot, has to justify a lot. Consequently, if someone gives in to the temptation - under the influence of his emotions - to exaggerate his claim and thus take an extreme position, he often becomes defenseless against the arguments of his opponent. There are at least three ways in which one can be pushed into an extreme corner without being aware of it:

10.3.1.1 Pushing someone into an extreme corner by fighting him violently/emotionally

Example 10.33. As chairman of a faculty meeting I was confronted with a colleague who evidently was lying repeatedly. Becoming more and more irritated by his lying I was led to say explicitly that he was a liar. Everyone in the faculty meeting was upset that I used these words and that I did not trust the words of my colleague. Consequently, the members at the meeting demanded that I offered my apologies; the truth or falsehood of the claims of my opponent was not further considered.

10.3.1.2 Pushing someone in an extreme corner by saddling him with more than he said

Example 10.34. 1. In a debate about immigration, a politician argues for restrictions on immigration. One of his opponents replies with: so, you want to deport all foreigners from the country.
2. One evening a husband came home and asked his wife whether she had been able to put a button on his jacket. The reaction of his wife was astonishing: You think I have nothing else to do than putting that button on your jacket! I worked all day, did shopping, had to prepare dinner, cleaned the house, etc.

The best reaction for the politician is to make clear that he did not claim the things his opponent said. The same holds for the husband. But - being irritated - the husband may be tempted to say that his wife with a little bit more efficiency would have been able to do what he hoped for, in which case the atmosphere in the family would only become worse.

10.3.1.3 Pushing someone into an extreme corner by drawing improper consequences from his statement

Example 10.35. In a discussion between a politician and businessmen, one of the businessmen was arguing for more roads because there are so many traffic jams. The reply of the politician was simply: sir, we cannot asphalt the whole country!

Clearly, the proposal of the businessman does not lead to the ultimate consequence that the whole country has to be asphalted. But the discussion was closed and the businessman gave up instead of making clear to the politician that his conclusion was inappropriate. Also nobody in the audience of more than one hundred people made any objection.

10.3.2 Straw man argument

By misrepresenting the position of a speaker, it becomes easy for the opponent to knock the speaker down. However, in fact the opponent does not refute the statement of the speaker, but he creates another and frequently much stronger statement which may easily be refuted, akin to the way that it is easy for a boxer to knock down a straw man. For this reason this unfair discussion method is also known as the *straw man argument*. The problem is that the position dismissed by the argument is not the real one, but only a caricature of the real position. In such cases the best strategy is to state explicitly: I did not say that.

Example 10.36 (Straw man argument).
A scientist submits a paper for publication in which an argument A is presented. The referee who has to judge whether the paper is suitable for publication, misinterprets the paper and believes that another argument B is presented. He then shows that argument B is incorrect or nonsense and subsequently recommends rejection of the submitted paper. In such a case the paper is rejected with a straw man argument.

Schopenhauer [4] calls this *extension*: carrying your opponent's proposition beyond its natural limits, so as to exaggerate it. He gives the following examples:

> I say that the English were supreme in drama. My opponent attempts to give an instance to the contrary, and replies that it is a well-known fact that in music, and consequently in opera, they could do nothing at all. I repel the attack by reminding him that music is not included in dramatic art, which includes tragedy and comedy alone. This he knew very well. What he did was try to generalize my proposition so that it would apply to all theatrical representations, and, consequently, to opera and then to music, in order to defeat me.

> Lamarck states that the polyp has no feeling, because it has no nerves. It is certain, however, that it has some sort of perception; for it advances towards light by moving in an ingenious fashion from branch to branch, and it seizes its prey. Hence it has been assumed that its nervous system is spread over the whole of its body in equal measure, as though it were blended with it; for it is obvious that the polyp possesses some faculty of perception without having any separate organs of sense. Since this assumption refutes Lamarck's position, he argues:
> *In that case all parts of its body must be capable of every kind of feeling, and also of motion,*

of will, of thought. The polyp would have all the organs of the most perfect animal in every point of its body; every point could see, smell, taste, hear, and so on; in fact, it could think, judge, and draw conclusions; every particle of its body would be a perfect animal, and it would stand higher than man, as every part of it would possess all the faculties which man possesses only in the whole of him. Further, there would be no reason for not extending what is true of the polyp to all monads, the most imperfect of all creatures, and ultimately to the plants, which are also alive, etc., etc.

By using dialectical tricks of this kind a writer betrays that he is secretly conscious of being in the wrong. Because it was said that the creature's whole body is sensitive to light, and therefore possessed of nerves, he makes out that its whole body is capable of thought. [Schopenhauer [4], Section 1]

10.3.3 Diversion maneuvers

In discussions it frequently happens that one tries to take someone away from his proposition, consciously or unconsciously, in a way similar to that of the young boy who came home with a great rip in his new pants and proudly showed to his mother the beautiful chestnuts which he had collected, hoping that she would not notice the rip. Below we present some of the methods used to embarrass someone.

10.3.3.1 Red herring argument: distracting someone from his original theme by moving the discussion unnoticed to another area

Changing the subject or diverting the argument from the real question at issue to some side-point is also known as a *red herring argument*. A red herring is a tactic to divert the opponent and/or audience from the relevant issue. A frequently heard example is this one: why should I pay for driving a few kilometers too fast; the police should chase dangerous criminals, not a decent tax payer like me.

Unlike the straw man argument, a red herring argument does not involve any misrepresentation of an opponent's position, but it concerns the introduction of a completely different issue which is not, or is only slightly, related to the real issue in question.

Example 10.37 (Red herring). (van Hoesel [2])
At a meeting of the elementary school board with the parents of the pupils, a mother asks one of the teachers about his opinion in the dispute between herself and her husband about beating their child because it had stolen some money. The teacher recognizes that the question is whether beating is admitted as a punishment. But instead of answering this question, he starts to talk about the punishment problem in more general terms, saying that the conscience of the child sometimes has to be corrected by punishment and that punishment is a translation from an ethical condemnation to empirical reality.

If the speaker continues to talk about this more general topic, illustrating more or less interesting aspects of the punishment problem, occasionally making a small joke, the woman in question will go home very satisfied and only realize later that the teacher in fact did not answer her question.

10.3 Unfair Discussion Methods

Example 10.38 (Red herring).
In a public debate with the mayor of the town the complaint is put forward that there is too much crime. The mayor then answers: well, this town has lots of problems, among which is also the housing shortage problem. But currently we are in conversation with cooperations to build new social housing. So, we are actually doing something about it.

Personally, I have experienced in many meetings of the faculty, the university and the city council that people frequently do not react to what might be strong arguments, they simply ignore them and pretend they did not hear them. This is usually a sign that they do not have appropriate counterarguments.

The College of Mayor and Aldermen is obliged to answer written questions of a council member within six weeks, and they do react within this period. However, frequently what they write is not an answer to the question! In such cases Schopenhauer [4] gives us in Section 34, Don't let him off the hook, the following advice:

> When you state a question or an argument and your opponent gives you no direct answer or reply, but evades it by a counter-question or an indirect answer (or some assertion which has no bearing on the matter, and, generally, tries to turn the subject), it is a sure sign that you have touched a weak spot, sometimes without knowing it. You have, as it were, reduced him to silence. You must, therefore, urge the point all the more, and not let your opponent evade it, even when you do not know where the weakness which you have hit upon really lies. [Schopenhauer [4], Section 34]

10.3.3.2 Distracting someone from his original theme by concentrating one's attack on one minor argument

If one has a number of arguments in favor of a certain proposition, one of the arguments may be a weaker one. Clever debaters may pick out this one weaker argument and with a great fanfare focus their attack on this minor argument. If they give a good show, they may achieve in this way that the strong arguments are forgotten and that they become the 'winner' of the discussion.

Example 10.39. (van Hoesel [2])
In a discussion about admitting or forbidding alcohol one of the participants brings in the following arguments against a total ban on alcohol:
1. Thousands of people would become unemployed;
2. It would mean an attack on the liberty of people;
3. Alcohol may have a positive influence on the health of people;
4. A total ban will encourage illegal trade and alcohol abuse;
5. Many people are *used to* alcohol, alcohol is like a friend which they do not want to miss.

One of the participants, strongly in favor of a total ban, focusses his attack on the last weaker argument as follows: Your son may *be used* to biting his nails, but you will not stop telling him he should not do so. You may *be used* to smoking a lot, but you keep trying to quit smoking. Your neighbor *is used* to throwing his garbage into your garden, but you will never accept this. Summarizing, let us remain sober (people laugh), that one *is used* to something does not mean that it is good and that one should not fight against it.

10.3.3.3 Distracting someone from his original theme by making an irrelevant objection

Example 10.40. (van Hoesel, [2])
A psychology professor has given a talk about the psychology of human resource management, in which he has emphasized the importance of showing respect and appreciation for the employees. Having given a number of good arguments to underpin this claim, he concludes with: summarizing, with one pat on the back you can achieve more than with thousand other measures. In the discussion following his presentation one of the attendees reacts as follows: mister chairman, I have not studied psychology, but I do not see myself walking through the factory giving pats on the back, taking my hat off for the employees, offering them cigars and cigarettes, bringing them coffee and tea in the morning and in the afternoon. (people laugh) Sorry, mister chairman, in this way one cannot run a company.

By taking the 'pat on the back' from the context and doing so in a humorous way, the attendee gets the laughers on his hand, but not the thinkers. This reminds us of Schopenhauer's [4] section 28: Persuade the audience, not the opponent, which was already mentioned in Section 10.2.3, Thinking simplistically.

10.3.3.4 Bluffing the community

Example 10.41. In the years 1970-1980 the idea emerged in the Netherlands that for students, from elementary school to university, it is social and emotional development that is most important; students may discover subject matters like number theory, language, history and geography themselves. Teachers who taught were in the way of both the emotional and the professional development of their pupils. The Dutch government from those days gave educational agencies plenty of room. These agencies sent out advisers on a large scale, who quickly spread the new insights. By working according to these new insights and the associated methods, the content level of education would improve.

An advisor explains to a group of teachers that explanations of any sort may last at most twenty minutes. The advisor himself takes more than one hour. A teacher asks for attention to the way in which the content of subjects can still be brought to the fore within the outlined framework. He expresses his serious concerns. The consultant blames the teacher for interfering with the process that his colleagues are going through. Also, this teacher apparently has no eye for the real interest of his students. Teachers like him are subject matter-oriented, while the proper attitude is student-oriented. Almost all colleagues remained silent, school directors almost always chose the side of the advisors. Impure methods like these have caused great suffering for many teachers (and students).

10.3.3.5 Distracting someone from his correct conclusion by pointing out a mistake in his argument

As we already know from Chapter 1 an invalid argument may have a true conclusion when its truth does not depend on the truth of the premises, but on other facts. So, it may happen that a speaker is drawing a right conclusion, but gives a wrong argument as in the following example (van Hoesel [2]).

Example 10.42. All planets are round. The earth is round. So, the earth is a planet. Every rectangle has four right angles. A square has four right angles. So, a square is a rectangle.

One may point out that the argument is invalid by remarking that a similar argument would be: all men have two eyes; an ape also has two eyes; so, an ape is a man. Nevertheless, the conclusion of the prior arguments is true, although its truth is not based on and independent of the given premises.

Example 10.43. (van Hoesel [2])
An engineer who just got a position at a certain firm concludes that he will belong to the management, because the managers have four weeks of vacation and he himself does too. His partner makes him doubt by pointing out that his argumentation is invalid; because a similar argument would be: the managers are wearing shoes and all employees are wearing shoes, so all employees are managers.

Again, the conclusion may be true, but if so, its truth does not depend on the given argument.

10.3.4 Suggestive Methods

There are three ways to bring people to accept our insights and objectives: by forcing them, by persuading them (but not by using good arguments) and by convincing them (by honest, proper and relevant arguments). The difference between being convinced and being persuaded is that in the first case one plays a more active role in the process (agreeing happily) than in the second case where one plays a more passive role. In this section we will analyse some discussion methods which have in common that the most important factor in bringing about an insight or opinion is not the quality of the argument used, but suggestive influence of one of the following kinds:
1. by using terms with tendentious emotional value or biased connotation;
2. by exploitation of certain thinking habits;
3. by abusing the analogy reasoning;
4. by all kinds of suggestive tricks.

10.3.4.1 Using terms with tendentious emotional value or biased connotation

Example 10.44 (Words with biased connotation).

protestants	heretics
alteration	innovation
existing order	antiquated prejudice
public worship	piety/godliness
system of religion	bigotry/superstition
the priests	the clergy
placing in safe custody	throwing into prison
an equivocal story	a bawdy story
religious zeal	fanaticism
through influence and connection	by bribery and nepotism

The difference between the objective and emotional meaning of a word becomes evident when one puts the words next to each other. For instance, in the sequence alcoholic – drunkard – boozer the meaning of the first word is a purely objective one, but the last word in addition expresses that the person who used it has already chosen a position.

Words with a tendentious emotional value can often be found in all kinds of political, moral and religious discussions.

Example 10.45 (Terms with biased connotation).
The city council was discussing building a new shopping mall at the border of the town and objections were raised that this might have disastrous consequences for the shopkeepers in the city center and hence for the city center itself, because many shops there would simply disappear. A representative of the labour party said that the shopkeepers are just tax evaders, so for him there was no problem at all.

Emotional words are frequently used in the political sphere. One can easily see this by reading how different newspapers report one and the same event. One newspaper calls a mistake of a minister in parliament a somewhat unfortunate mistake, while another newspaper calls it deliberate deception of the people.

Note that many initially completely neutral words can get an emotional connotation over time. Examples are the words workman and cleaning woman, who nowadays are called employee and interior caretaker, respectively.

In the public domain one really plays with words in order to make a positive impression. Since the word progressive for many people has a positive connotation, left-wing parties call themselves progressive, suggesting that they are focused on the future and go along with their time, thus ignoring the fact that one must keep the good things and only needs to correct or adapt what goes wrong.

If in a discussion many emotional terms are used, one should be careful: frequently these emotional terms are misused to mask bad argumentation. In such cases one should try to replace the emotional terms by more neutral expressions and see what remains of the argumentation. Van Hoesel [2] illustrates this with the following example of a discussion between a host and his guest.

10.3 Unfair Discussion Methods 517

Example 10.46 (Terms with biased connotation). (van Hoesel, [2])
Host: At Sundays I always like to drink a whisky before dinner; and I am fond of it.
Guest: Do you realize how much misery alcohol is causing to the world. Whole families and cultures have been destroyed by this poison. See how many human wrecks are walking in our big cities. Our psychiatric hospitals are overcrowded with victims of alcohol. Alcohol is causing a strong increase in criminality and sexual offences. I am deeply shocked by your statement that you enjoy your whisky so much.
Host: My dear friend, your words have impressed me. I will stop drinking.

In a less emotional and more business-like atmosphere this conversation would most likely have proceeded as follows:
Host: At Sundays I always like to drink a whisky before dinner; and I am fond of it.
Guest: You will have to admit that *misuse* of alcohol causes serious physical and mental problems.
Host: I fully agree! That is why I only take one.

Schopenhauer [4], section 32, points out that one may get rid of an assertion one does not like, or at any rate throw suspicion on it, by putting it into some odious category, even if the connection is only apparent or of a loose character. One might say for instance: that is Machiavellism, or Arianism, or Pantheism, or Atheism, or Spiritualism, or Ultra-Right, all words with a biased connotation. In making an objection of this kind, one essentially cries out 'Oh, I have heard that before' and one suggests that the system referred to has been entirely refuted and does not contain a word of truth.

10.3.4.2 Exploitation of certain thinking habits

It is not difficult to see that many of our thinking habits are based on incorrect and emotionally-based generalizations which we have already treated in Section 10.2.2. In this section we want to point out that our thinking habits may weaken our critical insight and make us vulnerable for suggestive influencing. For instance, we are used to talk about Russia as warlike and aggressive, which is exploited by our Western politicians without any scruples and without any attention for the way Russia looks at the West. A good example is the so called annexation of Crimea by Russia, where in fact the citizens of Crimea requested Russia to protect them against Ukraine, because they preferred to remain Russian. In addition, it is almost certain that if Russia had not taken Crimea, NATO would have built a naval base there.

Speakers in public - with the exception of a few good ones - rely more on the basis of our emotions and prejudices than on our common sense and critical insight. A smart speaker who, for example, wants the public to accept a dubious proposition, first formulates a number of propositions that are readily accepted by the public and only then presents his dubious proposition. As soon as used to nodding yes, chances are that they will not even think about the last questionable statement and nod again. For example, in a meeting of school teachers, the speaker may start by pointing out that the salaries have not been raised for many years, that the classrooms are getting bigger and bigger, that the pressure on the teachers is increasing and that their job is becoming more and more demanding, before eventually formulating his dubious proposition, like, for instance, that school teachers deserve a 10% salary increase. After saying so many things that the teachers can not disagree with, they will also be happy to accept his more dubious statement.

This technique is perfectly demonstrated by quacks at the market, for instance. They present a pseudo-scientific argument in which they formulate many propositions which are easily accepted by the general public. Since people are inclined to believe a person who proclaims their views, they will easily accept the dubious proposition at the end of the argument. Van Hoesel [2] gives the following example of such a quack.

Example 10.47 (Exploitation of certain thinking habits). (van Hoesel [2])
Ladies and gentlemen, we all know that the mind has a huge influence on the body. Did you have fear in the past? What did you feel? Precisely, that your heart beats faster. And what if you have suffered a great loss? Right, you start to cry, the tears come out. The mind affects the body. And perhaps you know someone who was paralyzed and could walk again under the influence of a strong emotion. The influence that body and mind have on each other is so strong. There are no physical illnesses and there are no mental illnesses, there are only sick people. Whether you suffer from nervous breakdowns, rheumatism, stomach- or head-aches, it really does not matter that much. Because in our laboratory - after many years of experimenting - we have discovered a method that can cure all your diseases, physically or mentally. Panasulfakin heals body and soul for the price of a doctor's visit. Thousands of fellow citizens owe their health to Panasulfakin.

10.3.4.3 Abusing the analogy reasoning

An analogy may be used to *clarify* something, like in the following example: The circulation of money for the well-being of the economy is like the circulation of the blood for the well-being of the body.

However, an analogy may also be misused when one tries to *prove* something. In such a case, one usually points out that two items have some properties in common and next one concludes that the second item has in addition another property of the first item.

Example 10.48 (Abusing the analogy reasoning). (van Hoesel [2])
Family doctor: You just said that your son already visited several doctors; nevertheless I advise you to consult a psychologist.
Father: No way! Look, I have a motorbike. If one mechanic pours Shell oil in it, a second one Renault oil and a third one again another oil, then my motor goes on the fritz. The more people mess with my son, the more they'll ruin him.

Although in this example a human being and a motorbike have some things in common, it goes too far to conclude that what is bad for the motorbike is also bad for a human being. The family doctor might have made clear that the analogy is inappropriate by suggesting the father to pour some oil in his son and to kick-start him.

10.3 Unfair Discussion Methods

Example 10.49 (Abusing the analogy reasoning). (van Hoesel [2])
Probation officer: Believe me, you will get a good craftsman.
Employer: maybe, but I am not inclined to employ someone who was in prison for theft. My father used to say: once a thief, always a thief.
Probation officer: Listen, your saying says nothing. On the contrary: no person wants to be more honest than the one who comes from jail for theft. Look: if you return from hospital after having fallen from the roof, would you climb on the roof again? No way!

In this example there is little analogy between thieving and falling from a roof that would allow one to draw any conclusion. Such forced analogies are frequently used in public or political speeches and in commercials, like in the following example.

Example 10.50 (Abusing the analogy reasoning). (van Hoesel [2])
A market vendor with a hoarse voice was trying to convince the public of the excellent qualities of his cough medicine. Colds, cough and bronchitis were according to him nothing else than dirt that had settled on the chest. In order to illustrate this he showed a glass of troubled water, and said that if he would not do anything, it will remain troubled forever. However, by pouring a bit of cleaning liquid in the glass, the water became crystal clear. He promised his audience that by taking three spoons of this liquid per day, their chest would become as clean as his glass of water.

That the reactions of a living being are very different from an anorganic reaction did not occur to his audience; the market vendor was doing good business.

Example 10.51 (Abusing the analogy reasoning). (van Hoesel [2])
A temperance advocate finished his speech by saying that liquor is not only bad for the mind, but also for the body, illustrating this by dropping a rain-worm in a glass of liquor. Indeed, the result was convincing, after a few seconds the rain-worm was as dead as a doornail. I cannot, he continued, give you a more convincing demonstration of the destructive effect of alcohol.

Of course, there is some similarity between a rain-worm and a human being: both are living beings. But this does not mean that what is bad for the rain-worm is also bad for a human being. In addition, the rain-worm was literally drowned, which would also have happened had the speaker used a glass of milk. One of the attendees was smart enough to realize these facts and drunk the glass of liquor with the excuse that he was troubled with worms.

Example 10.52 (Abusing the analogy reasoning).
Guns are like hammers: they are both tools with metal parts that could be used to kill someone. It would be ridiculous to restrict the purchase of hammers, so restrictions on purchasing guns are equally ridiculous.

Restrictions on the purchase of guns may be justified because they can easily be used to kill large numbers of people at a distance; this feature is not shared by hammers.

10.3.4.4 Suggestive tricks: using authority; suggestive influence of incomprehensible words; Argumentum ad Populum

A frequently used trick to suggest that a statement is true is to appeal to authority or prestige. This authority may be legitimate, but it may also be fictitious or pretended. When, for instance, a university professor in physics formulates a physical proposition, it is more than reasonable to accept its truth. However, when the same professor in physics formulates a proposition about some social problem, then we may attribute no more value to his claim than to the claims of other personalities of the same level and with the same level of information. The physician, the vicar, the pastor, the notary, to mention just a few examples, have for many people also authority on topics which have nothing to do with health, religion, morality and financial affairs, respectively. A similar thing holds for popstars when they make statements about political or social issues; their opinion has no more value than the ones uttered by arbitrary persons of the same intellectual level and competence.

Authority arguments are frequently used in practice, even in the scientific world.

Example 10.53 (Authority argument).
A PhD student had submitted a complaint to the national body of scientific integrity that the comments of a certain professor were inaccurate and careless. The professor in question replied in a letter to this body as follows: I want to draw your attention to the fact that I am the main editor of a journal of high reputation. Therefore you better take my opinion seriously.

When a person is an authority in a particular field he may also misuse this authority to intimidate others. Van Hoesel [2] gives an example of a university professor in psychology who gave a talk about the psychology of the factory girl. One of his students asks whether the factory girl does exist. To which he replies with 'I do not understand your question', making the student seem ridiculous. But the student is probably right that one cannot speak about *the* factory girl. By pretending he does not understand the student's question, the university professor insinuates to the bystanders, with whom he is in good repute, that what the student says is nonsense. The counter-trick for the student might be to admit that she might not have formulated her question clearly, but that when one compares a factory girl in a small bakery with a factory girl in a large Philips factory there may be more differences than similarities and that consequently it is unclear whether one can speak about the psychology of *the* factory girl. She might even add: with your intelligence it must be easy for you to understand this question.

Schopenhauer [4] gives another example:

> Thus, when Kant's Kritik appeared – or, rather, when it began to make a noise in the world – many professors of the old eclectic school declared that they failed to understand it, in the belief that their failure settled the business. But when the adherents of the new school proved to them that they were quite right, and had really failed to understand it, they were in a very bad temper. [Schopenhauer [4], section 31]

The suggestive effect of authority does not always have to be based on social position, title or the name of the speaker, but one may also successfully obtain authority

10.3 Unfair Discussion Methods

by using incomprehensible quasi-scientific terminology. It is amazing how many people consider incomprehensible and complicated terminology as scientific and interesting, while in fact it is only a mush of words. Some even claim that philosophers like Hegel and Heidegger are of this kind, but it may be that they did not spend enough time to study these authors properly. It is staggering to see how great the suggestive influence of incomprehensible words can be and how easily a belief in words arises. Management jargon, for instance, is an inexhaustible source of incomprehensible and quasi-scientific use of language.

Example 10.54 (Suggestive influence of incomprehensible words).
1. The unconscious Will of Nature eo ipso presupposes an unconscious idea as goal, content or object of itself. ... Instinct is defined as a purposive action without consciousness of the purpose. ... Instinct is conscious willing of the means to an unconsciously willed end. [Wilm, E.C., The Theories of Instinct. Yale University Press, 1925, pp. 135,139]
2. The prohibition on incest is in origin neither purely cultural nor purely natural, nor is it a composite mixture of elements from both nature and culture. It is the fundamental step because of which, by which, but above all in which, the transition from nature to culture is accomplished: the prohibition of incest is where nature transcends itself. [Lévi-Strauss, e.a., The elementary structure of kinship. Beacon Press, Boston, 1969, p. 24]

Sentences like these cannot be tested and have no clear meaning, which also means that no one can show that they are false. At the same time the authors of such sentences present themselves to be profound.

Example 10.55 (Suggestive influence of incomprehensible words). (van Hoesel [2])
A party-ideologist, at a party meeting at the end of his exposition about inflation, finishes enthusiastically with the words: We do not want *in*flation! We do not want *de*flation! But ... we want *re*flation!!! Followed by enthusiastic applause.

When someone after the meeting asked the speaker what he meant by reflation, his answer was: I do not know, but ask it to the people in the audience, because they seem to have understood it.

Evidently, in many cases people are satisfied with words they have gotten from persons with a certain authority. Incomprehensible secret language is one of the methods to seem important. D. Sperber calls this the Guru effect:

> All too often, what readers do, is judge profound what they have failed to grasp. Obscurity inspires awe, a fact I have been only too aware of, living as I have been in the Paris of Sartre, Lacan, Derrida and other famously hard to interpret maîtres à penser. ... Still the epidemiological mechanism I have briefly sketched, explains how many obscure texts and their authors come to be overestimated, often ridiculously so, not in spite but because of their very obscurity. [Sperber, D., The Guru effect. Review of Philosophy and Psychology 2010, pp. 583, 592]

Schopenhauer [4] points out that a universal prejudice may also be used as an authority. Using an appeal to popular assent is also called an *Argumentum ad Populum* (argument to the people). Such an appeal asserts that, since the majority of people believes an argument or chooses a particular course of action, the argument must be true or the course of action must be followed. Nowadays one sees this phenomenon in Western Europe, where wind-mills to generate electricity are built at a very large scale, although it is pretty clear that the enormous costs cannot outweigh the return.

> There is no opinion, however absurd, which men will not readily embrace as soon as they can be brought to the conviction that it is generally adopted. ... They are like sheep following the bellwether wherever he leads them. They would sooner die than think.
>
> It is very curious that the universality of an opinion should have so much weight with people. Their own experience might tell them that its acceptance is an entirely thoughtless and merely imitative process. But it tells them nothing of the kind, because they possess no self-knowledge whatever. ...
>
> To speak seriously, the universality of an opinion is no proof. In fact, it is not even a probability that the opinion is right. [For instance, almost all people once have thought planet earth was flat, but that majority's belief did not mean the earth really was flat.] ...
>
> When we come to look at the matter, so-called universal opinion is the opinion of two or three persons. We should be persuaded of this if we could see the way in which it really arises. ... [A few persons who select the news to be broadcasted and next more and more people are spreading the word.] ...
>
> When opinion reaches this stage [of universal acceptance], adhesion becomes a duty. Henceforward the few who are capable of forming a judgement hold their peace. Those who venture to speak are entirely incapable of forming any opinions or any judgement of their own, being merely the echo of other's opinions. Nevertheless, they defend them with all the greater zeal and intolerance. For what they hate in people who think differently is not so much the different opinions which they have as the presumption of wanting to form their own judgement. In short, there are very few who can think, but every man wants to have an opinion; and what remains but to take it ready-made from others, instead of forming opinions for himself. [Schopenhauer [4], section 30]

A particular type of argumentum ad populum does not assert that everybody is doing it, but rather that all the best people are doing it. For instance: any true intellectual would recognize the necessity for studying logical fallacies. The implication here is that anyone who fails to recognize the truth of this assertion is not an intellectual.

10.3.4.5 Suggestive tricks: repeating oneself, speaking confidently, suggestive questions

One would be surprised to realize how many of our ideas, views and convictions are in the end the result of commercials and propaganda. The media (TV and newspapers) get much of their information from the local and national governments, journalists have little or no time for research and almost everyone parrots what they have heard elsewhere. Consequently, many things are de facto not what they seem to be. For instance, religious Christian leaders in Syria give a completely different picture of the situation in their country than we are told by the mainstream media. In what follows, we discuss some of the more important tricks of persuasion.

Repeating oneself
We have the tendency to start to believe statements which are repeated again and again, either literally or with slight modifications. Repeating things is a well known

10.3 Unfair Discussion Methods

method to learn addition and multiplication, to learn French, but also to learn playing piano, etc. The speeches of Hitler, for instance, always had the same topics: the Jews, Gross Deutschland, die Partei, frequently presented in small variations. In the following example the speaker repeats several times more or less exactly the same thing without any convincing argument. Nevertheless, these repetitions suggest that what is said is absolutely true and that any further discussion is superfluous.

Example 10.56 (Suggestive repetitions). (van Hoesel [2])
Poverty is a lack of social adjustment. The economically weak are the ones who were not able to adjust to the social demands put on them. They are biologically less gifted than the working people, who were able to bring about such adjustment.

Speaking confidently Frequently people try to eliminate the critical attitude of their audience by speaking (very) confidently. A more modest speaker is frequently not taken very seriously, in particular if the audience is large. In political speeches, for instance, addressed to a large audience, the speaker will usually speak very confidently in order to prevent the audience from thinking that he has little or no knowledge or that his views are poorly substantiated.

Suggestive questions Questions are suggestive if they – by the way they are asked – actually suggest the answer.

Example 10.57 (Suggestive questions).
You certainly also buy a lottery ticket for the animal protection?
You will certainly agree with the usual 10% fee?
In a shop: you will certainly take it with you, madam? Instead of: do you want it to be delivered at home?

One may distinguish:
1. *The implying question* For instance: although the car which caused an accident, taken into custody by the police, does not have an antenna, the officer might ask: was the antenna of the car on the bonnet or on the roof?
2. *Question which contains a dilemma* For instance, although the car in question is green, the officer might ask: was the car black or red?
3. *Expectation question* For instance: he certainly drove too fast?
4. *Complex question* For instance: Did the driver give way, use his direction indicator and drive at the right side of the road, yes or no?

Another well-known example is the question: have you stopped beating your wife? Whether you answer this question with yes or no, in both cases you admit that you have beaten your wife before, because this question presupposes that you did so; see Section 7.11. In fact this question consists of two questions rolled into one: a) Did you beat your wife in the past? and b) If so, did you stop beating her?

Complex questions appear in written argument frequently. A student might write a bachelor thesis with the title 'Why is private development of resources so much more efficient than any public control?'. An observant reader may recognize that the prior implicit question, whether private development of resources really *is* more efficient in all cases, remains unaddressed.

10.3.5 Either/Or Fallacy

By the words we are using, we frequently make sharp distinctions which do not exist in reality. For instance, classifying people into rich and poor. However, when we would try to put ourselves into one of these two categories, many of us would notice that it is not really possible to do so. Similarly, in daily language we make sharp distinctions between beautiful and ugly, expensive and cheap, good and bad, intelligent and stupid, normal and abnormal. As already pointed out by the Dutch Significists, among them G. Mannoury and L.E.J. Brouwer, there are gradual transitions between these two extremes; see Section 7.3. Nevertheless, in discussions about a certain problem, people are frequently placed in front of a dilemma, while in fact there is no dilemma. In such a case, two extreme alternatives are offered to choose from, while in fact there is a whole range of possibilities. For instance: are you my friend or my enemy? Is he normal or abnormal? Are you healthy or sick?

Example 10.58 (Either/or fallacy).
Yesterday you criticized the Israeli government. But then you are an anti-semite. So, do you want to be an anti-semite or do you retract your comment?

The unfair element is that there is a whole range of possibilities between anti-semitism (hating *all* Jews) and disagreeing with *one* decision of the Israeli government.

Conversely, one may accentuate the gradual transition to explain away the difference between two different things.

Example 10.59. (van Hoesel [2])
Boss: John, you were ten minutes too late at work this morning.
John: If I would have been one minute too late, would you make a point of it?
Boss: Of course not.
John: And if I would have been two minutes too late?
Boss: I would not say anything.
John: And if these two minutes were three minutes?
Boss: Okay, I could live with that.
John might continue this way to conclude that there is no reason at all to blame him for anything. But the boss would nevertheless finish the conversation with: either you are on time or you are too late!

In a similar way one might try to explain away the difference between a small group of people and a crowd, by pointing out that one person more does not change a small group of people into a crowd. By this kind of reasoning one may cheat not only someone else, but also oneself.

10.3.6 The treacherous paradox

In this subsection we shall illustrate the disastrous influence that a paradox may exert on our critical thinking. In Section 10.3.4.3 we have already seen that using an analogy can have a paralyzing effect on our intellectual activity, probably because the analogy largely meets our laziness of thinking. No man is born as a good thinker, and without effort no one will probably ever learn to think well and clearly.

If one wants to sell a dubious position, one has to present it in the form of a paradox and one will notice that it is readily accepted.

Example 10.60 (Treacherous paradox). (van Hoesel [2])
A group of people discusses the education of children between say 15 and 20 years old. Some of them argue that one should give these children a lot of freedom, while others argue that too much freedom may have disastrous consequences. One of the participants, defending the larger freedom, summarizes the discussion in the following paradox: he who wants to hold his children must let them go.

Why is this paradoxical statement so convincing? First of all, because it suggests an (apparent) reconciliation between two different points of view, causing a kind of Eureka experience. In addition, since this paradoxical statement also seems to do justice to both points of view, everyone has the impression that his or her point of view has been taken into account. In the second place, this paradox suggests objectivity and distinction. Finally, the paradox caters to the laziness of thinking of the people involved.

Example 10.61 (Treacherous paradox).
A perfect organisation may be an organized chaos.
It takes a lot of reason to find something incomprehensible.
Strongly refusing outwardly means often accepting inwardly.
Less is more.
The voter is always right.

However, already the Roman writer Titus Livius ($\pm$ 10 CE) stated: but, as it mostly happens, the greater part overruled the better.

10.3.7 Ad Hominem Arguments

At the football field one sometimes hears fanatic supporters shout: first the man, next the ball. The reader may wonder what football has to do with argumentation. Well, there are many similarities: one sees many feints, tempers are often heated, the goal is often passed by, one does little with his head and cooperation is often lost. Similarly, in both cases one frequently gets personal.

When one has few or no arguments against a position defended by an opponent, one frequently jumps from the subject of discussion to the person in question, attacks him personally and tries to discredit him. This practice is fallacious because the personal character of an individual is irrelevant to the truth or falsity of the conclusion of the argument itself.

Example 10.62 (Ad hominem argument). (van Hoesel [2])
Mister X is in favor of Darwin's theory of evolution and mister Y opposes it, but

cannot find good counterarguments. So, he might ask the question: please tell me, do you descend from an ape from your grandmother's side or from your grandfather's side?

Another example of an ad hominem argument is: That plan cannot be good; he has not studied at a university. People, making remarks like this one, do not take the troubles to study the plan objectively and critically, again an indication of their laziness of thinking.

Example 10.63 (Ad hominem argument).
A local party LST in the Netherlands recently obtained the greatest number of seats in an election for the city council: 10 out of 45 seats, which means that almost 1 out of 4 voters had chosen for this party. Consequently, this party is entitled to form a coalition. However, one of the parties in the old coalition had – already before the election day – declared that they would not take part in a coalition with (the leader of the) LST, without giving any (good) reason. Interestingly, the party in question has the word democratic in its name! And since the other parties in the old coalition wanted to continue their cooperation, they did not want to form a coalition with LST either, in this way ignoring the votes of 22% of the citizens.

The same phenomenon occurred in several other cities in the Netherlands and also in the Dutch and Belgian parliaments, while anybody in any organization is supposed to cooperate with colleagues, even when they do not like each other very much.

Surprisingly, even in the academic world these ad hominem arguments are frequently used, in particular by referees of scientific journals and of proposed research projects.

Example 10.64 (Ad hominem argument).
This article might have been written by a beginning student.
The author of this PhD thesis is a charlatan.

Example 10.65 (Ad hominem argument).
A PhD candidate had written a thesis with a physical theory formulated in a logical mathematical language. Interestingly, this theory was inconsistent with the general theory of relativity. There was no claim at all that this theory was true. The thesis had been approved for defense by the PhD committee of the university. When the dean of the faculty learnt that this theory was inconsistent with general relativity, he sent the PhD thesis to a former classmate who had won a Nobel prize in physics with the request to have a look at it. Within a few hours his reply was there: *The idea of antimatter proposed in this thesis is inconsistent with the general theory of relativity, and in my opinion that can only mean that the PhD candidate has no clue whatsoever about what antimatter is; it would be a disgrace for the university to admit the candidate to the defense.* The dean decided to cancel the defense, even without consulting the two PhD supervisors, who spent weeks in order to be able to understand the formalism and the physical theory proposed.

Fortunately, later the PhD thesis was successfully defended at another university, the logical-mathematical part was published in a journal for logic and the physical part was published in a journal for physics, both of the highest level. The Nobel prize winner saved himself a lot of time by not having to look more carefully into the thesis.

Example 10.66 (Ad hominem argument).
A committee of the faculty, consisting of three professors, had to judge a number of research proposals which had been sent to its members quite in time. At the day of the meeting it turned out that one of the committee members had not looked at the proposal submitted by his colleague in the committee. So he asked to show him the research proposal in question. He looked at the title and after a few seconds said: that cannot be something interesting. The third committee member did not want to intervene and the research project was not granted without it having been studied properly.

Example 10.67 (Ad hominem argument). (van Hoesel [2])
A professor in psychology writes a book about the education of children. Without reading the book, someone might argue: that book cannot be good! Look at his own son; he is the terror of the neighborhood.

A man got the advice from his specialist not to smoke anymore. But he ignored the advice completely, for the specialist himself was smoking a big cigar when he gave his advice.

When one is confronted with such a personal attack, Schopenhauer [4] gives us the following advice:

> As soon as your opponent becomes personal, you quietly reply 'That has no bearing on the point in dispute' and immediately bring the conversation back to it, and continue to show him that he is wrong, without taking notice of his insults.

10.3.8 Argumentum ad baculum

This is an argument in which the opponent is physically or psychologically threatened, as it were with a stick (ad baculum).

Example 10.68 (Ad baculum).
Father made him an offer he could not refuse (Michael Corleone in *The Godfather*).
Your remarks smell of racism.

This argument prevents the opponent to speak freely. Frequently the threat is implicit. And when one makes the insinuation explicit, the other party has always the possibility to deny the insinuation. This makes this argument a very nasty one.

Of course, not all arguments ad baculum are fallacious. For instance, a policeman may threaten someone with a big fine if he does not respect the traffic lights.

10.3.9 Secrecy

By declaring a certain agreement to be secret, one may prevent critical questions or even hide that the agreement is illegal.

Example 10.69 (Secrecy).
The so called presidium of the city council, consisting of the chairmen of the different parties in the city council, had reached a majority decision that retired former members of the council would get half-pay during a certain period. It was known that this was illegal. For that reason the chairman of a local party announced that he would make this majority decision public. By this threat the presidium decided by majority to declare the agreement to be secret. Nevertheless, the party-leader made the decision public. He was arrested for violating secrecy, had to spend one day at the police office, his and his family's computers were taken into custody and he was sentenced to a fine of 350 euros.

The other members of the presidium, the mayor and the aldermen were not sentenced at all, although they knew that they had made an illegal decision.

Example 10.70 (Secrecy).
The mayor and aldermen of the city asked the city council for more money for transforming a former cinema to a theatre. Because the budget was already more than ten million euros, they knew that many members of the city council would be very critical, to say the least. In order to convince them still to make more funds available they declared that there was a contract with an entertainment company for making television programs in the new theatre. The leader of one of the parties in the city council asked whether he could see this contract. However this was refused with the argument that they could not make a trade secret public. Again the party leader asked: may I see this contract? Again the answer was: no, we are not allowed to make this trade secret public. Later it turned out that there was no contract at all, that there even had been no contacts with the entertainment company in question.

The mayor and some of the aldermen were dismissed by the city council. However, within half a year they all had new similar positions.

10.3.10 The Retirement Home's Discussion

Imagine two old men on a bench next to each other, talking alternately about the local football club and the youth of today. They do not listen to each other, but only are concerned with their own argument which they bring forward again and again, each time in a different form. They only listen to themselves, not to the other person. 'A debate is a generally heated conversation, in which two people talk to each other and listen to themselves' (Jean de Boisson). One might think or hope that such conversations do not occur in business or scientific discussions. Unfortunately, they do! Attend, for instance, a meeting of the local city council or of the parliament. It happens more than once that one speaker supports his position with various arguments, while his opponent restricts himself to repeating his own position without going into the arguments of the first speaker. In such a case the discussion leader, usually the mayor, should ask the 'old man' what he brings forward against the arguments of his opponent. Frequently it will turn out that he does not know them and/or that he will say: that may be true, but I stick to my point of view. By the way, there are mayors who do not care about the quality of the discussion and just wait till they are finished.

Another version of this phenomenon is *cherry picking*: only select evidence is presented in order to persuade the audience to accept a certain position, and evidence against this position is withheld. In other words, one picks the cherries one likes and ignores the cherries one does not like.

Example 10.71 (Cherry picking).
In the Netherlands there is an ongoing discussion about the future of the pension system, where it is difficult to find a balance between the interests of the younger people and those of the older people. Each group brings forward their favoured arguments, ignoring the arguments of the other group, even not mentioning them.

As we have already seen in Section 7.14 a statement may be true, but nevertheless not tell the whole truth and hence be misleading. For instance, if I answer your question whether I know a gas station because you are running out of gasoline and I answer 'yes, there is a gas station around the corner', I may be speaking the truth, but nevertheless be misleading if I know that the gas station is closed. The statement 'there is a gas station around the corner' together with simple *conversation rules*, like being relevant and maximally informative, *conversationally implicates* that the gas station is open.

As one may expect, politicians in particular are very good in telling truths that are misleading.

Example 10.72 (Cherry picking).
Politicians like to claim that they will solve a certain problem, for instance, great unemployment. But sometimes they forget to mention that they themselves were the ones who caused the problem in the first place.

10.4 Summary

One must keep in mind that our emotions, feelings and sentiments may have a strong negative influence on our thinking and that they can often overwhelm our critical thinking.

In the preceding sections we have treated a great number of mistakes which stand in the way of clear thinking and good discussion:
- An emotional thinker is frequently verbose, bombastic and theatrical, but at the same time inaccurate and vague.
- His words are tendentious, his definitions incoherent and meaningless.
- He simplifies the most difficult problems to meaningless formulas and he uses cliches as hand grenades.
- He starts with conclusions instead of finishing with them.
- He posits assumptions as established facts and he generalizes with the greatest ease on the basis of a few examples.
- He is a master in rationalizing his prejudices and he simply ignores evidence that

does not suit his purpose.
- He does not listen to the arguments of his opponent, but repeats his words again and again.
- He ascribes to his opponent assertions which he has never made.
- He draws extreme conclusions from moderate statements and creates dilemmas which do not exist.
- He camouflages his weak argumentation with a lot of words and he jumps from one subject to another.
- He makes objections that do not make sense and does everything to bluff to the audience.
- He poses suggestive questions and makes causal connections which are not realistic.
- He insinuates in a crude way and becomes all too easily personal.

All these fallacies and unfair discussion methods make us understand the complaint of Klemens von Metternich (1773-1859): Throughout my life I only knew ten or twelve people with whom it was pleasant to speak: who kept strictly to the subject, did not repeat themselves, did not speak about themselves, did not listen to their own words, were too civilized to lose themselves in commonplaces, and who had enough tact and good taste not to raise their own person above the subject.

Acknowledgements As is evident from this text, I owe a lot to the Dutch booklet Zindelijk Denken by A.F.G. van Hoesel. As far back as around the year 2000 I had tried to track down professor van Hoesel and his publisher in order to suggest to them to reprint this booklet. But I could not find any trace, neither of professor van Hoesel, nor of his publisher.

I am most grateful for many important and concrete suggestions made by Filip Buekens, Marcoen Cabbolet, Jan Cuijpers, Paul van Dongen and Michael Perrick, who commented on earlier drafts of this chapter. In particular, Sections 10.2.4 and 10.2.6, as well as several examples, are due to Marcoen Cabbolet. No less important is the correction of English in this Chapter, which was done by Naftali Weinberger.

References

1. Hamblin, C.L., *Fallacies*. Methuen & Co LTD, London, 1970.
2. Hoesel, A.F.G. van, *Zindelijk denken, Foutieve denkwijzen en oneerlijke discussiemethoden*. [Thinking clearly; fallacies and unfair discussion methods] (Out of print.) H. Nelissen, Baarn, NL, 1955, 1983.
3. Kahneman, D., *Thinking, Fast and Slow*. Penguin Books, 2011.
4. Schopenhauer, A., *The Art of always being right*. Gibson Square Books Ltd, London, 2005.
5. Tindale, C.W., *Fallacies and Argument Appraisal* (Critical Reasoning and Argumentation). Cambridge University Press, 2007.

Index

(v,w), 144
$2^{\mathbb{N}}$, 174
$A \Rightarrow B$, 313
$A \mapsto B$, 310
$A \models B$, 204
$A \to B$, 26
$A \rightleftarrows B$, 26
$A \vee B$, 26
$A \wedge B$, 26
$A \boxbox\to B$, 310
$A \diamond\to B$, 312
$A_1,\ldots,A_n \models B$, 38, 200, 284
$A_1,\ldots,A_n \models B$ in $K-$, 286
$A_1,\ldots,A_n \models_i B$, 395, 408
$A_1,\ldots,A_n \not\models B$, 39, 200
$A_1,\ldots,A_n \not\vdash B$, 66, 218
$A_1,\ldots,A_n \not\vdash' B$, 86, 224
$A_1,\ldots,A_n \not\vdash'_i B$, 389
$A_1,\ldots,A_n \not\vdash_i B$, 386, 405
$A_1,\ldots,A_n \vdash B$, 66, 218
$A_1,\ldots,A_n \vdash' B$, 86, 224
$A_1,\ldots,A_n \vdash' B$ in $K-$, 294
$A_1,\ldots,A_n \vdash'_i B$, 389, 406
$A_1,\ldots,A_n \vdash_i B$, 386, 405
$A_1,\ldots,A_n \models^* B$, 314
$A_1,\ldots,A_n \vdash^* B$, 313
CA, 303
$C(A)$, 214
EA, 303
GL, 318
Ind, 274
K, 280, 315
$K-$, 280
KT, 280, 315
$M \models A[v]$, 197
$M \models A$, 199, 283
$M \models_i A$, 395

$M \not\models A[v]$, 198
$M \not\models A$, 199, 283
$M \not\models_i A$, 395
$M,s \models_i A$, 394
$M,s \models_i A(a_1,\ldots,a_k)[n_1,\ldots,n_k]$, 407
$M,w \models A$, 283
$P(V)$, 138
$R;S, R \circ S$, 148
$S4$, 280, 315
$S5$, 280, 315
$Sk(A)$, 214
$V - W$, 135
$V <_1 W$, 169
$V \cap W$, 135
$V \cup W$, 134
$V \times W$, 145
V/R, 150
V^2, 146
V^n, 146
$W \not\subseteq V$, 137
$W \subset V$, 137
$W \subseteq V$, 137
W^V, 153
ZF, 132, 141
$[v]_R, v/R$, 149
$\Box A$, 280
$\Diamond A$, 280
$\Gamma \models B$, 38
$\Gamma \vdash B$, 66, 218
$\Gamma \vdash_i B$, 405
$\bigcup x$, 134
$\check{R}, R^{\mathsf{T}}$, 147
$\mathcal{N}$, 262
$\mathscr{P}\mathscr{A}$, 274
$\mathscr{P}$, 264, 265
$\emptyset$, 133
$\mathbb{N}$, 135

$\mathbb{Q}$, 162
$\mathbb{R}$, 162
$\mathbb{Z}$, 162
$\models B$, 284
$\models B$ in $K-$, 286
$\models_i A$, 395
$\neg A$, 26
σ_{01mon}, 413
σ_{01}, 413
σ_ω, 413
$\ulcorner A \urcorner$, 268, 270
$\vdash' B$ in $K-$, 294
$\vdash'_i B$, 390
$\vdash_i B$, 386
$\{v,w\}$, 133
$\{z \in x \mid A(z)\}$, 135
$f : V =_1 W$, 154
$f : V \leq_1 W$, 154
$f : V \to W$, 152
$f = g$, 155
$f \lceil V_0$, 156
$f(V')$, 152
$f(v)$, 151
$f^{-1}(W')$, 152
$f^{-1}(w)$, 155
$g \circ f$, 155
$v \in V$, 129
$v \notin V$, 129
$xRy, R(x,y)$, 147
$\mathscr{L}$, 262

a posteriori, 141, 243, 349
a priori, 141, 243, 349
absorption law, 136
accessibility relation, 112, 283, 368, 394
actually infinite, 162
ad baculum argument, 527
ad hominem argument, 525
agenda, 480
aleph-null-categorical, 272, 273
alethic operator, 278
alphabet, 434
alphabet of $\mathscr{L}$, 262, 263
alphabet of predicate logic, 190
alphabet of propositional logic, 25
always false, 34
always true, 33, 196, 199
analogy reasoning, 518
analytic, 141, 142, 243, 244, 299, 349, 350
anonymity, 469
anonymous variable, 445
antecedent, 26, 40
antecedent test, 354
anti-symmetric, 156

antinomies, 97, 98
antinomy of Berry, 99
antinomy of Russell, 99
any, 194
appeal to ignorance, 496
aprioristic reasoning, 504
argumentum ad ignorantiam, 496
argumentum ad populum, 521
arithmetic, 261
arithmetical completeness, 320
arithmetical soundness, 319
Arrow's impossibility theorem, 470
Arrow, K., 17
Artificial Intelligence, 15, 96, 445
assertion, 333
associative, 136
asymmetric, 158
atomic formula, 263, 435
atomic formula of predicate logic, 190
atomic formula of propositional logic, 25
atomic proposition, 22
attributive use, 297, 347
Aussonderungs Axiom, 130
automated theorem proving, 15, 96, 214
Automath, 71
autonymously used, 330
axiom schema, 65, 135
axiomatizable, 266
axioms for $\to$, 63
axioms for (classical) predicate logic, 216
axioms for (classical) propositional logic, 46, 64
axioms for (intuitionistic) propositional logic, 385
axioms of Peano, 265

backtracking, 429, 430
Balinski, M., 17
barber paradox, 99
Barcan formula, 316
basic term, 190
begging the question, 505
Begriffsschrift, 42, 106, 109
Bernays, P., 109
Berry's paradox, 344
Beth, E., 82
biased connotation, 516
bijection, 154
bijective, 154
binary connective, 52
binary spread, 413
body, 58, 428, 435
Borda Rule, 467
bound variable, 188

Index 533

Boyce-Codd Normal Form, 458
BR, 467
Brouwer, L.E.J., 162, 381
Brouwer-Kripke axiom, 416

calculemus, 41, 111, 368
Cantor, G., 129, 130, 169, 170, 174
Cartesian product, 145
causal chain, 297, 346
characteristic function, 154
cherry picking, 529
choice-law, 412, 413
choice-sequence, 413–415
Chomsky, N., 370
Church, A., 41, 111, 313, 368
circular reasoning, 504
classical logic, 29
clausal form, 214
clause, 57, 428, 436
cliché, 491
closed branch, 398, 399
closed formula, 192
closed interval, 171
closed search tree, 306, 399
closed tableau, 85, 224, 293, 389
closed tableau branch, 84, 223, 293, 389
closer to, 311
closure, 199
common knowledge, 300, 303
communicative act, 336
commutative, 136
compactness, 232, 240
compactness theorem, 35, 233
compatible, 459
complement, 136
complete, 64, 68
complete relation, 157
complete set of connectives, 52
completed branch, 84, 224, 293, 389
completed tableau, 85, 224, 293, 389
completeness, 93, 227, 240, 303, 316, 397, 409, 410, 432, 438, 469
completeness of intuitionistic propositional logic, 396
completeness theorem, 10, 68, 95, 227, 308, 401, 402
completion, 444, 448
composite formula, 25
composition of functions, 155
composition of relations, 148
compositionality, 333
concatenation, 237
conceptual content, 106
conclusion, 3, 40, 63

conditional, 33, 107, 309, 310, 313
Condorcet loser, 465
Condorcet paradox, 466
Condorcet portion, 466
Condorcet profile, 466
Condorcet winner, 465
confirmation principle, 245, 352
conjunctive normal form, 53
connected, 158
connective, 22
connotation, 332
consequentiae, 104
consistent, 33, 236, 266
constraints, 454
construction project, 381, 411
constructive, 402
content, 109
contingent, 34
continuity principle, 414
continuum, 174
Continuum Hypothesis, 174
contradictory, 34, 247
contraposition, 47
contrary, 247
conversational implicature, 107, 365
converse relation, 147
coordinated attack problem, 302
correct argument, 3, 22
correlation-law, 414
countable, 164
countably infinite, 164
counterexample, 49, 198, 199
counterfactual, 108, 310, 366
countermodel, 198, 199
cryptarithmetic puzzle, 446
cum hoc, ergo propter hoc, 509
Cusanus, 140
cut, 440

database-characterisation, 455
database-constraint, 455
databases, 455
de dicto, 299, 369
De Morgan, 47
de re, 299, 369
decidable, 40, 88, 95, 308, 392, 401
decision procedure, 40, 231
Dedekind infinite, 175
deducible, 11, 64, 66, 81, 218, 219, 221, 281, 405
deduction, 11, 64, 66, 217, 281, 386, 405
deduction theorem, 73, 220, 386
deductive databases, 445
deductive nomological explanations, 110

deep structure, 372
definite clause, 435
definite goal, 58, 435
definite program, 58, 435
definite program clause, 58, 435
denotation, 332
denumerable, 164
deontic logic, 284
deontic operator, 278
depth-first search, 433
derivable from, 371
derivation tree, 370
derived rule, 80
diagonal, 148
diagonal method, 169
diagonalisation, 169, 170
dictator, 469
Diodorus Cronus, 98, 103
direct reference, 297, 346
discursive paradox, 481
disjunction property, 296
disjunctive normal form, 53, 56
distributive law, 47, 136
district paradox, 480
Dom(R), 147
domain, 147, 194, 197, 211
domination, 471, 472
double negation elimination, 76
double standards, 505
dressed spread, 413
Duhem-Quine thesis, 51

either/or fallacy, 524
electric circuit, 53
elementary number theory, 261
elementary theory of fields, 241
elimination rule, 76, 220, 221
empty clause, 435
empty list, 439
empty relation, 148
empty set, 132, 133
entailment, 314
enthymemes, 18, 43, 489
enumerable, 164
enumeration, 164
epistemic logic, 287
epistemic operator, 278
equally great, 163
equipollent, 163
equivalence class, 149
equivalence relation, 149
Erathostenes, 164
essence, 296, 347
essentialism, 318, 359

Euclid, 164
everybody knows, 303
ex falso sequitur quod libet, 47
exclusive 'or', 37
exclusive negation, 339
existential quantifier, 23, 183
expert system, 15, 96, 447
extension, 332
extensionality axiom, 132
extensionally opaque, 317, 318

fact, 58, 246, 428, 430, 435
fallacy, 18, 489, 491
fallacy of incredulity, 500
falsidical paradoxes, 97, 99
falsification principle, 245, 352
finite, 164
first-order, 202
form, 69, 71
formal fallacy, 18, 490
formal language of predicate logic, 189
formal language of propositional logic, 26
formal number theory, 265
formalization, 265
formula, 280, 435
formula of predicate logic, 191, 211
formula of propositional logic, 6, 25
free for, 210
free variable, 188, 189
Frege's Begriffsschrift, 105
Frege, G., 42, 82, 106, 109, 142, 278, 331, 353
fulfilled, 84, 224, 293, 389
function, 151, 152
function symbols, 211
functionally dependent, 455

Gödel, K., 227
Gödel-number, 268
Gödel-numbering, 268
Gödel's logic, 318
general term, 332
Gentzen's elimination rules, 79
Gentzen's introduction rules, 79
Gentzen's system of natural deduction, 81, 221
Gentzen, G., 82
Gentzen-type rules, 82, 87
goal, 429
Goldbach's conjecture, 14, 382
grammar, 371
graph, 152
Grelling's paradox, 345
Grice, H.P., 47, 107, 108, 365, 367

head, 58, 428, 435, 439

Index   535

held constant, 217, 218, 405
Heyting, A., 385
Hilbert hotel, 165
Hilbert, D., 109
Hilbert-type, 62
homonyms, 502
Horn clause, 57, 58, 435
Hume, D., 352

idempotent, 136
identity relation, 148
ignorance, 496
IIA, 465
illocutionary act, 340
illocutionary force, 334
image, 151
immediate successor, 237
implicative verb, 355
impossibility theorem, 17, 469, 470
improper generalization, 493
incompleteness theorem, 12, 265, 266, 271
inconsistent, 34
incorrect argument, 5
incredulity, 499
independent, 62
Independent of Irrelevant Alternatives, 465, 470
indicative conditional, 107, 366
indirect reference, 334
individual constant, 185, 189
individual variable, 182, 185
induction basis, 27
induction hypothesis, 27
induction principle, 27, 192
inductive logic, 110
infinite, 164, 380
infinity, 135
infix notation, 451
informal fallacy, 18, 490
injection, 154
injective, 154
integer, 162
intended interpretation, 186, 262
intension, 332
intensional context, 278
interpolation theorem, 91
interpretation, 35, 194, 197, 211
interpreted sentence, 364
intersection, 135
interval, 171
introduction rule, 76, 220, 221
intuitionism, 14, 379
intuitionistic counterexample, 395
intuitionistic countermodel, 395

intuitionistic model, 395
intuitionistic propositional logic, 110, 385, 387
intuitionistically valid, 395
intuitionistically valid consequence, 393, 395
invalid, 4, 9
inverse function, 155
isomorphic, 159
isomorphism, 159

join, 459
judgment, 109
judgment aggregation, 481

König's Lemma, 237
Kant, I., 141, 243, 349
key, 455
killer, 491
knowledge base, 15, 57, 96, 246
knowledge-based-system, 447
Kripke model, 283, 394, 407
Kripke model for $K-$, 286
Kripke valid, 408
Kripke valid consequence, 408
Kripke, S., 279
Kleene, xiii
Löwenheim-Skolem theorem, 233, 235, 240
Löb's theorem, 270
Löwenheim's theorem, 232
language, 381
language generated by, 371
Laraki, R., 17
law of double negation, 47
law of excluded middle, 47
law of non-contradiction, 47
lawless sequence, 415
lawlike sequence, 415
Leibniz' law, 111, 316, 368
Leibniz, G., 111, 367
lexicographic ordering, 156
linear ordering, 157
lingua philosophica, 40, 111, 367
linguistic gradation, 338
list, 439
literal, 53, 55, 435
locution, 334
locutionary act, 340
logic of, 72
logic of provability, 318
logic program, 213, 215, 246
logical axioms, 62
logical consequence, 38, 196
logical positivism, 244, 352
logical symbols, 186, 264
logically complete, 304, 397

logically contingent, 141, 243, 349
logically necessary, 141, 243, 349
logicism, 142, 350

majority gauge, 475
majority grade, 471
Majority Judgment, 17, 462
Majority Rule, 17, 465, 466
majority-gauge rule, 475
Mannoury, G., 335
manufactured majority, 481
marriage theorem, 37
matching, 163, 430, 436
material implication, 25, 32, 33
mathematical induction, 27
mathematical language, 381
May's axioms, 469
meaning, 107, 333, 365
meaning as use, 359
merit-profile, 471, 472
meta-language, 7, 345
method of exhaustion, 337
method of transformation, 337
middle-most block, 472
MJ, 462
modal logic, 279
modal operator, 278
modality test, 354
model, 35, 195, 199, 233, 283, 313, 408
model theory, 199
modulo R, 149, 150
Modus Ponens, 5, 22, 63
Modus Tollens, 3, 5, 69
monadic formula, 237
monotonicity, 469
more than one meaning, 502
most general unifier, 438
most votes count, 464
MP, 22
MR, 466
muddy children puzzle, 288

naive comprehension principle, 130
name, 332
natural deduction, 11, 78, 79, 221
natural number, 130, 162
Negation as Failure, 444
negation of choice, 339
negation test, 354
negative introspection, 280, 320
negative literal, 57
negative test for validity, 231
neutrality, 469
node, 237

non-constructive, 403
non-intended interpretation, 186, 188
non-logical symbols, 186, 264
non-monotonic, 444
non-standard model of arithmetic, 273
non-terminals, 370, 371
normal database-type, 458
normal form theorem, 56
normal goal, 435
normal program, 435
normal program clause, 435
NP, 60
NP-complete, 60
numeral, 268

object-characterisation, 452, 453
object-language, 7, 190
oblique reference, 334
occur check, 437
one-one correspondence, 154, 163
one-sorted, 202
one-step expansion, 84, 224, 293, 389
ontology of mathematics, 141
open branch, 399
open formula, 192
open interval, 171
open search tree, 306, 399
open tableau, 85, 224, 293, 389
open tableau branch, 84, 224, 293, 389
opinion-profile, 471, 472
ordered n-tuple, 145
ordered pair, 144
ordering, 157
Organon, 247

pairing axiom, 133
pairwise comparison, 465, 466
paradox of the liar, 97
paradoxes, 97
paradoxes of material implication, 47, 108, 366
paradoxes of strict implication, 310
Pareto condition, 482
parsing, 446
partial function, 152
partial ordering, 156
partition, 150
path, 237
pattern of reasoning, 3
Peano, G., 264
perlocutionary act, 341
Philo of Megara, 103
Philosophical Investigations, 357
phrase marker, 370
picture theory of meaning, 358

plausible, 43
Plurality Rule, 17, 464
polarity element, 355
positive introspection, 280
positive literal, 57
positive test for validity, 231
possible world, 112, 279, 283, 368
post hoc, ergo propter hoc, 508
potentially infinite, 163
powerset, 138
PR, 464
pragmatics, 341, 364
precedence, 451
predicate, 183
predicate calculus, 182
predicate logic, 8, 181
predicate logic with equality, 237
predicate logic with function symbols, 211
predicate symbols, 187, 189
prefix, 237
premiss, 3, 40, 63
prenex (normal) form, 212
prenex formula, 212
prenex operation, 212
presupposition, 333, 353
Principia Mathematica, 109
principle of adequacy, 97
principle of extensionality, 112
probabilistic explanations, 110
procedure of searching for a tableau-deduction, 93, 228
procedure to construct a counterexample, 304, 397
production, 371
production rule, 370
projection, 458
projection problem, 356
Prolog, 57, 246, 428, 430, 433
proof, 66, 217
proof by cases, 51
proper subset, 137
property, 184
proposition, 21, 189, 333, 364
propositional calculus, 21, 24
propositional logic, 7, 21, 22
propositional operation, 22
provability predicate, 270
provable, 63, 66, 81, 217, 281

query language, 459
questionable analogy, 494
Quine, W.V., 142, 278, 329
quotient set, 150

Ran(R), 147
range, 147
rational number, 162
rationalizing, 506, 508
real number, 162
reciproca, 102
reckless, 14, 383
recursive definition, 431
red herring argument, 512
reductio ad absurdum, 51, 76
reference, 331, 333
referential use, 297, 347
referentially transparent, 317, 318
reflexive, 149, 156, 284, 394
regularity, 139
relation, 147, 184
relational database, 16, 445, 455
relative complement, 135
relevance logic, 72
relevant implication, 313
RelView, 149
replacement, 139
replacement rule, 50
replacement theorem, 50
representative, 149
resolution, 214
restriction of a function, 156
reversal portion, 465
Richard paradox, 346
rigid designator, 298, 348
Rosetta, 28
Ross's paradox, 285
Ross, A., 282
rule, 246, 428, 430, 435
rules of inference for (classical) predicate logic, 216
Russell's paradox, 130
Russell, B., 82, 109, 142, 242, 342

safeness condition, 448
same cardinality, 163
satisfiability problem, 59
satisfiable, 33, 35, 196, 199
satisfiable in, 232
scope, 188
search tree, 292, 294, 304, 306, 397, 399, 428, 432
second-order, 202
second-order logic, 234
secrecy, 527
selection rule, 430, 433
semantic tableaux rules, 82, 83, 290
semantics, 111, 341, 363
Sen's paradox, 482

sense, 331, 333
sentence, 192
sentential calculus, 21
separation axiom, 130, 135
sequence, 153
sequent, 82, 222, 290, 387
set, 129
Sextus Empiricus, 103
Sheffer stroke, 51
signed formula, 82, 222, 387
significs, 335, 501
simple consistency, 68
simplistic thinking, 494
simultaneously satisfiable in, 232
singleton, 133
singular term, 332
Skolem (normal) form, 213
Skolem's generalization, 232
Skolem's paradox, 235
slippery slope argument, 494
social choice theory, 16, 462
sophism, 491
sound, 43, 48, 67
soundness, 67, 219, 227, 239, 286, 314, 316, 396, 409, 410, 432, 438
speculative thinking, 497
speech act, 336, 340
split-rule, 290
spread, 381, 412
SQL, 459
square of opposition, 247
stable, 392
standard model, 262
start-symbol, 371
Stoic logic, 102
strategy-proof, 467, 473, 480
straw man argument, 511
strict associated ordering, 157
strict implication, 103, 278, 309, 310
structure, 158, 197
sub-contrary, 247
subaltern, 247
subformula, 49
subformula property, 85
subgoal, 58, 435
subject, 183
subjunctive conditional, 107, 366
subset, 137
subset-requirement, 456
substitution, 429, 430
substitution theorem, 45
substitutivity of identity, 112, 368
substitutivity of material equivalents, 112
succedent, 26, 40

successor function, 134
suggestive methods, 515
sum-set, 134
surface structure, 372
surjection, 154
surjective, 154
syllogism, 248
symmetric, 149, 286
synonyms, 502
syntactic, 86
syntax, 111, 363
synthetic, 141, 243, 349

table, 454
table-constraint, 454
tableau, 84, 85, 224, 293, 389
tableau branch, 84, 223, 293, 389
tableau-deducible, 86, 223, 224, 291, 294, 389, 390
tableau-deduction, 86, 223, 224, 294, 389
tableau-proof, 86, 223, 224, 294, 389
tableau-provable, 86, 223, 224, 294, 390
tableaux rules, 82, 223, 290, 387
tail, 439
Tarski's truth definition, 197
tautology, 33
tense operator, 278
term, 190, 211, 263, 434
terminals, 370, 371
ternary relation, 184
theory, 67
theory of types, 131
total ordering, 157
Tractatus, 357
transitive, 149, 156, 285, 394, 466
transitivity, 470
transposition, 147
Traveling Salesman Problem, 59
treacherous paradox, 524
tree, 237
true in, 199, 283, 395
truth of fact, 112, 368
truth of reason, 112, 368
truth table, 29, 30
truthfunctional, 30
tuple, 145, 453
tuple-constraint, 454
Turing, A., 41, 111, 368
two sorted, 202

undecidable, 230, 231, 266, 410
unfair discussion methods, 510
unification, 437
unification algorithm, 437

union, 133, 134
uniquely identifying, 455
unit clause, 58
universal closure, 199
universal quantifier, 23, 182
universal relation, 148
universal spread, 413
universe of discourse, 194, 197
unordered pair, 133
use and mention, 329

vague terms, 500
valid, 3, 9, 187, 238, 283, 284, 314
valid argument, 22
valid consequence, 38, 196, 200, 202, 238, 283, 284, 286, 314
valid consequence in $K-$, 286
valid formula, 33, 196, 199
valid in, 232

valid in $K-$, 286
valuation, 195, 197
variable general, 202
variable held constant, 201
veridical paradoxes, 97, 99
verification principle, 245, 352
Vienna Circle, 244, 352

weak implication, 313
weak negation elimination, 76
weak ordering, 157
well-ordering, 158
Whitehead, A., 109
Wittgenstein, L., 242, 342, 357

Zeno's paradox, 100
Zermelo, E., 129, 132
Zermelo-Fraenkel, 139
ZF, 139